T0343345

CNS Regeneration

CNS *Regeneration*

Basic Science and Clinical Advances

Edited by

Mark H. Tuszynski

Department of Neurosciences
University of California, San Diego
La Jolla, California

Jeffrey H. Kordower

Department of Neurological Science
Rush Presbyterian St. Luke's Medical Center
Chicago, Illinois

Academic Press

San Diego London Boston New York Sydney Tokyo Toronto

Cover photograph: Lesioned corticospinal axons [green, Alexa 488 (Molecular Probes)] within a field of extracellular matrix [red, Alexa 594 (Molecular Probes)] in the injured adult rat spinal cord. Photo courtesy of Raymond Grill.

This book is printed on acid-free paper. ♾

Academic Press
a division of Harcourt Brace & Company
525 B Street, Suite 1900, San Diego, California 92101-4495, USA
http://www.apnet.com

Academic Press
24-28 Oval Road, London NW1 7DX, UK
http://www.hbuk.co.uk/ap/

Library of Congress Catalog Card Number: 98-87916

International Standard Book Number: 0-12-705070-1

Printed and bound by CPI Group (UK) Ltd, Croydon, CR0 4YY

Transferred to Digital Print 2011

CONTENTS

PART II
Promoting CNS Recovery

PART III
Promoting Recovery in Neurological Disease

CONTRIBUTORS

Numbers in parentheses indicate the pages on which the authors' contributions begin.

PATRICK AEBISCHER (581), Gene Therapy Center and Division of Surgical Research, CHUV, Lausanne University Medical School, 1011 Lausanne, Switzerland

ANNE-CATHERINE BACHOUD-LÉVI (455), INSERM U421, IM3, Faculté de Médecine, Service de Neurologie, Hôpital Henri-Mondor, 94010 Crétil cedex, France

ROY A. E. BAKAY (389), Department of Neurological Surgery, Emory University School of Medicine, Atlanta, Georgia 30322

KRYZSTOF S. BANKIEWICZ (437), Molecular Therapeutics Section, Laboratory of Molecular Medicine and Neuroscience, National Institute of Neurological Disorders and Stroke, National Institutes of Health, Bethesda, Maryland 20892

ARMIN BLESCH (605), Department of Neurosciences, University of California, San Diego, La Jolla, California 92093

DIEGO BRAGUGLIA (581), Gene Therapy Center and Division of Surgical Research, CHUV, Lausanne University Medical School, 1011 Lausanne, Switzerland

XANDRA BREAKEFIELD (251), Molecular Neurogenetics Unit, Neurology Department, Massachusetts General Hospital and Harvard Medical School, Boston, Massachusetts 02114

PIERRE BRUGIÈRES (455), Service de Neuroradiologie, Hôpital Henri-Mondor, 94010 Crétil cedex, France

PATRIK BRUNDIN (299), Section on Neuronal Survival, Wallenberg Neuroscience Center, Department of Physiology and Neuroscience, Lund University, S-22362 Lund, Sweden

MARY BARTLETT BUNGE (631), The Chambers Family Electron Microscopy Laboratory, The Miami Project to Cure Paralysis, Departments of Neurological Surgery and Cell Biology and Anatomy, University of Miami School of Medicine, Miami, Florida 33101

WAYNE A. CASS (419), Department of Anatomy and Neurobiology, University of Kentucky, Lexington, Kentucky 40536

PIERRE CESARO (455), INSERM U421, IM3, Faculté de Médecine, Service de Neurologie, Hôpital Henri-Mondor, 94010 Crétil cedex, France

HAMED CHAKER (455), Hôpital Esquirol, 94413 Saint-Maurice, France

TIMOTHY J. COLLIER (321, 555), Research Center for Brain Repair and Department of Neurological Sciences, Rush Presbyterian St. Luke's Medical Center, Chicago, Illinois 60612

CARL W. COTMAN (529), Institute for Brain Aging and Dementia, University of California, Irvine, Irvine, California 92697

BRIAN J. CUMMINGS (529), Institute for Brain Aging and Dementia, University of California, Irvine, Irvine, California 92697

TERRENCE DEACON (365), Neuroregeneration Laboratory, McLean Hospital, Belmont, Massachusetts 02178

JAMIE L. EBERLING (437), Center for Functional Imaging, Lawrence Berkeley National Laboratory, University of California, Berkeley, California 94704

JOHN D. ELSWORTH (321), Departments of Psychiatry and Pharmacology, Yale Medical School, New Haven, Connecticut 06520

DWAINE F. EMERICH (477), Department of Neuroscience, Alkermes, Inc., Cambridge, Massachusetts 02903

MIA EMGÅRD (299), Section on Neuronal Survival, Wallenberg Neuroscience Center, Department of Physiology and Neuroscience, Lund University, S-22362 Lund, Sweden

MICHAEL T. FITCH (55), Department of Neurosciences, Case Western Reserve University School of Medicine, Cleveland, Ohio 44106

FRED H. GAGE (183, 505), Laboratory of Genetics, Salk Institute for Biological Studies, La Jolla, California 92037

DON M. GASH (419), Department of Anatomy and Neurobiology, University of Kentucky, Lexington, Kentucky 40536

GREG A. GERHARDT (419), Department of Pharmacology, University of Colorado Medical Center, Denver, Colorado 80262

RAYMOND J. GRILL (27, 605), Department of Neurosciences, University of California, San Diego, La Jolla, California 92093 and Department of Neurology, Veterans Affairs Medical Center, San Diego, California 92093

MICHAEL GRUNDMAN (647), Department of Neurosciences, University of California, San Diego, La Jolla, California 92093

BASSAM HADDAD (455), Centre Hospitalier Intercommunal de Crétil, 94010 Crétil cedex, France

PHILIPPE HANTRAYE (455), CEA-CNRS URA 2210, Service Hospitalier Frédéric Joliot,SHFJ, DRM, DSV, 91406 Orsay cedex, France

MELEIK A. HEBERT (419), Department of Pharmacology, University of Colorado Health Sciences Center, Denver, Colorado 80262

BARRY J. HOFFER (419), Department of Pharmacology, University of Colorado Medical Center, Denver, Colorado 80262

ALEXANDER F. HOFFMAN (419), Department of Pharmacology, University of Colorado Medical Center, Denver, Colorado 80262

BINGREN HU (89), Department of Neurosciences, University of California, San Diego, La Jolla, California 92093

OLE ISACSON (365, 477), Neuroregeneration Laboratory, Harvard Medical School and McLean Hospital, Belmont, Massachusetts 02178

ANDREAS JACOBS (251), Neuroscience Center, Massachusetts General Hospital, Molecular Neurogenetics Unit, Neurology Department, Massachusetts General Hospital and Harvard Medical School, Boston, Massachusetts 02114

NAOMI KLEITMAN (631), The Chambers Family Electron Microscopy Laboratory, The Miami Project to Cure Paralysis, Department of Cell Biology and Anatomy, University of Miami School of Medicine, Miami, Florida 33101

JEFFREY H. KORDOWER (159, 295, 389, 477, 503, 505, 555), Research Center for Brain Repair and Department of Neurological Sciences, Rush Presbyterian St. Luke's Medical Center, Chicago, Illinois 60612

EUGENE O. MAJOR (437), Center for Functional Imaging, Lawrence Berkeley National Laboratory, University of California, Berkeley, California 94704

ALBERTO MARTINEZ-SERRANO (203), Center of Molecular Biology "Severo Ochoa," Autonomous University of Madrid—CSIC, 28049 Madrid, Spain

ELLIOT J. MUFSON (505), Research Center for Brain Repair and Department of Neurological Sciences, Rush Presbyterian St. Luke's Medical Center, Chicago, Illinois 60612

ULRIKA MUNDT-PETERSEN (299), Sections on Neuronal Survival, Wallenberg Neuroscience Center, Department of Physiology and Neuroscience, Lund University, S-22362 Lund, Sweden

DEA NAGY (437), Center for Functional Imaging, Lawrence Berkeley National Laboratory, University of California, Berkeley, California 94704

JEAN-PAUL NGUYEN (455), INSERM U421, IM3, Faculté de Médecine, 94010 Crétil cedex, France

THEO D. PALMER (183), Laboratory of Genetics, Salk Institute, La Jolla, California 92037

MARC PESCHANSKI (455), INSERM U421, IM3, Faculté de Médecine, 94010 Crétil cedex, France

JASODHARA RAY (183), Laboratory of Genetics, Salk Institute, La Jolla, California 92037

D. EUGENE REDMOND, JR. (321), Departments of Psychiatry and Pharmacology, Yale Medical School, New Haven, Connecticut 06520

ROBERT H. ROTH (321), Departments of Psychiatry and Pharmacology, Yale Medical School, New Haven, Connecticut 06520

JAMES SCHUMACHER (365), Neurological Associates, Sarasota, Florida 34239

LAMYA S. SHIHABUDDIN (183), Laboratory of Genetics, Salk Institute, La Jolla, California 92037

JERRY SILVER (55), Department of Neurosciences, Case Western Reserve University School of Medicine, Cleveland, Ohio 44106

JOHN J. SLADEK, JR. (321), Department of Neurosciences, Chicago Medical School, North Chicago, Illinois 60064

EVAN Y. SNYDER (203), Departments of Neurology and Pediatrics, Harvard Medical School, Children's Hospital, Boston, Massachusetts 02115

MICHAEL V. SOFRONIEW (3), MRC Cambridge Centre for Brain Repair and Department of Anatomy, University of Cambridge, Cambridge CB2 2PY, United Kingdom

PHILIP A. STARR (389), University of California, San Francisco, San Francisco, California 94143

JANE R. TAYLOR (321), Departments of Psychiatry and Pharmacology, Yale Medical School, New Haven, Connecticut 06520

MARK H. TUSZYNSKI (27, 109, 159, 503, 505, 605, 647), Department of Neurosciences, University of California, San Diego, La Jolla, California 92093 and Veterans Affairs Medical Center, San Diego, California 92093

SAM WANG (251), Molecular Neurogenetics Unit, Neurology Department, Massachusetts General Hospital and Harvard Medical School, Boston, Massachusetts 02129

ZHIMING ZHANG (419), Department of Anatomy and Neurobiology, University of Kentucky, Lexington, Kentucky 40536

JUSTIN A. ZIVIN (89), Department of Neurosciences, University of California, San Diego, La Jolla, California 92093

INTRODUCTION

The dogma that regeneration in the central nervous system cannot occur is a myth. Over the past two decades it has become increasingly clear that the adult brain and spinal cord are responsive to signals provided by several classes of molecules that can promote neuronal survival, stimulate axon outgrowth, and even lead to self-replenishing sets of neural progenitor cells that form new neuronal and non-neuronal cells. This capacity for regeneration is rarely expressed in the central nervous system in the absence of specific intervention, but can be readily elicited by experimental manipulation. The culmination of these interventions is functional recovery, a phenomenon that has now been demonstrated in animal models of nervous system trauma, stroke, degeneration, and inherited degenerative disease. Based upon preclinical studies, several of these approaches are now undergoing clinical testing. We have entered a new era in the treatment of disease of the nervous system.

This book will focus on some of the leading current neurological disease models and methods for promoting nervous system regeneration. In Part I, responses of the nervous system to cellular and axonal insults will be described, identifying salient issues that are critical in mediating nervous system repair. Responses of neurons, axons, and glia to injury will be described, together with the deleterious effects on the nervous system of secondary damaging factors such as free radicals and excitotoxins.

In Part II, recent advances in neuroscience will be identified that offer the *tools* for promoting regeneration of the central nervous system. These tools include neurotrophic factors, fetal cell grafts, primary and immortalized progenitor cells, and genetic engineering.

In Part III, the application of regeneration-promoting strategies to specific animal models of human disease will be described. Means of promoting neural repair in models of Alzheimer's disease, Parkinson's disease, Huntington's disease and spinal cord injury will be described as prototypes for the treatment of neurological disease in general. Specific methods for treating these disorders, including the use of growth factors, gene therapy, and fetal grafting will be discussed. Finally, the testing of these strategies for the treatment of human disease will be discussed in a chapter on the design of clinical trials.

We are in the midst of a new era in the treatment of neurological disease. The focus of neurological therapy is shifting from *compensation* for damage already done to *rescue* and *regeneration* of existing pathways. No book on this topic can be complete as the field of regeneration expands with startling rapidity. Thus, the present offering is but a sampling of mechanistically enlightening and therapeutically promising progress in a field that has undergone remarkable evolution in a very short period of time.

Mark H. Tuszynski
Jeffrey H. Kordower

PART I

CNS Responses to Injury

CHAPTER 1

Neuronal Responses to Axotomy

MICHAEL V. SOFRONIEW

MRC Cambridge Centre for Brain Repair and Department of Anatomy, University of Cambridge, Cambridge CB2 2PY, United Kingdom

Neurons in the central and peripheral nervous system show various responses to axotomy, ranging from cell death or severe atrophy without axon regeneration to recovery with axon regeneration. The cellular and molecular mechanisms that underlie these responses and determine these differences are not well understood and are the subject of extensive investigation as reviewed in this chapter. Neurons that can be grouped together on the basis of topography, function, projection, and/or developmental origin, generally undergo similar reactions to axotomy, but different groups often react differently to each other. For example, motor neurons survive axotomy and regenerate, Purkinje cells survive but do not regenerate, and retinal ganglion cells die. Such differences are related to a combination of intrinsic properties and extracellular environmental factors. Current hypotheses regarding signals that mediate the neuronal response to axotomy include both the loss and the introduction of large molecules that signal via retrograde transport along the axon to the perikaryon, molecules that derive from glial cells around neuronal perikarya, and fast retrograde signals along the axon mediated by the influx of ions through membrane damage at the site of axonal injury. Understanding factors that regulate neuronal responses to axotomy will be fundamental to developing therapeutic strategies that induce axotomized neurons to survive and regenerate once their axons are provided with environmental cues permitting regrowth. In addition, understanding and preventing the tissue events that cause axonal discontinuity after blunt injury hold promise for therapeutic measures that markedly reduce delayed "secondary" axotomy after trauma to the central nervous system.

Axotomy resulting either from direct axonal transection or from indirect tissue reactions to injury is an important consequence of clinical conditions such as trauma, compression, and ischemia. Experimental studies on the neural responses to axotomy have a long history and were among the first attempts to investigate at the cellular level the effects of injury to the nervous system. Retrograde changes in neuronal cell bodies after axotomy were first recognized over 100 years ago (Nissl, 1892), and many of the basic degenerative and regenerative responses of neuronal peikarya and their axons to axotomy were well documented at the light microscopic level by the early part of this century (Cajal, 1928). Since that time, axotomy-induced changes have been characterized in many regions of the mammalian nervous system, in part because the analysis of retrograde or anterograde degeneration resulting from specifically placed lesions provided the first, and for many years the only, experimental means by which to study neuronal connectivity (Brodal, 1981).

There continues to be widespread interest in studies on neural responses to axotomizing injury. Although it is self-evident that neurons that have died after axotomy will not be available for therapeutic attempts at promoting regeneration, it also appears likely that neurons that have undergone severe retrograde atrophy will not regenerate well unless the atrophy can be prevented or reversed. In addition, circumstantial evidence suggests that the severity and extent of axotomy after central nervous system (CNS) injury are due not only to direct axonal transection, but also may be indirectly exacerbated by delayed cellular reactions at the injury site. Thus, efforts to understand and control both the retrograde reactions of neurons to axotomy and the local tissue responses around potentially axotomizing injury are essential to the development of therapies that facilitate neural regeneration and repair. Although the structural nuances of different types of degenerative and regenerative changes caused in neurons and their axons by axotomy have been described and reviewed extensively (Brodal, 1981; Cajal, 1928; Kreutzberg, 1995; Lieberman, 1971), the molecular and cellular mechanisms that underlie the different types of neuronal responses are not well understood and are the subject of much ongoing investigation. Information is becoming available on changes in molecular expression during neural reactions to axotomy, as well as on molecules that may trigger degenerative processes or protect neurons. This chapter will summarize and discuss: (1) factors that correlate with or appear to influence differences in the retrograde response of neurons to axotomy, (2) changes that characterize neurons that survive axotomy, (3) responses of glial cells around the site of axotomizing injuries and around the neuronal cell bodies of axotomized neurons, (4) molecular mechanisms that may regulate axotomy-induced degenerative or regenerative changes, (5) factors that may influence a transition from subtotal axonal injury to irreversible axotomy, and (6) strategies for facilitating regeneration following axonal injury in the CNS.

I. FACTORS THAT INFLUENCE NEURONAL DEATH OR SURVIVAL FOLLOWING AXOTOMY

The retrograde reaction of neurons to axotomy can vary enormously from transient changes followed by apparent recovery, to long-lasting cellular atrophy, to cell death (Brodal, 1981; Kreutzberg, 1982; Kreutzberg, 1995; Lieberman, 1971), as summarized in Fig. 1 (see color insert). Axotomy-induced cell death can have the characteristics of apoptosis (Berlelaar *et al.*, 1994; de Bilbao and Dubois-Dauphin, 1996; Quigley *et al.*, 1995; Rossiter *et al.*, 1996; Wilcox *et al.*, 1995) or of necrosis (Laiwand *et al.*, 1988). Retrograde degenerative changes can occur rapidly or very gradually over periods of many months. The cellular and molecular mechanisms that underlie the neuronal response to axotomy and determine these differences are not known. Identifying factors and conditions that correlate with, or are able to influence, the death or survival of different neurons in response to axotomy is an important first step in establishing the underlying molecular mechanisms.

A. Neuronal Groups

In most cases, similar reactions to axotomy are observed among neurons that can be grouped together on the basis of topography, function, projection, and/or developmental origin. For example, in adults, most motor neurons react similarly to axotomy, as do most cerebellar Purkinje neurons or most retinal ganglion cells. However, these different groups react quite differently to each other. Motor neurons survive axotomy and regenerate (Brodal, 1981), Purkinje cells survive but do not regenerate (Rossi *et al.*, 1995a), and retinal ganglion cells die (Bray *et al.*, 1991). The reasons for the similarity in reaction of specific groups of neurons are likely to be related both to common intrinsic properties and to similar extracellular environmental factors. Nevertheless, it is intriguing that a small percentage of neurons within a group often respond differently to axotomy than their neighbors and that cell groups that appear to be very similar can have different reactions. For example, basal forebrain cholinergic neurons constitute a large complex of contiguous cell groups that share a common developmental origin, chemistry, and appearance but are distinguished by projecting to different regions of the allo- and neocortex (Cuello and Sofroniew, 1984). Surprisingly, some of the subgroups respond differently to axotomy. Medial septal cholinergic neurons, which project to the hippocampus, rapidly lose their expression of choline acetyltransferase (ChAT) and high- and low-affinity NGF receptors and many die (O'Brien *et al.*, 1990; Sofroniew and Cooper, 1993; Tuszynski *et al.*, 1990), whereas basal nucleus cholinergic neurons, which project to the neocortex, do not lose ChAT or NGF receptor expression and do not die (Sofroniew *et al.*, 1983, 1987). The reasons underlying these different responses are not clear but may be related to target region or course of axonal projection.

B. Age of the Animal

Not only do different populations of neurons respond differently to axotomy, in some cases the same population of neurons will respond differently at different ages of the animal. For example, most motor neurons in the spinal cord and brain stem survive axotomy in adult vertebrates. In contrast, many if not most motor neurons die if they are axotomized during early stages of development. In mice and rats, the most vulnerable stage is during the first few days of postnatal developmental. If axotomized at this time, most motor neurons die, whereas by the time the animals are 2 or more weeks old, most motor neurons survive (de Bilbao and Dubois-Dauphin, 1996; Kreutzberg, 1995; Lieberman, 1971; Rossiter *et al.*, 1996). Axotomy of corticospinal neurons induces retrograde cell death before, but not after these neurons innervate their spinal targets (Merline and Kalil, 1990).

The molecular mechanisms that underlie the different responses of motor neurons to axotomy at different ages may involve changes in the local availability of growth factors or in the dependence of motor neurons on these factors. Considerable evidence indicates that target-derived growth factors such as GDNF, BDNF, and CNTF are made in various portions of the pathways and targets of motor neurons and regulate their survival during development. Some of these molecules are able to prevent the death of axotomized neonatal motor neurons (Oppenheim *et al.*, 1995; Sendtner *et al.*, 1990; Yan *et al.*, 1992). It is possible that the motor neurons undergo an intrinsic developmental change and become less dependent on growth factors for survival as they mature. Intrinsic factors such as the *bcl-2* family of genes that may regulate neuronal dependence on extrinsic growth factors have been identified. Some members of this family, *bcl-2* and *bcl-*x_L, promote neuronal survival, whereas others, *bax* and *bak*, promote cell death by apoptosis, and it appears that interactions between the various factors regulate an intrinsic cell death program (Korsmeyer, 1995). The targeted overexpression of *bcl-2* in motor neurons of transgenic mice protects these neurons from axotomy-induced apoptotic cell death in the postnatal period (Dubois-Dauphin *et al.*, 1994), suggesting that a developmental switch in the balance of *bcl-2* family-related gene expression could be responsible for the survival of mature motor neurons after axotomy. In other neuronal systems, both primary sensory neurons and basal forebrain cholinergic become less dependent on NGF (one of their principal target-derived growth factors) for survival as they mature (Barde *et al.*, 1980; Greene, 1977; Lindsay, 1988; Svendsen *et al.*, 1994; Svendsen and Sofroniew, 1995).

It deserves emphasis that not all populations of neurons show a developmental change in vulnerability to the effects of axotomy. Cerebellar Purkinje neurons appear to survive axotomy equally well in adult and young animals (Cajal, 1928; Dusart and Sotelo, 1994). The converse example also exists that similar numbers of septal cholinergic neurons die after axotomy in adult and neonatal rats (Cooper *et al.*, 1996).

C. Axonal Environment

The extracellular environment of the axon is one factor that may influence whether neurons survive or die following axotomy. Most neurons that send their axons into peripheral nerves survive axotomy in adults, perhaps because peripheral nerves provide an environment that is hospitable to axon regeneration (Lieberman, 1971). Notable exceptions are the dorsal motor neurons of the vagal nerve, which die slowly over a period of weeks to months after axotomy (Kreutzberg, 1982; Laiwand *et al.*, 1987, 1988). Whether this death is due to a unique property of the vagal neurons themselves or to the environment of the peripheral vagus nerve is not known. In contrast, far more, but by no means all, neurons whose projections lie entirely within the CNS die following axotomy in adults. It is not certain to what degree the difference between neurons with CNS and PNS projections is due to the inhospitable environment of the CNS to regeneration. Most retinal ganglion neurons die after axotomy, and survival is improved only partially when the axotomized neurons are presented with a peripheral nerve graft that sustains the regeneration of some of the neurons (Bray *et al.*, 1991). This observation suggests that although the environment around the transected axon can influence the survival of some neurons it is not the only factor.

Differences in the extracellular environments found within the CNS around axons may also play a role in neuronal responses to axotomy. Many neurons whose axons are transected in white matter tracts appear to die, including retinal ganglion cells axotomized in the optic nerve (Bray *et al.*, 1991), Clarke's column neurons after spinal cord white matter lesions (McBride *et al.*, 1988), septal cholinergic neurons after fornix transections (Fischer and Bjorklund, 1991; O'Brien *et al.*, 1990; Tuszynski *et al.*, 1990; Wilcox *et al.*, 1995), and substantia nigra neurons transected in the medial forebrain bundle (Reis *et al.*, 1978a). However, other CNS neurons appear to survive axotomy in the same white matter tracts, such as hippocampal formation neurons transected in the fornix or red nucleus and corticospinal neurons after spinal cord lesions (Merline and Kalil, 1990; Prendergast and Stelzner, 1976; Tetzlaff *et al.*, 1991). It will be important to investigate the role of the environment in the neuronal reaction to axotomy more systematically.

D. Distance of the Perikaryon from the Lesion

On the basis of silver-impregnated specimens, Cajal (1928) suggested that after axotomy in the spinal cord, the transected axons degenerate back to the first point of a collateral branch of sufficient caliber and terminal arborization to sustain the fiber. In the absence of such collaterals the axon might degenerate back all the way to the cell body and the neuron might die, unless it had another substantial axonal projection arising from or close to the cell body as in the case of dorsal root ganglion cells. The possibility that neuronal survival after axotomy depends heavily on remaining collateral branches has remained a popular hypothesis and has been supported by experimental evidence for some neurons (Fry and Cowan, 1972; O'Connor, 1947). Nevertheless, data from three well-studied systems, the septohippocampal cholinergic projection, the projection of retinal ganglion cells through the optic nerve, and cerebellar Purkinje cells, suggest that the presence of "sustaining collaterals" is not the only factor that influences whether neurons survive or die as the distance between the lesion and the neuronal perikaryon becomes smaller.

In rodents and nonhuman primates, axotomy of the septohippocampal pathway by complete transection of the fimbria–fornix results in the loss from the septum of about 75% of cholinergic neurons, as assessed by immunohistochemical staining for ChAT and $p75^{NTR}$ and histochemical staining for AChE (Lucidi-Phillipi and Gage, 1993; Sofroniew and Cooper, 1993). Progressively more septal cholinergic neurons are lost as the axotomizing lesion gets closer and closer to the neuronal perikarya in the septum, even though available evidence suggests that collateral branching does not occur in this region (Sofroniew and Isacson, 1988). This loss of immunohistochemically identified cholinergic neurons appears to represent the cell death of some but not all of these neurons as assessed by other experimental means: (1) some, but not all, septal neurons previously labeled with the retrograde fluorescent tracer are lost after a fimbria–fornix lesion (Fischer and Bjorklund, 1991; Naumann *et al.*, 1992; O'Brien *et al.*, 1990; Tuszynski *et al.*, 1990; Wilcox *et al.*, 1995), (2) large Nissl-stained neurons are lost (Daitz and Powell, 1954), and (3) TUNELstained or ultrastructural profiles of neurons undergoing cell death are present in the septal region between 1 and 4 weeks after fornix transection (Wilcox *et al.*, 1995, unpublished personal observations).

In the visual system, transection of the optic nerve leads to massive retrograde degeneration of ganglion neurons in the retina, as demonstrable by a variety of traditional histological as well as experimental techniques (Berlelaar *et al.*, 1994; Bray *et al.*, 1991; Quigley *et al.*, 1995). The severity of the degenerative reaction is directly proportional to the distance between the lesion and the neuronal cell body, such that the closer the lesion, the greater the number of retinal neurons die and the more rapid the time course of their degeneration (Berlelaar *et al.*, 1994; Villegas-Perez *et al.*, 1993). The time course of retrograde degeneration leading to neuron death in the retina following lesions in the distal part of the optic nerve can be quite prolonged over weeks to months and the transducing signal is not known. In this case it is clear that there are no collateral branches arising along the path of the axons in the optic nerve that might be involved in sustaining the retinal nerve cells.

In contrast to the two examples resulting in retrograde cell death, some neurons, such as Purkinje cells, survive axotomy even at very small distances from the cell body, as close as 100 μm (Dusart and Sotelo, 1994). These neurons have similar reactions to axotomy at varying distances from the cell body, and the reactions are independent of sites of emergence of any collateral branches.

E. Neuronal Activity

The potential effects that differences in neuronal activity might have on the death or survival of axotomized neurons have not been reported on extensively, perhaps in part because it is difficult to regulate experimentally the activity of an entire population of neurons. One system in which this can be done is that of the magnocellular neurohypophyseal neurons that secrete vasopressin under the regulation of serum osmolarity in a manner that correlates with neuronal activity. After axotomy, about two-thirds of these vasopressin neurons die under normonatremic conditions. Under hyponatremic conditions induced by administration of vasopressin or vasopressin analogues, almost all of the vasopressin neurons die after axotomy (Dohanics *et al.*, 1996; Herman *et al.*, 1987). Under hyponatremic conditions the activity of these neurons is substantially reduced, suggesting that reduced activity may make neurons more vulnerable to cell death after axotomy. Interestingly, hypernatremic conditions in which the activity of the vasopressin neurons would be increased did not significantly enhance the neuronal survival after axotomy (Dohanics *et al.*, 1996).

II. MOLECULAR, STRUCTURAL, AND FUNCTIONAL CHANGES IN NEURONS THAT SURVIVE AXOTOMY

The retrograde reaction of neurons that survive axotomy can vary from transient changes followed by restoration of a normal appearance to long-lasting changes in molecular expression, structural degeneration, and atrophy (Fig. 1).

A. Morphological Changes

The morphological changes that axotomy induces in the neuronal cell body have been referred to by terms such as "chromatolysis," "axon reaction," or "retrograde reaction" (Brodal, 1981; Kreutzberg, 1982; Lieberman, 1971). The acute phase of the retrograde reaction to axotomy is characterized by dispersal of the Nissl substance, displacement of the nucleus to the periphery of the cell, swelling of the cell body, and loss or "retraction" of synaptic terminals. This reaction is common to many neurons in the CNS and PNS, regardless of whether their axons are able to regenerate or not. In contrast, the long-term changes induced in neuronal perikarya by axotomy vary considerably. In neurons that are able to regenerate and are allowed to do so, such as motor neurons with axonal projections through peripheral nerves, the cell bodies remain hypertrophic during the period of regeneration, and there are increased levels of free ribosomes and other intracellular organelles associated with increased metabolism and protein synthesis (Kreutzberg, 1982; Lieberman, 1971). In motor neurons that are prevented from regenerating by tight ligation of the peripheral nerves, and in CNS neurons whose axons are in an environment that does not permit axon elongation, these hypertrophic changes do not occur. Instead, many neurons unable to regenerate slowly atrophy to a variable degree by reducing their cell volume and dendritic arborization, often persisting in this atrophic state for an indefinite period of time (Kreutzberg, 1982; Lieberman, 1971; McBride *et al.*, 1988; Sofroniew *et al.*, 1987). Nevertheless, some neurons, such as Purkinje cells, appear to change little after axotomy, even though they do not regenerate (Rossi *et al.*, 1995a).

B. Changes in Chemical Markers of Neurons: Transmitters, Receptors, Enzymes, etc.

The changes in biochemical or molecular markers induced by axotomy vary in different groups of neurons. Axotomized motor neurons that are allowed to regenerate show a rapid increase in oxygen utilization (Kreutzberg, 1982), upregulation of metabolic enzymes, and increased expression of microtubule-associated proteins, growth-associated proteins such as GAP-43 (Tetzlaff *et al.*, 1991, 1988a), and the p75 neurotrophin receptor (Ernfors *et al.*, 1989; Wood *et al.*, 1990). Many axotomized neurons change their expression of neuropeptides and transmitter-associated enzymes after axotomy. Sensory, sympathetic, and motor neurons do not normally express the neuropeptides galanin and vasoactive intestinal polypeptide, but begin to do so within a few days after axotomy, whereas the normal expression of substance P and neuropeptide Y by sensory and sympathetic neurons is downregulated (Zigmond *et al.*, 1996). Motor neurons and vagal neurons rapidly downregulate expression of ChAT after axotomy, and although both types of neurons survive for some time after the axotomy, ChAT levels recover in motor neurons but not in vagal neurons (Lams *et al.*, 1988), many of which go on to die (Laiwand *et al.*, 1987). Catecholaminergic neurons in the locus ceruleus downregulate levels of tyrosine hydroxylase after axotomy, but continue to survive (Reis *et al.*, 1978b). Some cholinergic medial septal neurons survive axotomy, downregulate ChAT, and recover ChAT expression after long survival times of 6 months or more (Naumann *et al.*, 1994). Not all axotomized neurons show such dramatic changes in transmitter-associated enzymes; cholinergic neurons of the basal nucleus that undergo severe shrinkage and atrophy after axotomy continue to express easily detectable levels of ChAT (Sofroniew *et al.*, 1983, 1987).

C. Electrophysiological Changes

Most axotomized neurons show an increased electrogenesis, apparently due to an increase in the number of functional Na^+ channels (Titmus and Faber, 1990). The functional significance of these changes is not clear. As discussed above, levels of neuronal activity appear able to influence whether neurons survive or die following axotomy of the same population of neurons. Axotomized vagal neurons that do not regenerate exhibit considerable changes in membrane properties that lead to increased firing duration and increased cytosolic Ca^{2+} concentrations that may be responsible for the gradual death of these neurons (Laiwand *et al.*, 1988). Some of these signals may derive from neuron–target interactions. Studies on peripheral neurons indicate that characteristic electrophysiological properties that are lost or altered by axotomy are recovered when regeneration of the axon restores connectivity of the neuron with target cells (Titmus and Faber, 1990). These observations suggest that ongoing neuron–target interaction may be required for the maintenance of distinctive electrophysiological properties of mature neurons. These neuron–

target interactions may in part involve retrograde signaling of neurotrophic factors. Studies on sympathetic neurons indicate that exogenous NGF can restore electrophysiological changes provoked by axotomy (Purves and Lichtman, 1978).

III. GLIAL CELL CHANGES AROUND THE PERIKARYA OF AXOTOMIZED NEURONS

The retrograde changes induced by axotomy involve not only the neurons themselves, but also the cellular elements of the local microenvironment of those neurons. In response to signals that are presumed to derive from the axotomized neurons, surrounding nonneuronal cells such as astrocytes and microglia undergo structural, biochemical, and proliferative changes.

A. Reactive Astrocytosis around Neuronal Perikarya

In response to axotomy of motor neurons, astrocytes local to the affected neuronal perikarya hypertrophy and upregulate their expression of the intermediate filament protein GFAP (Eddleston and Mucke, 1993; Eng and Ghirnikar, 1994; Graeber and Kreutzber, 1986). In cases in which the motor neurons are allowed to regenerate their axons, this astrocyte reaction resolves with time. In cases in which axon regeneration of the motor neurons is prevented or impeded by ligation, the astrocytic reaction persists indefinitely (Tetzlaff *et al.*, 1988b). The reaction of astrocytes in the vicinity of other axotomized neurons has not been described extensively, but a pronounced astrocyte reaction with cellular hypertrophy and upregulation of GFAP occurs rapidly around basal forebrain cholinergic neurons after ablation of cortical target neurons in the absence of axotomy. This astrocyte reaction persisted for over a year after the ablation of cortical neurons, the longest time examined, and was completely absent in animals in which grafts of embryonic cortical cells had been used to replace the ablated host neurons (Isacson and Sofroniew, 1992). These findings suggest that retrograde interactions between afferent neurons and cells in the target region may regulate astrocyte reactivity around the perikarya of the afferent neurons. The molecular nature of this interaction is not known.

The purpose of reactive astrocytosis around the cell bodies of neurons that had been axotomized or lost their target cells is not clear, but may be protective in nature or may help provide an environment conducive to regenerative attempts. Reactive astrocytes are able to secrete a variety of growth factors and cytokines (Eddleston and Mucke, 1993), which may promote survival or regeneration. Astrocytes are also thought to buffer extracellular ion concentrations and take up excess extracellular glutamate, which may be neurotoxic (Rothstein *et al.*, 1996; Tanaka *et al.*, 1997). Reactive astrocytosis may reflect increased activities of this kind.

B. Inflammation and Microglia

There is rapid proliferation of microglial cells around the neuronal perikarya of adult motor neurons that survive and regenerate after axotomy (Graeber *et al.*, 1988). Within several days these microglia have displaced synaptic terminals from the cell bodies and dendrites (Blinzinger and Kreutzberg, 1968) and may be involved in elimination of degenerating neurons (Thanos *et al.*, 1993). Under these conditions, the microglia also express major histocompatibility class (MHC) antigens (Streit *et al.*, 1989). After axotomy of retinal ganglion cells that results in substantial retrograde cell death, there are also a rapid proliferation and a hypertrophy of microglia that phagocytose degenerated neurons and may contribute to the process of neuronal degeneration (Thanos *et al.*, 1993).

IV. MOLECULAR TRANSDUCTION OF AXOTOMY-INDUCED CHANGES IN NEURONS

The molecular transduction of the local and retrograde effects of axotomy is likely to be complex and may vary in different regions of the nervous system and at different ages. Current hypotheses regarding signals that mediate the neuronal response to axotomy include the loss or introduction of large molecules that signal via retrograde transport along the axon to the perikaryon, molecules that derive from glial cells around neuronal perikarya, and fast retrograde signals along the axon mediated by the influx of ions through membrane damage at the site of axonal injury, as summarized in Fig. 1.

A. Loss of Target-Derived Factors

The hypothesis that retrograde changes induced in neurons by axotomy might be caused by the loss of retrograde signals derived from target cells has been popular for many years (Cragg, 1970). This hypothesis is a direct extension of observations that during critical periods of development, neuronal survival can be decreased by denying access to target cells or can be increased by increasing target size or by supplying target-derived factors. In agreement with this hypothesis, a number of target-derived growth factors that promote the survival of specific populations of developing neurons *in vitro* also prevent retrograde cell death of developing neurons after axotomy (Johnson *et al.*, 1986; Oppenheim *et al.*, 1995; Purves and Lichtman, 1978; Sendtner *et al.*, 1990; Yan *et al.*, 1992). This observation suggests that some of the cell death observed after axotomy of developing neurons, particularly in the PNS, may be due to the loss of retrograde signaling provided by target-derived factors.

The potential cause(s) of axotomy-induced retrograde cell death appears to be different in adult animals. Various lines of evidence suggest that many (if not most) adult neurons lose their dependence on target-derived growth factors for survival, perhaps due to shifts in the ratios of intracellular molecules that regulate programmed cell death (see Section I, B). The role of target-derived factors in the maintenance of afferent neurons in the adult has been investigated extensively *in vivo* in the basal forebrain cholinergic complex by the selective ablation of target cells. In this system NGF and other neurotrophins are produced in the target structures, and most if not all basal forebrain cholinergic neurons express high and low NGF receptors (Grimes *et al.*, 1993; Persson, 1993). NGF has been shown to regulate the survival of developing basal forebrain cholinergic neurons *in vitro* (Hartikka and Hefti, 1988; Kew *et al.*, 1996; Knusel *et al.*, 1990; Svendsen *et al.*, 1994). In adult animals, however, large lesions of target neurons that deplete neurotrophin levels in the absence of axotomy do not kill cholinergic neurons throughout the basal forebrain complex. Instead, the neurons atrophy (i.e., shrink) and downregulate ChAT, and these changes can be prevented or reversed by grafts of target neurons or NGF-producing cells (Isacson and Sofroniew, 1992; Maysinger *et al.*, 1995; Sofroniew *et al.*, 1990, 1993).

In some regions, such as the basal nucleus, the effects of axotomy are identical to those of loss of target neurons, and the cholinergic neurons atrophy to a more or less similar degree (Isacson and Sofroniew, 1992; Sofroniew and Pearson, 1985; Sofroniew *et al.*, 1983, 1987). In contrast, many cholinergic neurons in the medial septum die following axotomy, whereas most survive in an atrophic state following loss of target cells (Fischer and Bjorklund, 1991; Kordower *et al.*, 1993; O'Brien *et al.*, 1990; Sofroniew *et al.*, 1990, 1993; Sofroniew and Isacson, 1988; Tuszynski *et al.*, 1990; Wilcox *et al.*, 1995; Zhou *et al.*, 1998). Double-labeling tract-tracing experiments show that septal cholinergic neurons do not derive support from other targets via collateral projections (Sofroniew *et al.*, 1990), indicating that axotomy is lethal by having an effect other than, or in addition to, depriving them of target-derived factors. These observations call into question the role of neurotrophic factor withdrawal as the principal molecular trigger for precipitating retrograde neuronal death after axotomy in adults and suggest that the protective effect of NGF in maintaining the survival of these neurons following axotomy (Hefti, 1986;

Kromer, 1987; Williams *et al.*, 1986) is pharmacological, rather than replacement of a continuously required factor. This suggestion is supported by observations that: (1) NGF is required only for a number of weeks after the axotomy, after which it can be discontinued and the axotomized septal cholinergic neurons continue to survive (Montero and Hefti, 1988; Tuszynski and Gage, 1995), and (2) NGF and other growth factors can have pharmacological effects that protect neurons from cell death not caused by loss of those factors such as excitotoxicity and other forms of metabolic cell death (Matson *et al.*, 1993). Thus, in adult animals, loss of target-derived neurotrophic molecules may not directly cause cell death after axotomy, but may mediate other effects, such as cell atrophy or changes in molecular expression, which in turn may increase neuronal vulnerability to other insults (see Section IV, F).

Although loss of target-derived factors does not seem to be responsible for axotomy-induced retrograde cell death, experimentally depriving cells of target-derived factors is able to modulate the architecture of adult neuronal perikarya and dendrites and induce atrophic changes in many neurons in the PNS and CNS (Isacson and Sofroniew, 1992; Rossi *et al.*, 1995b; Sofroniew *et al.*, 1990, 1993; Voyvodic, 1989). Thus, interruption of a retrograde supply of target-derived factors may be responsible for the dendritic and cellular atrophy, cell shrinkage, and metabolic changes induced by axotomy.

B. Retrograde Transport of Novel Large Molecules Derived from the Site of Injury

Although considerably more attention has been focused on the *loss* of retrograde signals in regulating the neuronal response to axotomy, evidence also suggests that some changes after axotomy are caused by the *introduction* of novel retrogradely transported signals at the site of axonal injury. Many different cytokines and growth factors are produced in the tissue reaction that surrounds axotomizing injuries and many of these molecules have the capacity to signal retrogradely (Curtis *et al.*, 1993; Lindholm *et al.*, 1987). Damaged or transected axons have the capacity to take up and retrogradely transport macromolecules such as horseradish peroxidase (HRP) and wheat germ agglutinin (WGA) used in tract tracing or growth factors and cytokines (Curtis *et al.*, 1994; DiStefano *et al.*, 1992), which can have effects on the retrogradely transporting neurons. An example of how a change in molecular expression can be induced in this manner is the reexpression of low-affinity neurotrophin receptor $p75^{NTR}$ by axotomized and regenerating motor neurons. This molecule is expressed during development but not in normal circumstances in the adult (Ernfors *et al.*, 1989; Wood *et al.*, 1990) and its reexpression after axotomy is dependent on a retrograde signal deriving from the regenerating portion of the axon (Bussmann and Sofroniew, 1998). When retrograde transport from the regenerating portion of the nerve is blocked, the reexpression of $p75^{NTR}$ is prevented. These findings are consistent with observations in *Aplysia* neurons, in which retrograde transport of novel factors regulates neuronal plasticity and activates intrinsic molecular signals that stimulate regeneration after axotomy (Ambron *et al.*, 1996; Gunstream *et al.*, 1995; Provelones *et al.*, 1997).

C. Signals Derived from Local Nonneuronal Cells

Local glial cells represent a potential source of molecular signals to axotomized neurons. As discussed in Section III, glial cells around axotomized neurons undergo a variety of reactive changes, most likely in response to signals deriving from the neurons themselves. Reactive glial cells have been shown to produce a variety of growth factors (Eddleston and Mucke, 1993), and the glial reaction to axotomy may well include the production of growth factors or other signals that are able to influence the neuronal response to the axotomy. This possibility is supported by evidence that the growth factor LIF is produced by glial cells in sympathetic ganglia in response to axotomy of sympathetic neurons and that LIF is able to regulate peptide expression in adult sympathetic neurons (Sun *et al.*, 1996; Zigmond *et al.*, 1996).

D. Influx of Ions and Small Molecules at the Site of Axonal Injury

Axotomy is a physical injury to neurons that is accompanied by the intracellular influx of high concentrations of extracellular ions such as Na^{+} and Ca^{2+} through the membrane opened at the site of axonal damage (Borgens *et al.*, 1980; Emery *et al.*, 1991; George *et al.*, 1995; Meiri *et al.*, 1981; Strautman *et al.*, 1990; Titmus and Faber, 1990; Xie and Barrett, 1991; Ziv and Spira, 1997). This influx of ions can lead to prolonged retrograde depolarization, as well as to a retrograde wave of increased Ca^{2+} concentration that can reach the parent neuronal cell body (Borgens *et al.*, 1980; Emery *et al.*, 1991; Meiri *et al.*, 1981). This so-called retrograde "injury current" can trigger ultrastructural changes in the cell body (Emery *et al.*, 1991). Ca^{2+} entering at the site of membrane injury after axotomy also activates local phospholipase A2, which is involved in resealing the plasma membrane (Yawo and Kuno, 1983, 1985), and induces the dedifferentiation of axonal segments into growth cones (Ziv and Spira, 1997). Ca^{2+} reaching the perikaryon in this manner may act as a second messenger to influence various intracellular processes (Ambron and Walters,

1996). However, Ca^{2+} can also generate arachidonic acid and free radical toxicity (Duncan, 1988), and sustained abnormal elevation of cytosolic Ca^{2+} concentrations lead to cell death (Choi, 1995; Siesjo, 1981). Blockade of voltage-gated Ca^{2+} channels prevents axotomy-induced retrograde cell death of sympathetic neurons (Rich and Hoolowell, 1990). Sustained depolarization induced retrogradely by the injury current might also displace Mg^{2+} from the ion channel of NMDA receptors, rendering neurons more vulnerable to the potentially excitotoxic effects of glutamatergic inputs or trauma-induced elevations in extracellular glutamate. Blockade of NMDA channels reduces neuronal death after CNS injury (Faden *et al.*, 1989). Thus, the signaling effects of ions entering through damaged membrane at the axotomy site may help trigger regenerative responses, but may also contribute to degenerative changes and cell death.

E. Endogenous Molecules Toxic to Neurons

A number of endogenously produced molecules are toxic to neurons. These molecules include not only excitotoxic amino acids, free radicals, nitric oxide, intracellular Ca^{2+}, but also certain proteins such as β-amyloid and some cytokines. Many of these molecules are generated during neural responses to injury and may be involved in mediating some of the cell death that occurs following damage to neural tissue (Blumbergs *et al.*, 1995; Choi, 1995; Faden *et al.*, 1989; Matson *et al.*, 1993; Merrill and Benveniste, 1996). The potential roles of Ca^{2+} and excitotoxic amino acids have been discussed (Section I,E). The roles of large molecules like amyloid (Blumbergs *et al.*, 1995) and inflammatory cytokines (Merrill and Benveniste, 1996) in the degenerative or regenerative neural response to axotomy are less well understood, but represent important topics of future investigations. It has also been suggested that the degenerative effects of axotomy may be mediated in part by alterations in the metabolic state of the neuron in response to physical damage (Farel, 1989).

F. Endogenous Molecules Protective to Neurons

A number of endogenous molecules have been identified in particular growth factors such as the neurotrophins (NGF, BDNF, NT-3), GDNF, FGF, and others, that can prevent axotomy-induced retrograde neuronal death or reverse axotomy-induced neuronal atrophy in a wide variety of neuronal populations (Cuello, 1994; Kobayashi *et al.*, 1997; Lucidi-Phillipi and Gage, 1993; Peterson *et al.*, 1996; Svendsen and Sofroniew, 1995). The roles of these molecules in mediating the degenerative and regenerative responses to axotomy are not certain. It was originally assumed that these molecules given exogenously acted to replace endogenous target-derived factors upon which the neurons were dependent on for survival. Although this may be the case during development, as discussed in Section IV, A, it is less likely to be so in adult animals in which these molecules appear to be acting pharmacologically rather than replacing physiological requirements. Interestingly, various growth factors can protect neurons from excitotoxic and metabolic cell death (Lam *et al.*, 1998; Matson *et al.*, 1993), raising the possibility that some of the retrograde degeneration seen after axotomy results from the loss of protection provided by target-derived growth factors. In addition to being protective or reversing atrophy, some growth factors have the capacity to promote axon sprouting and elongation in specific populations of neurons (Campenot, 1977; Heisenberg *et al.*, 1994; Kawaja and Gage, 1991). Thus, although the precise mechanisms of action are not always known, a valuable storehouse of information is becoming available regarding the pharmacological effects of specific molecules that are able to provide protection from the lethal effects of axotomy, reverse the atrophic effects of axotomy, or promote axon elongation in specific groups neurons.

V. SECONDARY TISSUE EVENTS AFTER INJURY MAY LEAD TO DELAYED AXOTOMY

Most axotomizing injuries of clinical significance in the CNS are caused by blunt compression, crush, or shearing forces (Hilton, 1995). After a blunt injury, a core area is likely to suffer irreversible damage caused by direct mechanical axotomy. Neural tissue in a penumbra zone surrounding this core will also be damaged and become edematous, with high levels of extracellular K^+, glutamate, cytokines, and an inflammatory reaction (Fig. 1). Treatment with methylprednisolone and other anti-inflammatory agents such as indomethacin significantly improve clinical outcome, and reduce the amount of tissue damage and the number of axotomized neurons, apparently by blocking components of the inflammatory response (Bracken *et al.*, 1990; Guth *et al.*, 1994; Naso *et al.*, 1995; Young, 1993; Young *et al.*, 1994). These observations suggest that a component of the inflammatory response can cause delayed "secondary" axotomizing injuries resulting in a greater degree of axonal discontinuity than that caused by the original direct physical trauma. Characterizing the extent and importance of such secondary axotomy and identifying the cellular and molecular events that mediate this reaction, as well as additional ways to block it, represent important areas of future research. Given the many problems with achieving effective axonal regrowth in the adult mammalian

CNS, it seems likely that preventing delayed axotomy caused by secondary tissue events in damaged neural tissue has the potential to have a substantial clinical impact far sooner than attempts to promote axon sprouting and regrowth after irreversible axotomy has occurred.

VI. STRATEGIES FOR FACILITATING REGENERATION FOLLOWING AXONAL INJURY IN THE CNS: IMPORTANCE OF DIFFERENCES IN NEURAL RESPONSES TO AXOTOMY AND CAPACITY FOR AXON REGROWTH

As summarized in this chapter, the combined interactions of many different cellular and molecular events determine which neurons survive axotomy and go on to regenerate and which neurons die or survive in an atrophic state and do not regenerate. Some of these events may be governed by properties intrinsic to the particular neurons, and some may be determined by the extracellular environment around the injured axon or around the neuronal perikarya. Regardless of how the reactions are triggered, differences in the way individual neurons react to axotomy will influence their ability to regenerate if the axons are presented with an environment that allows growth and elongation. Neurons that have died will obviously not regenerate, but not all neurons that survive axotomy will be able to sustain axon elongation equally well simply because they are presented with a favorable environment. Although many mature CNS neurons will regenerate their axons into peripheral nerve or embryonic CNS grafts, others clearly will not do so without additional stimuli (Bray *et al.*, 1991; Dooley and Aguayo, 1982; Dusart and Sotelo, Rossi *et al.*, 1995a; Villegas-Perez *et al.*, 1988). Indeed, neurons able to regenerate can lose this capacity with prolonged time after axotomy in the absence of a substrate for regrowth, as shown for motor neurons (Fu and Gordon, 1995). On the bright side, although severe and prolonged retrograde atrophy is likely to compromise the capacity of axotomized neurons to regrow their axons, the capacity for regrowth may be stimulated by providing appropriate growth factors.

Obtaining detailed information about the factors that regulate the responses of different types of neurons to axotomy will be fundamental to developing effective therapeutic strategies to induce axotomized neurons to survive and regenerate once their axons are provided with environmental cues that permit regrowth, elongation, and guidance. Available evidence indicates that the factors required may vary considerably among different populations of neurons.

Only through extensive experimental characterization will the conditions that allow effective regeneration for specific groups of neurons be identified.

ACKNOWLEDGMENTS

The author's work is supported by The Wellcome Trust, Action Research and MRC. This article was written in part while the author was on sabbatical leave in the Department of Neurobiology, UCLA Medical Center, Los Angeles, California.

REFERENCES

Ambron, R. T., and Walters, E. T. (1996). "Priming events and retrograde injury signals: A new perspective on the cellular and molecular biology of nerve regeneration." *Mol Neurobiol* 13:61–79.

Ambron, R. T., Zhang, X., Gunstream, J. D., Povelones, M., and Walters, E. T. (1996). Intrinsic injury signals enhance growth, survival, and excitability of Aplysia neurons. *J Neurosci* 16:7469–7477.

Barde, Y.-A., Edgar, D., and Thoenen, H. (1980). Sensory neurons in culture: Changing requirements for survival factors during embryonic development. *Proc Natl Acad Sci USA* 77:1199–1203.

Berlelaar, M., Clarke, D. B., Wang, Y. C., Bray, G. M., and Aguayo, A. J. (1994). Axotomy results in delayed death and apoptosis of retinal ganglion cells in adult rats. *J Neurosci* 14:4368–4374.

Blinzinger, K., and Kreutzberg, G. W. (1968). Displacement of synaptic terminals from regenerating motoneurons by microglial cells. *Z Zellforsch* 85:145–157.

Blumbergs, P. C., Scott, G., Manavis, J., Wainwright, H., Simpson, D. A., and McLean, A. J. (1995). Topography of axonal injury as defined by amyloid precursor protein and the sector scoring method in mild and severe closed head injury. *J Neurotrauma* 12:565–572.

Borgens, R. B., Joffe, L. F., and Cohen, M. J. (1980). Large and persistent electrical currents enter the transected lamprey spinal cord. *Proc Natl Acad Sci USA* 77:1209–1213.

Bracken, M. B., Shepard, M. J., Collins, W. F., Holford, T. R., Young, W., Baskin, D. S., and Eisenberg, H. M. (1990). A randomized, controlled trial of methylprednisolone in the treatment of acute spinal injury. *New Engl J Med* 322:1405–1411.

Bray, G. M., Villegas-Perez, M. P., Vidal-Sanz, M., Carter, D. A., and Aguayo, A. J. (1991). Neuronal and nonneuronal influences on retinal ganglion cell survival, axonal regrowth, and connectivity after axotomy. *Ann NY Acad Sci* 633:214–228.

Brodal, A. (1981). *Neurological anatomy in relation to clinical medicine.* New York: Oxford University Press.

Bussmann, K. A. V., and Sofroniew, M. V. (1998). Re-expression of $p75^{NTR}$ by adult motor neurons after axotomy is triggered by retrograde transport of a positive signal after regenerating peripheral nerve. *Neuroscience,* in press.

Cajal, Y. R. (1928) Cajal's degeneration and regeneration of the nervous system. In: DeFilipe, J., and Jones, E. G., eds. *History of Neuroscience No. 5.* New York: Oxford University Press, 1991.

Campenot, R. B. (1977). Local control of neurite development by nerve growth factor. *Proc Natl Acad Sci USA* 74:4516–4519.

Choi, D. W. (1995). Calcium: Still center-stage in hypotoxic-ischemic neuronal death. *Trend Neurosci* 18:58–60.

Cooper, J. D., Skepper, J. N., Berzaghi, M. P., Lindholm, D., and Sofroniew, M. V. (1996). Delayed death of septal cholinergic neurons after excitotoxic ablation of hippocampal neurons during early postnatal development in the rat. *Exp Neurol* 139:143–155.

Cragg, B. G. (1970). What is the signal for chromatolysis? *Brain Res* 23:1–21.

Cuello, A. C. (1994). Trophic factor therapy in the adult CNS: Remodelling of injured basalo-cortical neurons. *Prog Brain Res* 100:213–221.

Cuello, A. C., and Sofroniew, M. V. (1984). The anatomy of the CNS cholinergic neurons. *Trends Neurosci* 7:74–78.

Curtis, R., Adryan, K. M., Zhu, Y., Harkness, P. J., Lindsay, R. M., and DiStefano, P. S. (1993). Retrograde axonal transport of ciliary neurotrophic factor is increased by peripheral nerve injury. *Nature* 365:253–255.

Curtis, R., Scherer, S. S., Somogyi, R., Adryan, K. M., Ip, N. Y., Zhu, Y., Lindsay, R. M., and DiStefano, P. S. (1994). Retrograde axonal transport of LIF is increased by peripheral nerve injury: Correlation with increased LIF expression in distal nerve. *Neuron* 12:191–204.

Daitz, H. M., and Powell, T. P. S. (1954). Studies on the connections of the fimbria fornix system. *J Neurol Neurosurg Psychiat* 17:75–82.

de Bilbao, F., and Dubois-Dauphin, M. (1996). Time course of axotomy-induced apoptotic cell death in facial motoneurons of neonatal wild type and bcl-2 transgenic mice. *Neuroscience* 71:1111–1119.

DiStefano, P. S., Friedman, B., Radziejewski, C., Alexander, C., Boland, P., Schick, C. M., Lindsay, R. M., and Wiegand, S. J. (1992). The neurotrophins BDNF, NT-3, and NGF display distinct patterns of retrograde axonal transport in peripheral and central neurons. *Neuron* 8:983–993.

Dohanics, J., Hoffman, G. E., and Verbalis, J. G. (1996). Chronic hyponatremia reduces survival of magnocellular vasopressin and oxytocin neurons after axonal injury. *J Neurosci* 16:2373–2380.

Dooley, J. M., and Aguayo, A. J. (1982). Axonal elongation from cerebellum into PNS grafts in the adult rat. *Ann Neurol* 12:221. Abstract.

Dubois-Dauphin, M., Frankowski, H., Tsujimoto, Y, Huarte, J., and Martinou, J.-C. (1994). Neonatal motoneurons overexpressing the *bcl-2* protooncogene in transgenic mice are protected from axotomy-induced cell death. *Proc Natl Acad Sci USA* 91:3309–3313.

Duncan, C. J. (1988). The role of phospholipase A2 in calcium-induced damage in cardiac and skeletal muscle. *Cell Tiss Res* 253:457–462.

Dusart, I., and Sotelo, C. (1994). Lack of Purkinje cell loss in adult rat cerebellum following protracted axotomy: Degenerative changes and regenerative attempts of the severed axons. *J Comp Neurol* 347:211–232.

Dusart, I., and Sotelo, C. (1997). Purkinje cell survival and axonal regeneration are age dependent: An *in vitro* study. *J Neurosci* 17:3710–3726.

Eddleston, M., and Mucke, L. (1993). Molecular profile of reactive astrocytes—implications for their role in neurological disease. *Neuroscience* 54:15–36.

Emery, D. G., Lucas, J. H., and Gross, G. W. (1991). Contributions of sodium and chloride to ultrastructural damage after dendrotomy. *Exp Brain Res* 86:60–72.

Eng, L. F., and Ghirnikar, R. S. (1994). GFAP and astrogliosis. *Brain Pathol* 4:229–237.

Ernfors, P., Henschen, A., Olson, L., and Persson, H. (1989). Expression of nerve growth factor receptor mRNA is developmentally regulated and increased after axotomy in rat spinal cord motoneurons. *Neuron* 2:1605–1613.

Faden, A. I., Demediuk, P., Scott Panter, S., and Vink, R. (1989). The role of excitatory amino acids and NMDA receptors in traumatic brain injury. *Science* 244:798–800.

Farel, P. B. (1989). Naturally occurring cell death and differentiation of developing spinal motoneurons following axotomy. *J Neurosci* 9:2103–2113.

Fischer, W., and Bjorklund, A. (1991). Loss of AChE- and NGFr-labeling precedes neuronal death of axotomized septal-diagonal band neurons: reversal by intraventricular NGF infusion. *Exp Neurol* 113:93–108.

Fry, F. J., and Cowan, W. M. (1972). A study of retrograde cell degeneration in the lateral mamillary nucleus of the cat, with special reference to the role of axonal branching in the preservation of the cell. *J Comp Neurol* 144:1–24.
Fu, S. Y., and Gordon, T. (1995). Contributing factors to poor functional recovery after delayed nerve repair: Prolonged axotomy. *J Neurosci* 15:3875–3876.
George, E. B., Glass, J. D., and Griffin, J. W. (1995). Axotomy-induced axonal degeneration is mediated by calcium influx through ion-specific channels. *J. Neurosci.* 15:6445–6452.
Graeber, M. B., and Kreutzber, G. W. (1986). Astrocytes increase in glial fibrillary acidic protein during retrograde changes of facial motor neurons. *J Neurocytol* 15:363–373.
Graeber, M. B., Tezlaff, W., Streit, W. J., and Kreutzberg, G. W. (1988). Microglial cells but not astrocytes undergo mitosis following rat facial nerve axotomy. *Neurosci Lett* 85:317–321.
Greene, L. A. (1977). Quantitative *in vitro* studies on the nerve growth factor (NGF) requirement of neurons. II. Sensory neurons. *Dev Biol* 58:106–113.
Grimes, M., Zhou, J., Li, Y., Holtzman, D., and Mobley, W. C. (1993). Neurotrophin signalling in the nervous system. *Semin Neurosci* 5:239–247.
Gunstream, J., Castro, G. A., and Walters, E. T. (1995). Retrograde transport of plasticity signals in Aplysia sensory neurons following axonal injury. *J Neurosci* 15:439–448.
Guth, L., Zhang, Z., and Roberts, E. (1994). Key role for pregnenolone in combination therapy that promotes recovery after spinal cord injury. *Proc Natl Acad Sci USA* 91:12308–12312.
Hartikka, J., and Hefti, F. (1988). Development of septal cholinergic neurons in culture: Plating density and glial cells modulate effects of NGF on survival, fiber growth, and expression of transmitter-specific enzymes. *J Neurosci* 8:2967–2985.
Hefti, F. (1986). Nerve growth factor promotes survival of septal cholinergic neurons after fimbrial transections. *J Neurosci* 6:2155–2162.
Heisenberg, C.-P., Cooper, J. D., Berke, J., and Sofroniew, M. V. (1994). NMDA potentiates NGF-induced sprouting of septal cholinergic neurons. *NeuroReport* 5:413–416.
Herman, J. P., Marciano, F. F., Wiegand, S. J., and Gash, D. M. (1987). Selelctive cell death of magnocellular vasopressin neurons in neurohypophysectomized rats following chronic administration of vasopressin. *J Neurosci* 7:2564–2575.
Hilton, G. (1995). Diffuse axonal injury. *J Trauma Nurs* 2:7–12.
Isacson, O., and Sofroniew, M. V. (1992). Effects of neuronal loss or replacement on the organization of intrinsic and afferent neural systems: Neurochemical and morphological evidence for plasticity in the injured adult cerebral neocortex. *Exp Neurol* 117:151–175.
Johnson, E. M., Rich, K. M., and Yip, H. K. (1986). The role of NGF in sensory neurons in vivo. *Trend Neurosci* 9:33–37.
Kawaja, M. D., and Gage, F. H. (1991). Reactive astrocytes are substrates for the growth of adult CNS axons in the presence of elevated levels of nerve growth factor. *Neuron* 7:1019–1030.
Kew, J. N. C., Smith, D. W., and Sofroniew, M. V. (1996). Nerve growth factor withdrawal induces the apoptotic death of developing septal cholinergic neurons *in vitro:* Protection by cAMP analogue and high K^+. *Neuroscience* 70:329–339.
Knusel, B., Michel, P. P., Schwaber, J. S., and Hefti, F. (1990). Selective and nonselective stimulation of central cholinergic and dopaminergic development *in vitro* by nerve growth factor, basic fibroblast growth factor, epidermal growth factor, insulin and the insulin-like growth factors I and II. *J Neurosci* 10:558–570.
Kobayashi, N. R., Fan, D.-P., Giehl, K. M., Bedard, A. M., Wiegand, S. J., and Tetzlaff, W. (1997). BDNF and NT-4/5 prevent atrophy of rat rubrospinal neurons after cervical axotomy, stimulate GAP-43 and Tα1 tubulin mRNA expression, and promote axonal regeneration. *J Neurosci* 17:9583–9595.
Kordower, J. H., Burke-Watson, M., Roback, J. D., and Wainer, B. H. (1993). Stability of septohippocampal neurons following excitotoxic lesions of the rat hippocampus. *Exp Neurol* 117:1–16.

Korsmeyer, S. J. (1995). Regulators of cell death. *Trends Genet* 11:101–105.

Kreutzberg, G. W. (1982). Acute neural reaction to injury. In Nicholls, J. G., ed. *Repair and regeneration of the nervous system*. Berlin: Springer, pp. 57–69.

Kreutzberg, G. W. (1995). Reaction of the neuronal cell body to axonal damage. In Waxman, S. G., Kocsis, J. D., and Stys, P. K., ed. *The axon: Structure, function, and pathology*. New York: Oxford University Press, pp. 355–374.

Kromer, L. F. (1987). Nerve growth factor treatment after brain injury prevents neuronal death. *Science* 235:214–216.

Laiwand, R., Werman, R., and Yarom, Y. (1987). Time course and distribution of motoneuronal loss in the dorsal motor vagal nucleus of guinea pig after cervical vagotomy. *J Comp Neurol* 256:527–537.

Laiwand, R., Werman, R., and Yarom, Y. (1988). Electrophysiology of degenerating neurones in the vagal motor nucleus of the guinea-pig following axotomy. *J Physiol* 404:749–766.

Lam, H. H. D., Horner, C. H., Berke, J., Cooper, J. D., and Sofroniew, M. V. (1998). Nerve growth factor protects basal forebrain cholinergic neurons from glutamate receptor-mediated toxicity. Submitted.

Lams, B. E., Isacson, O., and Sofroniew, M. V. (1988). Loss of transmitter-associated enzyme staining following axotomy does not indicate death of brainstem cholinergic neurons. *Brain Res* 475:401–406.

Lieberman, A. R. (1971). The axon reaction: A review of the principal features of perikaryal response to axonal injury. *Int Rev Neurobiol* 14:49–124.

Lindholm, D., Heumann, R., Meyer, M., and Thoenen, H. (1987). Interleukin-1 regulates synthesis of nerve growth factor in non-neuronal cells of rat sciatic nerve. *Nature* 330:658–659.

Lindsay, R. M. (1988). Nerve growth factors (NGF, BDNF) enhance axonal regeneration but are not required for survival of adult sensory neurons. *J Neurosci* 8:2394–2405.

Lucidi-Phillipi, C., and Gage, F. H. (1993). Functions and applications of neurotrophic molecules in the adult central nervous system. *Semin Neurosci* 5:269–277.

Matson, M. P., Cheng, B., and Smith-Swintosky, V. L. (1993). Neurotrophic factor mediated protection from excitotoxicity and disturbances in calcium and free radical metabolism. *Semin Neurosci* 5:295–307.

Maysinger, D., Piccardo, P., and Cuello, A. C. (1995). Microencapsulation and the grafting of genetically transformed cells as therapeutic strategies to rescue degenerating neurons of the CNS. *Rev Neurosci* 6:15–33.

McBride, R. L., Feringa, E. R., and Smith, B. E. (1988). The fate of prelabeled Clarke's column neurons after axotomy. *Exp Neurol* 102:236–243.

Meiri, H., Spira, M. E., and Parnas, I. (1981). Membrane conductance and action potentials of a regenerating axonal tip. *Science* 211:709–712.

Merline, M., and Kalil, K. (1990). Cell death of corticospinal neurons is induced by axotomy before but not after innervation of spinal targets. *J Comp Neurol* 296:506–516.

Merrill, J. E., and Benveniste, E. N. (1996). Cytokines in inflammatory brain lesions: Helpful and harmful. *Trends Neurosci* 19:311–338.

Montero, C. N., and Hefti, F. (1988). Rescue of lesioned septal cholinergic neurons by nerve growth factor: Specificity and requirements for chronic treatment. *J Neurosci* 8:2986–2999.

Naso, W. B., Perot, P. L. J., and Cox, R. D. (1995). The neuroprotective effect of methylprednisolone in rat spinal hemisection. *Neurosci Lett* 189:176–178.

Naumann, T., Kermer, P., and Frotscher, M. (1994). Fine structure of rat septohippocampal neurons. III. Recovery of choline acetyltransferase immunoreactivity after fimbria–fornix transection. *J Comp Neurol* 350:161–170.

Naumann, T., Peterson, G. M., and Frotscher, M. (1992). Fine structure of rat septohippocampal neurons: II. A time course analysis following axotomy. *J Comp Neurol* 325:219–242.

Nissl, F. (1892). Uber die veranderungen der ganglienzellen am facialiskern des kaninchens nach ausreissung der nerven. *Allg Z Psychiat* 48:197–198.
O'Brien, T. S., Svendsen, C. N., Isacson, O., and Sofroniew, M. V. (1990). Loss of true blue labelling from the medial septum following transection of the fimbria–fornix: Evidence for the death of cholinergic and noncholinergic neurons. *Brain Res* 508:249–256.
O'Connor, W. J. (1947). Atrophy of the supraoptic and paraventricular nuclei after interruption of the pituitary stalk in dogs. *O J Exp Physiol* 34:29–42.
Oppenheim, R. W., Houenou, L. J., Johnson, J. E., Lin, L. F., Lo, A. C., Newsome, A. L., Prevette, D. M., and Wang, S. (1995). Developing motor neurons rescued from programmed and axotomy-induced cell death by GDNF. *Nature* 373:344–346.
Persson, H. (1993). Neurotrophin production in the brain. *Semin Neurosci* 5:227–237.
Peterson, D. A., Lucidi-Phillipi, C. A., Murphy, D. P., Ray, J., and Gage, F. H. (1996). Fibroblast growth factor-2 protects entorhinal layer II glutamatergic neurons from axotomy-induced death. *J Neurosci* 16:886–898.
Prendergast, J., and Stelzner, D. J. (1976). Changes in the magnocellular portion of the red nucleus following thoracic hemisection in the neonatal and adult rat. *J Comp Neurol* 166:163–172.
Provelones, M., Tran, K., Thanos, D., and Ambron, R. T. (1997). An NF-κB-like transcription factor in axoplasm is rapidly inactivated after nerve injury in Aplysia. *J Neurosci* 17:4915–4920.
Purves, D., and Lichtman, J. W. (1978). Formation and maintenance of synaptic connections in autonomic ganglia. *Physiol Rev* 58:821–862.
Quigley, H. A., Nickells, R. W., Kerrigan, L. A., Pease, M. E., Thibault, D. J., and Zack, D. J. (1995). Retinal ganglion cell death in experimental glaucoma and after axotomy occurs by apoptosis. *Invest Ophthalmol Vis Sci* 36:774–786.
Reis, D. J., Gilad, G., Pickel, V. M., and Joh, T. H. (1978a). Reversible changes in the activities and amounts of tyrosine hydroxylase of dopamine neurons of the substantia nigra in response to axonal injury as studied by immunochemical and immunocytochemical methods. *Brain Res* 144:325–342.
Reis, D. J., Ross, R. A., Gilad, G., and Joh, T. H. (1978b). Reaction of central catecholaminergic neurons to injury: Model systems for studying the neurobiology of central regeneration and sprouting. In: Cotman, C. W., ed. *Neuronal plasticity*. New York: Raven Press, pp. 197–226.
Rich, K. M., and Hoolowell, J. P. (1990). Flunarizine protects neurons from death after axotomy or NGF deprivation. *Science* 248:1419–1421.
Rossi, F., Jankovski, A., and Sotelo, C. (1995a). Differential regenerative response of Purkinje cell and inferior olivary axons confronted with embryonic grafts: Environmental cues versus intrinsic neuronal determinants. *J Comp Neurol* 359:663–677.
Rossi, F., Jankovski, A., and Sotelo, C. (1995b). Target neuron controls the integrity of afferent axon phenotype: A study on the Purkinje cell-climbing fibre system in cerebellar mutant mice. *J Neurosci* 15:2040–2056.
Rossiter, J. P., Riopelle, R. J., and Bisby, M. A. (1996). Axotomy-induced apoptotic cell death of neonatal rat facial motoneurons: Time course analysis and relation to NADPH-diaphorase activity. *Exp Neurol* 138:33–44.
Rothstein, J. D., Dykes-Hoberg, M., Pardo, C. A., Bristol, L. A., Jin, L., Kunci, R. W., Kanai, Y., Hediger, M. A., Wang, Y., Schielke, J. P., and Welty, D. F. (1996). Knockout of glutamate transporters reveals a major role for astroglial transport in excitotoxicity and clearance of glutamate. *Neuron* 16:675–686.
Sendtner, M., Kreutzberg, G. W., and Thoenen, H. (1990). Ciliary neurotrophic factor prevents the degeneration of motor neurons after axotomy. *Nature* 345:440–441.
Siesjo, B. K. (1981). Cell damage in the brain: A speculative synthesis. *J Cereb Blood Flow Metab* 1:155–185.

Sofroniew, M. V., and Cooper, J. D. (1993). Neurotrophic mechanisms and neuronal degeneration. *Semin Neurosci* 5:285–294.

Sofroniew, M. V., Cooper, J. D., Svendsen, C. N., Crossman, P., Ip, N. Y., Lindsay, R. M., Zafra, F., and Lindholm, D. (1993). Atrophy but not death of adult septal cholinergic neurons after ablation of target capacity to produce mRNAs for NGF, BDNF and NT-3. *J Neurosci* 13:5263–5276.

Sofroniew, M. V., Galletly, N. P., Isacson, O., and Svendsen, C. N. (1990). Survival of adult basal forebrain cholinergic neurons after loss of target neurons. *Science* 247:338–342.

Sofroniew, M. V., and Isacson, O. (1988). Distribution of degeneration of cholinergic neurons in the septum following axotomy in different portions of the fimbria–fornix: A correlation between degree of cell loss and proximity of neuronal somata to the lesion. *J Chem Neuroanat* 1:327–337.

Sofroniew, M. V., and Pearson, R. C. A. (1985). Degeneration of cholinergic neurons in the basal nucleus following kainic or n-methyl-D-aspartic acid application to the cerebral cortex in the rat. *Brain Res.* 339:186–189.

Sofroniew, M. V., Pearson, R. C. A., Eckenstein, F., Cuello, A. C., and Powell, T. P. S. (1983). Retrograde changes in the basal forebrain of the rat following cortical damage. *Brain Res* 289:370–374.

Sofroniew, M. V., Pearson, R. C. A., and Powell, T. P. S. (1987). The cholinergic nuclei of the basal forebrain of the rat: Normal structure, development and experimentally induced degeneration. *Brain Res* 411:310–331.

Strautman, A. F., Cork, R. J., and Robinson, K. R. (1990). The distribution of free calcium in transected spinal axons and its modulation by applied electrical fields. *J Neurosci* 10:3564–3575.

Streit, W. J., Graeber, M. B., and Kreutzberg, G. W. (1989). Peripheral nerve lesion produces increased levels of major histocompability complex antigens in the central nervous system. *J Neuroimmunol* 21:117–123.

Sun, Y., Landis, S. C., and Zigmond, R. E. (1996). Signals triggering the induction of leukemia inhibitory factor in sympathetic superior cervical ganglia and their nerve trunks after axonal injury. *Mol Cell Neurosci* 7:152–163.

Svendsen, C. N., Kew, J. N. C., Staley, K., and Sofroniew, M. V. (1994). Death of developing septal cholinergic neurons following NGF withdrawal in vitro: Protection by protein synthesis inhibition. *J Neurosci* 14:75–87.

Svendsen, C. N., and Sofroniew, M. V. (1995). Do CNS neurons require target-derived neurotrophic support for survival throughout life and aging? *Perspect Dev Neurobiol* 3:131–140.

Tanaka, K., Watase, K., Manabe, T., Yamada, K., Watanabe, M., Takahashi, K., Iwama, H., Nishikawa, T., Ichihara, N., Kikuchi, T., Okuyama, S., Kawashima, N., Hori, S., Takimoto, M., and Wada, K. (1997). Epilepsy and exacerbation of brain injury in mice lacking the glutamate transporter GLT-1. *Science* 276:1699–1702.

Tetzlaff, W., Alexander, S. W., Miller, F. D., and Bisby, M. A. (1991). Response of facial and rubrospinal neurons to axotomy: Changes in mRNA expression for cytoskeletal proteins and GAP-43. *J Neurosci* 11:2528–2544.

Tetzlaff, W., Bisby, M. A., and Kreutzberg, G. W. (1988a). Changes in cytoskeletal proteins in the rat facial nucleus following axotomy. *J Neurosci* 8:3181–3189.

Tetzlaff, W., Graeber, M. B., Bisby, M. A., and Kreutzberg, G. W. (1988b). Increased glial fibrillary acidic protein synthesis in astrocytes during retrograde reaction of the rat facial nucleus. *Glia* 1:90–95.

Thanos, S., Mey, J., and Wild, M. (1993). Treatment of the adult retina with microglia-suppressing factors retards axotomy-induced neuronal degradation and enhances axonal regeneration *in vivo* and *in vitro*. *J Neurosci* 13:455–466.

Titmus, M. J., and Faber, D. S. (1990). Axotomy-induced alterations in the electrophysiological characteristics of neurons. *Progress in Neurobiology* 35:1–51.

Tuszynski, M. H., Armstrong, D. M., and Gage, F. H. (1990). Basal forebrain cell loss following fimbria/fornix transection. *Brain Res* 508:241–248.

Tuszynski, M. H., and Gage, F. H. (1995). Bridging grafts and transient nerve growth factor infusions promote long-term central nervous system neuronal rescue and partial functional recovery. *Proc Natl Acad Sci USA* 92:4621–4625.

Villegas-Perez, M. P., Vidal-Sanz, M., Bray, G. M., and Aguayo, A. J. (1988). Influences of peripheral nerve grafts on the survival and regrowth of axotomized retinal ganglion cells in adult rats. *J Neurosci* 8:265–280.

Villegas-Perez, M. P., Vidal-Sanz, M., Bray, G. M., Rasminsky, M., Bray, G. M., and Aguayo, A. J. (1993). Rapid and protracted phases of retinal ganglion cell loss follow axotomy in the optic nerve of adult rats. *J Neurobiol* 24:23–36.

Voyvodic, J. T. (1989). Peripheral target regulation of dendritic geometry in the rat superior cervical ganglion. *J Neurosci* 9:1997–2010.

Wilcox, B. J., Applegate, M. D., Portera-Cailliau, C., and Koliatsos, V. E. (1995). Nerve growth factor prevents apoptotic cell death in injured central cholinergic neurons. *J Comp Neurol* 359:573–585.

Williams, L. R., Varon, S., Peterson, G. M., Wictorin, K., Fischer, W., Bjorklund, A., and Gage, F. H. (1986). Continuous infusion of nerve growth factor prevents basal forebrain neuronal death after fimbria fornix transection. *Proc Natl Acad Sci USA* 83:9231–9235.

Wood, S. J., Pritchard, J., and Sofroniew, M. V. (1990). Re-expression of nerve growth factor receptor after axonal injury recapitulates a developmental event in motor neurons: Differential regulation when regeneration is allowed or prevented. *Eur J Neurosci* 2:650–657.

Xie, X.-Y., and Barrett, J. N. (1991). Membrane resealing in cultured rat septal neurons after neurite transection: Evidence for enhancement by $Ca2^{+}$-triggered protease activity and cytoskeletal disassembly. *J Neurosci* 11:3257–3267.

Yan, Q., Elliot, J., and Snider, W. D. (1992). Brain-derived neurotrophic factor rescues spinal motor neurons from axotomy-induced cell death. *Nature* 360:753–755.

Yawo, H., and Kuno, M. (1983). How a nerve fiber repairs its cut end: Involvement of phospholipase A2. *Science* 222:1351–1353.

Yawo, H., and Kuno, M. (1985). Calcium dependence of membrane resealing at the cut end of the cockroach giant axon. *J Neurosci* 5:1626–1632.

Young, W. (1993). Secondary injury mechanisms in acute spinal cord injury. *J Emerg Med* 11(Suppl. 1):13–22.

Young, W., Kume-Kick, J., and Constantini, S. (1994). Glucocorticoid therapy of spinal cord injury. *Ann NY Acad Sci* 743:241–263.

Zigmond, R. E., Hyatt-Sachs, H., Mohney, R. P., Schreiber, R. C., Shadiack, A. M., Sun, Y., and Vaccriello, S. A. (1996). Changes in neuropeptide phenotype after axotomy of adult peripheral neurons and the role of leukemia inhibitory factor. *Perspect Dev Neurobiol* 1:1–7.

Ziv, N. E., and Spira, M. E. (1997). Localized and transient elevations of intracellular Ca^{++} induce the dedifferentiation of axonal segments into growth cones. *J Neurosci* 17:3568–3579.

CHAPTER 2

Axonal Responses to Injury

RAYMOND J. GRILL AND MARK H. TUSZYNSKI

Department of Neurosciences, University of California, San Diego, La Jolla, California and Department of Neurology, Veterans Affairs Medical Center, San Diego, California

Injured neurons of the adult CNS exhibit little spontaneous capacity for regeneration compared to lesioned PNS axons. Differences both *intrinsic* and *extrinsic* to neurons may account for this dichotomy. The differential expression of growth-initiating genes, axonal structural genes, cell adhesion molecules, extracellular matrix molecules, neurotrophic factors, cytokines, and white matter-based inhibitors in specific combinations determines whether axonal growth is supported after injury. Growth is further modulated by responses of the local glial environment. In recent years it has been demonstrated that injured CNS axons are capable of growth when their environment is manipulated. Specifically, the provision of a permissive growth substrate in the form of peripheral nerve grafts, artificial matrices or neutralizers of white-matter based inhibitors augments the *distance* over which axons will grow, and the administration of neurotrophic factors increases the *number* of growing axons. These studies suggest that CNS neurons are not *intrinsically* incapable of new axonal growth after injury, but lack both the growth-inducing signals that are available to injured PNS axons, and that are handicapped by the presence of molecules in the local milieu that do not support new growth. This chapter will examine the nature of CNS axonal responses to injury, and will explore how the modification of the injured CNS environment may promote regrowth of injured axons.

I. INTRODUCTION

The axonal response to injury has been a subject of study for more than a century. The morphological features of axonal injury and degeneration were elegantly described by Ramon y Cajal in the early part of this century (Ramon y Cajal, 1928). More recent studies focus on molecular correlates of axonal injury, including the effects of injury on proteins of the axonal cytoskeleton (Banati *et al.*, 1993b; Banik *et al.*, 1997; Bignami *et al.*, 1981; Bisby and Tetzlaff, 1992; Geisert and Frankfurter, 1989; Goldstein *et al.*, 1988; Mikucki and Oblinger, 1991; Povlishock and Pettus, 1996; Schlaepfer and Hasler, 1979; Schlaepfer and Micko, 1978; Schlaepfer and Zimmerman, 1981; Trojanowski *et al.*, 1984) and interactions of injured axons with the surrounding cellular environment (Fitch and Silver, 1997a, 1997b; George and Griffin, 1994; La Fleur *et al.*, 1996; Liuzzi and Tedeschi, 1992; McKeon *et al.*, 1991; Pindzola *et al.*, 1993; Rudge and Silver, 1990; Savio and Schwab, 1989). Both early and more recent studies have attempted to elucidate why and how the ultimate outcome of axonal injury differs so markedly in the central nervous system (CNS) and the peripheral nervous system (PNS): axons in the PNS frequently exhibit efficient and successful regeneration after injury, whereas CNS axons abortively sprout but fail to regenerate after injury.

This dichotomy between the success of axonal regeneration in the adult CNS and the adult PNS may be explained by features both *intrinsic* and *extrinsic* to the axon. For example, peripheral neurons exhibit changes after injury in expression of developmentally associated axonal growth genes, upregulating the production of proteins such as tubulin isoforms, microtubule-associated proteins (MAPs) (Book *et al.*, 1996; Bush *et al.*, 1996; Chambers and Muma, 1997; Nunez and Fischer, 1997; Tonge *et al.*, 1996) and the growth-associated protein GAP-43 (Bisby and Tetzlaff, 1992; Geisert and Frankfurter, 1989; Kobayashi *et al.*, 1997; Mikucki and Oblinger, 1991). CNS axons demonstrate more attenuated changes in gene expression (Fournier and McKerracher, 1995; Kalil and Skene, 1986; Mikucki and Oblinger, 1991), suggesting that neurons of central origin may be *intrinsically* less capable of mounting successful regeneration. Alternatively, influences *extrinsic* to the axon may modulate axonal growth after injury including: (1) the molecular constituency of the extracellular matrix, which may be growth-supportive or inhibitory, (2) the presence of cell adhesion molecules on axons and cells of the surrounding environment with which injured axons may interact in a growth-promoting or a growth-inhibiting fashion, (3) the availability of neurotrophic factors to stimulate and guide axonal growth, (4) the influx of cytokines, inflammatory cells, and phagocytic cells, (5) the supportive or inhibitory nature of glia at the injury site, and (6) the growth-modulating effects of molecules associated with CNS white matter.

The response of the axon to injury is a vital area of research as new approaches are developed to enhance the regeneration of central neurons. This chapter will examine the axonal response to injury, beginning with a description of the classical morphological response to injury and followed by a description of the intracellular and extracellular events that influence axonal degeneration and modulate axonal regeneration.

II. AXONAL DEGENERATION

Ramon y Cajal described a set of degenerative events that occur in axons after injury in the PNS (Ramon y Cajal, 1928). Characteristic changes occur in the axon segment that is distal to the injury site, termed Wallerian or *anterograde degeneration*, and in the nerve segment proximal to the injury site that is still attached to the cell body.

A. Axotomy in the Peripheral Nervous System

1. Axonal Responses Distal to the Transection Site

Wallerian degeneration in the PNS is a rapid process in which the distal axonal and myelin segment, separated from the cell body, undergoes dissolution and absorption by the surrounding cellular environment (Figs. 1–3). Initial breakdown occurs in the axoplasm. This axonal breakdown is observed as a granular disintegration of the axonal cytoskeleton through disorder and disassembly of microtubule and neurofilament proteins, and is usually complete from 3 to 10 days post-injury (Bignami *et al.*, 1981; George and Griffin, 1994; Trojanowski *et al.*, 1984). Some neurofilament breakdown products can be observed up to 21 days posttransection (Trojanowski *et al.*, 1984). Within hours to days, the myelin sheath also swells, then retracts, becomes fragmented, and finally breaks down into fatty material (oily droplets) that is deposited in the space between axons 10 to 14 days post-injury (Lassmann *et al.*, 1978). Macrophages responsible for the phagocytosis of both the myelin breakdown products and the remaining axonal components appear in the distal axon as early as 1 to 3 days after injury (Avellino *et al.*, 1995) reviewed in (Perry and Brown, 1992). Schwann cells occupying the neurilemmal sheath divide and become hypertrophic, filling up the space formerly occupied by axons and contributing to phagocytosis of debris (Liu *et al.*, 1995; Reichert *et al.*, 1994). The infiltration of macrophages and other leukocytes reaches its peak between 12 to 21 days after injury (Avellino *et al.*, 1995; Ramon y Cajal, 1928), coinciding with the period of axonal breakdown and absorption. By the end of the

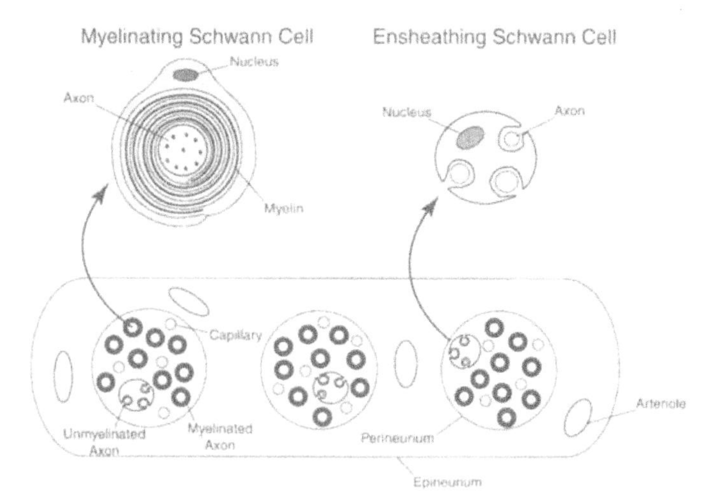

FIGURE 1 *Structure of the Peripheral Nerve.* Diagrammatic representation of a peripheral nerve cross-section discloses the regular structural elements of the epineurium, perineurium, and neurilemmal sheath. The entire peripheral nerve is covered by the epineurium, composed of collagenous connective tissue. Within the epineurium are hundreds of nerve fascicles, each encircled by a perineurium; the perineurial sheath is made up of tightly woven layers of collagen. Each axon is myelinated by a single Schwann cell; this axon/Schwann cell unit is contained within a basal lamina, forming a neurilemmal unit. More than one unmyelinated axons can be ensheathed (but not myelinated) by a single Schwann cell, and these units are also contained within a basal lamina.

second week, the majority of lipid fragments have been removed through both resorption and phagocytosis. By one month after injury, most distal axons and myelin have disappeared completely and the motor units that were formerly innervated have degenerated.

2. Axonal Responses Proximal to Transection

The axon proximal to a transection site in the PNS also undergoes a number of characteristic changes, including: (1) breakdown of the myelin sheath for up to several nodes of Ranvier proximal to the lesion site, (2) the development of granularity in injured axons that rapidly become necrotic, (3) early infiltration of phagocytic cells that engulf and digest the remains of necrotic axons, and (4) a variable degree of chromatolysis in the parent neuron, depending on the distance of axotomy from the cell body and the number of sustaining collaterals that are left unaffected by the lesion (Grafstein, 1975; Lieberman, 1971; Torvik, 1976) (see also Chapter 1). These events begin with the seepage of axoplasm from the proximal axonal segment. This seepage continues until

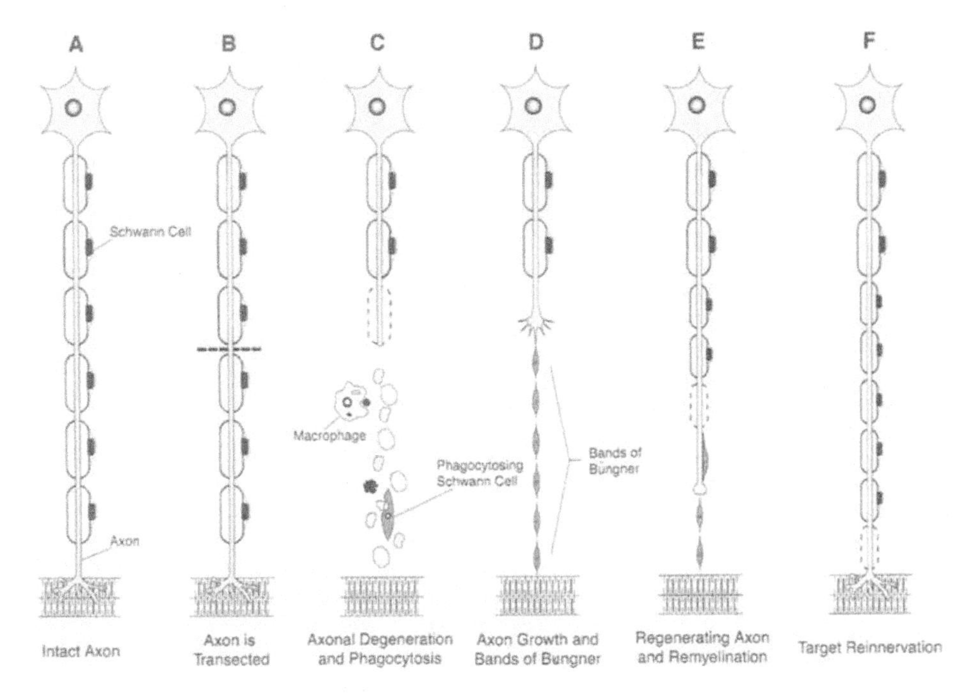

FIGURE 2 *Effects of Axotomy in the PNS.* (**A**) An intact motor axon is myelinated by several Schwann cells and projects to muscle. (**B**) The nerve is transected at a single site. (**C**) The proximal axonal segment undergoes limited degeneration to the previous node of Ranvier. Myelin breakdown occurs. The distal axonal segment begins to undergo Wallerian degeneration. The axon becomes granular, then disintegrates, and is phagocytosed by resident and recruited macrophages as well as activated Schwann cells. (**D**) Following Wallerian degeneration of the distal axonal segment, the remaining Schwann cells divide and align longitudinally within the basal lamina tubes. Growth cones from regenerating axons extend along the Schwann cell column, growing along the Schwann cell membranes and basal lamina. Schwann cells then begin to wrap regenerating axons as the growth cones extend along Schwann cell columns. (**E**) As the regenerating axons reinnervate the target, Schwann cells continue to ensheath or myelinate, forming substantially shorter internodal distances than found in the unlesioned state.

the severed end of the segment seals due to fusion of the damaged cell membrane. Within one day of injury, axons proximal to the site of injury begin to form bulbous swellings at their distal tips due to the accumulation of newly arrived axoplasm containing cytoskeletal components, mitochondria, and elements of smooth endoplasmic reticulum (Meller, 1987). The cytoskeletal components include neurofilaments, microtubules, and microfilaments synthesized in the cell body and transported down the axon by slow axonal transport (Fournier and McKerracher, 1995; Vallee and Bloom, 1991). Unable to move further down the axon, these proteins accumulate within the tips of the cut

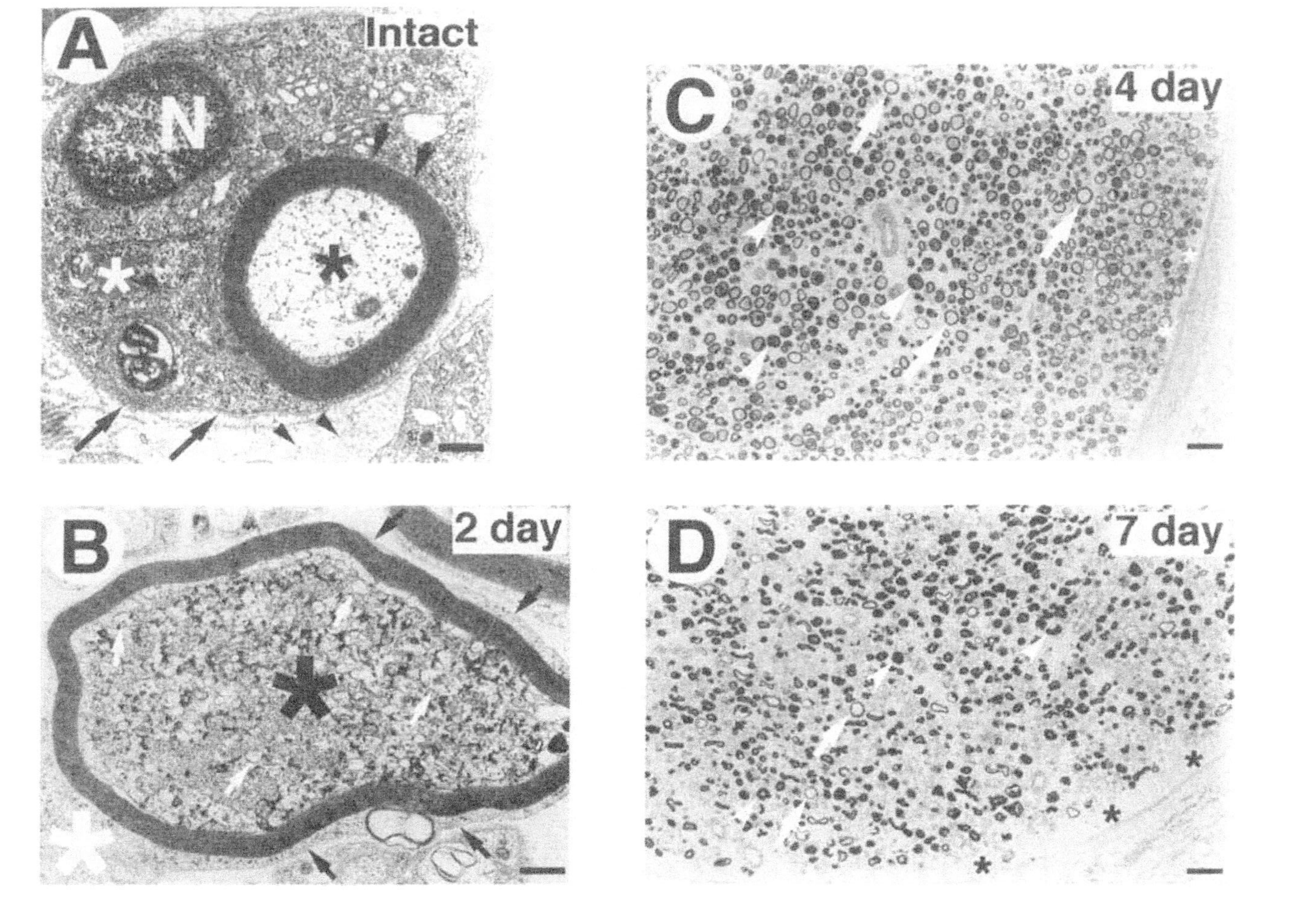
A
Intact
N
B
2 day
C
4 day
D
7 day

axons, forming clublike structures described by Ramon y Cajal (Ramon y Cajal, 1928) (Fig. 2). These give rise to growth cones that in some cases sustain axonal regeneration. These PNS growth conelike structures remain rich in mitochondria, smooth endoplasmic reticulum, and vesicles as they regenerate (Ohi *et al.*, 1989). As in the distal axon segment, removal of degenerated axon and myelin debris is rapid and efficient.

Thus, in the PNS rapid breakdown in axon and myelin components are met by equally rapid responses from Schwann cells and macrophages that clear debris and help to establish an environment in which regeneration can occur.

B. Axotomy in the Central Nervous System

A considerable difference between the CNS and PNS response to axotomy is the nature of the host cellular response. Whereas in the PNS an efficient mechanism exists to clear axon and myelin debris, these responses in the CNS are substantially slower and less extensive. George and Griffin (1994) compared the process of Wallerian degeneration in sensory axons within the spinal cord and sensory axons in a peripheral nerve environment. The rate at which sensory axons degenerate in the spinal cord and in the peripheral nerve follows virtually the same time course (George and Griffin, 1994) (Fig. 4). However, sensory axon and myelin debris are almost completely cleared from peripheral nerve roots by 30 days postlesion, whereas axon and myelin debris, including neurofilamentous accumulations in axons, are still visible 90 days after injury to sensory axons of the spinal cord (George and Griffin, 1994). In other CNS injury paradigms such as the lesioned optic nerve, Wallerian degeneration of

---

FIGURE 3 *Structural Axonal Alterations Following Peripheral Nerve Injury.* (**A**) Electron micrograph of a normal intact axon (black asterisk) and its associated Schwann cell (white asterisk) reveals ordered appearance of the axoplasm with intact cytoskeleton. Layers of myelin are present around axon (short arrows). A thin layer of basal lamina can be observed along the outer surface of the Schwann cell (long arrows). Collagen fibrils are present in the extracellular space (arrowheads). Scale bar = 0.83 μm (**B**) 48 hrs after chemically-induced injury, axoplasm (black asterisk) has become highly disordered with accumulation of swollen vesicles (white arrowheads) suggesting breakdown of axonal transport. A Schwann cell (white asterisk) with myelin sheath and a Schwann cell membrane (arrows) are relatively intact at this stage. Scale bar = 0.62 μm. (**C**) Semi-thin section of peripheral nerve 4 days post-injury. Degenerating axons are evident with dense inclusions and early myelin breakdown (white arrowheads). Some axons appear healthier at this early time point (white arrows). Scale bar = 16 μm. (**D**) Semi-thin section 7 days post-injury shows greater numbers of degenerating axons with dense cores (white arrowheads) and few remaining healthy axons (white arrows). Scale bar = 16 μm. (Micrographs courtesy H. Powell, Department of Pathology, University of California, San Diego.)

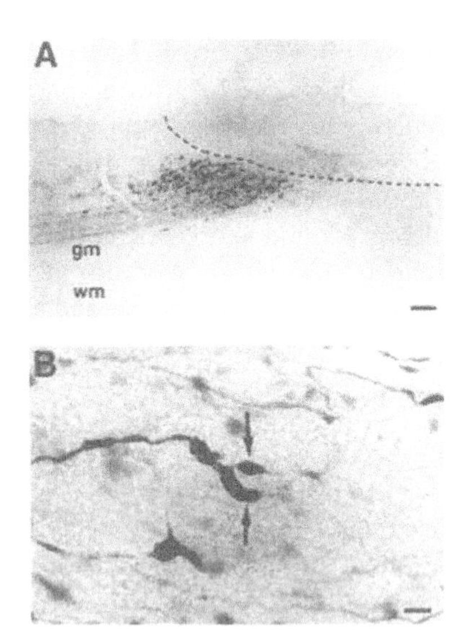

FIGURE 4 *Axonal Degeneration Following CNS Injury*. (**A**) Parasaggittal view of lesioned corticospinal tract in midthoracic spinal cord, anterogradely labeled with biotinylated dextran amine (BDA) 1 month after transection. Corticospinal axons are closely apposed to lesion site (dotted line). Degenerating axon tips appear as accumulation of black BDA reaction product. **gm**, gray matter; **wm**, white matter of spinal cord. Scale bar = 250 μm. (**B**) High magnification of clublike end bulbs (arrows) of lesioned corticospinal axons. Corticospinal axons are atrophic; there is no evidence of growth cone formation as seen in regenerating PNS axons. Scale bar = 10 μm.

the distal segment occurs over a time period extended by at least two weeks compared to lesioned peripheral nerve (Ludwin and Bisby, 1992).

The delay in clearance of axonal and myelin debris in the CNS may be attributable to at least two factors. First, CNS oligodendrocytes lack phagocytic properties. Unlike Schwann cells, whose phagocytic activity after axonal injury has been described both *in vitro* (Fernandez-Valle *et al.*, 1995) and *in vivo* (Liu *et al.*, 1995; Reichert *et al.*, 1994; Stoll *et al.*, 1989), oligodendrocytes do not exhibit phagocytic activity (Bignami and Ralston, 1969; Lampert and Cressman, 1966). Second, macrophages are rapidly recruited to peripheral but not central injury sites (George and Griffin, 1994; Perry *et al.*, 1987). Following injury to the PNS, resident populations of macrophages are activated, and additional blood-borne macrophages are recruited to the injury site. These "reactive" macrophages take on a rounded morphology that is observed even after they have engulfed damaged cellular debris. Activated macrophages can be observed within the injured nerve between 2 and 4 days after injury. In the CNS, microglial cells are the resident macrophage-lineage cell type. Following injury, microglial cells exhibit an early "activation" that is prominent by 3 days after injury, accompanied by an upregulation in major histocompatability complex II (MHC II) proteins and complement receptor-3 (CR3) proteins (George and Griffin, 1994). These markers decline sharply after 3 days, however, without further evidence of microglial–macrophage activity until 18 to 20 days after injury. By this time, approximately three weeks post-injury, cells with morphological features of macrophages are observed. The delay in the activation and transformation of microglia into actively phagocytosing cells may account for the delay in removal of myelin and axonal debris from the region of sensory axons in the dorsal columns following dorsal root axotomy (George and Griffin, 1994), thereby contributing to the failure of CNS axonal regeneration.

Thus, two principles emerge from the study of axotomy in the PNS and CNS: axons and myelin respond similarly to injury in the PNS and CNS, but the response of cells in the injured milieu is vastly different. In the PNS, Schwann cells and macrophages rapidly clear debris, whereas in the CNS products of axon and myelin breakdown are removed more slowly and less completely.

C. Changes in the Axonal Cytoskeleton after Injury

The mechanism of cytoskeletal disintegration of the axon has been studied in some detail in the CNS (Zimmerman and Schlaepfer, 1982). Calcium influx after injury activates intracellular proteases that dismantle the axon near the site of injury (Kampfl *et al.*, 1997; Schlaepfer and Zimmerman, 1981). One such protease, calpain, has an affinity for specific cytoskeletal components including neurofilaments (Pant, 1988), spectrin (Roberts-Lewis *et al.*, 1994), and microtubule-associated proteins (MAPs) (Johnson *et al.*, 1991). Following experimental brain injury, breakdown products of neurofilament and spectrin are found (Posmantur *et al.*, 1994). Further, in a traumatic model of brain damage, calpain inhibitors delivered directly into the blood stream diminish the accumulation of calpain breakdown products and maintain the integrity of axonal structures, reflected by a decrease in the number of cytoskeletally deranged axonal retraction balls (Posmantur *et al.*, 1997).

Activated microglia are another source of molecules that may contribute to axonal and myelin breakdown after injury. Microglia secrete the proteinases cathepsin B and cathepsin L (Banati *et al.*, 1993b; Banik *et al.*, 1997) reviewed in (Gehrmann *et al.*, 1995b), which can lead to dismantling of the cytoskeleton. In addition, microglia secrete reactive oxygen species (Banati *et al.*, 1991; Colton and Gilbert, 1987; Gehrmann *et al.*, 1995a; Woodroofe *et al.*, 1989) including nitric oxide (Banati *et al.*, 1993a; Murphy *et al.*, 1993; Thery *et al.*, 1993), which can induce neuronal breakdown and death *in vitro* (Thery *et al.*, 1991).

III. AXONAL REGENERATION

As noted previously, there is a vast difference in the regenerative capacity of lesioned central and peripheral axons that may be attributable to factors both intrinsic and extrinsic to the axon. "Intrinsic factors" refer to the innate ability of central or peripheral neurons to elaborate the growth of new axonal processes (growth cones), whereas "extrinsic factors" are elements of the extracellular environment that regulate growth potential by the provision of supportive or inhibitory growth signals. Axons located within the CNS generally show only abortive attempts at regeneration following injury, compared to consistent and functionally significant degrees of recovery that occur in axons located in the PNS. Indeed, the dissociation between the extent of regeneration observed if an axon of the same system is transected centrally or peripherally is profound, and propels forward the need to understand mechanisms that modulate axonal regeneration. The next section will examine elements that are intrinsic and extrinsic to the injured axon that may regulate regeneration after injury. These elements include the expression of growth-related genes, axonal transport mechanisms, neurotrophic factor availability, production of extracellular matrix and cell adhesion molecules, cytokine activation, and glial influences.

A. Intrinsic Elements in Axonal Regeneration

1. Retrograde Injury Signals

The initiation of a regenerative response appears to be linked to the retrograde transport of "axotomy signals" from the site of injury to the cell body (Fernandez *et al.*, 1981; Kenney and Kocsis, 1997, 1998; Singer *et al.*, 1982). In the PNS, there is a rapid upregulation of mRNA and protein for the immediate early gene *c-jun* in lumbar dorsal root ganglion (DRG) cell bodies following axotomy of the sciatic nerve (De Leon *et al.*, 1995; Herdegen *et al.*, 1992; Jenkins and Hunt, 1991; Kenney and Kocsis, 1997, 1998), suggesting that *c-jun* may influence post-injury events as well as the initiation of axonal regeneration. Upregulation of *c-jun* depends on the distance of axotomy from dorsal root ganglion cell bodies: activation of *c-jun* amino-terminal kinases (JNKs), the proteins responsible for activation of *c-jun* transcription, occurs within 30 minutes after a lesion located 1 cm from DRG cell bodies (Kenney and Kocsis, 1998), but is delayed by 3 hr if the lesion is located 4 cm from the cell body, suggesting that retrograde transport of an injury signal is responsible for JNK activation. The timing of JNK activation matches the rate of fast axonal transport (50–400 mm/day; (Jacob and McQuarrie, 1991).

The distance of axotomy from the cell body also affects retrograde transport in the CNS. Rubrospinal axons transected at the C3 level readily transport injected dyes in a retrograde direction, whereas rubrospinal axons lesioned at the more distal T10 level exhibit a marked decrease in retrograde transport (Tseng *et al.*, 1995), suggesting that distance of axotomy may impair the cellular response to injury by failing to retrogradely transport an injury-related signal to the cell soma. Supporting this possibility, an upregulation in expression of the growth-associated protein GAP-43 in the CNS depends on the distance of axotomy from the cell body. Corticospinal axons lesioned close to the cell body in the subcortical white matter upregulate GAP-43 expression (Kalil and Skene, 1986), whereas corticospinal axons lesioned more caudally at the level of the pyramidal decussation in the medulla fail to upregulate GAP-43 (reviewed in Bisby and Tetzlaff, 1992). The expression of GAP-43 positively correlates with successful axonal regeneration in the PNS (see the next section) (reviewed in Benowitz and Routtenberg, 1997).

Additional evidence supporting the importance of retrograde transport in initiating axon regeneration comes from studies in the hypoglossal system. Following transection of the peripheral hypoglossal nerve, one of the earliest events before the onset of regeneration is an increase in neuronal metabolism, identified by augmented 2-deoxyglucose uptake (Fernandez *et al.*, 1981; Singer *et al.*, 1982). Colchicine injections into the hypoglossal nerve proximal to the site of a lesion disrupt microtubular function and block injury-induced increases in 2-deoxyglucose. Retrograde transport of horseradish peroxidase injected into the proximal end of the transected hypoglossal nerve is also blocked by colchicine treatment. Blockade of axonal transport by application of vincristine *at* the lesion site (but not *distal* to the lesion site) prevents injury-induced upregulation of the low-affinity neurotrophin receptor (p75) in the soma of hypoglossal neurons, indicating that transport of an injury signal *from the lesion site* is required to elicit injury-related changes in gene expression (Hayes *et al.*, 1992). Collectively, these results suggest that the retrograde delivery of "axotomy signals" is required to enhance neuronal metabolism after injury and initiate regeneration (Fernandez *et al.*, 1981).

2. Synthesis of Cytoskeletal Components for Axonal Growth

The synthesis and transport of several cytoskeletal components are required to support the formation and extension of growth cones and axons during development. Actin, tubulin, and neurofilaments are all essential cytoskeletal components upon which axonal polarity, growth, and caliber during development are established (Caroni, 1997; Cleveland *et al.*, 1991; Edwards *et al.*, 1989; Gordon-Weeks, 1991a, 1991b, 1991c, 1993; Hoffman and Cleveland, 1988; Letourneau, 1996). In the adult, changes also occur in the synthesis and transport of essential cytoskeletal components after injury. Following a lesion to the adult facial nerve as it exits the stylomastoid foramen, motoneurons of the facial nucleus display massive increases in tubulin and actin synthesis and decreases in neurofilament protein production (Tetzlaff *et al.*, 1988). Similarly, thalamic neurons that have survived a lesion to the internal capsule upregulate the class III isoform of β-tubulin (Geisert and Frankfurter, 1989). Tubulin and actin upregulation appear to provide the cytoskeletal proteins necessary to support new axon growth. In particular, tubulin α-1, a developmentally regulated tubulin isoform not normally expressed in the adult, is upregulated following injury to the sciatic nerve (Gillen *et al.*, 1995; Mullins *et al.*, 1994) and facial nerve (Gloster *et al.*, 1994; Wu *et al.*, 1997). The concurrent downregulation in neurofilament synthesis following peripheral injury may relieve trafficking constraints on the transport mechanisms, allowing for increases in the rate of transport of other cytoskeletal proteins required for new axonal growth (Bisby and Tetzlaff, 1992).

In general, axotomized neurons in the CNS do not initiate the same degree of cytoskeletal protein synthesis that is observed in the lesioned PNS (reviewed in Bisby and Tetzlaff, 1992). For example, lesions of the rubrospinal tract result in rises of actin and tubulin (including tubulin-α1) expression that are similar to those that occur after peripheral nerve injury; however, these increases are only transient in the CNS. Downregulation of both tubulin and actin mRNA occurs between 8 to 14 days post-injury in the rubrospinal system, correlating with the appearance of atrophy in these neurons (Barron, 1983; Barron *et al.*, 1989; Tetzlaff *et al.*, 1991). During the period that tubulin and actin levels are increased, abortive sprouting of rubrospinal axons occurs, suggesting that a short-lived program of growth has been attempted (Barron *et al.*, 1989). Despite the subsequent downregulation of both total tubulin mRNA levels and actin in lesioned rubrospinal neurons, a specific tubulin isoform, tubulin-α1, actually exhibits a sustained upregulation in expression (Tetzlaff *et al.*, 1991). Tubulin-α1 is a cytoskeletal component that is expressed during axonal growth in development (Gloster *et al.*, 1994) and is also expressed following peripheral nerve injury (Gloster *et al.*, 1994; Matthew and Miller, 1993; Mohiuddin *et al.*, 1996; Wu *et al.*, 1997, 1993). Its expression remains elevated in the PNS after injury until contact is re-established with target tissue (Wu *et al.*, 1997). The injured corticospinal tract, on the other hand, exhibits an immediate drop in tubulin synthesis two days post-injury that fails to recover (Mikucki and Oblinger, 1991), and the corticospinal tract mounts little if any regenerative response after injury. Thus, the presence of new cytoskeletal protein synthesis correlates with attempted regeneration and is probably a requirement for axonal regrowth. It is unclear whether this failure to sustain synthesis of new cytoskeletal components after injury in the CNS is a primary *cause* of regeneration failure, or is merely a secondary and appropriate response to the failure of axonal regeneration for other reasons.

3. Anterograde Axonal Transport

The translocation of cytoskeletal proteins from the cell body to the site of axonal injury is crucial for supporting axonal regeneration (Fournier and McKerracher, 1995). In the adult nervous system, the movement of cytoskeletal proteins occurs through the process of slow axonal transport (Lorenz and Willard, 1978). There are two types of slow axonal transport in adult neurons, Slow component "a" (Sca) and Slow component "b" (Scb). In the CNS, Sca transports the major cytoskeletal proteins, including actin, α- and β-tubulins, and neurofilaments (Fournier and McKerracher, 1995; Vallee and Bloom, 1991). Scb transports a heterogeneous mixture of other proteins (Fournier and McKerracher, 1995). In the PNS, Sca transports actin and neurofilaments, whereas Scb transports a substantial amount of tubulin (Fournier and McKerracher, 1995). Following injury to peripheral nerves, Scb transport of tubulin *increases* to match the rate of axonal extension (Hoffman and Lasek, 1980; McKerracher and Hirscheimer, 1992), whereas in the CNS Sca transport of tubulin in crushed optic nerve *decreases* from a normal rate of 0.5 mm/day to 0.06 mm/day (McKerracher *et al.*, 1990b). This decrease in slow axonal transport in lesioned optic nerve results in a failure of both tubulin and neurofilament to reach nerve segments distal to the site of optic nerve crush. The failure to transport proteins that are integral to the construction of new growth correlates temporally with the diminished capacity of CNS axons to regenerate following injury. If a peripheral nerve graft is placed near the site of optic nerve axotomy, however, a small proportion of retinal ganglion cells will extend new processes into the graft (Fournier and McKerracher, 1997; McKerracher *et al.*, 1990a; So and Aguayo, 1985; Vidal-Sanz *et al.*, 1987). The retinal axons that successfully grow into peripheral nerve grafts increase their slow axonal transport to 1 mm/day (McKerracher *et al.*, 1990a), suggesting that recovery of slow transport can occur in successfully regenerating axons, regardless of central or peripheral location.

4. GAP-43 Expression

The exact role of GAP-43 in influencing axonal growth is incompletely understood. GAP-43 is a phosphorylated protein that is highly expressed in both axons and growth cones in the developing nervous system, and is re-expressed by regenerating PNS axons and sprouting CNS axons after injury in the adult (reviewed in Benowitz and Routtenberg, 1997; Oestreicher *et al.*, 1997). GAP-43 is a substrate of protein kinase C (PKC) (Akers and Routtenberg, 1985; Aloyo *et al.*, 1983) and may influence the reorganization of the axonal cytoskeleton (Benowitz and Routtenberg, 1997). More insight into the role of GAP-43 has been gained from the study of GAP-43 null mutation mice (Strittmatter *et al.*, 1995). Careful examination of the retinotectal system in GAP-43 −/− mice reveals that the extension of retinal axons becomes abnormally arrested at the level of the optic chiasm for a 6-day period during development; subsequently, axons extend past the optic chiasm to enter their appropriate tracts.

Additional information regarding the role of GAP-43 in axon growth is derived from transgenic mice that overexpress GAP-43. Normally, hippocampal mossy fibers arising from dentate granule cells project into the stratum lucidum of the CA3 region. However, mossy fibers in transgenic mice overexpressing GAP-43 project beyond their normal targets and into the stratum oriens (Aigner *et al.*, 1995). These findings support the hypothesis that GAP-43 transduces signals from the environment at the level of the growth cone, influencing axonal guidance and pathfinding during development (Benowitz and Routtenberg, 1997; Strittmatter *et al.*, 1995). GAP-43 may also enhance the capacity of neurons to sprout after injury. In the PNS, transgenic mice that overexpress GAP-43 exhibit spontaneous sprouting of motor neuron terminals at the neuromuscular junction (Aigner *et al.*, 1995). Following a sciatic nerve crush injury, these same GAP-43 overexpressors demonstrate early sprouting and augmented regeneration compared to wild type animals (Aigner *et al.*, 1995). However, GAP-43 overexpression in the CNS does not necessarily bring about regeneration. Overexpression of GAP-43 in the cerebellum using the Purkinje cell-specific L7 promoter enhances sprouting of Purkinje neurons after knife-cut injury (Buffo *et al.*, 1997), but does not promote axonal regeneration into growth-permissive substrates of embryonic neurons or Schwann cell grafts placed within the injury site (Buffo *et al.*, 1997).

Thus, several features *intrinsic* to the injured axon correlate with the occurrence of successful regeneration: the ability to retrogradely transport signals to the cell soma that initiate a growth state, the re-expression of genes coding for growth-related cytoskeletal components, the expression of GAP-43, and the ability to anterogradely transport materials down the axon that are required for growth (see Table I). Although axons of the adult CNS normally fail to express the full repertoire of these elements after injury, central axons placed in a experimentally manipulated milieu that is supportive of growth *can* extend axons, indicating that intrinsic signals can be induced to levels sufficient to

TABLE I Factors That Influence Axonal Regeneration

Intrinsic factors that may affect axonal growth
Retrograde transport of injury signals
• Efficient in PNS
• Inefficient in CNS
Synthesis of cytoskeletal components
• Efficient in PNS
• Inefficient in CNS
Anterograde axonal transport of cytoskeletal components
• Efficient in PNS
• Inefficient in CNS
GAP-43 Expression
• Regularly upregulated in PNS
• Inconsistently upregulated in CNS
Extrinsic factors that may affect axonal growth
Phagocytosis and removal of degenerating debris
• Blood-borne macrophages: Recruited to injury site in PNS and CNS
• Intrinsic phagocytic cells: Schwann cells and resident macrophages in PNS; microglia in CNS.
• Overall, more efficient in PNS
Growth factor production for cell survival and axon growth
• Schwann cells produce and spatially present growth factors in PNS
• Astrocytes produce some growth factors but without spatially specific presentation in CNS
Structural support and guidance of injured axons
• Bands of Büngner: Linear Schwann cell arrays and basal lamina in PNS provide spatial scaffold for axon growth
• Perineurium in PNS organizes axon fascicles
• No intrinsic structural support/guidance in CNS
Other extracellular molecules
• Growth-permissive substrates present in PNS and secreted by Schwann cells to be used as conduits by regenerating axons
• Inhibitory molecules present in CNS: Presence of myelin-based inhibitory molecules in adult CNS white matter tracts
• Upregulation of putatively nonpermissive growth molecules at and around the site of CNS injury, i.e., chondroitin sulfate proteoglycans.

support at least some degree of active regeneration. Indeed, the failure of central axons to recruit the full repertoire of intrinsic molecular and structural events after CNS injury may be the *result* of the axon's failure to enter a growth state, rather than *causing* growth failure.

B. Extrinsic Elements in Axonal Regeneration

1. The Milieu of the Injured Peripheral Axon

Following a *crush* injury to a peripheral nerve, axonal regeneration is typically successful and target reinnervation occurs despite the need to regrow and remyelinate crushed axons for distances that may exceed 50 cm. Axonal regeneration proceeds at a rate of 1–2 mm/day (Schaumberg *et al.*, 1992). Following a *transection* injury to a peripheral nerve, axonal regeneration will also occur if the gap between the proximal and distal nerve stumps is less than one centimeter and if the cut ends of the nerves are re-anastamosed in their original orientation (Schaumberg *et al.*, 1992). However, if the gap exceeds this distance or if the orientation of the nerve is disturbed, then axons may grow away from the basal lamina substrate into the collagenous connective tissue and form dense tangles of abortively sprouted nerve terminals ("neuromas") that spontaneously discharge and cause pain, or may simply fail to regenerate. Peripheral nerve regeneration therefore is highly dependent on the organization of the nerve.

The basis for the dependence of peripheral axon regeneration on the organization of the nerve is revealed by anatomical study. Axons within the peripheral nerve are organized into distinct bundles (fascicles) that are physically contained within three types of connective tissue sheaths: the epineurium, the perineurium, and the neurilemmal sheath (Fig. 1). The epineurium consists of connective tissue (primarily collagen) produced by fibroblasts (Thomas and Olsson, 1984) that surrounds the entire nerve and protects it to some extent from injury and stretch. The perineurium encloses bundles of dozens to hundreds of axons within layers of collagen, thereby dividing the nerve into multiple fascicles that project to related targets. The neurilemmal sheath consists of a basal lamina enclosing a Schwann cell and its ensheathed axons, or its single myelinated axon, providing a regular tube-like structure that extends the length of each individual axon.

After peripheral nerve injury, the distal axonal segment degenerates, leaving hollow tubes of basal lamina and the Schwann cells that formed them. These Schwann cells transiently proliferate, aligning themselves into columns known as the bands of Büngner (Fig. 2) (reviewed in Ide, 1996). When a regenerating axon encounters one of these Schwann cell–basal lamina tubes, they are guided down the distal axon to their terminal, and are remyelinated by the aligned resident Schwann cells that they encounter along their course. In addition, Schwann cells secrete neurotrophic factors (Ide, 1996), including nerve growth factor (NGF) (Funakoshi *et al.*, 1993; Heumann *et al.*, 1987; Lindholm *et al.*, 1987; Meyer *et al.*, 1992), brain-derived neurotrophic factor (BDNF) (Acheson *et al.*, 1991; Meyer *et al.*, 1992), and glial cell line-derived neurotrophic factor (GDNF) (Hammarberg *et al.*, 1996). These neurotrophic factors may both stimulate the growth of injured axons and guide axons down Schwann cells aligned in Bands of Büngner. This guidance may be achieved by the presentation of growth factors on the surface of Schwann cells. After injury, Schwann cells significantly upregulate expression of both NGF and the low-affinity neurotrophin receptor, p75 (Taniuchi *et al.*, 1986). Of note, NGF that is secreted by Schwann cells appears to bind to the p75 receptor on the Schwann cell surface, thereby presenting NGF to regenerating peripheral axons (Taniuchi *et al.*, 1988). Schwann cells do not express the trkA receptor for NGF, and the presence of p75 on the Schwann cell surface does not appear to modulate signal transduction within the Schwann cell. Thus, p75 could play a passive role on the Schwann cell surface by "capturing" NGF and presenting it to axons, thereby guiding regenerating axons down Bands of Büngner. When axons contact the surface of Schwann cells, growth factor production is subsequently downregulated (Taniuchi *et al.*, 1988). Glial cell line-derived neurotrophic factor (GDNF), a potent neurotrophic factor for motor neurons (Henderson *et al.*, 1995; Li *et al.*, 1995), may be presented to injured peripheral axons in a similar fashion (Trupp *et al.*, 1997). Schwann cells upregulate expression of both GDNF and the component of the GDNF receptor after injury, thereby presenting GDNF to regenerating axons (Trupp *et al.*, 1997). Besides promoting and guiding axon growth, growth factor secretion from Schwann cells could also prevent the death of the cells of origin of the injured axons.

Schwann cells additionally secrete extracellular matrix molecules such as collagen, laminin, and fibronectin (Wang *et al.*, 1992) (reviewed in Fawcett and Keynes, 1990; Martini, 1994), to which injured axons can attach and extend. Finally, Schwann cells produce and present the cell adhesion molecules L1 and N-CAM on their surfaces (Martini and Schachner, 1988), which may interact with regenerating axons to promote axonal attachment and propulsion through the peripheral nerve substrate (Martini, 1994). Both Schwann cells and resident macrophages in the peripheral nerve also assist in the rapid and efficient removal of degenerated axon and myelin debris from the site of injury so that regeneration can proceed (Avellino *et al.*, 1995; Holtzman and Novikoff, 1965; Lassmann *et al.*, 1978; Liu *et al.*, 1995; Reichert *et al.*, 1994).

Another potential molecular modulator of axonal growth after nervous system injury is the cytokine class of molecules. Following PNS injury, sources of cytokines can include resident and recruited macrophages, lymphocytes, mast cells, and Schwann cells (reviewed in Brecknell and Fawcett, 1996; Creange *et al.*, 1997; Griffin *et al.*, 1993). One cytokine, IL-1, may promote regeneration in injured peripheral nerves by upregulating neurotrophic factor expression (reviewed in Griffin *et al.*, 1993). For example, IL-1 promotes the expression of NGF in Schwann cells following peripheral nerve injury (Lindholm *et al.*, 1987). This increase in NGF correlates with recruitment of macrophages into the injured sciatic nerve (Heumann, 1987; Heumann *et al.*, 1987).

Thus it is clear that the spatial properties and the milieu of the peripheral nerve play a key role in bringing about axonal regeneration (see Table 1). The division of the peripheral nerve into guidance channels of neurilemmal sheaths, secretion of neurotrophic factors, production of extracellular matrix and cell adhesion molecules, and efficient phagocytosis of degenerating debris combine to create a new growth environment that is efficient and effective in promoting axonal regeneration.

2. The Milieu of the Injured Central Axon

The milieu of axons in the CNS differs markedly from that of the PNS. Unlike the peripheral nerve, axons in the CNS are not structurally separated by perineurial or neurilemmal sheaths. Rather, axons are organized into tracts of related function but without physical boundaries or divisions analogous to those created by perineurial or neurilemmal sheaths. There are no guidance tubes formed after injury that can support and guide axons through structures like bands of Büngner, and basal lamina is not deposited in a regular manner to delineate and separate guidance channels.

Glia are also organized differently in the CNS and PNS. In the PNS, one Schwann cell myelinates one axon or may ensheath several axons, but in the CNS a single oligodendrocyte myelinates several axons simultaneously and exhibits a less intimate relationship with its associated axon. Unlike Schwann cells, oligodendrocytes do not respond to injury by filling the injury site, aligning into regular longitudinal arrays, or assisting in phagocytosis of debris (Bignami and Ralston, 1969; Lampert and Cressman, 1966). Although CNS astrocytes secrete neurotrophic factors, this secretion is not known to be deposited in an organized manner that guides and supports regenerating axons. Whereas Schwann cells in the PNS help to reconstitute the normal extracellular matrix of the nerve, the most abundant extracellular matrix molecule in the CNS, hyaluronic acid, is not redeposited in sites of injury. Similarly, cell adhesion molecules such as L1 are not laid down in a regular array for presentation to injured central axons. Rather, astrogliosis occurs in sites of CNS injury. Fibroblasts and leptomeningeal cells may migrate into an injured zone and deposit extracellular collagen, which supports the growth of some but not all central axons (Grill *et al.*, 1997a, 1997b). As noted earlier, axon and myelin debris is not rapidly or efficiently cleared from CNS sites of injury, in part because microglial activation is delayed. Recent reports also suggest that inflammatory cells such as microglia–macrophages may deposit extracellular matrix molecules like chondroitin sulfate proteoglycans that inhibit axon growth (Davies *et al.*, 1997; Fitch and Silver, 1997a) (see also Chapter 3).

Changes in cytokine expression occur after injury in the CNS. Pro-inflammatory cytokine expression is upregulated after traumatic brain injury (Ghirnikar *et al.*, 1998; see also Ledeen and Chakraborty, 1998). Cytokines such as interleukin-1 (IL-1) and tumor necrosis factor-α (TNF-α) are also expressed in cells surrounding a spinal cord lesion (Bartholdi and Schwab, 1997) and may contribute both to the recruitment of inflammatory cells and to the modification of the host astroglial response (reviewed in Ghirnikar *et al.*, 1998). Following optic nerve crush, exogenous administration of two cytokines, TNF-α and colony stimulating factor-1 (CSF-1), increases the number of macrophages that migrate into the injured nerve (Lotan *et al.*, 1994). Sections of the crushed optic nerve treated with TNF-α exhibit an increase in PC12 cell adherence *in vitro*, suggesting that the activation of macrophages by some cytokines may increase the "adhesiveness" and may thereby improve the permissiveness of the CNS growth environment (Lotan *et al.*, 1994). Thus, cytokines influence the migration and adhesive properties of inflammatory cells in regions of CNS injury. The possible benefits or drawbacks of cytokines in modulating axonal responses to injury in both the CNS and PNS must, however, be balanced against the possibility that microglia and macrophages may secrete extracellular matrix molecules such as chondroitin sulfate proteoglycans, or upregulate expression of these molecules in astrocytes, which subsequently inhibit axonal growth (see Chapter 3) (Davies *et al.*, 1997; Fitch and Silver, 1997a). Overall, cytokines may perform functions and induce the expression and secretion of molecules that both promote and inhibit axonal growth after injury; ongoing studies continue to address these possibilities.

Finally, not only does the CNS lack many supportive elements of the peripheral nerve milieu, but molecular inhibitors of axon growth are present in the CNS, including myelin- or white-matter-associated inhibitory molecules (Bandtlow *et al.*, 1990; Bandtlow and Loeschinger, 1997; Cadelli *et al.*, 1992; Caroni and Schwab, 1993; Kapfhammer *et al.*, 1992; Rubin *et al.*, 1994; Schnell and Schwab, 1990; Schwab *et al.*, 1993; Schwab and Schnell, 1990) that further reduce the extent to which CNS axons can sprout and elongate.

IV. CONCLUSION

On balance, it appears that many factors extrinsic to the injured axon act to reduce the extent of axonal regeneration in the adult CNS. The environment of the injured peripheral axon is vastly different from that of the CNS and readily supports axonal growth at molecular, cellular, and tissue levels. This knowledge has been utilized to design strategies for promoting regeneration in the adult CNS, as described in detail in Chapters 24 and 25. Indeed, some of these strategies have met with clear success in several *in vitro* and *in vivo* models. The fact that alterations of the extracellular environment can elicit significant growth in the CNS suggests that adult axonal regeneration is primarily restrained by factors extrinsic rather than intrinsic to the axon. This knowledge will be valuable in propelling forward the search for new strategies for enhancing CNS axonal repair.

ACKNOWLEDGMENTS

We gratefully acknowledge the assistance of Dr. Henry Powell. This work was supported by the NIH (AG10435, R01NS37083), Veterans Affairs, the Hollfelder Foundation, the Paralyzed Veterans of America, and the National Paralysis Foundation.

REFERENCES

Acheson, A., Barker, P. A., Alderson, R. F., Miller, F. D., and Murphy, R. A. (1991). Detection of brain-derived neurotrophic factor-like activity in fibroblasts and Schwann cells: inhibition by antibodies to NGF. *Neuron* 7:265–275.

Aigner, L., Arber, S., Kapfammer, J. P., Laux, T., Schneider, C., Botteri, F., Brenner, H.-R., and Caroni, P. (1995). Overexpression of the neural growth-associated protein GAP-43 induces nerve sprouting in the adult nervous system of transgenic mice. *Cell* 83:269–278.

Akers, R. F., and Routtenberg, A. (1985). Protein kinase C phosphorylates a 47 Mr protein (F1) directly related to synaptic plasticity. *Brain Research* 334:147–151.

Aloyo, V. J., Zwiers, H., and Gispen, W. H. (1983). Phosphorylation of B-50 protein by calcium-activated, phospholipid-dependent protein kinase and B-50 protein kinase. *Journal of Neurochemistry* 41:649–653.

Avellino, A. M., Hart, D., Dailey, A. T., MacKinnon, M., Ellegala, D., and Kliot, M. (1995). Differential macrophage responses in the peripheral and central nervous system during Wallerian degeneration of axons. *Experimental Neurology* 136:183–198.

Banati, R. B., Gehrmann, J., Schubert, P., and Kreutzberg, G. W. (1993a). Cytotoxicity of microglia. *Glia* 7:111–118.

Banati, R. B., Rothe, G., Valet, G., and Kreutzberg, G. W. (1991). Respiratory burst activity in brain macrophages: A flow cytometric study on cultured rat microglia. *Neuropathology and Applied Neurobiology* 17:223–230.

Banati, R. B., Rothe, G., Valet, G., and Kreutzberg, G. W. (1993b). Detection of lysosomal cysteine proteinases in microglia: Flow cytometric measurement and histochemical localization of cathepsin B and L. *Glia* 7:183–191.

Bandtlow, C., Zachleder, T., and Schwab, M. E. (1990). Oligodendrocytes arrest neurite growth by contact inhibition. *Journal of Neuroscience* 10:3837–3848.

Bandtlow, C. E., and Loeschinger, J. (1997). Developmental changes in neuronal responsiveness to the CNS myelin-associated neurite growth inhibitor NI-35-250. *European Journal of Neuroscience* 9:2743–2752.

Banik, N. L., Matzelle, D., Terry, E., and Hogan, E. L. (1997). A new mechanism of methylprednisolone and other corticosteroids action demonstrated *in vitro:* Inhibition of a proteinase (calpain) prevents myelin and cytoskeletal protein degradation. *Brain Research* 748: 205–210.

Barron, K. D. (1983). *Comparative observations on the cytologic reactions of central and peripheral nerve cells to axotomy*. New York: Raven Press.

Barron, K. D., Banerjee, M., Dentinger, M. P., Scheibly, M. E., and Mankes, R. (1989). Cytological and cytochemical (RNA) studies on rubral neurons after unilateral rubrospinal tractotomy: The impact of GM1 ganglioside administration. *Journal of Neuroscience* 22:331–337.

Bartholdi, D., and Schwab, M. E. (1997). Expression of pro-inflammatory cytokine and chemokine mRNA upon experimental spinal cord injury in mouse: an in situ hybridization study. *European Journal of Neuroscience* 9:1422–1428.

Benowitz, L. I., and Routtenberg, A. (1997). GAP-43: An intrinsic determinant of neuronal development and plasticity. *Trends in Neurosciences* 20:84–91.

Bignami, A., Dahl, D., Nguyen, B. T., and Crosby, C. J. (1981). The fate of axonal debris in Wallerian degeneration of rat optic and sciatic nerves. Electron microscopy and immunofluorescence studies with neurofilament antisera. *Journal of Neuropathology and Experimental Neurology* 40:537–550.

Bignami, A., and Ralston, H. J. (1969). The cellular reaction to wallerian degeneration in the central nervous system of the cat. *Brain Research* 13:444–461.

Bisby, M. A., and Tetzlaff, W. (1992). Changes in cytoskeletal protein synthesis following axon injury and during axon regeneration. *Molecular Neurobiology* 6:107–123.

Book, A. A., Fischer, I., Yu, X. J., Iannuzzelli, P., and Murphy, E. H. (1996). Altered expression of microtubule-associated proteins in cat trochlear motoneurons after peripheral and central lesions of the trochlear nerve. *Experimental Neurology* 138:214–226.

Brecknell, J. E., and Fawcett, J. W. (1996). Axonal regeneration. *Biological Reviews* 71:227–255.

Buffo, A., Holtmaat, A. J. D. G., Savio, T., Verbeek, J. S., Oberdick, J., Oestreicher, A. B., Gispen, W. H., Verhaagen, J., Rossi, F., and Strata, P. (1997). Targeted overexpression of the neurite growth-associated protein B-50/GAP-43 in cerebellar purkinje cells induces sprouting after axotomy but not axon regeneration into growth-permissive transplants. *Journal of Neuroscience* 17:8778–8791.

Bush, M. S., Tonge, D. A., Woolf, C., and Gordon-Weeks, P. R. (1996). Expression of a developmentally regulated, phosphorylated isoform of microtubule-associated protein 1B in regenerating axons of the sciatic nerve. *Neuroscience* 73:553–563.

Cadelli, D. S., Bandtlow, C. E., and Schwab, M. E. (1992). Oligodendrocyte- and myelin-associated inhibitors of neurite outgrowth: Their involvement in the lack of CNS regeneration. *Experimental Neurology* 115:189–192.

Caroni, P. (1997). Intrinsic neuronal determinants that promote axonal sprouting and elongation. *Bioessays* 19:767–775.

Caroni, P., and Schwab, M. (1993). Oligodendrocyte- and myelin-associated inhibitors of neurite growth in the adult nervous system. *Advances in Neurology* 61.

Chambers, C. B., and Muma, N. A. (1997). Tau mRNA isoforms following sciatic nerve axotomy with and without regeneration. *Brain Research. Molecular Brain Research* 48:115–124.

Cleveland, D. W., Monteiro, M. J., Wong, P. C., Gill, S. R., Gearhart, J. D., and Hoffman, P. N. (1991). Involvement of neurofilaments in the radial growth of axons. *Journal of Cell Science. Supplement* 15:85–95.

Colton, C. A., and Gilbert, D. L. (1987). Production of superoxide anions by a CNS macrophage, the microglia. *Febs Letters* 223:284–288.

Creange, A., Barlovatz-Meimon, G., and Gherardi, R. K. (1997). Cytokines and peripheral nerve disorders. *European Cytokine Network* 8:145–151.

Davies, S. J., Fitch, M. T., Memberg, S. P., Hall, A. K., Raisman, G., and Silver, J. (1997). Regeneration of adult axons in white matter tracts of the central nervous system. *Nature* 390:680–683.

De Leon, M., Nahin, R. L., Molina, C. A., De Leon, D. D., and Ruda, M. A. (1995). Comparison of c-jun, junB, and junD mRNA expression and protein in the rat dorsal root ganglia following sciatic nerve transection. *Journal of Neuroscience Research* 42:391–401.

Edwards, M. A., Crandall, J. E., Wood, J. N., Tanaka, H., and Yamamoto, M. (1989). Early axonal differentiation in mouse CNS delineated by an antibody recognizing extracted neurofilaments. *Brain Research. Developmental Brain Research* 49:185–204.

Fawcett, J. W., and Keynes, R. J. (1990). Peripheral nerve regeneration. *Annual Review of Neurosciences* 13:43–60.

Fernandez, H. L., Singer, P. A., and Mehler, S. (1981). Retrograde axonal transport mediates the onset of regenerative changes in the hypoglossal nucleus. *Neuroscience Letters* 25:7–11.

Fernandez-Valle, C., Bunge, R. P., and Bunge, M. B. (1995). Schwann cells degrade myelin and proliferate in the absence of macrophages: Evidence from *in vitro* studies of Wallerian degeneration. *Journal of Neurocytology* 24:667–679.

Fitch, M. T., and Silver, J. (1997a). Activated macrophages and the blood-brain barrier: Inflammation after CNS injury leads to increases in putative inhibitory molecules. *Experimental Neurology* 148:587–603.

Fitch, M. T., and Silver, J. (1997b). Glial cell extracellular matrix: boundaries for axon growth in development and regeneration. *Cell and Tissue Research* 290:379–384.

Founier, A. E., and McKerracher, L. (1997). Tubulin expression and axonal transport in injured and regenerating neurons in the adult mammalian central nervous system. *Biochemistry and Cell Biology* 73:659-664.

Fournier, A. E., and McKerracher, L. (1995). Tubulin expression and axonal transport in injured and regenerating neurons in the adult mammalian central nervous system. *Biochemistry and Cell Biology* 73:659–664.

Funakoshi, H., Frisen, J., Barbany, G., Timmusk, T., Zachrisson, O., Verge, V. M., and Persson, H. (1993). Differential expression of mRNAs for neurotrophins and their receptors after axotomy of the sciatic nerve. *Journal of Cell Biology* 123:455–465.

Gehrmann, J., Banati, R. B., Wiessner, C., Hossman, K. A., and Kreutzberg, G. W. (1995a). Reactive microglia in cerebral ischaemia: An early mediator of tissue damage? *Neuropathology and Applied Neurobiology* 21:277–289.

Gehrmann, J., Matsumoto, Y., and Kreutzberg, G. W. (1995b). Microglia: Intrinsic immuneffector cell of the brain. *Brain Research. Brain Research Reviews* 20:269–287.

Geisert, E. E., Jr., and Frankfurter, A. (1989). The neuronal response to injury as visualized by immunostaining of class III beta-tubulin in the rat. *Neuroscience Letters* 102:137–141.

George, R., and Griffin, J. W. (1994). Delayed macrophage responses and myelin clearance during Wallerian degeneration in the central nervous system: The dorsal radiculotomy model. *Experimental Neurology* 129:225–236.

Ghirnikar, R. S., Lee, Y. L., and Eng, L. F. (1998). Inflammation in traumatic brain injury: role of cytokines and chemokines. *Neurochemical Research* 23:329–340.

Gillen, C., Gleichmann, M., Spreyer, P., and Müller, H. W. (1995). Differentially expressed genes after peripheral nerve injury. *Journal of Neuroscience Research* 42:159–171.

Gloster, A., Wu, W., Speelman, A., Weiss, S., Causing, C., Pozniak, C., Reynolds, B., Chang, E., Toma, J. G., and Miller, F. D. (1994). The T alpha 1 alpha-tubulin promoter specifies gene expression as a function of neuronal growth and regeneration in transgenic mice. *Journal of Neuroscience* 14:7319–7330.

Goldstein, M. E., Weiss, S. R., Lazzarini, R. A., Shneidman, P. S., Lees, J. F., and Schlaepfer, W. W. (1988). mRNA levels of all three neurofilament proteins decline following nerve transection. *Brain Research* 427:287–291.

Gordon-Weeks, P. R. (1991a). Control of microtubule assembly in growth cones. *Journal of Cell Science. Supplement* 15:45–49.

Gordon-Weeks, P. R. (1991b). Growth cones: The mechanism for neurite advance. *Bioessays* 13: 235–239.

Gordon-Weeks, P. R. (1991c). Microtubule organization in growth cones. *Biochemical Society Transactions* 19:1080–1085.

Gordon-Weeks, P. R. (1993). Organization of microtubules in axonal growth cones: A role for microtubule-associated protein MAP 1B. *Journal of Neurocytology* 22:717–725.

Grafstein, B. (1975). The nerve cell body response to axotomy. *Experimental Neurology* 48:32–51.

Griffin, J. W., George, R., and Ho, T. (1993). Macrophage systems in peripheral nerves. A review. *Journal of Neuropathology and Experimental Neurology* 52:553–560.

Grill, R., Murai, K., Blesch, A., Gage, F. H., and Tuszynski, M. H. (1997a). Cellular delivery of neurotrophin-3 promotes corticospinal axonal growth and partial functional recovery after spinal cord injury. *Journal of Neuroscience* 17:5560–5572.

Grill, R. J., Blesch, A., and Tuszynski, M. H. (1997b). Robust growth of chronically injured spinal cord axons induced by grafts of genetically modified NGF-secreting cells. *Experimental Neurology* 148:444–452.

Hammarberg, H., Piehl, F., Cullheim, S., Fjell, J., Hökfelt, T., and Fried, K. (1996). GDNF mRNA in Schwann cells and DRG satellite cells after chronic sciatic nerve injury. *Neuroreport* 7: 857–860.

Hayes, R. C., Wiley, R. G., and Armstrong, D. M. (1992). Induction of nerve growth factor receptor (p75NGFr) mRNA within hypoglossal motoneurons following axonal injury. *Brain Research. Molecular Brain Research* 15:291–297.

Henderson, C. E., Phillips, H. S., Pollock, R. A., Davies, A. M., Lemeulle, C., Armanini, M., Simmons, L., Moffet, B., Vandlen, R. A., and Simpson, L. C. (1995). GDNF: A potent survival factor for motoneurons present in peripheral nerve and muscle. *Science* 266:1062–1064.

Herdegen, T., Fiallos-Estrada, C. E., Schmid, W., Bravo, R., and Zimmermann, M. (1992). The transcription factors c-JUN, JUN D and CREB, but not FOS and KROX-24, are differentially regulated in axotomized neurons following transection of rat sciatic nerve. *Brain Research. Molecular Brain Research* 14:155–165.

Heumann, R. (1987). Regulation of the synthesis of nerve growth factor. *Journal of Experimental Biology* 132:133–150.

Heumann, R., Korsching, S., Bandtlow, C., and Thoenen, H. (1987). Changes of nerve growth factor synthesis in nonneuronal cells in response to sciatic nerve transection. *Journal of Cell Biology* 104:1623–1631.

Hoffman, P. N., and Cleveland, D. W. (1988). Neurofilament and tubulin expression recapitulates the developmental program during axonal regeneration: induction of a specific beta-tubulin isotype. *Proceedings of the National Academy of Sciences of the United States of America* 85: 4530–4533.

Hoffman, P. N., and Lasek, R. J. (1980). Axonal transport of the cytoskeleton in regenerating motor neurons: constancy and change. *Brain Research* 202:317–333.

Holtzman, E., and Novikoff, A. B. (1965). Lysosomes in the rat sciatic nerve following crush. *Journal of Cell Biology* 27:651–669.

Ide, C. (1996). Peripheral nerve regeneration. *Neuroscience Research* 25:101–121.

Jacob, J. M., and McQuarrie, I. G. (1991). Axotomy accelerates slow component b of axonal transport. *Journal of Neurobiology* 22:570–582.

Jenkins, R., and Hunt, S. P. (1991). Long-term increase in the levels of c-jun mRNA and jun protein-like immunoreactivity in motor and sensory neurons following axon damage. *Neuroscience Letters* 129:107–110.

Johnson, G. V., Litersky, J. M., and Whitaker, J. N. (1991). Proteolysis of microtubule-associated protein 2 and tubulin by cathepsin D. *Journal of Neurochemistry* 57:1577–1583.

Kalil, K., and Skene, J. H. (1986). Elevated synthesis of an axonally transported protein correlates with axon outgrowth in normal and injured pyramidal tracts. *Journal of Neuroscience* 6:2563–2570.

Kampfl, A., Posmantur, R. M., Zhao, X., Schmutzhard, E., Clifton, G. L., and Hayes, R. L. (1997). Mechanisms of calpain proteolysis following traumatic brain injury: Implications for pathology and therapy: implications for pathology and therapy: A review and update. *Journal of Neurotrauma* 14:121–134.

Kapfhammer, J. P., Schwab, M. E., and Schneider, G. E. (1992). Antibody neutralization of neurite growth inhibitors from oligodendrocytes results in expanded pattern of postnatally sprouting retinocollicular axons. *Journal of Neuroscience* 12:2112–2119.

Kenney, A. M., and Kocsis, J. D. (1997). Timing of c-jun protein induction in lumbar dorsal root ganglia after sciatic nerve transection varies with lesion distance. *Brain Research* 751:90–95.

Kenney, A. M., and Kocsis, J. D. (1998). Peripheral axotomy induces long-term c-jun amino-terminal kinase-1 activation and activator protein-1 binding activity by c-jun and junD in adult rat dorsal root ganglia In vivo. *Journal of Neuroscience* 18:1318–1328.

Kobayashi, N. R., Fan, D. P., Giehl, K. M., Bedard, A. M., Wiegand, S. J., and Tetzlaff, W. (1997). BDNF and NT-4/5 prevent atrophy of rat rubrospinal neurons after cervical axotomy, stimulate GAP-43 and Talpha1-tubulin mRNA expression, and promote axonal regeneration. *Journal of Neuroscience* 17:9583–9595.

La Fleur, M., Underwood, J. L., Rapolee, D. A., and Werb, Z. (1996). Basement membrane and repair of injury to peripheral nerve: defining a potential role for macrophages, matrix metalloproteinases, and tissue inhibitor of metalloproteinase-1. *Journal of Experimental Medicine* 184:2311–2326.

Lampert, P. W., and Cressman, M. R. (1966). Fine-structural changes of myelin sheaths after axonal degeneration in the spinal cord of rats. *American Journal of Pathology* 49:1139–1155.

Lassmann, H., Ammerer, H. P., Jurecka, W., and Kulnig, W. (1978). Ultrastructural sequence of myelin degradation. II. Wallerian degeneration in the rat femoral nerve. *Acta Neuropathologica* 44:103–109.

Ledeen, R. W., and Chakraborty, G. (1998). Cytokines, signal transduction, and inflammatory demyelination: Review and hypothesis. *Neurochemical Research* 23:277–289.

Letourneau, P. C. (1996). The cytoskeleton in nerve growth cone motility and axonal pathfinding. *Perspectives on Developmental Neurobiology* 4:111–123.

Li, L., Wu, W., Lin, L. F., Lei, M., Oppenheim, R. W., and Houenou, L. J. (1995). Rescue of adult mouse motoneurons from injury-induced cell death by glial cell line-derived neurotrophic factor. *Proceedings of the National Academy of Sciences* 92:9771–9775.

Lieberman, A. R. (1971). The axon reaction: A review of the principal features of perikaryal responses to axon injury. *International Review of Neurobiology* 14:49–124.

Lindholm, D., Heumann, R., Meyer, M., and Thoenen, H. (1987). Interleukin-1 regulates synthesis of nerve growth factor in non-neuronal cells of rat sciatic nerve. *Nature* 330:658–659.

Liu, H. M., Yang, L. H., and Yang, Y. J. (1995). Schwann cell properties: 3. C-fos expression, bFGF production, phagocytosis and proliferation during Wallerian degeneration. *Journal of Neuropathology and Experimental Neurology* 54:487–496.

Liuzzi, F. J., and Tedeschi, B. (1992). Axo-glial interactions at the dorsal root transitional zone regulate neurofilament protein synthesis in axotomized sensory neurons. *Journal of Neuroscience* 12:4783–4792.

Lorenz, T., and Willard, M. (1978). Subcellular fractionation of intra-axonally transport polypeptides in the rabbit visual system. *Proceedings of the National Academy of Sciences of the United States of America* 75:505–509.

Lotan, M., Solomon, A., Ben-Bassat, S., and Schwartz, M. (1994). Cytokines modulate the inflammatory response and change permissiveness to neuronal adhesion in injured mammalian central nervous system. *Experimental Neurology* 126:284–290.

Ludwin, S. K., and Bisby, M. A. (1992). Delayed Wallerian degeneration in the central nervous system of Ola mice: An ultrastructural study. *Journal of the Neurological Sciences* 109:140–147.

Martini, R. (1994). Expression and functional roles of neural cell surface molecules and extracellular matrix components during development and regeneration of peripheral nerves. *Journal of Neurocytology* 23:1–28.

Martini, R., and Schachner, M. (1988). Immunoelectron microscopic localization of neural cell adhesion molecules (L1, N-CAM, and myelin-associated glycoprotein) in regenerating adult mouse sciatic nerve. *Journal of Cell Biology* 106:1735–1746.

Matthew, T. C., and Miller, F. D. (1993). Induction of T alpha 1 alpha-tubulin mRNA during neuronal regeneration is a function of the amount of axon lost. *Developmental Biology* 158:467–474.

McKeon, R. J., Schreiber, R. C., Rudge, J. S., and Silver, J. (1991). Reduction of neurite outgrowth in a model of glial scarring following CNS injury is correlated with the expression of inhibitory molecules on reactive astrocytes. *Journal of Neuroscience* 11:3398–3411.

McKerracher, L., and Hirscheimer, A. (1992). Slow transport of the cytoskeleton after axonal injury. *Journal of Neurobiology* 23:568–578.

McKerracher, L., Vidal-Sanz, M., and Aguayo, A. J. (1990a). Slow transport rates of cytoskeletal proteins change during regeneration of axotomized retinal neurons in adult rats. *Journal of Neuroscience* 10:641–648.

McKerracher, L., Vidal-Sanz, M., Essagian, C., and Aguayo, A. J. (1990b). Selective impairment of slow axonal transport after optic nerve injury in adult rats. *Journal of Neuroscience* 10:2834–2841.

Meller, K. (1987). Early structural changes in the axoplasmic cytoskeleton after axotomy studied by cryofixation. *Cell and Tissue Research* 250:663–672.

Meyer, M., Matsuoka, I., Wetmore, C., Olson, L., and Thoenen, H. (1992). Enhanced synthesis of brain-derived neurotrophic factor in the lesioned peripheral nerve: Different mechanisms are responsible for the regulation of BDNF and NGF mRNA. *Journal of Cell Biology* 119:45–54.

Mikucki, S. A., and Oblinger, M. M. (1991). Corticospinal neurons exhibit a novel pattern of cytoskeletal gene expression after injury. *Journal of Neuroscience Research* 30:213–225.

Mohiuddin, L., Fernyhough, P., and Tomlinson, D. R. (1996). Acidic fibroblast growth factor enhances neurite outgrowth and stimulates expression of GAP-43 and T alpha 1 alpha-tubulin in cultured neurones from adult rat dorsal root ganglia. *Neuroscience Letters* 215:111–114.

Mullins, F. H., Hargreaves, A. J., Li, J. Y., Dahlström, A., and McLean, W. G. (1994). Tyrosination state of alpha-tubulin in regenerating peripheral nerve. *Journal of Neurochemistry* 62:227–234.

Murphy, S., Simmons, M. L., Agullo, L., Garcia, A., Feinstein, D. L., Galea, E., Reis, D. J., Minc-Golomb, D., and Schwartz, J. P. (1993). Synthesis of nitric oxide in CNS glial cells. *Trends in Neurosciences* 16:323–328.

Nunez, J., and Fischer, I. (1997). Microtubule-associated proteins (MAPs) in the peripheral nervous system during development and regeneration. *Journal of Molecular Neuroscience* 8:207–222.

Oestreicher, A. B., De Graan, P. N. E., Gispen, W. H., Verhaagen, J., and Schrama, L. H. (1997). B-50, the growth associated protein-43: Modulation of cell morphology and communication in the nervous system. *Progress in Neurobiology* 53:627–686.

Ohi, M., Ide, C., and Tohyama, K. (1989). Nerve regeneration in the dorsal funiculus of the rat spinal cord: A light and electron microscopic study. *Archives of Histology and Cytology* 52:373–386.

Pant, H. C. (1988). Dephosphorylation of neurofilament proteins enhances their susceptibility to degradation by calpain. *Biochemical Journal* 256:665–668.

Perry, V. H., and Brown, M. C. (1992). Role of macrophages in peripheral nerve degeneration and repair. *Bioessays* 14:401–406.

Perry, V. H., Brown, M. C., and Gordon, S. (1987). The macrophage response to central and peripheral nerve injury. A possible role for macrophages in regeneration. *Journal of Experimental Medicine* 165:1218–1223.

Pindzola, R. R., Doller, C., and Silver, J. (1993). Putative inhibitory extracellular matrix molecules at the dorsal root entry zone of the spinal cord during development and after root and sciatic nerve lesions. *Developmental Biology* 156:34–48.

Posmantur, R., Hayes, R. L., Dixon, C. E., and Taft, W. C. (1994). Neurofilament 68 and neurofilament 200 protein levels decrease after traumatic brain injury. *Journal of Neurotrauma* 11:533–545.

Posmantur, R., Kampfl, A., Siman, R., Liu, J., Zhao, X., Clifton, G. L., and Hayes, R. L. (1997). A calpain inhibitor attenuates cortical cytoskeletal protein loss after experimental traumatic brain injury in the rat. *Neuroscience* 77:875–888.

Povlishock, J. T., and Pettus, E. H. (1996). Traumatically induced axonal damage: Evidence for enduring changes in axolemmal permeability with associated cytoskeletal change. *Acta Neurochirurgica. Supplementum* 66:81–86.

Ramon y Cajal, S. (1928). *Degeneration and regeneration of the nervous system*. London: Oxford University Press.

Reichert, F., Saada, A., and Rotshenker, S. (1994). Peripheral nerve injury induces Schwann cells to express two macrophage phenotypes: Phagocytosis and the galactose-specific lectin MAC-2. *Journal of Neuroscience* 14:3231–3245.

Roberts-Lewis, J. M., Savage, M. J., Marcy, V. R., Pinsker, L. R., and Siman, R. (1994). Immunolocalization of calpain I-mediated spectrin degradation to vulnerable neurons in the ischemic gerbil brain. *Journal of Neuroscience* 14:3934–3944.

Rubin, B. P., Dusart, I., and Schwab, M. E. (1994). A monoclonal antibody (IN-1) which neutralizes neurite growth inhibitory proteins in the rat CNS recognizes antigens localized in CNS myelin. *Journal of Neurocytology* 23:209–217.

Rudge, J. S., and Silver, J. (1990). Inhibition of neurite outgrowth on astroglial scars *in vitro*. *Journal of Neuroscience* 10:3594–3603.

Savio, T., and Schwab, M. E. (1989). Rat CNS white matter, but not gray matter, is nonpermissive for neuronal cell adhesion and fiber outgrowth. *Journal of Neuroscience* 9:1126–1133.

Schaumberg, H. H., Berger, A. R., and Thomas, P. K. (1992). *Disorders of peripheral nerves*. Philadelphia: F.A. Davis.

Schlaepfer, W. W., and Hasler, M. B. (1979). Characterization of the calcium-induced disruption of neurofilaments in rat peripheral nerve. *Brain Research* 168:299–309.

Schlaepfer, W. W., and Micko, S. (1978). Chemical and structural changes of neurofilaments in transected rat sciatic nerve. *Journal of Cell Biology* 78:369–378.

Schlaepfer, W. W., and Zimmerman, U. P. (1981). Calcium-mediated breakdown of glial filaments and neurofilaments in rat optic nerve and spinal cord. *Neurochemical Research* 6:243–255.

Schnell, L., and Schwab, M. E. (1990). Axonal regeneration in the rat spinal cord produced by an antibody against myelin-associated neurite growth inhibitors. *Nature* 343:269–272.

Schwab, M., Kapfhammer, J. P., and Bandtlow, C. E. (1993). Inhibitors of neurite growth. *Annual Review of Neuroscience* 16:565–595.

Schwab, M. E., and Schnell, L. (1990). Channeling of developing rat corticospinal tract axons by myelin-associated neurite growth inhibitors. *Journal of Neuroscience* 11:709–721.

Singer, P. A., Mehler, S., and Fernandez, H. L. (1982). Blockade of retrograde axonal transport delays the onset of metabolic and morphologic changes induced by axotomy. *Journal of Neuroscience* 2:1299–1306.

So, K. F., and Aguayo, A. J. (1985). Lengthy regrowth of cut axons from ganglion cells after peripheral nerve transplantation into the retina of adult rats. *Brain Research* 328:349–354.

Stoll, G., Griffin, J. W., Li, C. Y., and Trapp, B. D. (1989). Wallerian degeneration in the peripheral nervous system: Participation of both Schwann cells and macrophages in myelin degradation. *Journal of Neurocytology* 18:671–683.

Strittmatter, S. M., Fankhauser, C., Huang, P. L., Mashimo, H., and Fishman, M. C. (1995). Neuronal pathfinding is abnormal in mice lacking the neuronal growth cone protein GAP-43. *Cell* 80:445–452.

Taniuchi, M., Clark, H. B., and Johnson, E. M. J. (1986). Induction of nerve growth factor receptor in Schwann cells after axotomy. *Proceedings of the National Academy of Sciences* 83:4094–4098.

Taniuchi, M., Clark, H. B., Schweitzer, J. B., and Johnson, E. M. J. (1988). Expression of nerve growth factor receptors by Schwann cells of axotomized peripheral nerves: Ultrastructural location, suppression by axonal contact, and binding properties. *Journal of Neurosciene* 8:664–681.

Tetzlaff, W., Alexander, S. W., Miller, F. D., and Bisby, M. A. (1991). Response of facial and rubrospinal neurons to axotomy: Changes in mRNA expression for cytoskeletal proteins and GAP-43. *Journal of Neuroscience* 11:2528–2544.

Tetzlaff, W., Bisby, M. A., and Kreutzberg, G. W. (1988). Changes in cytoskeletal proteins in the rat facial nucleus following axotomy. *Journal of Neuroscience* 8:3181–3189.

Thery, C., Chamak, B., and Mallat, M. (1991). Free radical killing of neurons. *European Journal of Neuroscience* 3:1155–1164.

Thery, C., Chamak, B., and Mallat, M. (1993). Neurotoxicity of brain macrophages. *Clinical Neuropathology* 12:288–290.

Thomas, P. K., and Olsson, Y. (1984). *Microscopic anatomy and function of the connective tissue components of peripheral nerve*. Philadelphia: W.B. Saunders.

Tonge, D. A., Golding, J. P., and Gordon-Weeks, P. R. (1996). Expression of a developmentally regulated, phosphorylated isoform of microtubule-associated protein 1B in sprouting and regenerating axons *in vitro*. *Neuroscience* 73:541–551.

Torvik, A. (1976). Central chromatolysis and the axon reaction: A reappraisal. *Neuropathology and Applied Neurobiology* 2:423–432.

Trojanowski, J. Q., Lee, V. M., and Schlaepfer, W. W. (1984). Neurofilament breakdown products in degenerating rat and human peripheral nerves. *Annals of Neurology* 16:349–355.

Trupp, M., Belluardo, N., Funakoshi, H., and Ibanez, C. F. (1997). Complementary and overlapping expression of glial cell line-derived neurotrophic factor (GDNF), c-ret proto-oncogene, and GDNF receptor-alpha indicates multiple mechanisms of trophic actions in the adult rat cns. *Journal of Neuroscience* 17:3554–3567.

Tseng, G. F., Shu, J., Huang, S. J., and Wang, Y. J. (1995). A time-dependent loss of retrograde transport ability in distally axotomized rubrospinal neurons. *Anatomy and Embryology* 191:243–249.

Vallee, R. B., and Bloom, G. S. (1991). Mechanisms of fast and slow axonal transport. *Annual Review of Neuroscience* 14:59–92.

Vidal-Sanz, M., Bray, G. M., Villegas-Pe;a;rez, M. P., Thanos, S., and Aguayo, A. J. (1987). Axonal regeneration and synapse formation in the superior colliculus by retinal ganglion cells in the adult rat. *Journal of Neuroscience* 7:2894–2909.

Wang, G. Y., Hirai, K., and Shimada, H. (1992). The role of laminin, a component of Schwann cell basal lamina, in rat sciatic nerve regeneration within antiserum-treated nerve grafts. *Brain Research* 570:116–125.

Woodroofe, M. N., Hayes, G. M., and Cuzner, M. L. (1989). Fc receptor density, MHC antigen expression and superoxide production are increased in interferon-gamma-treated microglia isolated from adult rat brain. *Immunology* 68:421–426.

Wu, W., Gloster, A., and Miller, F. D. (1997). Transcriptional repression of the growth-associated T alpha1 alpha-tubulin gene by target contact. *Journal of Neuroscience Research* 48:477–487.

Wu, W., Matthew, T. C., and Miller, F. D. (1993). Evidence that the loss of homeostatic signals induces regeneration-associated alterations in neuronal gene expression. *Developmental Biology* 158:456–466.

Zimmerman, U. J., and Schlaepfer, W. W. (1982). Characterization of a brain calcium-activated protease that degrades neurofilament proteins. *Biochemistry* 21:3977–3982.

CHAPTER 3

Beyond the Glial Scar

Cellular and Molecular Mechanisms by which Glial Cells Contribute to CNS Regenerative Failure

MICHAEL T. FITCH AND JERRY SILVER

Department of Neurosciences, Case Western Reserve University School of Medicine, Cleveland, Ohio

The responses of glial cells to injury in the adult central nervous system have long been considered to be detrimental to successful recovery. This chapter will provide a comprehensive review of selected aspects of gliosis at the tissue, cellular, and molecular levels as they relate to oligodendrocyte, astrocyte, and microglial–macrophage responses to trauma and the abortive attempts of neuronal regeneration. The historical perspective and modern approaches detailed in this chapter will demonstrate that the field of glial cell biology has allowed us to go beyond purely mechanical considerations of the glial scar, and in doing so has provided new insights into the complex reactions and interactions of glial cells following injury that generate the generally nonpermissive nature of lesion sites in the adult central nervous system.

I. INTRODUCTION

Glial cells of the central nervous system (CNS), as first described by Virchow (1846), were originally considered to be simply a mechanical framework to support neurons, as evidenced by their designation as "neuroglia," which literally translated means "nerve glue" (Weigert, 1895). In the years since their discovery, glial cells of the developing and mature CNS have been

CNS Regeneration

extensively studied and recognized as important functional components of the brain and spinal cord in addition to their role in the structural arrangement of the CNS. Oligodendrocytes are recognized as the cells responsible for the myelination of axons within the CNS (Bunge, 1968; Wood and Bunge, 1984), whereas astrocytes have been demonstrated to participate in formation and maintenance of the blood brain barrier (Wolburg and Risau, 1995), ion homeostasis (Walz, 1989), neurotransmitter transport (Schousboe and Westergaarde, 1995), production of extracellular matrix (Liesi *et al.*, 1983; Bernstein *et al.*, 1985; Liesi and Silver, 1988; Tomaselli *et al.*, 1988; Grierson *et al.*, 1990; McKeon *et al.*, 1991; Ard *et al.*, 1993; Smith-Thomas *et al.*, 1994; McKeon *et al.*, 1995; Canning *et al.*, 1996), and other functional roles. Microglial cells, widely believed to be related to cells of the peripheral macrophage and monocyte lineage (Ling and Wong, 1993; Theele and Streit, 1993), are the CNS resident phagocytic cells that maintain a resting phenotype during nonpathological conditions in the CNS (Banati and Graeber, 1994; Gehrmann *et al.*, 1995).

In the adult peripheral nervous system (PNS), it has been known for many years that traumatic crush injury leads to a cellular response that frequently results in successful regeneration of injured axons (for review see Guth, 1956). Unlike this robust regenerative response in the PNS, injury to the adult mammalian CNS leads, at best, to abortive regeneration within the long myelinated axon tracts resulting in permanent disability with little or no functional regeneration of injured axons (Ramon y Cajal, 1928). For many years, the hypothesis was maintained that regeneration failure following injury was due primarily to a structural barrier to axon growth, the so-called "glial scar," composed chiefly of glial cells (primarily astrocytes) and connective tissue elements (Windle and Chambers, 1950; Windle *et al.*, 1952; Clemente and Windle, 1954; Windle, 1956; Clemente, 1958).

The suggestion that the glial scar is an impenetrable mechanical barrier to regenerating axons was consistent with early histological observations and the classic idea that glial cells were merely structural components of the nervous system. However, in light of recent insights into the functional biology of glial cells, many investigators have reevaluated the role of the glial scar in preventing CNS regeneration following traumatic injury (for review see Reier *et al.*, 1983). Recent studies have taken the field of CNS regeneration beyond the physical ramifications of the glial scar, and this chapter is dedicated to a discussion of the cellular as well as molecular responses of oligodendrocytes, astrocytes, and microglial cells to traumatic injury and their proposed roles in the failure of functional regeneration of the adult mammalian CNS.

II. ARE ADULT CNS AXONS CAPABLE OF ROBUST REGENERATION?

Before addressing the issue of glial cell influences on CNS injury responses, the first, and perhaps simplest, explanation for CNS regenerative failure must be considered. One explanation for the failure of axon regeneration in the adult CNS is that mature adult mammalian neurons, unlike their immature counterparts, are incapable of long-distance regrowth of cut or damaged processes. Le Gros Clark (1942) suggested that adult CNS axons had a "feeble capacity" for axon growth as compared to PNS axons and even went so far as to claim that perhaps CNS neurons had no capacity to regenerate at all (Le Gros Clark, 1943). However, these suggestions were contradictory to the early observations that had been made by Ramon y Cajal (1928), which indicated that adult CNS axons could undergo limited growth and sprouting at the site of a lesion, but could not enter or cross the area of damage. Liu and Chambers (1958) demonstrated that axons of the spinal cord were capable of collateral sprouting, suggesting a plasticity and potential for growth in the adult CNS that had not been previously appreciated. Observations of limited growth by CNS axons on the proximal side of a lesion in the spinal cord (Guth *et al.*, 1981, 1985, 1986) also suggested that adult axons have at least a limited intrinsic ability to grow. Analyses of cut axon endings using orthograde labeling suggest that severed axons are, in fact, capable of dynamic albeit short-distance sprouting immediately following axotomy (Li and Raisman, 1994), but eventually the sprouting growth cones become quiescent and can persist for many months or even over a year without leaving the site of injury (Li and Raisman, 1995). Although Tello (1911a, 1911b) had reported growth of CNS axons within transplanted peripheral nerves, it was not until Aguayo and his colleagues demonstrated long-distance axon growth by intrinsic CNS neurons into grafts of peripheral nerve almost 70 years later that the robust intrinsic potential of certain adult CNS axons to regenerate *in vivo* became widely accepted (Richardson *et al.*, 1980; David and Aguayo, 1981). These studies have led to the notion that the environment of the PNS is in some respects more conducive to axon elongation than the environment of the adult CNS, and that adult CNS axons are capable of long-distance axon regeneration under certain conditions. However, the failure of large numbers of regenerating axons to successfully leave peripheral nerve grafts and regenerate long distances into the CNS highlights the transition between CNS and PNS environments as being a particularly important interface.

Further evidence that the interface between the adult CNS environment and the PNS is nonpermissive for axon regeneration comes from studies of injured dorsal roots of the spinal cord. The cell bodies of the sensory dorsal root ganglion (DRG) neurons are situated in the PNS and extend bipolar axons both peripherally and centrally. Crush injury to the peripheral branch of this axon leads to successful regeneration in the PNS, whereas similar injury to the central branch does not. However, careful analysis of the central branch of the axon after injury suggests that this portion of the axon does indeed possess regenerative potential because it can regenerate from the ganglion through the dorsal root right up to the surface of the spinal cord where it stops or turns abruptly at the dorsal root entry zone, the interface between PNS and CNS (Tower, 1931; Kimmel and Moyer, 1947; Perkins *et al.*, 1980; Kliot *et al.*, 1990). Thus, the axons of adult DRG neurons seem to be capable of regeneration, but regenerative failure occurs when these growing axons reach the interface between the cellular environments of the PNS and the CNS. Even axons from the ventral roots, known for their aggressive regenerative properties in the PNS, are unable to enter the CNS compartment when grafted onto injured dorsal roots (Carlstedt, 1983, 1985a, 1985b). It is important to note that in very young animals (less than one week of age) regenerating dorsal roots axons of the C-fiber system are capable of reentering the CNS compartment (Carlstedt *et al.*, 1987, 1988; Carlstedt, 1988), demonstrating *in vivo* that the interface between the CNS environment and the PNS environment in immature animals is permissive for axon navigation of this boundary, whereas this CNS/PNS interface in the adult is nonpermissive for axon regeneration into the CNS compartment.

Robust regeneration of adult DRG neurons up to 6–7 mm in just a few days has been recently demonstrated *in vivo* when these cells are transplanted directly into adult white matter tracts of the brain, minimizing trauma to CNS tissue and avoiding the PNS/CNS interface altogether (Davies *et al.*, 1997). This study demonstrates conclusively the intrinsic ability of at least one subset of adult neurons to regenerate long distances within the normal CNS environment. It is important to realize, however, that such intrinsic growth ability for adult cells may not be present among all neuron cell types and does not seem to equal the robust axon outgrowth capacity of transplanted DRG neurons or embryonic neurons during development. The subsets of adult neurons that elongate into PNS implants have been shown to dramatically increase expression of GAP-43, suggesting that the relative ability of neurons to upregulate GAP-43 may play an important role in determining which adult axons are most capable of robust axonal growth (Campbell *et al.*, 1991; Vaudano *et al.*, 1995), because upregulation of GAP-43 has been implicated as a necessary but not sufficient condition for neuron regeneration (Tetzlaff *et al.*, 1994). *In vitro* studies have demonstrated relatively slow growth for adult neurons by showing that various neuronal cell types grow at significantly faster rates at early ages (embryonic or postnatal) that adult neurons when presented with similar substrates, including superior cervical ganglion neurons (Argiro *et al.*, 1984), DRG neurons (Shewan *et al.*, 1995), and retinal ganglion neurons (Bahr *et al.*, 1995; Chen *et al.*, 1995), although these studies contrast with the *in vivo* documentation of adult DRG axon growth at the remarkable rate of 1–2 mm per day (Davies *et al.*, 1997). In addition, adult neurons may also respond differently to trophic factors, perhaps due to an intrinsic tendency to ramify. Thus, exogenous NGF leads to increased branching of adult DRG neurons as opposed to the increased axon length seen with NGF-treated neonatal DRG cells (Yasuda *et al.*, 1990). Such observations may help to explain why some adult axons will sprout and branch at the lesion site instead of extending axons linearly, which is more characteristic of embryonic neurite outgrowth. Therefore, improving the regeneration of adult axons may have to be done selectively for various neuronal subpopulations. Strategies to improve regeneration in the CNS may also need to consider incorporating molecules such as L1 that may play a role in increasing the rectilinear growth potential of different types of neurons (Brittis *et al.*, 1995), as well as making a cellular environment generally more conducive to axon growth.

III. DO GLIAL CELLS CONTRIBUTE TO CNS REGENERATIVE FAILURE?

A. Oligodendrocytes and Myelin

Oligodendrocytes, the cells responsible for the production of myelin in the CNS, have not been extensively studied in terms of their cellular responses to injury. They are not thought to play a major reactive role in the glial response to trauma. Although oligodendrocytes have been shown to proliferate following injury (Ludwin, 1984, 1985), this activity is found in the wound tract only transiently (Xie *et al.*, 1995), and no changes in the expression of myelin proteins have been demonstrated (Ludwin, 1985). Much attention, however, has been given to the hypothesis that the native cell membranes and myelin formed by mature oligodendrocytes throughout the CNS may be partially responsible for the failure of adult CNS axons to regenerate.

A number of studies have contributed evidence that suggests that the presence of myelin in the CNS may create an environment less conducive to axon growth (Caroni *et al.*, 1988; Savio and Schwab, 1990; Bastmeyer *et al.*, 1991; Stuermer *et al.*, 1992; Schwab *et al.*, 1993; Spillmann *et al.*, 1997). For example, the chicken nervous system is capable of complete recovery following spinal cord transection prior to myelin formation; but after the developmental onset of myelination, this capacity is lost (Keirstead *et al.*, 1992). Treatments that delay the onset of myelination in both chickens and rats lead to greater regrowth of cut axons at developmental time periods during which regeneration is normally absent (Savio and Schwab, 1990; Keirstead *et al.*, 1992), suggesting that the normal presence of myelin is at least associated with the time of the transition between a permissive and a nonpermissive environment for axon growth in the CNS. Some tissue culture studies have suggested that embryonic retinal neurons and certain peripheral neurons (such as early postnatal DRG or sympathetic ganglion neurons) avoid oligodendrocytes *in vitro* and have limited neurite outgrowth abilities on CNS myelin membranes when compared to myelin from the PNS (Caroni *et al.*, 1988; Fawcett *et al.*, 1989). Isolation of protein fractions from CNS myelin suggests that the nonpermissive nature of this substrate is due primarily to two main molecular weight fractions of 35 and 250 kD (Caroni and Schwab, 1988b). An antibody to these fractions (designated the IN-1 antibody) has been used to neutralize the inhibitory nature of CNS white matter in a variety of studies *in vitro* (Caroni and Schwab, 1988a; Savio and Schwab, 1989; Bandtlow *et al.*, 1990; Spillmann *et al.*, 1997), demonstrating that the inhibitory nature of CNS myelin may be modified to be more permissive to axonal outgrowth. The neutralizing IN-1 antibody has also been utilized *in vivo* in several studies and results in histological and behavioral recovery from injury (Schnell and Schwab, 1990; Bregman *et al.*, 1995). Myelin-associated glycoprotein (MAG) has also been suggested as an identified component of myelin that inhibits neurite outgrowth (McKerracher *et al.*, 1994), but studies using MAG deficient mice have demonstrated that the absence of MAG alone does not abolish inhibitory aspects of oligodendrocytes and myelin both *in vitro* and *in vivo* (Bartsch *et al.*, 1995; Bartsch, 1996). However other evidence suggests that MAG may in fact play some role in regenerative failure (Tang *et al.*, 1997). Until further studies identify MAG or other specific myelin components as instrumental in preventing axon regeneration *in vivo*, the investigations of myelin as an inhibitory component of the CNS suggest that the uncharacterized myelin antigens recognized by the IN-1 antibody may contribute to the regenerative failure seen in the adult CNS after injury and that modulation of these inhibitors may be therapeutic.

Although the histological and functional recovery demonstrated using the IN-1 antibodies *in vivo* suggests the importance of modifying myelin components following a CNS lesion, the specificity of this treatment as a therapeutic agent to allow axon regeneration remains to be further elucidated, particularly in light of studies demonstrating local neurite sprouting in response to IN-1 without long-distance regeneration through the lesion site (Guest *et al.*, 1997; Thallmair *et al.*, 1998; Z'Graggen *et al.*, 1998). Experiments using the IN-1 antibody *in vivo* utilized control antibodies raised against horseradish peroxidase (Schnell and Schwab, 1990; Bregman *et al.*, 1995; Guest *et al.*, 1997; Thallmair *et al.*, 1998; Z'Graggen *et al.*, 1998), an antigen not present within normal myelin structure where the IN-1 antigens are located (Rubin *et al.*,

1994). This raises the possibility that antibodies that recognize other myelin proteins may have similar beneficial effects as those demonstrated in these studies, not as a direct effect of blocking a specific inhibitory component of myelin, but as an indirect effect by causing a physical disruption to the myelin structure. In fact, experiments by other investigators have demonstrated regeneration of axons and electrophysiological recovery following disruption of the myelin structure via *in vivo* treatments with Gal-C or O_4 antibodies (against antigens found in normal myelin) combined with complement (Keirstead *et al.*, 1995). Such results demonstrate that less specific disruption of myelin is beneficial to axon regeneration, and that a component of the normal inflammatory response can mediate such alterations in myelin structure. Importantly, studies using the IN-1 antibodies as therapeutic agents have not addressed the possibility that antibodies that bind to oligodendrocytes and myelin membranes may, in turn, create a persistent low-grade inflammatory reaction that is secondary to antibody binding or myelin destruction. Such limited inflammatory responses have been associated previously with beneficial effects on axon regeneration *in vivo* (Windle and Chambers, 1950; Kliot *et al.*, 1990; Siegal *et al.*, 1990), perhaps through the direct effects of secreted inflammatory products on axon growth or indirectly via cytokine modulation of reactive astrocytes.

Although myelin-associated inhibitors have been demonstrated as one component of the CNS that may contribute to the failure of regeneration, the relative potency of their actions remains to be determined. A number of recent studies have questioned the role of myelin and myelin components as the sole inhibitory influences for axon growth. Tissue culture experiments have shown that although DRG neurons are inhibited by oligodendrocytes (Kobayashi *et al.*, 1995), retinal ganglion cell axons are not altered by contact with these myelinating cells (Ard *et al.*, 1991; Kobayashi *et al.*, 1995), observations that contrast with the initial experiments that established the inhibitory properties of oligodendrocytes (Caroni *et al.*, 1988). In addition, unmyelinated optic nerves from late-stage embryos have been shown to be nonpermissive for neonatal and adult axon growth *in vitro*, despite the lack of myelin components (Shewan *et al.*, 1993), whereas embryonic DRG neurons can grow on both unmyelinated neonatal or fully myelinated adult optic nerves (Shewan *et al.*, 1995), suggesting that issues of neuronal age and subtype may be critical to consider before proposing global inhibitory effects for myelin substrates. The Bowman-Wyse mutant rat has areas of the optic nerve devoid of myelin and oligodendrocytes that still demonstrate a lack of regenerative ability *in vivo* (Berry *et al.*, 1992), suggesting that the absence of myelin and oligodendrocytes is not sufficient to allow regeneration to occur. Embryonic neurons, transplanted gently into the adult nervous system, have the ability to extend long axons through the heavily myelinated environment of an adult white matter axon tract *in vivo* (Davies *et al.*, 1993; Li and Raisman, 1993). Adult axons in the cingulum have also been observed growing distances of up to 200 micrometers within a myelinated white matter tract following axotomy, although such growth was not through the lesion site but rather back toward the neuron cell body (Davies *et al.*, 1996). A study by Berry and associates (1996) suggests that myelin components do not necessarily need to be neutralized to get regeneration in the acutely lesioned optic nerve, as trophic factors administered to the retinal ganglion cell bodies via a crude PNS graft can partially overcome any inhibitory factors that may be acting on the severed axons distally in the injured optic nerve. The most convincing evidence to date that the presence of myelin in adult CNS white matter tracts is not sufficient to inhibit the regeneration of adult CNS axons is the series of *in vivo* experiments that demonstrate robust regeneration of transplanted DRG cell axons across the corpus callosum into the contralateral hemisphere (Davies *et al.*, 1997). These recent studies suggest that the inhibitory nature of myelin is not the universal factor that leads to regenerative failure in the CNS. It remains to be determined under what circumstances myelin inhibitors may play a role as a component of regenerative failure, and where factors other than myelin work to establish a nonpermissive environment for regeneration of the adult CNS.

B. Astrocytes and the Glial Scar

1. Astrogliosis

The astrocytic cell responses to injury have been extensively studied in a variety of experimental models, and the terms "gliosis" and "astrogliosis" are often used to describe the astrocyte reactions to injury. Astrocyte cellular hypertrophy, hyperplasia, and increased production of intermediate filaments characterize astrocyte gliosis, and cells responding in these ways to injury are often referred to as "reactive astrocytes." Astocytes are easily identified by immunocytochemical methods directed toward the astrocyte specific glial fibrillary acidic protein (GFAP) (Bignami *et al.*, 1972; Bignami and Dahl, 1974; Eng, 1985), and astrocyte hypertrophy and increased GFAP following injury have been demonstrated using these techniques to label reactive astrocytes (Bignami and Dahl, 1976). The issue of astrocyte cell division following CNS injury is less clearly recognized, as various studies have determined that some reactive astrocytes divide after injury but the majority of them do not (Norton *et al.*, 1992). A number of investigators have suggested that there are relatively few proliferating astrocytes (ranging from 1–2% to 5–6%), that they are restricted to the immediate area of the wound, and that the apparent increase in numbers of reactive astrocytes is not primarily due to proliferation, but rather to migration of cells and the enhanced ability to visualize them with

GFAP antibodies (Adrian and Williams, 1973; Murray and Walker, 1973; Latov *et al.*, 1979; Ludwin, 1985; Miyake *et al.*, 1988; Takamiya *et al.*, 1988; Murray *et al.*, 1990; Hatten *et al.*, 1991; Miyake *et al.*, 1992; Amat *et al.*, 1996). Thus, astrocytic hypertrophy and increased expression of GFAP are widely recognized as characteristics of reactive astrocytes and are often used as markers for gliosis. The previously described cellular hyperplasia as a component of gliosis is predominantly due to increases in inflammatory cells within the lesion.

2. Astrocytes as a Mechanical Barrier

As briefly discussed in the introduction to this chapter, the idea that the astrocytes that comprise the glial scar serve primarily as a mechanical barrier to regeneration has been discounted in recent years (for review see Reier *et al.*, 1983). Although this hypothesis was originally widely supported (Windle and Chambers, 1950; Windle *et al.*, 1952; Clemente and Windle, 1954; Windle, 1956; Clemente, 1958), recent work suggests that the astrocytic scar that forms after traumatic injury does not prevent axon growth simply via a mechanical mechanism. Spinal cord injury in hibernating squirrels does not lead to formation of a dense glial scar, but regeneration still does not occur in the absence of a histologically apparent barrier (Guth *et al.*, 1986). Relatively normal histological tissue architecture is restored within days of a microlesion injury to the cingulum that transects axons without severely disrupting the glial framework, and despite the absence of a histological glial scar, regenerating axons do not cross the lesion site even though some turn and elongate proximally for distances that would have carried them across the lesion site had they been growing distally (Davies *et al.*, 1996). These experiments demonstrate that the lack of a histologically dense and mechanically obstructive glial scar does not in itself lead to successful regeneration. Current theories of how astrocytes may contribute to the lack of CNS regeneration include: (1) failure to provide a proper cellular substrate for axon growth, (2) failure to produce molecules that support axons, and (3) production of molecules that actively inhibit axon elongation.

3. Astrocytes as Substrates for Axon Elongation

Astrocytes have been implicated as negative components of the mature mammalian CNS response to injury that may be responsible in part for the failure of regeneration, and much attention has been given to the role of astrocytes as substrates for neuron outgrowth. Glial cells of the astrocyte lineage play an integral role during development of the nervous system as a substrate for neuronal migration and axon elongation *in vivo* (Silver *et al.*, 1993; Rakic, 1995). Neonatal astrocytes have been shown to be a supportive substrate for axon growth *in vitro* (Fallon, 1985b, 1985a; Rudge *et al.*, 1989; Ard *et al.*, 1991; Bahr *et al.*, 1995), whereas reactive astrocytes are a nonpermissive substrate for axon growth (Rudge *et al.*, 1989; Smith *et al.*, 1990; Geisert and Stewart, 1991; Bahr *et al.*, 1995; Le Roux and Reh, 1996), suggesting that the reactive astrocytes present following a traumatic injury to the CNS may contribute to an inhibitory cellular environment. However, it is interesting to note that although axon elongation is severely limited, dendritic growth is much less altered by reactive astrocytes (Le Roux and Reh, 1996). This observation suggests that perhaps some of the limited neurite growth and abortive sprouting seen in the vicinity of wounds in grey matter are dendritic processes instead of short axonal projections. The lack of axonal outgrowth on adult reactive astrocytes *in vitro* suggests that these cells may contribute to CNS regenerative failure either by actively inhibiting or by not providing an appropriately supportive environment.

Although reactive astrocytes may not be favorable substrates for axon elongation, the presence of a cellular terrain for axons to grow on is certainly essential (Guth *et al.*, 1981, 1985), especially considering that axons will not traverse a purely fluid environment (Harrison, 1910, 1914). A problem that has plagued the field of CNS injury research for many years is the progressive necrosis and development of cavities or cysts as secondary events following trauma to the CNS. Such cavitation can develop from a small initial lesion that progresses to a large cavity extending far rostral and caudal to the original area of injury (Balentine, 1978). Although investigators have hypothesized that cavitation and central necrosis are related to ischemic injury (Balentine, 1978), hemorrhage (Ducker *et al.*, 1971; Wallace *et al.*, 1987), neuronal lysozyme activity (Kao *et al.*, 1977a; Kao *et al.*, 1977b), macrophage infiltration and inflammation (Blight, 1994; Fujiki *et al.*, 1996; Szczepanik *et al.*, 1996; Zhang *et al.*, 1996; Fitch and Silver, 1997a), or leakage of serum proteins across the blood brain barrier (Fitch and Silver, 1997a), the underlying causes of progressive necrosis leading to cyst formation are still under active investigation. This is an important topic for further study, because an acellular cyst obviously lacks the appropriate cellular substrates for axon regeneration to occur.

4. Astrocyte Production of Axon Supportive Molecules

Another possible way in which astrocytes could play a role in the failure of regeneration is to fail to produce appropriate factors that are supportive for axon growth. The developing nervous system contains a number of molecules produced by astrocytes or their precursors that encourage axon growth (Tomaselli *et al.*, 1988; Smith *et al.*, 1990), but until recently it was believed that adult astrocytes did not produce the same types of supportive molecules. It is now recognized that adult astrocytes can produce laminin *in vitro* and *in vivo* (Liesi *et al.*, 1984; Bernstein *et al.*, 1985; Liesi and Silver, 1988; Frisen *et al.*, 1995), and laminin has been described as a component of glial scars *in vivo* (Bernstein *et al.*, 1985; McKeon *et al.*, 1991, 1995; Risling *et al.*, 1993; Frisen *et al.*, 1995). Many other molecules thought to promote axon regeneration have been shown to be produced by reactive astrocytes (for review see Eddleston and Mucke, 1993). However, questions remain that may prove to be important in determining the regenerative response of injured axons in relation to other molecules of the CNS. Thus, although astrocytes certainly do produce axon supportive molecules in the vicinity of CNS lesions, it is not known whether sufficient quantities, correct temporal or spatial sequences, or combinations of these axon growth supportive molecules are produced. It is also not well appreciated whether axotomized neurons reexpress the appropriate receptor molecules that would enable regeneration to occur. It is likely that interventions designed to increase concentrations or combinations of trophic and/or tropic molecules may be beneficial at certain time points following a traumatic injury, and active investigations into these issues are ongoing.

For example, some evidence indicates that growing adult axons can associate with reactive astrocytes after injury when supplied with exogenous trophic molecules such as NGF, although the temporal sequence of these cellular events has not been completely determined. Implantation of nitrocellulose filters containing NGF along with fetal tissue into the lesioned spiral cord leads to increases in axon growth from re-implanted cut dorsal roots, perhaps due to effects of the NGF on the glial cells that organize around the implant (Houle, 1992). Fibroblasts that are genetically engineered to secrete NGF have been implanted into the sites of CNS damage, and this treatment encourages large numbers of axons to regenerate into the usually nonpermissive central region of injury (Kawaja and Gage, 1991; Kawaja *et al.*, 1992; Tuszynski *et al.*, 1994, 1996, 1997). However, although these approaches have shown the ability to stimulate sprouting into the lesion site, the difficulty appears to be in encouraging the axons to leave the area of trophic support and to regenerate out of the immediate site of injury toward their proper functional connections. This highlights a major dilemma inherent in any repair strategy that uses local application of trophic molecules at the cut ends of axons to stimulate regeneration. Unfortunately, the axons that are stimulated to grow toward the site of factor release are then unable to leave the "trophic oasis" that is present at the source of molecules and, thus, simply remain indefinitely in the immediate vicinity of the exogenous factors. A successful modulation of trophic factors must not only encourage axons to grow following injury, but must also provide a stimulus for elongation through the lesion site and back into the CNS parenchyma. Alternatively, delivery of such potential therapeutic agents could encourage growth of axons without attracting regeneration directly into the cellular graft, as demonstrated with grafts of NT-3 secreting cells leading to partial functional recovery and growth of corticospinal axons in adjacent grey matter (Grill *et al.*, 1997). Perhaps the addition of trophic factors combined with other therapeutic agents may provide such a stimulus, as demonstrated by Schnell and colleagues (1994) using NT-3 combined with an antibody to neutralize myelin inhibitors.

Another intriguing issue concerning the availability of trophic factors in the area of a CNS wound has been suggested by Frisen and associates (1993) and Fryer and colleagues (1997). A truncated form of the neurotrophin TrkB receptor that lacks the catalytic domain has been shown to be expressed by astrocytes and oligodendrocytes and is upregulated in the glial scar following injury (Frisen *et al.*, 1993; Fryer *et al.*, 1997). The hypothesis has been suggested that neurotrophins bound to the surface of glial cells via the truncated TrkB receptors may be used as a growth-promoting substrate for sprouting axons at the site of a lesion (Frisen *et al.*, 1993). However, this theory fails to explain why axon regrowth does not progress beyond minimal sprouting. Perhaps a more satisfying hypothesis is supported by *in vitro* experiments that demonstrate that cells expressing the truncated TrkB receptor fail to support neurite outgrowth from BDNF responsive neurons even in the presence of exogenous BDNF (Fryer *et al.*, 1997), suggesting that the truncated receptors may be binding and internalizing neurotrophins. Such a situation could, in a sense, form a "molecular sponge" that acts to soak up and remove trophic factors that may be required by growing axons. Thus, factors bound to truncated TrkB receptors (or other similar "molecular sponges") may be unavailable for growing axons, which could help to explain the lack of functional regeneration even in wounds in which abundant trophic molecules have been demonstrated. Astrocytes in the immediate vicinity of an injury may be inhibitory to axon regeneration in part by virtue of removing functional access to essential growth factors in the lesion area.

5. Production of Inhibitory Molecules by Astrocytes

A number of studies have suggested that astrocyte production of axon inhibitory molecules may explain several aspects of CNS regenerative failure (for reviews see McKeon and Silver, 1995; Hoke and Silver, 1996; Fitch and Silver, 1997b). A particular emphasis has been placed on extracellular matrix molecules produced by reactive astrocytes that are upregulated following injury, a notable example being tenascin (Laywell and Steindler, 1991; McKeon *et al.*, 1991, 1995; Laywell *et al.*, 1992; Brodkey *et al.*, 1995; Lips *et al.*, 1995; Zhang *et al.*, 1995). Some forms of tenascin have been implicated as negative influences of axon growth on astrocytes (Grierson *et al.*, 1990; Ard *et al.*, 1993; Smith-Thomas *et al.*, 1994; Chiquet-Ehrismann *et al.*, 1995; Gates *et al.*, 1997), and tenascin has been demonstrated as a component of glial scars extracted from adult brain that are poorly supportive of axon growth (McKeon *et al.*, 1991, 1995).

Proteoglycans are another class of molecules produced by reactive astrocytes that are suggested to play a role in the modulation of axon growth and regeneration. Proteoglycans are molecules consisting of a protein core with attached sugar moieties called glycosaminoglycans (GAGs) and are characterized by their GAG compositions as chondroitin sulfate, heparan sulfate, keratan sulfate, and dermatan sulfate (for a review of nervous tissue proteoglycans see Margolis and Margolis, 1993). The upregulation of such proteoglycans is found in many tissues throughout the body in pathological conditions. Changes in proteoglycans are found in regenerating skeletal muscle (Carrino *et al.*, 1988), arterial injury (Richardson and Hatton, 1993; Nikkari *et al.*, 1994; Jain *et al.*, 1996), atherogenesis (Srinivasan *et al.*, 1995), and corneal injury (Brown *et al.*, 1995). The nervous system is no exception, because increases in proteoglycans have also been demonstrated *in vivo* following trauma to the adult CNS. For example, the NG2 proteoglycan is increased transiently after injury to the CNS (Levine, 1994), and phosphacan is increased in glial scars (McKeon *et al.*, 1995). Chondroitin sulfate proteoglycans are increased and have been shown to persist in the extracellular matrix of the CNS following injury, including the spinal cord following dorsal root injury (Pindzola *et al.*, 1993), in the fornix following transection (Lips *et al.*, 1995), in the brain following a stab wound (Fitch and Silver, 1997a), in explants of wounded striatum (Gates *et al.*, 1997), and in the spinal cord following penetrating crush injury (Fitch and Silver, 1997a). In addition, chondroitin sulfate proteoglycans are closely associated with the abortive regeneration and turning of the axons in this model in the absence of physical glial scarring (S. J. A. Davies, M. T. Fitch, G. Raisman, and J. Silver, unpublished observations). The presence of these putative inhibitory proteoglycans *in vivo* following injury suggests a role for these molecules in contributing to the nonpermissive environment encountered in the CNS. Along these lines, studies have shown that changes in the ratios of chondroitin sulfate and heparan sulfate proteoglycans in the developing chicken spinal cord occur precisely at the transition between the permissive and nonpermissive environments for spinal cord regeneration (Dow *et al.*, 1994). Proteoglycans and other inhibitory molecules have been implicated in the failure of the adult CNS environment to support robust axon growth following traumatic injury, and some recent experiments are the first to demonstrate the association of these molecules in reactive glial extracellular matrix with the failure of adult axon regeneration (Davies *et al.*, 1997).

In support of these ideas, astrocytes *in vitro* have been demonstrated to produce proteoglycans and associated glycosaminoglycans (Norling *et al.*, 1984; Gallo *et al.*, 1987; Gallo and Bertolotto, 1990; Johnson-Green *et al.*,

1991; Geisert *et al.*, 1992; Smith-Thomas *et al.*, 1994), and proteoglycans associated with reactive astrocytes have been demonstrated to inhibit neurite outgrowth (Snow *et al.*, 1990; McKeon *et al.*, 1991, 1995; Snow *et al.*, 1991; Canning *et al.*, 1993, 1996; Dou and Levine, 1994). Such proteoglycans have been shown to inhibit neurite outgrowth as a result of their GAG chains *in vitro* (Snow *et al.*, 1990, 1991; Cole and McCabe, 1991; Fichard *et al.*, 1991; Bovolenta *et al.*, 1993) and sometimes as a function of their core proteins (Oohira *et al.*, 1991; Geisert and Bidanset, 1993; Dou and Levine, 1994). It has also been suggested that, in certain situations, proteoglycans may positively influence neuronal survival, attachment, and/or axon growth as well (Iijima *et al.*, 1991; Maeda *et al.*, 1995; Challacombe and Elam, 1997; Gates *et al.*, 1997; Kappler *et al.*, 1997). However, an intriguing observation made recently is that the increased neurite outgrowth seen on certain proteoglycans is often dendritic rather than axonal (Maeda and Noda, 1996), which is consistent with observations that neurite outgrowth demonstrated on reactive astrocyte substrates is also largely dendritic (Le Roux and Reh, 1996). The growth of axons on glial scar tissue is inhibited by proteoglycans present in the extracellular matrix, and this inhibition can be partially reversed by treatment of the scar tissue with enzymes to remove specific sugar moieties from the proteoglycan molecules (McKeon *et al.*, 1991, 1995). Furthermore, neutralizing antibodies directed against a heparan/chondroitin sulphate proteoglycan expressed after brain injury block the inhibition of neurite outgrowth and growth cone collapse activity *in vitro* (Bovolenta *et al.*, 1997). Although such studies provide evidence for a direct effect of proteoglycans on neurite extension, a recent report suggests that in some situations chondroitin sulfate may regulate and organize other extracellular matrix-associated molecules, perhaps by directly binding to various undefined growth promoting or inhibiting factors (Emerling and Lander, 1996). Such findings indicate that proteoglycan molecules present in glial scars may play a direct and/or indirect role in the lack of regeneration in the injured adult CNS, and that modification of the inhibitory effects of post-injury CNS scar tissue may be possible.

C. The Function of Inhibitory Molecules

The function of increased levels of proteoglycans or other inhibitory molecules following injury to the adult CNS remains unclear. Normal environmental factors of the adult CNS may favor the inhibition of axon growth in an attempt by the body to maintain normal synaptic connections (Hockfield *et al.*, 1990), and after injury the CNS cellular environment may become even more inhibitory in an effort to prevent the aberrant growth of axons and the formation of inappropriate connections. Alternatively, the upregulation of proteoglycans may be part of the protective CNS response to injury much as it is for other body tissues, and the functional aspects of the increases in these molecules may be unrelated to axon growth and regeneration. In other words, the negative effects of proteoglycan upregulation or other inhibitors on axon growth may simply be an unfortunate side effect of the normal wound-healing response by an injured tissue.

Thus, although it is clear that one effect of proteoglycan upregulation following injury may be the inhibition of functional regeneration, the question still remains about nature's intended function for these molecules in the CNS response to injury. Proteoglycans have been suggested to play a role in modulating growth factors (Ruoslahti and Yamaguchi, 1991), a function that could help regeneration by making appropriate growth factors available, or alternatively could hinder axon growth by binding and functionally removing important growth signals from the injury site. Further investigations are required to determine whether modulation of growth factors by proteoglycans has any effect on the regenerative responses of CNS axons. Proteoglycans have also been implicated as modulators of cell adhesion or migration (Grumet *et al.*, 1993), and have been shown to lead to the increased migration of astrocytes in response to wound-associated factors (Faber-Elman *et al.*, 1996), another role that could certainly exert a positive influence on the CNS repair response by directing the migration of astrocytes into areas of damage to stabilize the tissue structure. Unfortunately, such astrocyte migration often occurs in the opposite direction, as the necrotic cyst characteristic of many CNS wounds develops into a large cavity devoid of astrocytes (Mathewson and Berry, 1985; Fujiki *et al.*, 1996; Zhang *et al.*, 1996; Fitch and Silver, 1997a; Zhang *et al.*, 1997).

Proteoglycans and other astrocyte extracellular matrix products have been implicated in the formation of cystic cavities in the CNS. Astrocytes normally produce a basal lamina (which contains proteoglycans) at the pial surface of the CNS, and the astrocyte is polarized with respect to the production of this basal lamina that is produced on one side of the cell only (Kusaka *et al.*, 1985). Following traumatic injury to the CNS, astrocytes produce ectopic basal lamina components at the borders of the cut edges of the injured tissue, again in a polarized fashion on the side facing the lesion cavity (Lawrence *et al.*, 1984; Bernstein *et al.*, 1985). Chondroitin sulfate proteoglycans have been described at the interface between developing cavities and the surrounding viable tissue (MacLaren, 1996; Fitch and Silver, 1997a), perhaps as a component of an ectopic basal lamina secreted by reactive astrocytes. The function of extracellular matrix products surrounding a necrotic cavity of the CNS is open for speculation, and it is an intriguing possibility that components such as proteoglycans may play a role in "walling off" the injured tissue in an effort to protect the surrounding viable cellular environment from further damage. In fact, proteoglycans have been demonstrated to inhibit phagocytosis and destruction of β-amyloid protein by macrophages (Shaffer *et al.*, 1995). Thus, it is possible that the CNS uses proteoglycans as a molecular protectant of tissue destruction by degradative enzymes or secondary tissue damage by inflammatory cells following a traumatic injury.

IV. WHAT CAUSES ASTROCYTE GLIOSIS AND INCREASES IN INHIBITORY MOLECULES?

A. Triggers of Astrocyte Gliosis

Although astrocyte gliosis in response to injury to the CNS has been studied for many years, the specific causes and mechanisms leading to astrocyte hypertrophy, hyperplasia, and increased production of GFAP remain unknown. Several investigators have suggested that astrocyte gliosis can be triggered by the degeneration of severed axon tracts, perhaps secondary to Wallerian degeneration or retrograde axon changes, based on observations of the spread of GFAP immunoreactivity along white matter tracts far from the site of trauma to the spinal cord or dorsal roots (Barrett *et al.*, 1981, 1984; Murray *et al.*, 1990) or the delayed increases in GFAP mRNA in mutant animals with delayed Wallerian degeneration (Steward and Trimmer, 1997). Lesions to one side of the brain lead to gliosis throughout the ipsilateral hemisphere (Berry *et al.*, 1983; Mathewson and Berry, 1985), and sometimes gliosis is seen to extend to areas of the contralateral hemisphere as well (Amaducci *et al.*, 1981; Ludwin, 1985; Schiffer *et al.*, 1986; Takamiya *et al.*, 1988; Xie *et al.*, 1995; Fitch and Silver, 1997a), suggesting that axons that span the brain hemispheres may contribute to the glial responses on the side contralateral to the injury. Various cytokines and other molecules have been implicated as possible triggers of astrocyte gliosis, including IL-1 (Giulian and Lachman, 1985; Giulian *et al.*, 1988; Rostworowski *et al.*, 1997), IL-6 (Chiang *et al.*, 1994; Klein *et al.*, 1997), thrombin (Nishino *et al.*, 1993, 1994), CNTF (Kahn *et al.*, 1995, 1997; Winter *et al.*, 1995), endothelin-1 (Hama *et al.*, 1997), and TNF alpha (Rostworowski *et al.*, 1997). Such factors that potentially play a role in astrocyte gliosis may derive from the injured astrocytes themselves, endogenous serum factors, activated microglial cells, or the invading inflammatory cells from the periphery.

Although it is certainly important to identify molecules that may play a role in inducing astrocyte gliosis, it is becoming increasingly apparent that using GFAP upregulation and cellular hypertrophy as markers for gliosis (as most studies have done) may be of limited benefit. Observations by several investigators suggest that astrocyte gliosis may be heterogeneous (Miller *et al.*,

1986; Hill *et al.*, 1996), particularly in its association with the production of inhibitory molecules (McKeon *et al.*, 1991, 1995; Fitch and Silver, 1997a). Although some reactive astrocytes, particularly those near the wound epicenter, have been associated with the production of boundary molecules, it is certainly apparent that not all astrocytes with the morphological characteristics of reactive astrocytes (i.e., increased GFAP) are present in areas with increased levels of extracellular matrix molecules (McKeon *et al.*, 1991, 1995; Pindzola *et al.*, 1993; Levine, 1994; Fitch and Silver, 1997a). These observations suggest that perhaps not all astrocytes that react to injury play a role in the failure of CNS regeneration, and that only those astrocytes associated with inhibitory molecules are detrimental to axon growth whereas those further away from the lesion may be more conducive to neurite sprouting, functional plasticity, and long-distance regeneration.

B. Triggers for Inhibitory Molecules

Therefore, the pivotal question that remains is: What molecular triggers are responsible for the production of astroglial inhibitory extracellular matrix? A series of studies have identified β-amyloid protein as one trigger of reactive astrogliosis that leads to increases in the production of inhibitory molecules (Canning *et al.*, 1993, 1996; Hoke *et al.*, 1994; Hoke and Silver, 1996). Because it is unlikely that β-amyloid is a trigger for the production of inhibitory molecules following traumatic injury to the CNS, it is important to begin considering what other factors may play such a role. Using lesion models of the brain and spinal cord, recent work suggests that the increases in one family of inhibitory molecules (chondroitin sulfate proteoglycans) are associated with a breakdown of the blood brain barrier and infiltrating macrophages present within the lesion site (Fitch and Silver, 1997a) (Fig. 1, see color insert). These observations suggest that either leakage of serum proteins or infiltrating inflammatory cells (or both) could influence the production of extracellular matrix molecules in the vicinity of a CNS wound. In fact, cytokines associated with inflammatory infiltrates have been shown to modulate extracellular matrix production by astrocytes *in vitro* (DiProspero *et al.*, 1997). These hypotheses are in agreement with previous studies concerning general triggers of astrocyte gliosis (Giulian *et al.*, 1988; Nishino *et al.*, 1993, 1994), but work remains to be done to confirm whether known triggers of widespread gliosis also induce inhibitory molecules in the discrete region surrounding the wound. The role of degenerating axons as a trigger for gliosis is unclear at least *in vitro,* because one study claims that dying axons are not sufficient to signal matrix production by astrocytes (Ard *et al.*, 1993), and another study indicates that degenerating axons can lead to increases in extracellular matrix production by astrocytes

(Guenard *et al.*, 1996). However, observations *in vivo* illustrate that the degeneration of injured axons is not sufficient to induce the upregulation of inhibitory molecules outside of the immediate vicinity of a CNS wound (Fitch and Silver, 1997a). It should be noted, however, that degenerating axons could indirectly lead to slow increases in inhibitory molecules, because the presence of dying axons has been demonstrated to lead to inflammation within the CNS (Zhang *et al.*, 1996), and such inflammatory cells may in turn trigger the upregulation of extracellular matrix molecules.

V. WHAT IS THE ROLE OF INFLAMMATION IN CNS INJURY?

A. Macrophages and Microglia

The inflammatory response in the CNS following injury is composed primarily of two components: activation of intrinsic microglial cells and recruitment of bone marrow-derived inflammatory cells from the peripheral bloodstream (for review see Perry *et al.*, 1993). Chemical injuries to the brain appear to lead to a predominantly microglial cell inflammatory response, whereas direct stab wounds and injections are composed mostly of peripheral monocytes (Murabe *et al.*, 1982; Riva-Depaty *et al.*, 1994). However, it is generally accepted that both microglia and peripherally derived macrophages respond to injury in various proportions depending on the type and severity of the lesion (Giulian *et al.*, 1989). Some authors have suggested that the limited and delayed recruitment of macrophages into a CNS lesion may explain the lack of efficient myelin clearance seen after such injury (Perry *et al.*, 1987; George and Griffin, 1994), and the persistence of myelin has been implicated as one component that may explain some aspects of CNS regenerative failure as discussed previously in this chapter (for review see Schwab *et al.*, 1993).

The inflammatory response to injury is thought by some investigators to contribute to secondary tissue damage within the CNS (Blight, 1994). Microglial cytokines have been suggested as possible sources of nervous system impairment following injury (Giulian *et al.*, 1989), and neutrophilic leukocytes may augment necrosis and inflammation following a CNS wound (Means and Anderson, 1983). Microgial cells are capable of releasing cytotoxic factors that can kill neurons (Banati *et al.*, 1993; Giulian, 1993), and have been suggested to play a role in disconnecting existing neuronal connections and destroying neurons surounding areas of injury (Giulian *et al.*, 1994a, 1994b). Many authors have advocated the use of therapeutic agents to modify the secretory activity of microglia and macrophages as a way to limit secondary damage to the CNS (Giulian and Lachman, 1985; Giulian *et al.*, 1989; Banati *et al.*, 1993; Guth *et al.*, 1994a, 1994b; Zhang *et al.*, 1997).

Perhaps a bit paradoxically, the inflammatory response within the CNS has also been suggested to have positive effects on the healing of nervous system wounds (Lotan *et al.*, 1994; Klusman and Schwab, 1997). Experiments that demonstrated regeneration of PNS axons into the CNS environment noted the presence of a mild inflammatory reaction at the site of axon entry into the CNS, suggesting a positive role for regeneration (Le Gros Clark, 1943; Kliot *et al.*, 1990; Siegal *et al.*, 1990), and the use of certain pro-inflammatory agents was demonstrated to have positive effects on axon growth (Windle and Chambers, 1950; Windle *et al.*, 1952; Clemente and Windle, 1954; Windle, 1956; Clemente, 1958; Guth 1994a, 1994b). Macrophages secrete factors that can promote axon growth, such as NGF, NT-3 (Elkabes *et al.*, 1996), thrombospondin (Chamak *et al.*, 1994), and IL-1 (Giulian *et al.*, 1994a, 1994b), suggesting that appropriate secretory activity by inflammatory cells may indeed promote regeneration of axons. Studies using transplantation of activated macrophages indicate that increasing the phagocytic activity of macrophages within the wound can have beneficial effects for regeneration (Lazarov-Spiegler *et al.*, 1996). In terms of the resident functions of brain macrophages, Banati and Graeber (1994) describe microglial cells as "sensors of pathology" and maintain that there is little evidence that activation of microglia alone is harmful for the CNS. In fact, one study suggests that activated microglial cells may actually support axon regeneration via the production of growth promoting molecules (Rabchevsky and Streit, 1997).

B. Macrophages, Microglia, and Inhibitory Molecules

The relationship between inflammatory macrophages, activated microglia, and proteoglycan upregulation following trauma to the CNS has been discussed previously in this chapter (see Fig. 1). In addition to their potential role as a trigger for astrocyte production of extracellular matrix molecules, microglia have been suggested as a possible source of proteoglycans with neurite inhibitory properties (Bovolenta *et al.*, 1993). In fact, peripheral macrophages have been shown to produce cell surface and secreted proteoglycans and their associated glycosaminoglycans in tissue culture experiments (Uhlin-Hansen and Kolset, 1987, 1988; Kolset *et al.*, 1986, 1988, 1996; Kolset and Larsen, 1988; Uhlin-Hansen *et al.*, 1989, 1993; Petricevich and Michelacci, 1990; Owens and Wagner, 1992; Yeaman and Rapraeger, 1993a, 1993b; Haidl and Jefferies, 1996). However, the majority of proteoglycan production by macrophages appears to be secreted into the fluid media compartment *in vitro*,

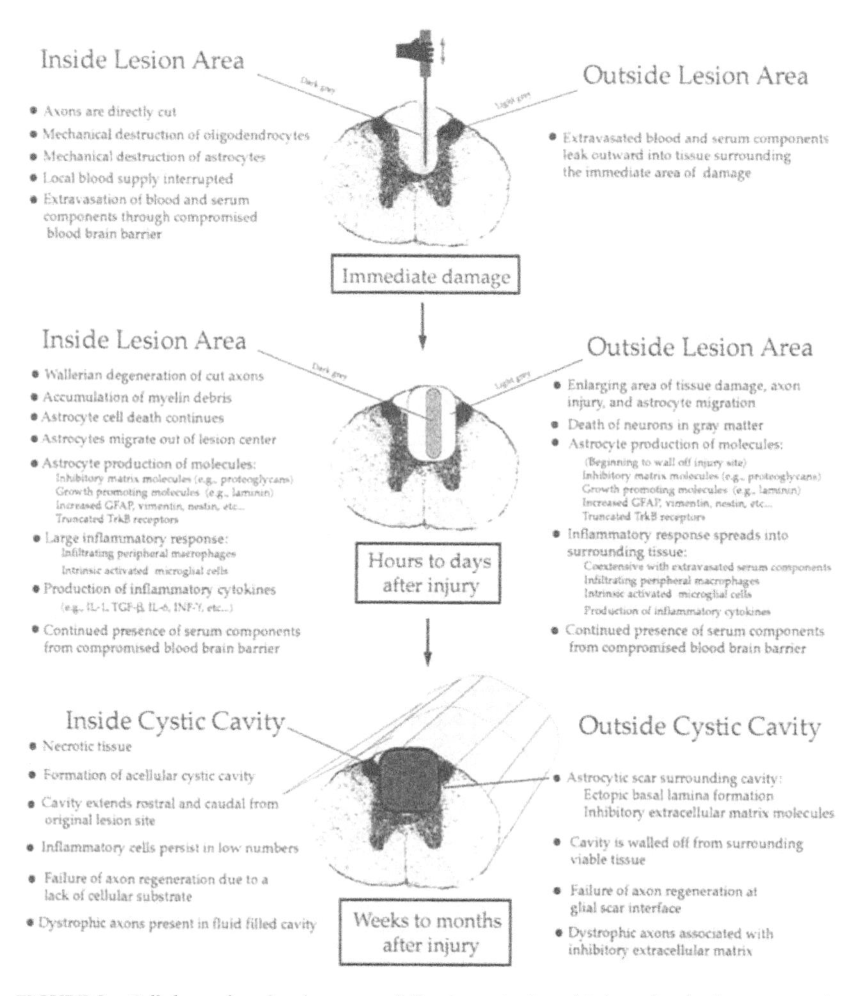

FIGURE 2 Cellular and molecular events following spinal cord injury that lead to progressive necrosis and cystic cavitation. Schematic illustration summarizing some of the major events that occur in three general phases of the response to spinal cord injury: immediate cellular and tissue damage, events occurring hours to days after injury, and the resolution of the wound weeks to months after injury. Although this figure uses a spinal cord stab injury as a model for central nervous system wound healing, the general features of the tissue responses are relevant to trauma to both brain and spinal cord by various mechanisms. Note especially the enlarging acellular cavity that can extend laterally as well as rostral and caudal to the lesion site, and the astrocytic response to this progressive necrosis, which includes the production of inhibitory molecules as a way to "wall off" the injury site. This figure demonstrates that leakage of serum factors through a compromised blood brain barrier and subsequent inflammatory infiltrates interact to influence the astrocyte reaction to CNS damage that ultimately leads to increased cell death and regenerative failure.

unlike the culture experiments in which astrocytes deposited quantities of proteoglycans onto culture substrates (McKeon *et al.*, 1991, 1995; Canning *et al.*, 1993, 1996; Hoke *et al.*, 1994; McKeon and Silver, 1995; Hoke and Silver, 1996). In addition to production, macrophages are also active in degrading and inducing other cells to degrade proteoglycan molecules (Laub *et al.*, 1982). The proteoglycans secreted by macrophages have been suggested to protect themselves from their own degradative enzymes (Kolset and Larsen, 1988), or perhaps to regulate the inflammatory response itself (Kolset *et al.*, 1996). The possible roles of macrophage-derived proteoglycans in the CNS injury response remain to be elucidated.

VI. CAN GLIAL RESPONSES BE MODIFIED TO ENHANCE REGENERATION?

This chapter has been dedicated to a discussion of the glial cell responses to CNS injury and their possible roles in contributing to regenerative failure (see summary Fig. 2). Oligodendrocytes and myelin, astrocytes and extracellular matrix, and microglia–macrophages and inflammation have all been examined as potential players in the abortive regenerative responses seen in adult CNS lesions. Many potential repair strategies have been attempted in the past, and several recent studies that have shown encouraging results suggest that perhaps the use of a combination of approaches will be beneficial in improving the CNS regenerative response. Combinations of therapeutic agents designed to modify inflammatory responses (Bracken *et al.*, 1990, 1992, 1997; Guth *et al.*, 1994a, 1994b; Taoka, 1997), and peripheral nerve bridges supplemented with acidic fibroblast growth factor combined with expert surgical technique (Cheng *et al.*, 1996) are several examples of such potential approaches to these problems. It is likely that no single response to injury will fully explain the complex pathology observed after CNS trauma. Instead, it is highly probable that a plethora of factors including those discussed in this chapter are at work in creating a local CNS environment that is not conducive to axonal regeneration, and the major challenge that remains is to find ways to assist injured axons at the site of a lesion to grow beyond the glial scar. Further research should be directed toward repair strategies that modify the responses of glial cells in the immediate vicinity of trauma, in addition to providing molecules to enhance the regenerative response by adult neurons, to provide a truly interdisciplinary approach to solve the problem of CNS injury and the resulting permanent disability.

REFERENCES

Adrian, E. J., and Williams, M. G. (1973). Cell proliferation of injured spinal cord. An electron microscopic study. *J. Comp. Neurol* 151:1–24.

Amaducci, L., Forno, K. I., and Eng. L. F. (1981). Glial fibrillary acidic protein in cryogenic lesions of the rat brain. *Neurosci Lett* 21:27–32.

Amat, J. A., Ishiguro, H., Nakamura, K., and Norton, W. T. (1996). Phenotypic diversity and kinetics of proliferating microglia and astrocytes following cortical stab wounds. *Glia* 16:368–382.

Ard, M. D., Bunge, M. B., Wood, P. M., Schachner, M., and Bunge, R. P. (1991). Retinal neurite growth on astrocytes is not modified by extracellular matrix, anti-L1 antibody, or oligodendrocytes. *Glia* 4:70–82.

Ard, M. D., Schachner, M., Rapp, J. T., and Faissner, A. (1993). Growth and degeneration of axons on astrocyte surfaces: effects on extracellular matrix and on later axonal growth. *Glia* 9:248–259.

Argiro, V., Bunge, M. B., and Johnson, M. I. (1984). Correlation between growth form and movement and their dependence on neuronal age. *J Neurosci* 4:3051–3062.

Bahr, M., Przyrembel, C., and Bastmeyer, M. (1995). Astrocytes from adult rat optic nerves are nonpermissive for regenerating retinal ganglion cell axons. *Exp Neurol* 131:211–220.

Balentine, J. D. (1978). Pathology of experimental spinal cord trauma. I. The necrotic lesion as a function of vascular injury. *Lab Invest* 39:236–253.

Banati, R. B., Gehrmann, J., Schubert, P., and Kreutzberg, G. W. (1993). Cytotoxicity of microglia. *Glia* 7:111–118.

Banati, R. B., and Graeber, M. B. (1994). Surveillance, intervention and cytotoxicity: Is there a protective role of microglia? *Dev Neurosci* 16:114–127.

Bandtlow, C., Zachleder, T., and Schwab, M. E. (1990). Oligodendrocytes arrest neurite growth by contact inhibition. *J Neurosci* 10:3837–3848.

Barrett, C. P., Donati, E. J., and Guth, L. (1984). Differences between adult and neonatal rats in their astroglial response to spinal injury. *Exp Neurol* 84:374–385.

Barrett, C. P., Guth, L., Donati, E. J., and Krikorian, J. G. (1981). Astroglial reaction in the gray matter lumbar segments after midthoracic transection of the adult rat spinal cord. *Exp Neurol* 73:365–377.

Bartsch, U. (1996). Myelination and axonal regeneration in the central nervous system of mice deficient in the myelin-associated glycoprotein. *J Neurocytology* 25:303–313.

Bartsch, U., Bandtlow, C. E., Schnell, L., Bartsch, S., Spillmann, A. A., Rubin, B. P., Hillenbrand, R., Montag, D., Schwab, M. E., and Schachner, M. (1995). Lack of evidence that myelin-associated glycoprotein is a major inhibitor of axonal regeneration in the CNS. *Neuron* 15:1375–1381.

Bastmeyer, M., Beckmann, M., Schwab, M. E., and Stuermer, C. A. (1991). Growth of regenerating goldfish axons is inhibited by rat oligodendrocytes and CNS myelin but not by goldfish optic nerve tract oligodendrocytelike cells and fish CNS myelin. *J Neurosci* 11:626–640.

Bernstein, J. J., Getz, R., Jefferson, M., and Kelemen, M. (1985). Astrocytes secrete basal lamina after hemisection of rat spinal cord. *Brain Res* 327:135–141.

Berry, M., Carlile, J., and Hunter, A. (1996). Peripheral nerve explants grafted into the vitreous body of the eye promote the regeneration of retinal ganglion cell axons severed in the optic nerve. *J Neurocytol* 25:147–170.

Berry, M., Hall, S., Rees, L., Carlile, J., and Wyse, J. P. (1992). Regeneration of axons in the optic nerve of the adult Browman-Wyse (BW) mutant rat. *J Neurocytol* 21:426–448.

Berry, M., Maxwell, W. L., Logan, A., Mathewson, A., McConnell, P., Ashhurst, D. E., and Thomas, G. H. (1983). Deposition of scar tissue in the central nervous system. *Acta Neurochir Suppl Wien* 32:31–53.

Bignami, A., and Dahl, D. (1974). Astrocyte-specific protein and neuroglial differentiation. An immunofluorescence study with antibodies to the glial fibrillary acidic protein. *J Comp Neurol* 153:27–38.

Bignami, A., and Dahl, D. (1976). The astroglial response to stabbing. Immunofluorescence studies with antibodies to astrocyte-specific protein (GFA) in mammalian and submammalian vertebrates. *Neuropathol Appl Neurobiol* 2:99–110.

Bignami, A., Eng, L. F., Dahl, D., and Uyeda, C. T. (1972). Localization of the glial fibrillary acidic protein in astrocytes by immunofluorescence. *Brain Res* 43:429–435.

Blight, A. R. (1994). Effects of silica on the outcome from experimental spinal cord injury: Implication of macrophages in secondary tissue damage. *Neuroscience* 60:263–273.

Bovolenta, P., Fernaud-Espinosa, I., Mendez-Otero, R., and Nieto-Sampedro, M. (1997). Neurite outgrowth inhibitor of gliotic brain tissue. Mode of action and cellular localization, studied with specific monoclonal antibodies. *Eur J Neurosci* 9(5):977–989.

Bovolenta, P., Wandosell, F., and Nieto, S. M. (1993). Neurite outgrowth inhibitors associated with glial cells and glial cell lines. *Neuroreport* 5:345–348.

Bracken, M. B., Shepard, M. J., Collins, W. J., Holford, T. R., Baskin, D. S., Eisenberg, H. M., Flamm, E., Leo, S. L., Maroon, J. C., and Marshall, L. F. (1992). Methylprednisolone or naloxone treatment after acute spinal cord injury: 1-year follow-up data. Results of the second National Acute Spinal Cord Injury Study. *J Neurosurg* 76:23–31.

Bracken, M. B., Shepard, M. J., Collins, W. F., Holford, T. R., Young, W., Baskin, D. S., Eisenberg, H. M., Flamm, E., Leo, S. L., and Maroon, J. (1990). A randomized, controlled trial of methylprednisolone or naloxone in the treatment of acute spinal-cord injury. Results of the second National Acute Spinal Cord Injury Study. *N Engl J Med* 322:1405–1411.

Bracken, M. B., Shepard, M. J., Holford, T. R., Leo, S. L., Aldrich, E. F., Fazl, M., Fehlings, M., Herr, D. L., Hitchon, P. W., Marshall, L. F., Nockels, R. P., Pascale, V., Perot, Pl Jr, Piepmeier, J., Sonntag, V. K., Wagner, F., Wilberger, J. E., Winn, H. R., and Young, W. (1997). Administration of methylprednisolone for 24 or 48 hours or tirilazad mesylate for 48 hours in the treatment of acute spinal cord injury. Results of the Third National Acute Spinal Cord Injury Randomized Controlled Trial. National Acute Spinal Cord Injury Study. *JAMA* 277:1597–1604.

Bregman, B. S., Kunkel, B. E., Schnell, L., Dai, H. N., Gao, D., and Schwab, M. E. (1995). Recovery from spinal cord injury mediated by antibodies to neurite growth inhibitors. *Nature* 378:498–501.

Brittis, P. A., Lemmon, V., Rutishauser, U., and Silver, J. (1995). Unique changes of ganglion cell growth cone behavior following cell adhesion molecule perturbations: A time lapse study of the living retina. *Mol Cell Neurosci* 6:433–449.

Brodkey, J. A., Laywell, E. D., O'Brien, T. F., Faissner, A., Stefansson, K., Dorries, H. U., Schachner, M., and Steindler, D. A. (1995). Local brain injury and upregulation of a developmentally regulated extracellular matrix protein. *J Neurosci* 82:106–121.

Brown, C. T., Applebaum, E., Banwatt, R., and Trinkaus, R. V. (1995). Synthesis of stromal glycosaminoglycans in response to injury. *J Cell Biochem* 59:57–68.

Bunge, R. P. (1968). Glial cells and the central myelin sheath. *Phys Reviews* 48:197.

Campbell, G., Anderson, P. N., Turmaine, M., and Lieberman, A. R. (1991). GAP-43 in the axons of mammalian CNS neurons regenerating into peripheral nerve grafts. *Exp Brain Res* 87:67–74.

Canning, D. R., Hoke, A., Malemud, C. J., and Silver, J. (1996). A potent inhibitor of neurite outgrowth that predominates in the extracellular matrix of reactive astrocytes. *Int. J. Devl. Neuroscience* 14:153–175.

Canning, D. R., McKeon, R. J., DeWitt, D. A., Perry, G., Wujek, J. R., Frederickson, R. C., and Silver, J. (1993). Beta-amyloid of Alzheimer's disease induces reactive gliosis that inhibits axonal outgrowth. *Exp Neurol* 124:289–298.

Carlstedt, T. (1983). Regrowth of anastomosed ventral root nerve fibers in the dorsal root of rats. *Brain Res* 272:162–165.

Carlstedt, T. (1985a). Regenerating axons form nerve terminals at astrocytes. *Brain Res* 347:188–191.

Carlstedt, T. (1985b). Regrowth of cholinergic and catecholaminergic neurons along a peripheral and central nervous pathway. *Neuroscience* 15:507–518.

Carlstedt, T. (1988). Reinnervation of the mammalian spinal cord after neonatal dorsal root crush. *J Neurocytol* 17:335–350.

Carlstedt, T., Cullheim, S., Risling, M., and Ulfhake, B. (1988). Mammalian root-spinal cord regeneration. *Prog Brain Res* 78:225–229.

Carlstedt, T., Dalsgaard, C. J., and Molander, C. (1987). Regrowth of lesioned dorsal root nerve fibers into the spinal cord of neonatal rats. *Neurosci Lett* 74:14–18.

Caroni, P., Savio, T., and Schwab, M. E. (1988). Central nervous system regeneration: Oligodendrocytes and myelin as non-permissive substrates for neurite growth. *Prog Brain Res* 78:363–370.

Caroni, P., and Schwab, M. E. (1988a). Antibody against myelin-associated inhibitor of neurite growth neutralizes nonpermissive substrate properties of CNS white matter. *Neuron* 1:85–96.

Caroni, P., and Schwab, M. E. (1988b). Two membrane protein fractions from rat central myelin with inhibitory properties for neurite growth and fibroblast spreading. *J Cell Biol* 106:1281–1288.

Carrino, D. A., Oron, U., Pechak, D. G., and Caplan, A. I. (1988). Reinitiation of chondroitin sulphate proteoglycan synthesis in regenerating skeletal muscle. *Development* 103:641–656.

Challacombe, J. F., and Elam, J. S. (1997). Chondroitin 4-sulfate stimulates regeneration of goldfish retinal axons. *Exp Neurol* 143:10–47.

Chamak, B., Morandi, V., and Mallat, M. (1994). Brain macrophages stimulate neurite growth and regeneration by secreting thrombospondin. *J Neurosci Res* 38:221–233.

Chen, D. F., Jhaveri, S., and Schneider, G. E. (1995). Intrinsic changes in developing retinal neurons result in regenerative failure of their axons. *Proc Natl Acad Sci USA* 92:7287–7291.

Cheng, H., Cao, Y., and Olson, L. (1996). Spinal cord repair in adult paraplegic rats: Partial restoration of hind limb function. *Science* 273:510–513.

Chiang, C. S., Stalder, A., Samimi, A., and Campbell, I. L. (1994). Reactive gliosis as a consequence of interleukin-6 expression in the brain: Studies in transgenic mice. *Dev Neurosci* 16:212–221.

Chiquet-Ehrismann, R., Hagios, C., and Schenk, S. (1995). The complexity in regulating the expression of tenascins. *Bioessays* 17:873–887.

Clemente, C. D. (1958). The regeneration of peripheral nerves inserted into the cerebral cortex and the healing of cerebral lesions. *J Comp Neurol* 109:123–143.

Clemente, C. D., and Windle, W. F. (1954). Regeneration of severed nerve fibers in the spinal cord of the adult cat. *J Comp Neurol* 101:691–731.

Cole, G. J., and McCabe, C. F. (1991). Identification of a developmentally regulated keratan sulfate proteoglycan that inhibits cell adhesion and neurite outgrowth. *Neuron* 7:1007–1018.

David, S., and Aguayo, A. J. (1981). Axonal elongation into peripheral nervous system "bridges" after central nervous system injury in adult rats. *Science* 214:931–933.

Davies, S. J. A., Field, P. M., and Raisman, G. (1993). Long fibre growth by axons of embryonic mouse hippocampal neurons microtransplanted into the adult rat fimbria. *Eur J Neurosci* 5:95–106.

Davies, S. J. A., Field, P. M., and Raisman, G. (1996). Regeneration of cut adult axons fails even in the presence of continuous aligned glial pathways. *Exp Neurol* 142:203–216.

Davies, S. J. A., Fitch, M. T., Memberg, S. P., Hall, A. K., Raisman, G., and Silver, J. (1997). Regeneration of adult axons in white matter tracts of the central nervous system. *Nature* 390:680–683.

DiProspero, N. A., Meiners, S., and Geller, H. M. (1997). Inflammatory cytokines interact to modulate extracellular matrix and astrocytic support of neurite outgrowth. *Exp Neurol* 148:628–639.

Dou, C.-L., and Levine, J. M. (1994). Inhibition of neurite growth by the NG2 chondroitin sulfate proteoglycan. *J Neurosci* 14:7616–7628.

Dow, K. E., Ethell, D. W., Steeves, J. D., and Riopelle, R. J. (1994). Molecular correlates of spinal cord repair in the embryonic chick: Heparan sulfate and chondroitin sulfate proteoglycans. *Exp Neurol* 128:233–238.

Ducker, T. B., Kindt, G. W., and Kempe, L. G. (1971). Pathological findings in acute experimental spinal cord trauma. *J Neurosurg* 35:700–707.

Eddleston, M., and Mucke, L. (1993). Molecular profile of reactive astrocytes—implications for their role in neurologic disease. *Neuroscience* 54:15–36.

Elkabes, S., DiCicco, B. E., and Black, I. B. (1996). Brain microglia/macrophages express neurotrophins that selectively regulate microglial proliferation and function. *J Neurosci* 16:2508–2521.

Emerling, D. E., and Lander, A. D. (1996). Inhibitors and promoters of thalamic neuron adhesion and outgrowth in embryonic neocortex: Functional association with chondroitin sulfate. *Neuron* 17:1089–1100.

Eng, L. F. (1985). Glial fibrillary acidic protein (GFAP): The major protein of glial intermediate filaments in a differentiated astrocytes. *J Neuroimmunol* 8:203–214.

Faber-Elman, A., Solomon, A., Abraham, J. A., Marikovsky, M., and Schwartz, M. (1996). Involvement of wound-associated factors in rat brain astrocyte migratory response to axonal injury: *in vitro* simulation. *J Clin Invest* 97:162–171.

Fallon, J. R. (1985a). Neurite guidance by non-neuronal cells in culture: Preferential outgrowth of peripheral neurites on glial as compared to nonglial cell surfaces. *J Neurosci* 5:3169–3177.

Fallon, J. R. (1985b). Preferential outgrowth of central nervous system neurites on astrocytes and schwann cells as compared with nonglial cells *in vitro*. *J Cell Biol* 100:198–207.

Fawcett, J. W., Rokos, J., and Bakst, I. (1989). Oligodendrocytes repel axons and cause axonal growth cone collapse. *J Cell Sci* 92:93–100.

Fichard, A., Verna, J. M., Olivares, J., and Saxod, R. (1991). Involvement of a chondroitin sulfate proteoglycan in the avoidance of chick epidermis by dorsal root ganglia fibers: A study using beta-D-xyloside. *Dev Biol* 148:1–9.

Fitch, M. T., and Silver, J. (1997a). Activated macrophages and the blood brain barrier: Inflammation after CNS injury leads to increases in putative inhibitory molecules. *Exp Neurol* 148:587–603

Fitch, M. T., and Silver, J. (1997b). Glial cell extracellular matrix: Boundaries for axon growth in development and regeneration. *Cell Tis Res* 290:379–384.

Frisen, J., Haegerstrand, A., Risling, M., Fried, K., Johansson, C. B., Hammarberg, H., Elde, R., Hokfelt, T., and Cullheim, S. (1995). Spinal axons in central nervous system scar tissue are closely related to laminin-immunoreactive astrocytes. *Neuroscience* 65:293–304.

Frisen, J., Verge, V. M., Fried, K., Risling, M., Persson, H., Trotter, J., Hokfelt, T., and Lindholm, D. (1993). Characterization of glial trkB receptors: Differential response to injury in the central and peripheral nervous system. *Proc Natl Acad Sci USA* 90:4971–4975.

Fryer, R. H., Kaplan, D. R., and Kromer, L. F. (1997). Truncated TrkB receptors on non-neuronal cells inhibit BDNF induced neurite outgrowth *in vitro*. *Exp Neurol* 148:616–627.

Fujiki, M., Zhang, Z., Guth, L., and Steward, O. (1996). Genetic influences on cellular reactions to spinal cord injury: Activation of macrophages/microglia and astrocytes is delayed in mice carrying a mutation (Wld) that causes delayed wallerian degeneration. *J Comp Neurol* 371:469–484.

Gallo, V., and Bertolotto, A. (1990). Extracellular matrix of cultured glial cells: Selective expression of chondroitin sulfate by type-2 astrocytes and progenitors. *Exp Cell Res* 187:211–223.

Gallo, V., Bertolotto, A., and Levi, G. (1987). The proteoglycan chondroitin sulfate is present in a subpopulation of cultured astrocytes and in their precursors. *Dev Biol* 123:282–285.

Gates, M. A., Fillmore, H., and Steindler, D. A. (1997). Chondroitin sulfate proteoglycan and tenascin in the wounded adult mouse neostriatum *in vitro*: Dopamine neuron attachment and process outgrowth. J Neurosci 16:8005–8018.

Gehrmann, J., Matsumoto, Y., and Kreutzberg, G. W. (1995). Microglia: Intrinsic immuneffector cell of the brain. *Brain Res Reviews* 20:269–287.

Geisert, E. J., and Bidanset, D. J. (1993). A central nervous system keratan sulfate proteoglycan: Localization to boundaries in the neonatal rat brain. *Brain Res Dev Brain Res* 75:163–173.

Geisert, E. E., and Stewart, A. M. (1991). Changing interactions between astrocytes and neurons during CNS maturation. *Dev Biol* 143:335–345.

Geisert, E. J., Williams, R. C., and Bidanset, D. J. (1992). A CNS specific proteoglycan associated with astrocytes in rat optic nerve. *Brain Res* 571:165–168.

George, R., and Griffin, J. W. (1994). Delayed macrophage responses and myelin clearance during Wallerian degeneration in the central nervous system: The dorsal radiculotomy model. *Exp Neurol* 129:225–236.

Giulian, D. (1993). Reactive glia as rivals in regulating neuronal survival. *Glia* 7:102–110.

Giulian, D., Chen, J., Ingeman, J. E., George, J. K., and Noponen, M. (1989). The role of mononuclear phagocytes in wound healing after traumatic injury to adult mammalian brain. *J Neurosci* 9:4416–4429.

Giulian, D., and Lachman, L. B. (1985). Interleukin-1 stimulation of astroglial proliferation after brain injury. *Science* 228:497–499.

Giulian, D., Li, J., Leara, B., and Keenen, C. (1994a). Phagocytic microglia release cytokines and cytotoxins that regulate the survival of astrocytes and neurons in culture. *Neurochem Int* 25:227–233.

Giulian, D., Li, J., Li, X., George, J., and Rutecki, P. A. (1994b). The impact of microglia-derived cytokines upon gliosis in the CNS. *Dev Neurosci* 16:128–136.

Giulian, D., Woodward, J., Young, D. G., Krebs, J. F., and Lachman, L. B. (1988). Interleukin-1 injected into mammalian brain stimulates astrogliosis and neovascularization. *J Neurosci* 8:2485–2490.

Grierson, J. P., Petroski, R. E., Ling, D. S., and Geller, H. M. (1990). Astrocyte topography and tenascin cytotactin expression: Correlation with the ability to support neuritic outgrowth. *Brain Res Dev Brain Res* 55:11–19.

Grill, R., Murai, K., Blesch, A., Gage, F. H., and Tuszynski, M. H. (1997). Cellular delivery of neurotrophin-3 promotes corticospinal axonal growth and partial functional recovery after spinal cord injury. *J Neurosci* 17:5560–5572.

Grumet, M., Flaccus, A., and Margolis, R. U. (1993). Functional characterization of chondroitin sulfate proteoglycans of brain: Interactions with neurons and neural cell adhesion molecules. *J Cell Biol* 120:815–824.

Guenard, V., Frisch, G., and Wood, P. M. (1996). Effects of axonal injury on astrocyte proliferation and morphology *in vitro:* Implications for astrogliosis. *Exp Neurol* 137:175–190.

Guest, J. D., Hesse, D., Schnell, L., Schwab, M. E., Bunge, M. B., and Bunge, R. P. (1997). Influence of IN-1 antibody and acidic FGF–fibrin glue on the response of injured corticospinal tract axons to human Schwann cell grafts. *J Neurosci Res* 50:888–905.

Guth, L. (1956). Regeneration in the mammalian peripheral nervous system. *Physiological Reviews* 36:441–478.

Guth, L., Barrett, C. P., and Donati, E. J. (1986). Histological factors influencing the growth of axons into lesions of the mammalian spinal cord. *Exp Brain Res Suppl* 13:271–282.

Guth, L., Barrett, C. P., Donati, E. J., Anderson, F. D., Smith, M. V., and Lifson, M. (1985). Essentiality of a specific cellular terrain for growth of axons into a spinal cord lesion. *Exp Neurol* 88:1–12.

Guth, L., Barrett, C. P., Donati, E. J., Deshpande, S. S., and Albuquerque, E. X. (1981). Histopathological reactions on axonal regeneration in the transected spinal cord of hibernating squirrels. *J Comp Neurol* 203:297–308.

Guth, L., Zhang, Z., DiProspero, N. A., Joubin, K., and Fitch, M. T. (1994a). Spinal cord injury in the rat: Treatment with bacterial lipopolysaccharide and indomethacin enhances cellular repair and locomotor function. *Exp Neurol* 126:76–87.

Guth, L., Zhang, Z., and Roberts, E. (1994b). Key role for pregnenolone in combination therapy that promotes recovery after spinal cord injury. *Proc Natl Acad Sci USA* 91:12308–12312.

Haidl, I. D., and Jefferies, W. A. (1996). The macrophage cell surface glycoprotein F4/80 is a highly glycosylated proteoglycan. *European Journal of Immunology* 26:1139–1146.

Hama, H., Kasuya, Y., Sakurai, T., Yamada, G., Suzuki, N., Masaki, T., and Goto, K. (1997). Role of endothelin-1 in astrocyte responses after acute brain damage. *J Neurosci Res* 47:590–602.

Harrison, R. G. (1910). The outgrowth of the nerve fiber as a mode of protoplasmic movement. *J Exp Zool* 9:787–848.

Harrison, R. G. (1914). The reaction of embryonic cells to sold structures. *J Exp Zool* 17:521–544.

Hatten, M. E., Liem, R. K., Shelanski, M. L., and Mason, C. A. (1991). Astroglia in CNS injury. *Glia* 4:233–243.

Hill, S. J., Barbarese, E., and McIntosh, T. K. (1996). Regional heterogeneity in the response of astrocytes following traumatic brain injury in the adult rat. *J Neuropathol Exp Neurol* 55:1221–1229.

Hockfield, S., Kalb, R. G., Zaremba, S., and Fryer, H. J. (1990). Expression of neural proteoglycan correlates with the acquisition of mature neuronal properties in the mammalian brain. *Cold Spring Harbor Symp Quant Biol* 55:505–514.

Hoke, A., Canning, D. R., Malemud, C. J., and Silver, J. (1994). Regional differences in reactive gliosis induced by substrate-bound beta-amyloid. *Exp Neurol* 130:56–66.

Hoke, A., and Silver, J. (1996). Proteoglycans and other repulsive molecules in glial boundaries during development and regeneration of the nervous system. *Prog Br Res* 108:149–163.

Houle, J. D. (1992). Regeneration of dorsal root axons is related to specific non-neuronal cells lining NGF-treated intraspinal nitrocellulose implants. *Exp Neurol* 118:133–142.

Iijima, N., Oohira, A., Mori, T., Kitabatake, K., and Kohsaka, S. (1991). Core protein of chondroitin sulfate proteoglycan promotes neurite outgrowth from cultured neocortical neurons. *J Neurochem* 56:706–708.

Jain, M., He, Q., Lee, W. S., Kashiki, S., Foster, L. C., Tsai, J. C., Lee, M. E., and Haber, E. (1996). Role of CD44 in the reaction of vascular smooth muscle cells to arterial wall injury. *J Clin Invest* 97:596–603.

Johnson-Green, P., Dow, K. E., and Riopelle, R. J. (1991). Characterization of glycosaminoglycans produced by primary astrocytes in vitro. *Glia* 4:314–321.

Kahn, M. A., Ellison, J. A., Speight, G. J., and de, V. J. (1995). CNTF regulation of astrogliosis and the activation of microglia in the developing rat central nervous system. *Brain Res* 685:55–67.

Kahn, M. A., Huang, C. J., Caruso, A., Barresi, V., Nazarian, R., Condorelli, D. F., and de Vellis, J. (1997). Ciliary neurotrophic factor activates JAK/Stat signal transduction cascade and induces transcriptional expression of glial fibrillary acidic protein in glial cells. *J Neurochem* 68:1413–1423.

Kao, C. C., Chang, L. W., and Bloodworth, J. J. (1977a). Axonal regeneration across transected mammalian spinal cords: An electron microscopic study of delayed microsurgical nerve grafting. *Exp Neurol* 54:591–615.

Kao, C. C., Chang, L. W., and Bloodworth, J. M. (1977b). The mechanism of spinal cord cavitation following spinal cord transection. Part 3: Delayed grafting with and without spinal cord retransection. *J Neurosurg* 46:757–766.

Kappler, J., Junghans, U., Koops, A., Stichel, C. C., Hausser, H. J., Kresse, H., and Muller, H. W. (1997). Chondroitin/dermatan sulphate promotes the survival of neurons from rat embryonic neocortex. *Eur J Neurobiol* 9:306–318.

Kawaja, M. D., and Gage, F. H. (1991). Reactive astrocytes are substrates for the growth of adult CNS axons in the presence of elevated levels of nerve growth factor. *Neuron* 7:1019–1030.

Kawaja, M. D., Rosenberg, M. B., Yoshida, K., and Gage, F. H. (1992). Somatic gene transfer of nerve growth factor promotes the survival of axotomized septal neurons and the regeneration of their axons in adult rats. *J Neurosci* 12:2849–2864.

Keirstead, H. S., Dyer, J. K., Sholomenko, G. N., McGraw, J., Delaney, K. R., and Steeves, J. D. (1995). Axonal regeneration and physiological activity following transection and immunological disruption of myelin within the hatchling chick spinal cord. *J Neurosci* 15:6963–6974.

Keirstead, H. S., Hasan, S. J., Muir, G. D., and Steeves, J. D. (1992). Suppression of the onset of myelination extends the permissive period for the functional repair of embryonic spinal cord. *Proc Natl Acad Sci USA* 89:11664–11668.

Kimmel, D. L., and Moyer, E. K. (1947). Dorsal roots following anastomosis of the central stumps. *J Comp Neurol* 87:289–321.

Klein, M. A., Moller, J. C., Jones, L. L., Bluethmann, H., Kreutzberg, G. W., and Raivich, G. (1997). Impaired neuroglial activation in interleukin-6 deficient mice. *Glia* 19:227–233.

Kliot, M., Smith, G. M., Siegal, J. D., and Silver, J. (1990). Astrocyte-polymer implants promote regeneration of dorsal root fibers into the adult mammalian spinal cord. *Exp Neurol* 109:57–69.

Klusman, I., and Schwab, M. E. (1997). Effects of pro-inflammatory cytokines in experimental spinal cord injury. *Brain Res* 762:173–184.

Kobayashi, H., Watanabe, E., and Murikami, F. (1995). Growth cones of dorsal root ganglion but not retina collapse and avoid oligodendrocytes in culture. *Dev Biol* 168:383–394.

Kolset, S. O., Ehlorsson, J., Kjellen, L., and Lindahl, U. (1986). Effect of benzyl beta-D-xyloside on the biosynthesis of chondroitin sulphate proteoglycan in cultured human monocytes. *Biochemical Journal* 238:209–216.

Kolset, S. O., Ivhed, I., Overvatn, A., and Nilsson, K. (1988). Differentiation-associated changes in the expression of chondroitin sulfate proteoglycan in induced U-937 cells. *Cancer Research* 48:6103–6108.

Kolset, S. O., and Larsen, T. (1988). Sulfur-containing macromolecules in cultured monocyte-like cells. *Acta Histochemica* 84:67–75.

Kolset, S. O., Mann, D. M., Uhlin, H. L., Winberg, J. O., and Ruoslahti, E. (1996). Serglycin-binding proteins in activated macrophages and platelets. *Journal of Leukocyte Biology* 59:545–554.

Kusaka, H., Hirano, A., Bornstein, M. B., and Raine, C. S. (1985). Basal lamina formation by astrocytes in organotypic cultures of mouse spinal cord tissue. *J Neuropathol Exp Neurol* 44:295–303.

Latov, N., Nilaver, G., Zimmerman, E. A., Johnson, W. G., Silverman, A. J., Defendini, R., and Cote, L. (1979). Fibrillary astrocytes proliferate in response to brain injury: A study combining immunoperoxidase technique for glial fibrillary acidic protein and radioautography of tritiated thymidine. *Dev Biol* 72:381–384.

Laub, R., Huybrechts, G. G., Peeters, J. C., and Vaes, G. (1982). Degradation of collagen and proteoglycan by macrophages and fibroblasts. Individual potentialities of each cell type and cooperative effects through the activation of fibroblasts by macrophages. *Biochimica et Biophysica Acta* 721:425–433.

Lawrence, J. M., Huang, S. K., and Raisman, G. (1984). Vascular and astrocytic reactions during establishment of hippocampal transplants in adult host brain. *Neuroscience* 12:745–760.

Laywell, E. D., Dorries, U., Bartsch, U., Faissner, A., Schachner, M., and Steindler, D. A. (1992). Enhanced expression of the developmentally regulated extracellular matrix molecule tenascin following adult brain injury. *Proc Natl Acad Sci USA* 89:2634–2638.

Laywell, E. D., and Steindler, D. A. (1991). Boundaries and wounds, glia and glycoconjugates. Cellular and molecular analyses of developmental partitions and adult brain lesions. *Ann N Y Acad Sci* 633:122–141.

Lazarov-Spiegler, O., Solomon, A. S., Zeev-Brann, A. B., Hirschberg, D. L., Lavie, V., and Schwartz, M. (1996). Transplantation of activated macrophages overcomes central nervous system regrowth failure. *FASEB J* 10:1296–1302.

Le Gros Clark, W. E. (1942). The problem of neuronal regeneration in the central nervous system. I. The influence of spinal ganglia and nerve fragments grafted in the brain. *J Anatomy* 77:20–45.

Le Gros Clark, W. E. (1943). The problem of neuronal regeneration in the central nervous system. II. The insertion of peripheral nerve stumps into the brain. *J Anatomy* 77:251–259.

Le Roux, P. D., and Reh, T. A. (1996). Reactive astroglia support primary dendritic but not axonal outgrowth from mouse cortical neurons in vitro. *Exp Neurol* 137:49–65.

Levine, J. M. (1994). Increased expression of the NG2 chondroitin-sulfate proteoglycan after brain injury. *J Neurosci* 14:4716–4730.

Li, Y., and Raisman, G. (1993). Long axon growth from embryonic neurons transplanted into myelinated tracts of the adult rat spinal cord. *Brain Res* 629:115–127.

Li, Y., and Raisman, G. (1994). Schwann cells induce sprouting in motor and sensory axons in the adult rat spinal cord. *J Neurosci* 14:4050–4063.

Li, Y., and Raisman, G. (1995). Sprouts from cut corticospinal axons persist in the presence of astrocytic scarring in long-term lesions of the adult rat spinal cord. *Exp Neurol* 134:102–111.

Liesi, P., Dahl, D., and Veheri, A. (1983). Laminin is produced by early rat astrocytes in primary culture. *J Cell Biol* 96:920–924.

Liesi, P., Kaakkola, S., Dahl, D., and Vaheri, A. (1984). Laminin is induced in astrocytes of adult brain by injury. *EMBO J* 3:683–686.

Liesi, P., and Silver, J. (1988). Is astrocyte laminin involved in axon guidance in the mammalian CNS? *Dev Biol* 130:774–785.

Ling, E. A., and Wong, W. C. (1993). The origin and nature of ramified and amoeboid microglia: a historical review and current concepts. *Glia* 7:9–18.

Lips, K., Stichel, C. C., and Muller, H. W. (1995). Restricted appearance of tenascin and chondroitin sulphate proteoglycans after transection and sprouting of adult rat postcommissural fornix. *J Neurocytol* 24:449–464.

Liu, C.-N., and Chambers, W. W. (1958). Intraspinal sprouting of dorsal root axons. *Arch Neurol Psychiat* 79:46–61.

Lotan, M., and Schwartz, M. (1994). Cross talk between the immune system and the nervous system in response to injury: Implications for regeneration. *Faseb J* 8:1026–1033.

Ludwin, S. K. (1984). Proliferation of mature oligodendrocytes after trauma to the central nervous system. *Nature* 308:274–275.

Ludwin, S. K. (1985). Reaction of oligodendrocytes and astrocytes to trauma and implantation. A combined autoradiographic and immunohistochemical study. *Lab Invest* 52:20–30.

MacLaren, R. E. (1996). Development and role of retinal glia in regeneration of ganglion cells following retinal injury. *Br J Opthalmol* 80:458–464.

Maeda, N., Hamanaka, H., Oohira, A., and Noda, M. (1995). Purification, characterization and developmental expression of a brain-specific chondroitin sulfate proteoglycan, 6B4 proteoglycan/phosphacan. *Neuroscience* 67:23–35.

Maeda, N., and Noda, M. (1996). 6B4 proteoglycan/phosphacan is a repulsive substratum but promotes morphological differentiation of cortical neurons. *Development* 122:647–658.

Margolis, R. K., and Margolis, R. U. (1993). Nervous tissue proteoglycans. *Experientia* 49:429–446.

Mathewson, A. J., and Berry, M. (1985). Observations on the astrocyte response to a cerebral stab wound in adult rats. *Brain Res* 327:61–69.

McKeon, R. J., Hoke, A., and Silver, J. (1995). Injury-induced proteoglycans inhibit the potential for laminin-mediated axon growth on astrocytic scars. *Exp Neurol* 136:32–43.

McKeon, R. J., Schreiber, R. C., Rudge, J. S., and Silver, J. (1991). Reduction of neurite outgrowth in a model of glial scarring following CNS injury is correlated with the expression of inhibitory molecules on reactive astrocytes. *J Neurosci* 11:3398–3411.

McKeon, R. J., and Silver, J. (1995). Functional significance of glial-derived matrix during development and regeneration. In Kettenmann, H., and Ransom, B. R., eds., *Neuroglia* New York: Oxford University Press, pp. 398–409.

McKerracher, L., David, S., Jackson, D. L., Kottis, V., Dunn, R. J., and Braun, P. E. (1994). Identification of myelin-associated glycoprotein as a major myelin-derived inhibitor of neurite growth. *Neuron* 13:805–811.

Means, E. D., and Anderson, D. K. (1983). Neuronophagia by leukocytes in experimental spinal cord injury. *J Neuropathol Exp Neurol* 42:707–719.

Miller, R. H., Abney, E. R., David, S., Ffrench, C. C., Lindsay, R., Patel, R., Stone, J., and Raff, M. C. (1986). Is reactive gliosis a property of a distinct subpopulation of astrocytes? *J Neurosci* 6:22–29.

Miyake, T., Hattori, T., Fukuda, M., Kitamura, T., and Fujita, S. (1988). Quantitative studies on proliferative changes of reactive astrocytes in mouse cerebral cortex. *Brain Res* 451:133–138.

Miyake, T., and Okada, M., and Kitamura, T. (1992). Reactive proliferation of astrocytes studied by immunohistochemistry for proliferating cell nuclear antigen. *Brain Res* 590:300–302.

Murabe, Y., Ibata, Y., and Sano, Y. (1982). Morphological studies on neuroglia III. Macrophage response and "microgliocytosis" in kainic acid-induced lesions. *Cell Tissue Res* 218:75–86.

Murray, H. M., and Walker, B. F. (1973). Comparative study of astrocytes and mononuclear leukocytes reacting to brain trauma in mice. *Exp Neurol* 41:290–302.

Murray, M., Wang, S. D., Goldberger, M. E., and Levitt, P. (1990). Modification of astrocytes in the spinal cord following dorsal root or peripheral nerve lesions. *Exp Neurol* 110:248–257.

Nikkari, S. T., Jarvelainen, H. T., Wright, T. N., Ferguson, M., and Clowes, A. W. (1994). Smooth muscle cell expression of extracellular matrix genes after arterial injury. *Am J Pathol* 144:1348–1356.

Nishino, A., Suzuki, M., Ohtani, H., Motohashi, O., Umezawa, K., Nagura, H., and Yoshimoto, T. (1993). Thrombin may contribute to the pathophysiology of central nervous system injury. *J Neurotrauma* 10:167–179.

Nishino, A., Suzuki, M., Yoshimoto, T., Otani, H., and Nagura, H. (1994). A novel aspect of thrombin in the tissue reaction following central nervous system injury. *Acta Neurochir Suppl Wien* 60:86–88.

Norling, B., Glimelius, B., and Wasteson, A. (1984). A chondroitin sulphate proteoglycan from human cultured glial and glioma cells. *Biochem J* 221:845–853.

Norton, W. T., Aquino, D. A., Hozumi, I., Chiu, F. C., and Brosnan, C. F. (1992). Quantitative aspects of reactive gliosis: a review. *Neurochem Res* 17:877–885.

Oohira, A., Matsui, F., and Katoh, S. R. (1991). Inhibitory effects of brain chondroitin sulfate proteoglycans on neurite outgrowth from PC12D cells. *J Neurosci* 11:822–827.

Owens, R. T., and Wagner, W. D. (1992). Chondroitin sulfate proteoglycan and heparan sulfate proteoglycan production by cultured pigeon peritoneal macrophages. *Journal of Leukocyte Biology* 51:626–633.

Perkins, S., Carlstedt, T., Mizuno, K., and Aguayo, A. J. (1980). Failure of regenerating dorsal root axons to regrow into the spinal cord. *Canad J Neurol Sci* 7:323.

Perry, V. H., Andersson, P. B., and Gordon, S. (1993). Macrophages and inflammation in the central nervous system. *Trends Neurosci* 16:268–273.

Perry, V. H., Brown, M. C., and Gordon, S. (1987). The macrophage response to central and peripheral nerve injury. A possible role for macrophages in regeneration. *J Exp Med* 165:1218–1223.

Petricevich, V. L., and Michelacci, Y. M. (1990). Proteoglycans synthesized in vitro by nude and normal mouse peritoneal macrophages. *Biochimica et Biophysica Acta* 1053:135–143.

Pindzola, R. R., Doller, C., and Silver, J. (1993). Putative inhibitory extracellular matrix molecules at the dorsal root entry zone of the spinal cord during development and after root and sciatic nerve lesions. *Dev Biol* 156:34–48.

Rabchevsky, A. G., and Streit, W. J. (1997). Grafting of cultured microglial cells into the lesioned spinal cord of adult rats enhances neurite outgrowth. *J Neurosci Res* 47:34–48.

Rakic, P. (1995). Radial glial cells: Scaffolding for brain construction. In Kettenmann, H., and Ransom, B. R., eds., *Neuroglia.* New York: Oxford University, pp. 746–762.

Ramon y Cajal, S. (1928). Degeneration and regeneration of the nervous system. (May, R. M., trans.) London: Oxford University Press.

Reier, P. J., Stensaas, L. J., and Guth, L. (1983). The astrocytic scar as an impediment to regeneration in the central nervous system. In: Bunge, R. P., Kao, C. C., and Reier, P. J., eds.), New York: Raven Press, pp. 163–195.

Richardson, M., and Hatton, M. W. (1993). Transient morphological and biochemical alterations of arterial proteoglycan during early wound healing. *Exp Mol Pathol* 58:77–95.

Richardson, P. M., McGuinness, U. M., and Aguayo, A. J. (1980). Axons from CNS neurons regenerate into PNS grafts. *Nature* 284:264–265.

Risling, M., Fried, K., Linda, H., Carlstedt, T., and Cullheim, S. (1993). Regrowth of motor axons following spinal cord lesions: Distribution of laminin and collagen in the CNs scar tissue. *Brain Res Bull* 30:405–414.

Riva-Depaty, I., Fardeau, C., Mariani, J., Bouchaud, C., and Delhaye, B. N. (1994). Contribution of peripheral macrophages and microglia to the cellular reaction after mechanical or neurotoxin-induced lesions of rat brain. *Exp Neurol* 128:77–87.

Rostworowski, M., Balasingam, V., Chabot, S., Owens, T., and Yong, V. W. (1997). Astrogliosis in the neonatal and adult murine brain post-trauma: Elevation of inflammatory cytokines and the lack of requirement for endogenous interferon-gamma. *J Neurosci* 17:3664–3674.

Rubin, B. P., Dusart, I., and Schwab, M. E. (1994). A monoclonal antibody (IN-1) which neutralizes neurite growth inhibitory proteins in the rat CNS recognizes antigens localized in CNS myelin. *J Neurocytol* 23:209–217.

Rudge, J. S., Smith, G. M., and Silver, J. (1989). An in vitro model of wound healing of the CNS: analysis of cell reaction and interaction at different ages. *Exp Neurol* 103:1–16.

Ruoslahti, E., and Yamaguchi, Y. (1991). Proteoglycans as modulators of growth factor activities. *Cell* 64:867–869.

Savio, T., and Schwab, M. E. (1989). Rat CNS white matter, but not gray matter, is nonpermissive for neuronal cell adhesion and fiber outgrowth. *J Neurosci* 9:1126–1133.

Savio, T., and Schwab, M. E. (1990). Lesioned corticospinal tract axons regenerate in myelin-free rat spinal cord. *Proc Natl Acad Sci USA* 87:4130–4133.

Schiffer, D., Giordana, M. T., Migheli, A., Giaccone, G., Pezzotta, S., and Mauro, A. (1986). Glial fibrillary acidic protein and vimentin in the experimental glial reaction of the rat brain. *Brain Res* 374:110–118.

Schnell, L., Schneider, R., Kolbeck, R., Barde, Y. A., and Schwab, M. E. (1994). Neurotrophin-3 enhances sprouting of corticospinal tract during development and after adult spinal cord lesion. *Nature* 367:170–173.

Schnell, L., and Schwab, M. E. (1990). Axonal regeneration in the rat spinal cord produced by an antibody against myelin-associated neurite growth inhibitors. *Nature* 343:269–272.

Schousboe, A., and Westergaarde, N. (1995). Transport of neuroactive amino acids in astrocytes. In: Kettenmann, H., and Ransom, B. R., eds., *Neuroglia.* Oxford University Press, New York: pp. 246–258.

Schwab, M. E., Kapfhammer, J. P., and Bandtlow, C. E. (1993). Inhibitors of neurite growth. *Annu Rev Neurosci* 16:565–595.

Shaffer, L. M., Dority, M. D., Gupta, B. R., Frederickson, R. C., Younkin, S. G., and Brunden, K. R. (1995). Amyloid beta protein (A beta) removal by neuroglial cells in culture. *Neurobiology of Aging* 16:737–745.

Shewan, D., Berry, M., Bedi, K., and Cohen, J. (1993). Embryonic optic nerve tissue fails to support neurite outgrowth by central and peripheral neurons *in vitro*. *Eur J Neurosci* 5:809–817.
Shewan, D., Berry, M., and Cohen, J. (1995). Extensive regeneration *in vitro* by early embryonic neurons on immature and adult CNS tissue. *J Neurosci* 15:2057–2062.
Siegal, J. D., Kliot, M., Smith, G. M., and Silver, J. (1990). A comparison of the regeneration potential of dorsal root fibers into gray or white matter of the adult rat spinal cord. *Exp Neurol* 109:90–97.
Silver, J., Edwards, M. A., and Levitt, P. (1993). Immunocytochemical demonstration of early appearing astroglial structures that form boundaries and pathways along axon tracts in the fetal brain. *J Comp Neurol* 328:415–436.
Smith, G. M., Rutishauser, U., Silver, J., and Miller, R. H. (1990). Maturation of astrocytes in vitro alters the extent and molecular basis of neurite outgrowth. *Dev Biol* 138:377–390.
Smith-Thomas, L., Fok, S. J., Stevens, J., Du, J. S., Muir, E., Faissner, A., Geller, H. M., Rogers, J. H., and Fawcett, J. W. (1994). An inhibitor of neurite outgrowth produced by astrocytes. *J Cell Sci* 107:1687–1695.
Snow, D. M., Lemmon, V., Carrino, D. A., Caplan, A. I., and Silver, J. (1990). Sulfated proteoglycans in astroglial barriers inhibit neurite outgrowth *in vitro*. *Exp Neurol* 109:111–130.
Snow, D. M., Watanabe, M., Letourneau, P. C., and Silver, J. (1991). A chondroitin sulfate proteoglycan may influence the direction of retinal ganglion cell outgrowth. *Development* 113:1473–1485.
Spillmann, A. A., Amberger, V. R., Schwab, M. E. (1997). High molecular weight protein of human central nervous system myelin inhibits neurite outgrowth: An effect which can be neutralized by the monoclonal antibody IN-1. *Eur J Neurosci* 9:549–555.
Srinivasan, S. R., Xu, J. H., Vijayagopal, P., Radhakrishnamurthy, B., and Berenson, G. S. (1995). Low-density lipoprotein binding affinity of arterial chondroitin sulfate proteoglycan variants modulates cholesteryl ester accumulation in macrophages. *Biochim Biophys Acta* 1272:61–67.
Steward, O., and Trimmer, P. A. (1997). Genetic influences on cellular reactions to CNS injury: The reactive response of astrocytes in denervated neuropil regions in mice carrying a mutation (Wld(S)) that causes delayed Wallerian degeneration. *J Comp Neurol* 380:70–81.
Stuermer, C. A. O., Bastmeyer, M., Bahr, M., Strobel, G., and Paschke, K. (1992). Trying to understand axonal regeneration in the CNS of fish. *J Neurobiol* 23:537–550.
Szczepanik, A. M., Fishkin, R. J., Rush, D. K., and Wilmot, C. A. (1996). Effects of chronic intrahippocampal infusion of lipopolysaccharide in the rat. *Neuroscience* 70:57–65.
Takamiya, Y., Kohsaka, S., Toya, S., Otani, M., and Tsukada, Y. (1988). Immunohistochemical studies on the proliferation of reactive astrocytes and the expression of cytoskeletal proteins following brain injury in rats. *Brain Res* 466:201–210.
Tang, S., Woodhall, R. W., Shen, Y. J., deBallard, M. E., Saffell, J. L., Doherty, P., Walsh, F. S., and Filbin, M. T. (1997). Soluble myelin-associated glycoprotein (MAG) found *in vivo* inhibits axonal regeneration. *Mol Cell Neurosci* 9:333–346.
Taoka, Y., Okajima, K., Uchiba, M., Murakami, K., Harada, N., Johno, M., Naruo, M., Okabe, H., and Takatsuki, K. (1997). Reduction of spinal cord injury by administration of iloprost, a stable prostacyclin analog. *J Neurosurg* 86:1007–1011.
Tello, F. (1911a). La influencia del neurotropismo en la regeneracion de los centros nerviosos. *Trab Lab Invest Univ Madrid* 9:123–159.
Tello, F. (1911b). Un experimento sobre las influencia del neurotropismo en la regeneracion de la corteza cerebral. *Rev Clin Madrid* 5:292–294.
Tetzlaff, W., Kobayashi, N. R., Giehl, K. M. G., Tsui, B. J., Cassar, S. L., and Bedard, A. M. (1994). Response of rubrospinal and corticospinal neurons to injury and neurotrophins. *Prog Brain Res* 103:271–286.

Thallmair, M., Metz, G. A. S., Z'Graggen, W. J., Raineteau, O., Kartje, G. L., and Schwab, M. E. (1998). Neurite growth inhibitory restrict plasticity and functional recovery following corticospinal tract lesions. *Nature Neurosci* 1:124–131.

Theele, D. P., and Streit, W. J. (1993). A chronicle of microglial ontogeny. *Glia* 7:5–8.

Tomaselli, K. J., Neugebauer, K. M., Bixby, J. L., Lilien, J., and Reichardt, L. F. (1988). N-cadherin and integrins: Two receptor systems that mediate neuronal process outgrowth on astrocyte surfaces. *Neuron* 1:33–43.

Tower, S. (1931). A search for trophic influence of posterior spinal roots on skeletal muscle, with a note on the nerve fibers found in the proximal stumps of the roots after excision of the root ganglia. *Brain* 54:99–110.

Tuszynski, M. H., Gabriel, K., Gage, F. H., Suhr, S., Meyer, S., and Rosetti, A. (1996). Nerve growth factor delivery by gene transfer induces differential outgrowth of sensory, motor, and noradrenergic neurites after adult spinal cord injury. *Exp Neurol* 137:157–173.

Tuszynski, M. H., Murai, K., Blesch, A., Grill, R., and Miller, I. (1997). Functional characterization of NGF-secreting cell grafts to the acutely injured spinal cord. *Cell Transplant* 6:361–368.

Tuszynski, M. H., Peterson, D. A., Ray, J., Baird, A., Nakahara, Y., and Gage, F. H. (1994). Fibroblasts genetically modified to produce nerve growth factor induce robust neuritic ingrowth after grafting to the spinal cord. *Exp Neurol* 126:1–14.

Uhlin-Hansen, L., Eskeland, T., and Kolset, S. O. (1989). Modulation of the expression of chondroitin sulfate proteoglycan in stimulated human monocytes. *Journal of Biological Chemistry* 264:14916–14922.

Uhlin-Hansen, L., and Kolset, S. O. (1987). Proteoglycan biosynthesis in relation to differentiation of cord blood monocytes in vitro. *Cell Differentiation* 21:189–197.

Uhlin-Hansen, L., and Kolset, S. O. (1988). Cell density-dependent expression of chondroitin sulfate proteoglycan in cultured human monocytes. *Journal of Biological Chemistry* 263:2526–2531.

Uhlin-Hansen, L., Wik, T., Kjellen, L., Berg, E., Forsdahl, F., and Kolset, S. O. (1993). Proteoglycan metabolism in normal and inflammatory human macrophages. *Blood* 82:2880–2889.

Vaudano, E., Campbell, G., Anderson, P. N., Davies, A. P., Woolhead, C., Schreyer, D. J., and Lieberman, A. R. (1995). The effects of a lesion or a peripheral nerve graft on GAP-43 upregulation of the adult rat brain: An *in situ* hybridization and immunocytochemical study. *J Neurosci* 15:3594–3611.

Virchow, R. (1846). Ueber gas granulierte Ansehen der Wandungen der Gerhirnventrikel. *Allg Z Psychiatr Psych-Gerichtl Med* 3:424–450.

Wallace, M. C., Tator, C. H., and Lewis, A. J. (1987). Chronic regenerative changes in the spinal cord after cord compression injury in rats. *Surg Neurol* 27:209–219.

Walz, W. (1989). Role of glial cells in the regulation of the brain ion microenvironment. *Prog Neurobiol* 33:309–333.

Weigert, C. (1895). Beitrage zur Kenntnis der normalen menshlichen neuroglia. Moritz Diesterweg, Frankfurt aum Maim, Weisbrod.

Windle, W. F. (1956). Regeneration of axons in the vertebrate central nervous system. *Phys Rev* 36:427–439.

Windle, W. F., and Chambers, W. W. (1950). Regeneration in the spinal cord of the cat and dog. *J Comp Neurol* 93:241–258.

Windle, W. F., Clemente, C. D., and Chambers, W. W. (1952). Inhibition of formation of a glial barrier as a means of permitting a peripheral nerve to grow into the brain. *J Comp Neurol* 96:359–369.

Winter, C. G., Saotome, Y., Levison, S. W., and Hirsh, D. (1995). A role for ciliary neurotrophic factor as an inducer of reactive gliosis, the glial response to central nervous system injury. *Proc Natl Acad Sci USA* 92:5865–5869.

Wolburg, H., and Risau, W. (1995). Formation of the blood-brain-barrier. In Kettenmann, H., and Ransom, B. R., eds., *Neuroglia.* New York: Oxford University Press, pp. 763–776.

Wood, P. and Bunge, R. P. (1984). The biology of the oligodendrocyte. In Norton, W. T., ed., *Oligodendroglia.* New York: Plenum Press, pp. 1–46.

Xie, D., Schultz, R. L., and Whitter, E. F. (1995). The oligodendroglial reaction to brain stab wounds: An immunohistochemical study. *J Neurocytol* 24:435–484.

Yasuda, T., Sobue, G., Ito, T., Mitsuma, T., and Takahashi, A. (1990). Nerve growth factor enhances neurite arborization of adult sensory neurons; a study in single-cell culture. *Brain Res* 524:54–63.

Yeaman, C., and Rapraeger, A. C. (1993a). Membrane-anchored proteoglycans of mouse macrophages: P388D1 cells express a syndecan-4-like heparan sulfate proteoglycan and a distinct chondroitin sulfate form. *Journal of Cellular Physiology* 157:413–425.

Yeaman, C., and Rapraeger, A. C. (1993b). Post-transcriptional regulation of syndecan-1 expression by cAMP in peritoneal macrophages. *Journal of Cell Biology* 122:941–950.

Z'Graggen, W. J., Metz, G. A. S., Kartje, G. L., Thallmair, M., and Schwab, M. E. (1998). Functional recovery and enhanced corticofugal plasticity after unilateral pyramidal tract lesion and blockage of myelin-associated neurite growth inhibitors in adult rats. *J Neurosci* 18:4744–4757.

Zhang, Y., Anderson, P. N., Campbell, G., Mohajeri, H., Schachner, M., and Lieberman, A. R. (1995). Tenascin-C expression by neurons and glial cells in the rat spinal cord: Changes during postnatal development and after dorsal root or sciatic nerve injury. *J Neurocytol* 24:585–601.

Zhang, Z., Krebs, C. J., and Guth, L. (1997). Experimental analysis of progressive necrosis after spinal cord trauma in the rat: Etiological role of the inflammatory response. *Exp Neurol* 143:141–152.

Zhang, Z., Fujiki, M., Guth, L., and Steward, O. (1996). Genetic influences on cellular reactions to spinal cord injury: A wound-healing response present in normal mice is impaired in mice carrying a mutation (Wld) that causes delayed wallerian degeneration. *J Comp Neurol* 371:485–495.

CHAPTER 4

Excitotoxins and Free Radicals

BINGREN HU AND JUSTIN A. ZIVIN
Department of Neurosciences, University of California, San Diego, La Jolla, California

Excitotoxins and free radicals are important contributors to cell dysfunction and loss in several diseases of the nervous system. Glutamate systems are a primary pathway mediating excitotoxic damage to the nervous system, with several toxic effects being mediated through ionotropic glutamate receptors. Given the fact that several glutamate receptor subtypes have been cloned and their ligand sensitivity has been linked to protein kinase activity, new drugs are under development that target glutamate receptor subtypes or protein kinases to reduce neuronal damage. Free radical formation also contributes substantially to cell damage in neurological disease. Various approaches including free radical scavengers, and genetic manipulation of enzymes in free radical metabolic pathways have been shown to provide neuroprotection in models of nervous system disease. The implications of these findings will be discussed in the context of brain ischemia, trauma, and neurodegenerative disorders including Huntington's, Parkinson's, and Alzheimer's disease.

I. INTRODUCTION

Excitotoxins and free radicals have been widely considered to be involved in tissue damage caused by brain ischemia, trauma, and epilepsy. Both excitotoxins and free radicals are also implicated in mechanisms of neurodegenerative diseases including Huntington's, Parkinson's, and Alzheimer's diseases. Knowledge of glutamate receptors, glutamate receptor-mediated intracellular signaling and free radical metabolic pathways have accumulated in recent years. The diversity of glutamate receptors provides a basis for the development of specific and effective glutamate receptor antagonists. Several free radical scavengers are neuroprotective in animal models of focal ischemia and trauma. Recently, studies using transgenic mice have provided evidence that free radicals are involved in brain injury. The objective of this chapter is to discuss the roles of excitotoxins and free radicals in the mechanisms of acute brain damage, mainly ischemic injury.

II. EXCITOTOXICITY

Excessive activation of neuronal glutamate receptors by a group of structural analogs of L-glutamate damages neurons. This action is termed excitotoxicity. Discovery of this phenomenon stems from pioneering studies demonstrating that exposure of the central nervous system (CNS) to dicarboxylic acid structural analogs such as L-glutamate (Glu), L-aspartate (Asp), kainate (KA), ibotanate or quinolinic acid (QUIN) are able to produce seizures and degeneration of groups of neurons (Hayashi, 1954; Lucas and Newhouse, 1957; Curtis and Watkins, 1963; Olney, 1969; Olney and Ho, 1970). These compounds are often referred to as excitotoxins. However, the concept of excitotoxicity was not established until it was shown that neuronal damage can be induced by L-glutamate analogs through glutamate receptors (Olney *et al.*, 1971; Watkins, 1978). This theory was further supported by discoveries that excitotoxicity is mediated by depolarization-induced cell swelling in the early phase of damage and was due to calcium-mediated events by overactivation of glutamate receptors in the late phase of neuronal death in cultured neurons (Rothman and Olney, 1985; Choi, 1988).

Excitatory amino acid receptors and excitotoxicity have been a focus of attention in neuroscience research in recent years. This is (1) because glutamate is a major neurotransmitter in the CNS and plays an important role in synaptic plasticity, learning, and memory (Bliss and Collingridge 1993; Linden, 1994); and (2) because excessive activation of glutamate receptors has been implicated in the mechanisms of acute brain insults such as brain ischemia, hypoglycemia, seizure, trauma (Choi, 1988), and several chronic neurodegenerative disorders (Ikonomidou and Turski, 1995).

Glutamate receptors can be classified into ionotropic and metabotropic receptors based on their pharmacological, physiological, and biochemical properties. The ionotropic receptors are named after their preferred agonists. They consist of N-methyl-D-aspartate (NMDA), α-amino-3-hydroxy-5-methyl-4-isoxazole propionate (AMPA), and kainate (KA) receptors. These receptors are directly coupled to ion channels permeable to cations (Seeburg, 1993; Mori and Mishina, 1995). The metabotropic glutamate receptors mediate their biological functions through G-protein-coupled cascades (Nikanishi, 1992). Among the ionotropic receptors, the NMDA receptor channel is gated in a unique use- and voltage-dependent manner (Mori and Mishina, 1995). At resting cell membrane potential, NMDA receptor channels are blocked by Mg^{2+}. Upon ligand binding and depolarization, the Mg^{2+} blockade is removed from the channel, which permits a substantial conductance of Ca^{2+}. Glycine is a co-agonist and polyamine is a positive modulator of the NMDA receptor-gated channels. Both the AMPA and KA (AMPA–KA) receptor-gated ion channels are permeable to Na^+ and K^+. To date, two gene families that encode NMDA receptors have been cloned and termed NR1 and NR2. The NR1 family has eight splicing variants, and the NR2 family is comprised of NR2A, NR2B, NR2C, and NR2D. Transfection studies have demonstrated that NMDA receptors are probably pentamers composed of channel subunits of NR1 and at least one of the regulatory subunits of NR2 (Mori and Mishina, 1995). AMPA receptors have four subunits named GluR1–GluR4, and each subunit has two forms, flip and flop, created by alternative splicing (Seeburg, 1993; Wisden and Seeburg, 1993). Recombinant AMPA receptors lacking a subunit of GluR2 are permeable to Ca^{2+} (Burnashev *et al.*, 1992). GluR5, GluR6, GluR7, and KA1 and KA2 are subunits for the kainate receptors (Seeburg, 1993; Wisden and Seeburg, 1993). Different combinations and uneven regional distributions of these receptor subunits in the CNS create a functional diversity of NMDA or AMPA–KA receptors that may have clinical significance. This is because these different combinations of the receptor subunits have different affinities for agonists and antagonists, different phosphorylation-induced modulatory activities and different ion conductance kinetics (Seeburg, 1993; Wisden and Seeburg, 1993; Buller *et al.*, 1994; Petralia *et al.*, 1994; Zukin and Bennett, 1995). Phosphorylation of glutamate receptors by protein kinases is an important regulatory mechanism for receptor function (Roche *et al.*, 1994).

Excitotoxicity has been extensively studied in cultured neurons and ischemic brains *in vivo*. Global brain ischemia leads to a selective delayed neuronal death in some population of neurons such as in hippocampal CA1 pyramidal neurons. During the delay (two or three days), neurons destined for death not only functionally recover, but also look normal under the light microscope (Kirino, 1982; Pulsinelli *et al.*, 1982; Smith *et al.*, 1984). Delayed neuronal death is also implicated in several other types of brain injuries including focal ischemia, trauma, seizure, and chronic neurodegenerative diseases. Mechanisms underlying this phenomenon are not completely understood. Among many hypotheses, glutamate-mediated excitotoxicity through an overload of intracellular calcium ([Ca^{2+}]i) has been considered to be an important contributor to ischemic neuronal damage (Siesjö, 1988; Choi, 1988; Rothman and Olney, 1995). Three lines of evidence support the concept that glutamate receptors, particularly the NMDA and AMPA subtypes, are involved in ischemia-induced neuronal damage. First, neurons can be killed by exposure to glutamate structural analogs either in culture or *in vivo* (Choi, 1988; Jhamadas and Boegman, 1994). Second, using the microdialysis technique, it was found that extracellular glutamate is significantly increased, about 100-fold, during ischemia (Benveniste *et al.*, 1984). Third, ionotropic glutamate receptor blockade has been shown to be neuroprotective in brain ischemic and traumatic injuries (for review see Lipton, 1993).

Several issues remain controversial. The increase in extracellular glutamate present at the onset of ischemia rapidly returns to control values after an ischemic episode (Benveniste *et al.*, 1984), whereas the neuroprotective effects of glutamate receptor antagonists have been reported when these agents are given during reperfusion phases up to several hours following an ischemic episode (Buchan *et al.*, 1991; Li and Buchan, 1993; Xue *et al.*, 1994). This paradox suggests that there are additional factors in the postischemic brain participating in glutamate-mediated neuronal damage.

The synapse, consisting of the presynaptic membrane, the synaptic cleft, and the specialized postsynaptic membrane containing the postsynaptic density (PSD), is the basic neurotransmission unit in the CNS (Kennedy, 1994). Glutamate receptors are highly concentrated in the postsynaptic density (Petralia *et al.*, 1994, Hu *et al.*, 1998). We recently found evidence that transient cerebral ischemia induces a marked accumulation of protein kinases in PSDs including calcium–calmodulin-dependent kinase II (Fig. 1, CaMKII) and protein kinase C (Fig. 1, PKC-γ) (Hu *et al.*, 1998). Tyrosine phosphorylation of PSD-proteins including the NMDA receptor (Fig. 1, arrow) is substantially increased in the reperfusion phases after an ischemic period (Fig. 1, Ptyr) (Hu and Wieloch, 1994, 1995a, 1995b). The conclusion drawn from these studies is that these kinases are persistently increased in the PSDs of neurons destined to die, but only transiently increased in neurons that are relatively resistant to ischemic insults. Consistent with these synaptic biochemical alterations, the 2- and 3-dimensional ultrastructures of synapses selectively stained with ethanolic phosphotungstic acid (Hu *et al.*, 1998) are also markedly modified in the postischemic phase (Fig. 2A and B). These changes in synaptic biochemistry and ultrastructure may, at least in part, result from excitotoxicity that follows transient cerebral ischemia (Hu *et al.*, 1995a, 1995b, 1998).

Excitotoxin-induced neuronal damage has been proposed to be mediated by several Ca^{2+} mediated pathways (Fig. 3) (Choi, 1988; Siesjö, 1988). Activation of proteases (calpain and caspases) and production of free radicals have all received attention. It has been demonstrated that spectrin and microtubule-

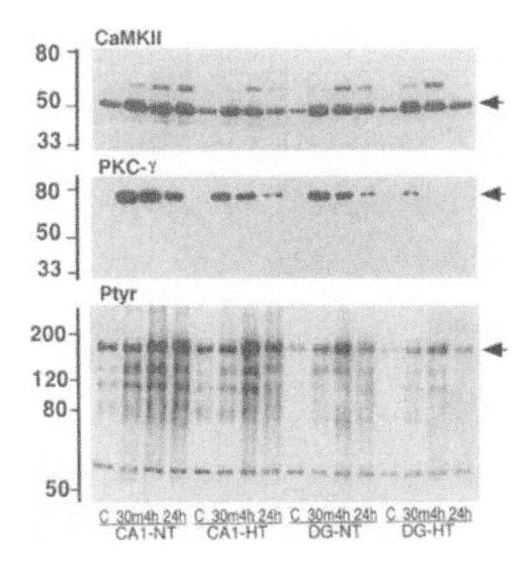

FIGURE 1 Ischemia-induced translocation of Ca^{2+}–calmodulin-dependent protein kinase II (CaM-kinase II), protein kinase C (PKC) and increase in tyrosine phosphorylation (Ptyr) of NMDA receptor in crude postsynaptic density fractions after normothermic (NT) and hypothermic (HT) ischemia. Samples from sham-operated rats and rats subjected to 15 min ischemia, followed by 30 min (30m), 4 hr (4h) and 24 hr (24h) of reperfusion, were separated on SDS-PAGE, transferred onto immobilon-P, labeled with antibodies against the α-subunit of CaM-kinase II (Hu *et al.*, 1995), protein kinase C-γ and phosphotyrosine proteins (Hu *et al.*, 1994) and developed with the Amersham ECL system. CA1 = samples from hippocampal CA1 region; DG = sample from hippocampal dentate gyrus area. Molecular weight standards are indicated on the left. Results demonstrate that CaM-kinase II, protein kinase C-γ and tyrosine phosphorylation of NMDA receptor-2 (NR2) (arrow) markedly increase in postsynaptic density fractions after transient cerebral ischemia. The increases are more persistent in the CA1 region than in the DG area. Intra-ischemic hypothermia (HT) reduces these changes. (The original studies were conduced in the Laboratory for Exp. Brain Res., Lund, Sweden.) Thus, changes in calcium and second messenger systems occur after ischemia are regionally expressed, and can be modified by hypothermia.

associated protein-2 (MAP-2) are degraded long before delayed neuronal death in the postischemic brain (Roberts-Lewis *et al.*, 1994). Several free radical scavengers (see next section) and protease inhibitors have been shown to be neuroprotective in brain ischemia models *in vivo* (Lee *et al.*, 1991; Rami and Krieglstein, 1993; Gottron *et al.*, 1997; Hara *et al.*, 1997). This evidence suggests that free radical-induced damage and proteolysis may contribute to excitotoxin-induced neuronal damage.

III. FREE RADICALS

Any chemical substance possessing unpaired electrons is a free radical. The most important free radicals in biological systems are oxygen radicals and

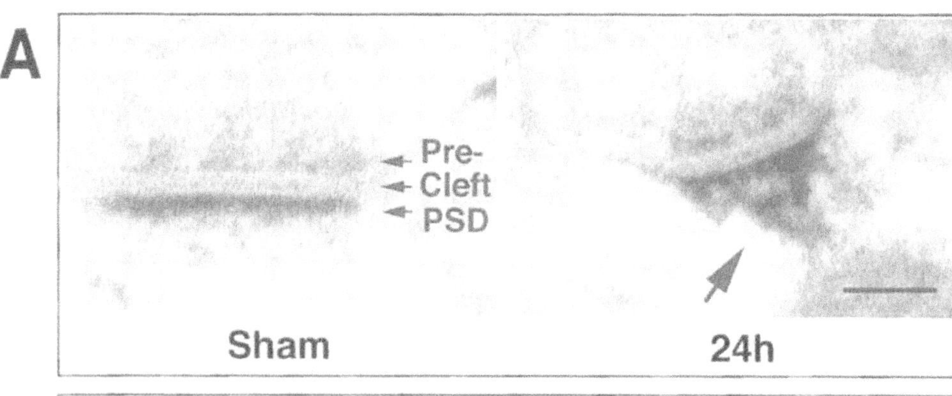

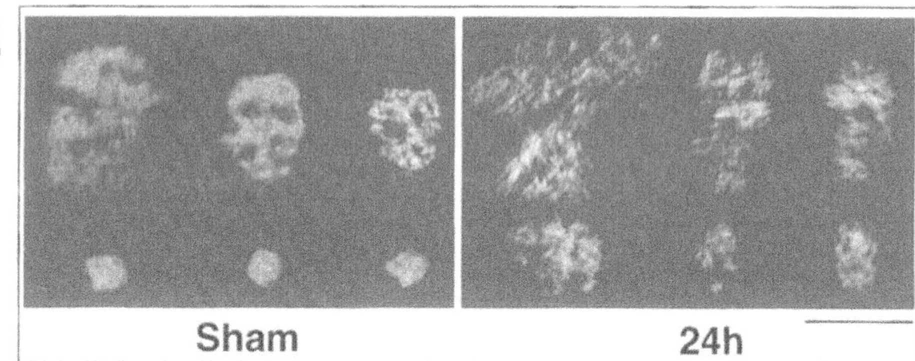

FIGURE 2 (A) Electron micrographs of ethanolic phosphatungstic acid (E-PTA) stained synapses in hippocampal CA1 regions from sham-operated rats (sham) and rats at 24 hr reperfusion following 15 min cerebral ischemia. Examples of postsynaptic density (PSDs) are shown. In comparison with the control synapses (sham), postischemic PSDs were generally curved and surrounded by fluffy E-PTA stained materials in the PSDs. Scale bar = 0.1 μm. Pre = pre-synaptic specialization: cleft = synaptic cleft; PSD = postsynaptic density. (B) Three-dimensional images of E-PTA stained PSDs from tomographic volumes of control (sham) and 24 hr of reperfusion (24h) following 15 min cerebral ischemia in hippocampal CA1 region. E-PTA stained thick sections (1 μm) were examined by a JEOL 4000EX intermediate high-voltage electron microscope at an accelerating voltage of 400 keV. Data for tomographic reconstruction was acquired using the single axis tilt method. A series of images was obtained over a range of $+/- 60°$ at $2°$ increments using film as the specimen and was tilted in equal angular increments and reconstructed into three-dimensional images of the synapses. Additional details of the tomographic method used can be found in Soto et al. (1994) and Perkins (1997). Each PSD was rotated individually to appear in *en face* configuration. The control PSDs were compact, disklike structures with generally round edges and discrete perforations, whereas the postischemic PSDs were more loosely configured, netlike structures with irregular edges and less well-defined perforations. Scale bar = 0.5 μm (Submitted data). Thus, ischemia disrupts synaptic architecture.

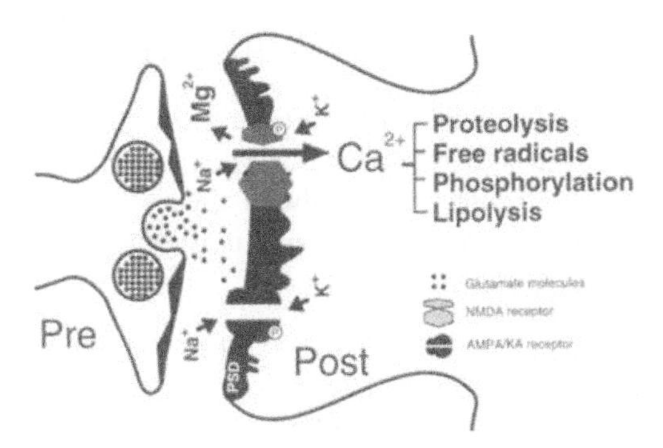

FIGURE 3 A hypothetical model of excitotoxicity. Glutamate is released from glutamatergic nerve terminals during brain injuries. It binds to postsynaptic glutamate receptors, induces depolarization through AMPA receptor channels, and releases the Mg^{2+} block in NMDA receptor channels of postsynaptic neurons. Calcium influx into postsynaptic neurons occurs through NMDA receptor channels. High levels of intracellular calcium can overactivate proteases, NO synthases, protein kinases, and lipases, and lead to neuronal damage. Pre: presynaptic terminal. Post: postsynaptic spine. PSD: postsynaptic density.

their derivatives. Free radicals are so highly reactive and short-lived that it is extremely difficult to prove their existence directly *in vivo*. This leads to some concerns about the importance of free radicals in brain injury. Nevertheless, there is a wide array of evidence from studies of free radical production, endogenous metabolic enzymes, scavengers, and genomic modifications that suggest the importance of free radicals in brain pathology. Free radicals can be produced from the following sources during brain injury: (1) leakage from mitochondria in the process of electron transfer, (2) nitric oxide-related reactions, (3) deliberate release from phagocytes during inflammation, (4) enzyme-catalyzed oxidization reactions, and (5) lipid peroxidation.

A. The Iron-Catalyzed Haber-Weiss Reaction (Fenton Reaction)

Superoxide can be produced by the transfer of a single electron to oxygen in the process of mitochondrial electron transfer and in enzyme-catalyzed oxidization reactions (see below).

$$O^2 + e^- \longrightarrow O_2^{-\bullet}$$

Two superoxides can react with each other to form hydrogen peroxide and this reaction can be accelerated by superoxide dismutases (SODs, including Cu-Zn-SOD in the cytoplasm and Mn-SOD in mitochondria).

$$2\,O_2^{-\bullet} + 2H^+ \xrightarrow{\text{SODs}} H_2O_2 + O_2$$

Hydrogen peroxide can be removed from tissues by catalases or glutathione peroxidases (GSHPxs). It has been shown that selenium-dependent glutathione peroxidase is a major enzyme that catalyzes the reaction for clearance of hydrogen peroxide at the expense of reduced glutathione (GSH) in brain tissue (Jain *et al.*, 1991).

$$H_2O_2 \xrightarrow[\text{2GSH} \quad \text{GSSG}]{\text{GSHPxs}} 2H_2O$$

Hydrogen peroxide, unlike superoxides, can easily cross cell membranes. If hydrogen peroxide is not cleared and trace amounts of free Fe^{2+} are present, hydrogen peroxide can produce the most reactive and damaging oxygen free radical, the hydroxyl radical ($^{\bullet}OH$).

$$H_2O_2 + Fe^{2+} \longrightarrow {}^{\bullet}OH + OH^- + Fe^{3+}$$

$$Fe^{3+} + O_2^{-\bullet} \longrightarrow Fe^{2+} + O_2$$

The overall result of these two reactions is often referred to as the iron-catalyzed Haber-Weiss reaction (Fenton reaction).

$$H_2O_2 + O_2^{-\bullet} \xrightarrow{Fe^{2+}/\,Fe^{3+}} {}^{\bullet}OH + OH^- + O_2$$

Under normal circumstances, free iron in the brain is maintained at an extremely low level by iron-binding proteins such as plasmic transferrin and intracellular ferritin. However, when acidosis occurs during and after brain insults ($pH < 6.0$), iron can be released from these binding proteins and actively catalyze oxygen free radical reactions. Hemoglobin from an intracerebral hemorrhage is another source of iron for catalyzing oxygen free radical reactions. The hemoglobin released from brain hemorrhage has been described as a "dangerous protein" in brain injuries (Halliwell, 1992).

From the above discussion, it is clear that the key players in the Fenton reaction are oxygen itself (O_2), superoxide ($O_2^{-\bullet}$), hydrogen peroxide (H_2O_2), the hydroxyl radical ($^{\bullet}OH$), and transition metals such as irons (i.e., Fe^{2+} and Fe^{3+}). Among these molecules, superoxide and hydrogen peroxide themselves are not particularly damaging species, but they are sources of hydroxyl radicals, the most reactive oxidizing radical. The hydroxyl radical has an extremely short half-life that limits its damaging distance in tissue. Nevertheless, the hydroxyl radical is able to strongly react with all cellular molecules including lipids, DNA, and proteins, and may cause great damage at its production site.

B. Nitric Oxide

Nitric oxide (NO) was discovered to be a vasodilator in vascular endothelial cells (Moncada *et al.*, 1987). This simple molecule has been intensively investigated recently due to its involvement in neurotransmission, learning and memory (Garthwaite, 1991), brain injuries (Chan, 1996; Beckman, 1994), and phagocytic activity (Rosen *et al.*, 1995). Production of NO is catalyzed by NO synthase (NOS) with L-arginine as a precursor (Moncada *et al.*, 1987). There are three different isoforms of NOS, neuronal (NOS-I or nNOS), inducible (NOS-II or iNOS), and endothelial (NOS-III or eNOS) (Garthwaite, 1991; Moncada and Higgs, 1993). Neuronal and endothelial NOS, constitutively expressed in neurons and endothelial cells respectively, are often referred to as constitutive NOS (cNOS), and they are abundant in the CNS. Both neuronal and endothelial NOS require Ca^{2+}–calmodulin as a cofactor for their activities (Masters *et al.*, 1996), and these enzymes produce small quantities of NO under physiological conditions. The inducible NOS usually is not expressed, but it can be induced in macrophages and neutrophils by tumor necrosis factor-α (TNF-α) and interleukin-1β (Xie *et al.*, 1992). Unlike neuronal and endothelial NOS, iNOS does not require Ca^{2+}–calmodulin for its activity and can produce large and potentially toxic quantities of NO in inflammatory cells recruited to an injury site (Moncada and Higgs, 1993; De Groote and Fang, 1995). It has been shown that iNOS is also expressed in microglia and astrocytes in the CNS (Moncada and Higgs, 1993).

Nitric oxide produced by NOSs is able to react with superoxide to form peroxynitrite ($ONOO^-$) at a faster rate than SOD-catalyzed reactions (Beckman, 1994).

$$NO + O_2^{-\bullet} \longrightarrow ONOO^-$$

$$ONOO^- + H^+ \longrightarrow ONOOH$$

$$ONOOH \longrightarrow {}^\bullet OH + NO_2$$

The peroxynitrite itself is a strong oxidant, and the hydroxyl radical formed from the decomposition of the protonated form of peroxynitrite is also extremely reactive as previously mentioned.

NOS activity can be regulated by multiple phosphorylation sites. Neuronal NOS can be phosphorylated by protein kinase A (Dinerman and Synder, 1994), protein kinase C (Bredt *et al.*, 1992), and CaM-kinase II (Bredt *et al.*, 1992) and dephosphorylated by calcineurin, a Ca^{2+}–calmodulin-dependent phosphatase. In general, phosphorylation of NOS by all these kinases decreases its activity and dephosphorylation increases its activity. It has been reported that

FK506 is neuroprotective in cultured neurons exposed to glutamate and in an ischemic stroke model (Dawson, 1995; Zhang and Steiner, 1995). This neuroprotective effect has been explained by the formation of a complex of FK506 and its binding protein (FKBP), which binds to calcineurin and inhibits its activity, thereby decreasing NO production (Dawson, 1995).

C. Deliberate Release of Free Radicals from Phagocytes During Inflammation

Phagocytes recruited to an injury site can produce toxic amounts of nitric oxide and other free radicals through the following reactions.

$$NADPH + O_2 \xrightarrow{\text{NADH oxidase}} O_2^{-\bullet} + NADP^+ + H^+$$

$$O_2^{-\bullet} + O_2^{-\bullet} + 2\,H^+ \xrightarrow{\text{SOD}} H_2O_2 + O_2$$

$$H_2O_2 + Cl^- \xrightarrow{\text{Myeloperoxidase}} H_2O_2 + HOCl$$

These processes are known as an "oxidative burst" because oxygen consumption in these cells are markedly increased during the reactions. Hypochlorous acid (HOCl) and its unprotonated anion (OCl^-) are very powerful oxidizing agents that will react at multiple sites on many cellular molecules (Rosen *et al.*, 1995).

D. The Oxidases

Xanthine oxidase and monoamine oxidase are two important enzymes potentially causing free radical-induced brain pathology. Xanthine oxidase is synthesized as xanthine dehydrogenase, which transfers electrons to the electron carrier NAD^+ (nicotinamide adenine dinucleotide). In ischemic conditions, a calcium-dependent protease converts this dehydrogenase to its oxidase form, which transfers electrons to oxygen instead of to NAD^+ and produces superoxide (Morris *et al.*, 1995). Dopamine turnover in dopamine nerve terminals may generate excessive amounts of H_2O_2 by monoamine oxidase, which has been hypothesized as a possible cause of neurodegeneration in Parkinson's disease (Cohen, 1988).

E. Lipid Peroxidation

Polyunsaturated fatty acids of membrane phospholipids can be attacked by free radicals to cause lipid peroxidation. The process is initiated when a radical

(R) attacks and removes an allylic hydrogen from an unsaturated fatty acid (LH). The initiation of this reaction can result in a chain reaction (Hall and Braughler, 1993).

$$LH + R^{\bullet} \longrightarrow L^{\bullet} + RH$$

$$L^{\bullet} + O_2 \longrightarrow LOO^{\bullet}$$

$$LOO^{\bullet} + LH \longrightarrow LOOH + L^{\bullet}$$

Once iron is involved, it reacts with lipid hydroperoxides and decomposes them to peroxyl radicals ($LOO^{\bullet}$) and alkoxyl radicals ($LO^{\bullet}$).

$$LOOH + Fe^{2+} \longrightarrow LO^{\bullet} + OH^{-} + Fe^{3+}$$

$$LOOH + Fe^{3+} \longrightarrow LOO^{\bullet} + Fe^{2+}$$

$$LO^{\bullet} + LH \longrightarrow LOH + L^{\bullet}$$

Lipid peroxidation can result in membrane dysfunctions such as membrane leakage or damage to neurotransmitter receptors or cause microvascular dysfunction that triggers postinjury hypoperfusion (Hall and Braughler, 1993).

IV. IMPLICATION OF FREE RADICALS IN BRAIN INJURY

Although free radicals are damaging chemical substances, biological systems have the machinery to defend against this type of injury. Defenses include iron-binding proteins, the metabolic enzymes such as SOD, catalases, or glutathione peroxidases. Under pathological conditions, however; either an increase in free radical production or a decrease in free radical clearance may result in free radical-induced tissue damage. This has been hypothesized to be an important contributor to age-related neurodegeneration including Parkinson's and Alzheimer's diseases and to acute brain injuries (Williams, 1995; Chan, 1996).

Because free radicals are so short-lived that it is extremely difficult to detect them directly, several methods have been developed to measure free radical reaction products *in vivo*. These include the salicylate-trapping method in conjunction with microdialysis and HPLC (high-performance liquid chromatography) to detect hydroxyl radicals (Oliver *et al.*, 1990; Globus *et al.*, 1995), a porphyrinic microsensor for NO (Malinski *et al.*, 1993), hydroethine imaging (Bindokas *et al.*, 1996), and spin trap reagents in combination with electron paramagnetic resonance (Tominaga *et al.*, 1993) for superoxide.

Free radical-induced tissue damage in the CNS has been extensively studied in animal models of ischemic and traumatic brain injuries (Chan, 1996). Available evidence suggests that free radicals are transiently increased in brain tissue subjected to ischemia–reperfusion and trauma (Oliver *et al.*, 1990; Malinski *et al.*, 1993; Tominaga *et al.*, 1993; Hall and Braughler, 1993; Globus *et al.*, 1995). Free radicals may also be produced in cultured neurons or brain slices exposed to glutamate and its analogs (Bindokas *et al.*, 1996) and in cultured neurons deprived of neurotrophic factor (Greenlund *et al.*, 1995). Several free radical scavengers have been tested for their neuroprotective effects in ischemic and traumatic brain injuries. Steroid antioxidants, such as 21-aminosteroid and its derivatives, are potent inhibitors of lipid peroxidation, but they have generally failed to prevent selective and delayed neuronal damage in global ischemia models (Carlsson *et al.*, 1994; Hall *et al.*, 1996) or in clinical trials in stroke patients. However, this group of free radical scavengers has been shown to be effective in some models of brain trauma and focal ischemic injuries (Park and Hall, 1994; Sterz *et al.*, 1996; Hall *et al.*, 1996).

Inhibitors of NOSs have been studied in animal models of brain injuries. Inconsistent results have been obtained. The inconsistency may be due to the diverse isoforms of NOS in the CNS (Huang *et al.*, 1994, 1996). For example, it has been suggested that activation of neuronal NOS by glutamate is detrimental but activation of endothelial NOS is beneficial in ischemic and traumatic brains (Huang *et al.*, 1994). The neurotoxic action of NO has been hypothesized to be mediated by peroxynitrite ($ONOO^-$), a reaction product of NO and the superoxide anion (Beckman, 1994; Dawson and Dawson, 1996; Kamii *et al.*, 1996). Inducible NOS is highly expressed in ischemia-affected vessels and in the neutrophils that infiltrate into the parenchyma after ischemic and traumatic brain injuries (Iadecola *et al.*, 1996; Clark *et al.*, 1996).

Recent studies have shown that antileukocyte treatments using antibodies against ICAM-1 and CD11–CD18 are effective in several animal models of brain injury (Matsuo *et al.*, 1994; Chopp, 1995; Bowes *et al.*, 1993, 1996). This may indicate that free radicals released by leukocytes recruited to the injury site may exacerbate brain injuries (Matsuo *et al.*, 1995). Spin trap agents have been used as free radical scavengers in several brain injury models (Carney and Floyd, 1991). Several studies have demonstrated that α-phenyl-N-tert-butyl nitrone (PBN) reduces infarction size in rats when given up to 12 hours after permanent middle cerebral artery (MCA) occlusion (Cao and Phillis, 1994) and up to 3 hours after the initiation of recirculation subsequent to 2 hours of MCA occlusion (Zhao *et al.*, 1994). The protective effects of PBN have been attributed to prevention of microvascular dysfunction or polymorphonuclear leukocyte accumulation because PBN has no protective effect in a model of transient forebrain ischemia (Siesjö *et al.*, 1995).

Genomic manipulation studies support the free radical theory in brain injury (Kinouchi *et al.*, 1991; Chan, 1996). It was observed that the total infarct volume was reduced by 36% at 24 hours after MCA occlusion in mice overexpressing SOD-1 (CuZn-SOD). Brain edema was also reduced in this study

(Kinouchi *et al.*, 1991). The same group of investigators also showed that infarct volume was significantly reduced in an ischemia–reperfusion MCA occlusion model (Kamii *et al.*, 1995), but infarct volume was unchanged in permanent MCA occlusion in the transgenic mice (Chan *et al.*, 1993). In another study, a significant decrease in infarction size was observed in a middle cerebral artery occlusion model using neuronal NOS knockout mice (Huang *et al.*, 1994).

Many sources of free radicals, including mitochondrial leakage, oxidase catalyzed reactions, NO-related products and release from inflammatory cells, may contribute to CNS damage after injury. However, it is unclear which source plays a major role in CNS injury. Several free radical scavengers and antileukocyte treatments can protect the brain from focal ischemic and traumatic injury but they have no effect on global ischemic neuronal damage. This may indicate that brain microvascular dysfunction caused by free radicals during inflammation is an important factor in acute brain injury.

V. CONCLUSION

Knowledge about excitotoxins and free radicals in brain injury has accumulated in recent years although important details about intracellular mechanisms remain elusive. A diversity of functional subtypes of glutamate receptors and free radical production pathways in the CNS provides a challenge for developing anti-excitotoxin and anti-free radical drugs. More information about the composition, regulation and antagonists of glutamate receptors, sources of damaging free radicals, and their intracellular signaling pathways leading to cell death and and survival after brain insults is required for further understanding excitotoxin- and free radical-induced brain injury.

ACKNOWLEDGMENTS

The authors thank Dr. Maryann Martone for critical reading of the manuscript.

REFERENCES

Beckman, J. S. (1994). Peroxynitrite versus hydroxyl radical: The role of nitric oxide in superoxide-dependent cerebral injury. *Ann NY Acad Sci* 738:69–75.

Benveniste, H., Drejer, J., Schousboe, A., and Diemer, N. H. (1984). Elevation of the extracellular concentrations of glutamate and aspartate in rat hippocampus following transient cerebral ischemia monitored by intracerebral microdialysis. *J Neurochem* 43:1369–1374.

Bindokas, V. P., Jordan, J., Lee, C. C., and Miller, R. J. (1996). Superoxide production in rat hippocampal neurons: Selective imaging with hydroethine. *J Neurosci* 16:1324–1336.

Bliss, T. V. P., and Collingridge, G. L. (1993). Synaptic model memory: Long-term potentiation in the hippocampus. *Nature* 361:31–39.

Bowes, M. P., Rothlein, R., Fagan, S. C., and Zivin, J. A. (1996). Monoclonal antibodies preventing leukocyte activation reduce experimental neurological injury and enhance efficacy of thrombolytic therapy. *Neurology* 45:815–819.

Bowes, M. P., Zivin, J. A., *et al.* (1993). Monoclonal antibody to the ICAM-1 adhesion site reduces neurological damage in a rabbit cerebral embolism stroke model. *Exp Neurol* 119:215–219.

Bredt, D. S., Derris, C. D., and Synder, S. H. (1992). Nitric oxide synthase regulatory sites. *J Biol Chem* 267:10976–10981.

Buchan, A. M., Xue, D., Huang, Z. G., Smith, K. H., and Howard, L. (1991). Delayed AMPA receptor blockade reduces cerebral infarction induced by focal ischemia. Neuroreport 2:473–476.

Buller, A. L., Larson, H. C., Schneider, B. E., Beaton, J. A., Morrisett, R. A., and Monaghan, D. T. (1994). The molecular basis of NMDA receptor subtypes: Native receptor diversity is predicted by subunit composition. *J Neurosci* 14:5171–5484.

Burnashev, N., Monyer, H., Seeburg, P. H., and Sakmann, B. (1992). Divalent ion permeability of AMPA receptor channels is dominated by the edited form of a single subunit. *Neuron* 8:189–198.

Cao, X., and Phillis, J. (1994). α-phenyl-N-tert-butyl nitrone (PBN) reduces cortical infarct and edema in rats subjected to focal ischemia. *Brain Res* 644:267–272.

Carlsson, B. R., Loberg, E. M., Grogaard, B., and Steen, P. A. (1994). The antioxidant tirilazad does not affect cerebral blood flow or histopathologic changes following severe ischemia in rats. *Act Neurol Scand* 90:256–262.

Carney, J. M., and Floyd, R. A. (1991). Protection against oxidative damage to CNS by α-phenyl-N-tert-butyl nitrone (PBN) and other spin trap agents: A novel series of nonlipid free radical scavengers. *J Mol Neurosci* 3:47–57.

Chan, P. H. (1996). Role of oxidants in ischemic brain damage. *Stroke* 27:1124–1129.

Choi, D. W. (1988). Glutamate neurotoxicity and diseases of the nervous system. *Neuron* 1:623–634.

Chopp, M. (1995). Anti-adhesion molecule antibody reduces ischemic cell damage after transient MCAO in the rat. *J Cereb Blood Flow Metab* 15 (Suppl 1):S57.

Clark, R. S., Kochasnek, P. M., Schwarz, M. A., Schiding, J. K., Turner, D. S., Chen, M., Carls, M., and Watkins, S. C. (1996). Inducible nitric oxide synthase expression in cerebral vascular smooth muscle and neutrophils after traumatic brain injury in immature rats. *Pediatric Res* 39:784–790.

Cohen, G. (1988). Oxygen free radical and Parkinson's disease. In: Halliwell, B., ed., *Oxygen free radical and tissue injury*. pp. 130–135.

Curtis, D. R., and Watkins, J. C. (1963). Acidic amino acids with strong excitatory actions on mammalian neurons. *J Physiol* 166:1–14.

Dawson, V. L. (1995). Nitric oxide: Role in neurotoxicity. *Clin Exp Pharmacol Physiol* 22:305–308.

Dawson, V. L., and Dawson, T. M. (1996). Nitric oxide neurotoxicity. *J Chem Neuroanatomy* 10:179–190.

De Groote, M. A., and Fang, F. C. (1995). NO inhibition: Antimicrobial properties of nitric oxide. *Clin Infect Diseases* 219 (Suppl.):S162–S165.

Dinerman, J. L., and Synder, S. H. (1994). Cyclic nucleotide dependent phosphorylation of nitric oxide synthase inhibit catalytic activity. *Neuropharm* 267:1245–1251.

Garthwaite, J. (1991). Glutamate, nitric oxide and cell-cell signalling in the nervous system. *TINS* 14:67.

Globus, M. Y., Alonso, O., Detrich, W. D., Busto, R., and Ginsberg, M. D. (1995). Glutamate release and free radical production following brain injury: Effects of posttraumatic hypothermia. *J Neurochem* 65:1704–1711.

Gottron, F. J., Ying, H. S., and Choi, D. W. (1997). Caspase inhibition selectively reduces the apoptotic component of oxygen-glucose deprivation-induced cortical neuronal death. *Molecular and Cellular Neurosci* 9(3):159–169.

Greenlund, L. J. S., Deckwerth, T. L., and Jonson, E. M., Jr. (1995). Superoxide dimutase delays neuronal apoptosis: A role for reactive oxygen species in programmed neuronal death. *Neuron* 14:303–315.

Hall, E. D., and Braughler, J. M. (1993). Free radicals in CNS injury. In: Waxman, S. G., *Molecular and cellular approaches to the treatment of neurological disease.* New York: Raven Press Ltd., pp. 81–105.

Hall, E. D., Andrus, P. K., Smith, S. L., *et al.* (1996). Neuroprotective efficacy of microvascularly-localized versus brain-penetrating antioxidants. *Act Neurochirurgica* 66 (Suppl.):107–113.

Halliwell, B. (1992). Reactive oxygen species and the central nervous system. *J Neurochem* 59:1609–1623.

Hara, H., Friedlander, R. M., Gagliardini, V., Ayata, C., Fink, K., Huang, Z., Shimizu-Sasamata, M., Yuan, J., and Moskowitz, M. D. (1997). Inhibition of interleukin 1beta converting enzyme family proteases reduces ischemic and excitotoxic neuronal damage. *Proc Natl Acad Sci USA* 94(5):2007–2012.

Hayashi, T. (1954). Effects of sodium glutamate on the nervous system. *Keio J Med* 3:183–192.

Hu, B. R., and Wieloch, T. (1994). Tyrosine phosphorylation and activation of mitogen-activated protein kinase in the rat brain following transient cerebral ischemia. *J Neurochem* 62:1357–1367.

Hu, B. R., Kamme, F., and Wieloch, T. (1995a). Alterations of Ca^{2+} calmodulin-dependent protein kinase II and its messenger RNA in the rat hippocampus following normal and hypothermic ischemia. *Neurosci* 68:1003–1016.

Hu, B. R., and Wieloch, T. (1995b). Persistent translocation of Ca^{2+}/calmodulin-dependent protein kinases II to synaptic junctions in the vulnerable hippocampal CA1 region following transient cerebral ischemia. *J Neurochem* 64:277–284.

Hu, B. R., Park, M., Martone, M. E., Fischer, W. H., Ellisman, M. H., and Zivin, J. A. (1998). Assembly of proteins to postsynaptic densities after transient cerebral ischemia. *J Neurosci* 18(2):625–633.

Huang, Z., Huang, P. L., Fishman, M. C., and Moskowitz, M. A. (1996). Focal cerebral ischemia in mice deficient in either endothelial (eNOS) or neuronal nitric oxide (nNOS) synthase. *Stroke* 27:173 (Abstract).

Huang, Z., Huang, P. L., Panahian, N., Dalkara, T., Fishman, M. E., and Moskowitz, M. (1994). Effects of cerebral ischemia in mice deficient in neuronal nitric oxide synthase. *Science* 265:1883–1885.

Iadecola, C., Zhang, F., Casey, R., Clark, H. B., and Rose, M. E. (1996). Inducible nitric oxide synthase gene expression in vascular cells after transient focal cerebral ischemia. *Stroke* 27:1373–1380.

Ikonomidou, C., and Turski, L. (1995). Excitotoxicity and neurodegenerative diseases. *Curr Opin Neurol* 8:487–497.

Jain, A., Martensson, J., Stole, B. E., Auld, P. A. M., and Meister, A. (1991). Glutathione deficiency leads to mitochondrial damage in brain. *Proc Natl Acad Sci USA* 88:1913–1917.

Jhamadas, K. H., and Boegman, R. J. (1994). Quinolinic acid induced brain neurotransmitter deficits: Modulation by endogenous excitotoxin antagonists. *Can J Physical Pharmacol* 72:1473–1482.

Kamii, H., Kinouchi, H., Chen, S. F., Sharp, F. R., Epstein, C. J., and Chan, P. H. (1995). SOD-1 transgenic mice: an application to the study of ischemic brain injury. In: Ohnishi, S. T., ed., *Membrane-linked diseases,* Vol. 4. CRC Press, pp. 423–431.

Kamii, H., MIkawa, S., Murakami, K., Kinouchi, H., Yoshimoto, T., Reola, L., Carlson, E., Epstein, C. J., and Chan, P. H. (1996). Effects of nitric oxide synthase inhibition on brain infarction in SOD-1-transgenic mice following transient focal cerebral ischemia. *J Cereb Blood Flow Metab* 16:1153–1157.

Kennedy, M. B. (1994). The biochemistry of synaptic regulation in the central nervous system. *Ann Rev Biochem* 63:571–600.

Kinouchi, H., Epstein, C. J., Mizui, T., Carlson, S. J., Chen, S. F., and Chan, P. H. (1991). Attenuation of focal cerebral ischemic injury in transgenic mice overexpressing CuZn superoxide dismutase. *Proc Natl Acad Sci USA* 88:11158–11162.

Kirino, T. (1982). Delayed neuronal death in the gerbil hippocampus following ischemia. *Brain Res* 239:57–69.

Lee, K. S., Frank, S., Vanderklish, P., Arai, A., and Lynch, G. (1991). Inhibition of proteolysis protects hippocampal neurons from ischemia. *Proc Natl Acad Sci USA* 88:7233–7237.

Li, H., and Buchan, A. M. (1993). Treatment with an AMPA antagonist 12 hours following severe normothermic forebrain ischemia prevents CA1 neuronal injury. *J Cereb Blood Flow Metab* 13:933–939.

Linden, D. J. (1994). Long-term depression in the mammalian brain. *Neuron* 12:457–472.

Lipton, S. A. (1993). Prospective for clinically tolerated NMDA antagonists: Open-channel blockers and alternative redox states of nitric oxide. *TINS* 16:527–532.

Lucas, D. R., and Newhouse, J. P. (1957). The toxic effects of sodium L-glutamate on the inner layers of the retina. *AMA Ophthalmal* 58:193–201.

Malinski, T., Bailey, F., Zhang, Z. G., and Chopp, M. (1993). Nitric oxide measured by a porphyrinic microsensor in rat brain after middle cerebral artery occlusion. *J Cereb Blood Flow Metab* 13:355–358.

Masters, B. S. S., McMillan, K., Sheta, E. A., Nishimura, J. S., Roman, L. J., and Martasek, P. (1996). Neuronal nitric synthase, a modular enzyme formed by convergent evolution. *FASEB* 10:552–558.

Matsuo, Y., Kihara, T., Ikeda, M., Ninomiya, M., Onodera, H., and Kogure, K. (1995). Role of neutrophils in radical production during ischemia and reperfusion of the rat brain: Effect of neutrophil depletion on extracellular ascorbyl radical formation. *J Cereb Blood Flow Metab* 15:941–947.

Matsuo, Y., Onodera, H., Shiga, Y., *et al.* (1994). Role of cell adhesive molecules in brain injury after transient middle cerebral artery occlusion in the rat. *Brain Res* 656:344–352.

Moncada, S., and Higgs, A. (1993). The L-arginine-nitric oxide pathway. *N Engl J Med* 329:2002–2012.

Moncada, S. N., Herman, A. G., and Vanhoutte, P. M. (1987). Endothelin-derived relaxing factor is identified as nitric oxide. *Trends Pharmacol Sci* 8:326–329.

Mori, H., and Mishina, M. (1995). Structure and function of the NMDA receptor channel. *Neuropharmacol* 34:1219–1237.

Morris, C. J., Earl, J. R., Trenam, C. W., and Blake, D. R. (1995). Reactive oxygen species and iron—a dangerous partnership in inflammation. *Int J Biochem Cell Biol* 27:109–122.

Nikanishi, S. (1992). Molecular diversity of glutamate receptors and implications for brain function. *Science* 258:597–603.

Oliver, C. N., Starke-Reed, P. E., Stadman, E. R., Liu, G. L., Carney, J. M., and Floyd, R. A. (1990). Oxidative damage to brain proteins, loss of glutamine synthetase activity and production of free radicals during ischemia/reperfusion-induced injury to gerbil brain. *Proc Natl Acad Sci* 87:5144–5147.

Olney, J. W. (1969). Brain lesion, obesity and other disturbances in mice treated with monosodium glutamate. *Science* 164:719–721.

Olney, J. W., and Ho, O. L. (1970). Brain damage in infant mice treated with monosodium glutamate. *Nature* 227:609–610.

Olney, J W., Ho, O. L., and Rhee, V. (1971). Cytotoxic effects of acidic and sulphur-containing amino acids on the infant mouse central nervous system. *Exp Brain Res* 14:61–76.

Park, C. K., and Hall, E. D. (1994). Dose-dependent analysis of the effect of 21-aminosteroid tirilazad mesylate (U-74006F) upon neurological outcome and ischemic brain damage in permanent focal cerebral ischemia. *Brain Res* 645:157–163.

Perkins, G., Renken, C., Martone, M. E., Young, S. J., Ellisman, M., and Frey, T. (1997). Electron tomography of neuronal mitochondria: Three-dimensional structure and organization of cristae and membrane contacts. *J Struct Biol* 119:260–272.

Petralia, R. S., Yokotani, N., and Wenthold, R. J. (1994). Light and electron microscope distribution of the NMDA receptor subunit NMDARI in the rat nervous system using a selective anti-peptide antibody. *J Neurosci* 14:667–696.

Pulsinelli, W. A., Brierley, J. B., and Plum, F. (1982). Temporal profile of neuronal damage in a model of transient forebrain ischemia. *Ann Neurol* 11:491–499.

Rami, A., and Krieglstein, J. (1993). Protective effects of calpain inhibitors against neuronal damage in a model of transient forebrain ischemia. *Ann Neurol* 11:491–498.

Roberts-Lewis, J. M., Savage, M. J., Marcy, V. R., Pinsker, L. R., and Siman, R. (1994). Immunolocalization of calpain I-mediated spectrin degradation to vulnerable neurons in the ischemic gerbil brain. *J Neurosci* 14:3934–3944.

Roche, K. W., Tingley, W. G., and Huganir, R. L. (1994). Glutamate receptor phosphorylation and synaptic plasticity. *Curr Opin Neurobiol* 4:383–388.

Rosen, G. M., Pou, S., Ramos, C. L., Cohen, M. S., and Britigan, B. E. (1995). Free radical and phagocytic cells. *FASEB J* 9:200–209.

Rothman, S. M., and Olney, J. W. (1985). Neurotoxicity of excitatory amino acids is produced by passive chloride influx. *J Neurosci* 5:1483–1489.

Rothman, S. M., and Olney, J. W. (1995). Excitotoxicity and the NMDA receptor—still lethal after eight years. *TINS* 18:57–58.

Seeburg, P. H. (1993). The molecular biology of mammalian glutamate receptor channels. *Trends Neurosci* 16:359–365.

Siesjö, B. K. (1988). Historical overview. Calcium, ischemia, and death of brain cell. *Ann NY Acad Sci* 522:638–661.

Siesjö, B. K., Zhao, Q., Pahlmark, K., Siesjö, P., Katsura, K., and Folbergrova, J. (1995). Glutamate, Calcium, and free radicals as mediators of ischemic brain damage. *Ann Thorac Surg* 59:1316–1320.

Smith, M. L., Bendek, G., Dahlgren, N., Rosen, I., Wieloch, T., and Siesjö, B. K. (1984). Models for studying long-term recovery following forebrain ischemia in the rat. A 2-vessel occlusion model. *Acta Neurol Scand* 69:385–401.

Soto, G. E., Young, S. J., Martone, M. E., Deerinck, T. J., Lamont, S., Carragher, B. O., Hama, K., and Ellisman, M. H. (1994). Serial section electron tomography: a method for three-dimensional reconstruction of large structures. *NeuroImage* 1:230–243.

Sterz, F., Janata, K., Kurkciyan, I., Mullner, M., Malzer, R., and Schreiber, W. (1996). Possibilities of brain protection with tirilazad after cardiac arrest. *Seminars in Throm Hemostasis* 22:105–112.

Tominaga, T., Sato, S., Ohnishi, J., and Ohnishi, S. T. (1993). Potentiation of nitric oxide formation following bilateral carotid occlusion and focal cerebral ischemia in the rat: *In vivo* detection of the nitric oxide radical by electron paramagnetic spin trapping. *Brain Res* 414:342–346.

Watkins, J. C. (1978). Excitatory amino acids. In: McGeer, E., Olney, J. W., and McGeer, P., eds., *Kainic acid as a tool in neurobiology*. New York: Raven Press, pp. 37–69.

Williams, L. R. (1995). Oxidative stress, age-related neurodegeneration, and the potential for neurotrophic treatment. *Cerebrovascular and Brain Metab Rev* 7:55–73.

Wisden, W., and Seeburg, P. H. (1993). Mammalian ionotropic glutamate receptors. *Curr Opin Neurobiol* 3:291–298.

Xie, Q. W., Cho, H. J., Calaycay, J., Mumford, R. A., Swiderek, K. M., Lee, T. D., Ding, A., Troso, T., and Nathan, C. (1992). Cloning and characterization of inducible nitric oxide synthase from mouse macrophage. *Science* 256:225–228.

Xue, D., Huang, Z. G., Barnes, K., Lesiuk, H. J., Smith, K. E., and Bachan, A. M. (1994). Delayed treatment with AMPA, but not NMDA, antagonists reduces neocortical infarction. *J Cereb Blood Flow Metab* 14:251–261.

Zhang, J. and Steiner, J. P. (1995). Nitric oxide synthase, immunophilins and poly(ADP-ribose) synthetase: Novel targets from the development of neuroprotective drugs. *Neurol Res* 17: 285–288.

Zhao, Q., Pahlmark, K., Smith, M. L., and Siesjö, B. (1994). Delayed treatment with the spin trap α-phenyl-n-tert-butyl nitrone (PBN) reduces infarct size following transient middle cerebral artery occlusion in the rats. *Acta Physiol Scand* 152:349–350.

Zukin, R. S., and Bennett, M. V. L. (1995). Alternative spliced isoforms of the NMDAR1 receptor subunit. *TINS* 18:306–313.

PART II

Promoting CNS Recovery

CHAPTER 5

Neurotrophic Factors

MARK H. TUSZYNSKI

Department of Neurosciences, University of California, San Diego, La Jolla, California, and Department of Neurology, Veterans Affairs Medical Center, San Diego, California

Neurotrophic factors are proteins that regulate neuronal survival, axonal growth, synaptic plasticity, and neurotransmission. Different populations of neurons respond to specific neurotrophic factors, thereby providing potentially powerful tools for treating various neurological disorders. In animal models of human diseases, including Alzheimer's disease, Parkinson's disease, Huntington's disease, amyotrophic lateral sclerosis, stroke, spinal cord injury, and peripheral neuropathy, neurotrophic factors have exhibited an ability to promote neuronal rescue, elicit axonal growth, and generate functional recovery. Yet clinical trials of neurotrophic factors have yielded little success to date, demonstrating preliminary efficacy for the treatment of peripheral neuropathy but failing to show benefit in several other neurological diseases. An emerging concept in the use of this class of molecules for the treatment of nervous system disease is that their mode of delivery to the nervous system is critical: neurotrophic factors must be delivered *specifically* and in a *regionally restricted* fashion to neuronal targets. Nonetheless, neurotrophic factors are likely to become a mainstay of neurological therapy in the future, providing not only a means of *compensating* for nervous system disease once it has occurred, but for the first time, *preventing* or *reducing* the rate of disease progression.

CNS Regeneration

I. INTRODUCTION

The existence of nervous system growth factors has been known for several decades. In 1951 the first nervous system growth factor, nerve growth factor (NGF), was serendipitously discovered and found to promote the survival of peripheral sensory and sympathetic neurons during nervous system development (Levi-Montalcini and Hamburger, 1951, 1953; Cohen and Levi-Montalcini, 1957; Levi-Montalcini, 1987). For the following three decades it was believed that the role of growth factors in the nervous system was restricted to modulating neuronal survival during *development;* the *adult* brain and spinal cord were not thought to be capable of responding to growth factors or to other molecules that influence neuronal survival and growth. Clearly, this perception has been dramatically altered in the last 10 to 15 years by the discovery that neurotrophic factors are present in the adult CNS, are capable of completely preventing the death of specific populations of neurons after injury in the adult (Hefti, 1986; Kromer, 1987), and are capable of modulating a number of nervous system functions including synaptic efficacy and neurotransmitter turnover (see Levi-Montalcini, 1987; Thoenen, 1995; Barde, 1990; Bothwell, 1995; Davies, 1994). These findings have broad implications for neural repair strategies, creating for the first time the possibility of not merely *compensating* for neural degeneration after it has occurred, but of intervening *early* in the course of a disease to prevent or reduce neuronal degeneration. In addition, the ability of neurotrophic factors to stimulate axonal growth both naturally and after experimental manipulation suggests a potential role for neurotrophic factors in promoting axonal regeneration and reconstruction of host circuitry after injury in the adult CNS.

Currently, more than three dozen putative nervous system growth factors have been identified (Table I). These growth factors possess a wide range of structures, receptor signaling mechanisms, responsive neurons, and *in vivo* biological effects. The actions of a specific growth factor are determined by the location and extent of expression of the growth factor itself, and by the expression of an appropriate receptor for the factor on a target cell. The spatial and temporal balance between neurotrophic factor and receptor expression thus determines the spectrum of biological activity. Based upon these properties, different growth factors may be useful for treating of a variety of neurological diseases.

The known neurotrophic factors can be divided into several families based upon shared coding sequences and structures.

II. THE CLASSIC NEUROTROPHIN FAMILY

The first neurotrophic factor family to be identified was the "classic" neurotrophin family consisting of nerve growth factor (NGF) (Levi-Montalcini and

Hamburger, 1953; Levi-Montalcini, 1987; Thoenen and Barde, 1980), brain-derived neurotrophic factor (BDNF; Barde *et al.*, 1982), neurotrophin-3 (NT-3; Maisonpierre *et al.*, 1990; Rosenthal *et al.*, 1990), neurotrophin-4/5 (NT-4/5; Ip *et al.*, 1992a; Berkmeier *et al.*, 1991), and the nonmammalian growth factors neurotrophin-6 (NT-6; Gotz *et al.*, 1994), and neurotrophin-7 (NT-7; Nilsson *et al.*, 1998). NGF was the first known neurotrophic factor, discovered serendipitously 50 years ago in the course of examining properties of a sarcoma cell line *in vitro* (Levi-Montalcini and Hamburger, 1951; Levi-Montalcini *et al.*, 1954). A role for NGF as an essential survival factor for peripheral sensory and sympathetic neurons during development of the nervous system was soon described (Levi-Montalcini and Angeletti, 1963; Levi-Montalcini, 1987). However, the presence and importance of NGF in the adult nervous system was not appreciated until nearly 35 years later. Beginning in 1979, it became clear that NGF is transported (Schwab *et al.*, 1979; Seiler and Schwab, 1984) and produced (Johnson *et al.*, 1971; Korsching *et al.*, 1985) in the adult CNS. Several years later, the potent neuroprotective effects of NGF on injured adult neurons of the cholinergic basal forebrain were described (Hefti, 1986; Kromer, 1987; Gage *et al.*, 1988), revolutionizing the perception that the adult CNS was an inflexible, nonplastic structure. These findings helped usher in the modern era of research directed at enhancing CNS plasticity, preventing neuronal loss, and promoting CNS regeneration.

In 1983, NGF became the first neurotrophin to be cloned and sequenced (Scott *et al.*, 1983). In humans, the NGF gene is located on the short arm of chromosome 1. Studies regarding the biochemistry of NGF have been performed extensively on material isolated from the adult male mouse submaxillary gland because it contains the highest natural levels of NGF in mammals (Petrides and Shooter, 1986; Varon *et al.*, 1967b). NGF produced in the submaxillary gland exists as a protein pentamer consisting of two alpha, one beta, and two gamma subunits with a total molecular weight of 130 kD and a sedimentation rate of 7S (Varon *et al.*, 1967b, 1967a). The biological activity of the NGF molecule is wholly contained within the β-subunit, a 118 amino acid polypeptide with a molecular weight of 12,250 daltons and a sedimentation rate of 2.5S (Angeletti *et al.*, 1973). NGF is stable in the pH range of 5 to 8 (Thoenen and Barde, 1980). The gamma subunit of the NGF pentamer is a peptidase involved in cleavage of the prepro molecule to release the active β-fragment, rendering NGF available as a diffusible molecule. Due to the fact that NGF is a very basic molecule with an isoelectric point greater than 10, it is also possible that NGF becomes bound to charged molecules on the extracellular membrane following secretion from the cell (Blochl and Thoenen, 1996). Within the mammalian brain it is not clear how neurotrophin precursors are cleaved into mature, active peptides although recent studies suggest that this is accomplished by specific prohormone convertases (Seidah *et al.*, 1996a,

TABLE I Known Neurotrophic Factors and their Targets in Mammalian Systems

Name of growth factor	Target*
Nerve Growth Factor (NGF)	BFC, Striatal, Nociceptive Sensory, Sympathetic
Brain-Derived Neurotrophic Factor (BDNF)	BFC, HPC, Cortex, Mechanorec. Sensory, Motor (Alpha), Vestibular, Auditory, Retinal Ggl.
Neurotrophic-3 (NT-3)	BFC, HPC, Cortex, Proprio. Sensory, Muscle Spindle, Auditory, Oligodendrocytes
Neurotrophin-4/5 (NT-4/5)	Motor (Alpha), Retinal Ganglion, Sensory
Neurotrophin-6 (NT-6)	Nonmammalian
Neurotrophin-7 (NT-7)	Nonmammalian
Ciliary Neurotrophic Factor (CNTF)	Motor (Alpha- & Cortex), Striatal, Parasympathetic, Sensory, HPC, BFC, 02A
Leukemia Inhibitory Factor (LIF) or Cholinergic Differentiation Factor (CDF)	Motor (Alpha- & Cortex), Glia, Cortical
Cardiotrophin-1	Motor (Alpha)
Transforming growth factors β1, β2, β3 (TFGβ1-3)	Motor (Alpha), Sensory
Glial Cell-Line Derived Neurotrophic Factor (GDNF)	DA neurons, Motor (Alpha- & Cortex), Sensory
Neurturin	DA neurons, Motor (Alpha- & Cortex), Sensory
Persephin	DA neurons, Motor (Alpha- & Cortex)
Activin	Motor (Alpha)
Bone morphogenetic proteins (BMPs)	CNS 7 PNS pattern formation during development
Fibroblast Growth Factor-1 (acidic FGF)	Multiple cord & brain stem; Cortex
Fibroblast Growth Factor-2 (basic FGF)	BFC, Cortex, DA neuron, Retinal ganglion, Stem cells
Fibroblast Growth Factors 2-15	
Insulin-like Growth Factor I (IGF-I)	Motor (alpha), Sensory, Multiple Brain, Oligodendrocyte

Insulin-like Growth Factor II (IGF-II)	Sensory, Sympathetic
Epidermal Growth Factor (EGF)	DA neurons, HPC, Cortex, Stem cells
Transforming growth factor-α (TGF-α)	DA neurons, HPC, Cortex (in vitro)
Neuregulins	Maturation of glial cells and synapses
NTAK	
Hepatocyte Growth Factor	HPC, Motor (alpha), Sensory, Schwann cells
Immunophilins	Sensory, Brain stem
Midkine	Photoreceptors
Erythropoietin	BFC
Interleukins 1,2,3,6 (IL-1-6)	BFC
Heparin-binding neurotrophic factor	
Platelet Derived Growth Factor (PDGF)	
Axon ligand-1 (Al-1)	
elf-1	
ehkl-L	
LERK2	

* The listing of targets is based upon findings of both *in vivo* and *in vitro* studies.
BFC: Basal forebrain cholinergic neurons
HPC: Hippocampus
DA neurons: Dopaminergic neurons
02A: Oligodendrocyte/astrocyte progenitor cells

1996b). It is also unknown whether NGF in the CNS is stored in a complex similar to the pentamer present in the mouse submaxillary gland. The function of the NGF α subunit is unknown.

The second neurotrophic factor to be identified, brain-derived neurotrophic factor (BDNF), was isolated in 1982 from pooled extracts of porcine brain (Barde *et al.*, 1982). The BDNF gene, subsequently cloned and sequenced (Leibrock *et al.*, 1989), consists of a 120 amino acid protein with a molecular weight of 12,300 daltons. Like NGF, the BDNF gene codes for a large prepro molecule with a secretory signal peptide that presents BDNF as an extracellular factor.

The finding that NGF and BDNF share approximately 50% homology for base pair codons suggested that a family of related proteins might exist possessing similar neurotrophic properties. This led to a search for additional molecules. Using hybridization probes to homologous NGF and BDNF coding sequences, NT-3 (Hohn *et al.*, 1990; Maisonpierre *et al.*, 1990; Rosenthal *et al.*, 1990), NT-4/5 (Ip *et al.*, 1992a; Berkmeier *et al.*, 1991), NT-6 (Gotz *et al.*, 1994), and NT-7 (Nilsson *et al.*, 1998) have all been discovered since 1990. NT-6 and NT-7 are only present in nonmammalian species, however. Like NGF and BDNF, NT-3 and NT-4/5 are naturally expressed in the nervous system during development and in adulthood. The regional patterns of expression of neurotrophins in the adult brain exhibit both specificity and overlap, suggesting that each neurotrophic factor plays a distinct role in modulating neuronal function. Thus, each neurotrophin may possess potential for treating degeneration of specific neuronal populations, and combinations of trophic factors could act synergistically to promote the survival of single or multiple neuronal populations in nervous system disease (Mitsumoto *et al.*, 1994b).

The neurotrophins exert their actions on subsets of neurons by binding to specific transmembrane receptors. Two distinct classes of neurotrophin receptors have been identified: a low-affinity receptor termed p75, to which all neurotrophins can bind, and a tyrosine kinase-linked (trk) family of receptors that bind different neurotrophins with unique and specific affinities.

p75 is a glycosylated transmembrane protein that is a member of the tumor necrosis factor-α (TNF-α) receptor family (Carter *et al.*, 1996). p75 appears to serve two roles: it modulates neurotrophin interactions with trk receptors, and it influences cellular function directly via the p75 intracellular domain (Majdan *et al.*, 1997; Rao *et al.*, 1995; Chao, 1994; Rabizadeh *et al.*, 1993; Van der Zee *et al.*, 1996; Dobrowsky *et al.*, 1994). p75 is capable of associating with trk receptors to accelerate NGF binding, leading to a high-affinity receptor state (Mahadeo *et al.*, 1994; Hempstead *et al.*, 1991). Thus, neurons expressing both p75 and the NGF-specific receptor, trkA, can respond to lower concentrations of NGF than neurons expressing trkA alone (Chao, 1994). p75 expression may also enhance the ability to which specific trk receptors discriminate between their neurotrophin ligands (Lewin and Barde, 1996; Clary and Reichardt, 1994; Benedetti *et al.*, 1993). Independently of trk interactions, p75 also promotes *apoptosis,* or programmed cell death, during development (Rabizadeh *et al.*, 1993; Van der Zee *et al.*, 1996; Yeo *et al.*, 1997; but see also Peterson *et al.*, 1997), and can modulate sphingomyelin metabolism (Dobrowsky *et al.*, 1994).

Most effects of neurotrophins on neurons appear to be mediated by trk receptors. Biological responses mediated by trk receptors include neuronal survival, neuronal differentiation, axon outgrowth, synaptic plasticity, and neurotransmitter expression. Three forms of the trk receptor have been characterized and are designated trkA (Kaplan *et al.*, 1991; Cordon-Cardo *et al.*, 1991; Klein *et al.*, 1991), trkB (Squinto *et al.*, 1991), and trkC (Lamballe *et al.*, 1991). Each trk receptor possesses specific ligands: NGF binds to *trkA,* BDNF and NT-4/5 bind to *trkB,* and NT-3 binds primarily to *trkC* (see Chao, 1992; Barbacid, 1994 for review). Trk receptors are transmembrane glycoproteins with intracellular tyrosine kinase elements. Binding of a neurotrophin induces dimerization of its trk receptor, thereby activating the tyrosine kinase and inducing autophosphorylation of specific trk residues, which then bind and activate other messenger systems including phospholipase C-γ and phosphatidylinositol 3-kinase (Saltiel and Ohmichi, 1993; Grimes *et al.*, 1993). Other kinases and second messengers subsequently become activated, eliciting multiple biological responses within the cell. Thus, neurons exhibit responses to a given neurotrophic factor by signaling through distinct and specific receptors. Interestingly, trk receptors are evolutionarily conserved, with 100% amino acid homology in most species thus far examined.

The biological effects of the neurotrophins in the developing and adult nervous system have been described extensively (see Levi-Montalcini, 1987; Thoenen *et al.*, 1985; Thoenen, 1995; Yuen and Mobley, 1996; Lewin and Barde, 1996 for reviews). Although originally thought to function primarily as cell survival factors, neurotrophic factors in general and the neurotrophin family in particular have now been shown to be capable of modulating several aspects of nervous system development and function, including cellular differentiation, axon growth, axon target finding, synaptic efficacy, synaptic plasticity, and neurotransmission–neurotransmitter turnover (Fig. 1).

A. Neurotrophin Roles in the PNS

In the PNS, neurotrophic factors are survival factors for several classes of neurons. The availability of limited quantities of NGF in a target-derived manner is essential for the survival of sympathetic neurons of the paravertebral ganglia, and for 70 to 80% of sensory neurons (nociceptors) in the dorsal root

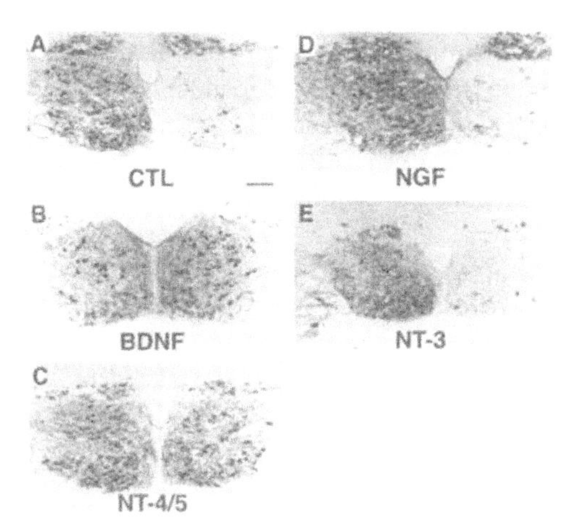

FIGURE 1 Neurotrophins prevent injury-induced alterations in the neuronal phenotype. (**A**) After hypoglossal nerve transection in adult rats, motor neurons ipsilateral to the lesioned side (right) in the medulla downregulate expression of choline acetyltransferase (ChAT). ChAT-labeled neurons on the left side of the brain are unaffected by the lesion. Intracerebroventricular infusions of BDNF (**B**) and NT-4/5 (**C**) completely prevent downregulation of the cholinergic phenotype, whereas infusions of NGF (**D**) and NT-3 (**E**) have no effect on motor neurons. Scale bar = 40 μm. Reprinted from *Neuroscience* 1996, 71:761–771.

ganglia (Levi-Montalcini, 1987; Johnson *et al.*, 1980; Pearson *et al.*, 1983; Lewin and Mendell, 1993; Crowley *et al.*, 1994; Smeyne *et al.*, 1994; Gorin and Johnson, 1980; Fagan *et al.*, 1996; Silos-Santiago *et al.*, 1995). Specific subpopulations of other PNS neurons exhibit dependence on either BDNF (Liu *et al.*, 1995a; Ernfors *et al.*, 1994a; Jones *et al.*, 1994), NT-3 (Ernfors *et al.*, 1994b; Farinas *et al.*, 1994), NT-4/5 (Liu *et al.*, 1995a), or the cytokine GDNF (see the following discussion) during development.

Once a neuron has successfully survived the period of developmental cell death, it may subsequently downregulate expression of a trophic factor receptor and may fail to respond to the trophic factor even if it is exogenously delivered. Cells may also upregulate expression of a different neurotrophic factor receptor to modulate another aspect of cell function. For example, a subpopulation of dorsal root ganglion nociceptive sensory neurons requires NGF for survival during development and expresses trkA prior to embryonic day 15 (E15). After E15, however, these neurons downregulate trkA expression and upregulate expression of the GDNF receptor RET, a switch in neurotrophin receptor expression that temporally correlates with a switch in the phenotype of the neuron (Molliver *et al.*, 1997). Other nociceptive neurons continue to express trkA throughout life although they do not depend upon NGF for survival; rather, NGF serves a neuroregulatory role in these neurons by altering neuropeptide–neurotransmitter levels, including substance P and calcitonin gene-related peptide (CGRP) levels (Otten, 1984; Lindsay and Harmar, 1989). In contrast to sensory neurons, sympathetic neurons continue to require NGF for survival into adulthood (Levi-Montalcini, 1987; Gorin and Johnson, 1980). Thus, three different classes of neurons that require NGF as a survival factor during development diverge in the nature of their subsequent NGF-dependence: one class permanently downregulates NGF receptor expression and responds to a different growth factor, GDNF; a second class switches from NGF-dependence for survival to NGF-modulation of transmitter expression and function; and a third class retains NGF-dependence for survival through adulthood.

In addition to regulating neuronal survival and function, neurotrophins modulate neuronal differentiation and synapse function during PNS development. Thus, the administration of anti-NGF immunoglobulins during development leads to a loss of nociceptive neurons in the dorsal root ganglia (DRG) and their replacement with D-hair afferent neurons (Ritter *et al.*, 1991; Lewin *et al.*, 1992), indicating that NGF modulates neuronal differentiation. NGF also modulates neuronal morphology, evidenced by an increase in the number of dendrites on sympathetic neurons treated with NGF (Ruit *et al.*, 1990). Neurotrophin effects at the synaptic level are exhibited by BDNF and NT-3, which potentiate synapse function at the neuromuscular junction (Lohof *et al.*, 1993). Thus, neurotrophins exert effects on cell survival, differentiation, synapse function, and neurotransmitter regulation in the PNS. In many cases these functions continue into adulthood: sensory neurons in the adult bind and retrogradely transport NGF, BDNF, NT-3, and NT-4/5 (DiStefano *et al.*, 1992). Schwann cells, a source of NGF, BDNF, NT-4/5, GDNF, and CNTF, produce augmented amounts of neurotrophic factors after peripheral nerve injury and likely contribute both to the survival and regrowth of injured neurons in the adult (Meyer *et al.*, 1992; Funakoshi *et al.*, 1993).

B. Neurotrophin Roles in the CNS

Neurotrophins modulate several neuronal systems in the CNS. During development, BDNF is required for the survival of vestibular neurons (Ernfors *et al.*, 1995) and NT-3 is required for the survival of most cochlear neurons (Ernfors *et al.*, 1995). In the brain proper, however, there has been surprisingly little evidence that neurotrophins are required for neuronal survival during development. For example, despite the demonstration that NGF is capable of completely preventing the death of basal forebrain cholinergic neurons after lesions in the adult, mice with targeted homozygous (−/−) (Crowley *et al.*, 1994) or heterozygous (+/−) (Chen *et al.*, 1997) mutations of NGF or trkA (Fagan *et *al.*, 1996; Silos-Santiago *et al.*, 1995) do not exhibit a loss of basal forebrain cholinergic neurons during development. Rather, NGF influences the location and density but not the number of basal forebrain cholinergic inputs to the hippocampus (Chen *et al.*, 1997). Additional evidence also supports a role for NGF in modulating neuronal size, dendritic structure and complexity (Purves, 1988), and synapse number and size in the brain (Garofalo *et al.*, 1992).

BDNF null mutant mice and NT-3 null mutant mice also exhibit relatively modest reductions in neuronal numbers in the CNS. Little if any reduction in numbers of cortical and hippocampal neurons are observed in these null mutants, despite the abundant developmental expression of trkB and trkC receptors in cortical and hippocampal regions (Ernfors *et al.*, 1994a). In the spinal cord, BDNF null mutants exhibit no loss of motor neurons (Ernfors *et al.*, 1994a; Jones *et al.*, 1994; Conover *et al.*, 1995; Liu *et al.*, 1995a), despite the fact that experiments in neonatal animals demonstrate that BDNF can prevent injury-induced death of motor neurons (Sendtner *et al.*, 1992a; Yan *et al.*, 1992; Clatterbuck *et al.*, 1994; Koliatsos *et al.*, 1993). NT-3 null mutant mice exhibit a modest reduction in motor neuron numbers, accounted for by a reduction in the number of muscle spindle afferent Ia motor neurons (Klein *et al.*, 1994; Kucera *et al.*, 1995a, 1995b). trkB null mutant mice exhibit reductions in numbers of motor neurons in the facial nucleus and spinal cord (Klein *et al.*, 1993), although these changes are inconsistently noted in different strains of trkB −/− mice. trkC null mutants support findings from NT-3 null mutant mice by exhibiting fewer muscle spindle afferent motor neurons, therefore indicating that these neurons require NT-3 for survival during development (Klein *et al.*, 1994; Kucera *et al.*, 1995a, 1995b).

Evidence also exists for neurotrophin-mediated modulation of target finding and synapse formation during CNS development. In the developing visual system, injections of NGF antibodies extend the developmental period over which ocular dominance column plasticity can be elicited (Domenici *et al.*, 1995). Further, injections of BDNF, NT-4/5, or antagonists to trkB receptors disrupt normal formation of ocular dominance columns in the visual cortex (Cabelli *et al.*, 1995, 1997). Injections of NGF, BDNF, NT-3, or NT-4 also alter dendritic pattern formation in the visual cortex (McAllister *et al.*, 1995); indeed, formation of normal dendrites requires both neurotrophins and neuronal activity in the developing visual cortex (McAllister *et al.*, 1996). Supporting the importance of neurotrophins in the CNS, mice with targeted mutations of BDNF (−/−) exhibit abnormal dendritic branching and pattern formation in the cerebellum (Schwartz *et al.*, 1997). Thus, neurotrophins modulate neuronal differentiation, axon target finding, and synapse function in the CNS; unlike the PNS, there is less evidence for a role of the classic neurotrophins in modulating CNS neuronal survival.

Several studies demonstrate an additional role for neurotrophins in modulating synaptic function and neurotransmitter availability in the CNS. BDNF null mutant mice exhibit impaired hippocampal long-term potentiation (LTP; Korte *et al.*, 1995), which can be restored by application of exogenous BDNF. Conversely, the induction of LTP in normal mice alters BDNF, NT-3, and trk expression (Patterson *et al.*, 1992; Castren *et al.*, 1993; Nawa *et al.*, 1995), supporting an interrelation between neurotrophin production and synaptic mechanisms. Indeed, release of neurotrophins from neurons may be a function of activity in ensembles of neurons (Lewin and Barde, 1996). Neurotrophin levels in the brain appear to be regulated by events that lead to synaptic plasticity, and neurotrophins in turn can have effects on the strength of synaptic connections. If neurotrophins indeed contribute substantially to short-term neuronal signaling and synaptic plasticity, then mechanisms would need to exist for the rapid and efficient removal of neurotrophins from the extracellular space, as suggested by Lewin and Barde (1996). This function may be served in the adult CNS by splice variants of the trk receptors, which are truncated forms of full-length receptors (Barbacid, 1994). Truncated trk receptors lack the tyrosine kinase domains that normally mediate signal transudation after neurotrophin binding, and could eliminate neurotrophins by binding and internalizing them without stimulating signal transduction. This elimination of neurotrophins would be analogous to uptake mechanisms that rapidly remove neurotransmitters from the extracellular space (Lewin and Barde, 1996).

Expanding beyond the traditional concept that BDNF is a target-derived growth factor, recent findings demonstrate that BDNF is anterogradely transported down axons in the CNS, indicating that BDNF compartmentalization and availability at axonal projection sites may modulate the nature and strength of some projecting systems in an anterograde manner (Conner *et al.*, 1997, 1998). Indeed, *in vitro* preparations indicate that BDNF is present in dendrites of hippocampal neurons and is released upon neuronal depolarization, suggesting that local BDNF availability can modulate synaptic function.

Neurotrophins have further been shown to modulate synaptic efficacy over very short time scales of seconds and minutes, suggesting the existence of non-trk/non-p75 signaling mechanisms (Lewin and Barde, 1996). For example, NT-3 and BDNF increase spontaneous and evoked synaptic activity over minutes or less (Lohof *et al.*, 1993; LeBmann *et al.*, 1994; Kang and Schuman, 1995; Knipper *et al.*, 1994). These effects occur at the presynaptic level and do not require kinase activation. Within seconds, BDNF and NT-3 can increase intracellular calcium in cultured hippocampal neurons (Berninger *et al.*, 1993), which may lead to increased transmitter release. NT-3 also increases activity of hippocampal neurons in culture, possibly by decreasing GABAergic transmission (Kin *et al.*, 1994). In the intact adult brain, NGF, BDNF, NT-3, or

NT-4/5 elicits rapid effects on whole neuronal systems (Berzaghi *et al.*, 1995; see also Lewin and Barde, 1996).

Thus, the family of "classic neurotrophins" modulates a number of functions in the intact CNS and PNS. In the PNS, neurotrophins act as survival, differentiation, and synaptic plasticity factors. In the CNS, neurotrophin effects on cell survival are less clear, but other neuronal functions are extensively modulated by neurotrophins, including axonal and/or dendritic target finding, neuritic complexity, neuronal phenotype, neurotransmitter expression, and modulation of synaptic function.

C. Cellular Targets and Effects of the Classic Neurotrophins in the Injured Nervous System

After injury, several neurotrophins have been shown to protect injured or degenerating adult nervous system neurons and axons. These survival and growth-promoting capabilities of the neurotrophins are described extensively in subsequent chapters that review specific animal models of human diseases, and are briefly summarized in this chapter.

NGF has been shown to influence the survival and/or growth of several neuronal populations in the injured adult nervous system. NGF prevents the degeneration of basal forebrain cholinergic neurons after injury and as a result of normal aging in rats and primates (Hefti, 1986; Kromer, 1987; Fischer *et al.*, 1987; Nebes and Brady, 1989; Koliatsos *et al.*, 1990; Barnett *et al.*, 1990; Tuszynski *et al.*, 1991, 1996c; Rosenberg *et al.*, 1988). For these reasons, NGF is being considered as a therapeutic agent for treating the cholinergic component of neuronal loss in Alzheimer's disease (see Chapter 20). However, significant problems related to the mode of administration of NGF into the CNS have arisen (Williams, 1991; Winkler *et al.*, 1996b), generating a need for an intraparenchymal, accurately targeted means of delivering NGF to the brain that limits the extent of diffusion from the delivery site. NGF also reduces cholinergic and noncholinergic neuronal degeneration in the striatum of rodents with quinolinic acid lesions (Kordower *et al.*, 1996), suggesting its potential use in Huntington's disease. In the spinal cord, cells genetically modified to produce and secrete NGF induce extensive growth of supraspinal coerulospinal axons and the central processes of nociceptive sensory axons in acute and chronic spinal cord injury (Tuszynski *et al.*, 1994, 1996a; Mufson *et al.*, 1989) (see Chapter 24). In the PNS, NGF ameliorates several parameters of nerve dysfunction in rodent models of diabetic neuropathy (Apfel, 1997; Apfel *et al.*, 1994) and taxol-induced (Apfel *et al.*, 1991) neuropathy. Based upon these findings, phase I trials (Petty *et al.*, 1994) and phase II trials of

NGF in diabetic neuropathy have been conducted, and phase III trials are in progress. To date, NGF has exhibited safety (Petty *et al.*, 1994) and preliminary efficacy in these human trials, representing the first successful use of a neurotrophic factor to treat a human disease.

BDNF influences several types of neurons *in vitro,* including sensory neurons, basal forebrain cholinergic neurons, GABAergic neurons of the ventral mesencephalon, as well as retinal ganglion cells (see Lewin and Barde, 1996; Hefti, 1997; Hyman *et al.*, 1994; Ip *et al.*, 1993). Messenger RNA (mRNA) for BDNF is diversely expressed in the nervous system in the hippocampus, cortex, hypothalamus, brain stem, and cerebellum (see Schecterson and Bothwell, 1992; Lewin and Barde, 1996). TrkB mRNA is also diversely expressed in the hippocampus, cortex, basal forebrain, ventral mesencephalon, and spinal cord. When delivered *in vivo* to the injured nervous system, BDNF rescues basal forebrain cholinergic and GABAergic neurons (Widmer *et al.*, 1993), although it is less potent on cholinergic neurons of this system than NGF. BDNF also prevents the degeneration of brain stem neurons after axotomy in spinal cord injury models (Kobayashi *et al.*, 1997). BDNF prevents the death of injured primary motor neurons in neonatal animals (Sendtner *et al.*, 1992a; Yan *et al.*, 1992; Clatterbuck *et al.*, 1994; Koliatsos *et al.*, 1993), and prevents the downregulation of the cholinergic neuronal phenotype after motor neuron injury in the adult (Fig. 1) (Yan *et al.*, 1994; Tuszynski *et al.*, 1996b; Chiu *et al.*, 1994; Clatterbuck *et al.*, 1994; Friedman *et al.*, 1995; Xu *et al.*, 1995). BDNF also reduces motor neuron degeneration and improves function in models of mouse mutant neuropathy, a correlational model of the human disease amyotrophic lateral sclerosis (ALS, or Lou Gherig's disease; see Chapter 23). Notably, BDNF rescues neurons in this motor neuron disease model if administered to adult (Mitsumoto *et al.*, 1994b; Ikeda *et al.*, 1995) but not neonatal animals (Blondet *et al.*, 1997). Based upon the effects of BDNF on motor neuron degeneration in animal models, a clinical trial of BDNF in ALS is in progress in humans. In this trial BDNF is injected subcutaneously; questions arise regarding the ability of BDNF to access degenerating motor neurons in sufficient concentrations to achieve neuroprotection in the spinal cord with this delivery method. BDNF is retrogradely transported to spinal motor neurons when injected into the neuromuscular junction or intravascularly (e.g., Yan *et al.*, 1994), but it is unclear that pharmacological levels of BDNF can be attained in the spinal cord by this route of administration. BDNF also influences the survival and function of a subset of peripheral sensory axons (Conover *et al.*, 1995; Hofer and Barde, 1988; Acheson *et al.*, 1995), although this finding has not been the basis for clinical trials in neuronopathy to date. Finally, BDNF promotes the survival of vestibular neurons and a subset of auditory neurons, suggesting a potential future use in treating disorders of the middle ear (Ernfors *et al.*, 1995).

In vitro, NT-3 influences the survival of neurons of the dorsal root ganglion (Kucera *et al.*, 1995b; Ernfors *et al.*, 1994b), sympathetic ganglia, hippocampus, ventral mesencephalon (dopaminergic and GABAergic neurons), and a subpopulation of spinal cord motor neurons (muscle spindle afferents; Tessarollo *et al.*, 1994; Kucera *et al.*, 1995a; reviewed in Lewin and Barde, 1996). NT-3 mRNA is present in muscle, hippocampus, neocortex, cerebellum, olfactory bulb, diencephalon, midbrain, and spinal cord. mRNA for trkC is present in the hippocampus, neocortex, cerebellum, ventral mesencephalon, and motoneurons (Lamballe *et al.*, 1991, 1994; Lewin and Barde, 1996). After injury in the adult, NT-3 can augment the function of large caliber, proprioceptive peripheral sensory axons (Helgren *et al.*, 1997; Gao *et al.*, 1995); on this basis, NT-3 is currently undergoing clinical trials in neuropathy. NT-3 also promotes the growth of corticospinal motor axons after spinal cord injury (Schnell *et al.*, 1994; Grill *et al.*, 1997), and partially ameliorates motor deficits (Grill *et al.*, 1997) (see Chapter 24), although these findings at present are not the basis for clinical trials. NT-3 is also a survival factor for the majority of cochlear neurons (Cabelli *et al.*, 1995), indicating potential uses for hearing disorders. Finally, NT-3 appears to modulate oligodendrocyte function (Cohen *et al.*, 1996), raising the possibility that it could be used to treat demyelinating conditions.

The influences of NT-4/5 on neuronal populations show overlap with BDNF because the two neurotrophins share signal transduction through the trkB receptor. *In vitro,* NT-4/5 promotes the survival of neurons of the dorsal root ganglia, sympathetic ganglia, nodose ganglion, hippocampus, basal forebrain, ventral mesencephalon (dopamine and GABAergic neurons), motor neurons, and retinal ganglion cells (see Lewin and Barde, 1996). mRNA for NT-4/5 is present in muscle, the pons, medulla, hypothalamus, thalamus, and cerebellum. In the adult, NT-4/5 ameliorates motor neuron dysfunction (Hughes *et al.*, 1993; Koliatsos *et al.*, 1994; Friedman *et al.*, 1995; Tuszynski *et al.*, 1996b), reduces brain stem neuronal loss after spinal cord injury, and promotes axonal growth of supraspinal brain stem projections to the spinal cord (Kobayashi *et al.*, 1997). NT-4/5 has not yet undergone clinical testing.

Thus, the first and most thoroughly studied family of growth factors illustrates fundamental features of neurotrophic factor biology that are common to the other families as well. Neurotrophins are diversely expressed during development and in adulthood in specific distributions that may overlap with other trophic factors. Most trophic factor actions are paracrine in nature, although autocrine effects are also observed (Acheson *et al.*, 1995). Signaling occurs through specific receptors that modulate the action of receptor-linked tyrosine kinases or other intracellular transduction proteins. Both neurons and glia are affected by trophic mechanisms. A number of cellular functions are influenced by growth factors, including survival, differentiation, process outgrowth, target finding, maintenance of the cellular phenotype, modulation of transmitter levels, and synaptic plasticity.

III. THE CYTOKINE GROWTH FACTORS

A second class of nervous system growth factors is the *cytokine* family, consisting of leukemia inhibitory factory (LIF; also known as cholinergic differentiation factor) (Patterson, 1994; Taupin *et al.*, 1998), ciliary neurotrophic factor (CNTF; Barbin *et al.*, 1984; Lin *et al.*, 1989; Sendtner *et al.*, 1994), and cardiotrophin-1 (Herrmann *et al.*, 1993; Arce *et al.*, 1998). The cytokines differ from the neurotrophins not only in molecular structure, but also in fundamental features of receptor physiology. CNTF is a protein of 200 amino acids (molecular weight 22,700 daltons) that interacts with a tripartite receptor complex including a specific CNTF receptor α, an LIF receptor β, and a gp130 receptor (Ip *et al.*, 1992b). Binding of CNTF activates the Janus tyrosine kinase (JAK), which phosphorylates members of the STAT family of transcription factors. The phosphorylated STAT proteins then translocate to the nucleus where they activate CNTF-responsive genes (Segal and Greenberg, 1996). Unlike the classic neurotrophins, CNTF lacks a signal peptide for secretion from the cell. Because CNTF is produced by both Schwann cells and astrocytes, it has been suggested that CNTF may act as a factor that is released from cells after injury in the CNS or PNS; however, its precise role remains to be established.

In vitro, CNTF can act as a survival factor for motor neurons, and can prevent motor neuronal degeneration after neonatal axotomy and as a consequence of spontaneous degeneration in adult mice with mutant motor neuronopathy (Arakawa *et al.*, 1990; Sendtner *et al.*, 1990, 1992b; Lindsay, 1996; Mitsumoto *et al.*, 1994a). Mouse mutant motor neuronopathy resembles, in some respects, the spontaneous degeneration of motor neurons that occurs in amyotrophic lateral sclerosis. Of note, the neuroprotection resulting from combined treatment with both BDNF and CNTF in mouse mutant motor neuronopathy is greater than the protection resulting from delivery of either growth factor alone, suggesting that neurotrophins can act synergistically to promote neuronal protection even if targeting a single neuronal population (Mitsumoto *et al.*, 1994b). Expression of CNTF is naturally upregulated in peripheral nerves following axotomy, and may support motor and sensory neuronal survival and axonal regeneration after injury (Apfel *et al.*, 1993). CNTF has also been reported to protect basal forebrain cholinergic neurons (Hagg *et al.*, 1992) and substantia nigra dopaminergic neurons (Hagg and Varon, 1993) from lesion-induced degeneration without restoring the transmitter phenotype of the injured neurons. CNTF also rescues striatal interneurons after excitotoxic lesions (Emerich *et al.*, 1996). Further, in many instances neurons responding to CNTF do not express any components of the tripartite receptor complex, suggesting that CNTF may act by inducing the release of neuroprotective agents from neighboring neurons or glia. Additional neurons that have been reported to exhibit CNTF responses include ciliary ganglion neurons, hippocampal neurons, neurons of the spinal nucleus of the bulbocavernosus (Forger *et al.*, 1997), retinal photoreceptors (Ezzeddine *et al.*, 1997), corticospinal neurons (Dale *et al.*, 1995), and progenitor cells that can differentiate into oligodendrocytes or type II astrocytes (O2A progenitor cells). CNTF has also been reported to ameliorate neuronal loss and improve functional outcomes in rats after cerebral ischemia (Kumon *et al.*, 1996). Despite this extensive characterization of neuronal responses to CNTF, its role in normal development remains unclear because CNTF null mutant mice exhibit few changes in the nervous system until postnatal ages, when motor neuronal loss becomes evident (see Sendtner *et al.*, 1994; Richardson, 1994 for review). Of note, and possibly of concern with regard to the treatment of neurological disease, CNTF induces several components of reactive gliosis in the brain (Levison *et al.*, 1996).

CNTF was one of the first neurotrophic factors to undergo large-scale clinical trials for the treatment of a human neurological disease, amyotrophic lateral sclerosis (ALS CNTF Treatment Study Group, 1996). This disease was targeted for neurotrophic factor treatment because CNTF prevented progressive motor neuronal degeneration in mouse mutant neuronopathy (Sendtner *et al.*, 1992b), an effect on motor neurons that was subsequently supported by the finding that targeted mutation of the CNTF gene induced motor neuron degeneration during development (Masu *et al.*, 1993). CNTF was administered subcutaneously in this trial, thereby depending on uptake and retrograde transport of CNTF from the neuromuscular junction to rescue spinal motor neurons. CNTF failed to demonstrate efficacy in humans (ALS CNTF Treatment Study Group, 1996). Reasons for this failure might have included (1) the extremely short half-life of CNTF after peripheral injection (minutes), (2) the possibility that subcutaneous injections of CNTF failed to achieve concentrations in the spinal cord sufficient to rescue motor neurons, and (3) the absence of preclinical efficacy data from a species larger than the mouse; if negative outcomes had resulted from experiments in larger mammals, human trials might have been designed differently or postponed until superior efficacy data and delivery methods had been developed. It is also possible, of course, that CNTF would fail to exhibit efficacy even in the presence of adequate central delivery of the growth factor because the underlying mechanism of neuronal toxicity in ALS cannot be affected by neurotrophic factors. An alternative approach for treating ALS that is currently undergoing clinical trials is the combined delivery of CNTF and BDNF. This approach is based upon mouse data showing that the two growth factors act synergistically to reduce motor neuronal degeneration (Mitsumoto *et al.*, 1994b). Clearly, the adequate delivery of growth factors to their intended therapeutic targets remains an important factor that limits satisfactory clinical testing for efficacy. An attempt to deliver CNTF intrathecally in patients with ALS has been conducted by placing encapsulated cells genetically modified to produce CNTF into the thecal space of the spinal cord (see Chapter 23); a Phase I study with this approach reported safety, but efficacy has not yet been addressed (Aebischer *et al.*, 1996). A problem that may be difficult to surmount in the therapeutic use of CNTF is the occurrence of cachexia (weight loss) after systemic administration (Henderson *et al.*, 1994).

The cytokine growth factor LIF also influences several classes of neurons. *In vitro*, LIF has been reported to promote survival of motor neurons (Patterson, 1994), cortical neurons, and glia (Taupin *et al.*, 1998). In the injured adult peripheral nerve, LIF supports motor and sensory axonal regeneration (Arakawa *et al.*, 1990; Cheema *et al.*, 1994a, 1994b), a function that may be mediated in part by production of LIF in Schwann cells (Banner and Patterson, 1994). Recently, cellular LIF delivery has been found to upregulate NT-3 expression in the injured spinal cord and to augment growth of corticospinal axons (Blesch, Patterson, and Tuszynski, unpublished observations). LIF is the first neurotrophic factor outside of the "classic neurotrophin family" that has been shown to upregulate expression of another trophic factor in the CNS after *in vivo* injury, a finding that correlates with augmented recovery. Thus, neurotrophins influence CNS function at yet another level by modulating production of growth factors by other cells. Similar findings have been reported in *in vitro* preparations (Thoenen, 1995) and in the PNS, in which NGF delivery augments BDNF expression in motor neurons (Apfel *et al.*, 1996).

Cardiotrophin-1 (CT-1) is a newer member of the cytokine family of growth factors that also promotes motor neuron survival during development (Pennica *et al.*, 1996; Arce *et al.*, 1998). Interestingly, motoneurons express receptors for both GDNF and CT-1, and both growth factors may be required to support motor neuron survival during development (Arce *et al.*, 1998). GDNF is produced by Schwann cells whereas CT-1 is produced by muscle cells, and distinct signals from both of these growth factors may determine the developmental fate of motor neurons. CT-1 could be useful for the treatment of motor neuron disease.

IV. THE FIBROBLAST GROWTH FACTORS

A third class of nervous system growth factors consists of the fibroblast growth factors (FGF; Bohlen *et al.*, 1984; Thomas *et al.*, 1984; Eckenstein, 1994;

Unsicker *et al.*, 1991). The most thoroughly characterized members of this family are FGF-1 and FGF-2 (also known as acidic FGF and basic FGF, respectively), which share 55% sequence homology. At least 17 members of this family have now been identified (Hoshikawa *et al.*, 1998). Most FGF family members consist of 150 to 250 amino acids, and most bind to heparin with high affinity. Like selected members of the cytokine family of growth factors, some fibroblast growth factors, including FGF-1 and FGF-2, lack signal sequences and do not appear to undergo secretion from cells. The lack of a secretory signal indicates that these factors may play a role as molecules that are released at the time of injury to modulate CNS plasticity, rather than undergoing release in a target-derived manner to influence neuronal survival and pattern formation during development (Eckenstein, 1994). Other members of the FGF family do contain classic secretory signals, and may mediate a classic paracrine neurotrophic function in modulating cell survival and growth during development (Hoshikawa *et al.*, 1998). In addition, some FGF-1 and FGF-2 may be presented to the extracellular surface of the cell membrane by associating with heparin (Vlodavsky *et al.*, 1991).

Most studies of the distribution and function of FGFs in the nervous system have centered on FGF-1 and FGF-2. Studies on the temporal and spatial distribution of mRNA and protein of these growth factors are subject to the caveat of cross-reactivity between various immunolabels and probes because of the structural similarities between the 17 FGF family members (see Eckenstein, 1994). Both FGF-1 and FGF-2 are primarily expressed postnatally. FGF-1 is localized within neurons of the cholinergic basal forebrain, substantia nigra, sensory systems, motor neurons, and other subcortical populations (Eckenstein, 1994). FGF-2 is present in pyramidal neurons of the hippocampal CA2 region and in astrocytes. Overall levels of FGF protein in the adult CNS are very high compared to amounts of the classic neurotrophins (Eckenstein *et al.*, 1991), suggesting that the FGFs play a major role in modulating function in the adult CNS.

Actions of FGFs are modulated by binding to a family of transmembrane tyrosine kinase receptors (see Rapraeger *et al.*, 1991; Yayon *et al.*, 1991). Several genes code for different types of FGF receptors, and some FGF receptor genes are alternately spliced resulting in expression of a number of different FGF receptors that exhibit overlapping or restricted specificities for different FGF family ligands. Members of the FGF family also interact with a low-affinity heparan sulfate proteoglycan receptor that appears to enhance FGF binding (Wolff *et al.*, 1987). Removal of heparan sulfate from the cell surface abolishes FGF's ability to signal through transmembrane FGF receptors, indicating the importance of interactions with proteoglycans in mediating FGF signal transduction (Rapraeger *et al.*, 1991; Yayon *et al.*, 1991). It is hypothesized that FGFs are typically in an inactive conformation, but binding with a cell surface heparan proteoglycan may induce a conformational change and activate FGF, rendering it capable of successful binding and signal transduction with the transmembrane FGF receptor (see Eckenstein, 1994). Thus, the biological activity of FGFs are modulated by a combination of their regional expression, expression of their receptors, and interactions with heparan sulfate proteoglycans (Eckenstein, 1994).

FGFs can elicit several types of responses from their target cells, including direct activation of growth or survival-related gene expression, altered expression of cell adhesion molecules (Meiners *et al.*, 1993), modulation of expression of other growth factors *in vitro* (Yoshida and Gage, 1992), modulation of neuronal electrophysiology (Terlau and Seifert, 1990), or reduction–blockade in production of toxic molecules such as nitric oxide (Maiese *et al.*, 1993) or excitotoxins (Mattson *et al.*, 1989). FGF-1 and FGF-2 act as mitogens or differentiation factors for oligodendrocytes, Schwann cells, and astrocytes (Genburger *et al.*, 1987; Stemple *et al.*, 1988; Kalcheim, 1989; Murphy *et al.*, 1990), and promote the *in vitro* survival of cortical, hippocampal, motor, sensory, sympathetic, and parasympathetic neurons (see Eckenstein, 1994; Unsicker *et al.*, 1991; Grothe, Wewetzer, 1996 and the following discussion). FGF-5 has also been reported to promote motor neuron survival *in vitro,* and FGF-9 acts as a mitogen for cultured astrocytes.

When examined *in vivo,* FGF-1 reduces neuronal loss after hippocampal ischemia (Sasaki *et al.*, 1992; Mac Millan *et al.*, 1993), and is currently being examined in clinical trials for stroke. The receptor for FGF-1 is widely distributed in the CNS, and its delivery has also been reported to promote the growth of axons from several brain stem neurons after spinal cord injury (Cheng *et al.*, 1996) and to promote peripheral nerve regeneration (Cordiero *et al.*, 1989).

FGF-2 has been reported to promote the survival of layer II to III neurons of the entorhinal cortex following perforant path lesions *in vivo* (Gomez-Pinilla *et al.*, 1992; Peterson *et al.*, 1996) (see also Chapter 21). This population of neurons degenerates both in normal aging and in Alzheimer's disease. This survival effect may related to the ability of FGF-2 to enhance the calcium buffering capacity of vulnerable neurons (Peterson *et al.*, 1996). FGF-2 has also been reported to protect basal forebrain cholinergic neurons from injury-induced degeneration (Anderson *et al.*, 1988; Otto *et al.*, 1989; Gomez-Pinilla *et al.*, 1992), another cell population that degenerates in normal aging and in Alzheimer's disease. However, this protective effect may be indirectly mediated by activation of neurotrophic factor production in astrocytes. FGF-2 also reduces MPTP-induced or lesion-induced loss of substantia nigra neurons (Otto and Unsicker, 1990), and reduces retinal ganglion cell loss after optic nerve transection (Sievers *et al.*, 1987).

Importantly, FGFs have also emerged as potent molecules for establishing and propagating multipotent neural stem cells and progenitor cells (Ray and

Gage, 1994; Mayer-Proschel *et al.*, 1997) (see Chapters 7 and 8) in mammalian systems. Provided either solely or in combination with other growth factors such as epidermal growth factor (Suhonen *et al.*, 1996; Gage *et al.*, 1995; Ray and Gage, 1994), FGFs can generate pluripotent cells capable of differentiation into glial or neuronal lineages. Such cells could be used in strategies to replace lost host cells.

V. THE INSULIN-LIKE GROWTH FACTORS

Insulin and the insulin-like growth factors (IGF-I and IGF-II) constitute another family of molecules (reviewed in Roberts and LeRoith, 1988; Folli *et al.*, 1996) with growth-promoting qualities (Bothwell, 1982; Mill *et al.*, 1985; LeRoith *et al.*, 1993; Ishii, 1993). Most studies of this family in the nervous system have focused on IGF-I and IGF-II, which are broadly expressed in adulthood and development. Patterns of IGF-I and IGF-II expression in the nervous system are regionally specific and temporally regulated (LeRoith *et al.*, 1993; Ishii, 1993; Folli *et al.*, 1996). In particular, IGF-I mRNA is expressed in neurons and glia, and IGF-II mRNA is expressed in cells of the choroid plexus, the leptomeninges, and to a lesser extent on glia (LeRoith *et al.*, 1993). Although insulin and insulin-like growth factors modulate glucose utilization by many cell types, there are suggestions that these molecules may also act as specific, localized signals in the nervous system that modulate cell survival (DiCicco-Bloom and Black, 1988; Recio-Pinto *et al.*, 1986; Beck *et al.*, 1993; McMorris and Dubois-Dalcq, 1988; Knusel *et al.*, 1990), growth, differentiation (McMorris and Dubois-Dalcq, 1988; McMorris *et al.*, 1986; Bondy *et al.*, 1990), synapse function (Ishii, 1989), and neurotransmitter release (Dahmer and Perlman, 1988; see LeRoith *et al.*, 1993; Ishii, 1993 for general reviews).

The IGF receptor is a tetramer consisting of two α and two β subunits (LeRoith *et al.*, 1993) linked by disulfide units. The α subunits lie outside the cell and bind IGF ligands at cysteine-rich domains. The β subunits are receptors containing a short extracellular domain, a transmembrane domain, and a cytoplasmic domain; the latter contains an ATP-binding site and a tyrosine kinase domain. IGF family ligands bind to the α subunit, which in turn binds to β subunits. The binding of ligand-occupied α to β induces a conformational change in β that activates the receptor tyrosine kinase and induces autophosphorylation, leading to phosphorylation of additional substrates and cascades that mediate biological responses. The IGF-II receptor is distinct from the IGF-I receptor: it consists of a single protein with a large extracellular and a short intracellular domain, and signal transduction occurs through activation of glycine. A family of insulin-like growth factor binding proteins also exists, presumably modulating the binding of IGF-I and IGF-II to the IGF receptor (Yee *et al.*, 1987).

mRNA for IGF-I is detectable and is differentially regulated during development in several regions of the nervous system (LeRoith *et al.*, 1993). These regions include trigeminal and sympathetic neurons, sensory and cerebellar relay systems, nonpyramidal cells of the cortex and hippocampus, Purkinje cells of the cerebellum, and muscle (LeRoith *et al.*, 1993; Ishii, 1993). In the postnatal and mature brain, IGF-I mRNA is present in frontal cortex, the hippocampus, amygdala, suprachiasmatic nucleus, and cerebellar and sensory relay systems. Expression of IGF-I receptors peaks during development and is found in cerebellum, neocortex, olfactory bulb, retina, and choroid plexus (LeRoith *et al.*, 1993). *In vitro,* IGF-I promotes the survival or proliferation of cortical neurons (Aizenman and de Vellis, 1987), basal forebrain neurons (Knusel *et al.*, 1990), ventral mesencephalon neurons (Knusel *et al.*, 1990; Beck *et al.*, 1993), primary motor neurons (Yin *et al.*, 1994), sympathetic neurons (Zackenfels *et al.*, 1995), and oligodendrocytes (Aizenman *et al.*, 1986), and induces neurite outgrowth from motor (Caroni and Grandes, 1990; Caroni *et al.*, 1994), sensory (Bothwell, 1982), sympathetic (Zackenfels *et al.*, 1995), and cortical neurons (Aizenman and de Vellis, 1987). It may also influence oligodendrocyte differentiation and myelination (LeRoith *et al.*, 1993). Overexpression of IGF-I in transgenic mice during development results in a disturbance in the formation of normal barrel fields in primary sensory cortex: an increase is observed in the number of neurons, neuronal area, and volume per barrel, but there is a decrease in total neuronal density (Gutierrez-Ospina *et al.*, 1996). There is also an increase in myelin synthesis in IGF-I-overexpressing mice (Carson *et al.*, 1993). These data support a role for IGF-I in modulating neuronal number and nervous system patterning.

The effects of IGF-I in *in vivo* models of nervous system disease have not been the subject of extensive study. IGF-I immunoreactivity increases in Schwann cells after nerve crush (Hansson *et al.*, 1986). Injections of IGF-I or IGF-II into rat muscle induces sprouting of motor neurons (Caroni and Grandes, 1990), and IGF-I infusions into the lesioned peripheral nerve or dorsal root ganglion improve sensitivity to sensory stimuli post-injury, an effect that is blocked by injection of anti-IGF-I (Kanje *et al.*, 1989). Systemic administration of IGF-I improves the extent of functional recovery in the hindlimbs of mice after sciatic nerve lesions (Contreras *et al.*, 1995). In addition, IGF-I reduces the number and extent of demyelination sites in mice with experimental autoimmune encephalomyelitis, a model of multiple sclerosis, and improves their functional outcome (Yao *et al.*, 1995; Liu *et al.*, 1995b).

On the basis of IGF-I effects on motor neurons, clinical trials of IGF-I on patients with ALS were conducted in a prospective, randomized, controlled manner. At its conclusion, the trial showed a slight IGF-I effect on prolongation of life and symptom severity, but the overall magnitude of the effect was disappointing. Once again, IGF was administered peripherally in this trial, raising questions regarding access to IGF by degenerating motor neurons in quantities sufficient to influence motor neuron survival and function. Central and targeted delivery of the growth factor to motor neurons might be required to ultimately determine the value of IGF-I in treating ALS. The potential use of IGFs to treat multiple sclerosis and peripheral neuropathy is also being considered.

The distribution of IGF-II mRNA is less extensive than that of IGF-I in the CNS and is associated primarily with the choroid plexus, leptomeninges, and glia. *In vitro,* IGF-II promotes survival and neurite extension from sensory and sympathetic neurons (Recio-Pinto *et al.,* 1986; Zackenfels *et al.,* 1995). IGF-I and IGF-II also stimulate proliferation of rat inner ear epithelial cells (Zheng *et al.,* 1997). These limited findings have not served as the basis for clinical trials.

VI. THE TRANSFORMING FACTOR-β FAMILY OF GROWTH FACTORS

Another recently discovered neurotrophic factor, glial cell line-derived neurotrophic factor (GDNF), is a member of the transforming growth factor-β (TGF-β) superfamily (Lin *et al.,* 1993). Subsequently, three new members of this family, neurturin, persephin, and the bone morphogenic proteins (BMPs), that possess neurotrophic activities have been discovered. This section will focus specifically on these molecules.

GDNF is a secreted polypeptide growth factor of 134 amino acids. Although GDNF and neurturin are related to other members of the TGF-β superfamily, they mediate intracellular signal transduction by unique interactions with a multicomponent receptor system. The receptor components consist of (1) a receptor tyrosine kinase called RET (Durbec *et al.,* 1996; Worby *et al.,* 1996), which interacts with (2) a glycosyl-phosphatidylinositol (GPI)-linked receptor that specifically binds either GDNF (termed GDNF receptor α-1) or neurturin (termed GDNF receptor α-2; see below). GDNF also appears capable of interacting with RET directly (Treanor *et al.,* 1996).

GDNF acts as a survival factor for dopaminergic neurons of the substantia nigra, both supporting neuronal survival *in vitro* (Lin *et al.,* 1993) and preventing axotomy-induced degeneration of these neurons in rodents (Beck *et al.,* 1996). GDNF also rescues degenerating neurons and improves symptoms in primates with MPTP-induced parkinsonism (Gash *et al.,* 1996). Interestingly, short-term administration of GDNF elicits long-term effects on lesioned dopaminergic mesencephalic neurons (Winkler *et al.,* 1996a). GDNF is also a particularly potent growth factor for motor neurons (Hendersen *et al.,* 1994;

Yan *et al.*, 1995; Oppenheim *et al.*, 1995). In addition, GDNF rescues ischemic cortical neurons (Wang *et al.*, 1997), lesioned neurons of the locus coeruleus (Arenas *et al.*, 1995), and axotomized neurons in primary motor cortex (Giehl *et al.*, 1997), the cholinergic basal forebrain (Williams *et al.*, 1996), and the retina (retinal ganglion cells) (Klocker *et al.*, 1997). GDNF null mutant mice exhibit a striking agenesis of the kidneys and defective enteric innervation (Pichel *et al.*, 1996; Moore *et al.*, 1996), and also exhibit defects in the development of neurons of the dorsal root ganglion, sympathetic neurons, and nodose ganglion neurons (Moore *et al.*, 1996). Surprisingly, GDNF null mutant mice do not exhibit abnormalities of catecholaminergic neurons or motor neurons.

Based upon the ability of GDNF to rescue dopaminergic neurons in rodent and primate models of Parkinson's disease, effects of GDNF are currently being examined in a clinical trial for Parkinson's disease. In this study, patients are receiving intermittent intracerebroventricular injections of GDNF, thereby delivering high doses of GDNF into the cerebrospinal fluid (CSF) system. There are two potential drawbacks to this delivery method. First, it is not clear that GDNF will diffuse across the ependymal lining of the brain in quantities sufficient to reach and activate dopaminergic neurons of the substantia nigra. Second, exogenous GDNF is diffusely distributed throughout the CSF space, and may elicit undesired effects from stimulation of GDNF-sensitive, nondopaminergic neurons. Once again, a means of accurately targeting and regionally restricting delivery of the growth factor to nigral neurons is required to optimally test the therapeutic efficacy of GDNF in Parkinson's disease. One potential means of achieving targeted delivery is gene therapy (see Chapter 20); in rodent models of Parkinson's disease, *in vivo* gene delivery of GDNF reduces dopaminergic neuronal degeneration and improves functional outcomes (Bilang-Bleuel *et al.*, 1997).

Neurturin (Kotzbauer *et al.*, 1996; Heuckeroth *et al.*, 1997; Widenfalk *et al.*, 1997) and persephin (Milbrandt *et al.*, 1998) are two recently discovered members of the TGF-β superfamily. Neurturin is a neurotrophic factor with 40% amino acid homology to GDNF, and is expressed during both development and in adulthood (Widenfalk *et al.*, 1997). It is encoded as a prepro molecule by two exons on human chromosome 19 (Heuckeroth *et al.*, 1997). Like GDNF, neurturin activates the MAP kinase-signaling pathway. Signaling occurs through a multicomponent receptor similar to that employed by GDNF, consisting of the RET receptor tyrosine kinase and a distinct glycosylphosphatidylinositol (GPI)-linked receptor termed GDNF receptor α-2 (GDNFR-α_2) (Klein *et al.*, 1997; Buj-Bello *et al.*, 1997; Sanicola *et al.*, 1997; Baloh *et al.*, 1997; Widenfalk *et al.*, 1997). GDNFR-α_2 (also referred to as TrnR2, RETL2, NTNR-α, and GDNFR-β) shares approximately 50% sequence homology with GDNFR-α_1. Recently, a third specific GPI-linked receptor termed GDNFR-α_3 was identified (Jing *et al.*, 1997; Worby *et al.*, 1998; Naveilhan *et al.*, 1998; Masure *et al.*, 1998) with approximately 35% homology to GDNFR-$\alpha_{1\&2}$; its ligand remains to be identified. Neurturin is expressed in the postnatal brain in the cortex, striatum, and several brain stem areas and in the pineal gland (Widenfalk *et al.*, 1997). Its receptor is also widely expressed in the cortex, cerebellum, thalamus, hypothalamus, zona incerta, brain stem, spinal cord, and in subpopulations of sensory neurons (Widenfalk *et al.*, 1997). During development neurturin promotes the survival of sympathetic, nodose, and dorsal root ganglion sensory neurons (Heuckeroth *et al.*, 1997; Kotzbauer *et al.*, 1996). *In vivo* effects of neurturin in models of adult CNS injury have not yet been published.

Persephin is a peptide growth factor that was identified by degenerate polymerase chain reaction and shares 40% base pair homology with GDNF and neurturin (Milbrandt *et al.*, 1998). Persephin appears to signal through a receptor type that is distinct from GDNF and neurturin. *In vitro,* persephin supports the survival of ventral midbrain dopaminergic neurons and motor neurons (like GDNF and neurturin; Milbrandt *et al.*, 1998), but has no effect on sensory neurons (unlike GDNF and neurturin). *In vivo,* persephin prevents the loss of dopaminergic neurons of the substantia nigra after 6-hydroxydopamine lesions, and supports the survival of spinal cord motor neurons after sciatic nerve lesions (Milbrandt *et al.*, 1998). Thus, persephin may be of potential use in the therapy in Parkinson's disease or motor neuron disease.

An extensive subfamily of the TGF-β superfamily is the group of Bone Morphogenetic Proteins (BMPs; Mehler *et al.*, 1997; Shah *et al.*, 1996). Described relatively recently, the BMPs are a group of cytokines with several members that transduce signals by acting upon serine-threonine kinase receptor subunits. These molecules and their receptors are found in discrete regions throughout the developing neuraxis, and appear to modulate several functions (Mehler *et al.*, 1997). Acting in the earliest stages of neural development, BMPs inhibit formation of neuroectoderm during gastrulation. In the neural tube, BMPs act as gradient morphogens that promote differentiation of dorsal cell types and intermediate cell types. At later stages in development, BMPs act in the CNS on subventricular zone progenitor cells to promote astroglial differentiation and to suppress differentiation of these cells into neurons and oligodendroglia, while also acting on nonsubventricular zone progenitor cells to promote neuronal differentiation and survival. BMPs also promote the *death* of rhombencephalonic neural crest cells. In the PNS, specific BMPs promote the differentiation of neural crest stem cells to a neuronal lineage (Shah *et al.*, 1996; Mehler *et al.*, 1997). Thus, BMPs exert a number of influences on neural development including the promotion of cell survival, cell death, cell differentiation, and pattern formation at various stages (Mehler *et al.*, 1997). Autocrine, paracrine and cooperative signaling loops are involved in these processes. The presence and influence of BMPs in the adult nervous system has not yet been the subject of extensive study.

VII. EPIDERMAL GROWTH FACTOR FAMILY

Members of this family include epidermal growth factor (EGF), transforming growth factor-α (TGF-α), the newly discovered class of molecules referred to as the neuregulins, and NTAK (Neural- and Thymus-derived Activator for ErbB Kinases). These molecules act upon tyrosine kinase-linked EGF-like receptors and influence several different classes of cells in the nervous system.

EGF (Morrison *et al.*, 1988) and transforming growth factor-α (TGF-α) act by binding to the same EGF tyrosine kinase receptor (Alexi and Hefti, 1993; Plata-Salaman, 1991). *In vitro,* several nervous system cells respond to these factors, including hippocampal neurons, mesencephalic dopaminergic neurons, retinal cells, and Schwann cells (Alexi and Hefti, 1993; Peng *et al.*, 1998; Lillien, 1995; Toma *et al.*, 1992; Maiese *et al.*, 1993; Hadjiconstantinou *et al.*, 1991; Plata-Salaman, 1991; Casper *et al.*, 1991). In primary cultures of hippocampal neurons, EGF improves neuronal survival and neurite outgrowth, and blocks damage caused by hydroxyl radicals and lipid peroxidation (Peng *et al.*, 1998). *In vivo,* intracerebroventricular infusions of EGF prevent neuronal degeneration, increase synapse number, and improve functional outcomes after cerebral ischemia in gerbils (Peng *et al.*, 1998). Intracerebroventricular injections of EGF also protect dopaminergic neurons of the substantia nigra from MPTP-induced toxicity, and increase tyrosine hydroxylase labeling of these same neurons (Hadjiconstantinou *et al.*, 1991). EGF receptor numbers increase in peripheral nerve following axotomy (Toma *et al.*, 1992), primarily in Schwann cells and fibroblasts, indicating that EGF may play a role in peripheral nerve regeneration. EGF was also the first molecule reported to elicit growth of neural progenitor cells isolated from the adult mammalian CNS (Reynolds *et al.*, 1992). Further, in combination with FGF, EGF promotes the isolation of neural progenitor cells from the adult spinal cord (Weiss *et al.*, 1996). Clinical trials with EGF or TGF-α have not been conducted.

The neuregulins are a family of closely related proteins with important roles in neural and cardiac development (Gassman and Lemke, 1997; Rosen *et al.*, 1996; Jo *et al.*, 1995). Neuregulins and their receptors influence the survival, growth, differentiation, and fate of several cell types, including glia, and play an important role in modulating synapse development, particularly at the neuromuscular junction. Neuregulins also play an important role in pattern formation in the CNS. There are two sets of neuregulin-like growth factors, termed neuregulin-1 and neuregulin-2. Each set is encoded by its own single gene, and each gene generates several different protein products (as many as

15) resulting from alternative splicing of single transcripts (Gassman and Lemke, 1997). Members of the neuregulin-1 family include NDF (Neu differentiation factor), GGF (glial growth factor), ARIA (acetylcholine receptor-inducing agent), and heregulin (Falls *et al.*, 1993; Marchionni *et al.*, 1993; Holmes *et al.*, 1992; Ozaki *et al.*, 1997). Members of the neuregulin-2 family are currently being identified. Each neuregulin exhibits a distinct expression pattern in the developing and adult peripheral and/or central nervous systems.

Molecules of the neuregulin-1 and neuregulin-2 family possess EGF-like domains that bind to ErbB3 and ErbB4 receptor tyrosine kinases. Neuregulin-1 and neuregulin-2 molecules stimulate ErbB-receptor tyrosine phosphorylation profiles that are distinct, thereby mediating different and specific biochemical responses in cells at distinct tissue sites (Carraway *et al.*, 1997). In transgenic Erb3 null mutant mice, Schwann cells are absent in peripheral nerves leading to the death of motor and sensory neurons (Riethmacher *et al.*, 1997). These findings indicate that motor and sensory neurons require the presence of Schwann cells to survive during development and that Erb3 functions as a major modulator of Schwann cell survival or differentiation.

During development and regeneration in the PNS, neuregulins secreted from peripheral neuronal terminals are important determinants of reciprocal interactions between neurons and glia. Neuregulins stimulate transcription of genes encoding acetylcholine receptors at the developing neuromuscular junction, thereby inducing neuromuscular synapse differentiation (Ozaki *et al.*, 1997), an increase in synthesis of acetylcholine receptors, and an increase in Na^+ channels (see Loeb and Fishbach, 1997). ARIA (AChR-inducing activity), another neuregulin present at motor endplates (Falls *et al.*, 1993), stimulates acetylcholine receptor synthesis in cultured myotubes. Schwann cells possess neuregulin receptors (Rosenbaum *et al.*, 1997), and yet another neuregulin, glial growth factor (GGF), regulates Schwann cell survival and mitogenesis at the developing neuromuscular junction (Trachtenberg and Thompson, 1996; Rosenbaum *et al.*, 1997). Glial growth factor 2 (GGF2) stimulates myogenesis, and may be involved in skeletal muscle maintenance and nerve growth (Florini *et al.*, 1996). In the CNS, neuregulins also determine the phenotype of receptors for neurotransmitters in some maturing synapses, including the establishment of NMDA receptors in the cerebellum (Ozaki *et al.*, 1997). GGF is produced by granular neurons in the developing cerebellum and induces radial glia formation, which subsequently supports migration of granule cells along radial glial fibers. GGF also mediates interactions between migrating neurons and radial glial cells in the developing cortex (Anton *et al.*, 1997), signaling via ErbB2. Finally, GGF also regulates oligodendrocyte progenitor cell proliferation, survival, and differentiation (Canoll *et al.*, 1996). Heregulin, $\beta1$, another neuregulin isoform, is also mitogenic.

Neuregulin mRNA expression is induced by the neurotrophic factors BDNF, NT-3, NT-4/5, and GDNF (Loeb and Fishbach, 1997), indicating that neuregulin expression by neurotrophic factors may constitute a reciprocal, regulatory loop that promotes peripheral or central nervous system development (Loeb and Fishbach, 1997). Indeed, the density of some receptors such as acetylcholine receptors may be determined throughout life by ongoing expression of neuregulin (Sandrock *et al.*, 1997). Conversely, during development neuroblasts produce neuregulins that in turn stimulate nonneuronal cells to make NT-3 (Verdi *et al.*, 1996).

Overall, neuregulins are primarily produced by neuronal cells in the PNS and CNS, and determine glial cell survival and differentiation, as well as development and maturation of several types of synapses. Their effects in *in vivo* models of nervous system injury are currently under study.

NTAK is present in the developing and adult brain and in the adult thymus (Higashiyama *et al.*, 1997). It is structurally similar to the neuregulin molecules in its immunoglobulin-like, EGF-like, and hydrophobic domains. NTAK binds to two isoforms of the EGF receptor, ErbB3, and ErbB4. *In vitro,* NTAK stimulates the differentiation of MDA-MB-453 cells and competitively inhibits binding of neuregulin (Higashiyama *et al.*, 1997), but its specific *in vitro* and *in vivo* targets require additional characterization.

VIII. OTHER GROWTH FACTORS

Several other molecules with putative nervous system growth factor properties have been identified. Some of these factors were identified on the basis of searches for ligands to "orphan" tyrosine kinase receptors; that is, receptors that are known to exist in the developing nervous system at times and in locations where they might influence cellular survival, growth, and differentiation. For example, ligands to the erbB family of tyrosine kinase receptors led to discovery of the large class of neuregulins, previously described. Growth-associated protein-6 (GAS-6) was discovered as a ligand to the orphan tyrosine kinase receptor Rse (Godowski *et al.*, 1995).

Additional molecules with growth-promoting activities have been identified, including platelet-derived growth factor (PDGF), the interleukins, the immunophilin class of compounds, midkine, heparin-binding neurotrophic factors (pleiotrophin), erythropoietin, granulocyte-macrophage colony stimulating factor, and others. The actions of some of these factors have been partially characterized. For example, the erythrogeneis-promoting factor erythropoietin has been reported to support the survival of basal forebrain cholinergic neurons after lesions (Konishi *et al.*, 1993). The nature and spectra of effects of several other recently discovered growth factors remain to be more fully characterized, as the number of such molecules steadily grows. These proteins exhibit, singly or in combination, effects on neuronal and glial survival, growth, differentiation, and terminal morphology, and several examples will be discussed in more detail.

Hepatocyte growth factor (HGF) is widely expressed in mammals. HGF exhibits mitogenic and differentiation effects on cells of the liver, kidney, lung, and nervous system, and mediates signal transduction through a transmembrane tyrosine kinase receptor encoded by the proto-oncogene *c-met* . In the nervous system, HGF promotes the survival and growth of several types of cells, including motor, sensory and hippocampal neurons, muscle cells, and Schwann cells (Wong *et al.*, 1997; Ebens *et al.*, 1996; Miyazawa *et al.*, 1998; Maina *et al.*, 1997). HGF stimulates ChAT activity in motor neurons of the developing spinal cord. Expression of HGF is upregulated in muscle after injury (Di Renzo *et al.*, 1993), suggesting a role as a modulator of motor neuron or muscle function after injury. HGF also acts synergistically with CNTF to promote motor neuronal survival (Wong *et al.*, 1997). Mutations of the HGF receptor (*met* tyrosine kinase) cause reductions in the number of peripheral sensory nerves (Maina *et al.*, 1997). *In vivo* administration of HGF reduces the extent of ischemic neuronal death in the hippocampus, suggesting that HGF may have a future role in the therapy of stroke (Miyazawa *et al.*, 1998).

Immunophilins are molecules with growth-related potential that were discovered in the course of a search for new immunosuppressant agents (Gold, 1997; Snyder and Sabatini, 1995; Steiner *et al.*, 1997). Immunophilins are a family of receptor proteins of two classes: the FK506-binding proteins (FKBPs), which bind the immunosupressant molecules FK506 and rapamycin, and the cyclophilins, which bind cyclosporin A. The binding of either FK506 or cyclosporin A to its immunophilin receptor inhibits the phosphatase calcineurin, leading to immunosuppression. Rapamycin binds to FKBP and this complex then binds to *R*apamycin *A*nd *F*KBP-12 *T*arget (RAFT), which subsequently inhibits protein translation. In addition, immunophilins regulate calcium flux by association with the ryanodine and inositol 1,4,5-triphosphate (IP3) receptors, and reduce nitric oxide formation by stimulating phosphorylation of neuronal nitric oxide synthase. Immunophilin levels are higher in nervous system tissue than in the immune system. FKBP12 levels increase in injured peripheral nerve in parallel with GAP-43 (Sabatani *et al.*, 1997). Immunophilin analogs that lack immunosupressant properties have been generated, and have exhibited neuronal growth and neuronal protection properties *in vitro* and *in vivo* (Steiner *et al.*, 1997). Immunophilin compounds reportedly promote neurite outgrowth from sensory neuron cultures, and increase axonal regeneration after sciatic nerve lesions. In the brain, the immunophilin analog GPI-1046 reportedly promotes sprouting of lesioned axons of the substantia nigra in rodents and improves functional outcomes (Steiner *et al.*, 1997). GPI-

1046 also promotes sprouting of lesioned serotonergic axons in somatosensory cortex. Based on these findings, the immunophilins are being developed as potential therapeutic agents for nervous system disease.

Midkine is a heparin-binding growth factor that promotes differentiation of the neuromuscular junction by inducing clustering of acetylcholine receptors (Zhou *et al.*, 1997); in addition to this developmental role, midkine also acts as a neurotrophic factor by protecting photoreceptors in the adult rat retina from light-induced injury (Unoki *et al.*, 1994).

Several molecules have been identified that are ligands to eph-type tyrosine kinase receptors and promote axon guidance and target-finding in the retinotectal pathway (Winslow *et al.*, 1995; Drescher *et al.*, 1995; Cheng *et al.*, 1995; Xu *et al.*, 1996). These molecules include AL-1 (axon ligand 1), RAGS, ELF-1, and *rtk1*. Other molecules have been identified as putative trophic factors based upon their ability to promote neurite outgrowth in *in vitro* neuronal assays, including heparin-binding neurotrophic factor (or pleiotrophin; Nakagawara *et al.*, 1995; Fabri *et al.*, 1992), granulocyte-macrophage colony stimulating factor (Konishi *et al.*, 1993), and the brain interleukins 1, 2, 3 and 6 (Spranger *et al.*, 1990; Bandtlow *et al.*, 1990). The potential impact of these molecules on survival and growth of cells in the adult nervous system remains to be established.

Synthetic molecules have been generated that are peptide analogs to naturally occurring neurotrophic factors (see Hefti, 1997). These molecules are relatively short peptides of 9 to 20 amino acids that contain putative active sites of their larger parent neurotrophic factors, and possess the advantage that they can cross the blood–brain barrier after peripheral administration. Neuroprotective or growth effects have yet to be reported *in vivo* models of nervous system disease with peripherally administered peptide analogs, however. These analogs also are subject to the hypothetical disadvantage that their delivery would be nontargeted and therefore subject to the risk of eliciting adverse growth effects from nontargeted neural systems.

Another strategy for delivering neurotrophic molecules across the blood–brain barrier is to link a trophic factor to a molecule that is actively transported into the CNS (Friden *et al.*, 1993). Using this type of approach, NGF has been linked to transferrin receptor antibodies and has been reported to cross the blood–brain barrier. In a quinolinic acid lesion model of Huntington's disease in rodents, striatal neuronal protection has been reported following peripheral administration of an NGF-transferrin receptor antibody conjugate in rodent (Charles *et al.*, 1996). Similarly, NGF-transferrin receptor antibody conjugates have been reported to protect basal forebrain cholinergic neurons from retrograde degeneration following excitotoxic lesions of the cortex (Kordower *et al.*, 1994). Clinical trials with these conjugates have not been conducted to date.

Finally, pan-neurotrophins and neurotrophic factor "fusion" proteins are synthetic peptides that contain active sites of more than one neurotrophic factor within a single molecule (Snyder and Sabatini, 1995). Pan-neurotrophin-1 is a fusion of NGF, BDNF and NT-3, and exhibits transport properties and neurotrophic properties *in vitro* that are consistent with the known activities of its three trophic factor components independently (Snyder and Sabatini, 1995). Pan-neurotrophic factors are of potential use in neurological disorders characterized by the degeneration of multiple types of neurons, including stroke, trauma, and Alzheimer's disease. However, convincing evidence regarding the efficacy of these molecules in *in vivo* models of nervous system disease remains to be established.

IX. SUMMARY OF GROWTH FACTOR EFFECTS AND POTENTIAL USES

From the preceding summary of growth factor effects in the nervous system, it is evident that different neurotrophic proteins influence nervous system function at many structural and functional levels, and at diverse times during life. Growth factors are retrogradely or anterogradely transported down axons, and can act in either an autocrine or paracrine manners. Under physiological conditions and in a therapeutic context, growth factors provide a means of preventing cell loss, promoting axonal growth, influencing synaptic transmission, modulating injury responses, and generating pluripotent stem cells for neural repair.

Some neurotrophic factors exhibit exquisite localization in their sites of production and biological effects, whereas other growth factors or their receptors are broadly expressed throughout the neuraxis. In fact, it remains to be definitively proven that some of the more globally expressed growth factors, such as the insulin-like growth factors, are true neurotrophic factors rather than indirect modulators of growth through augmentation of cellular (e.g., glucose) metabolism.

Perhaps the most straightforward clinical target for growth factor therapy is the prevention of neuronal loss. Applications to progressive neurodegenerative disorders in which cell loss occurs include Alzheimer's disease, Parkinson's disease, Huntington's disease, ALS, and multiple sclerosis. A cell rescue strategy is also logical for disorders such as stroke and neural trauma. Based on the results of preclinical studies, candidate growth factors for each of these disorders have been identified.

Another logical target of growth factor therapy is the maintenance or regrowth of axons. Peripheral neuropathies provide an attractive target for therapy because of the relative safety and ease of growth factor delivery, and indeed clinical trials of NGF for diabetic neuropathy may provide the first instance of a human disease successfully treated with neurotrophic factors. Neuropathies provide a model for the treatment of neurological disease with more than one growth factor, because neurons of the dorsal root ganglia exhibit multiple phenotypes and functions, and respond to different growth factors. Once again, the identification of appropriate growth factors for the treatment of peripheral neuropathy can be based upon extensive preclinical studies that have examined the normal role of growth factors in peripheral nervous system development and function, and the effects of growth factor delivery using *in vivo* models of peripheral nerve damage.

Growth factors might also find application for the treatment of other types of nervous system disease. For example, neurotrophic factors are capable of modulating neurotransmission and synaptic efficacy, and thus might be used to treat memory disorders by enhancing synaptic efficacy in the hippocampus. Applying trophic factor therapy to such conditions is a topic requiring further exploration.

X. CONSIDERATIONS AND FUTURE DIRECTIONS

The abundant data summarized above generally provides a strong basis for continuing to pursue neurotrophic factors for neuroprotective and regenerative therapy of neurological disease. It is critical, however, that the results of these early studies play a vital role in directing future avenues in trophic factor research. Issues that have been brought into focus by experiments performed to date include: (1) the importance of understanding the pharmacokinetics and bioavailability of trophic factors after various routes of delivery, (2) restricting neurotrophic factor actions to targeted cell populations, and (3) developing effective delivery systems for introducing trophic agents to humans.

Ensuring that exogenously delivered trophic factors remain biologically active is a primary concern in designing successful trophic factor therapy. Most neurotrophic factors are natural proteins and various elements can affect their bioavailability following systemic administration in pharmacological doses. Proteolytic cleavage by nonspecific peptidases may deactivate the trophic factor molecule, a possibility that may partially account for the negative outcome of CNTF in ALS trials (ALS CNTF Treatment Study Group, 1996). In addition, various factors may bind to exogenously delivered trophic agents rendering them unavailable to neurons. For instance, serum-born alpha-2 macroglobulin may bind circulating NGF and diminish its bioavailability (Hefti, 1997). The binding of trophic factors to endogenous receptors on nontargeted cell populations should also be taken into account when evaluating pharmacokinetics, because such interactions would further reduce bioavailability for interacting with targeted cells.

Perhaps one of the most crucial lessons gleaned from clinical trials of neurotrophic factors for nervous system disease thus far is the importance of restricting the biological actions of exogenously delivered trophic factors to the cell population targeted for therapy, as indicated in the preceding sections. The widespread presence of trophic factor receptors on various neuronal and nonneuronal cells throughout the body makes the goal of restricting trophic factor actions a challenging one. Other trophic factor mediated actions, such as axonal sprouting, alterations in neurotransmitter systems, and modulation of synaptic function (Lewin and Barde, 1996), heighten concerns regarding nonspecific trophic factor delivery. In addition, NGF affects the immune system by stimulating mast cell migration and histamine release, and can influence cells of the reproductive system (Rohrer and Sommer, 1983) The functional impact of long-term trophic factor administration on these responsive systems remains to be fully evaluated.

Thus, the development of techniques for delivering trophic factors in a restricted manner to targeted cell populations will play a key role in determining the success of neurotrophic factor therapy as a clinical tool. One means of achieving targeted and regionally restricted growth factor delivery in the CNS is intraparenchymal growth factor infusion. NGF infusions to the brain in rats can elevate NGF levels in tissues near the infusion site sufficiently to rescue cholinergic neurons (Hu *et al.,* 1997) but do not result in the widespread diffusion of NGF throughout the CNS (J. Conner and M. Tuszynski, unpublished observations). Alternatively, gene therapy is another means of specifically targeting trophic factor to localized neuronal populations, as noted above and in subsequent chapters.

In conclusion, neurotrophic factors are an extremely promising group of biological compounds with potential for treating a variety of neurological diseases for which therapy is currently nonexistent or suboptimal. A focused effort to bring these pharmacological agents to clinical application will require additional basic research into mechanisms of biological action, a more thorough understanding of the spectra of neuronal responses to different factors, and targeted and safe delivery methods. Once these goals are achieved, neurotrophic factors are likely to radically alter the landscape of neurological therapy.

ACKNOWLEDGMENTS

Supported by the NIH (NINDS 1R01NS37083, NIA 2P01AG10435), American Academy of Neurology Research Foundation, Veterans Affairs, and the Hollfelder Foundation.

REFERENCES

Acheson, A., Conover, J. C., Fandl, J. P., DeChiara, T. M., Russell, M., et al. (1995). A BDNF autocrine loop in adult sensory neurons prevents cell death. *Nature* 374:450–453.

Aebischer, P., Schluep, M., Deglon, N., Joseph, J. M., Hirt, L., Heyd, B., Goddard, M., Hammang, J. P., Zurn, A. D., Kato, A. C., et al. (1996). Intrathecal delivery of CNTF using encapsulated genetically modified xenogenic cells in amyotrophic lateral sclerosis patients. *Nature Med* 2:1041.

Aizenman, Y., Weischel, M. E., and De Vellis, K. (1986). Changes in insulin and transferrin requirements of pure brain neuronal cultures during embryonic development. *Proc Natl Acad Sci USA* 83:2263–2266.

Aizenman, Y., and de Vellis, J. (1987). Brain neurons develop in a serum and glial free environment: Effects of transferrin, insulin, insulin-like growth factor-I and thyroid hormone on neuronal survival, growth and differentiation. *Brain Res* 406:32–42.

Alexi, T., and Hefti, F. (1993). Trophic actions of transforming growth factor alpha on mesencephalic dopaminergic neurons developing in culture. *Neurosci* 55:803–808.

ALS CNTF Treatment Study Group. (1996). A double-blind placebo-controlled clinical trial of subcutaneous recombinant human ciliary neurotrophic factor (rHCNTF) in amyotrophic lateral sclerosis. *Neurology* 46:1244—1249.

Anderson, K. J., Dam, D., Lee, S., Cotman, C. W. (1988). Basic fibroblast growth factor prevents death of lesioned cholinergic neurons in vivo. *Nature* 332:360–362.

Angeletti, R. H., Hermodson, M. A., and Bradshaw, R. A. (1973). Amino acid sequences of mouse 2.5S nerve growth factor. II. Isolation and characterization of the thermolytic and peptic peptides and the complete covalent structure. *Biochem* 12:100–115.

Anton, E. S., Marchionni, M. A., Lee, K. F., and Rakic, P. (1997). Role of GGF/neuregulin signaling interactions between migrating neurons and radial glia in the developing cerebral cortex. *Development* 124:3501–3510.

Apfel, S. (1997). Treatment of peripheral nervous system disorders. In: Apfel, S., ed., *Clinical applications of neurotrophic factors.* Philadelphia: Lippincott-Raven, pp. 63–81.

Apfel, S. C., Lipton, R. B., Arezzo, J. C., and Kessler, J. A. (1991). Nerve growth factor prevents toxic neuropathy in mice. *Ann Neurol* 29:87–90.

Apfel, S. C., Arezzo, J. C., Moran, M., and Kessler, J. A. (1993). Effects of administration of ciliary neurotrophic factor on normal motor and sensory peripheral nerves *in vivo. Brain Res* 604:1–6.

Apfel, S. C., Arezzo, J. C., Brownlee, M., Federoff, H., and Kessler, J. A. (1994). Nerve growth factor administration protects against experimental diabetic sensory neuropathy. *Brain Res* 634:7–12.

Apfel, S. Y., Wright, D. E., Wiideman, A. M., Dormia, C., Snider, W. D., and Kessler, J. A. (1996). Nerve growth factor upregulates expression of brain-derived neurotrophic factor in the peripheral nervous system. *Molec Cell Neurosci* 7:134–142.

Arakawa, Y., Sendtner, M., and Thoenen, H. (1990). Survival effect of ciliary neurotrophic factor on chick embryonic motoneurons in culture: Comparison with other neurotrophic factors and cytokines. *J Neurosci* 10:3507–3515.

Arce, V., Pollock, R. A., Philippe, J. M., Pennica, D., Henderson, C. E., and deLapeyiere, O. (1998). Synergistic effects of Schwann- and muscle-derived factors on motoneuron survival involve GDNF and cardiotropin-1 (CT-1). *J Neurosci* 18:1440–1448.

Arenas, E., Trupp, M., Akerud, P., and Ibanez, C. F. (1995). GDNF prevents degeneration and promotes the phenotype of brain noradrenergic neurons *in vivo. Neuron* 15:1465–1473.

Baloh, R. H., Tansey, M. G., Golden, J. P., Creedon, D. J., Heuckeroth, R. O., Keck, C. L., Zimonjic, D. B., Popescu, N. C., Johnson, E. M. J., and Milbrandt, J. (1997). TrnR2, a novel receptor that mediates neurturin and GDNF signaling through Ret. *Neuron* 18:793–802.

Bandtlow, C. E., Meyer, M., Lindholm, D., Spranger, M., Heumann, R., and Thoenen, H. (1990). Regional and cellular codistribution of interleukin 1 beta and nerve growth factor mRNA in the adult rat brain: Possible relationship to the regulation of nerve growth factor synthesis. *J Cell Biol* 111:1701–1711.

Banner, L. R., and Patterson, P. H. (1994). Major changes in the expression of the mRNAs for cholinergic differentiation factor/leukemia inhibitory factor and its receptor after injury to adult peripheral nerves and ganglia. *Proc Natl Acad Sci USA* 91:7109–7913.

Barbacid, M. (1994). The trk family of neurotrophin receptors. *J Neurobiol* 25:1386–1403.

Barbin, G., Manthorpe, M., and Varon, S. (1984). Purification of the chick eye ciliary neurotrophic factor. *J Neurochem* 43:1468–1478.

Barde, Y. A. (1990). The nerve growth factor family. *Progr Growth Factor Res* 2:237–248.

Barde, Y. A., Edgar, D., and Thoenen, H. (1982) Purification of a new neurotrophic factor from mammalian brain. *EMBO* 1:549–553.

Barnett, J., Baecker, P., Routledge-Ward, C., Bursztyn-Pettegrew, H., Chow, J., Nguyen, B., Bach, C., Chan, H., Tuszynski, M. H., Yoshida, K., Rubalcava, R., and Gage, F. H. (1990). Human beta nerve growth factor obtained from a baculovirus expression system has potent *in vitro* and *in vivo* neurotrophic activity. *Exp Neurol* 110:11–24.

Beck, K. D., Knusel, B., and Hefti, F. (1993). The nature of the trophic action of brain-derived neurotrophic factor, des(1-3)-insulin-like growth factor-1, and basic fibroblast growth factor on mesencephalic dopaminergic neurons developing in culture. *Neurosci* 52:855–866.

Beck, K. D., Irwin, I., Valverde, J., Brennan, T. J., Langston, J. W., and Hefti, F. (1996). GDNF induces a dystonia-like state in neonatal rats and stimulates dopamine and serotonin synthesis. *Neuron* 16:665–673.

Benedetti, M., Levi, A., and Chao, M. V. (1993). Differential expression of nerve growth factor receptors leads to altered binding affinity and neurotrophin responsiveness. *Proc Natl Acad Sci USA* 90:7859–7863.

Berkmeier, L. R., Winslow, J. W., Kaplan, D. R., Nikolics, K., Goeddel, D. V., and Rosenthal, A. (1991). Neurotrophin-5: A novel neurotrophic factor that activates trk and trkB. *Neuron* 7:857–866.

Berninger, B., Garcia, D. E., Inagaki, N., Hahnel, C., and Lindholm, D. (1993). BDNF and NT-3 induce intracellular calcium elevation in hippocampal neurones. *NeuroReport* 4:1303–1306.

Berzaghi, M. P., Gutierrez, R., Heinemann, U., Lindholm, D., and Thoenen, H. (1995). Neurotrophins induce acute transmitter-mediated changes in brain electrical activity. *Abstr Soc Neurosci* 21:226.3.

Bilang-Bleuel, A., Revah, F., Colin, P., Locquet, I., Robert, J. J., Mallet, J., and Horellou, P. (1997). Intrastriatal injection of an adenoviral vector expressing glial-cell-line-derived neurotrophic factor prevents dopaminergic neuron degeneration and behavioral impairment in a rat model of Parkinson disease. *Proc Natl Acad Sci USA* 94:8818–8823.

Blochl, A., and Thoenen, H. (1996). Localization of cellular storage compartments and sites of constitutive and activity-dependent release of nerve growth factor (NGF) in primary cultures of hippocampal neurons. *Molec Cell Neurosci* 7:173–190.

Blondet, B., Murawsky, M., Houenou, L. J., Li, L., Ait-Ikhlef, A., Yan, Q., and Rieger, F. (1997). Brain-derived neurotrophic factor fails to arrest neuromuscular disorders in the paralysed mouse mutant, a model of motoneuron disease. *J Neurol Sci* 153:20–24.

Bohlen, P., Baird, A., Esch, F., Ling, N., and Gospodarowicz, D. (1984). Isolation and partial molecular characterization of pituitary fibroblast growth factor. *Proc Natl Acad Sci USA* 81:5364–5368.

Bondy, C. A., Werner, H., Roberts, C. T. J., and LeRoith, D. (1990). Cellular pattern of insulin-like growth factor-I (IGF-I) and type I IGF receptor gene expression in early organogenesis: Comparison with IGF-II gene expression. *Mol Endocrinol* 4:1386–1398.

Bothwell, M. (1982). Insulin and somatomedin MSA promote nerve growth factor-independent neurite formation by cultured chick dorsal root ganglionic sensory neurons. *J Neurosci Res* 8:225–231.

Bothwell, M. (1995). Functional interactions of neurotrophins and neurotrophin receptors. *Annu Rev Neurosci* 18:223–253.

Buj-Bello, A., Adu, J., Pinon, L. G., Horton, A., Thompson, J., Rosenthal, A., Chinchetru, M., Buchman, V. L., and Davies, A. M. (1997). Neurturin responsiveness requires a GPI-linked receptor and the Ret receptor tyrosine kinase. *Nature* 387:721–724.

Cabelli, R. J., Hohn, A., and Shatz, C. J. (1995). Inhibition of ocular dominance column formation by infusion of NT-4/5 or BDNF. *Science* 267:1662–1666.

Cabelli, R. J., Shelton, D. L., Segal, R. A., and Shatz, C. J. (1997). Blockade of endogenous ligands of trkB inhibits formation of ocular dominance columns. *Neuron* 19:63–76.

Canoll, P. D., Musacchio, J. M., Hardy, R., Reynolds, R., Marchionni, M. A., and Salzer, J. L. (1996). GGF/neuregulin is a neuronal signal that promotes the proliferation and survival and inhibits the differentiation of oligodendrocyte progenitors. *Neuron* 17:229–243.

Caroni, P., and Grandes, P. (1990). Nerve sprouting in innervated adult skeletal muscle induced by exposure to elevated levels of insulin-like growth factors. *J Cell Biol* 110:1307–1317.

Caroni, P., Schneider, C., Kiefer, M. C., and Zapf, J. (1994). Role of insulin-like growth factors in nerve sprouting: Suppression of terminal sprouting in paralyzed muscles by IGF-binding protein 4. *J Cell Biol* 125:893–902.

Carraway, K. L., Weber, J. L., Unger, M. J., Ledesma, J., Yu, N., Gassman, M., and Lai, C. (1997). Neuregulin-2, a new ligand of ErbB3/ErbB4-receptor tyrosine kinases. *Nature* 387:512–516.

Carson, M. J., Behringer, R. R., Brinster, R. L., and McMorris, F. A. (1993). Insulin-like growth factor increases brain growth and central nervous system myelination in transgenic mice. *Neuron* 10:729–740.

Carter, B. D., Kaltschmidt, C., Kaltschmidt, B., Offenhauser, N., Bohm-Matthaei, R., Baeverle, P. A., Barde, Y. A. (1996). Selective activation of NF-kB by nerve growth factor through the neurotrophin receptor p75. *Science* 272:542–545.

Casper, D., Mytilineou, C., and Blum, M. (1991). EGF enhances the survival of dopamine neurons in rat embryonic mesencephalon primary cell culture. *J Neurosci Res* 30:372–381.

Castren, E., Pitkanen, M., Sirvio, J., Parasadanian, A., Lindholm, D., et al. (1993). The induction of LTP increases BDNF and NGF mRNA but decreases NT-3 mRNA in the dentate gyrus. *NeuroReport* 4:895–898.

Chao, M. V. (1992). Neurotrophin receptotrs: A window into neuronal differentiation. *Neuron* 9:583–593.

Chao, M. V. (1994). The p75 neurotrophin receptor. *J Neurobiol* 25:1373–1385.

Charles, V., Mufson, E. J., Friden, P. M., Bartus, R. T., and Kordower, J. H. (1996). Atrophy of cholinergic basal forebrain neurons following excitotoxic cortical lesions is reversed by intravenous administration of an NGF conjugate. *Brain Res* 728:193–203.

Cheema, S. S., Richards, L., Murphy, M., and Bartlett, P. F. (1994a). Leukemia inhibitory factor prevents the death of axotomised sensory neurons in the dorsal root ganglia of the neonatal rat. *J Neurosci Res* 37:213–218.

Cheema, S. S., Richards, L. J., Murphy, M., and Bartlett, P. F. (1994b). Leukaemia inhibitory factor rescues motoneurones from axotomy-induced cell death. *NeuroReport* 5:989–992.

Chen, K. S., Nishimura, M. C., Armanini, M. P., Crowley, C., Spencer, S. D., and Phillips, H. S. (1997). Disruption of a single allele of the nerve growth factor gene results in atrophy of basal forebrain cholinergic neurons and memory deficits. *J Neurosci* 17:7288–7296.

Cheng, H.-J., Nakamoto, M., Bergemann, A., and Flanagan, J. C. (1995). Complementary gradients in expression and binding of ELF-1 and Mek4 in development of the topographic and retinotectal projection map. *Cell* 82:371.

Cheng, H., Yihai, C., and Olson, L. (1996). Spinal cord repair in adult paraplegic rats: Partial restoration of hind limb function. *Science* 273:510–513.

Chiu, A. Y., Chen, E. W., and Loera, S. (1994). Distinct neurotrophic responses of axotomized motor neurons to BDNF and CNTF in adult rats. *Neuroreport* 5:693–696.

Clary, D. O., and Reichardt, L. F. (1994). An alternatively spliced form of the nerve growth factor receptor TrkA confers an enhanced response to neurotrophin-3. *Proc Natl Acad Sci USA* 91:11133–11137.

Clatterbuck, R. E., Price, D. L., and Koliatsos, V. E. (1994). Further characterization of the effects of brain-derived neurotrophic factor and ciliary neurotrophic factor on axotomized neonatal and adult mammalian motor neurons. *J Comp Neurol* 342:45–56.

Cohen, R. I., Marmur, R., Norton, W. T., Mehler, M. F., and Kessler, J. A. (1996). Nerve growth factor and neurotrophin-3 regulate the proliferation and survival of developing rat brain oligodendrocytes. *J Neurosci* 16:6433–6442.

Cohen, S., and Levi-Montalcini, R. (1957). Purification and properties of a nerve growth-promoting factor isolated from mouse sarcoma. *Cancer Res* 17:15–20.

Conner, J. M., Irvine, Person, and Varon, S. (1997). Localization of BDNF in the rat brain. *J Neurosci* x:x.

Conner, J. M., Lauterborn, J. C., and Gall, C. M. (1998). Anterograde transport of neurotrophin proteins in the CNS: a reassessment of the neurotrophin hypothesis. *Rev Neurosci* In press.

Conover, J. C., Erickson, J. T., Katz, D. M., Bianchi, L. M., Poueymirou, W. T., McClain, J., Pan, L., Helgren, M., Ip, N. Y., Boland, P., Friedman, B., Wiegand, S., Vejsada, R., Kato, A. C., DeChiara, T. M., and Yancopoulos, G. D. (1995). Neuronal deficits, not involving motor neurons, in mice lacking BDNF and/or NT4. *Nature* 375:235–238.

Contreras, P. C., Steffler, C., Yu, E., Callison, K., Stong, D., and Vaught, J. L. (1995). Systemic administration of rhIGF-1 enhanced regeneration after sciatic nerve crush in mice. *J Pharmacol Exp Ther* 274:1443–1449.

Cordiero, P. G., Seckel, P. R., Lipton, S. A., D'Amore, P. A., Wagner, J., and Madison, R. (1989). Acidic fibroblast growth factor enhances peripheral nerve regeneration *in vivo*. *Plast Reconstr Surg* 83:1013–1019.

Cordon-Cardo, C., Tapley, P., Jing, S., Nanduri, V., O'Rourke, E., Lamballe, F., Kovary, K., Klein, R., Jones, K. R., Reichart, L. F., and Barbacid, M. (1991). The trk tyrosine protein kinase mediates the mitogenic properties of nerve growth factor and neurotrophin-3. *Cell* 66:173–183.

Crowley, C., Spencer, S. D., Nishimura, M. C., Chen, K. S., Pitts-Meeks, et al. (1994). Mice lacking nerve growth factor display perinatal loss of sensory and sympathetic neurons yet develop basal forebrain cholinergic neurons. *Cell* 76:1001–1011.

Dahmer, M. K., and Perlman, R. L. (1988). Bovine chromaffin cells have insulin-like growth factor-I (IGF-I) receptors: IGF-I enhances catecholamine secretion. *J Neurochem* 51:321–323.

Dale, S. M., Kuang, R. Z., Wei, X., and Varon, S. (1995). Corticospinal motor neurons in the adult rat: degeneration after intracortical axotomy and protection by ciliary neurotrophic factor. *Experimental Neurology* 135:67–73.

Davies, A. M. (1994). The role of neurotrophins in the developing nervous system. *J Neurobiol* 25:1334–1348.

Di Renzo, M. F., Bertolotto, A., Olivero, M., Putzolu, P., Crepaldi, T., Schiffer, D., Pagni, C. A., and Comoglio, P. M. (1993). *Oncogene* 8:219–222.

DiCicco-Bloom, E., and Black, I. B. (1988). Insulin growth factors regulate the mitotic cycle in cultured rat sympathetic neuroblasts. *Proc Natl Acad Sci USA* 85:4066–4070.

DiStefano, P. S., Freidman, B., Radziejewski, C., Alexander, C., Boland, P., Schick, C.M., Lindsay, R. M., and Wiegand, S. J. (1992). The neurotrophins BDNF, NT-3 and NGF display distinct patterns of retrograde axonal transport in peripheral and central neurons. *Neuron* 8:983–993.

Dobrowsky, R. T., Werner, M. H., Castellino, A. M., Chao, M. V., and Hannun, Y. A. (1994). Activation of the sphingomyelin cycle through the low-affinity neurotrophin receptor. *Science* 265:1596–1599.

Domenici, L., Cellerino, A., Berardi, N., Cattaeno, A., and Maffei, L. (1995). Antibodies to nerve growth factor (NGF) prolong the sensitive period for monocular deprivation in the rat. *NeuroReport* 5:2041–2044.

Drescher, U., Kremoser, C., Handwerker, C., Loschinger, J., Noda, M., and Bonhoeffer, F. (1995). *In vitro* guidance of retinal ganglion cell axons by RAGS, a 25 kDa tectal protein related to ligands for Eph receptor tyrosine kinases. *Cell* 82:359–370.

Durbec, P., Marcos-Gutierrez, C. V., Kilkenny, C., Grigoriou, M., Wartiowaara, K., Suvanto, P., Smith, D., Ponder, B., Costantini, F., Saarma, M., et al. (1996). GDNF signaling through the Ret receptor tyrosine kinase. *Nature* 381:789—793.

Ebens, A., Brose, K., Leonardo, D., Hanson, M. G., Bladt, F., Birchmeier, C., Barres, B. A., and Tessier-Lavinge, M. (1996). *Neuron* 17:1157–1172.

Eckenstein, F. P., Shipley, G. D., and Nishi, R. (1991). Acidic and basic fibroblast growth factors in the nervous system: Distribution and differential alteration of levels after injury of central vs. peripheral nerve. *J Neurosci* 11:412–419.

Eckenstein, F. P. (1994). Fibroblast growth factors in the nervous system. *J Neurobiol* 11:1467–1480.

Emerich, D. F., Lindner, M. D., Winn, S. R., Chen, E. Y., Frydel, B. R., and Kordower, J. H. (1996). Implants of encapsulated human CNTF-producing fibroblasts prevent behavioral deficits and striatal degeneration in a rodent model of Huntington's disease. *J Neurosci* 16:5168–5181.

Ernfors, P., Lee, K.-F., and Jaenisch, R. (1994a). Mice lacking brain-derived neurotrophic factor develop with sensory deficits. *Nature* 368:147–150.

Ernfors, P., Lee, K.-F., Kucera, J., and Jaenisch, R. (1994b). Lack of neurotrophin-3 leads to deficiencies in the peripheral nervous system and loss of limb proprioception. *Cell* 77:503–512.

Ernfors, P., Van de Water, T., Loring, J., and Jaenisch, R. (1995). Complementary roles of BDNF and NT-3 in vestibular and auditory development. *Neuron* 14:1153–1164.

Ezzeddine, Z. D., Yang, X., DeChiara, T., Yancopoulos, G., and Cepko, C. L. (1997). Postmitotic cells fated to become photoreceptors can be respecified by CNTF treatment of the retina. *Development* 124:1055–1067.

Fabri, L., Nice, E. C., Ward, L. D., Maruta, H., Burgess, A. W., and Simpson, R. J. (1992). Characterization of bovine heparin-binding neurotrophic factor (HBNF): assignment of disulfide bonds. *Biochem Intl* 28:1–9.

Fagan, A. M., Zhang, H., Landis, S., Smeyne, R. J., Silos-Santiago, I., and Barbacid, M. (1996). TrkA, but not TrkC, receptors are essential for survival of sympathetic neurons *in vivo*. *J Neurosci* 16:6208–6218.

Falls, D. L., Rosen, K. M., Corfas, G., Lane, W. S., and Fischbach, G. D. (1993). ARIA, a protein that stiulates acetylcholine receptor synthesis, is member of the neu ligand family. *Cell* 72:801–815.

Farinas, I., Jones, K. R., Backus, C., Wang, X.-Y., and Reichardt, L. F. (1994). Severe sensory and sympathetic deficits in mice lacking neurotrophin-3. *Nature* 369:658–661.

Fischer, W., Wictorin, K., Bjorklund, A., Williams, L. R., Varon, S., and Gage, F. H. (1987). Amelioration of cholinergic neuron atrophy and spatial memory impairment in aged rats by nerve growth factor. *Nature* 329:65–68.

Florini, J. R., Samuel, D. S., Ewton, D. Z., Kirk, C., and Sklar, R. M. (1996). Stimulation of myogenic differentiation by a neuregulin, glial growth factor 2. Are neuregulins the long-sought muscle trophic factors secreted by nerves? *J Biol Chem* 271:12699–12702.

Folli, F., Ghidella, S., Bonfanti, L., Kahn, C. R., and Merighi, A. (1996). The early intracellular signaling pathway for the insulin/insulin-like growth factor receptor family in the mammalian central nervous system. *Molec Neurobiol* 13:155–183.

Forger, N. G., Howell, M. L., Bengston, L., MacKenzie, L., DeChiara, T. M., and Yancopoulos, G. D. (1997). Sexual dimorphism in the spinal cord is absent in mice lacking the ciliary neurotrophic factor receptor. *J Neurosci* 17:9605–9612.

Friden, P. M., Walus, L. R., Watson, P., Doctrow, S. R., Kozarich, J. W., Backman, C., Hoffer, B., Bloom, F., and Granholm, A. C. (1993). Blood–brain barrier penetration and *in vivo* activity of an NGF conjugate. *Science* 259:373–377.

Friedman, B., Kleinfeld, D., Ip, N. Y., Verge, V. M., Moulton, R., Boland, P., Zlotchenko, E., Lindsay, R. M., and Liu, L. (1995). BDNF and NT-4/5 exert neurotrophic influences on injured adult spinal motor neurons. *J Neurosci* 15:1044–1056.

Funakoshi, H., Frisen, J., Barbany, G., Timmusk, T., Zachrisson, O., Verge, V. M., and Persson, H. (1993). Differential expression of mRNAs from neurotrophins and their receptors after axotomy of the sciatic nerve. *J Cell Biol* 123:455–465.

Gage, F. H., Armstrong, D. M., Williams, L. R., and Varon, S. (1988). Morphologic response of axotomized septal neurons to nerve growth factor. *J Comp Neurol* 269:147–155.

Gage, F. H., Coates, P. W., Palmer, T. D., Kuhn, H. G., Fisher, L. J., Suhonen, J. O., Peterson, D. A., Suhr, S. T., and Ray, J. (1995). Survival and differentiation of adult neuronal progenitor cells transplanted to the adult brain. *Proc Natl Acad Sci U S A* 92:11879–11883.

Gao, W. Q., Dybdal, N., Shinsky, N., Murnane, A., Schmelzer, C., Siegel, M., Keller, G., Hefti, F., Phillips, H. S., and Winslow, J. W. (1995). Neurotrophin-3 reverses experimental cisplatin-induced peripheral sensory neuropathy. *Ann Neurol* 38:30–37.

Garofalo, L., Ribeiro-da-Silva, A., and Cuello, A. C. (1992). Nerve growth factor-induced synaptogenesis and hypertrophy of cortical cholinergic terminals. *Proc Natl Acad Sci USA* 89:2639–2643.

Gash, D. M., Zhang, Z., Ovadia, A., Cass, W. A., Yi, A., Simmerman, L., Russell, D., Martin, D., Lapchak, P. A., Collins, F., Hoffer, B. J., and Gerhardt, G. A. (1996). Functional recovery in parkinsonian monkeys treated with GDNF. *Nature* 380:252–255.

Gassman, M., and Lemke, G. (1997). Neuregulins and neuregulin receptors in neural development. *Curr Opin Neurobiol* 7:87–92.

Genburger, C., Lebourdette, G., and Sensenbrenner, M. (1987). Brain basic fibroblast growth factor stimulates the proliferation of rat neuronal precursor cells *in vitro*. *Febs Lett* 217:1–5.

Giehl, K. M., Schacht, C. M., Yan, Q., and Mestres, P. (1997). GDNF is a trophic factor for adult rat corticospinal neurons and promotes their long-term survival after axotomy *in vivo*. *Eur J Neurosci* 9:2479–2488.

Godowski, P. J., Mark, M. R., Chen, J., Sadick, M. D., Raab, H., and Hammonds, R. G. (1995). Reevaluation of the roles of protein S and Gas6 as ligands for the receptor tyrosine kinase Rse-Tyro 3. *Cell* 82:355–358.

Gold, B. G. (1997). FK506 and the role of immunophilins in nerve regeneration. *Molec Neurobiol* 15:285–306.

Gomez-Pinilla, F., Lee, J. W., Cotman, C. W. (1992). Basic fibroblast growth factor in adult rat brain: Cell distribution and response to entorhinal lesion and fimbria–fornix transection. *J Neurosci* 12:345–355.

Gorin, P. D., and Johnson, E. M. (1980). Effects of long term nerve growth factor deprivation on the nervous system of the adult rat: an autoimmune approach. *Brain Res* 198:27–42.

Gotz, R., Koster, R., Winkler, C., Raulf, F., Lottspeich, F., Schartl, M., and Thoenen, H. (1994). Neurotrophin-6 is a new member of the nerve growth factor family. *Nature* 372:266–269.

Grill, R., Murai, K., Blesch, A., Gage, F. H., and Tuszynski, M. H. (1997). Cellular delivery of neurotrophin-3 promotes corticospinal axonal growth and partial functional recovery after spinal cord injury. *J Neurosci* 17:5560–5572.

Grimes, M., Zhou, J., Li, Y., Holtzman, D., and Mobley, W. C. (1993). Neurotrophin signaling in the nervous system. *Semin Neurosci* 5:239–247.

Grothe, C., and Wewetzer, K. (1996). Fibroblast growth factor and its implications for developing and regenerating neurons. *Int J Dev Biol* 40:403–410.

Gutierrez-Ospina, G., Calikoglu, A. S., Ye, P., and D'Ecoule, A. J. (1996) *In vivo* effects of insulin-like growth factor-I on the development of sensory pathways: Analysis of the primary somatic sensory cortex (S1) of transgenic mice. *Endocrinology* 137:5484–5492.

Hadjiconstantinou, M., Fitkin, J. G., Dalia, A., and Neff, N. H. (1991). Epidermal growth factor enhances striatal dopaminergic parameters in the 1-methyl-4-phenyl-1,2,3,6-tetrahydropyridine-treated mouse. *J Neurochem* 57:479–482.

Hagg, T., Quon, D., Higaki, J., and Varon, S. (1992). Ciliary neurotrophic factor prevents neuronal degeneration and promotes low affinity NGF receptor expression in the adult rat CNS. *Neuron* 8:145–158.

Hagg, T., and Varon, S. (1993). Ciliary neurotrophic factor prevents neuronal degeneration of adult rat substantia nigra dopaminergic neurons *in vivo*. *Proc Natl Acad Sci USA* 90:6315–6319.

Hansson, H.-A., Dahlin, L. B., Danielson, N., Fryklund, L., Nachemson, A. K., Polleryd, P., Rozell, B., Skottner, A., Stemme, S., and Lundborg, G. (1986). Evidence indicating trophic importance of IGF-I in regenerating peripheral nerves. *Acta Physiol Scand* 126:609–614.

Hefti, F. (1986). Nerve growth factor (NGF) promotes survival of septal cholinergic neurons after fimbrial transection. *J Neurosci* 6:2155–2162.

Hefti, F. (1997). Pharmacology of neurotrophic factors. *Annu Rev Pharmacol Toxicol* 37: 239–267.

Helgren, M. E., Cliffer, K. D., Torrento, K., Cavnor, C., Curtis, R., DiStefano, P. S., Wiegand, S. J., and Lindsay, R. M. (1997). Neurotrophin-3 administration attenuates deficits of pyridoxine-induced large-fiber sensory neuropathy. *J Neurosci* 17:372–382.

Hempstead, B. L., Martin-Zanca, D., Kaplan, D. R., Parada, L. F., and Chao, M. V. (1991). High affinity NGF binding requires coexpression of the trk proto-oncogene and the low-affinity NGF receptor. *Nature* 350:678–682.

Hendersen, C. E., Phillips, H. S., Pollock, R. A., Davies, A. M., Lemeulle, C., Armanini, M., Simpson, L.C., Moffet, B., Vandlen, R. A., Koliatsos, V. E., and Rosenthal, A. (1994). GDNF: A potent survival factor for motoneurons present in peripheral nerve and muscle. *Science* 266:1062–1064.

Henderson, J. T., Seniuk, N. A., Richardson, P. M., Gauldie, J., and Roder, C. (1994). Systemic administration of ciliary neurotrophic factor induces cachexia in rodents. *J Clin Invest* 93:2632–2638.

Herrmann, J. L., Menter, D. G., Hamada, J., Marchetti, D., Nakajima, M., and Nicolson, G. L. (1993). Mediation of NGF-stimulated extracellular matrix invasion by the human melanoma low-affinity p75 neurotrophin receptor: Melanoma p75 functions independently of trkA. *Mol Biol Cell* 4:1205–1216.

Heuckeroth, R. O., Kotzbauer, P., Copeland, N. G., Gilbert, D. J., Jenkins, N. A., Zimonjic, D. B., Popescu, N. C., Johnson, E. M. J., and Milbrandt, J. (1997). Neurturin, a novel neurotrophic factor, is localized to mouse chromosome 17 and human chromosome 19p13.3. *Genomics* 44:137–140.

Higashiyama, S., Horikawa, M., Yamada, K., Ichino, N., Nakano, N., Nakagawa, T., Miyagawa, J., Matsushita, N., Nagatsu, T., Taniguchi, N., et al. (1997). A novel brain-derived member of the epidermal growth factor family that interacts with ErbB3 and ErbB4. *J Biochem* 122:675–680.

Hofer, M. M., and Barde, Y. A. (1988). Brain-derived neurotrophic factor prevents neuronal death *in vivo*. *Nature* 331:261–262.

Hohn, A., Leibrock, J., Bailey, K., and Barde, Y. (1990). Identification and characterization of a novel member of the nerve growth factor/brain derived neurotrophic factor family. *Nature* 344:339–341.

Holmes, W. E., Slowkowski, M. X., Akita, R. W., Henzel, W. J., Lee, J., et al. (1992). Identification of heregulin, a specific activator of p185erbB2. *Science* 256:1205–1208.

Hoshikawa, M., Ohbayashi, N., Yonamine, A., Konishi, M., Ozaki, K., Fukui, S., and Itoh, N. (1998). Structure and expression of a novel fibroblast growth factor, FGF-17, preferentially expressed in the embryonic brain. *Biochem Biophys Res Commun* 244:187–191.

Hu, L., Cote, S. L., and Cuello, A. C. (1997). Differential modulation of the cholinergic phenotype of the nucleus basalis magnocellularis neurons by applying NGF at the cell body or cortical terminal fields. *Exp Neurol* 143:162–171.

Hughes, R. A., Sendtner, M., and Thoenen, H. (1993). Members of several gene families influence survival of rat motoneurons *in vitro* and *in vivo*. *J Neurosci Res* 36:663–671.

Hyman, C., Juhasz, M., Jackson, C., et al. (1994). Overlapping and distinct actions of the neurotrophins BDNF, NT-3 and NT-4/5 on cultured dopaminergic and GABAergic neurons of the ventral mesencephalon. *J Neurosci* 360:735–755.

Ikeda, K., Klinkosz, B., Greene, T., Cedarbaum, J. M., Wong, V., Lindsay, R. M., and Mitsumoto, H. (1995). Effects of brain-derived neurotrophic factor on motor dysfunction in wobbler mouse motor neuron disease. *Ann Neurol* 37:505–511.

Ip, N. Y., Ibanez, C. F., Nye, S. H., McClain, J., Jones, P. F., Gies, D. R., Belluscio, L., LeBeau, M. M., Espinosa, R., Squinto, S. P., Persson, H., and Yancopoulos, G. D. (1992a). Mammalian neurotrophin-4: Structure, chromosonal localization, tissue distribution and receptor specificity. *Proc Natl Acad Sci USA* 89:3060–3064.

Ip, N. Y., Nye, S. H., Boulton, T. G., Davis, S., Taga, T., Li, Y., Birren, S. J., Yasukawa, K., Kishimoto, T., and Anderson, D. J. (1992b). CNTF and LIF act on neuronal cells via shared signalling pathways that involve IL-6 signal transducing receptor component gp130. *Cell* 69:1121–1132.

Ip, N. Y., Li, Y., Yancopoulos, G. D., and Lindsay, R. M. (1993). Cultured hippocampal neurons show response to BDNF, NT-3 and NT-4, but not NGF. *J Neurosci* 13:3394–3405.

Ishii, D. N. (1989). Relationship of insulin-like growth factor II gene expression in muscle to synaptogenesis. *Proc Natl Acad Sci U SA* 86:2898–2902.

Ishii, D. N. (1993). Neurobiology of insulin and insulin-like growth factors. In: Loughlin, S. E. and Fallon, J. H., eds., *Neurotrophic factors*. San Diego: Academic Press, pp. 415–442.

Jing, S., Yu, Y., Fang, M., Hu, Z., Holst, P. L., Boone, T., Delaney, J., Schultz, H., Zhou, R., and Fox, G. M. (1997). GFRalpha-2 and GFRalpha-3 are two new receptors for ligands of the GDNF family. *J Biol Chem* 272:33111–33117.

Jo, S. A., Zhu, X., Marchionni, M. A., and Burden, S. J. (1995). Neuregulins are concentrated at nerve-muscle synapses and activate ACh-receptor gene expression. *Nature* 373:158–161.

Johnson, D. G., Gorden, P., and Kopin, I. J. (1971). A sensitive radioimmunoassay for p75 nerve growth factor antigen in serum and tissues. *J Neurochem* 18:2355–2362.

Johnson, E. M., Jr., Gorin, P. D., Brandeis, L. D., and Pearson, J. (1980). Dorsal root ganglion neurons are destroyed by exposure in utero to maternal antibody to nerve growth factor. *Science* 210:916–918.

Jones, K. R., Farinas, I., Backus, C., Reichardt, L. F. (1994). Targeted disruption of the BDNF gene perturbs brain and sensory neuron development but not motor neuron development. *Cell* 76:989–999.

Kalcheim, C. (1989). Basic fibroblast growth factor stimulates survival of nonneuronal cells developing from trunk neural crest. *Dev Biol* 134:1–10.

Kang, H. J., and Schuman, E. M. (1995). Long lasting neurotrophin-induced enhancement of synaptic transmission in the adult hippocampus. *Science* 267:1658–1662.

Kanje, M., Skottner, A., Sjoberg, J., and Lundborg, G. (1989). Insulin-like growth factor I (IGF-I) stimulates regeneration of the rat sciatic nerve. *Brain Res* 486:396–398.

Kaplan, D. R., Hempstead, B. L., Martin-Zanca, D., Chao, M. V., and Parada, L. F. (1991). The trk proto-oncogene product: a signal transducing receptor for nerve growth factor. *Science* 252:554–560.

Kin, H. G., Wang, T., Olafsson, P., and Lu, B. (1994). Neurotrophin 3 potentiates neuronal activity and inhibits gamma-aminobutyratergic synaptic transmission in cortical neurons. *Proc Natl Acad Sci USA* 91:12341–12345.

Klein, R., Jing, S., Nanduri, V., O'Rourke, E., and Barabcid, M. (1991). The trk proto-oncogene encodes a receptor for nerve growth factor. *Cell* 65:189–197.

Klein, R., Smeyne, R. J., Wurst, W., Long, L. K., Auerbach, B. A., Joyner, A. L., and Barbacid, M. (1993). Targeted disruption of the trkB neurotrophin receptor gene results in nervous system lesions and neonatal death. *Cell* 75:113–122.

Klein, R., Silos-Santiago, I., Smeyne, R. J., Lira, S. A., Brambilla, R., Bryant, S., Zhang, L., Snider, W. D., and Barbacid, M. (1994). Disruption of the neurotrophin-3 receptor gene trkC eliminates Ia muscle afferents and results in abnormal movements. *Nature* 368:249–251.

Klein, R. D., Sherman, D., Ho, W. H., Stone, D., Bennett, G. L., Moffat, B., Vandlen, R., Simmons, L., Gu, Q., Hongo, J. A., Devaux, B., Poulsen, K., Armanini, M., Nozaki, C., Asai, N., Goddard, A., Phillips, H., Henderson, C. E., Takahashi, M., and Rosenthal, A. (1997). A GPI-linked protein that interacts with Ret to form a candidate neurturin receptor. *Nature* 387:717–721.

Klocker, N., Braunling, F., Isenmann, S., and Bahr, M. (1997). *In vivo* neurotrophic effects of GDNF on axotomized retinal ganglion cells. *NeuroReport* 8:3439–3442.

Knipper, M., Leung, L. S., Zhao, D., and Rylett, R. J. (1994). Short-term modulation of glutamatergic synapses in adult rat hippocampus by NGF. *NeuroReport* 5:2433–2436.

Knusel, B., Michel, P. P., Schwaber, J. S., and Hefti, F. (1990). Selective and nonselective stimulation of central cholinergic and dopaminergic development *in vitro* by nerve growth factor, basic fibroblast growth factor, epidermal growth factor, insulin and the insulin-like growth factors I and II. *J Neurosci* 10:558–570.

Kobayashi, N. R., Fan, D.-P., Giehl, K. M., Bedard, A. M., Wiegand, S. J., and Tetzlaff, W. (1997). BDNF and NT-4/5 prevent atrophy of rat subrospinal neurons after cervical axotomy, stimulate GAP-43 and Alpha-1-tubulin mRNA expression, and promote axonal regeneration. *J Neurosci* 17:9583–9595.

Koliatsos, V. E., Nauta, H. J., Clatterbuck, R. E., Holtzman, D. M., Mobley, W. C., and Price, D. L. (1990). Mouse nerve growth factor prevents degeneration of axotomized basal forebrain cholinergic neurons in the monkey. *J Neurosci* 10:3801–3813.

Koliatsos, V. E., Clatterbuck, R. E., Winslow, J. W., Cayouette, M. H., and Price, D. L. (1993). Evidence that brain-derived neurotrophic factor is a trophic factor for motor neurons *in vivo*. *Neuron* 10:359–367.

Koliatsos, V. E., Cayouette, M. H., Berkemeier, L. R., Clatterbuck, R. E., Price, D. L., and Rosenthal, A. (1994). Neurotrophin 4/5 is a trophic factor for mammalian facial motor neurons. *Proc Natl Acad Sci USA* 91:3304–3308.

Konishi, Y., Chui, D. H., Hirose, H., Kunishita, T., and Tabira, T. (1993). Trophic effect of erythropoietin and other hematopoietic factors on central cholinergic neurons *in vitro* and *in vivo*. *Brain Res* 609:29–35.

Kordower, J. H., Charles, V., Bayer, R., Bartus, R. T., Putney, S., Walus, L. R., and Friden, P. M. (1994). Intravenous administration of a transferrin receptor antibody-nerve growth factor conjugate prevents the degeneration of cholinergic striatal neurons in a model of Huntington's disease. *Proc Natl Acad Sci USA* 91:9077–9080.

Kordower, J. H., Chen, E. Y., Mufson, E. J., Winn, S. R., and Emerich, D. F. (1996). Intrastriatal implants of polymer encapsulated cells genetically modified to secrete human nerve growth factor: trophic effects upon cholinergic and noncholinergic striatal neurons. *Neuroscience* 72:63–77.

Korsching, S., Auburger, G., Heumann, R., Scott, J., and Thoenen, H. (1985). Levels of nerve growth factor and its mRNA in the central nervous system of the rat correlate with cholinergic innervation. *EMBO J* 4:1389–1393.

Korte, M., Carroll, P., Wolfe, E., Brem, G., Thoenen, H., et al. (1995). Hippocampal long-term potentiation is impaired in mice lacking brain-derived neurotrophic factor. *Proc Natl Acad Sci USA* 192:8856–8860.

Kotzbauer, P. T., Lampe, P. A., Heuckeroth, R. O., Golden, J. P., Creedon, D. J., Johnson, E. M. J., and Milbrandt, J. (1996). Neurturin, a relative of glial-cell-line-derived neurotrophic factor. *Nature* 384:467–470.

Kromer, L. F. (1987). Nerve growth factor treatment after brain injury prevents neuronal death. *Science* 235:214–216.

Kucera, J., Ernfors, P., Walro, J., and Jaenisch, R. (1995a). Reduction in the number of spinal motor neurons in neurotrophin-3 deficient mice. *Neurosci* 69:321–330.

Kucera, J., Fan, G., Jaenisch, R., Linnarsson, S., and Ernfors, P. (1995b). Dependence of developing group Ia afferents on neurotrophin-3. *J Comp Neurol* 363:307–320.

Kumon, Y., Sakaki, S., Watanabe, H., Nakano, K., Ohta, S., Matsuda, S., Yoshimura, H., and Sakanaka, M. (1996). Ciliary neurotrophic factor attenuates spatial cognition impairment, cortical infarction and thalamic degeneration in spontaneously hypertensive rats with focal cerebral ischemia. *Neurosci Lett* 206:141–144.

Lamballe, F., Klein, R., and Barabcid, M. (1991). TrkC, a new member of the trk family of tyrosine protein kinases, is a receptor for neurotrophin-3. *Cell* 66:967–979.

Lamballe, F., Smeyne, R. J., and Barbacid, M. (1994). Developmental expression of trkC, the neurotrophin-3 receptor, in the mammalian nervous system. *J Neurosci* 14:14–28.

LeBmann, V., Gottman, K., and Heumann, R. (1994). BDNF and NT-4/5 enhance glutamatergic synaptic transmission in cultured hippocampal neurones. *NeuroReport* 6:21–25.

Leibrock, J., Lottspeich, F., Hohn, A., Hofer, M., Hengerer, B., Masiokowski, P., Thoenen, H., and Barde, Y. A. (1989). Molecular cloning and expression of brain derived neurotrophic factor. *Nature* 341:149–152.

LeRoith, D., Roberts, C. T. J., Werner, H. and et al. (1993). Insulin-like growth factors in the brain. In: Loughlin, S., and Fallon, J., eds., *Neurotrophic factors*. San Diego: Academic Press, pp. 391–414.

Levi-Montalcini, R., Meyer, H., and Hamburger, V. (1954). *In vitro* experiments on the effects of mouse sarcoma 180 and 37 on the spinal and sympathetic ganglia of the chick embryo. *Cancer Res* 14:49–57.

Levi-Montalcini, R., and Angeletti, P. U. (1963). Essential role of the nerve growth factor on the survival and maintenance of dissociated sensory and sympathetic embryonic nerve cells *in vitro*. *Dev Biol* 7:653–659.

Levi-Montalcini, R. (1987). The nerve growth factor 35 years later. *Science* 237:1154–1162.

Levi-Montalcini, R., and Hamburger, V. (1951). Selective growth stimulating effects of mouse sarcoma on the sensory and sympathetic nervous system of the chick embryo. *J Exp Zool* 116:321–362.

Levi-Montalcini, R., and Hamburger, Y. (1953). A diffusible agent of mouse sarcoma, producing hyperplasia of sympathetic ganglia and hyperneurotization of viscera in the chick embryo. *J Exp Zool* 123:233–288.

Levison, S. W., Ducceschi, M. H., Young, G. M., and Wood, T. L. (1996). Acute exposure to CNTF in vivo induces multiple components of reactive gliosis. *Exp Neurol* 141:256–268.

Lewin, G. R., Ritter, A. M., and Mendell, L. M. (1992). On the role of nerve growth factor in the development of myelinated nociceptors. *J Neurosci* 12:1896–1905.

Lewin, G. R., and Mendell, L. M. (1993). Nerve growth factor and nociception. *Trends Neurosci* 16:353–359.

Lewin, G. R., and Barde, Y.-A. (1996). Physiology of the neurotrophins. *Annu Rev Neuroscience* 19:289–317.

Lillien, L. (1995). Changes in retinal cell fate induced by overexpression of EGF receptor. *Nature* 377:158–162.

Lin, L. F., Mismer, D., Lile, J. D., Armes, L. G., Butler, E. T., Vanice, J. L., and Collins, F. (1989). Purification, cloning and expression of ciliary neurotrophic factor (CNTF). *Science* 246:1023–1025.

Lin, L. F., Doherty, D. H., Lile, J. D., Bektesh, S., and Collins, F. (1993). GDNF: A glial cell-line derived neurotrophic factor for midbrain dopaminergic neurons. *Science* 260:1130–1132.

Lindsay, R. M., and Harmar, A. J. (1989). Nerve growth factor regulates expression of neuropeptide genes in adult sensory neurons. *Nature* 337:362–364.

Lindsay, R. M. (1996). Therapeutic potential of the neurotrophins and neurotrophin-CNTF combinations in peripheral neuropathies and motor neuron diseases. *Ciba Found Symp* 196:39–48; discuss.

Liu, X., Ernfors, P., Wu, H., and Jaenisch, R. (1995a). Sensory but not motor neuron deficits in mice lacking NT4 and BDNF. *Nature* 375:238–241.

Liu, X., Yao, D-L., and Webster, H. D. (1995b). Insulin-like growth factor I treatment reduces clinical deficits and lesion severity in acute demyelinating experimental autoimmune encephalomyelitis. *Multiple Sclerosis* 1:2–9.

Loeb, J. A., and Fishbach, G. D. (1997). Neurotrophic factors increase neuregulin expression in embyronic ventral spinal cord neurons. *J Neurosci* 17:1416–1624.

Lohof, A. M., Ip, N. Y., and Poo, M. (1993). Potentiation of developing neuromuscular synpases by the neurotrophins NT-3 and BDNF. *Nature* 363:350–353.

Mac Millan, V., Walton, R. K., and Davis, J. (1993). Acidic fibroblast growth factor infusion reduces ischemic CA1 hippocampal damage in the gerbil. *Can J Neurol Sci* 20:37–40.

Mahadeo, D., Kaplan, L., Chao, M. V., and Hempstead, B. L. (1994). High affinity nerve growth factor binding displays a faster rate of association than p140trk binding. Implications for multisubunit polypedtide receptors. *J Biol Chem* 269:6884–6891.

Maiese, K., Boniece, I., De, M. D., and Wagner, J. A. (1993). Peptide growth factors protect against ischemia in culture by preventing nitric oxide toxicity. *J Neurosci* 13:3034–3040.

Maina, F., Hilton, M. C., Ponzetto, C., Davies, A. M., and Klein, R. (1997). Met receptor signaling is required for sensory nerve development and HGF promotes axonal growth and survival of sensory neurons. *Genes and Development* 11:3341–3350.

Maisonpierre, P. C., Belluscio, L., Squinto, S., Ip, N. Y., Furth, M. E., Lindsay, R. M., and Yancopoulos, G. D. (1990). Neurotrophin-3: A neurotrophic factor related to NGF and BDNF. *Science* 247:1446–1451.

Majdan, M., Lachance, C., Gloster, A., Aloyz, R., Zeindler, C., Bamji, S., Bhakar, A., Belliveau, D., Fawcett, J., Miller, F. D., et al. (1977). Transgenic mice expressing the intracellular domain of the p75 neurotrophin receptor undergo neuronal apoptosis. *J Neurosci* 17:6988–6998.

Marchionni, M. A., Goodearl, A. D. J., Chen, M. S., Bermingham-McDonogh, O., Kirk, C., et al. (1993). Glial growth fators are alternatively spliced erbB2 ligands expressed in the nervous system. *Nature* 362:312–316.

Masu, Y., Wolf, E., Holtmann, B., Sendtner, M., Brem, G., and Thoenen, H. (1993). Disruption of the CNTF gene results in motor neuron degeneration. *Nature* 365:27–32.

Masure, S., Cik, M., Pangalos, M. N., Bonaventure, P., Verhasselt, P., Lesage, A. S., Leysen, J. E., and Gordon, R. D. (1998). Molecular cloning, expression and tissue distribution of glial-cell-line-derived neurotrophic factor family receptor alpha-3 (GFRalpha-3). *Eur J Biochem* 251:622–630.

Mattson, M. P., Murrain, M., Guthrie, P. B., and Kater, S. B. (1989). Fibroblast growth factor and glutamate: Opposing roles in the generation and degeneration of hippocampal neuroarchitecture. *J Neurosci* 9:3728–3740.

Mayer-Proschel, M., Kalyani, A. J., Mujtaba, T., and Rao, M. S. (1997). Isolation of lineage-restricted neuronal precursors from multipotent neuroepithelial stem cells. *Neuron* 19:773–785.

McAllister, A. K., Lo, D. C., and Katz, L. C. (1995). Neurotrophins regulate dendritic growth in developing visual cortex. *Neuron* 15:791–803.

McAllister, A. K., Katz, L. C., and Lo, D. C. (1996) Dendritic regulation of cortical dendritic growth requires activity. *Neuron* 17:1057–1064.

McMorris, F. A., Smith, T.M., DeSalvo, S., and Furlanetto, R. W. (1986). Insulin-like growth factor I/somatomedin C: A potent inducer of oligodendrocyte development. *Proc Natl Acad Sci USA* 83:822–826.

McMorris, F.A., and Dubois-Dalcq, M. (1988). Insulin-like growth factor I promotes cell proliferation and oligodendroglial commitment in rat glial progenitor cells developing *in vitro*. *J Neurosci Res* 21:199–209.

Mehler, M. F., Mabie, P. C., Zhang, D., and Kessler, J. A. (1997). Bone morphogenetic proteins in the nervous system. *TINS* 20:309–317.

Meiners, S., Marone, M., Rittenhouse, J. L., and Geller, H. M. (1993). Regulation of astrocytic tenascin by basic fibroblast growth factor. *Dev Biol* 160:480–493.

Meyer, M., Matsuoka, I., Wetmore, C., Olson, L., and Thoenen, H. (1992). Enhanced synthesis of brain-derived neurotrophic factor in the lesioned peripheral nerve: Different mechanisms are responsible for the regulation of BDNF and NGF mRNA. *J Cell Biol* 119:45–54.

Milbrandt, J., de Sauvage, F. J., Fahrner, T. J., Baloh, R. H., Leitner, M. L., Tansey, M. G., Lampe, P. A., Neuckeroth, R. O., Kotzbauer, P. T., Simburger, K. S., et al. (1998). Persephin, a novel neurotrophic factor related to GDNF and neurturin. *Neuron* 20:245–253.

Mill, J. F., Chao, M. V., and Ishii, D. N. (1985). Insulin, insulin-like growth factor II, and nerve growth factor effects on tubulin mRNA levels and neurite formation. *PNAS* 82:7126–7130.

Mitsumoto, H., Ikeda, K., Holmlund, T., et al. (1994a). The effects of ciliary neurotrophic factor (CNTF) on motor neuron dysfunction in Wobbler mouse motor neuron disease. *Ann Neurol* 36:142–148.

Mitsumoto, H., Ikeda, K., Klinkosz, B., Cedarbaum, J., Wong, V., and Lindsay, R. (1994b). Arrest of motor neuron disease in wobbler mice cotreated with CNTF and BDNF. *Science* 265:1107–1110.

Miyazawa, T., Matsumoto, K., Ohmichi, H., Katoh, H., Yamashima, T., and Nakamura, T. (1998). Protection of hippocampal neurons from ishemia-induced neuronal death by hepatocyte growth factor: A novel neurotrophic factor. *J Cerebral Blood Flow Metab* 18:345–348.

Molliver, D. C., Wright, D. E., Leitner, M. L., Parsadanian, A. S., Doster, K., Wen, D., Yan, Q., and Snider, W. D. (1997). IB4-binding DRG neurons switch from NGF to GDNF dependence in early postnatal life. *Neuron* 19:849–861.

Moore, M. W., Klein, R. D., Farinas, I., Sauer, H., Armanini, M., Phillips, H., Reichardt, L. F., Ryan, A. M., Carver-Moore, J., and Rosenthal, A. (1996). Renal and neuronal abnormalities in mice lacking GDNF. *Nature* 382:76–79.

Morrison, R. S., Keating, R. F., and Moskal, J. R. (1988). Basic fibroblast growth factor and epidermal growth factor exert differential trophic effects on CNS neurons. *J Neurosci Res.* 21:71–79.

Mufson, E. J., Bothwell, M., Hersh, L. B., and Kordower, J. H. (1989). Nerve growth factor receptor immunoreactive profiles in the normal, aged human basal forebrain: Colocalization with cholinergic neurons. *J Comp Neurol* 285:196–217.

Murphy, M., Drago, J., and Bartlett, P. F. (1990). Fibroblast growth factor stimulates the proliferation and differentiation of neural precursor cells *in vitro*. *J Neurosci Res* 25:463–475.

Nakagawara, A., Milbrandt, J., Muramatsu, T., Deuel, T. F., Zhao, H., et al. (1995). Differential expression of pleiotrophin and midkine in advanced neuroblastomas. *Cancer Res* 55:1792–1797.

Naveilhan, P., Baudet, C., Mikaels, A., Shen, L., Westphal, H., and Ernfors, P. (1998). Expression and regulation of GFRalpha3, a glial cell line-derived neurotrophic factor family receptor. *Proc Natl Acad Sci USA* 95:1295–1300.

Nawa, H., Carnahan, J., and Gall, C. (1995). BDNF protein measured by a novel enzyme immunoassay in normal brain and after seizure: Partial disagreement with mRNA levels. *Eur J Neurosci* 7:1527–1536.

Nebes, R. D., and Brady, C. B. (1989). Focused and divided attention in Alzheimer's disease. *Cortex* 25:305–315.

Nilsson, A. S., Fainziliber, M., Falck, P., and Ibanez, C. F. (1998). Neurotrophin-7: A novel member of the neurotrophin family from the zebrafish. *FEBS Lett* 424:285–290.

Oppenheim, R. W., Houenou, L. J., Johnson, J. E., Lin, L. F., Li, L., Lo, A. C., Newsome, A. L., Prevette, D. M., and Wang, S. (1995). Developing motor neurons rescued from programmed and axotomy-induced cell death by GDNF. *Nature* 373:344–346.

Otten, U. (1984). Nerve growth factor and the peptidergic sensory neurons. *Trends in Pharmacol Sci* 5:307–310.

Otto, D., Frotscher, M., and Unsicker, K. (1989). Basic fibroblast growth factor and nerve growth factor administered in gelfoam rescue medial septal neurons after fimbria fornix transection. *J Neurosci Res* 22:83–91.

Otto, D., and Unsicker, K. (1990). Basic FGF reverses chemical and morphological deficits in the nigrostriatal system of MPTP-treated mice. *J Neurosci* 10:1912–1921.

Ozaki, M., Sasner, M., Yano, R., Lu, H. S., and Buonanno, A. (1997). Neuregulin-beta induces expression of an NMDA-receptor subunit. *Nature* 390:691–694.

Patterson, P. H. (1994). Leukemia inhibitory factor, a cytokine at the interface between neurobiology and immunology. *Proc Natl Acad Sci USA* 91:7833–7835.

Patterson, S. L., Grover, L. M., Schwartzkroin, P. A., and Bothwell, M. (1992). Neurotrophin expression in rat hippocampal slices: A stimulus paradigm inducing LTP in CA1 evokes increases in BDNF and NT-3 mRNAs. *Neuron* 9:1081–1088.

Pearson, J., Johnson, E. M., and Brandeis, L. (1983). Effects of antibodies to nerve growth factor on intrauterine development of derivatives of cranial neural crest and placode in the guinea pig. *Dev Biol* 96:32–36.

Peng, H., Wen, T. C., Tanaka, J., Maeda, N., Matsuda, S., Desaki, J., Sudo, S., Zhang, B., and Sakanaka, M. (1998). Epidermal growth factor protects neuronal cells *in vivo* and *in vitro* against transient forebrain ischemia- and free radical-induced injuries. *J Cerebral Blood Flow Metab* 18:349–360.

Pennica, D., Arce, V., Swanson, T. A., Vejsada, R., Pollock, R. A., Armanini, M., Dudley, K., Phillips, H. S., Rosenthal, A., Kato, A. C., et al. (1996). Cardiotrophin-1, a cytokine present in embryonic muscle, supports long-term survival of spinal motoneurons. *Neuron* 17:63–74.

Peterson, D. A., Lucidi-Phillipi, C. A., Murphy, D. P., Ray, J., and Gage, F. H. (1996). Fibroblast growth factor-2 protects entorhinal layer II glutamatergic neurons from axotomy-induced death. *J Neurosci* 16:886–898.

Peterson, D. A., Leppert, J. T., Lee, K. F., and Gage, F. H. (1997). Basal forebrain neuronal loss in mice lacking neurotrophin receptor p75 [letter; comment]. *Science* 277:837–839.

Petrides, P. E., and Shooter, E. M. (1986). Rapid isolation of the 7S-nerve growth factor complex and its subunits from murine submaxillary glands and saliva. *J Neurochem* 46:721–725.

Petty, B. G., Cornblath, D. R., Adornato, B. T., Chaudhry, V., Flexner, C., Wachsman, M., Sinicropi, D., Burton, L. E., and Peroutka, S. J. (1994). The effect of systemically administered recombinant human nerve growth factor in healthy human subjects. *Ann Neurol* 36:244–246.

Pichel, J. G., Shen, L., Sheng, H. Z., Granholm, A. C., Drago, J., Grinberg, A., Lee, E. J., Huang, S. P., Saarma, M., Hoffer, B. J., et al. (1996). Defects in enteric innervation and kidney development in mice lacking GDNF. *Nature* 382:73–76.

Plata-Salaman, C. R. (1991). Epidermal growth factor and the nervous system. *Peptides* 12:653–663.

Purves, D. (1988). *Body and brain: A trophic theory of neural connections.* Cambridge, Mass.: Harvard University Press.

Rabizadeh, S., Oh, J., Zhong, L. T., Yang, J., Bitler, C. M., Butcher, L. L., and Bredesen, D. E. (1993). Induction of apoptosis by the low-affinity NGF receptor. *Science* 261:345–348.

Rao, P., Hsu, K. C., and Chao, M. V. (1995). Upregulation of NF-kappa B-dependent gene expression mediated by the p75 tumor necrosis factor receptor. *J Interferon Cytokine Res* 15:171–177.

Rapraeger, A. C., Krufka, A., and Olwin, B. B. (1991). Requirement of heparan sulfate for bFGF-mediated fibroblast growth and myoblast differentiation. *Science* 252:1705–1708.

Ray, J., and Gage, F. H. (1994). Spinal cord neuroblasts proliferate in response to basic fibroblast growth factor. *J Neurosci* 14:3548–3564.

Recio-Pinto, E., Rechler, M. M., and Ishii, D. N. (1986). Effects of insulin, insulin-like growth factor-II, and nerve growth factor on neurite formation and survival in cultured sympathetic and sensory neurons. *J Neurosci* 6:1211–1219.

Reynolds, B. A., Tetzlaff, W., and Weiss, S. (1992). A multipotent EGF-responsive striatal embryonic progenitor cell produces neurons and astrocytes. *J Neurosci* 12:4565–4574.

Richardson, P. M. (1994). Ciliary neurotrophic factor: A review. *Pharmac Ther* 63:187–198.

Riethmacher, D., Sonnenberg-Riethmacher, E., Brinkmann, V., Yamaai, T., Lewin, G. R., and Birchmeier, C. (1997). Severe neuropathies in mice with targeted mutations in the ErbB3 receptor. *Nature* 389:725–730.

Ritter, A. M., Lewin, G. R., Kremer, N. E., and Mendell, L. M. (1991). Requirement for nerve growth factor in the development of myelinated nociceptors *in vivo*. *Nature* 350:500–502.

Roberts, C. T., and LeRoith, D. (1988). Molecular aspects of the insulin-like growth factors, their binding proteins and receptors. *Balliere's Clin Endocrinol Metab* 2:1069–1085.

Rohrer, H., and Sommer, I. (1983). Simultaneous expression of neuronal and glial properties by chick ciliary ganglion cells during development. *J Neurosci* 3:1683–1699.

Rosen, K. M., Sandrock, A. W., Goodearl, A. D., Loeb, J. A., and Fishbach, G. D. (1996). The role of neuregulin (ARIA) at the neuromuscular junction. *Cold Spring Harbor Symposia on Quantitative Biology* 61:427–434.

Rosenbaum, C., Karyala, S., Marchionni, M. A., Kim, H. A., Krasnoselsky, A. L., Happel, B., Isaacs, I., Brackenbury, R., and Ratner, N. (1997). Schwann cells express NDF and SMDF/n-ARIA mRNAs, secrete neuregulin, and show constitutive activation of erbB3 receptors: evidence for a neuregulin autocrine loop. *Exp Neurol* 148:604–615.

Rosenberg, M. B., Friedmann, T., Robertson, R. C., Tuszynski, M., Wolff, J. A., Breakefield, X. O., and Gage, F. H. (1988). Grafting genetically modified cells to the damaged brain: restorative effects of NGF expression. *Science* 242:1575–1578.

Rosenthal, A., Goeddel, D. V., Nguyen, T., Lewis, M., Shih, A., Laramee, G. R. Nikolics, K., and Winslow, J. W. (1990). Primary structure and biological activity of a novel human neurotrophic factor. *Neuron* 4:767–773.

Ruit, K. G., Osborne, P. A., Schmidt, R. E., Yan, Q., Snider, W. D., et al. (1990). Nerve growth factor regulates sympathetic ganglion cell morphology and survival in the adult mouse. *J Neurosci* 10:2412–2419.

Sabatani, D. M., Lai, M. M., and Snyder, S. H. (1997). Neural roles of immunophilins and their ligands. *Molec Neurobiol* 15:223–239.

Saltiel, A. R., and Ohmichi, M. (1993). Pleiotropic signaling from receptor tyrosine kinases. *Curr Opin Neurobiol* 3:352–359.

Sandrock, A. W., Dryer, S. E., Rosen, K. M., Gozani, S. N., Kramer, R., Theill, L. E., and Fishbach, G. D. (1997). Maintenance of acetylcholine receptor number by neuregulins at the neuromuscular junction *in vivo*. *Science* 276:599–603.

Sanicola, M., Hession, C., Worley, D., Carmillo, P., Ehrenfels, C., Walus, L., Robinson, S., Jaworski, G., Wei, H., Tizard, R., Whitty, A., Pepinsky, R. B., and Cate, R. L. (1997). Glial cell line-derived neurotrophic factor-dependent RET activation can be mediated by two different cell-surface accessory proteins. *Proc Natl Acad Sci USA* 94:6238–6243.

Sasaki, K., Oomura, Y., Suzuki, K., Hanai, K., and Yagi, H. (1992). Acidic fibroblast growth factor prevents death of hippocampal CA1 pyramidal cells following ischemia. *Neurochem Int* 21:397–402.

Schecterson, L. C., and Bothwell, M. (1992). Novel roles for neurotrophins are suggested by BDNF and NT-3 mRNA expression in developing neurons. *Neuron* 9:449–463.

Schnell, L., Schneider, R., Kolbeck, R., Barde, Y. A., and Schwab, M. E. (1994). Neurotrophin-3 enhances sprouting of corticospinal tract during development and after adult spinal cord lesion. *Nature* 367:170–173.

Schwab, M. E., Otten, U., Agid, Y., and Thoenen, H. (1979). Nerve growth factor (NGF) in the rat CNS: Absence of specific retrograde axonal transport and tyrosine hydroxylase induction in locus coeruleus and substantia nigra. *Brain. Res* 168:473–483.

Schwartz, P. M., Borghesani, P. R., Levy, R. L., and Pomeroy, S. L. (1997). Abnormal cerebellar development and foliation in BDNF −/− mice reveals a role for neurotrophins in CNS patterning. *Neuron* 19:269–281.

Scott, J., Selby, M., Urdea, M., Quiroga, M., Bell, G. I., and Rutter, W. J. (1983). Isolation and nucleotide sequence of a cDNA encoding the precursor of mouse nerve growth factor. *Nature* 302:538–540.

Segal, R. A., and Greenberg, M. E. (1996). Intracellular signalling pathways activated by multiple neurotrophic factors. *Ann Rev Neurosci* 45:463–489.

Seidah, N. G., Benjannet, S., Pareek, S., Chretien, M., and Murphy, R. A. (1996a). Cellular processing of the neurotrophin precursors of NT-3 and BDNF by the mammalian proprotein convertases. *FEBS Lett* 379:247–250.

Seidah, N. G., Benjannet, S., Pareek, S., Savaria, D., Hamelin, J., Goulet, B., Laliberte, J., Lazure, C., Chretien, M., and Murphy, R. A. (1996b) Cellular processing of the nerve growth factor prescursor by the mammalian pro-protein convertases. *Biochem J* 314:951–960.

Seiler, M., and Schwab, M. E. (1984). Specific retrograde transport of nerve growth factor (NGF) from cortex to nucleus basalis in the rat. *Brain Res* 300:33–39.

Sendtner, M., Kreutzberg, G. W., and Thoenen, H. (1990). Ciliary neurotrophic factor prevents the degeneration of motor neurons after axotomy. *Nature* 345:440.

Sendtner, M., Holtmann, B., Kolbeck, R., Thoenen, H., and Barde, Y.-A. (1992a). Brain-derived neurotrophic factor prevents the death of motoneurons in newborn rats after nerve section. *Nature* 360:757–759.

Sendtner, M., Schmalbruch, H., Stockli, K. A., Carrol, P., Kreutzberg, G. W., and Thoenen, H. (1992b). Ciliary neurotrophic factor prevents degeneration of motor neurons in mouse mutant neuronopathy. *Nature* 358:502–504.

Sendtner, M., Carroll, P., Holtmann, B., Hughes, R. A., and Thoenen, H. (1994). Ciliary neurotrophic factor. *J Neurobiol* 25:1436–1453.

Shah, N. M., Groves, A. K., and Anderson, D. J. (1996). Alternative neural crest cell fates are instructively promoted by TGF beta superfamily members. *Cell* 85:331–343.

Sievers, J., Hausmann, B., Unsicker, K., and Berry, M. (1987). Fibroblast growth factors promote the survival of adult rat retinal ganglion cells after transection of the optic nerve. *Neurosci Lett* 76:157.

Silos-Santiago, I., Molliver, D. C., Ozaki, S., Smeyne, R. J., Fagan, A. M., Barbacid, M., and Snider, W. D. (1995). Non-TrkA-expressing small DRG neurons are lost in TrkA deficient mice. *J Neurosci* 15:5929–5942.

Smeyne, R. J., Klein, R., Schnapp, A., Long, L. K., Bryant, S., et al. (1994). Severe sensory and sympathetic neuropathies in mice carrying a disrupted Trk/NGF receptor gene. *Nature* 368:246–249.

Snyder, S. H., and Sabatini, D. M. (1995). Immunophilins and the nervous system. *Nature Med* 1:32–37.

Spranger, M., Lindholm, D., Bandtlow, C., Gnahn, H., Nahar-Noe, M., and Thoenen, H. (1990). Regulation of nerve growth factor (NGF) synthesis in the rat central nervous system: Comparison between the effects of interleukin-1 and various growth factors in astrocyte cultures and *in vivo*. *Eur J Neurosci* 2:69–76.

Squinto, S. P., Stitt, S. T., Aldrich, T. H., Davis, S., Bianco, S. M., Radziejewski, C., Glass, D. J., Masiakowski, P., Furth, M. E., Valenzuela, D. M., Distefano, D. S., and Yancopoulos, G. D. (1991). Trk B encodes a functional receptor for brain-derived neurotrophic factor and neurotrophin-3 but not nerve growth factor. *Cell* 65:885–893.

Steiner, J. P., Hamilton, G. S., Ross, D. T., Valentine, H. L., Guo, H., Connolly, M. A., Liang, S., Ramsey, C., Li, J. H., Huang, W., et al. (1997). Neurotrophic immunophilin ligands stimulate structural and functional recovery in neurodegenerative animal models. *Proc Natl Acad Sci USA* 94:2019–2024.

Stemple, D. L., Mahanthappa, N. K., and Anderson, D. J. (1988). Basic FGF induces neuronal differentiation, cell division, and NGF-dependence in chromaffin cells: A sequence of events in sympathetic development. *Neuron* 1:517–525.

Suhonen, J. O., Peterson, D. A., Ray, J., and Gage, F. H. (1996). Differentiation of adult hippocampus-derived progenitors into olfactory neurons *in vivo*. *Nature* 383:624–627.

Taupin, J. L., Pitard, V., Dechanet, J., Miossec, V., Gulade, N., and Moreau, J. F. (1998). Leukemia inhibitory factor: Part of a large ingathering family. *Int Rev Immunology* 16:397–426.

Terlau, H., and Seifert, W. (1990). Fibroblast growth factor enhances long-term potentiation in the hippocampal slice. *Eur J Neurosci* 2:973–977.

Tessarollo, L., Vogel, K. S., Palko, M. E., Reid, S. W., and Parada, L. F. (1994). Targeted mutation in the neurotrophin-3 gene results in loss of muscle sensory neurons. *Proc Natl Acad Sci USA* 91:11844–11848.

Thoenen, H., Korsching, S., Heumann, R., and Acheson, A. (1985). Nerve growth factor. In: *Growth factors in biology and medicine*, London (Ciba Foundation Symposium 116): Pitman, pp. 113–128.

Thoenen, H. (1995). Neurotrophins and neuronal plasticity. *Science* 270:593–598.

Thoenen, H., and Barde, Y. A. (1980). Physiology of nerve growth factor. *Physiol Rev* 60:1284–1335.

Thomas, K. A., Rios, C. M., and Fitzpatrick, S. (1984). Purification and characterization of acidic fibroblast growth factor from bovine brain. *Proc Natl Acad Sci USA* 81:357–361.

Toma, J. G., Pareek, S., Barker, P., Mathew, T. C., Murphy, R. A., Acheson, A., and Miller, F. D. (1992). Spatiotemporal increase in epidermal growth factor receptors following peripheral nerve injury. *J Neurosci* 12:2504–2514.

Trachtenberg, J. T., and Thompson, W. J. (1996). Schwann cell apoptosis at developing neuromuscular junction is regulated by glial growth factor. *Nature* 379:174–177.

Treanor, J. J., Goodman, L., de Sauvage, F., Stone, D. M., Poulsen, K. T., Beck, C. D., Gray, C., Armanini, M. P., Pollock, R. A., Hefti, F., et al. (1996). Characterization of a multicomponent receptor for GDNF. *Nature* 382:80–83.

Tuszynski, M. H., U, H.-S., and Gage, F. H. (1991). Recombinant human nerve growth factor infusions prevent cholinergic neuronal degeneration in the adult primate brain. *Ann Neurol* 30:625–636.

Tuszynski, M. H., Peterson, D. A., Ray, J., Baird, A., Nakahara, Y., and Gage, F. H. (1994). Fibroblasts genetically modified to produce nerve growth factor induce robust neuritic ingrowth after grafting to the spinal cord. *Exp Neurol* 126:1–14.

Tuszynski, M. H., Gabriel, K., Gage, F. H., Suhr, S., Meyer, S., and Rosetti, A. (1996a). Nerve growth factor delivery by gene transfer induces differential outgrowth of sensory, motor and noradrenergic neurites after adult spinal cord injury. *Exp Neurol* 137:157–173.

Tuszynski, M. H., Mafong, E., and Meyer, S. (1996b). BDNF and NT-4/5 prevent injury-induced motor neuron degeneration in the adult central nervous system. *Neurosci* 71:761–771.

Tuszynski, M. H., Roberts, J., Senut, M. C., U, H.-S., and Gage, F. H. (1996c). Gene therapy in the adult primate brain: Intraparenchymal grafts of cells genetically modified to produce nerve growth factor prevent cholinergic neuronal degeneration. *Gene Therapy* 3:305–314.

Unoki, K., Ohba, N., Arimura, H., Muramatsu, H., and Muramatsu, T. (1994). Rescue of photoreceptors from the damaging effects of constant light by midkine, a retinoic acid-responsive gene product. *Invest. Ophth Visual Sci* 35:4063–4068.

Unsicker, K., Grothe, C., Otto, D., and Westermann, R. (1991). Basic fibroblast growth factor in neurons and its putative functions. *Ann NY Acad Sci* 638:300–305.

Van der Zee, C. E., Ross, G. M., Riopelle, R. J., and Hagg, T. (1996). Survival of cholinergic forebrain neurons in developing p75NGFR-deficient mice [see comments]. *Science* 274:1729–1732.

Varon, S., Nomura, J., and Shooter, E. M. (1967a). The isolation of the mouse nerve growth factor protein in a high molecular weight form. *Biochem* 6:2202–2209.

Varon, S., Nomura, J., and Shooter, E. M. (1967b). Subunit structure of a high molecular weight form of the nerve growth factor from mouse submaxillary gland. *Proc Natl Acad Sci USA* 57:1782–1789.

Verdi, J. M., Groves, A. K., Farinas, I., Jones, K., Marchionni, M. A., Reichardt, L. F., and Anderson, D. J. (1996). A reciprocal cell-cell interaction mediated by NT-3 and neuregulins controls the early survival and development of sympathetic neuroblasts. *Neuron* 16:515–527.

Vlodavsky, I., Bashkin, P., Ishai-Michaeli, R., Chajek, T., Bar-Shavit, R., Haimovitz-Friedman, A., Klagsburn, M., and Fuks, Z. (1991). Sequestration and release of basic fibroblast growth factor. *Ann NY Acad Sci* 638:207–220.

Wang, Y., Lin, S. Z., Chiou, A. L., Williams, L. R., and Hoffer, B. J. (1997). Glial cell line derived neurotrophic factor protects against ischemia-induced injury in the cerebral cortex. *J Neurosci* 17:4341–4348.

Weiss, S., Dunne, C., Hewson, J., Wohl, C., Wheatley, M., Peterson, A. C., and Reynolds, B. A. (1996). Multipotent CNS stem cells are present in the adult mammalian spinal cord and ventricular neuraxis. *J Neurosci* 16:7599–7609.

Widenfalk, J., Nosrat, C., Tomac, A., Westphal, H., Hoffer, B., and Olson, L. (1997). Neurturin and glial cell line-derived neurotrophic factor receptor-beta (GDNFR-beta), novel proteins related to GDNF and GDNFR-alpha with specific cellular patterns of expression suggesting roles in the developing and adult nervous system and in peripheral organs. *J Neurosci* 17:8506–8519.

Widmer, H. R., Knusel, B., and Hefti, F. (1993). BDNF protection of basal forebrain cholinergic neurons after axotomy: Complete protection of p75 NGFR-positive cells. *NeuroReport* 4:363–366.

Williams, L. R. (1991). Hypophagia is induced by intracerebroventricular administration of nerve growth factor. *Exp Neurol* 113:31–37.

Williams, L. R., Inouye, G., Cummins, V., Pelleymounter, M. A. (1996). Glial cell line-derived neurotrophic factor sustains axotomized basal forebrain cholinergic neurons *in vivo*: Dose-response comparison to nerve growth factor and brain-derived neurotrophic factor. *J Pharmacol Exp Ther* 277:1140–1151.

Winkler, C., Sauer, H., Lee, C. S., and Bjorklund, A. (1996a). Short-term GDNF treatment provides long-term rescue of lesioned nigral dopaminergic neurons in a rat model of Parkinson's disease. *J Neurosci* 16:7206–7215.

Winkler, J., Ramirez, G. A., Kuhn, H. G., Peterson, D. A., Day-Lollini, P. A., Stewart, G. R., Tuszynski, M. H., Gage, F. H., and Thal, L. J. (1996b). Reversible induction of Schwann cell hyperplasia and sprouting of sensory and sympathetic neurites *in vivo* after continuous intracerebroventricular administration of nerve growth factor. *Ann Neurol* 40:128–139.

Winslow, J. W., Moran, P., Valverde, J., Shih, A., Yuan, J. Q, et.al. (1995). Coling of AL-1, a ligand for an eph-related tyrosine kinase receptor involved in axon bundle formation. *Neuron* 14:973–981.

Wolff, J. A., Yee, J. K., Skelly, H. F., Morres, J. C., Respess, J. G., Friedmann, T., and Leffert, H. (1987). Expression of retrovirally transduced genes in primary cultures of adult rat heptocytes. *PNAS* 84:3344–3348.

Wong, V., Glass, D. J., Arriaga, R., Yancopoulos, G. D., Lindsay, R. M., and Conn, G. (1997). Hepatocyte growth factor promotes motor neuron survival and synergizes with ciliary neurotrophic factor. *J Biol Chem* 272:5187–5191.

Worby, C. A., Vega, Q. C., Zhao, Y., Chao, H. H. J., Seasholtz, A. F., and Dixon, J. E. (1996). Glial cell-line derived neurotrophic factor signals through the RET receptor and activates mitogen-activated protein kinase. *J Biol Chem* 271:23619–23622.

Worby, C. A., Vega, Q. C., Chao, H. H., Seasholtz, A. F., Thompson, R. C., and Dixon, J. E. (1998). Identification and characterization of GFRalpha-3, a novel Co-receptor belonging to the glial cell line-derived neurotrophic receptor family. *J Biol Chem* 273:3502–3508.

Xu, Q., Alldus, G., MacDonald, R., Wilkinson, D. G., and Holder, N. (1996). Function of the Eph-related kinase rtk1 in patterning of the zebrafish forebrain. *Nature* 381:319.

Xu, X. M., Guenard, V., Kleitman, N., Aebischer, P., and Bunge, M. B. (1995). A combination of BDNF and NT3 promotes supraspinal axonal regeneration in Schwann cell grafts in adult rat thoracic spinal cord. *Exp Neurol* 134:261–272.

Yan, Q., Elliot, J., and Snider, W. D. (1992). Brain-derived neurotrophic factor rescues spinal motor neurons from axotomy-induced cell death. *Nature* 24:753–755.

Yan, Q., Matheson, C., Lopez, O., and Miller, J. A. (1994). The biological responses of axotomized motoneurons to brain-derived neurotrophic factor. *J Neurosci* 14:5281–5291.

Yan, Q., Matheson, C., and Lopez, O. T. (1995). *In vivo* neurotrophic effects of GDNF on neonatal and adult facial motor neurons. *Nature* 373:341–344.

Yao, D. L., Liu, X., Hudson, L. D., and Webster, H. D. (1995). Insulin-like growth factor I treatment reduces demyelination and upregulates gene expression of myelin-related proteins in experimental autoimmune encephalomyelitis. *Proc Natl Acad Sci USA* 92:6190–6194.

Yayon, A., Klagsburn, M., Esko, J. D., Leder, P., and Ornitz, D. M. (1991). Cell surface, heparin-like molecules are required for binding of basic fibroblast growth factor to its high affinity receptor. *Cell* 64:841–848.

Yee, J. K., Jolly, D. J., Miller, A. D., Willis, R. C., Wolff, J. A., and Friedmann, T. (1987). Epitope insertion into the human hypoxanthine phosphoribosyltransferase protein and detection of the mutant protein by an anti-peptide antibody. *Gene* 53:97–104.

Yeo, T. T., Chua-Couzens, J., Butcher, L. L., Bredesen, D. E., Cooper, J. D., Valletta, J. S., Mobley, W. C., and Longo, F. M. (1997). Absence of p75NTR causes increased basal forebrain cholinergic neuron size, choline acetyltransferase activity, and target innervation. *J Neurosci* 17:7594–7605.

Yin, Q. W., Johnson, J., Prevette, D., and Oppenheim, R. W. (1994). Cell death of spinal motoneurons in the chick embryo following deafferentation: Rescue effects of tissue extracts, soluble proteins, and neurotrophic agents. *J Neurosci* 14:7629–7640.

Yoshida, K., and Gage, F. H. (1992). Cooperative regulation of nerve growth factor synthesis and secretion in fibroblasts and astrocytes by fibroblast growth factor and other cytokines. *Brain Res* 569:14–25.

Yuen, E. C., and Mobley, W. C. (1996). Therapeutic potential of neurotrophic factors for neurological disorders. *Ann Neurol* 40:346–354.

Zackenfels, K., Oppenheim, R. W., and Rohrer, H. (1995). Evidence for an important tole of IGF-I and IGF-II for the early development of chick sympathetic neurons. *Neuron* 14:731–741.

Zheng, J. L., Helbig, C., and Gao, W. Q. (1997). Induction of cell proliferation by fibroblast and insulin-like growth factors in pure rat inner ear epithelial cell cultures. *J Neurosci* 17:216–226.

Zhou, H., Muramatsu, T., Halfter, W., Tsim, K. W., and Peng, H. B. (1997). A role of midkine in the development of the neuromuscular junction. *Molec Cell Neurosc* 10:56–70.

CHAPTER 6

Fetal Neuronal Grafting for CNS Regeneration

General Principles

JEFFREY H. KORDOWER* AND MARK H. TUSZYNSKI†,**

*Research Center for Brain Repair and Department of Neurological Sciences, Rush Presbyterian St. Luke's Medical Center, Chicago, Illinois, †Department of Neurosciences, University of California, San Diego, La Jolla, California, and **Department of Neurology, Veterans Affairs Medical Center, San Diego, California

Studies using transplantation of fetal cells were among the first to demonstrate regeneration in the injured or degenerating adult central nervous system. Beginning with studies from the early part of the twentieth century, transplantation of fetal nervous system tissue to the brain and spinal cord has been a useful tool for elucidating both the regenerative potential of the central nervous system and normal mechanisms of nervous system development and function. Various studies in models of neuronal and axonal injury have demonstrated that fetal neuronal grafts can, to varying degrees, ameliorate anatomical, neurochemical, electrophysiological, and functional abnormalities in the CNS. Practical limitations in fetal cell availability and societal acceptance, as well as the recent development of alternative cell replacement strategies such as neural stem cells, may limit the extent to which fetal-grafting strategies will eventually be used to treat human disease. Nonetheless, trials of fetal grafting are currently underway in patients with Parkinson's disease, Huntington's disease, and spinal cord injury. Recent reports regarding xenotransplantation (cross-species grafting) suggest a potentially interesting alternative to the use of human fetal tissue for neural repair strategies.

I. INTRODUCTION

The first studies to examine transplantation paradigms in the central nervous system were conducted nearly 100 years ago. Work in the ensuing decades explored the potential of a variety of cell types placed in the CNS to influence development and regeneration (Thompson, 1890; Forssman, 1898, 1900; Del Conte, 1907; Le Gros Clark, 1940; Lugaro, 1906; Sugar and Gerard, 1940; Tello, 1911a, 1911b; Tidd, 1932; Ransom, 1909). Collectively, this work laid the foundation for the modern era of fetal nervous system transplantation that began in the 1960s and continues to the present time. Fetal nervous system transplantation studies have significantly contributed to our understanding of CNS function in three ways: (1) they have provided insight into features of nervous system development; (2) they have defined the nature of *normal* nervous system function in the neonatal and adult animal, and (3) they have generated substantial information regarding the extent to which the *adult* nervous system remains responsive to growth signals, leading to promising strategies for generating neural repair. The present chapter will focus on fetal grafting as a means of promoting CNS repair (see Björklund and Stenevi, 1985; Fisher and Gage, 1993; Das, 1990; Bakay and Sladek, 1993 for more comprehensive reviews).

Fetal transplants can promote CNS regeneration in several ways. First, transplants can replace lost or degenerating host neurons, a neuronal *replacement* strategy. In this case, the intent of the grafted neurons is to integrate into host neural circuitry and reinnervate brain regions that have lost normal innervation patterns following experimental lesions or genetic mutations. In replacement strategies, grafted neurons secrete a desired neurotransmitter and ameliorate lost functions by directly assuming the role of host neurons. This mechanism contrasts with that of grafting strategies that use non-fetal donor sources, such as adrenal medullary transplants or genetically modified cells, which do not rely on specific *patterned* reinnervation and *regulatable* neurotransmitter release, but rather function by releasing transmitters from a cellular point source in a paracrine fashion. Second, fetal grafts can act as *bridges* for *host* axonal regeneration. The fetal environment is highly conducive to axonal growth, and fetal transplant bridging strategies can promote reconstruction of connections between host neural circuits. Last, fetal grafts can *prevent the degeneration* of host systems.

These strategies will be reviewed in the following sections. It should be noted that, when properly performed, grafts of fetal tissue can survive and innervate virtually all regions of the neuraxis, and a vast literature has emerged employing numerous animal models (e.g., Björklund and Stenevi, 1985). For brevity, this chapter will review three types of animal models as exemplars for the potential uses of fetal-grafting strategies. First, grafting strategies to *replace specific neural circuits* (Fig. 1A, see color insert) will be illustrated in striatal injury (models of Parkinson's disease and Huntington's disease) and hippocampal injury (a model for memory loss). Second, the use of fetal cell transplant as *bridges* to reconnect *host* circuitry (Fig. 1B, see color insert) will be illustrated in a septohippocampal model of memory loss. Third, fetal grafting to replace *endocrine cells* by forming graft connections to blood vessels (*neurohemal* mechanisms; Fig. 1C, see color insert) will be illustrated in models of hypothalamic dysfunction.

This chapter will begin by examining practical aspects in the use of fetal nervous system grafting to enhance CNS repair. These aspects include the choice and condition of donor tissue; the choice of site for graft placement in the host; conditions in the host that may influence graft survival; graft vascularization; and immunological considerations. Most of the studies described in this section utilized within-species grafts (allografts). More recent developments using cross-species grafting (xenotransplantation) of fetal tissue will also be briefly described (see also Chapter 13).

II. PRACTICAL CONSIDERATIONS IN FETAL GRAFTING FOR CENTRAL NERVOUS SYSTEM REPAIR

Findings from the body of fetal transplantation work suggest that several factors must be taken into account in the implementation of fetal transplantation for promoting neural repair.

A. Donor Tissue

The age at which donor tissue is harvested is critical. Neurons in an active stage of neurogenesis appear to comprise the optimal stage for grafting (Stenevi *et al.*, 1976; Olson *et al.*, 1983; Das *et al.*, 1980) and it is clear that donor cells obtained outside of a regionally specific gestational window will not survive transplantation. Because different nuclei across the neuraxis exhibit distinct time courses of development *in vivo*, the optimal age for obtaining host tissue will differ, depending on which CNS region is being harvested. For example, spinal cord tissue is optimally harvested in the rat from embryonic day 11 to 15 (ED11–15) (Altman and Bayer, 1984), whereas hippocampal tissue is optimally harvested from E13–E18 (Altman and Das, 1965; Bayer, 1980; Lauder and Bloom, 1974). For human grafting studies, fetal dopamine cells are best obtained between 6.5 and 9 weeks postconception. In general, fetal tissue should be harvested at a developmental stage in which some neuronal differentiation has occurred, but prior to terminal differentiation and the extension of axons from the fetal donor nucleus to its targets.

A second consideration in the procurement and processing of donor tissue is the handling of cells. Grafts are removed from the host as solid blocks of tissue and can then be placed immediately into host tissue, or can be gently processed into a cell suspension using mechanical dissociation and trypsinization. The initial dogma was that suspension grafts were superior to solid grafts. However, graft viability similar to that of cell suspension grafts has been achieved with solid grafts of human fetal dopaminergic tissue, and this preparation procedure provides the added benefit of extending the window of gestation from which successful grafts can be obtained. For bridging strategies, block grafts are preferred because they are more structurally integrated into the host when placed in cavities and will not diffuse away from the graft site.

Another practical aspect of fetal grafting for clinical use is tissue storage. This is an important parameter to allow time for the collection of multiple fetuses and for the performance of critical serological tests to ensure the safety of the procedure. Some studies have cryopreserved fetal tissue prior to transplantation (Collier *et al.*, 1987). Although this procedure has had some success in animal models, there is a significant drop-off in graft viability. Furthermore, fetal grafts consisting of cryopreserved dopamine cells have never been demonstrated to survive in patients with Parkinson's disease as assessed via postmortem examination or determined *in vivo* with PET scans. An alternative to cryopreservation is cool storage (Nikkah *et al.*, 1995). Fetal dopaminergic grafts can be stored for up to 4 days in culture in a cool "hibernation medium." Graft survival is often as good as fresh grafts after 4 days of storage.

B. Host Considerations

Several studies have addressed the importance of host age on the survival of grafted fetal donor neurons. In general, fetal transplants survive and integrate well in brain parenchyma regardless of host age; however, the extension of neuritic processes from grafts appears to be superior in younger as opposed to aged host tissue. Reasons for this may include an inherently greater degree of neuroplasticity in younger as opposed to aged tissue, modulated by the presence of neurotrophic factors or other substances (Gage *et al.*, 1983; Crutcher, 1990). Graft survival is only adversely affected when fetal tissue is grafted into extremely aged subjects.

The timing of transplantation surgery after lesion placement is also critical for optimizing graft survival and function. Nieto-Sampedro and associates

(1983) reported several years ago that brain lesions induce a transient, time-dependent increase in neurotrophic factors that aid in graft viability. Other lines of evidence support this concept. Studies of peripheral nerve injury document an upregulation of growth factor expression by glial cells that occurs within hours and persists for days after injury. Thus, lesions may "prime" host tissue for grafting and regeneration by upregulating expression of neurotrophic factors and other substances. For this reason, a number of paradigms graft fetal tissue 7 to 10 days after placing a CNS lesion, a time point at which the level of neurotrophic factors created by the lesion appears to be highest.

Another consideration in neuronal transplantation paradigms is the optimal choice of the host target site to receive the fetal graft. The adult CNS environment is generally nonconducive to new axonal growth. Thus neuronal replacement strategies that attempt to achieve integration of graft circuitry with the host require fetal cell placement in the host brain region that contains the target of lesioned or degenerated inputs. For example, nigral allografts show optimal host connectivity when placed directly into the striatum rather than into the host substantia nigra. Additionally, particular loci within a single region may need to be specifically targeted. For example, in Parkinson's disease the postcommissural putamen rather than the anterior putamen or caudate nucleus is the principal target for grafting, because the former region displays the greatest dopaminergic denervation and is most intimately connected with motor circuitry. Furthermore, the positioning of a graft in a particular locus may be critical. In rats receiving neurohypophysectomies, anterior hypothalamic transplants impair water balance and fluid osmolarity when they are located in the dorsal aspect of the third ventricle, but restore fluid homeostasis when placed in the ventral aspect of the third ventricle in such a position as to contact the median eminance (Marciano and Gash, 1986).

C. Graft Vascularization

Several studies have reported that fetal graft survival is optimal if grafts become well vascularized (Dunn, 1917; Glees, 1940; Stenevi *et al.*, 1976; Krum and Rosenstein, 1988; Kromer *et al.*, 1979; Finger and Dunnett, 1989). In general, the extent to which grafts become vascularized correlates with graft survival and size. In addition, some strategies have attempted to enhance graft vascularization using the drug nimodipine (Finger and Dunnett, 1989; Zhou *et al.*, 1991). The type of vasculature that invests grafted tissue is dependent upon the source of the donor tissue and the site of grafting into the host. Normally, the blood supply of the CNS is nonfenestrated and contains tight junctions, except in the circumventricular organs that lack a blood–brain barrier. If fetal tissue is placed within brain parenchyma, nonfenestrated vasculature forms in the graft similar to that which is observed within the normal CNS. In contrast, if fetal tissue is placed into regions of the host brain that are devoid of a normal blood–brain barrier (e.g., within the third ventricle, dorsal to the median eminence), vessels that grow into the transplant from the *ventral* surface of the brain are fenestrated and the grafted tissue has access to factors from the peripheral circulation. In contrast, vessels growing into a third ventricular graft along its *lateral* aspects are of the parenchymal CNS type and therefore contain tight junctions. Neural crest-derived tissues placed into the brain form fenestrated capillary plexuses regardless of location of graft placement.

D. Immunological Considerations

The central nervous system has classically been considered an "immunologically privileged" site. This tenet is based both upon empirical observations and studies of central immune activation. Further, the CNS lacks a lymphatic drainage system. Although graft rejection inside the central nervous system occurs after a longer latency compared to outside the CNS, antigen-presenting cells exist in the adult CNS, major histocompatability (MHC) class antigens are expressed, and immunologically mediated graft rejection occurs (see Streilein, 1988; Widner and Brundin, 1988; Sachs and Bach, 1990 for reviews). Marked differences in the immune systems of various species exist, leading to some confusion in comparing results of grafting studies in rodents, nonhuman primates, and humans. Immune responses in primate systems can result in earlier and more marked graft rejection than in rodent systems.

Another factor that may influence fetal graft rejection in the adult brain is the time course of expression of fetal cell antigens. As fetal cells survive over time in the host brain, they presumably mature and begin to express new antigens. If the blood–brain barrier re-forms within several weeks after grafting, then newly expressed antigens may be partially sheltered from the host systemic immune system. As noted earlier, this sheltering will not prevent, but may delay, the development of an immune response. On the other hand, a premature cessation of immunosuppression may leave the host immune system fully competent to respond to fetal cell antigens expressed later in the course of neural differentiation in the host.

Thus, the need for immunosuppression of fetal grafts in the human brain is currently unresolved. In some clinical protocols of fetal grafting, human patients receive immunosuppression with cyclosporin A for a period of 3 to 6 months and drugs are then discontinued. Freed and coworkers (1980) report superior functional recovery in patients who do not receive cyclosporin at all. In contrast, two human patients in whom immunosuppression was discontinued 6 months after fetal striatal grafting showed surviving healthy fetal neurons that extended processes into the adult striatum for up to 18 months after the original surgery (Kordower *et al.*, 1995, 1996). In these human transplant cases, T-cells and B-cells were observed within fetal nigral grafts in spite of robust graft survival, excellent graft integration, and progressive clinical improvement (Kordower *et al.*, 1997). Additional studies are needed in primate systems to more clearly evaluate the role of immunological factors in fetal transplantation.

A surprising set of findings highlights the extent to which CNS immunology remains enigmatic. Outside the nervous system, xenografts rarely survive for prolonged periods even when immunosuppressive agents are used (Steele and Auchincloss, 1995; Sachs and Bach, 1990). (Galpern and colleagues 1996) transplanted fetal porcine substantia nigra into the brains of immunosuppressed rats with nigrostriatal lesions. Despite the fact that these were xenografts, implanted neurons survived in the host striatum. Based upon these porcine-to-rodent transplantation studies, fetal pig substantia nigra was recently transplanted to immunosuppressed patients with Parkinson's disease. One patient in this series later died of causes unrelated to Parkinson's disease; examination of the brain demonstrated survival of a few grafted dopaminergic neurons with limited innervation of the *human* striatum (Deacon *et al.*, 1997).

E. Other Factors That Influence Neuronal Grafting to the Nervous System

Donor graft survival is an important consideration in fetal-grafting studies. One means of potentially augmenting donor graft survival is the concurrent administration of neurotrophic factors into host brain regions that receive fetal grafts. Different trophic factors influence specific populations of CNS neurons. Clearly, NGF can support the viability and phenotypic expression of cholinergic basal forebrain neurons. Glial-derived neurotrophic factor (GDNF) is a potent neurotrophic factor for fetal dopaminergic neurons. Both of these trophic factors have been demonstrated to enhance the survival of their respective cell type following transplantation. Other growth factors support the survival of neurons from various brain regions. In addition, epidermal growth factor (EGF) and fibroblast growth factor-2 (FGF-2) promote the growth of neural stem cells in the CNS (Reynolds *et al.*, 1992; Ray and Gage, 1994), a finding of some interest with regard to applications in fetal grafting. Thus, supplementing fetal grafts with neurotrophic factors may augment their survival and efficacy. Experiments aiming to augment fetal graft survival using agents that reduce cell death, including calcium channel antagonists, free radical scavengers, and anti-apoptotic genes, are also the subject of current study.

Thus, the extensive body of work on fetal transplantation indicates that several factors influence the survival and efficacy of fetal neuronal grafts in the nervous system. The potential value of fetal transplants in enhancing CNS repair will now be illustrated using selected *in vivo* models as exemplars that target either (1) *neuronal replacement,* (2) *bridging* to reconnect host projections, or (3) *endocrine replacement* as *a priori* experimental strategies.

III. FETAL GRAFTS FOR NEURONAL REPLACEMENT

A. Parkinson's Disease (PD) and Huntington's Disease (HD) Models

In PD, progressive neuronal degeneration occurs in melatonin-containing neurons of the pars compacta of the substantia nigra, causing a loss of dopaminergic inputs to the striatum (caudate and putamen). In most cases this cell loss is idiopathic, although several human cases have resulted from the illicit ingestion of the neurotoxin n-methy-4 phenyl 1,2,3,6 tetrahydropyridine (MPTP). Patients with Parkinson's disease develop a paucity of movement (bradykinesia), rigidity, and a characteristic resting tremor (Yahr, 1984). The changes are progressive over several years (see Chapters 11, 12, and 14). To model PD in animals, injections of dopaminergic neurotoxins such as MPTP or 6-hydroxydopamine (6-OHDA) are made to cause selective degeneration of substantia nigra neurons (see Chapter 11).

In HD, the primary pathological abnormality occurs within intrinsic neurons of the striatum. GABAergic medium spiny interneurons of the caudate-putamen are selectively vulnerable and undergo progressive degeneration due to a mutation in the *Huntingtin* gene (Li *et al.*, 1995; Norremolle *et al.*, 1993). Other cell types such as cholinergic interneurons degenerate later in the disease process. These losses lead to a progressive movement disorder with severe cognitive disturbances (see Chapters 17 and 18). To model HD, intrastriatal injections of excitotoxins such as quinolinic acid or the administration of mitochondrial toxins such as 3-nitroproprionic acid are used, resulting in degeneration of GABAergic medium spiny neurons (Davies and Roberts, 1988) (see Chapter 18).

Thus, abnormalities of striatal function occur in both PD and HD, but in PD the abnormality results from a loss of *extrinsic* inputs to the striatum whereas in HD the abnormality results from a loss of *intrinsic* neurons. The usefulness of the animal models results from their reproduction of a broad spectrum of morphological, neurochemical, and functional abnormalities that occur in the human diseases. Studies of fetal nervous system grafting in striatal models have elucidated fundamental features of CNS plasticity, and have also helped to resolve fundamental questions related to fetal grafting procedures such as: (1) Should fetal neurons be placed into the region where host neurons are lost or into the targets to which these neurons normally project to optimally restore function? (2) Do multiple grafts of dissociated fetal cell *suspensions* into the degenerating striatum result in superior survival and functional restoration compared to fewer grafts of larger *blocks* of fetal tissue? (3) What degree of graft *integration* and *connectivity* with the host is required to support functional recovery?

Animal models of PD were most appropriate for addressing optimal locations of graft placement for restoring function. Following lesions of the nigrostriatal projection, fetal nigral allografts were placed either into the substantia nigra the source of striatal inputs, or into the striatum, the target of nigral projections (Perlow *et al.*, 1979; Björklund and Stenevi, 1979; Björklund *et al.*, 1980a; Dunnett *et al.*, 1988). Most fetal grafts survived, but grafts placed into the striatum generally elicited superior functional recovery compared to grafts placed in the nigra on tasks that assessed spontaneous and drug-induced motor behaviors. Grafts into the striatum resulted in superior degrees of reciprocal connections between graft and host as evidenced by tracing experiments and immunocytochemical labeling. These initial studies also demonstrated that superior functional recovery occurred in animals receiving cell *suspension* grafts into multiple striatal locations rather than solid block grafts into fewer sites. However, neither graft type elicited complete functional recovery. Although rodent nigral allografts appeared to survive best as suspensions, it is unclear whether similar grafts of *human* nigral anlagen are superior to solid grafts when placed into the striatum of nigrostriatal lesioned rats. Under these xenograft conditions, comparable survival and neurite outgrowth was seen when comparing suspensions and solid implants (Freeman *et al.*, 1995). In fact, the gestational window for achieving successful transplants was broadened when solid implants were employed. Unfortunately, functional studies were not performed in this experiment to confirm that similar structural findings translated into equivalent functional recovery.

Another important feature of fetal grafting is the specificity of functional recovery relative to graft location. Grafts placed into the dorsal striatum primarily ameliorated rotational abnormalities (Perlow *et al.*, 1979; Björklund *et al.*, 1980a, 1980b; Björklund and Stenevi, 1979; Dunnett *et al.*, 1983, 1988; Freed *et al.*, 1980), whereas grafts placed in the ventrolateral regions of the striatum exhibited greater effects on sensorimotor abnormalities and bradykinesia (Dunnett *et al.*, 1983, 1981; Mandel *et al.*, 1990). *In vivo* microdialysis of heterotypic grafts showed release of dopamine in the striatum (Strecker *et al.*, 1987; Zetterström *et al.*, 1986). Of note, some *control* animals in these experiments also exhibited functional recovery and sprouting of lesioned systems, although the extent of recovery was generally superior in grafted subjects. Subsequently, elegant studies demonstrated that the act of simply placing lesions in the striatum-induced sprouting of host neural systems (Fiandaca *et al.*, 1988; Plunkett *et al.*, 1990), possibly as a result of upregulated expression of neurotrophic factors. This highlights the importance of using appropriate lesion controls in CNS-grafting experiments.

Animal models of HD confirmed many of these findings. Grafts of fetal *striatal* tissue taken from E13–17 rats survived in the quinolinic acid-lesioned or ibotenic acid-lesioned adult striatum (McGeer and McGeer, 1976; Coyle and Schwartz, 1976) and established reciprocal connections between graft and host tissue. Processes extended from fetal striatal grafts into the host striatum, globus pallidus, substantia nigra pars reticulata, thalamus, cerebral cortex, and putatively even as far as the brain stem raphae nucleus (Levivier *et al.*, 1995; Pearlman *et al.*, 1991; Giordano *et al.*, 1990; Wictorin and Björklund, 1989; Wictorin *et al.*, 1988; Isacson *et al.*, 1987). Thus, appropriate projections extended from grafts, although some inappropriate connections were also made (Wictorin *et al.*, 1989). At the ultrastructural level, several types of synapses within striatal grafts were identified, some of which putatively arose from host projections. Grafts released the appropriate neurotransmitter gamma amino butyric acid (GABA) into the globus pallidus and substantia nigra (Sirinathsinghji *et al.*, 1988). Partial restoration of electrophysiological potentials was observed, although some novel electrical properties were also detected (Walsh *et al.*, 1988; Xu *et al.*, 1991). Functional amelioration was reported on tasks measuring hyperactivity, locational asymmetries, skilled coordinated use of forepaws, and some features of cognitive performance (Isacson *et al.*, 1986; Dunnett *et al.*, 1988; Deckel *et al.*, 1986; Sanberg *et al.*, 1986).

Thus, findings from striatal models confirmed that fetal neuronal grafts survived in the host brain and were capable of restoring specific physiological and functional features of the normal CNS. Fetal grafts survived and functioned best when grafted into natural targets of innervation, and when grafted as cell suspensions into multiple target locations.

B. Memory Models: Hippocampal Denervation

In another animal model of neuronal replacement, fetal basal forebrain tissue has been grafted to the denervated hippocampus to promote mnemonic recovery. Hippocampal inputs are lesioned either by transecting the fimbria–fornix, thereby depriving the hippocampus of several inputs from the basal forebrain and hindbrain, or by transecting the perforant path, thereby interrupting entorhinal cortical inputs to the hippocampus. This section will focus primarily on findings from fimbria–fornix lesion studies.

Rats that undergo fimbria–fornix lesions exhibit comprehensive reductions in hippocampal acetylcholine, serotonin, and norepinephrine. They also exhibit significant and long-lasting deficits on tasks related to spatial memory. To determine whether grafts of fetal tissue can compensate for these losses, E16–18 basal forebrain tissue has been grafted into the hippocampus. These grafts contain cholinergic neurons but few if any serotonergic or noradrenergic neurons; grafts to repopulate the latter transmitter systems would also require fetal tissue from the hindbrain raphae nucleus (serotonin) and locus coeruleus (norepinephrine). Cholingeric basal forebrain grafts survive in the hippocampus and reinnervate it in an organotypic pattern. In contrast, cholinergic neurons derived from the striatum also survive when grafted to the hippocampus but do not reinnervate it in a pattern that resembles normal inputs. Graft-derived axons from fetal basal forebrain implants into the hippocampus make synaptic contacts with host hippocampal neurons (Clark *et al.*, 1985; Segal *et al.*, 1985; Gage *et al.*, 1987) and release acetylcholine (Nilsson *et al.*, 1990b); see also Chapter 22). Of note, graft-derived axonal contacts with host neurons make a disproportionately large number of somatic contacts onto granule cell somata relative to dendrites, a reversal of normal afferent topography. Thus, projection patterns of grafted fetal neurons are not necessarily normal. Nonetheless, extensive functional recovery on memory tasks has been reported following grafts of fetal forebrain cells to the hippocampus (Dunnett, 1990; Björklund and Stenevi, 1985; Dunnett *et al.*, 1982b). Animals receiving striatal grafts to the hippocampus display only attenuated functional recovery, highlighting the importance of using fetal tissue derived from the normal source of hippocampal inputs. Interestingly, more extensive functional recovery occurs when co-grafts of acetylcholine-containing and serotonin-containing cells are placed in the hippocampus compared to the degree of recovery observed with either graft type alone (Nilsson *et al.*, 1990a). In some cases, stimulation of host brain regions that project to the hippocampus results in acetylcholine release from grafts, indicating that grafts can be regulated to some degree by host circuitry (Nilsson *et al.*, 1990b). Thus, fetal-grafting studies to the hippocampus demonstrate that grafts into a region of relatively complex neural circuitry can restore function, but recovery is optimal when donor tissue is obtained from a homotypic brain region affected by a lesion.

In summary, these models illustrate the ability of fetal neurons to replace cells that have been lost due to experimental lesions that mimic those seen in neurodegenerative disease. These cells survive and innervate normal targets in an organotypic manner and, for the most part, display features indicative of normal *in situ* cells.

IV. FETAL GRAFTS AS "BRIDGES" FOR RECONNECTING HOST CIRCUITS

Pioneering studies performed by Aguayo and colleagues in the 1970s and 1980s demonstrated that injured adult CNS axons are capable of regeneration when provided with appropriate substrates for growth (Richardson *et al.*, 1980; David and Aguayo, 1981). In these studies, adult peripheral nerve segments were placed into the injured spinal cord or brain stem to serve as "bridges" for host axonal growth. Indeed, host CNS axons penetrated the nerve grafts and extended for distances of up to several centimeters. In subsequent optic nerve lesion paradigms, regenerating retinal ganglion cell axons reinnervated their targets in the superior colliculus and reestablished electrophysiological transmission (Aguayo *et al.*, 1990; Carter *et al.*, 1989). These findings demonstrated that adult CNS axons are capable of regeneration if provided with an appropriate growth environment.

Extensive axonal growth also occurs in the developing nervous system, suggesting that grafts of fetal nervous system tissue to the lesioned adult brain might provide a supportive mileu for axonal regeneration. In this type of paradigm, fetal transplantation is not used as a means of *replacing* host neurons. Rather, fetal transplants provide an environment rich in extracellular matrix molecules, cell adhesion molecules, cellular substrates, and trophic factors that can augment *host* axonal regeneration (Segal *et al.*, 1981; Kromer *et al.*, 1981; Tuszynski *et al.*, 1990; Tuszynski and Gage, 1995). Using the septohippocampal axotomy paradigm to eliminate cholinergic inputs to the dorsal hippocampus, Björklund and co-workers grafted solid blocks of E18 fetal hippocampal tissue were grafted into lesion cavities of adult rats (Segal *et al.*, 1981; Kromer *et al.*, 1981) (e.g., see Fig. 2). The fetal grafts provided a physical bridge of tissue spanning the cholinergic basal forebrain to the hippocampus (Fig. 2). In control animals lacking bridges, and in animals containing fetal bridges that did not fully span the lesion cavity, no cholinergic axons reinnervated the host hippocampus. In contrast, grafts that completely spanned the lesion cavity contained host cholinergic axons that partially reinnervated the host dorsal hippocampus. Injections of retrogradely transported labels into the hippocampus demonstrated that host axons extended from the cholinergic basal forebrain, through the graft, and into the host hippocampus (Tuszynski and Gage, 1995). Although several regenerating host cholinergic axons terminated within the fetal hippocampal bridge itself, many continued to grow completely through the graft and into the host target. Such grafts were subsequently shown to restore specific features of hippocampal electrophysiology (Tuszynski *et al.*, 1990). Expanding upon the bridging concept, grafts of fetal hippocampal bridges were combined with infusions of nerve growth factor (NGF) in rats with

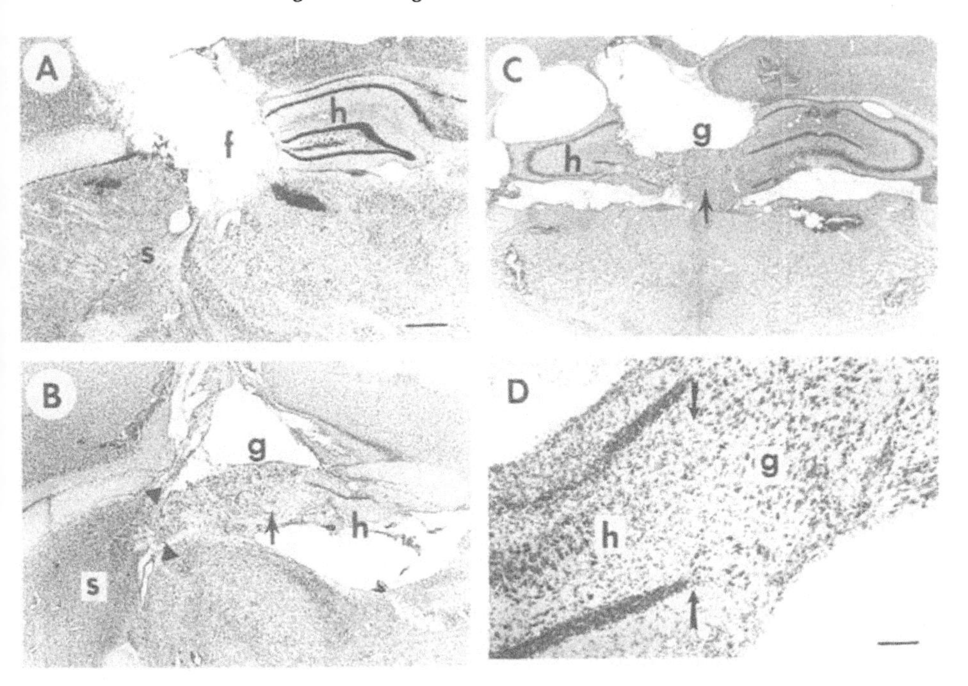

FIGURE 2 Fetal neuronal grafting to the adult CNS. (A) Following lesion-aspiration of the fimbria–fornix (f), axons projecting from the host septal nucleus (s) are disconnected from their normal target, the host hippocampus (h). Sagittal section; rostral left, caudal right. Nissl stains in A–D; scale bar = 1400 μm in A–C. (B) A graft (g and arrow) of fetal hippocampal tissue has been placed in the fimbria–fornix lesion cavity, thereby reestablishing physical continuity between the host septal nucleus and host hippocampus. There is excellent graft integration in at the interface between the host septum and the graft (arrowheads), and host hippocampus and graft. Sagittal section. (C) Coronal section illustrating another example of excellent graft integration into host hippocampal region. (D) At higher magnification, the interface (arrows) from fetal hippocampal graft (g) to host hippocampus (h) is without glial scar or boundary. Scale bar = 15 μm.

fimbria–fornix lesions (Figs. 2–4). NGF infusions enhanced cholinergic neuronal survival while the bridging fetal grafts supported the regeneration of host axons into the denervated hippocampus. As a result, lesioned rats showed recovery on a simple memory task (Tuszynski and Gage, 1995). These findings demonstrated for the first time that functional losses in the lesioned adult CNS could be ameliorated by reconstruction of the *host* neural circuitry.

These studies support the notion that fetal nervous system grafts can provide a milieu that supports regeneration of lesioned adult axons. The specific components of fetal tissue that supports regeneration require further characterization, and are likely to consist of neurotrophic factors and extracellular matrix mole-

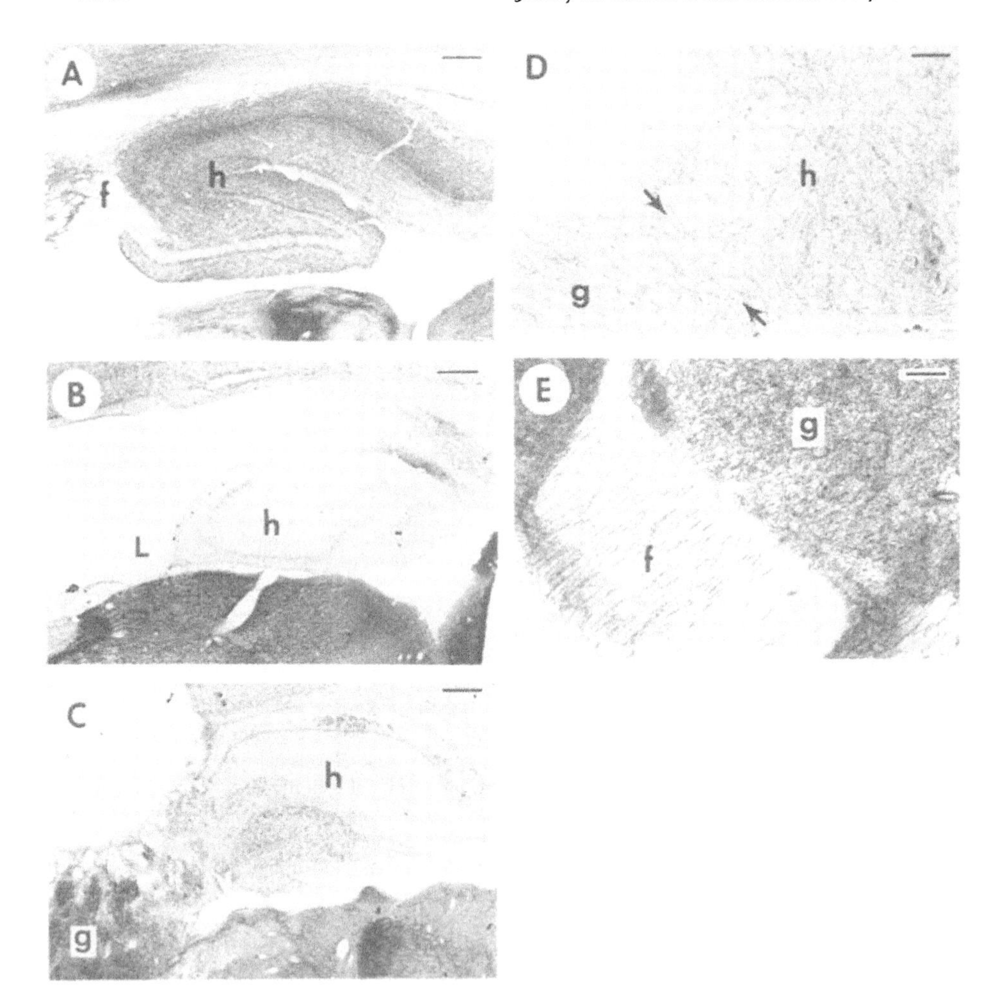

FIGURE 3 Acetylcholine esterase (AChE) stain demonstrates regeneration of host basal forebrain cholinergic axons across fetal hippocampal bridge placed in fimbria fornix lesion cavity. (A) The normal cholinergic innervation of the intact adult hippocampus (h) is shown. Cholinergic axons penetrate the hippocampus from the host fimbria–fornix (f). Scale bar = 70 μm in A–C. (B) After fimbria–fornix lesions, the host hippocampus loses nearly all cholinergic innervation. L, lesion cavity. (C) The placement of a hippocampal fetal graft (g) in the lesion cavity allows the regeneration of host axons from the septal nucleus to the fetal hippocampal graft; many host axons continue to extend all the way through the graft to repenetrate the host hippocampus (h). Cholinergic axons reinnervating the host hippocampus are not derived from the fetal hippocampal graft itself because the hippocampus possess very few cholinergic neurons. (D) Higher magnification of the graft-host interface (arrows) at the indicates abundant crossing of cholinergic axons from the graft (g) into the host hippocampus (h). Scale bar = 8 μm in D,E. (E) Many host cholinergic axons in the fornix (f) enter the hippocampal graft (g).

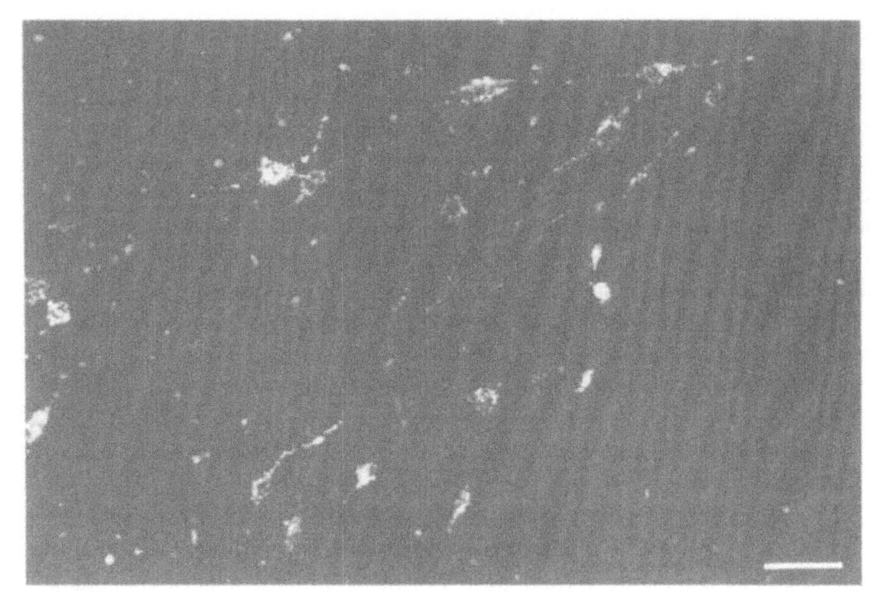

FIGURE 4 Host septal basal forebrain neurons are labeled by retrogradely transported fluorescent rhodamine microspheres injected into the host hippocampus after fetal grafting to a fimbria–fornix lesion cavity, reflecting the regeneration of host cholinergic axons. Scale bar = 12 μm.

cules that provide a conducive growth scaffold and favorable cell–axon adhesion molecules. The characterization of these substances would permit their specific application to the injured adult CNS in optimal, potentially synthetic combinations, thereby eliminating the need to use fetal tissue.

V. FETAL GRAFTS FOR ENDOCRINE REPLACEMENT

Some of the earliest modern studies of fetal transplantation examined the hypothesis that grafts of fetal neurons could compensate for the loss of endocrine function in the brain. Particularly valuable findings came from models of hypothalamic dysfunction.

A. Hypogonadal Mice

In models of mouse hypogonadism (abortive development of the sexual system), the value of fetal neuronal grafting in restoring anatomical, neurochemical, and functional features of the brain have been described. In 1977 Cattanach and associates (1977) described a mutant mouse strain, the *hypogonadal mouse*, with a deficiency in gonadotrophin-releasing hormone (GnRH) secretion that resulted in hypogonadism. In *hypogonadal* mice, gonadotrophin-releasing hormone cells of the preoptic area of the hypothalamus fail to secrete GnRH, causing a secondary reduction in the production of luteinizing hormone (LH) and follicle-stimulating hormone (FSH) from the pituitary gland. As a result, mice show a failure to develop secondary sex characteristics and are incapable of reproduction. A series of pioneering studies by Krieger and colleagues (1982) reported that fetal preoptic area hypothalamic tissue derived from wild-type embryonic day 16 to 18 (E16–18) mice survived when grafted into the third ventricle of hypogonadal mutant mice and restored serum LH and FSH levels. In addition, mutant mice became capable of reproduction, demonstrating that fetal neuronal replacement grafts could generate functional recovery.

Not all functions in hypogonadal mice were restored to normal. Testicular weight recovered to only one-fourth of wild-type levels, although this amount was sufficient to generate normal spermatogenesis. Mutant female hypogonadal mice developed normal uteri and ovaries, but normal cyclical menses were not reestablished. Nonetheless, grafted female rats became capable of reproduction. Morphological analysis showed that many grafted neurons extended GnRH immunoreactive processes to the median eminence, the vascular structure into which GnRH is normally secreted from the hypothalamus to enter the hypothalamic portal circulation and directly access gonadotrophin-secreting cells in the pituitary gland. Evidence from Kokoris and associates suggested that termination of grafted GnRH processes on vascular cells of the median eminence was required for restoration of normal function (Kokoris *et al.*, 1988, 1987).

B. Vasopressin-Deficient Rats

Studies in another hypothalamic mutant model also demonstrated that fetal nervous system grafting could correct functional deficits in the brain. *Brattleboro* rats exhibit deficiencies in vasopressin (or antidiuretic hormone, ADH) secretion from the hypothalamus secondary to a genetic mutation. As a result of this hormonal deficiency, rats fail to reabsorb water in the kidneys and suffer excessive water loss. *Brattleboro* rats exhibit low urine osmolality and high urine flow rates, causing them to drink large amounts of water (polydipsia). When *Brattleboro* rats received transplants of E17–19 anterior hypothalamus into the third ventricle, water-drinking behavior and urine osmolality were normalized in a subset (20%) of graft recipients (Gash *et al.*, 1982; Sladek *et al.*, 1982). Restoration of the normal phenotype correlated with the extension of neuritic processes from the grafted neurons into blood vessels of the median eminence and the establishment of adequate local vascular circulation in the graft (Scharoun *et al.*, 1984). The abnormal behavioral phenotype was ameliorated despite only partial restoration of serum vasopressin levels. Of note, nearly all grafted rats showed *transient* behavioral recovery, but only rats with grafts that were in physical contact with the median eminence and that extended vasopressin-immunoreactive processes into the median eminence exhibited long-term functional recovery. Rats with transient but not long-term recovery may have had damage to the adjacent paraventricular preoptic hypothalamus, a nucleus that normally stimulates drinking. This highlights an important principle that should be considered in all grafting studies; appropriate control groups must be included to ensure that outcomes are specifically graft-derived rather than a consequence of lesion-induced changes in the host brain caused by the grafting procedure.

Other models of vasopressin deficiency have been employed in grafting studies that demonstrate the diverse ways in which fetal implants can restore function in the CNS. *Brattleboro* rats show deficient synthesis of vasopressin, but hypothalamic circuitry is normal. Thus the functional recovery seen following anterior hypothalamic grafting appears to be mediated by vasopressin that is synthesized and secreted by grafted neurons, which make neurohemal (axonal-vascular) contacts with host vessels within the median eminence. In contrast, a similar syndrome of polydipsia and polyurea is seen in normal rats following selective lesions of the posterior lobe of the pituitary gland. In this model system, functional recovery is also achieved following anterior hypothalamic transplants; however, these grafts induce functional recovery by preventing host hypothalamic-hypophyseal neurons from degenerating rather than by secreting vasopressin (Marciano *et al.*, 1989).

Thus, hypothalamic grafting studies demonstrate that fetal grafts can restore endocrine production and correct functional disturbances in the adult brain. These studies also indicate that grafts in some instances require integration with host tissue in specific regions to elicit recovery.

The preceding models of neuronal replacement, bridging, and endocrine replacement are a sampling of fetal grafting studies that reflect the potential of undifferentiated neural cells to replace or promote the regeneration of injured adult neural systems. For more complete reviews that encompass these paradigms and others in greater depth, refer to Chapters 11, 12, 21, 24, and 25, and to additional references (Björklund and Stenevi, 1985; Das, 1990; Fisher and Gage, 1993, 1994; Bakay and Sladek, 1993; Dunnett *et al.*, 1982a; Dunnett, 1990).

VI. CONCLUSION

Fetal neuronal grafting in animal models has served as a useful tool for understanding features of nervous system development, CNS plasticity, and adult CNS repair. In several models there is compelling evidence that fetal grafting could be used to reduce disease severity and perhaps slow the rate of decline. However, practical considerations in the routine extension of fetal neuronal transplantation for the treatment of human neurological disease are formidable. First, it is unlikely that sufficient amounts of fetal tissue would ever be available to treat human neurological diseases that are potential targets of grafting, including Parkinson's disease, Huntington's disease, Alzheimer's disease, spinal cord injury, and other forms of CNS trauma and neurodegeneration. Second, there is a potential risk of transmission of infectious disease from fetal tissue to the host if the donor tissue is not carefully screened, a consideration of more importance in Third-World transplantation studies. Third, there is the risk that unexpected complications could arise from the practice of neural transplantation in inexperienced hands (Folkerth and Durso, 1996). Fourth, there is a lack of societal consensus on the ethical acceptability of fetal transplantation.

On the other hand, in the clinical realm there may be a role for fetal-grafting studies in identifying human disorders that can benefit from neural repair strategies. For example, if controlled, prospective clinical trials of fetal grafting for Parkinson's disease that are now in progress prove successful, then neural replacement will presumably have a role in the treatment of this disease; studies of neural stem cell therapies should then be pursued with renewed vigor. Other molecules in the fetal environment, such as neurotrophic factors, extracellular matrix molecules, and cell adhesion molecules, may also contribute to neural repair, and positive findings from future clinical trials employing fetal grafts in bridging strategies could lead to the precise characterization of molecules in the fetal environment, other than neurons, that augment neural plasticity. Of course, one might argue that trials in humans are not necessary to spur investigators on to discover molecules that promote neural plasticity. It is likely, however, that positive findings in human trials would generate an exceptional level of interest and activity in these investigations.

Thus the ultimate value of fetal transplantation studies may lie in generating sufficient evidence to prove that neural repair is achievable in a specific disease. This in turn would enhance the search for basic cellular and molecular mechanisms underlying the beneficial effects of fetal cells, leading in turn to the development of new and, hopefully, broadly applicable treatments for disease. Many of the key "plastic" elements of the fetal environment have already been identified, isolated, and synthesized: neural stem cells, neurotrophic molecules, extracellular matrix molecules, and cell adhesion molecules. Unique combinations of these factors in specific models of neural damage can eventually lead to the development of effective therapies for neurological disease.

ACKNOWLEDGMENTS

Supported by the NIH (NINDS 1R01NS37083 1R01NS35785 NIA 2P01AG10435), American Academy of Neurology Research Foundation, Veterans Affairs, and the Hollfelder Foundation.

REFERENCES

Aguayo, A. J., Bray, G. M., Rasminsky, M., Zwimpfer, T., Carter, D., and Vidal-Sanz, M. (1990). Synaptic connections made by axons regenerating in the central nervous system of adult mammals. *J Exp Biol* 153:199–224.

Altman, J., and Bayer, S. A. (1984). The development of the rat spinal cord. *Adv Anat Embryol Cell Biol* 85:1–168.

Altman, J., and Das, G. (1965). Autoradiographic and histological evidence of postnatal hippocampal neurogenesis in rats. *J Comp Neurol* 124:319–336.

Bakay, R. A., and Sladek, J. R. J. (1993). Fetal tissue grafting into the central nervous system: yesterday, today, and tomorrow [see comments]. *Neurosurgery* 33:645–647.

Bayer, S. A. (1980). The development of the hippocampal region in the rat. I. Neurogenesis examined with [3H]thymidine autoradiography. *J Comp Neurol* 190:87–114.

Björklund, A., Dunnett, S. B., Stenevi, U., Lewis, M. E., and Iversen, S. D. (1980a). Reinnervation of the denervated striatum by substantia nigra transplants: Functional consequences as revealed by pharmacological and sensorimotor testing. *Brain Res* 199(2):307–333.

Björklund, A., Schmidt, R. H., and Stenevi, U. (1980b). Functional reinnervation of the neostriatum in the adult rat by use of intraparenchymal grafting of dissociated cell suspensions from the substantia nigra. *Cell Tissue Res* 212(1):39–45.

Björklund, A., and Stenevi, U. (1979). Reconstruction of the nigrostriatal dopamine pathway by intracerebral nigral transplants. *Brain Res* 177:555–560.

Björklund, A., and Stenevi, U. (1985). Neural grafting in the mammalian CNS. Amsterdam: Elsevier.

Carter, D., Bray, G. M., and Aguayo, A. J. (1989). Regenerated retinal ganglion cells axons can form well-differentiated synapses in the superior colliculus of adult hamsters. *J Neurosci* 9:4042–4050.

Cattanach, B. M., Iddon, C. A., Charlton, H. M., Ciappa, S. A., and Fink, G. (1977). Gonadotropin-releasing hormone deficiency in a mutant mouse with hypogonadism. *Nature* 269:338–340.

Clark, D. J., Gage, F. H., and Björklund, A. (1985). Formation of cholinergic synpases by intrahippocampal septal grafts as revealed by choline acetyltransferase immunohistochemistry. *Brain Res* 369:151–162.

Collier, T. J., Redmond, D. E., Sladek, C. D., Gallagher, M. J., Roth, R. H., and Sladek, J. R. (1987). Intracerebral grafting and culture of cryopreserved primate dopamine neurons. *Brain Res* 374:363–366.

Coyle, J. T., and Schwartz, R. (1976). Lesion of striatal neurons with kainic acid provides a model for Huntington's chorea. *Nature* 263:244–246.

Crutcher, K. A. (1990). Age-related decrease on sympathetic spouting is primarily due to decreased target receptivity: Implications for understanding brain aging. *Neurobiol Aging* 11:175–183.

Das, G. D. (1990). Neural transplantation: An historical perspective. *Neurosci Biobehav Rev* 14:389–401.
Das, G. D., Hallas, B. H., and Das, K. G. (1980). Transplantation of brain tissue in the brain of rat. I. Growth characteristics of neocortical transplants of embryos of different ages. *Am J Anat* 158(2):135–145.
David, S., and Aguayo, A. (1981). Axonal elongation into peripheral nervous system "bridges" after central nervous system injury in adult rats. *Science* 214:931–933.
Davies, S. W., and Roberts, P. J. (1988). Model of Huntington's disease. *Science* 241:474–475.
Deacon, T., Schumacher, J., Dinsmore, J., Thomas, C., Palmer, P., Kott, S., Edge, A., Penney, D., Kassissieh, S., Dempsey, P., and Isacson, O. (1997). Histological evidence of fetal pig neural cell survival after transplantation into a patient with Parkinson's disease. *Nature Med* 3:350–353.
Deckel, A. W., Moran, T. H., and Robinson, R. G. (1986). Behavioral recovery following kainic acid lesions and fetal implants of the striatum occurs independent of dopaminergic mechanisms. *Brain Res* 363:383–385.
Del Conte, G. (1907). Einpflanzungen von embryohalem Gewebe ins Gehirn. *Beitrage zur Path Anat* 42:193–202.
Dunn, E. H. (1917). Primary and secondary findings in a series of attempts to transplant cerebral cortex in the albino rat. *J Comp Neurol* 27:565–582.
Dunnett, S. B. (1990). Neural transplantation in animal models of dementia. *Eur J Neurosci* 2:567–587.
Dunnett, S. B., Björklund, A., Schmidt, R. H., Stenevi, U., and Iversen, S. D. (1983). Intracerebral grafting of neuronal cell suspensions. V. Behavioural recovery in rats with bilateral 6-OHDA lesions following implantation of nigral cell suspensions. *Acta Physiol Scand Suppl* 522:39–47.
Dunnett, S. B., Björklund, A., Stenevi, U., and Iversen, S. D. (1981). Grafts of embryonic substantia nigra reinnervating the ventrolateral striatum ameliorate sensorimotor impairments and akinesia in rats with 6-OHDA lesions of the nigrostriatal pathway. *Brain Res* 229:209–217.
Dunnett, S. B., Björklund, A., Stenevi, U., and Iversen, S. D. (1982a). CNS transplantation: Structural and functional recovery from brain damage. *Prog Brain Res* 55:431–443.
Dunnett, S. B., Hernandez, T. D., Summerfield, A., Jones, G. H., and Arbuthnott, G. (1988). Graft-derived recovery from 6-OHDA lesions: Specificity of ventral mesencephalic graft tissues. *Exp Brain Res* 71:411–424.
Dunnett, S. B., Low, W. C., Iversen, S. D., Stenevi, U., and Björklund, A. (1982b). Septal transplants restore maze learning in rats with fornix–fimbria lesions. *Brain Res* 251(2):335–348.
Elliot, E. J., and Muller, K. J. (1982). Synapses between neurons regenerate accurately after destruction of ensheathing glial cells in the leech. *Science* 215:1260–1262.
Fiandaca, M. S., Kordower, J. H., Hansen, J. T., Jiao, S. S., and Gash, D. M. (1988). Adrenal medullary autografts into the basal ganglia of Cebus monkeys: Injury-induced regeneration. *Exp Neurol* 102:76–91.
Finger, S., and Dunnett, S. B. (1989). Nimodipine enhances growth and vascularization of neural grafts. *Exp Neurol* 104:1–9.
Fisher, L. J., and Gage, F. H. (1993). Grafting in the mammalian central nervous system. *Physiol Rev* 73:583–616.
Fisher, L. J., and Gage, F. H. (1994). Intracerebral transplantation: Basic and clinical applications to the CNS. *FASEB J* 8:489–496.
Folkerth, R. D., and Durso, R. (1996). Survival and proliferation of nonneural tissues, with obstruction of cerebral ventricles, in a parkinsonian patient treated with fetal allografts. *Neurology* 46:1219–1225.
Forssman, J. (1898). Ueber die Ursachen, welche die Wachsthumsrichtung der peripheren Nervenfasern bei der Regeneration bestimmen. Inaugural-Dissertation, University of Lund, pp. 3–47.

Forssman, J. (1900). Zur Kenntniss des Neurotropismus. *Ziegler:Beitrage zur Patologischen Anatomie* 27:407–430.

Freed, W. J., Perlow, M. J., Karoum, F., Seiger, A., Olson, L., Hoffer, B. J., and Wyatt, R. J. (1980). Restoration of dopaminergic function by grafting of fetal rat substantia nigra to the caudate nucleus: Long-term behavioral, biochemical, and histochemical studies. *Ann Neurol* 8(5):510–519.

Freeman, T. B., Sandberg, P. R., Nauert, G. M., Borlongan, C., Liu, E. Z., Boss, B. D., Spector, D., Olanow, C. W., and Kordower, J. H. (1995). The influence of donor age on the survival of solid and suspension intraparenchymal human embryonic nigral grafts. *Cell Transpl* 4:141–154.

Gage, F. H., Björklund, A., Stenevi, U., and Dunnett, S. B. (1983). Intracerebral grafting of neuronal cell suspensions. VIII. Survival and growth of implants of nigral and septal cell suspensions in intact brains of aged rats. *Acta Physiol Scand Suppl* 522:67–75.

Gage, F. H., Buzsaki, G., Nilsson, O. G., and Björklund, A. (1987). Grafts of fetal cholinergic neurons to the deafferented hippocampus. *Prog Brain Res* 71:335–347.

Galpern, W. R., Burns, L. H., Deacon, T. W., Dinsmore, J., and Isacson, O. (1996). Xenotransplantation of porcine fetal ventral mesencephalon in a rat model of Parkinson's disease: Functional recovery and graft morphology. *Exp Neurol* 140:1–13.

Gash, D. M., Warren, P. H., Dick, L. B., Sladek, J. R., Jr., and Ison, J. R. (1982). Behavioral modification in *Brattleboro* rats due to vasopressin administration and neural transplantation. *Ann NY Acad Sci* 394:672–688.

Giordano, M., Ford, L. M., Shipley, M. T., and Sandberg, P. R. (1990). Neural grafts and pharmacological intervention in a model of Huntington's disease. *Brain Res Bull* 25:453–465.

Glees, P. (1940). The differentiation of the brain and other tissues in an implanted portion of embryonic head. *J Anat* 75:239–247.

Isacson, O., Dawbarn, D., Brundin, P., Gage, F. H., Emson, P. C., and Björklund, A. (1987). Neural grafting in a rat model of Huntington's disease: Striosomal-like organization of striatal grafts as revealed by acetylcholinesterase histochemistry, immunocytochemistry and receptor autoradiography. *Neuroscience* 22:481–497.

Isacson, O., Dunnett, S. B., and Björklund, A. (1986). Graft-induced behavioral recovery in an animal model of Huntington disease. *Proc Natl Acad Sci USA* 83:2728–2732.

Kokoris, G. J., Lam, N. Y., Ferin, M., Silverman, A. J., and Gibson, M. J. (1988). Transplanted gonadotropin-releasing hormone neurons promote pulsatile luteinizing hormone secretion in congenitally hypogonadal (hpg) male mice. *Neuroendocrinology* 48:45–52.

Kokoris, G. J., Silverman, A. J., Zimmerman, E. A., Perlow, M. J., and Gibson, M. J. (1987). Implantation of fetal preoptic area into the lateral ventricle of adult hypogonadal mutant mice: The pattern of gonadotropin-releasing hormone axonal outgrowth into the host brain. *Neurosci* 22:159–167.

Kordower, J. H., Freeman, T. B., Snow, B. J., Vingerhoets, F. J., Mufson, E. J., Sanberg, P. R., Hauser, R. A., Smith, D. A., Nauert, G. M., Perl, D. P., *et al.* (1995). Neuropathological evidence of graft survival and striatal reinnervation after the transplantation of fetal mesencephalic tissue in a patient with Parkinson's disease. *N Engl J Med* 332:1118–1124.

Kordower, J. H., Rosenstein, J. M., Collier, T. J., Burke, M. A., Chen, E. Y., Li, J. M., Martel, L., Levey, A. E., Mufson, E. J., Freeman, T. B., et al. (1996). Functional fetal nigral grafts in a patient with Parkinson's disease: Chemoanatomic, ultrastructural, and metabolic studies. *J Comp Neurol* 370:203–230.

Kordower, J. H., Goetz, C. G., Freeman, T. B., and Olanow, C. W. (1997). Dopaminergic transplants in patients with Parkinson's disease. *Exp Neurol* 144:41–46.

Krieger, D. T., Perlow, M. J., Gibson, M. J., Davies, T. F., Zimmerman, E. A., Ferin, M., and Charlton, H. M. (1982). Brain grafts reverse hypogonadism of gonadotropin releasing hormone deficiency. *Nature* 298(5873):468–471.

Kromer, L. F., Björklund, A., and Stenevi, U. (1979). Intracephalic implants: A technique for studying neuronal interactions. *Science* 204(4397):1117–1119.

Kromer, L. F., Björklund, A., and Stenevi, U. (1981). Regeneration of the septohippocampal pathways in adult rats is promoted by utilizing embryonic hippocampal implants as bridges. *Brain Res* 210:172–200.

Krum, J. M., and Rosenstein, J. M. (1988). Patterns of angiogenesis in neural transplant models: II. Fetal neocortical transplants. *J Comp Neurol* 271:331–345.

Lauder, J. M., and Bloom, F. E. (1974). Ontogeny of monoamine neurons in the locus ceruleus, raphae nuclei and substantia nigra of the rat. *J Comp Neurol* 155:469–482.

Le Gros Clark, W. E. (1940). Neuronal differentiation in implanted foetal cortical tissue. *J Neurol Psychiat* 3:263–272.

Levivier, M., Gash, D. M., and Przedborski, S. (1995). Time course of neuroprotective effect of transplantation on quinolinic acid-induced lesions of the striatum. *Neurosci* 69:43–50.

Li, X. J., Li, S. H., Sharp, A. H., Nucifora, F. C., Schilling, G., Lanahan, A., Worley, P., Snyder, S. H., and Ross, C. A. (1995). A huntingtin-associated protein enriched in brain with implications for pathology. *Nature* 378:398–402.

Lugaro, E. (1906). Sul neurotropismo e suitrapanti dei nervi. *Riv path nerve mentale* 11:320–327.

Mandel, R. J., Brundin, P., and Björklund, A. (1990). The importance of graft placement and task complexity for transplant-induced recovery on simple and complex sensorimotor deficits in dopamine denervated rats. *Eur J Neurosci* 2:888–894.

Marciano, F. F., and Gash, D. M. (1986). Structural and functional relationships of grafted vasopressin neurons. *Brain Res* 370:338–342.

Marciano, F. F., Wiegand, S. J., Sladek, J. R., Jr., and Gash, D. M. (1989). Fetal hypothalamic transplants promote survival and functional regeneration of axotomized adult supraoptic magnocellular neurons. *Brain Res* 483:135–142.

McGeer, P. L., and McGeer, E. G. (1976). Duplication of biochemical changes of Huntington's chorea by intrastriatal injections of glutamic and kainic acids. *Nature* 263:517–519.

Nieto-Sampedro, M., Manthorpe, M., Barbin, G., Varon, S., and Cotman, C. W. (1983). Injury-induced neuronotrophic activity in adult rat brain: Correlation with survival delayed implants in the wound cavity. *J Neurosci* 3:2219–2229.

Nikkah, G., Eberhard, J., Olsson, M., and Björklund, A. (1995). Preservation of fetal ventral mesencephalic cells by cool storage: *In vitro* viability and TH-positive neuron survival after microtransplantation to the striatum. *Brain Res* 687:22–34.

Nilsson, O. G., Brundin, P., and Björklund, A. (1990a). Amelioration of spatial memory impairment by intrahippocampal grafts of mixed septal and raphae tissue in rats with combined cholinergic and serotonergic denervation of the forebrain. *Brain Res* 515:193–206.

Nilsson, O. G., Kalen, P., Rosengren, E., and Björklund, A. (1990b). Acetylcholine release from intrahippocampal septal grafts is under the control of the host brain. *Proc Natl Acad Sci USA* 87:2647–2651.

Norremolle, A., Riess, O., Epplen, J. T., Fenger, K., Hasholt, L., and Sorensen, S. A. (1993). Trinucleotide repeat elongation in the Huntingtin gene in Huntington disease patients from 71 Danish families. *Human Molec Genet* 2:1475–1476.

Olson, L., Björklund, H., Freedman, R., *et al.* (1983). Intrinsic and extrinsic determinants of brain development as evidenced by grafting of fetal brain tissue. In: *Development and regenerating vertebrate nervous systems*. New York: Liss, pp. 121–150.

Pearlman, S. H., Levivier, M., Collier, T. J., Sladek, J. R., and Gash, D. M. (1991). Striatal implants protect the host striatum against quinolinic acid toxicity. *Exp Brain Res* 84:303–310.

Perlow, M. J., Freed, W. J., Hoffer, B. J., Seiger, A., Olson, L., and Wyatt, R. J. (1979). Brain grafts reduce motor abnormalities produced by destruction of nigrostriatal dopamine system. *Science* 204:643–647.

Plunkett, R. J., Bankiewicz, K. S., Cummins, A. C., Miletich, R. S., Schwartz, J. P., and Oldfield, E. H. (1990). Long-term evaluation of hemiparkinsonian monkeys after adrenal autografting or cavitation alone. *J Neurosurg* 73:918–926.

Ransom, S. W. (1909). Transplantation of the spinal ganglion into the brain. *Q Bull Northwest Univ Med Sch* 11:176–178.

Ray, J., and Gage, F. H. (1994). Spinal cord neuroblasts proliferate in response to basic fibroblast growth factor. *J Neurosci* 14:3548–3564.

Reynolds, B. A., Tetzlaff, W., and Weiss, S. (1992). A multipotent EGF-responsive striatal embryonic progenitor cell produces neurons and astrocytes. *J Neurosci* 12:4565–4574.

Richardson, P. M., McGuiness, U. M., and Aguayo, A. J. (1980). Axons from CNS neurons regenerate into PNS grafts. *Nature* 284:264–265.

Sachs, D. H., and Bach, F. H. (1990). Immunology of xenograft rejection. *Hum Immuno* 28:245–251.

Sanberg, P. R., Henault, M. A., and Deckel, A. W. (1986). Locomotor hyperactivity: Effects of multiple striatal transplants in an animal model of Huntington's disease. *Pharmacol Biochem Behav* 25:297–300.

Scharoun, S. L., Gash, D. M., and Notter, M. F. (1984). *In vitro* and *in vivo* studies on development and regeneration of vasopressin neurons. *Peptides* 5:157–167.

Segal, M., Björklund, A., and Gage, F. H. (1985). Transplanted septal neurons make viable cholinergic synapses with a host hippocampus. *Brain Res* 336:302–307.

Segal, M., Stenevi, U., and Björklund, U. (1981). Reformation in adult rats of functional septo-hippocampal connections by septal neurons: Regeneration across an embryonic hippocampal tissue bridge. *Neurosci Lett* 27:7–12.

Sirinathsinghji, D. J., Dunnett, S. B., Isacson, O., Clarke, D. J., Kendrick, K., and Björklund, A. (1988). Striatal grafts in rats with unilateral neostriatal lesions. II. *In vivo* monitoring of GABA release in globus pallidus and substantia nigra. *Neuroscience* 24:803–811.

Sladek, J. R., Jr., Scholer, J., Notter, M. D., and Gash, D. M. (1982). Immunohistochemical analysis of vasopressin neurons transplanted into Brattleboro rat. *Ann NY Acad Sci* 394:102–115.

Steele, D. J., and Auchincloss, H. J. (1995). Xenotransplantation. *Annu Rev Med* 46:345–360.

Stenevi, U., Björklund, A., and Svendgaard, N. A. (1976). Transplantation of central and peripheral monoamine neurons to the adult rat brain: Techniques and conditions for survival. *Brain Res* 114:1–20.

Strecker, R. E., Sharp, T., Brundin, P., Zetterström, T., Ungerstedt, U., and Björklund, A. (1987). Autoregulation of dopamine release and metabolism by intrastriatal nigral grafts as revealed by intracerebral dialysis. *Neuroscience* 22:169–178.

Streilein, J. W. (1988). Transplantation immunobiology in relation to neural grafting: Lessons learned from immunologic privilege in the eye. *Int J Dev Neurosci* 6:497–511.

Sugar, O., and Gerard, R. W. (1940). Spinal cord regeneration on the rat. *J Neurophysiol* 3:1–19.

Tello, F. (1911a). La influencia del neurotropismo en la regeneracion de las centros nerviosos. *Trab Lab Invest Biol Univ Madr* 9:123–159.

Tello, F. (1911b). Un experimento sobre la influencia del neurotropismo en la regeneracion de la corteza cerebral. *Rev Clin Madr* 5:292–294.

Thompson, W. G. (1890). Successful brain grafting. *New York Medical Journal* 51:701–702.

Tidd, C. W. (1932). The transplantation of spinal glanglia in the white rat. A study of the morphological changes in surviving cells. *J Comp Neurol* 55:531–543.

Tuszynski, M. H., Buzsaki, G., and Gage, F. H. (1990). NGF infusions combined with fetal hippocampal grafts enhance reconstruction of the lesioned septo-hippocampal projection. *Neurosci* 36:33–44.

Tuszynski, M. H., and Gage, F. H. (1995). Bridging grafts and transient NGF infusions promote long-term CNS neuronal rescue and partial functional recovery. *PNAS* 92:4621–4625.

Walsh, J. P., Zhou, F. C., Hull, C. D., Fisher, R. S., Levine, M. S., and Buchwald, N. A. (1988). Physiological and morphological characterization of striatal neurons transplanted into the striatum of adult rats. *Synapse* 2:37–44.

Wictorin, K., and Björklund, A. (1989). Connectivity of striatal grafts implanted into the ibotenic acid-lesioned striatum. II. Cortical afferents. *Neuroscience* 30:297–311.

Wictorin, K., Isacson, O., Fischer, W., Nothias, F., Peschanski, M., and Björklund, A. (1988). Studies on host afferent inputs to fetal striatal transplants in the excitotoxically lesioned striatum. *Prog Brain Res* 78:55–60.

Wictorin, K., Simerly, R. B., Isacson, O., Swanson, L. W., and Björklund, A. (1989). Connectivity of striatal grafts implanted into the ibotenic acid-lesioned striatum. III. Efferent projecting graft neurons and their relation to host afferents within the grafts. *Neuroscience* 30:313–330.

Widner, H., and Brundin, P. (1988). Immunological aspects of grafting in the mammalian central nervous system. A review and speculative synthesis. *Brain Research Reviews* 13:287–324.

Xu, Z. C., Wilson, C. J., and Emson, P. C. (1991). Synaptic potentials evoked in spiny neurons in rat neostriatal grafts by cortical and thalamic stimulation. *J Neurophysiol* 65:477–493.

Yahr, M. D. (1984). The parkinsonian syndrome. In: Merritt, H., ed., *A Textbook of Neurology*. Philadelphia: Lee & Febiger, pp. 526–537.

Zetterström, T., Brundin, P., Gage, F. H., Sharp, T., Isacson, O., Dunnett, S. B., Ungerstedt, U., and Björklund, A. (1986). *In vivo* measurement of spontaneous release and metabolism of dopamine from intrastriatal nigral grafts using intracerebral dialysis. *Brain Res* 362:344–349.

Zhou, F. C., Pu, C. F., and Finger, S. (1991). Nimopidine-enhanced survival of suboptimal neural grafts. *Restor Neurol Neurosci* 3:211–215.

CHAPTER 7

The Use of Neural Progenitor Cells for Therapy in the CNS Disorders

JASODHARA RAY, THEO D. PALMER, LAMYA S. SHIHABUDDIN, AND FRED H. GAGE
Laboratory of Genetics, Salk Institute, La Jolla, California

In recent years a significant number of neurological diseases have been defined at the molecular level. Somatic gene therapy using genetically modified nonneuronal cells expressing therapeutic factors have been successfully used in animal models of neurodegenerative diseases. Ability to grow central nervous system (CNS)-derived neural progenitor cells has proven to be extremely useful to study a diverse phenomenon including the fate choice, differentiation, and synaptic maturation of cells. Immortal or perpetual cultures of neural progenitor cells implanted into the rodent brain survive, migrate, and integrate in the host cytoarchitecture. These cells can be genetically modified to express therapeutic gene products. The ability of the implanted cells to integrate in the host brain and express transgene products *in situ* offer potential approaches for gene therapy in certain CNS diseases. The utility of this approach has already been explored in animal models of neurodegenerative diseases. This chapter reviews the recent advances made in understanding the nature and potentiality of neural progenitor cells *in vitro* and *in vivo* as well as their possible use for cell replacement and gene therapy.

I. INTRODUCTION

Neurons in the adult central nervous system are terminally differentiated and most exist throughout the life of the animal but unlike other organ systems, the CNS is unique in its inability to replace the dead or dying neurons following damage. Regardless of the cause of the neuronal cell loss, the cellular deficit leads to lost motor and/or cognitive functions in the affected individual. There are several factors that lead to cell loss. The most common of these are neurodegenerative diseases, the normal aging process, or physical injury. Neurodegenerative diseases are characterized by the progressive and selective degeneration or loss of neurons in specific brain loci. In many cases, the exact cause is unknown but may be due to the absence of metabolic enzymes, neurotransmitters, trophic factors, or cofactors (for review see Suhr and Gage, 1993; Martin, 1995; Rosenberg and Iannaccone, 1995; Fisher, 1995; Snyder and Fisher, 1996). Pharmacological agents often provide transient or erratic benefit, and may cause undesirable side effects. In addition, the limitations imposed by the blood–brain barrier prevent the peripheral administration of enzymes and other macromolecules required in the CNS. Given this inaccessibility, the treatment of disorders in the brain requires special approaches that specifically target the delivery of therapeutic products to the CNS.

II. APPROACHES TO THERAPY

Treatment of degenerative disease or injury could focus on one or more of the steps leading to cell loss. In concept, one of the most effective strategies would be to interrupt the ongoing process of neuronal cell death and thus prevent progressive degeneration. Developmental studies suggest that the establishment and maintenance of each neuronal population depend on a complex interaction of intrinsic, extrinsic, and target specific cues. These signals include a variety of classical growth or trophic factor interactions as well as the more subtle effects induced by establishing and maintaining appropriate synaptic connections (Patterson, 1990; Heumann, 1994; Jelsma and Aguayo, 1994; Barbacid, 1995; Calof, 1995; Temple and Qian, 1995).

A number of novel approaches have been developed to deliver therapeutic agents directly to an affected region of the brain. The most promising of these include fetal tissue grafting and somatic gene therapy (for review see Fisher and Gage, 1993; Suhr and Gage, 1993; Martin, 1995; Fisher, 1995; Crystal, 1995). Fetal tissue grafting has been used in a variety of experimental paradigms to explore developmental functions in the brain and more recently, to examine the potential for replacing dying or dead cells in specific areas of the CNS (Fisher and Gage, 1993). Though shown to be effective in several animal models and in limited human applications, fetal tissue grafting raises several technical and ethical issues related to acquiring and using aborted human tissues. As an alternative, somatic cell-mediated gene therapy, i.e., the grafting of genetically engineered autologous tissues, provides a plausible alternative for delivery of therapeutic factors to the brain (Suhr and Gage, 1993; Martin, 1995; Fisher, 1995; Crystal, 1995).

Three diseases that have received the majority of attention in gene therapy research are Alzheimer's, Parkinson's, and Huntington's diseases. Each disease has well-documented pathologies that involve cholinergic, dopaminergic, or GABA-ergic systems, respectively. And, each disease can be effectively modeled by inducing discreet lesions in a number of laboratory animal systems. In the case of Alzheimer's or Parkinson's disease, simple replacement of acetylcholine or dopamine within the affected region appears to alleviate some of the more acute symptoms of the disease. These observations have led to two parallel strategies in gene therapy: First, to prevent further degeneration of host circuitry by delivery of neurotrophic and/or growth factors and, second, to supplement the diminishing supply of neurotransmitter by grafting transmitter-synthesizing cells to areas normally innervated by the missing cells (Fisher and Gage, 1993; Suhr and Gage, 1993; Fisher, 1995; Martin, 1995).

In Alzheimer's disease, a loss of cholinergic neurons in the basal forebrain may likely lead to dementia and memory impairment. Approaches to gene therapy involve *in vivo* delivery of nerve growth factor (NGF) that prevents cholinergic cell loss, or direct delivery of choline acetyltransferase (ChAT) needed for the production of acetylcholine (Fisher, 1995). The dopaminergic neurons of the substantia nigra pars compacta that project to the striatum are lost in Parkinson's disease causing motor impairment. This model system focuses on a dopamine-replacement strategy by using genetically modified cells to deliver tyrosine hydroxylase (TH) or dopa decarboxylase, enzymes involved in the conversion of tyrosine to dopamine (Fisher and Gage, 1995). Huntington's disease is linked to selective loss of GABA-ergic interneurons in the striatum, the primary cause of choreaic movement associated with the disease (Dunnett and Svendsen, 1993; Martin, 1995). Gene therapy approaches for treatment of this disease are based on the delivery of the GABA by fibroblasts modified to produce the GABA synthesizing enzyme, glutamic acid decarboxylase (GAD) (Dunnett and Svendsen, 1993; Martin, 1995; Peschanski *et al.*, 1995).

Genetically modified skin fibroblasts have been successfully used to deliver transgenes in the CNS; the ectopically grafted cells survive as compact collagen-ensheathed deposits and are unable to make synaptic connection with host brain cells. Cells derived from CNS offer advantages over fibroblasts and new developments in the culture of neural progenitors from both fetal and adult brain tissues offer a unique alternative to acutely isolated fetal tissue. The concept of growing replacement neurons from small tissue biopsies is attractive but there are substantial advances needed to make neuronal replacement practical. If progenitors are to be used to restore neuronal function in the adult brain, it will be necessary to generate cells that function like the missing cells.

For each disease, transmitter phenotype, interconnection, and responses to the local environment must be appropriate for the cells being replaced. Given the complexity of developmental cues involved in establishing CNS structure or function, it may prove difficult to accurately mimic the developmental processes necessary to generate authentic circuits in the adult. These issues will undoubtedly be resolved but there are potential applications that may be within our current abilities.

III. WHAT DO WE KNOW ABOUT NEURAL PROGENITORS OR STEM CELLS OF THE CNS?

During development, multipotent neural stem cells proliferate to generate all of the neuronal or glial precursors that eventually make up the mature structures of the brain. Immature cells capable of self-renewal as well as making neuronal and glial progeny are termed stem cells. Immature cells in general are interchangeably called precursors or progenitors, though some tend to use one or the other term to indicate cells committed to either neuronal or glial lineage. For simplicity, in this chapter, we will group immature cells capable of proliferating *in vitro* (including multipotent stem cells and committed precursors) into the single category of progenitor. Although there are no markers that directly identify neural progenitors, their presence has been documented in a variety of ways. Transplantation of discreet cell populations between embryos, retroviral marking, or isolation of cells *in vitro* (for review see Gage *et al.*, 1995a; Temple and Qian, 1995, 1996; Kilpatrick *et al.*, 1995; Weiss *et al.*, 1996; Luskin *et al.*, 1997; Mckay, 1997; Ray *et al.*, 1997) all provide methods by which neural progenitors have been identified and evaluated for the ability to differentiate into glia and/or neurons of a specific type. These studies suggest several basic concepts. A given structure is formed by the proliferation of immature multipotent cells, the progeny of which migrate and differentiate into neurons and glia. Just as individual structures become specialized, the progenitors in each region become more restricted in the types of cells they generate. Although the adult brain remains competent to generate new glia, neurogenesis ultimately declines until it ceases in the young adult mammalian brain. There are, however, two locations in the rodent brain where neurogenesis continues well into adult life. These are the olfactory bulb and the granule cell layer of the hippocampus.

In the adult olfactory bulb, new neurons are generated from progenitors that originate in the subventricular zone (SVZ) of the lateral ventricle. These cells migrate from the lateral ventricle along a rostral migratory stream and eventually differentiate into neurons in the olfactory bulb (Luskin, 1993; Lois and Alvarez-Buylla, 1993; Lois *et al.*, 1996). In the subgranule cell layer of the hippocampal dentate gyrus, proliferative progenitors migrate a short distance into the granule cell layer where they differentiate into hippocampal granule cells (Altman and Das, 1966; Bayer, 1982; Kaplan and Bell, 1984; Kuhn *et al.*, 1996). Recent studies have shown that progenitors isolated from the adult mouse subventricular zone or the adult rat hippocampus are multipotent stem cells, i.e., cells capable of proliferating as immature progenitors that can differentiate into either neurons or glia (Gritti *et al.*, 1996; Gage *et al.*, 1995b; Palmer *et al.*, 1995). This indicates that even the adult brain may provide a source of cells that could be recruited *in vitro* or *in situ* as a source for new neurons and glia.

IV. IMMORTALIZED PROGENITORS CAN BE ISOLATED AND MAINTAINED IN VITRO

Until recently it has been difficult to maintain the long-term survival and/or proliferation of primary neural progenitors *in vitro*. Among approaches that have been utilized to obtain progenitors capable of proliferation and subsequent differentiation *in vitro* or *in vivo* involves immortalization of cells from developing brain by retroviral transduction of conditional oncogenes. Immortalized neural cell lines have been derived from different regions of mouse and rat developing brain (Snyder, 1994; Whittmore and Snyder, 1996; Snyder *et al.*, 1997; McKay, 1997). The majority of these neural cell lines were multipotent in culture retaining the potential to differentiate either along a neuronal or glial lineage (Bartlett *et al.*, 1988; Bernard *et al.*, 1989; Frederiksen *et al.*, 1988; Ryder *et al.*, 1990). Some immortalized clonal cell lines did not generate multiple phenotypes but differentiated to express exclusively glial or neuronal properties (Evard *et al.*, 1990; Frederiksen and McKay, 1988; Redies *et al.*, 1991; Whittemore and White, 1993; Hoshimaru *et al.*, 1996).

V. PRIMARY PROGENITORS CAN BE ISOLATED AND MAINTAINED IN VITRO

In the last several years, a number of investigators have utilized various cytokines, trophic factors, and/or astrocyte-conditioned medium to propagate nontransformed neural progenitor cells. Progenitors responsive to epidermal growth factor (EGF) have been isolated and cultured from mouse embryonic and adult striatum (Reynolds *et al.*, 1992; Reynolds and Weiss, 1992). Another growth factor, FGF-2, has been successfully used to isolate and culture progenitor cells from embryonic mouse cerebral cortex (Kilpatrick and Bartlett, 1995), mesencephalon and telencephalon (Murphy *et al.*, 1990; Kilpatrick and Bartlett,

1993), embryonic rat striatum (Catteneo and McKay, 1990), spinal cord (Deloulme *et al.*, 1991; Ray and Gage, 1994), and hippocampus (Ray *et al.*, 1993; Vicario-Abejon *et al.*, 1995). FGF-2-responsive progenitors have also been isolated from adult rat hippocampus (Gage *et al.*, 1995b), septum, and striatum (Palmer *et al.*, 1995) and adult mouse striatum (Gritti *et al.*, 1996). It has been suggested and demonstrated (Morshead *et al.*, 1994; Lois and Alvarez-Buylla, 1993) that EGF-responsive multipotential progenitors may be located predominantly in the subventricular zone, especially in the region surrounding the lateral ventricles. Alternatively, FGF-2 appears to be able to recruit similar populations of progenitors from both subventricular zone and underlying quiescent parenchyma of the adult rat brain (Richards *et al.*, 1992; Palmer *et al.*, 1995).

It has been reported that some neural progenitors that proliferate in response to FGF-2 generate progeny that predominantly express neuronal markers (Catteno and McKay, 1990; Ray *et al.*, 1993; Ray and Gage, 1994), suggesting that they represent a neuroblast population. However, the mitogenic action of FGF-2 is not restricted to neuroblasts; FGF-2 is also involved in the stimulation of multipotent precursors (Kilpatrick and Bartlett, 1995; Gage *et al.*, 1995b). Some of the multipotent FGF-2 responsive cells were completely unaffected by EGF (Kilpatrick and Bartlett, 1993). A recent study has reported that both EGF and FGF-2 are needed to recruit and culture progenitor cells from adult spinal cord (Weiss *et al.*, 1997). Thus, EGF and FGF-2 have the potential to induce proliferation of some population of cells from both fetal and adult CNS. However, at this point it is unclear whether the two growth factors act on the same or different progenitor cell populations.

VI. PROPERTIES OF CULTURED PRIMARY PROGENITORS

Clonal analysis of EGF and FGF-responsive cells demonstrated that some of these progenitors were capable of self-renewal and thus possess some of the properties of a stem cell. Individual cells from embryonic rat or mouse CNS (Kilpatrick and Bartlett, 1995; Reynolds and Weiss, 1996), adult mouse striatum (Gritti *et al.*, 1996), or adult rat hippocampus (Palmer *et al.*, 1997) (an intermediate filament protein expressed by progenitor cells) produced populations of nestin-immunoreactive cells that could differentiate into neurons, astrocytes, or oligodendrocytes.

Several studies show that a variety of trophic factors, growth factors, and hormones can be used to influence the differentiation of these cultured progenitors into neurons and/or glia. Factors such as serum, retinoic acid, FGF-2, brain-derived neurotrophic factor (BDNF), or neurotrophin 3 (NT-3), act on neural progenitor to enhance their differentiation into mature neurons or glia

(Vescovi, *et al.*, 1993; Kilpatrick and Bartlett, 1993; Ahmed *et al.*, 1995; Palmer *et al.*, 1997). In some experiments, serum-enriched or serum-free conditions as well as conditioned medium collected from astrocyte cell line Ast-1 have been used in combination with growth factors to influence the proliferation and differentiation of progenitors (Kilpatrick *et al.*, 1995; Gage *et al.*, 1995a; Ray *et al.*, 1997; Palmer *et al.*, 1995, 1997). As in development, a complex array of factors will likely be required to direct cell fate into a desired phenotype and a temporal cascade of growth factors may differentially influence the proliferation and differentiation of these multipotent stem cells.

VII. GRAFTING PROGENITORS TO THE FETAL AND ADULT BRAIN

The generation of immortalized neural cell lines and the ability to culture and expand primary neural progenitor cells *in vitro* have raised the questions: How do these cultured cells compare to their *in vivo* counterparts? Can these cells survive and integrate into normal cytoarchitectures of the intact adult brain in a functionally meaningful way? Can the local environment influence the fate of grafted cells? If these cells behave similarly to endogenous cells *in situ* or grafted fetal tissues, then it is very likely that they will have great potential for cell replacement and gene therapy purposes.

A. Grafting of Immortalized Progenitors in the Intact or Lesioned CNS

A number of studies have reported that immortalized progenitor cells, when grafted back into embryonic, neonatal, or adult brain, migrate from the site of implantation and differentiate into cells appropriate for their location (for review see Snyder, 1994; Whittemore and Snyder, 1996; Snyder *et al.*, 1997). Upon grafting into developing hippocampus and cerebellum, embryonic hippocampus-derived immortalized HiB5 cells integrate into host parenchyma and differentiate into cells with morphologies consistent with those of endogenous hippocampal dentate gyrus granule neurons and cerebellar granule neurons, respectively (Renfranz *et al.*, 1991). A neonatal cerebellum-derived cell line (C17-2) grafted in neonatal cerebellum differentiated into region specific neurons (granule and basket cells), astrocytes, and oligodendrocytes (Snyder *et al.*, 1992). Similarly, when the raphe-derived neuronally restricted cell line RN33B was grafted into neonatal cerebral cortex and hippocampus, cells differentiated in a site-specific manner (Shihabuddin *et al.*, 1995, 1996). Morphologies of cells grafted in the hippocampus were identical to endogenous CA1 and CA3 pyramidal neurons and dentate gyrus neurons. In the cerebral cortex, grafted cells differentiated into cortical pyramidal and stellate neurons, expressed mature pyramidal cell protein, and made synaptic connections with host axons. The above studies raised a crucial question: Can the adult CNS similarly support the differentiation of the implanted progenitor cells? A number of researchers have addressed this issue by grafting immortalized embryonic cell lines into different CNS regions of the adult. RN33B differentiated into region-specific neurons when grafted into adult cerebral cortex, hippocampus, and spinal cord (Shihabuddin *et al.*, 1995, 1996; Onifer *et al.*, 1993) but RN46A, another raphe-derived cell line, did not exhibit extensive morphological differentiation and survived for only a short period of time (Whittemore *et al.*, 1995). Hippocampus-derived HiB5 and the striatum-derived ST14A cells preferentially differentiated into glia-like cells and were found primarily within the injection site (Lundberg and Björklund, 1996). It is clear from these studies that although adult CNS retains the potential for directing integration and differentiation of the grafted cells, the environment is restrictive, and proliferation and specific differentiation of all cell lines are not supported similarly.

B. Grafting of Cultured Fetal Progenitors in the Intact or Lesioned Adult Brain

The ability to isolate and culture fetal and adult brain-derived primary neural progenitor cells has raised the question whether the cultured primary cells grafted in the adult brain behave similarly to the immortalized cells and undergo differentiation in a site- and age-specific manner. FGF-2 responsive progenitor cells isolated and cultured from fetal hippocampus have been grafted homotypically (hippocampus) and heterotypically (striatum) in the adult rat brain. In each case grafted cells migrated from the site of implantation and differentiated into neurons and astrocytes (Ray *et al.*, 1997). In the hippocampus, eight weeks post-grafting, cells migrated to the dorsal leaf (where there was damage due to injection of the fluid) and differentiated into astrocytes. Cells migrating to the granule cell layer of the dentate gyrus differentiated into neurons. Progenitor cells cultured from embryonic spinal cord behaved similarly to hippocampus-derived cells when implanted into adult rat spinal cord and gave rise to both neurons and astrocytes (Ray *et al.*, 1997). These findings are different from studies that implanted EGF-responsive progenitor cells cultured from embryonic rat and human brains into different regions of the adult rat brain (Svendsen *et al.*, 1996). Cultured EGF-responsive progenitor cells from embryonic striatum or mesencephalon grafted into the striatum with ibotenic acid or nigrostriatal lesions formed small sparse grafts containing only a very few surviving cells, which remained confined mostly to the injection site.

There was little evidence of migration of the grafted cells. EGF-responsive progenitors from mesencephalon grafted into the striatum did not give rise to any graft-derived TH-positive cells and no behavioral improvement was seen. In the same study, EGF-responsive fetal human progenitor cells were grafted in immunosuppressed rats with nigrostriatal lesions. These grafts behave similarly to grafts of cells derived from fetal rat brain. These studies showed that although both EGF and FGF-2-responsive progenitors from fetal brain can survive in adult brain after transplantation, they behave differently *in vivo*. FGF-2-responsive fetal rat brain-derived progenitors survive well, migrate extensively, and differentiate into neurons and glia, whereas EGF-responsive progenitors showed poor survival, and little or no migration and differentiation. These differences may be due to the inherent properties of EGF- and FGF-2-responsive progenitors or due to lack of signals in lesioned brains where EGF-generated progenitors were grafted. Future studies involving the grafting of both EGF and FGF-2-generated progenitors in the same experimental paradigm may address these issues. Interestingly, when a nonpassaged culture containing a mixture of primary cells and EGF-responsive progenitors from human mesencephalon were implanted in lesioned adult rat striatum, cells gave rise to large grafts containing many small undifferentiated cells (Svendsen *et al.*, 1996). These grafts had the appearance of primary human mesencephalic tissue grafts. Implantation of nonpassaged FGF-2 responsive human fetal progenitors (presumably containing a mixture of primary and expanded population of cells) in intact adult rat striatum showed clusters of surviving cells around the injection site only in animals implanted with one million cells or more (Sabate *et al.*, 1995). Morphology of most of the surviving cells were neuroblast-like, and glial cells represented a minority population. These studies indicate that primary cells grafted along with progenitors may be providing a microenvironment in which the immature progenitors can survive better and form larger grafts.

C. Grafting of Adult-Derived Progenitors in the Adult CNS

Although progenitors exist in the adult rodent brain, there is considerably less information regarding the potential utility of using grafts of adult-derived progenitor cultures to restore function. Two studies provide some insight on how the local cues presented to the progenitors may have a significant impact on graft outcome. Adult-derived hippocampal progenitors were implanted into the normal adult hippocampus, the rostral migratory stream leading to the olfactory bulb (Fig. 1, see color insert) or into the cerebellum (Gage *et al.*, 1995b; Suhonen *et al.*, 1996). Though cells survived in each location, there were striking differences in the phenotypes of the surviving cells. The rostral migratory stream and the hippocampal granule cell layer provide cues that direct endogenous progenitors to migrate and differentiate into site-specific neurons (Fig. 2, see color insert). These studies showed that the grafts of cultured progenitors migrate from the site of implantation and, within neurogenic zones, differentiate into neurons. Cells found outside of these areas or within the area of injury caused by the injection itself differentiated into glia. This dependence on local cues suggests that simply injecting immature progenitors into an injured area of the adult brain may not yield significant repair. Effective therapy may require manipulation of the cells or host prior to injection and, as part of this manipulation, it may be necessary to recapitulate the developmental cues (if possible) required for generating appropriate cells or connections.

D. Cross-Species Grafting of Fetal Tissues

An alternative to human tissue grafts is the use of fetal tissues from other species. To examine whether host brain signals have influence on the donor tissues, dissociated fetal cells have been transplanted heterotypically (Campbell *et al.*, 1995; Isacson *et al.*, 1995). These studies have shown that cells grafted cross-species survived and exhibited distinct patterns of integration in the host brain. Neural cells from embryonic mouse lateral ganglionic eminence (LGE) grafted into embryonic rat striatum incorporated into the host striatum and differentiated into characteristic striatal projection neurons. Cells from medial ganglionic eminence or ventral mesencephalon did not show such preference (Campbell *et al.*, 1995). To explore the possible use of xenotransplantation in human diseases, Isacson and colleagues (1995) grafted embryonic pig ventral mesencephalon cells into striatum or into the dorsal mesencephalon of adult rats with nigrostriatal lesions (Parkinson's model) or LGE cells in neuron-depleted striata (Huntington's model). Examination of patterns and specificity of target-directed long-distance axon and glia fiber growth showed the presence of donor axons in host white and gray matters. However, glial fibers grew nonspecifically into host white matter, but did not penetrate into any gray matter zone. These observations suggest that cell migration, differentiation, and donor cell axonal growth are regulated by the species-common environmental signals and not by the intrinsic program of the cells. Thus adult host brain may well be able to direct the growth and development of the transplanted xenotypic cells.

E. Grafting of Oligodendrocyte Precursors

Rather than approach the most difficult problems first, i.e., restoration of complex neural circuits, the initial utility of progenitors that can differentiate into neurons or glia may lie in treating disorders that are primarily glial in nature. The most relevant of these may be inherited or injury-induced demyelination. In theory, remyelination may be as simple as injecting oligodendrocyte progenitors into a demyelinating tract (Duncan, 1996). Many of the issues regarding the repair of long tract injuries such as those seen in spinal trauma are largely problems of oligodendrocyte biology.

The ease with which oligodendrocytes can be generated from progenitors suggests that they could be used for remyelination. One of the areas receiving the most attention is the spinal cord, where in oligodendrocyte replacement could be useful in demyelinating syndromes or in the repair of spinal cord injury. Even in advanced demyelinating disease or injury, there are often large tracts of descending projections that could be rescued if endogenous or exogenously introduced progenitors were induced to form mature oligodendrocytes. The potential utility of such an approach is demonstrated in several experiments in rats and dogs (Tontsch *et al.*, 1994; Duncan and Milward, 1995; Duncan, 1996). The striking remyelination following grafting of normal oligodendrocyte precursors into the diseased cord provides a therapeutic approach for the repair of myelin in demyelinating diseases.

VIII. GENETIC MANIPULATION AND PROGENITOR GRAFTS

The feasibility of using neural progenitors to deliver therapeutic factors in the brain as integral members of the brain structures with minimal interference with the normal biological processes has been tested by number of groups (Table I). Because the immortalized or primary neural stem cells divide in culture, they can be stably infected with viral vectors to express a gene of interest. Snyder and colleagues (1995) have used genetically modified immortalized cerebellar cell line C17-2 to deliver β-glucuronidase (GUSB) in an animal model of genetic neurodegenerative lysosomal storage disease mucopolysaccharidosis type VII (MPS VII). Diffused grafting of C17-2 cells stably expressing GUSB into the cerebral ventricles of MPS VII transgenic newborn mice corrected lysosomal storage throughout the brains of homozygous mutants devoid of the secreted enzyme.

Immortalized stem cells expressing a-subunit of β-hexosaminidase (HexA) have also been used to explore their potential use in Tay-Sachs disease (Lacorazza *et al.*, 1996). In this disease a mutation in HexA a-subunit causes an accumulation of GM_2 ganglioside leading to a severe neurological disorder. Genetically engineered cerebellar cell lines expressing human HexA a-subunit were grafted in normal fetal and newborn mice. Engrafted brains, analyzed at various time points after transplantation, showed the presence of both tran-

TABLE I Studies with Genetically Modified Neural Cells in Animal Models of Neurodegenerative Diseases

Model system	Cell type	Transgene	Graft site/lesion	Differentiation *in vivo*	Length of study/transgene expression *in vivo*	*In vivo* measure	References
Lysosomal storage (Sly) disease	C17-2	β-glucuronidase (GUSB)	Newborn mucopolysaccharidosis VII mouse cerebral ventricles	Neurons/Glia	12 weeks	Correction of lysosomal storage in neurons, and glia	Snyder et al., 1995
Tay-Sachs disease	C17-2	β-hexosaminidase a-subunit (HexA)	Normal fetal and newborn mouse	Neurons/Glia	8 weeks	Expression of enzyme	Lacorazza *et al.*, 1995
Learning and memory	HiB5	NGF	Adult rat septum/septohippocampal lesion	Glia	2 weeks	Septal cholinergic cell savings	Martinez-Serrano *et al.*, 1995
Age-induced memory impairments	HiB5	NGF	Age-impaired rat medial septum and nucleus basalis magnocellularis	Glia	10 weeks	Functional recovery of cognitive function	Martinez-Serrano *et al.*, 1996
Huntington's disease	HiB5	NGF/BDNF	Adult rat striatum/excitotoxic lesions	Glia	4 weeks	Rescue of striatal projecting and cholinergic neurons by NGF; No effect of BDNF	Martinez-Serrano and Björklund, 1996

script and protein of HexA a-subunit throughout the brain at a level proposed to be therapeutic in Tay-Sachs disease.

Immortalized cell lines expressing trophic factors have also been used in animal models of memory impairment and Huntington's disease. A conditionally immortalized hippocampal cell line (HiB5) secreting NGF was grafted in the striatum of adult intact rats and it showed that the cells migrated from the site of injection to scatter in the host brain parenchyma around the injection site and differentiate into cells with glial morphology (Martinez-Serrano *et al.*, 1995). In addition, there was an accumulation of low-affinity NGF receptor (p75) expressing fibers around the grafted cells that expressed NGF *in vivo* for 2 weeks posttransplantation. When grafted into the septum of adult rats with complete fimbria–fornix transection, NGF-secreting cells prevented over 90% of the cholinergic cell loss, and cell size of the rescued cells was similar to those on the contralateral side (Martinez-Serrano *et al.*, 1995). The functional efficacy of these cells was demonstrated by improved spatial learning of rats with NGF-secreting grafts in nucleus basalis magnocellularis (NBM) alone or in both NBM and medial septum (MS). When NGF-secreting cells were grafted in the basal forebrain of learning-impaired aged rats, the long-term recovery of spatial learning and memory was observed in a 10-week testing paradigm (Martinez-Serrano *et al.*, 1996). NGF-producing grafts induced a hypertrophic response of cholinergic neurons in the target-grafted regions of medial septum. In contrast, BDNF-producing grafts showed no such effects.

In a rat model of Huntington's disease based on the excitotoxic lesion of the striatum with quinolonic acid, NGF-producing HiB5 cells rescued striatal cholinergic and spiny projection neurons and reduced the astroglial and microglial reactions to the excitotoxic lesion in the striata (Martinez-Serrano and Björklund, 1996). BDNF-producing HiB5 cells showed only marginal effect. These results indicate that genetically modified neural stem cells represent a highly effective way to deliver therapeutic factors to counteract neurodegeneration due to inherited diseases, trauma, or excitotoxic damages.

IX. CONCLUSION

Emerging data suggests that CNS-derived neural progenitor cells, as either immortalized or primary propagated cultures, will be very useful for understanding issues related to neuronal plasticity and the extracellular signals and genes involved in cell fate decisions and differentiation. The plasticity of the multipotent progenitors and the influence of site-specific signals have been examined by grafting the cultured cells into different regions of the adult CNS (Gage *et al.*, 1995b; Suhonen *et al.*, 1996). While showing a promising trend to form cells appropriate for the graft location, grafts into the adult brain are highly sensitive to the presence or absence of local cues that direct cell fate. Adult rat hippocampus-derived progenitor cells grafted in intact hippocampus or olfactory system migrate and undergo differentiation at neurogenic areas to generate site-specific neurons. However, outside of these areas, the cues direct primarily glial differentiation. These results indicate that the repair of many areas of the adult brain will require considerable outside influence to achieve appropriate cell types. Some of this manipulation could take the form of genetic manipulation of either host or graft, a possibility proven effective in several rodent models.

Most of the neural progenitor grafting studies reported so far have used immortalized or primary mouse- or rat-CNS-derived progenitors cells. However, for application in human diseases it will be necessary to show that progenitor cells exist in the fetal and adult human brain and that they have properties similar to those of rodent cells. To this extent, clonal cultures of progenitor cells from the fetal human brain have been shown to be multipotent in that they give rise to cells expressing either neuronal or glial markers after treatments with appropriate differentiating factors (Sah *et al.*, 1997). Properties of these cells are similar to those of rat CNS-derived progenitor cells. When fetal human brain-derived progenitor cells genetically modified to express *E. coli* LacZ gene were grafted into rat brain, a small number of cells expressed the transgene *in vivo* for 2 to 3 weeks (Sabate *et al.*, 1995). In short-term cultures, cells isolated from adult human forebrain expressed neuronal markers (Kirschenbaum *et al.*, 1994). However, only a very small number of neurons were actually derived from proliferating precursor cells indicating that although precursor cells are present in the adult human brain, their numbers may be extremely low and the recovery of useful numbers of neuronal precursors may be limiting.

Because most neurological diseases are characterized by multifocal or widespread pathology, human neural progenitors (in spite of their apparent rarity) have inherent biological advantages over nonneural tissues in that grafted cells can migrate and integrate into the site of damage where they could differentiate to replace the dying neurons and glia. This intrinsic capability and the potential to produce transgene factors that would be beneficial to the host cells and that may aid in the regeneration of neural circuitry *in situ* make autologous grafts very intriguing. On the other hand, autologous cells carrying an inherited genetic abnormality would, in theory, require some form of genetic modification prior to grafting. Rodent models have established that progenitor cells can be genetically modified to produce factors that affect the local tissues following grafting. In the treatment of inherited disease, gene therapy could be combined with progenitor culture and grafting to generate cells ideally suited for the delivery of therapeutic molecules to halt the degenerating process and promote regeneration and the establishment of functional connections.

REFERENCES

Ahmed, S., Reynolds, B. A., and Weiss, S. (1995). BDNF enhances the differentiation but not the survival of CNS stem cell-derived neuronal precursors. *J Neurosci* 15:5765–5778.

Altman, J., and Das, G. D. (1966). Autoradiographic and histological studies of postnatal neurogenesis. I. A longitudinal investigation of the kinetics, migration and transformation of cells incorporating tritiated thymidine. *J Comp Neurol* 126:337–390.

Barbacid, M. (1995). Neurotrophic factors and their receptors. *Curr Opin Cell Biol* 7:148–155.

Bartlett, P. F., Reid, H. H., Bailey, K. A., and Bernard, O. (1988). Immortalization of mouse neural precursor cells by the c-myc oncogene. *Proc Natl Acad Sci USA* 85:3255–3259.

Bayer, S. A. (1982). Changes in the total number of dentate granule cells in juvenile and adult rats: a correlated volumetric and ^{3}H-thymidine autoradiographic study. *Exp Brain Res* 46:315–323.

Bernard, O., Reid, H. H., and Bartlett, P. F. (1989). Role of c-myc and the N-myc proto-oncogenes in the immortalization of neural precursors. *J Neurosci Res* 24:9–20.

Calof, A. L. (1995). Intrinsic and extrinsic factors regulating vertebrate neurogenesis. *Curr Opin Neurobiol* 5:19–27.

Campbell, K., Olsson, M., and Björklund, A. (1995). Regional incorporation and site-specific differentiation of striatal precursors transplanted to the embryonic forebrain ventricle. *Neuron* 15:1259–1273.

Catteneo, E., and McKay, R. (1990). Proliferation and differentiation of neuronal stem cells regulated by nerve growth factor. *Nature* 347:762–765.

Crystal, R. G. (1995). Transfer of genes to humans: Early lessons and obstacles to success. *Science* 270:404–410.

Deloulme, J. C., Baudier, J., and Sensenbrenner, M. (1991). Establishment of pure neuronal cultures from fetal rat spinal cord and proliferation of the neuronal precursor cells in the presence of fibroblast growth factor. *J Neurosci Res* 29:499–509.

Duncan, I. D. (1996). Glial cell transplantation and remyelination of the central nervous system. *Neuropathol Appl Neurobiol* 22:87–100.

Duncan, I. D., and Milward, E. A. (1995). Glial cell transplants: Experimental therapies of myelin diseases. *Brain Pathol* 5:301–310.

Dunnett, S. B., and Svendsen, S. N. (1993). Huntington's disease: Animal models and transplantation repair. *Curr Opin Neurobiol* 3:790–796.

Evard, C., Borde, I., Martin, P., Galiana, S. E., Premont, J., Gros, F., and Rouget, P. (1990). Immortalization of bipotential and plastic glio-neuronal precursor cells. *Proc Natl Acad Sci USA* 87:3062–3066.

Fisher, L. J. (1995). Engineered cells: A promising therapeutic approach for neural disease. *Resto Neurol Neurosci* 8:49–57.

Fisher, L. J., and Gage, F. H. (1993). Grafting in the mammalian central nervous system. *Physiol Rev* 73:583–616.

Fisher, L. J., and Gage, F. H. (1995). Novel therapeutic directions for Parkinson's disease. *Mol Med Today*. Alsevier Trends Journal, pp. 181–187.

Frederiksen, K., Jat, P. S., Valtz, N., Levy, D., and McKay, R. (1988). Immortalization of precursor cells from the mammalian CNS. *Neuron* 1:439–448.

Frederiksen, K., and McKay, R. D. G. (1988). Proliferation and differentiation of rat neuroepithelial precursor cells *in vivo*. *J Neurosci* 8:1144–1151.

Gage, F. H., Coates, P. W., Palmer, T. D., Kuhn, H. G., Fisher, L. J., Suhonen, J. O., Peterson, D. A., Suhr, S. T., and Ray, J. (1995b). Survival and differentiation of adult neuronal progenitor cells transplanted to the adult brain. *Proc Natl Acad Sci USA* 92:11879–11883.

Gage, F. H., Ray, J., and Fisher, L. J. (1995a). Isolation, characterization and use of stem cells from the CNS. *Ann Rev Neurosci* 18:159–192.

Gritti, A., Parati, E. A., Cova, L., Frolichsthal, P., Galli, R., Wanke, E., Faravelli, L., Morassutti, D. J., Roisen, F., Nickel, D. D., and Vescovi, A. L. (1996). Mutipotential stem cells from the adult mouse brain proliferate and self-renew in response to basic fibroblast growth factor. *J Neurosci* 16:1091–1100.

Heumann, R. (1994). Neurotrophin signalling. *Curr Opin Neurobiol* 4:668–679.

Hoshimaru, M., Ray, J., Sah, D. W. Y., and Gage, F. H. (1996). Differentiation of immortalized adult neuronal progenitor cell ine HC2S2 into neurons by regulatable suppression of v-myc oncogene. *Proc Natl Acad Sci USA* 93:1518–1523.

Isacson, O., Deacon, T. W., Pakzaban, P., Galpern, W. R., Dinsmore, J., and Burns, L. H. (1995). Transplanted xenogeneic neural cells in neurodegenerative disease models exhibit remarkable axonal target specificity and distinct growth patterns of glial and axonal fibers. *Nature Med* 1:1189–1194.

Jelsma, T. N., and Aguayo, A. J. (1994). Trophic factors. *Curr Opin Neurobiol* 4:717–725.

Kaplan, M. S., and Bell, D. H. (1984). Mitotic neuroblasts in the 9-day-old and 11-month-old rodent hippocampus. *J Neurosci* 4:1429–1441.

Kilpatrick, T. J., and Bartlett, P. F. (1993). Cloning and multipotential neural precursors: Requirements for proliferation and differentiation. *Neuron* 10:255–265.

Kilpatrick, T. J., and Bartlett, P. F. (1995). Cloned multipotential precursors from the mouse cerebrum require FGF-2, whereas glial restricted precursors are stimulated with either FGF-2 or EGF. *J Neurosci* 15:3563–3661.

Kilpatrick, T. J., Richards, L. J., and Barlett, P. F. (1995). The regulation of neural precursor cells within the mammalian brain. *Mol Cell Neurosci* 6:2–15.

Kirschenbaum, B., Nedergaard, M., Preuss, A., Barami, K., Fraser, R., and Goldman, S. (1994). *In vitro* neuronal production by precursor cells derived from the adult human brain. *Cereb Cortex* 4:576–589.

Kuhn, H. G., Dickinson-Anson, H., and Gage, F. H. (1996). Neurogenesis in the dentate gyrus of the adult rat: Age-related decrease of neuronal progenitor proliferation. *J Neurosci* 16:2027–2033.

Lacorazza, H. D., Flax, J. D., Snyder, E. Y., and Jendoubi, M. (1996). Expression of human β-hexosaminidase a-subunit gene (the gene defect of Tay-Sachs disease) in mouse brains upon engraftment of transduced progenitor cells. *Nature Med* 2:424–429.

Lois, C., and Alvarez-Buylla, A. (1993). Proliferating subventricular zone cells in the adult mammalian forebrain can differentiate into neurons and glia. *Proc Natl Acad Sci USA* 90:2074–2077.

Lois, C., Garcia-Verdugo, J. M., and Alvarez-Buylla, A. (1996). Chain migration of neuronal precursors. *Science* 264:1145–1148.

Lundberg, C., and Björklund, A. (1996). Host regulation of glial markers in intrastriatal grafts of conditionally immortalized neural stem cell lines. *NeuroReport* 7:847–852.

Luskin, M. B. (1993). Restricted proliferation and migration of postnatally generated neurons derived from the forebrain subventricular zone. *Neuron* 11:173–189.

Luskin, M. B., Zigova, T., Betarbet, R., and Soteres, B. J. (1997). Characterization of neuronal progenitor cells of the neonatal forebrain. In: Gage, F. H., and Christen, Y., eds., *Research and perspective in neurosciences. Isolation, characterization and utilization of CNS stem cells.* Heidelberg: Fondation IPSEN, Springer, pp. 67–86.

Martin, J. B. (1995). Gene therapy and pharmacological treatment of inherited neurological disorders. *Trends Biotechnol* 13:28–35.

Martinez-Serrano, A., Lundberg, C., Horellou, P., Fischer, W., Bentlage, C., Campbell, K., McKay, R. D. G., Mallet, J., and Björklund, A. (1995). CNS-derived neural progenitor cells for gene transfer of nerve growth factor to the adult rat brain: Complete rescue of axotomized cholinergic neurons after transplantation into the septum. *J Neurosci* 15:5668–5680.

Martinez-Serrano, A., and Björklund, A. (1996). Protection of the neostriatum against excitotoxic damage by neurotrophin-producing, genetically modified neural stem cells. *J Neurosci* 16:4604–4616.

Martinez-Serrano, A., Fischer, W., Soderstrom, S., Ebendal, T., and Björklund, A. (1996). Long-term functional recovery from age-induced spatial memory impairments by nerve growth factor gene transfer to the rat basal forebrain. *Proc Natl Acad Sci USA* 93:6355–6360.

McKay, R. (1997). Stem cells in the central nervous system. *Science* 276:66–71.

Morshead, C. M., Reynolds, B. A., Craig, C. G., Mcburney, M. W., Staines, W. A., Morassutti, D., Weiss, S., and van der Kooy, D. (1994). Neural stem cells in the adult mammalian forebrain: A relatively quiescent subpopulation of subependymal cells. *Neuron* 13:1071–1082.

Murphy, M., Drago, J., and Bartlett, P. F. (1990). Fibroblast growth factor stimulates the proliferation and differentiation of neural precursor cells *in vitro*. *J Neurosci Res* 25:463–475.

Onifer, S. M., Whittemore, S. R., and Holets, V. R. (1993). Variable morphological differentiation of raphe-derived neuronal cell line following transplantation into the adult rat CNS. *Exp Neurol* 122:130–142.

Palmer, T. D., Ray, J., and Gage, F. H. (1995). FGF-2-responsive neuronal progenitors reside in proliferative and quiescent regions of the adult rodent brain. *Mol Cell Neurosci* 6:474–486.

Palmer, T. D., Takahashi, J., and Gage, F. H. (1997). The adult rat hippocampus contains primordial neural stem cells. *Mol Cell Neurosci* 8:389–404.

Patterson, P. H. (1990). Control of cell fate in a vertebrate neurogenic lineage. *Cell* 62:1035–1038.

Peschanski, M., Cesaro, P., and Hantraye, P. (1995). Rational for intrastriatal grafting of striatal neuroblasts in patients with Huntington's disease. *Neurosci* 68:273–285.

Ray, J., and Gage, F. H. (1994). Spinal cord neuroblasts proliferate in response to basic fibroblast growth factor. *J Neurosci* 14:3548–3564.

Ray, J., Palmer, T. D., Suhonen, J. O., Takahasi, J., and Gage, F. H. (1997). Neurogenesis in the adult brain: Lessons learned from the studies of progenitor cells from embryonic and adult central nervous system. In: Gage, F. H., and Christen, Y. eds., *Research and perspective in neurosciences. Isolation, characterization and utilization of CNS stem cells.* Heidelberg: Fondation IPSEN, Springer, pp. 129–149.

Ray, J., Peterson, D. A., Schinstine, M., and Gage, F. H. (1993). Proliferation, differentiation, and long-term culture of primary hippocampal neurons. *Proc Natl Acad Sci USA* 90:3602–3606.

Redies, C., Lendahl, U., and Mckay, R. D. G. (1991). Differentiation and heterogeneity in T-antigen immortalized precursor cell lines from mouse cerebellum. *J Neurosci Res* 30:601–615.

Renfranz, P. J., Cunningham, M. G., and Mckay, R. D. G. (1991). Region-specific differentiation of the hippocampal stem cell line HiB5 upon implantation into the developing mammalian brain. *Cell* 66:713–729.

Reynolds, B. A., Tetzlaff, W., and Weiss, S. (1992). A multipotent progenitor cell produces neurons and astrocytes. *J Neurosci* 12:4565–4574.

Reynolds, B. A., and Weiss, S. (1992). Generation of neurons and astrocytes from isolated cells of the adult mammalian central nervous system. *Science* 255:1707–1710.

Reynolds, B. A., and Weiss, S. (1996). Clonal and population analyses demonstrate that an EGF-responsive mammalian embryonic CNS precursor is a stem cell. *Develop Biol* 175:1–13.

Richards, L. J., Kilpatrick, T. J., and Bartlett, P. F. (1992). *De novo* generation of neuronal cells from the adult mouse brain. *Proc Natl Acad Sci USA* 89:8591–8595.

Rosenberg, R. N., and Iannaccone, S. T. (1995). The prevention of neurogenetic disease. *Arch Neurol* 52:356–362.

Ryder, E. F., Snyder, E. Y., and Cepko, C. L. (1990). Establishment and characterization of multipotent neural cell lines using retrovirus vector-mediated oncogene transfer. *J Neurobiol* 21:356–375.

Sabata, O., Horellou, P., Vigne, E., Colin, P., Perricaudet, M., Buc-Caron. M.-H., and Mallet, J. (1995). Transplantation to the rat brain of human neural progenitors that were genetically modified using adenovirus. *Nature Genet* 9:256–260.

Sah, D. W. Y., Ray, J., and Gage, F. H. (1997). Bipotent progenitor cell lines from the human CNS. *Nature Biotech* 15 (in press).

Shihabuddin, L. S., Hertz, J. A., Holets, V. R., and Whittemore, S. R. (1995). The adult CNS retains the potential to direct region-specific differentiation of a transplanted neuronal precursor cell line. *J Neurosci* 15:6666–6678.

Shihabuddin, L. S., Brunschwig, J.-P., Holets, V. R., Bunge, M. B., and Whittemore, S. R. (1996). Induction of mature neuronal properties in immortalized neuronal precursor cells following transplantation in the neonatal CNS. *J Neurocyto* 125:101–111.

Snyder, E. Y. (1994). Grafting immortalized neurons to the CNS. *Curr Opin Neurobiol* 4:742–751.

Snyder, E. Y., Deitcher, D. L., Walsh, C., Arnold-Aldea, S., Hatweig, E. A., and Cepko, C. L. (1992). Multipotent neural cell lines can engraft and participate in development of mouse cerebellum. *Cell* 66:33–51.

Snyder, E. Y., and Fisher, L. J. (1996). Gene therapy in neurology. *Curr Opin Ped* 8:558–568.

Snyder, E. Y., Flax, B. D., Yandava, K. I., Park, S., Liu, C. M., and Aurora, S. (1997). Transplantation and differentiation of neural "stem-like" cells: Possible insights into development and therapeutic potentials. In: Gage, F. H. and Christen, Y. eds., *Research and perspective in neurosciences. Isolation, characterization and utilization of CNS stem cells.* Heidelberg: Fondation IPSEN, Springer, pp. 173–196.

Snyder, E. Y., Taylor, R. M., and Wolfe, J. H. (1995). Neural progenitor cell engraftment corrects lysosomal storage throughout the MPS VII mouse brain. *Nature* 374:367–370.

Suhonen, J. O., Peterson, D. A., Ray, J., and Gage, F. H. (1996). Differentiation of adult-derived hippocampal progenitor cells into olfactory bulb neurons. *Nature* 382:624–627.

Suhr, S., and Gage, F. H. (1993). Gene therapy for neurologic disease. *Arch Neurol* 50:1252–1268.

Svendsen, C. N., Clarke, D. J., Rosser, A. E., and Dunnett, S. B. (1996). Survival and differentiation of rat and human epidermal growth factor-responsive precursor cells following grafting into the lesioned adult central nervous system. *Exp Neurol* 137:376–388.

Temple, S., and Qian, X. (1995). bFGF, neurotrophins, and the control of cortical neurogenesis. *Neuron* 15:249–252.

Temple, S., and Qian, X. (1996). Vertebrate neural progenitor cells: Subtypes and regulation. *Curr Opin Neurobiol* 6:11–17.

Tontsch, U., Archer, D. R., Dubois-Dalcq, M., and Duncan, I. D. (1994). Transplantation of an oligodendrocyte cell line leading to extensive myelination. *Proc Natl Acad Sci USA* 91::11616–11620.

Vescovi, A. L., Reynolds, B. A., Fraser, D. D., and Weiss, S. (1993). bFGF regulates the proliferative fate of unipotent (neuronal) and bipotent (neuronal/astroglial) EGF-generated CNS progenitor cells. *Neuron* 11:951–966.

Vicario-Abejon, C., Johe, K. K., Hazel, T. G., Collazo, D., and McKay, R. D. G. (1996). Functions of basic fibroblast growth factor and neurotrophins in the differentiation of hippocampal neurons. *Neuron* 15:105–114.

Weiss, S., Dunne, C., Hewson, J., Wohl, C., Wheatley, M., Peterson, A. C., and Reynolds, B. A. (1997). Multipotent CNS stem cells are present in the adult mammalian spinal cord and ventricular neuroaxis. *J Neurosci* 16:7599–7609.

Weiss, S., Reynolds, B. A., Vescovi, A. L., Morshead, C., Craig, C. G., and van der Kooy, D. (1996). Is there a neural stem cell in the mammalian forebrain? *Trends Neurosci* 19:387–393.

Whittemore, S. R., and Snyder, E. Y. (1996). Physiological relevance and functional potential of central nervous system-derived cell lines. *Mol Neurobiol* 12:13–38.

Whittemore, S. R., and White, L. A. (1993). Target regulation of neuronal differentiation in a temperature-sensitive cell line derived from medullary raphe. *Brain Res* 615:27–40.

Whittemore, S. R., White, L. A., Shihabuddin, L. S., and Eaton, M. J. (1995). Phenotypic diversity in neuronal cell lines derived from raphe nucleus by retroviral transduction. *Methods: Companion Meth Enzymol* 7:285–293.

CHAPTER 8

Neural Stem Cell Lines for CNS Repair

ALBERTO MARTÍNEZ-SERRANO* AND EVAN Y. SNYDER†

*Center of Molecular Biology "Severo Ochoa," Autonomous University of Madrid–CSIC, Campus Cantoblanco, Madrid, Spain, †Departments of Neurology (Division of Neuroscience) and Pediatrics (Division of Newborn Medicine), Harvard Medical School, Children's Hospital, Boston, Massachussetts

The establishment and use of stable, engraftable, clonal, multipotent neural stem cell lines have recently added an exciting new dimension to strategies for cell replacement and gene transfer to the diseased mammalian CNS. These neural stem cell clones can serve as convenient, well-controlled models for the *in vivo* study of CNS development and regeneration; can constitute readily available, well-characterized, safe sources of graft material for the replacement of multiple types of degenerated neural cells; and can provide excellent vehicles for the transfer of genes encoding diffusible and nondiffusible factors directly to the CNS. By exploiting their basic biologic properties, these cells are be able to bypass restrictions imposed by the blood–brain barrier to deliver therapeutic gene products in a sustained, direct, and perhaps regulated fashion throughout the CNS (either because they intrinsically produce these substances or because they have been genetically engineered *ex vivo* to do so). Furthermore, although they may disseminate these gene products throughout the brain, they nevertheless restrict that distribution to *only* the CNS. In addition, they may replace dysfunctional neural cells in both a site-specific and more global manner (circumventing the concern that in many alternative gene transfer techniques "new" genetic information is supplied to "old" neural circuits, many of which may have degenerated). Neural stem cell clones may be used for neurodegenerative conditions that occur both during development and in the mature brain. In fact, they appear to be capable of altering their migration and differentiation in response to certain as yet unspecified signals elaborated during active neurodegeneration. Thus, these vehicles may overcome many of the limitations of viral and nonneural cellular vectors, as well as pharmacologic and genetic interventions. A

CNS Regeneration

growing body of evidence has, indeed, affirmed the efficacy of a neural stem cell-based strategy for the replacement of defective or absent genes and cells, and has suggested that repopulation of the diseased or injured CNS with such cells may promote both anatomic and behavioral recovery in animal models of neurodegenerative conditions. These recent experiments with clones of rodent neural progenitor and stem cells are bringing us rapidly closer to the challenge of repairing the CNS in genuine clinical settings using similarly well-characterized, fully controlled, multifaceted cellular tools of *human* origin.

I. INTRODUCTION

With the earliest recognition that multipotent neural progenitors and stem cells, propagated in culture, could be reimplanted into mammalian brain where they could reintegrate appropriately and stably express foreign genes [Snyder *et al.*, 1992; Renfranz *et al.*, 1991], restorative neurobiologists began to speculate how such a phenomenon might be harnessed for therapeutic advantage. These, and the studies they spawned (reviewed briefly in this chapter), provide hope that the use of such cells might circumvent some of the limitations of presently available graft material for cell replacement (e.g., primary fetal tissue) and some of the difficulties of extant gene transfer vehicles (e.g., viral vectors, bone marrow transplanatation, pharmacologic agents, nonneural cells, synthetic pumps).

Multipotent neural progenitors and stem cells are immature, uncommitted cells that exist in the developing and even adult nervous system and are responsible for giving rise to the vast array of more specialized cells of the mature CNS. These cells, which are self-maintaining, self-renewing, and multipotent (the operational definition of a stem cell), are extremely plastic and may be "molded" by their environment. They can be isolated from the nervous system and propagated by a number of strategies (both by genetic manipulation or by exposure to mitogens) in culture where they can be characterized, cloned, passaged, frozen and stored, yet can be thawed and reimplanted into the developing and adult CNS in which they integrate nondisruptively throughout the CNS, respect appropriate neurogenic cues, and differentiate into multiple appropriate types of neural cells of both neuronal and glial lineages as integral members of normal structures, without disturbing other neurobiological processes.

Not only can they become multiple neural cell types, but they can also be easily engineered before transplantation ("*ex vivo* genetic manipulation") to carry and import exogenous genes directly into the CNS where they may serve as "factories" or "pumps" for the sustained delivery of such gene products.

The same cells can be isolated once, for example, from a mouse, expanded into multiple vials, cryopreserved, and subsequently used for numerous mouse hosts from multiple mouse strains (there appears to be no immunologic barrier) with numerous and varying types of neurologic disease. One neural stem cell clone can, therefore, serve multiple functions; it can be a ready and inexhaustible source of plastic and adaptable neural cells for mutiple needs.

Although the initial assumption was that such cells might simply substitute for primary fetal neural tissue as a graft material in classic experimental transplantation paradigms in which implants are placed in a circumscribed locus [reviewed in Bjorklund and Dunnett, 1995; Fisher and Gage, 1995], the somewhat surprising observation that neural progenitor or stem cells might integrate so robustly and seamlessly *throughout* the mammalian CNS and so effectively express exogenous, transgenic proteins (creating virtually "chimeric" brains or regions of brain, e.g., Fig. 1, see color insert [Snyder *et al.*, 1995; Lacorraza *et al.*, 1996]), suggested both a new role for neural transplantation, addressing disorders not typically regarded as within its purview, as well as possible approaches to the CNS manifestations of diseases heretofore refractory to intervention [Table I]. If this same adaptability applies to human cells, enormous progress could be made in the treatment of multiple human afflictions through the availability of such a therapeutic resource of human origin.

These abilities, it should be emphasized, are predicated on harnessing the basic biologic principles and plasticity inherent in this class of heretofore underrecognized, underappreciated, and underexplored type of neural cell. Therefore, the use of neural progenitors and/or stem cells for repair of the dysfunctional CNS is an area of investigation that may not only yield novel strategies against neurologic disability, but may also offer insights into both stem cell biology and the pathobiology underlying certain types of neurodegeneration. The pursuit of one goal typically nurtures the other. Several recent

TABLE I Neural Stemlike Cells May Serve As Cellular Vehicles for the Widespread CNS Distribution of the Following:

- Enzymes (*e.g., for enzyme replacement*)
- Cells (*e.g., for neural cell replacement*)
- Neutrophic agents and diffusible factors
- Myelin and nondiffusible factors
- Extracellular matrix molecules (*e.g., reelin*)
- Putative pathogenic agents (*e.g., to create models of disease*)
- Viruses (*e.g., to test neurovirulence*)
- Viral vectors (*e.g., as producer/packaging cells*)
- Substances predominatly within regions of pathology (*e.g., tumor, infarct, cell death*)

reviews have dealt extensively with basic aspects of CNS stem cell biology [Snyder 1994, 1998; Gage *et al.*, 1995; Gage and Christen, 1997; Brüstle and McKay, 1996; Weiss *et al.*, 1996; Whittemore and Snyder, 1996; Stemple and Mahantharppa, 1997; Morrison *et al.*, 1997; McKay, 1997]. In this chapter, therefore, we will concentrate on summarizing investigations in which neural stem cells have shown particular promise for the development of neuroregenerative strategies.

II. CNS STEM CELLS AND PROGENITORS IN A CULTURE DISH: "FREEZING" A DEVELOPMENTAL PROGRAM?

As mentioned in the introduction, it was recognized that the opportunity to transplant well-characterized, homogenous cells that self-renew in culture but, *in vivo*, integrate appropriately throughout the neuraxis and express exogenous (therapeutic) genes directly to the CNS without disrupting other neurobiologic processes could potentially circumvent some of the limitations of extant techniques for CNS gene therapy and repair. However, neural stem cells removed from the brain do not normally remain in a proliferative or uncommitted state *in vitro* [Davis and Temple, 1994; Ghosh and Greenberg, 1995; Qian *et al.*, 1997]; after 1 or 2 mitoses, if any, they typically cease dividing and differentiate. To maintain them in a proliferative, immature, "stemlike" state requires an intervention. Interventions that have successfully circumvented these limitations have included both a number of genetic and epigenetic means: transduction of propagation or immortalizing genes into neural progenitors (i.e., "internal commands"); chronic exposure of progenitors to mitogenic cytokines (i.e., "external commands"); and culture of progenitors on such substrates as astroglial membrane homogenates or in cellular aggregates [Davis and Temple, 1994; Quian *et al.*, 1997; all reviewed by Whittemore and Snyder, 1996]. Evidence is mounting, in fact, that external growth stimulatory (epigenetic) signals such as cytokines and the various genetic means for propagating cells actually impinge on many of the same final common pathways influencing the expression of various cell cycle regulatory proteins (Nourse *et al.*, 1994; Steiner *et al.*, 1995; Galaktionov *et al.*, 1996; Coats *et al.*, 1996; Iavarone and Massague, 1997; Leone *et al.*, 1997). For neural stem cells, the goal of "immortalization" is simply to suspend differentiation and hold the often rapid progression of a developmental program in abeyance by inducing the cells to remain in the cell cycle in culture. From a practical point of view, although mitogens are clearly effective (Weiss *et al.*, 1996; McKay, 1997; Gage *et al.*, 1995; Suhonen *et al.*, 1996), the most effective, efficient, convenient, and safest procedure to date has been the introduction of genes that encode for proteins that interact with cell cycle proteins ("immortalizing" or "propagation" genes) (Frederiksen *et al.*, 1988; Bartlett *et al.*, 1988; Ryder *et al.*, 1990]. The impetus to "cascade" through a particular developmental program is temporarily overridden in culture, probably by the action of these genes in combination with the particular *in vitro* environmental conditions and the elimination of signals normally present *in vivo*. However, it is important to recognize that, while maintaining neural stem cells in a proliferative state *in vitro* "suspends" progression through a developmental program (and what might ordinarily be a rapid narrowing of a stem cell's phenotypic options), it does not "subvert" these programs or the ability of these cells to respond to normal microenvironmental cues (e.g., withdraw from the cell cycle, interact with host cells, differentiate) both *in vitro* and *in vivo* (Snyder *et al.*, 1997).

Different isoforms of myc, neu, p53, adenoviral E1A, and SV40 large T-antigen (T-ag) have all been used to different extents for the perpetuation of cells from very different brain regions and donor ages (see Table II, and references therein) (Frederiksen *et al.*, 1988; Bartlett *et al.*, 1988; Ryder *et al.*, 1990; Renfranz *et al.*, 1991; Snyder *et al.*, 1992; White and Whittemore, 1992; Cattaneo *et al.*, 1993, 1994; Whittemore and White, 1993; White *et al.*, 1994; Lundberg *et al.*, 1997). The genes used most extensively and successfully to date have been *vmyc* and a mutated allele of *T-ag*, the *tsA58* temperature-sensitive form of the protein. This latter gene product is stable in cells cultured at the permissive temperature (33°C) and induces cells to progress through the cell cycle, but is degraded when the temperature is raised to 37°C in culture or, for instance, after grafting into the rodent brain where the temperature reaches 39°C. For neural cells transduced with *vmyc*, stable integration in the brain seems to be accompanied by a constituitive and spontaneous loss of *vmyc* immunoreactivity, suggesting auto-controlled downregulation or degradation of the immortalizing gene product. A common and appealing feature for all these perpetuated cells is that they behave like established lines, but do *not* show signs of transformation either *in vitro* or *in vivo* (as discussed by Whittemore and Snyder, 1996). Though the number of cells often seen at maturity suggests 0 to 3 mitoses posttransplant and prior to end differentiation–in fact, donor progenitors may *need* to be initially plastic and mitotic for optimal engraftment–no tumors are *ever* seen (Cattaneo *et al.*, 1994, Lundberg *et al.*, 1997; Snyder, 1998; Park *et al.*, 1995). This fact is consistent with the observation that, when recipient animals are "pulsed" with bromodeoxyuridine (BrdU) at various times following transplantation, the proportion of donor cells that are still mitotic falls to 0 by 48 to 72 hrs post engraftment (a finding that mirrors their behavior in culture with contact inhibition) (Snyder, 1998; Park *et al.*, 1995, 1999). Transplanted mice exhibit no neurologic dysfunction. Structures that receive contributions from donor cells develop normally.

Among the numerous clonal cell lines generated recently, one can find not only cells with stemlike characteristics and a multipotency for generating both neuronal and glial progeny, but also cells with a more limited differentiation potential, behaving as progenitors for only neurons or glia (Tables II and III). Precisely why some lines differ remains an active area of investigation. It is unclear whether these differences reflect differences in the developmental stage of the cells prior to their perpetuation or the means by which the cells were and are manipulated.

The cell lines listed in Table II are those that have been experimentally tested in neuroregeneration paradigms as discussed next in this chapter. (When compared with the lines listed in Table III, one notes that these tend to be less restricted in their differentiation potential.) These lines share a number of advantageous characteristics for intracerebral transplantation and gene transfer studies (Table IV). (1) They are "self-maintaining" and "self-renewing," meaning that they can be indefinitely expanded and passaged in culture and that, even though a cell may give rise to differentiated progeny as some of these exit the cell cycle (e.g., upon contact inhibition or encountering differentiating cues), other progeny will renew the dish's stem cell pool by retaining such features as remaining immature, uncommitted, proliferative, and multipotent. (2) Their ease of handling and mitotic properties has allowed the introduction of reporter and therapeutic genes by most viral- and non-viral-mediated transfection strategies as in any cell line. (3) They have been isolated as single clones, making their genetic background homogeneous, as opposed to cultures of primary tissue or even most growth factor-expanded cultures of mixed and heterogeneous stem cells and progenitors. (4) They are all of CNS origin, permitting the generation of cells with neural characteristics that can integrate in a recipient host brain after transplantation. (5) The perpetuation process seems not to have altered their stemlike or progenitor properties; they have been consistently reported to differentiate into appropriate glia and/or neurons, both *in vitro* and *in vivo*.

III. "DEFROSTING" THE PROGRAM: SURVIVAL AND INTEGRATION OF PROPAGATED NEURAL STEM CELLS IN A RECIPIENT BRAIN

Perhaps the most remarkable property of these perpetuated stem cell clones is the last one enumerated above: their apparently seamless integration into the surrounding parenchyma of a recipient brain after transplantation (e.g., Fig. 1) and their differentiation into ostensibly mature neural cells of various phenotypes. In contrast, for instance, to primary fetal neural tissue, fibroblasts, or muscle cells, neural stem cell clones soon migrate away from the implanta-

TABLE II Progenitor & Stem Cell Lines Obtained from the Mammalian CNS That Have Been Used in Neuroregeneration and Gene Transfer Experiments *In Vivo*

Cell line	Species	Source	Propagating gene	Differentiation potential	Representative references
HiB5	rat	E16 hippocampus[a]	tsA58/U19 T-ag	multipotent	Renfranze *et al.*, 1991
C17.2, C27.3, C36.4	mouse	PN cerebellar EGL[c]	v-myc	multipotent	Ryder *et al.*, 1990 Snyder *et al.*, 1992, 1995, 1997 Lacorraza *et al.*, 1996 Rosario *et al.*, 1997
ST14A, ST79-13A, ST86[b]	rat	E14 striatum	tsA58/U19 T-ag	multipotent	Cattaneo *et al.*, 1993, 1994 Lundberg *et al.*, 1997
RN33B, RN46A	rat	E13 medullary raphe	tsA58 T-ag	neurons	White and Whittemore, 1992 Whittemore and White, 1993 White *et al.*, 1994
CSM 14.1.4	rat	E14 ventral mesenc.	tsA58 T-ag	?	Anton *et al.*, 1994

[a] E = embryonic.
[b] These cell lines are collectively termed in the text as "ST cell lines".
[c] PN = postnatal; EGL = external germinal layer.

TABLE III Other Examples of Progenitor Cell Lines Derived from the CNS

Cell line	Species	Source	Propagating gene	Differentiation potential	Representative refs. & notes
2.3D	mouse	E10 mesencephalon	c-myc	multipotent	a
Several cell lines	rat	PN2 cerebellum	tsA58 t-ag, v-myc, neu	variable	Frederiksen et al., 1988
Several cell lines	rat	E17-18 hippocampus	tsA58 T-ag	multi- & uni-potent	b
MK31 and other lines	mouse	E17 hippocampus	tsA58 T-ag	neurons	c,d
Astrocyte progenitor	rat	PN1 cortex, diencephalon	wt T-ag	type 1 astrocytes	e
Oligodendrocyte progenitor	rat	PN3 optic nerve	tsA58/U19 T-ag	oligodendrocyte	f
Striatal neuroblasts	rat	E14 striatal primordium	tsA58 T-ag	GABAergic neurons	g
Striatal neuroblasts	mouse	E18 striatal primordium	none	cholinergic	h
HC2S2	mouse	adult hippocampus	v-myc	neurons	i
Cb-E1A	rat	PN7 cerebellum	adenoviral E1A	neurons	j
Neuroblast HMR 10-3	rat	E17 hippocampus	tsA58 T-ag	neurons	k

[a] Bartlett, P. F. *et al.* (1988). *Proc Natl Acad Sci USA* 85:3255–3259. Multipotent *in vitro*. The cell line became transformed.
[b] Eves, *et al.* (1992). *Proc Natl Acad Sci USA 89:4373–4377.*
[c] Mehler, M. F. *et al.* (1993). *Nature* 362:62–65. Cytokine regulation of neuronal differentiation.
[d] Mehler, *et al.* (1995). Int J Dev Neurosci 13:213–240.
[e] Radany, E. H. *et al.* (1992). *Proc Natl Acad Sci USA 89:6467–6471. T-ag expressed from the GFAP promoter.*
[f] Almazan, G., and McKay, R. D. G. (1992). *Brain Res* 579:234–245. Improved immortalization with the U19 T-ag mutation.
[g] Giordano, M. *et al.* (1993). *Exp Neurol* 124:395–400. GAD-positive and GABA releasing neurons are obtained.
[h] Wainwright, M. S. *et al.* (1995). *J Neuroscience* 15:676–688. Cholinergic phenotype and dopaminergic receptors.
[i] Hoshimaru, M. *et al.* (1996). *Proc Natl Acad Sci USA* 93:1518–1523. Conditional immortalization with v-myc (tet-regulation).
[j] Seigel, G. M. *et al.* (1996). *Neuroscience* 74:511–518.
[k] Eves, E. M. *et al.* (1994). *Brain Res* 656:396–404. Reversible neuronal differentiation.

TABLE IV Properties of Progenitors and Stem Cell Lines That Make Them Ideally Suited For CNS Repair and *Ex Vivo* Gene Therapy

In vitro	In vivo
Unlimited proliferation	Non-unitotic; harmless to the recipient brain; Nondisruptive integration
Predictable and abundant availability	Avoids systemic side effects
Homogeneous genetic background; clonal	Engraftment at any age (embryos to aged subjects)
Ease of handling, drug-selection, manipulation, and cloning in culture	Nontumorigenic *in vivo*
Amenable to genetic modification	Nonimmunogenic
Allows for reproducible gene transfer	Long-term survival
Controllable material	Allows for long-term gene transfer *in vivo*
Reduces the need for fetal tissue or primary cultures	Targeted incorporation in predictable brain regions possible
Low-cost (regular cell culture equipment)	Focal, circumscribed delivery of transgenic proteins
Nontransformed *in vitro*	Widespread integration possible by delivery to germinal zones
Minimal variation	Delivery of transgenic proteins to large brain areas possible
From experiment to experiment	*In vivo* differentiation into and/or replacement of multiple neural cell types
	Site-specific differentiation
	Reconstruction of neural structures and synaptic routes
	Regulation by host physiology
	Functional integration

tion site (often moving from appropriate germinal zones and traveling along well-established migratory pathways) and integrate in a cytoarchitecturally proper manner.

These observations were initially made for the C17.2, C27.3, and HiB5 stem cell lines after grafting to the neonatal brain (Snyder *et al.*, 1992; Renfranz *et al.*, 1991). In these studies, performed to analyze the developmental potential of these cells, the grafted cells were found to integrate well, and differentiate into both neurons and glia in the recipient brain, resembling the host cell types found at their implantation site (hippocampus or cerebellum).

Subsequently, HiB5 cells, the series of striatum (ST)-derived cell lines (ST14A, ST79-13A, ST86), and clone C17.2 stem cells have been successfully grafted into multiple CNS regions of embryos, newborns, and adult or aged recipients, showing similarly appropriate integration and differentiation (Table II; Cattaneo *et al.*, 1994; Martínez-Serrano *et al.*, 1995a; Lundberg and Björklund, 1996, Lundberg *et al.*, 1997; Snyder *et al.*, 1995, 1997; Lacorraza *et al.*, 1996; Snyder, 1998). Engrafted cells become integral members of the cytoarchitecture: donor-derived neurons receive appropriate synapses and possess appropriate ion channels; the blood–brain barrier (BBB) remains intact where donor-derived astroglia put foot processes onto cerebral vasculature; donor-derived oligodendroglia express myelin basic protein (MBP) and myelinate neuronal processes (Snyder *et al.*, 1993, 1997; Lynch *et al.*, 1996]. That such cells may functionally integrate into host neural circuitry was suggested by the observation that HiB5 cells incorporated in the hippocampal dentate gyrus granule cell layer upregulated their *cfos* expression in a manner similar to host interneurons in response to an experimentally induced seizure (McKay, 1992). Interestingly, some donor-derived cells remain integrated as simply immature, undifferentiated but quiescent progenitors or stem cells (Cattaneo *et al.*, 1994; Lundberg *et al.*, 1996a, 1997; Snyder *et al.*, 1997). These often reside intimately associated with blood vessels, resembling pericytes, or are positioned in the subventricular zone (SVZ) (a postnatal germinal zone thought to harbor quiescent, endogenous stem cells), or are even interspersed with other more differentiated cells within the CNS parenchyma.

What is the differentiation potential of a given grafted stem cell in a given region? Donor stem cells appear to intermingle nondisruptively with local endogenous progenitors/stem cells and to respond to the same spatial and temporal cues in the same manner as host cells. Accordingly, they give rise to multiple and varied cell types in these different regions, but differentiate only into the types of neurons and glia expected for a respective region at the particular developmental stage of the transplant. In other words, in regions where neurogenesis is ongoing, they become the appropriate neuronal cell type; in regions where neurogenesis has ceased and gliogenesis is the predominant developmental process, they appropriately become glia. For instance, C17.2 stem cells differentiate into pyramidal neurons in neocortex when implanted into the ventricular zone (VZ) of a mid-embryonic mouse brain during normal corticogenesis, but do not undergo neuronal differentiation in the neocortex at a later stage when neurogenesis is normally completed and gliogenesis predominates (Snyder, 1994, 1998; Snyder and Macklis, 1996; Snyder *et al.*, 1997). In the cerebellum at embryonic day 12 to 14 (E12-14) mice, when Purkinje cell (PC) neurons are born, C17.2 cells can give rise to PCs. At a later stage in cerebellar development, however, they yield only the cell types normally born then–small interneurons and glia. In the adult brain, most neurogenesis has ceased with the exception of that in the olfactory bulb (OB) (in addition to the hippocampus) to which endogenous progenitors in the

TABLE V Plasticity of Progenitor/Stem Cell Lines for Phenotypical Differentiation *In Vivo*

Cell lines	Host age/species	Condition/placement	Neurons	Glia	Immature cells	Reference
HIB5, St	E15 rat	intact-ventricles	+	+	+	*a*
	newborn rat	cerebellum, hippocampus	+	+	nd	Renfranz *et al.*, 1991
	adult rat	intact striatum, septum, NBM	+	+	+	Renfranz *et al.*, 1994 Martínez-Serrano *et al.*, 1995a Lundberg et al., 1997
	aged rat	aged-septum, NBM	+	+	nd	Martínez-Serrano et al., 1995b, 1996a
	adult rat	striatum, pre-graft IA lesion	−	+	+	Lundberg et al., 1997
	adult rat	striatum, post-graft IA lesion	nd	+++	+	Lundberg and Björklund, 1996
	adult rat	striatum, post-graft 6-OH-DA lesion	nd	+++	+	Lundberg and Björklund, 1996
C17.2	E14.5 mouse	intact-ventricles	++	++	nd	Lacorazza *et al.*, 1996
	newborn mouse	intact-ventricles, cerebellum	++	+	+	Snyder *et al.*, 1992, Lacorazza *et al.*, 1996
	newborn mouse	MPS VII-ventricles	+	+	+	Snyder *et al.*, 1995
	newborn mouse	meander tail, cerebellum	++	+	+	Rosario *et al.*, 1997
	newborn mouse	shiverer-ventricles	+	++	+	Yandava *et al.*, 1995
	newborn mouse	reeler, cerebellum	++	+	+	Auguste *et al.*, 1996
	neonatal mouse	hypoxia-ischemia	++	++	++	Park *et al.*, 1995, 1997, 1998
		spinal cord axotomy	++	+	+	Flax & Snyder, 1995
	adult mouse	intact and KA-lesioned cortex	−	++	+	Snyder *et al.*, 1997
	adult mouse	photolytic lesion-cortex	++	+	+	Snyder *et al.*, 1997
	adult mouse	intact-SVZ	+	+	+	Snyder, 1994, 1998, Snyder, 1997
	adult mouse	intact-OB	++	+	+	Snyder, 1994, 1998, Snyder *et al.*, 1997
RN33B	newborn rat	intact-raphe, hipp., spinal cord	+++	−	+	Onifer *et al.*, 1993
	adult rat	cortex, hippocampus	++	−	++	Shihabuddin *et al.*, 1995
	adult rat	KA/colchicine lesion hipp.	+/++	−	++/+	Shihabuddin *et al.*, 1996
	newborn rat	intact striatum	++	+	++	Lundberge *et al.*, 1996c
	adult rat	intact-striatum	++	+	++	Lundberge *et al.*, 1996c
	adult rat	KA-lesioned striatum	+/−	+	+++	Lundberge *et al.*, 1996c
RN46A-(BDNF)	adult rat	intact-hippocampus and cortex	++	−	+	Eaton and Whittemore, 1996

Notes and Abbreviations:

a, Martínez-Serrano A, Olsson M, Gates M and Björklund A, (1998).

Hipp, hippocampus; IA, ibotenic acid; KA, kainic acid; NBM, nucleus basalis magnocellularis; nd, non-determined; OB, olfactory bulb; SVZ, subventricular zone; 6-OH-DA, 6-hydroxy-dopamine.

SVZ will migrate and become granule neurons. Accordingly, when C17.2 cells are implanted into the ventricles of an adult mouse, allowing them access to the SVZ, they intermingle with host SVZ cells, and migrate with them to the OB via the rostral migratory stream after which they uniformly differentiate into granule neurons. Interestingly, if the same donor cells enter the adult OB through an alternative route, they become only glia. Similarly, C17.2 cells that migrate from the adult ventricles into neocortex (and into most other regions of the adult brain, where gliogenesis predominates) produce glia exclusively (Snyder *et al.*, 1995; Snyder, 1998]. These are the very same cells that, if similarly implanted in the ventricles of a fetus, can still give rise to the aforementioned pyramidal neurons. This multipotency, plasticity, and responsiveness to environmental cues is a remarkable characteristic of these cells.

In general, and not surprisingly given the developmental profiles of the normal developing rodent brain (Bayer and Altman, 1998), although glia and immature progenitors are generated at all ages of engraftment, more neurons are produced in brains of hosts that are at an earlier (embryos, newborns) rather than a later (adult or aged) developmental stage (Renfranz *et al.*, 1991, Snyder *et al.*, 1992; Rosario *et al.*, 1997; Martínez-Serrano *et al.*, 1995a, 1995b; Lundberg *et al.*, 1996a, 1997). Furthermore, the migration of engrafted stem cells is not as extensive in intact adult or aged animals as it is during the perinatal period.

Although this chapter deals primarily with lines of multipotent stem cells and not with lines of more restricted neural lineage, certain cell lines with more restricted phenotypes warrant mention because they present interesting similarities and contrasts. RN33B and RN46A, initially derived from the raphe nucleus, and perpetuated by transduction of *ts T-Ag*, appear to be a neuroblast cell lines in that their differentiation programs seem to be restricted to the generation of solely neurons (summarized in Whittemore and Snyder, 1996). Nevertheless there appears to be a plasticity in the way at least one of these lines responds to *in vivo* microenvironmental cues following transplantation, much in the way we have seen stem cells accommodate. RN33B, even when implanted in the adult brain, will give rise to neurons that mimic the phenotypical characteristics of the targeted brain areas (Onifer *et al.*, 1993; Shihabuddin *et al.*, 1995, 1996; Lundberg *et al.*, 1996c). For example, in the neocortex and CA1-3 fields of the hippocampus, the line will generate cells with the appearance of pyramidal neurons; in the hippocampal dentate gyrus, granule cells; in the hilar region of the hippocampal formation, polymorphic cells. Further consistent with a site-specific differentiation, the progeny of RN33B cells in the striatum, but not in the cortex of adult rats, will express typical striatal neuronal markers, and will project to appropriate target territories like the globus pallidus (Lundberg *et al.*, 1996c). (Interestingly, the inferences from these experiments might be of greatest importance not for what they say about the cells themselves, but rather for what they suggest about the residual flexibility of the adult CNS–that even the adult brain retains an ability to mold neuronal phenotypes based on environmental cues.) Yet, even though these cells possess a plasticity and a responsiveness to microenvironmental cues, they appear also to possess an autonomy that stem cells may not possess. In fact, a sister line RN46A, which ostensibly should be identical in its behavior to RN33B, produces only serotonergic neurons under *all* circumstances *in vitro* and *in vivo,* even when implanted into the hippocampus or cortex. This autonomy, which, for example, allows for the appearance of donor-derived neurons even though the intact adult neocortex does not normally give rise to neurons, might make such lines more useful for certain applications than stem cell lines. (Interestingly, however as will be explored in the following paragraphs, the scenarios might be quite different in the face of injury or neurodegeneration.)

For stem cells, survival and integration following transplantation appears to be stable and lifelong. Donor cells do not appear to induce host-mediated immune rejection in spite of their prolonged cell culture history and various genetic modifications. C17.2 cells have been identified in transplanted animals at least 2 years after grafting (Snyder *et al.*, 1992). Similarly, the HiB5 and ST cell lines survive for prolonged periods in young animals; after an initial phase of mitosis (~2–3 divisions during a 7-day period of active migration), the cells remain in the host brain for at least 6 months with no marked reduction in number (Lundberg *et al.*, 1997). HiB5 cells have been detected 9 months after grafting into the brains of middle-aged rats when the animals had now reached an advanced age (Martínez-Serrano and Björklund, 1998). Intriguingly (and in contrast to these data from stem cell lines), long-term survival of cells from the neuroblastic cell lines RN33B and RN46A may somehow be compromised. In long-term experiments, the number of identified grafted cells from these lines decline several-fold in a period of 10 weeks (Eaton and Whittemore, 1996; Lundberg *et al.*, 1996c). Might this imply that donor cells that more closely approximate an end-differentiated, postmitotic stage at the time of engraftment are also more likely to complete their life cycle and undergo apoptosis?

IV. *EX VIVO* GENE TRANSFER TO THE BRAIN

The idea of transferring therapeutic genes (or their end metabolic products) to the mammalian brain has attracted considerable attention over the past decade. Many methods of gene transfer to the brain are under study for the treatment of the CNS manifestations of a number of diseases. One strategy employs vectors–typically genetically altered viruses–for the delivery of exogenous genes directly to a host's brain cells *in situ*. (This topic, beyond the scope of this chapter, is reviewed elsewhere; e.g., Kaplit and Lowey, 1995; Snyder and Fisher, 1996). Alternatively, transplanting cells–either of neural or nonneural origin–that intrinsically secrete missing or therapeutic gene products, or that are genetically engineered *ex vivo* to do so, may provide another strategy. Nonmigratory cells implanted into small, discrete anatomic sites are well suited for disease processes whose pathology is very focal. Neural cells with the capacity to migrate in the CNS provide more powerful vehicles for delivering factors to diseases with pathology that is more widely disseminated, extensive, multifocal, or even global (e.g., many neurogenetic diseases; Snyder *et al.*, 1995). Both neural and nonneural cell types have been successfully engineered to produce a variety of neurotransmitter synthetic enzymes, metabolic enzymes, and neurotrophic factors (reviewed in Björklund, 1991; Gage *et al.*, 1991, 1995; Gage and Fisher, 1991; Snyder and Wolfe, 1995; Martínez-Serrano and Björklund, 1996b; Fisher, 1997; Snyder and Senut, 1997). The use of nonneuronal cells has recently been reviewed elsewhere (Snyder and Senut, 1997). In this chapter, we focus on the use of stable, clonal neural stem cell lines as vehicles for gene transfer. Impressive results with such cells in animal models of CNS disease encourages the continued development of this gene therapy strategy for treating neural dysfunction.

The earliest "gene therapy" of sorts was actually the transplantation of primary fetal tissue in diseases such as parkinsonism (Dunnett and Bjorklund, 1994). The expectation was that engrafted exogenous substantia nigral tissue from a fetus would constitutively produce a neurotransmitter of interest, in this case dopamine, to the targeted dopamine-deficient basal ganglia of a transplanted parkinsonian patient. However, should the supply of dopamine prove insufficient, primary fetal tissue cannot be easily genetically manipulated to overproduce it. What if the supply of fetal substantia nigra is limited? Also, what might one do for other types of therapeutic proteins whose source in the fetal brain cannot be as easily identified as dopamine? Therefore, for both biologic and ethical reasons, there has been an active search for ultimate substitutes for primary fetal tissue as a graft material for brain repair.

"*Ex vivo*" or "somatic cell gene therapy" relies on the use of genetically modified cultured cells that are transplanted to the living organism to transfer a gene, and ultimately a protein, with a function of interest. The earliest experiments of this sort were performed with transformed cell lines of fibroblastic, neural, or endocrine origin. However, because these cells formed tumors in the brain, efforts rapidly turned toward the use of primary dissociated cultures of fibroblasts and astrocytes (reviewed in Snyder and Senut, 1997). Such primary tissue is generally nontumorgenic, but has the disadvantage of requiring that cell preparations be generated, genetically modified, and characterized individually for each surgical session. In addition to being laborious and time-consuming, significant problems of inter-experimental reproducibility and control are posed, all of which could eventually compromise practical clinical application (discussed by Martínez-Serrano and Björklund, 1996b).

Table VI summarizes the requirements for safety, efficiency, reproducibility, and control that any gene transfer strategy to the brain should ideally fulfill. It appeared that neural stem cell lines could ably meet these stringent demands and overcome most of the limitations of other cell types and gene therapy methods currently available (Tables IV and VI; Björklund, 1993; Snyder, 1994, 1995, 1998; Gage *et al.*, 1995; Martínez-Serrano and Björklund, 1996b; Snyder and Wolfe, 1995; Snyder and Fisher, 1996; Fisher, 1997; McKay, 1997). Stable and passageable neural stem cell clones provide the researcher with a fully defined cellular "reagent" for *ex vivo* gene transfer, the cellular and molecular properties of which can be readily studied and often modified under well-controlled conditions. Furthermore, such an approach to gene transfer is unlikely to provoke "functional or metabolic chaos" in the recipient brain. Following transplantation, the genetic content of cells in the recipient brain remains essentially unaltered, unlike following direct ("*in vivo*") gene transfer procedures with viral vectors, thus minimizing the risk of functional disturbances and the emergence of unanticipated, noncoherent or aberrant genotypes and phenotypes (neurons expressing functionally opposed neurotransmitters might be such an example.)

Through the use of neural progenitor and stem cells, a number of reporter and therapeutic genes have been successfully transferred to the rodent and primate brain (Table VII), correcting a variety of metabolic and functional deficits. In some cases, part of the correction or cross-correction in the treated brains was mediated by the intrinsic capacity of the cells to produce the necessary protein (Snyder *et al.*, 1995); in other cases the cells were genetically manipulated to overproduce these proteins Snyder *et al.*, 1995; Lacorazza *et al.*, 1996; Park *et al.*, 1997) or to produce gene products not inherent to the cell transplanted (Martínez-Serrano *et al.*, 1995a, 1995b, 1996a]. Although most studies to date have employed retroviral-mediated *ex vivo* transduction of the cDNA of interest into the stem cell (probably the most efficient gene transfer technique for these mitotic cells in culture), other gene transfer methods such as lipofection, calcium-phosphate precipitation, and electroporation, as well as adenoviral-, adeno-associated viral-, and herpesviral-mediated approaches are also successful, although their value for *in vivo* experiments has not yet been clarified.

V. MARKER AND REPORTER GENES

Typical for the field of gene transfer technology, the earliest tests of the feasibility of neural stem cells as gene delivery vehicles was based on the

TABLE VI Neural Stem Cell Lines Satisfy the Stringent Demands for Gene Transfer to the Brain

Harmless to the receipient brain and not interfering with brain function.
Any gene transfer procedure should be completely innocuous for the recipient organism, both at the histological and functional level. Special caution must be taken not to disturb other functions apart from the one to be targeted. Inappropriate signaling to healthy cells, or mixing of incompatible genetic information in the host brain cells may result in poor therapeutic outcome if other brain functions are altered.

Stability
Because, in most cases, the transferred protein needs to solve a protracted problem, sometimes for the life of the individual, any gene transfer vehicle should be designed for permanent residence in the brain to avoid repeated surgical interventions. Genetically modified cells must remain stably integrated, whether they are host origin (following direct gene transfer) or the product of a transplant (*ex vivo* gene transfer).

Silent to the immune system.
Gene transfer should not trigger an immune reaction which might not only eliminate the cells of interest, but can also cause considerable elaboration of cytokines which could disturb delicate balances in the brain and adversely influence transgene transfer and expression.

Long-term efficiency of gene expression.
Any gene transfer protocol should ensure long-term production of the transgenic protein and stability of its functional effect in the diseased organism.

Distribution of the transgenic protein in the host brain.
Focal delivery is sometimes required to avoid disturbing an equilibrium among different, often functionally unrelated brain structures. In other cases, widespread gene transfer is necessary, particularly to correct inherited genetic, metabolic, toxic, traumatic, or immunologic defects.

Controllable.
When dealing with the transfer of genetic information to a living organism it is always desirable to keep control of that process; the degree of control will determine the degree of reproducibility.

Regulatable.
Ideally, gene expression by the gene delivery vehicle within the brain should be capable of control from the outside, e.g., pharmacologic regulation of a regulatable promoter. This would allow gene expression to be shut off if problems arise or pathology changes. Even more desirable would be to have aspects of brain physiology regulating foreign protein production in a feedback regulated manner.

Ethics.
Because a large number of individuals may require treatment, the gene transfer vehicle should be capable of being generated on a large scale. The procedure must be regarded as an ethical research and clinical tool worldwide. Costs should be affordable and capable of being integrated into the research and public health programs of most societies for distribution to all patients in need, without hindering efficiency and overall quality.

expression of "reporter genes," genes that produce proteins that are functionally irrelevant but are detectable histologically, immunocytochemically, or biochemically, such as E. coli β-galalactosidase (β-gal) (the *lacZ* gene), placental alkaline phosphatase, firefly luciferase, jellyfish green fluorescent protein (GFP). These genes have typically, but not exclusively, been transduced via retroviral vectors. The rationale has been that, if these genes can be transferred efficiently, or if one studies the variables influencing their expression, then principles may be extrapolated to the transfer of more biologically relevant genes. Furthermore, having a reporter gene incorporated into the cell facilitates identification of grafted material in the brain, particularly in the case of neural stem cells where a distinct border between graft and host does not exist; individual donor cells intermingle imperceptibly with host cells (e.g., Fig. 1).

The C17.2 stem cell line, for example, was found to be capable of stable, long-term production (over 2 years) of β-gal *in vivo* (Snyder *et al.*, 1992, 1995, 1997; Rosario *et al.*, 1997), suggesting that foreign gene expression by stem cells may be expected to be permanent (if so desired by the demands of the clinical situation). GFP has also been expressed in C17.2 cells to study their potential for gene delivery to intracerebral tumors (Aboody-Guterman *et al.*, 1996, 1998). GFP has, in fact, been expressed under the control of a cell-type specific promoter, the neuron-specific enolase promoter; GFP is produced only when the stem cell assumes a neuronal phenotype, hence signaling that fate choice and providing a useful assay for developmental studies *in vitro* and *in vivo* (Nissim *et al.*, 1997). *Luciferase*, under the transcriptional control of a tetracycline-regulatable promoter, has been used to assess the feasibility of regulatable gene expression *in vivo* by stem cell lines (Corti *et al.*, 1996). In this study, luciferase production by subclones of the ST14A cell line grafted to the rat striatum could be successfully altered (at least in short-term experiments–2 to 6 days post grafting) by systemic administration of oxytetracycline.

In some situations and for some cell lines, however, loss of transgene expression can be quite problematic. For example, in the neuroblastic cell line RN33B, *lacZ* seems to be downregulated in a large percentage of the grafted cells (Onifer *et al.*, 1993; Lundberg *et al.*, 1996c). Also, there appear to be to date no stable subclones of the HiB5 or ST lines that express *lacZ* or *alkaline phosphatase in vivo* (Renfranz *et al.*, 1991; Martínez-Serrano, unpublished). It is not unusual to observe that, despite a reporter gene's having been transferred to ostensibly identical cells of the same clone using the same retroviral vector with the same promoter, its expression may nevertheless vary from subclone to subclone *in vivo*.

TABLE VII *Ex Vivo* **Gene Transfer Experiments Performed Using Genetically Modified Neural Progenitor and Stem Cell Lines***

Gene	Model[a]	Age	Carrier cell	Expression	Reference
LacZ	rat Cbl	newborn	HiB5	nd	Renfranz *et al.*, 1991
	mouse Cbl, crb	newborn, fetus	C17.2	22 mo.	Snyder *et al.*, 1992, 1993, 1995; Lacorazza *et al.*, 1996
	neuron-def. *mea* Cbl	newborn	C17.2	1 mo.	Rosario *et al.*, 1997
(Reelin)	neuron-def. *reeler* Cbl	newborn	C17.2	1 mo.	Auguste *et al.*, 1996
(MBP)	olig-impaired *shi* crb	newborn	C17.2	2 mo.	Yandava *et al.*, 1995
	neuronal degen. mouse ctx	adult	C17.2	3 mo.	Snyder *et al.*, 1997
	asphyxiated mouse crb	juvenile, adult	C17.2	3 mo.	Park *et al.*, 1995, 1997
	axotomized mouse SC	juvenile, adult	C17.2	3 mo.	Flax and Snyder, 1995
	rat	newborn, adult	RN33B	15 d	Onifer *et al.*, 1993
	rat	adult	RN46A	2 wk	Eaton and Whittemore, 1996
tet-regulated luciferase	rat intact striatum	adult	ST14A	6 d	Cortie *et al.*, 1996
m-NGF	rat striatum	adult	HiB5	2 wk	Martínez-Serrano *et al.*, 1995a, b, 1996a
	rat FF transection	adult	HiB5	2 wk	Martínez-Serrano *et al.*, 1995a
	rat NBM	adult	HiB5	10 wk	Martínez-Serrano *et al.*, 1995b
	rat NBM	aged	HiB5	9 mo.	Martínez-Serrano *et al.*, 1995b, 1996a, (1998)
	at HD	adult	HiB5	nd	Martínez-Serrano and Björklund, 1996a
	rat MCAO	adult	HiB5	nd	Andsberg *et al.*, 1998

h-BDNF	rat striatum	adult	HiB5	nd	Martínez-Serrano *et al.*, 1996b
	HD	adult	HiB5	nd	Martínez-Serrano and Björklund, 1996a
r-BDNF	rat Ctx, Hipp.	adult	RN46A	2 wk	Eaton and Whittemore, 1996
r-NT-3	asphyxiated mouse crb	juvenile, adult	C17.2	3 mo.	Park *et al.*, 1997
	hemi-sectioned SC	adult	C17.2	2 mo.	Himes *et al.*, 1995
h-GUSB	MPS VII mouse	newborn	C17.2	8 mo.	Snyder *et al.*, 1995
h-Hex	normal	E14.5 embryo, newborn	C17.2, C27.3	3-8 wks	Lacorazza *et al.*, 1996
b- & r-TH	6-OH-DA rat and primate	adult	CSM 14.1.4	2 mo	Anton *et al.*, 1994

Notes and Abbreviations:

* See text for a discussion of the most prominent therapeutic findings

[a], unless otherwise specified, the animals were not lesioned.

b, bovine; Cbl, cerebellum; crb, cerebrum; ctx, cortex; SC, spinal cord; MBP, myelin basic protein; neuron degen., selective apoptotic neuronal degeneration in adult neocortex via targed toxic chromophore-mediated photolytic neuronal degeneration [as per Sheen & Macklis, 1995]; CSM 14.1.4, E14 rat Ventral Mesencephalon immortalized progenitors; Ctx, cerebral cortex; FF, fimbria-fornix; h, human; hBDNF, human brain-derived neurotrophic factor; NT-3, Neurotrophin-3; HD, quinolinic acid lesion rat striatum as a rodent model of Huntington's disease; Hipp. Hippocampus; m, mouse; MCAO, transient middle cerebral artery occlusion; mNGF, mouse nerve growth factor; GUSB, β-glucuronidase; MPS VII, mucopolysaccharidosis type VII; Hex, β-hexosaminidase; NBM, nucleus basalis magnocellularis; nd, nondetermined; r, rat; TH, tyrosine hydroxylase; 6-OH-DA, 6-hydroxydopamine

Loss of transgene expression is the bane of all gene therapy. The variables that influence the downregulation of foreign genes are poorly understood in general and, for somatic cell-mediated gene therapy to the brain, virtually unknown.

Because the magnitude of the problem had never actually been systematically quantified for neural transplantation nor had the circumstances been specified under which it occurs, Snyder and associates (Tsai *et al.*, 1999) sought to measure the actual efficiency of foreign gene expression *in vivo* by *individual* neural stem cells. The technique entailed combining fluorescent *in situ* hybridization (FISH) (for detecting the transgene in a given engrafted stem cell) with immunocytochemistry (ICC) (for detecting the product of the specific gene in the same cell) (as per Gussoni *et al.*, 1996, 1997). By calculating the discordance between presence of the reporter gene (FISH+ for *lacZ*) and presence of the gene product (ICC+ for βgal), the efficiency and time course of gene expression by donor neural stem cells could be traced. They affirmed that neural stem cells are, indeed, capable of robust expression of foreign genes ($\geq$80%) for at least 2 months and that even after 1 year post-engraftment, the level never falls to 0, suggesting that therapeutic levels (even if reduced) of a foreign gene *do* persist lifelong. They determined that the retroviral insertion site in the host cell's genome (which is random) does influence the efficiency and duration of transgene expression (certain sites promote better expression than others, as determined by the variable time course of stem cell clones that varied only in insertion site). They found that transducing a given stem cell multiple times with a retroviral vector, thus inserting multiple copies of a therapeutic transgene within the same cell, is a simple, easy, and immediately available method for blunting decrements in transgene expression. This intervention may be effective because (1) there is more gene product produced by the cell; (2) it provides a given cell with more copies of the gene, thus limiting the likelihood that all copies are inactivated simultaneously; and/or (3) it increases the likelihood of a gene achieving a favorable insertion site in the host cell genome. (Not yet specifically explored with this technique but probably another important variable, in addition to retroviral integration site, is the differentiation state of the engrafted cell.)

Because the problem of foreign gene nonexpression cannot be solved until it has been adequately defined, studies such as these, as well as the immediate practical solutions offered for overcoming these limitations (until more elegant interventions are devised), should begin to benefit stem cell-mediated CNS gene therapy.

It should be mentioned that donor stem cells can be tagged by *nongenetic* means, as well (detailed in Cadusseau and Peschansli, 1992; Onifer *et al.*, 1993; Lundberg *et al.*, 1996c, 1997). These alternative methods, which typically depend on the incorporation of a detectable foreign compound in the donor cells prior to transplantation, have included: fluorescent molecules or beads sequestered in the nucleus, cytosol, or plasma membrane (e.g., DiI, Hoechst blue, or PKH); and nucleotide analogues that incorporate into the DNA of donor cells while they are dividing in culture (e.g., bromodeoxyuridine [BrdU] or ^{3}H-thymidine). The pitfalls of some of these techniques are that, over extended periods (sometimes as short as a week), some dyes might diffuse (mislabeling host cells) or might dissipate (precluding their use for analysis following long intervals post-transplant). Grafted cells may die and be phagocytized by host cells leading to their misidentification as donor cells. Should a donor cell undergo extensive mitosis *in vivo,* a marker may be at risk for becoming diluted among the progeny making their detection difficult. Finally, there is the need to insure that the markers are not toxic or antimitotic.

If cells from one species are transplanted into hosts of another species, then antibodies to species-specific cell markers may be used to identify donor cells. The mouse-specific M6 cell surface antibody marker has been used to identify engrafted mouse cells in the rat host brain (Campell *et al.*, 1995). The concern with this approach is that cross-species transplantation may not be as robust or reliable as transplants within the same species; graft rejection may compromise the experiments. A similar approach that circumvents the concerns of xenotransplantation may be the implantation of cells of one sex into hosts of the other. For example, clone C17.2 mouse stem cells, which are male, have been readily and efficiently detected with a cDNA probe to the Y-chromosome following engraftment into female mouse host brains (Snyder, E. Y. and Harvey, A. R., unpublished).

Although identification of engrafted donor cells by virtue of a retrovirally transduced reporter gene is probably the best labeling technique, a few potential pitfalls must still be anticipated and avoided here. The recrudescence of replication-competent virus by recombination (i.e., "helper virus") must be ruled out prior to transplantation (Ryder *et al.*, 1990). The presence of helper virus might mean that the retroviral genome encoding the reporter gene could escape and infect host cells, leading to their misidentification as donor cells. A "helper test," which should be performed for each new donor line generated, can be easily performed by simply demonstrating that the supernatant from a neural cell line is incapable of transmitting the reporter gene to naive cells (Ryder *et al.*, 1990; Snyder *et al.*, 1992). Some early studies in the field went so far as demonstrating that βgal+ cells within a recipient brain actually shared the same unique retroviral insertion site as the donor clonal cell line in the culture dish (Snyder *et al.*, 1992). In offering unambiguous proof that βgal+-cells were of donor origin, these studies affirmed the reliability of a simple helper test for future studies.

VI. THERAPEUTIC GENES

A. Neurotransmitter Synthesis-Related Enzymes: Tyrosine Hydroxylase and the Generation of DOPA/Dopamine-Secreting Cells

On the basis of numerous studies demonstrating the efficacy of primary fetal tissue transplants in experimental models and clinical cases of Parkinson's disease [Björklund, 1991; Olanow *et al.*, 1996), the transfer of gene-encoding enzymes for the synthesis of neurotransmitters has long been a focus for researchers interested in the gene therapy of neurodegenerative disorders. Following the assumption that neurotransmitter replacement in a denervated target brain structure may help reverse some functional disabilities, cells of various types, genetically modified to "pump out" neurotransmitters, have been transplanted into those targets to compensate for the lack of neuronal input. Early experiments for the gene transfer of tyrosine hydroxylase (TH) (the rate limiting enzyme for dopamine synthesis) (Björklund, 1991; Gage *et al.*, 1991; Gage and Fisher, 1991) or choline acetyl transferase (ChAT) (pivotal to acetylcholine synthesis) (Winkler *et al.*, 1995), using a variety of cellular vehicles (e.g., tumorgenic cell lines, primary fibroblasts, or astrocytes (Lundberg *et al.*, 1996b), have supported this assumption, showing to some extent an impact on motor or cognitive behaviors (reviewed by Snyder and Fisher, 1996; Martínez-Serrano and Björklund, 1996b).

Anton and colleagues (1994) used the *tsA58 T-ag* gene to immortalize cells (albeit not stem cells) obtained from the ventral mesencephalic region of E14 rat brain [Tables II and VII]. These cells (designated "CSM 14.1.4") were subjected to several genetic modifications before transplantation. The mix of cells that was ultimately grafted into hemiparkinsonian animals expressed rat TH from the cell's own genetic information, plus bovine and rat TH from 3 different vectors (1 expression plasmid and 2 retroviral vectors). The cells synthesized up to 65 ng L-DOPA per million cells (in cell lysates). Prior to grafting, the cells were treated with a mix of factors to enhance neuronal differentiation (bFGF plus retinoic acid) and supplemented with external BH_4. When pre- and post-grafting values were compared in transplanted animals, abnormal apomorphine-induced rotation appeared to be reduced. Interestingly, however, these animals had the same absolute score as sham- or control-grafted animals after transplantation, a phenomenon that requires further investigation. When a similar experiment was performed in 2 MPTP-lesioned monkeys, a reduction of about 50% in their apomorphine-induced rotation was observed. Although the precise amount of DOPA/dopamine replacement in the implanted striata remains to be determined, it is encouraging that immortalized neural cells were effective in promoting sustained (2 month) expression of immunocytochemically detectable TH.

Newer insights from gene transfer studies combined with older knowledge from fetal transplant studies have served to derive a set of minimum criteria that genetically modified cells should probably meet in order to foster functional recovery in models of parkinsonism: (1) The cells should be equipped with a complete metabolic pathway for DOPA production rather than only expressing TH (Kang *et al.*, 1993; Bencsics *et al.*, 1996). (2) The amount of DOPA released *in vitro* should be on the order of 1 nmol/hr/10^6 cells and the ratio of DOPA-released-to-DOPA-stored should be high. (3) After grafting to the striatum, there should be some degree of migration by donor cells from the implantation site to provide an even source of DOPA/dopamine. (4) There should be biochemical documentation of sustained expression of transgenes *in vivo* at a biologically relevant level. (5) It should be possible to study the correlation between cell survival, biochemical compensation, and behavioral recovery. In order to assign any behavioral effect to the gene transfer procedure itself, it would be desirable to test for the reversion of any behavioral improvement after experimentally downregulating TH expression or killing the grafted cells.

B. Nerve Growth Factor and the Cholinergic System

We will next summarize work developed by one of us at A. Björklund's laboratory in Lund (Sweden) for the gene transfer of neurotrophic factors using neural stem cell lines. Although these studies largely focus on the prototypic neurotrophin, nerve growth factor (NGF), the approach is rapidly expanding to include other neurotrophic agents beneficial for various neuronal populations in the mammalian brain. The therapeutic approach (reviewed in Björklund, 1991; Gage *et al.*, 1991; Gage and Fisher, 1991) is to provide the brain directly with a supply of these highly active peptides, which normally do not cross the BBB, in order to halt or diminish neurodegeneration and/or to stimulate regeneration of injured and/or diseased neurons with the ultimate goal of restoring lost function.

NGF has been transferred to the rat brain using a derivative of the stem cell line HiB5, originally of hippocampal origin. After repeated retroviral infection (using a vector designed to avoid internal promoter elements to ensure long-term expression *in vivo*), a stable NGF-secreting clone, HiB5-NGF (E8), was isolated (Martínez-Serrano *et al.*, 1995a). It produced 2 ng/hr/10^5cells of NGF in culture not only in dividing cells but also in cells that had exited the cell cycle, simulating an end-differentiated phenotype in the brain. Initial short-term experiments revealed that the cells could express NGF message *in vivo* and had neurotrophic effects on cholinergic neurons, stimulating sprouting of axon collaterals from nucleus basalis neurons toward the source of NGF (Martínez-Serrano *et al.*, 1995a). What was more interesting, however, was that transplants of the NGF-secreting cells could block degeneration of medial septum cholinergic neurons after axotomy (the complete unilateral fimbria–fornix transection model) (Martínez-Serrano and Björklund, 1996b).

Taking advantage of these properties, experiments were next aimed at correcting the cholinergic deficits that normally appear with aging in the mammalian brain. The aged rat offers a natural neurodegeneration model in which atrophy of the cholinergic system is paralleled by deficits in the ability of the animals to perform certain cognitive tasks. As a first step, the NGF-producing cells were shown to induce a trophic response in intact, healthy nucleus basalis cholinergic neurons in young animals (Martínez-Serrano *et al.*, 1995b). These results, combined with results from previous studies in which NGF infused by means of osmotic mini-pumps, could correct neuronal atrophy and reverse deficits in learning and memory in aged rats (Fischer *et al.*, 1987), then prompted an investigation of the utility of these engineered stem cells in this older animal population. The behavioral results of this (and subsequent) experiments are illustrated in Fig. 2. In summary, previously memory-impaired, aged animals receiving transplants of NGF-secreting HiB5 cells improved their performance in the task (both acquisition and retention) to a level similar to that seen in young animals; the atrophy of cholinergic neurons in the nucleus basalis and medial septum was corrected; and expression and production of NGF-like neurotrophic activity *in vivo* was demonstratable at least 10 weeks post-grafting. Stem-cell-mediated gene transfer to the aged rat brain could increase levels of NGF protein in grafted tissue to values similar to those seen in young animals (Martínez-Serrano *et al.*, 1996a). Subsequent studies of longer duration (Martínez-Serrano *et al.*, 1996a) demonstrated that cognitive improvement could extend beyond just acquisition and retention, to the even more complex demands of consecutively learning different tasks. These experiments represented the first indication that a gene therapy protocol could influence a spontaneous behavior in mammals, and one of high cognitive function at that. These observations were substantiated by a concurrent similar study using primary fibroblasts engineered *ex vivo* to transfer the NGF gene (Chen and Gage, 1995), although behavioral improvement may actually have been more profound using the stem cells.

These studies in aged animals illustrated the potential of acutely administered trophin-secreting neural stem cells to reverse established behavioral deficits in a neurodegeneration condition. In more recent studies, it was learned that the development of both age-induced cholinergic neuronal atrophy and learning deficits might, in fact, be prevented in engrafted middle-aged rats

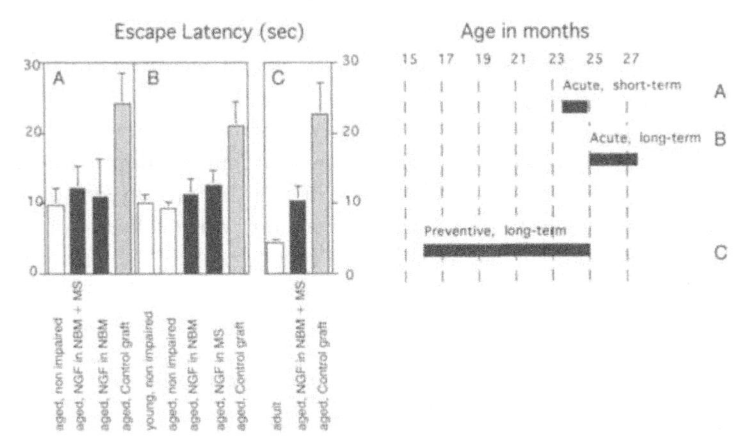

FIGURE 2 Behavioral effects of NGF gene transfer by neural stem cell lines. The diagram summarizes the behavioral results obtained in three independent studies (**A**, **B**, **C**) aimed at inducing functional recovery of cognitively impaired aged animals (**A**, **B**) or prevention of the development of these deficits (**C**). Rats' ability to learn and perform a cognitive task was evaluated in the Morris watermaze test. The left-hand diagram shows the escape latency score of the different groups of animals (time required to find a submerged platform in an opaque water-filled circular pool, on the basis of extra-maze, external spatial cues). As illustrated, animals receiving NGF-producing transplants in the nucleus basalis (NBM) or medial septum (MS) show a low score in this type of task, similar to the nonimpaired young, adult, or aged animals, whereas rats receiving control cells remained impaired (high score). The time window in the animal's life when the equipment was conducted (from transplantation to testing) is indicated in the right panel. [Data from Martínez-Serrano *et al.*, 1995b (A), 1996a (B), and Martínez-Serrano and Björklund, 1998 (C)].

who were then examined when they had reached advanced age, 9 months post-transplant (Martínez-Serrano and Björklund, 1998). Such long-term protection against neurodegeneration through the administration of neurotrophic factors by stem-cell-mediated gene transfer illustrates its relevance also as a more prophylactic treatment for slowing down or forestalling the progression of the behavioral and structural changes of some neurodegenerative processes.

These experiments (Martínez-Serrano *et al.*, 1995b, 1996a) also afforded an opportunity to study the distribution of an exogenous secreted protein delivered to the adult or aged brain (Table IV). By transplanting the NGF-producing cells in discrete locations (medial septum or nucleus basalis) it could be demonstrated that neurotrophic effects on cholinergic neurons were confined to the targeted region, as expected from a graft that migrates moderately (1–1.5 mm) around the implantation site. This property suggests that, under certain circumstances, integrated neural stem cells can supply trophic factors in close vicinity to those neurons that require it without affecting other cell groups at more remote locations for which the molecule might be problematic. It is instructive to realize, however (and will be explored in more detail in subsequent sections of this chapter), that neural stem cells are so multifaceted that simply altering their mode of administration in the brain (e.g., to a germinal zone) may allow them to deliver therapeutic gene products and cells in a more disseminated, brain-wide manner when a disease of a more global nature demands such an intervention (Snyder *et al.*, 1995; Lacorraza *et al.*, 1996; Yandava and Snyder, 1995].

C. Neuroprotection Against Excitotoxic and Ischemic Insults by NGF Gene Transfer

Ex vivo gene transfer of NGF has contributed to our understanding of neurotrophic actions in the mammalian brain, sometimes providing unexpected insights. For example, when NGF is delivered to the brain by genetically modified cells, it appears to exert neurotrophic actions on medium-size spiny projection striatal neurons, cells that one would not expect to be sensitive to this neurotrophin given that they lack $p75^{NTR}$ or Trk A NGF receptors. These neurons can, in fact, be protected against quinolinic acid-induced lesions (an experimental model of Huntington's disease in rodents) by grafts of genetically engineered NGF-producing primary fibroblasts overlying the corpus callosum (Schumacher *et al.*, 1991; Frim *et al.*, 1993a, 1993b). Intriguingly, infused NGF protein is not known to have similar effects in the rodent brain. Martínez-Serrano and Björklund (1996a) subsequently analyzed the neuroprotective abilities of NGF-secreting neural stem cells in this model for both striatal cholinergic interneurons and GABA-ergic projection neurons. The degree of innervation of target territories of spiny projection neurons was assessed as a measure of the preservation of endogenous circuitry in the brains of animals engrafted with NGF-secreting HiB5 one week before quinolinic-acid lesioning. One month after the lesion (five weeks post-grafting), striatal degeneration, lesion size, and loss of spiny and cholinergic neurons were noted to be reduced. The precise mechanism for these neuroprotective effects on striatal projection neurons is not well understood, although it seems to be mediated by amelioration of the oxidative stress that accompanies the excitotoxic lesion (Martínez-Serrano and Björklund, 1996a; Galpern *et al.*, 1996). A similar experiment, using cultures of EGF-expanded progenitors derived from a transgenic mouse expressing human NGF under control of the GFAP promoter has essentially reproduced these results (Carpenter *et al.*, 1996). As an extension of these data, it appears that the engrafted NGF-secreting stem cell line may also protect striatal neurons from ischemic injury following experimental transient occlusion of the middle cerebral artery in the rat. Infarct size and neuronal loss appear to be reduced at 2 days of reperfusion time (Andsberg *et al.*, 1998).

D. Other Neurotrophic Factors

Genetically modified neural stem cells have proven effective for delivering other neurotrophic trophic agents and cytokines to the brain in various neurodegeneration paradigms. BDNF, NT-3, NT-4/5, CNTF, and GDNF are examples of bioactive peptides whose direct intracranial infusion has been demonstrated to be neuroregenerative *in vivo* in various experimental lesion models.

Brain-derived neurotrophic factor (BDNF) is a well-known neurotrophin for cholinergic, dopaminergic, and GABA-ergic neurons from a variety of brain regions. BDNF is a good example of a factor that does not diffuse well through the brain parenchyma; therefore, local delivery by genetically engineered grafted cells may have advantages over conventional infusion by pump. In a manner similar to that for NGF previously described, a BDNF-secreting subclone of the HiB5 cell line was generated (initially producing 0.28ng/h/10^5cells). Its transplantation into the striatum and into the nucleus basalis resulted in increased neuronal soma size and upregulated cholinergic markers (e.g., ChAT) (Martínez-Serrano *et al.*, 1996b). Interestingly, over many passages, this line, for unclear reasons, has seemed to lose its ability to secrete BDNF. These data contrast with those obtained following genetic modification of the neuroblastic cell line RN46A, which not only appears to produce BDNF without difficulty, but whose BDNF production seems to enhance its otherwise poor survival *in vivo* (Eaton and Whittemore, 1996).

C17.2 neural stem cells, which intrinsically express low levels of neurotrophin-3 (NT-3), have been successfully engineered to stably overproduced this neurotrophic agent. When implanted into the hemisectioned rodent spinal cord, they appear to protect host neurons in Clarke's nucleus from cell death (Himes *et al.*, 1995), and when implanted into ischemic areas of mouse brain, they appear to increase significantly the number of donor-derived neurons replacing degenerated host neurons (Park *et al.*, 1997). (These data will be discussed in more detail later.)

Such variable findings might suggest that not all growth factors or cytokines can be stably expressed with equal efficiency by stem cells or progenitors at all developmental stages. Furthermore, because some factors are known to influence the development and differentiation of neural stem cells (Johe *et al.*, 1996), one may need to be cognizant of the effects such trophic agents might have on their neural stem cell vehicles, working in an autocrine or paracrine fashion and influencing their stability and survival. In other words, for some of these factors, stem cells are not simply passive delivery vehicles.

This realization underscores the point that each new gene requires rigorous individual study, analysis, and assessment.

E. Delivery of Metabolic Enzymes and the Correction of Genetic Neurodegenerative Defects

In the various experimental paradigms described in previous sections, in which cells expressing various foreign genes were implanted into discrete neural regions (e.g., hippocampus, striatum, septum), it could be argued that grafts of neural stem cells offer no significant advantage over grafts of primary nonneural cells (such as fibroblasts), which have the added appeal of being easily biopsied from the host into whose brain they will be reimplanted. Indeed, even synthetic pumps could be efficacious. In fact, the inherent biology of neural stem cells endows them with capabilities that circumvent many of the limitations of such vehicles. Furthermore, these properties enable a variety of novel therapeutic strategies heretofore regarded as beyond the purview of neural transplantation and consigned solely (albeit ineffectually) to pharmacologic and genetic interventions or to the transplantation of hematopoietic cells administered via bone marrow transplantation (BMT). In fact, the first report of neural stem cells for gene therapy actually also served to enunciate the uniqueness of this cell type as a vehicle for gene transfer (Snyder *et al.*, 1995).

The majority of neurodegenerative diseases–particularly those of genetic, infectious, immunologic, metabolic, traumatic, ischemic, or toxic etiology–are characterized not by focal but rather by extensive, multifocal, or even global neuropathology. Such widely disseminated CNS lesions have not been traditionally regarded as amendable to neural transplantation, particularly by local CNS grafts of nonmigratory nonneural cells or even primary fetal neural tissue; these materials have been reserved for focal or regionally restricted neurologic diseases (such as Parkinsonism). The mainstays for such neurogenetic diseases (particularly of a metabolic nature)–enzyme replacement, drug treatment, dietary manipulations, BMT–have generally not been successful in reversing or forestalling damage to the CNS, presumably because of restrictions imposed by the BBB to the sustained entrance of corrective molecules. However, it was hypothesized that the very ability of neural stem cells to engraft diffusely and to become integral members of structures (particularly developing structures) throughout the host CNS would permit these cells to address the challenge of widespread neuropathology. The potentially and uniquely appealing characteristics of neural stem cells, it was reasoned, could best be exploited experimentally in genetic mouse models of neurodegeneration requiring disseminated enzyme and/or neural cell replacement. Therefore, the feasibility of a stem-cell-based strategy for the delivery of absent gene products was actually first affirmed by correcting the widespread neuropathology of a murine model of a genetic neurodegenerative lysosomal storage disease called *mucopolysaccharidosis type VII* (MPS VII). Caused by a frameshift mutation of the *β-glucuronidase* (GUSB) gene, this heritable condition causes progressive mental retardation in humans and inexorable fatal neurodegeneration in mice. GUSB is a secreted enzyme required for the degradation of glycosaminoglycans (GAGs); the enzymatic deficiency results in lysosomal accumulation of undegraded GAGs in the brain (neurons and glia) and other tissues (spleen, liver, kidney, cornea, skeleton). Treatments for MPS VII and most other lysosomal storage disorders are designed to provide a source of normal enzyme for uptake by diseased cells, a process termed "cross-correction." C17.2 neural stem cells were genetically modified with a retrovirus encoding human GUSB to augment the mouse GUSB constitutively secreted by these cells. Transplantation of these GUSB-overexpressing stem cells into the cerebroventricular system of newborn MPS VII mice (a rapid, safe, and minimally intrusive technique devised to permit the cells access to germinal zones lining the ventricles) resulted in profuse incorporation of donor-derived cells throughout the mutant neuraxis (Fig. 1). This brain-wide distribution of engrafted GUSB-secreting stem cells corresponded to the distribution of corrective levels of GUSB throughout the mutant brains devoid of that enzyme (Snyder *et al.*, 1995). This diffuse GUSB expression in turn resulted in widespread permanent cross-correction of lysosomal storage in mutant neurons and glia, including the neocortex, compared to age-matched untreated MPS VII controls. Subjective observation of cage behavior suggested that MPS VII recipients were more active and alert. The neonatal transplantation of GUSB-producing neural stem cells produced dramatic long-term improvements in the brain, as neonatal BMT and peripheral enzyme replacement did for the skeletal and visceral disease. Although MPS VII may be regarded as "uncommon," the broad category of diseases it models (i.e., neurogenetic degenerative diseases) afflicts as many as 1 in 1500 persons, a significant challenge to society's human and economic resources. These observations suggested a strategy for CNS gene therapy of this class of diseases, which heretofore have not been adequately treated. This work also represented the first report that neural stem cells could deliver an absent gene product or therapeutic molecule throughout and directly to the CNS.

This therapeutic approach is now being extended to other untreatable neurodegenerative diseases characterized by an absence of discrete gene products and/or the accumulation of toxic metabolites. For example, this strategy is being tested in two recently generated mouse models of the gangliosidoses: Tay–Sachs (TSD) and Sandhoff diseases (SD). When transduced with a retroviral vector encoding the α-subunit of β-hexosaminidase (Hex) (a mutation of which leads to accumulation of GM_2 ganglioside), genetically engineered neural stem cells, implanted into the cerebroventricles, increased whole brain Hex levels of wildtype fetal and neonatal mice by 14 to 76%, an amount sufficient to restore normal ganglioside turnover (Lacorazza *et al.*, 1996). In preliminary studies, transplants into mutant brains suggest that Hex may be boosted to at least 10% of normal, an amount that should prove therapeutic. *In vitro,* the secretory products of these neural stem cells are cross-corrective of TSD and SD neurons and glia (Aurora *et al.*, 1997).

In approaching metabolic diseases such as those above, there is an interesting point to be made that might be applicable to *ex vivo* gene therapy in general. In almost all cases, neural stem cells, because they are normal CNS cells, constitutively express normal amounts of the particular enzyme in question. The extent to which this amount needs to be augmented by genetic engineering may vary from model to model and enzyme to enzyme, and, of course, needs to be studied individually. Reassuringly, in most inherited metabolic diseases and in most neurologic diseases in general, the amount of enzyme required to restore normal metabolism and forestall CNS disease may be quite small. That low levels of neural cell products may already be expressed intrinsically by neural stem cells may actually enhance their value over vectors in gene therapy paradigms. For example, although downregulation of neural-specific gene expression in engineered nonneural cells may leave them "incapacitated," neural stem cells, which originate from brain and which may intrinsically produce low levels of various CNS factors, might allow correction to proceed despite inactivation of the introduced gene. (For example, in addition to most lysosomal enzymes, it is known that C17.2 stem cells inherently express low levels of NT-3, GDNF, IGF-1, syndecan-3, and L1, to name a few.) Finally, donor tissue originating from the brain may sustain expression of introduced neural genes longer than nonneural cells.

Such studies in genetic diseases have helped to establish the paradigm of using genetically modified neural stem cells for the transfer of a range of factors (including other types of genes, extracellular matrix, myelin, even cells) of therapeutic or developmental interest throughout the CNS both in mature brains (as described in previous sections), but especially in developing brains, where the cells can actually contribute to organogenesis of multiple structures in the CNS (Table II). Because neural stem cells can populate widely disseminated developing or degenerating CNS regions with cells of multiple lineages, their use as graft material in the brain can be considered almost analogous to hematopoietic stem-cell-mediated reconstitution and gene transfer. Therefore, neural stem cells have opened a new role for neural transplantation and a possible strategy for addressing the CNS manifestations of diseases that heretofore had been refractory to intervention but now appear amenable to cell-based genetic interventions (see the following discussion and Table II).

VII. USING NEURAL STEM CELLS TO REPLACE MORE THAN GENES: CELLS, MYELIN, EXTRACELLULAR MATRIX IN MODELS OF DEGENERATION AND ABNORMAL DEVELOPMENT

Many neurologic diseases, even those that can be ameliorated by replacement of an identified gene product, are characterized by the degeneration of specific neural cell types or circuits. These structures may need to be replaced in a functional manner, and may need to be resistant to ongoing or residual metabolic toxic processes. One reassuring insight from the classic fetal transplant literature is that even modest anatomical reconstruction may sometimes have an unexpectedly beneficial behavioral or functional effect (Fisher and Gage, 1993). It has long been known that, in spite of the intricate circuitry of the mammalian brain, fetal CNS grafts can integrate focally into lesioned homotopic regions of rodent brain (reviewed in Fisher and Gage, 1994). Neural stem cells may also be able to replace some degenerated or dysfunctional cells. Experiments with prototypical clone C17.2 neural stem cells in various mouse mutants and injury models have, indeed, provided the first evidence affirming the validity of this assumption and the feasibility of this approach (Snyder *et al.*, 1997; Rosario *et al.*, 1997; Yandava and Snyder, 1995). These and other representative studies illustrating the plasticity of cell lines for generating various neural cell types *in vivo* and responding to different environmental cues are summarized in Table V; some are also described in the following sections. These data highlight an additional advantage that cells may have over other vehicles for gene therapy. Viral vectors, synthetic pumps, and nonneural cells depend on relaying new genetic information through established neural circuits and substrates, which may, in fact, have degenerated and require replacement. Neural stem cells may provide that replacement, in addition to supplying the missing gene (e.g., Park *et al.*, 1997; Auguste *et al.*, 1996).

First, we will discuss the replacement of some cellular and extracellular components in models of abnormal neural development. Later we will discuss cell replacement in injury models.

Through studying mouse mutants, it has been learned that neural stem cells may replace some degenerated or dysfunctional neural cells. In the *meander tail* (*mea*) mutant, which is characterized by failure of sufficient granule cell (GC) neurons to develop in the cerebellum (particularly in the anterior lobe), neural stem cells (clone C17.2), implanted at birth, were capable of "repopulating" large portions of the internal granular layer by differentiating into GCs and replenishing that largely absent neuronal population (Fig. 3, see color insert; Rosario *et al.*, 1997). A pivotal observation to emerge from this work, with implications for fundamental stem cell biology, was that cells with the potential for multiple fates might "shift" their differentiation to compensate for a deficiency in a particular cell type. As compared with their differentiation fate in normal cerebella, a large majority of these donor stem cells in *mea* regions deficient in GCs pursued a GC phenotype in preference to other potential phenotypes, suggesting a "push" on undifferentiated, multipotent cells toward repletion of the inadequately developed cell type (perhaps "pressured" by cues from a microenvironment deficient in a particular cell type; Fig. 4). This presents a possible developmental mechanism with therapeutic value.

Given that many neurologic diseases, particularly those of neurogenetic etiology, are characterized by global degeneration or dysfunction, other experiments sought to test whether neural stem cells could address this type of therapeutic challenge (not traditionally viewed as attainable by neural transplantation). It was hypothesized that the techniques devised for diffuse engraftment of enzyme-expressing stem cells to treat global metabolic lesions by gene product replacement (e.g., intracerebroventricular injection) could also be

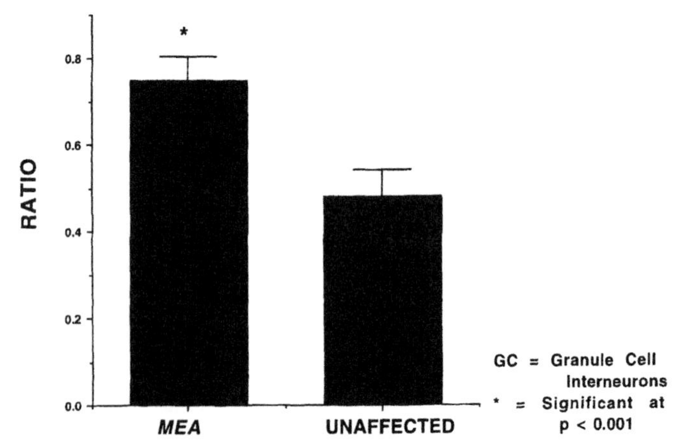

FIGURE 4 Comparison of proportion of all donor-derived cells that became granule cell neurons (CGs) in transplanted *meander tail (mea)* anterior lobes (ALs) with the same ratio in unaffected ALs. There was significantly more neuronal differentiation in *mea* mutant cerebellar ALs than in unaffected ALs (*$p<0.001$). The *mea* AL is dramatically diminished in GCs. These data suggest that, although there is no significant difference in the degree of engraftment between normal and mutant ALs, *differentiation toward a neuronal phenotype* is favored in the granule neuron-deficient regions of the *mea* cerebellum. (See text for further discussion.) (Figure reproduced with permission from Rosario *et al.*, 1997)

applied to the remediation of other types of diffuse neuropathology (Table I), including those requiring widespread cell replacement (Fig. 1). Mouse mutants characterized by CNS-wide white matter disease provided an ideal model for testing this hypothesis. The oligodendroglia of the dysmyelinated *shiverer* (*shi*) mouse are dysfunctional because they lack MBP essential for effective myelination. Therapy, therefore, requires widespread replacement with MBP-expressing oligodendrocytes. Neural stem cells (clone C17.2), transplanted at birth, resulted in widespread engraftment throughout the *shi* brain with repletion of significant amounts of MBP (Yandava and Snyder, 1995). Accordingly, of the many donor cells that differentiated into oligodendroglia, a subgroup myelinated up to 40% of host neuronal processes with better compacted myelin that contained major dense lines and that had a thickness and periodicity approximating those of normal myelin. Some recipient animals appeared to show a decrement in their symptomatic tremor. Therefore, global cell replacement seems feasible for some pathologies if cells with stemlike features are employed. This approach is being extended to other poorly myelinated mutants, such as the *twitcher* mouse model of *Krabbe's disease* in which preliminary results suggest successful engraftment by C17.2 cells and probable remyelination (Taylor and Snyder, 1997). The ability of stem cells to remyelinate is of particular importance because dysmyelination plays an important role in many genetic and acquired (injury, infection) neurodegenerative diseases.

Work in another mutant, the *reeler* mouse, has suggested that stem cells may not only replace developmentally impaired neural cells, but may also help to correct certain aspects of abnormal cytoarchitecture in models of dysfunctional lamination, particularly those characterized by deficiencies in extracellular matrix (ECM) (Auguste *et al.*, 1996). The laminar positioning of neurons in the *reeler* mouse brain is profoundly abnormal due, most likely, to a mutation in a gene encoding the secreted ECM molecule, Reelin. Neural stem cells (clone C17.2), implanted at birth into the defective developing *reeler* cerebellum, appeared not only to replace missing GCs in the *reeler* cerebellum, but also to promote a more normally laminated cytoarchitectural appearance in engrafted regions by "rescuing" aspects of the abnormal migration, positioning, and survival of neurons. Not only did donor-derived GCs migrate to the appropriate layer themselves, but transplanted stem cells also seemed to influence neighboring mutant neurons: *reeler* PCs were more appropriately aligned in a monolayer in engrafted regions; *reeler* GCs, their survival enhanced, also migrated appropriately in greater numbers deeply below the PCs into the deeper strata of the internal granular layer. It appeared that donor stem cells may have helped to promote normal layering and survival by providing molecules (including Reelin), which work at the cell surface to guide proper migration and positioning of neurons during histogenesis. These findings, therefore, suggested a possible stem-cell-based strategy for the treatment of

CNS diseases characterized by abnormal cellular migration, lamination, and cytoarchitectural arrangement.

More broadly, complementation studies with mutants such as those described in this chapter help to support such an approach–whether with exogenous stem cells or with appropriately mobilized endogenous stem cells–for compensating for neurodevelopmental problems of many etiologies.

VIII. MIGHT NEURAL STEM CELLS BE RESPONSIVE TO AND THEREFORE BE IDEALLY SUITED FOR NEURODEGENERATION?

An intriguing phenomenon with possible therapeutic dividends has begun to emerge from observations of the behavior of neural stem clones in various mouse models of acquired CNS injury. During phases of active neurodegeneration, as-yet-unidentified factors seem to be transiently elaborated to which neural stem cells may respond by migrating to degenerating regions and differentiating toward replacement of dying neural cells. In other words, neural stem cells, we believe, have the capacity to respond to neurogenic cues *in vivo* not only when they occur during their normal developmental expression but even when induced or "reactivated" at later stages, for example, by certain types of cell death.

This phenomenon was first observed in a model of experimentally induced apoptosis of selectively targeted pyramidal neurons in the adult neocortex (Snyder *et al.*, 1997). When transplanted into this neuron-specific apoptotic degenerative environment in the adult neocortex (Sheen and Macklis, 1995), though not when implanted into intact or even necrotically lesioned adult neocortex, 15% of transplanted neural stem cells (clone C17.2) differentiated specifically into that type of degenerating neuron, partially replacing that lost neuronal population (Snyder *et al.*, 1997; Fig. 5, see color insert). (In intact neocortex, donor cells differentiated exclusively into glia, the expected finding given that gliogenesis is the predominant differentiative process normally ongoing in adult neocortex.) Some replacement neurons send axons across the corpus callosum to appropriate targets in the contralateral hemisphere (Leavitt *et al.*, 1996). The neuronal phenotype assumed by these stem cells within regions of apoptosis suggested that this form of degeneration created a microenvironment permissive or instructive for a neuronal fate choice, perhaps through reactivation of signals ordinarily available only during embryonic corticogenesis.

That the type and time course of the cell death encountered by neural stem cells may actually be quite pivotal is reinforced by observations with the HiB5 and ST stem cell lines grafted into the ibotenic acid (IA) lesioned adult striatum, an environment characterized ultimately by the development of a good deal of astrogliotic scarring. When pre-engrafted animals were subsequently subjected to IA lesioning, both host- and donor-derived cells underwent an astrogliotic reaction; the number of GFAP+ donor-derived cells (astrocytes) increased from a baseline of 10 to 40% (paralleled by an increase in vimentin+, probably immature, donor cells) (Lundberg and Björklund, 1996). If the cells were transplanted into the striatum following the IA lesion, when astrogliosis had ensued, the proportion of donor-derived astrocytes remained at baseline levels, but, because the donor cells divided once, the absolute number of donor-derived cells, and hence astrocytes, was at least doubled; graft-derived neurons were virtually undetectable (<0.1%) (Lundberg and Björklund, 1996). In experiments in which the adult striatum was denervated by intra-nigral 6-hydroxy-dopamine, grafted cells in the striatum similarly yielded astrocytes (Lundberg and Björklund, 1996). These kinds of neurodegenerative environments, during the periods examined, appeared to elaborate signals for astrogliogenesis (rather than neuronogenesis), perhaps to provide a protective astrogliotic reaction; both host and grafted cells responded similarly. In a converse yet complementary line of evidence, even the oligo-potent neuroblastic RN33B line, whose cells give rise *only* to neurons, seems to require as-yet-unspecified instructive signals present in areas of neurodegeneration to achieve its differentiated neuronal phenotype; in intact areas, the grafted cells remain undifferentiated (Lundberg *et al.*, 1996c; Shihabuddin *et al.*, 1996].

Apoptosis (at least at particular critical phases) is becoming implicated in a growing number of both neurodegenerative and normal developmental processes (reviewed in Bredesen, 1995). It has been hypothesized that neurodegeneration–perhaps those aspects mediated by apoptosis–may create a *milieu* that recapitulates normal embryonic developmental cues (particularly for neuronogenesis) to which neural stem cells can respond to therapeutic advantage (Snyder and Macklis, 1996). In other words, acute CNS injury or degeneration (of certain types and/or at critical times) might positively influence the migration, proliferation, and differentiation of neural stem cells, both of host and donor origin. Preliminary evidence in two experimental mouse models in which apoptosis is known to be contributory suggests that this may, indeed, be feasible. The first is a model of hypoxic-ischemic (HI) brain injury, in which neural tissue is infarcted throughout the cerebrum ipsilateral to a unilateral carotid ligation combined with reduced ambient oxygen. The second is a model of lower motoneuron (LMN) degeneration within spinal cord (SC) segments receiving contributions from, and ipsilateral to, a unilaterally axotomized sciatic nerve. Axotomy at birth results in a selective permanent degeneration of 75% of LMNs in the fourth through sixth lumbar spinal segments (L_{4-6}) in the adult mouse.

When C17.2 stem cells, in preliminary studies, are transplanted into brains of postnatal mice subjected to unilateral HI brain injury (optimally 3–7 days following HI), donor-derived cells migrate preferentially to and integrate extensively within the large ischemic areas that typically span the injured hemisphere (Fig. 6, see color insert). Even donor cells implanted in more distant locations (including the intact contralateral hemisphere) migrate toward the regions of HI injury. A subpopulation of donor stem cells, particularly in the penumbra of the infarct, "shift" their differentiation fate towards the repletion of neurons and oligodendrocytes, the neural cell types typically damaged following asphyxia/stroke and the cell types least likely to regenerate spontaneously in the postnatal CNS (Park *et al.*, 1995, 1999). In fact, consistent with our previous findings, no neurons and only a small percent of oligodendrocytes are derived from stem cells in *intact* postnatal neocortex (Snyder *et al.*, 1997). Clearly, as in the targeted apoptosis model, novel signals appear to be transiently elaborated following HI to which stem cells might respond. (Interestingly, the behaviors revealed in these studies are what might be expected from a classic, quiescent stem cell population "attempting" to repopulate and reconstitute a damaged region in response to injury. Yet, even though this is *de rigor* in the hematopoietic system, this has never been shown to occur in the CNS.) Because engrafted stem cells continued to express *lacZ*, it appeared feasible that neuronal differentiation of both host- and donor-derived cells might be enhanced if donor cells were genetically manipulated *ex vivo* prior to transplantation to express certain neurotrophic agents. In fact, in preliminary studies when subclones of C17.2 engineered to overexpress NT-3 were implanted into asphyxiated mouse brain, the proportion of donor-derived neurons was increased up to 20% in the infarction cavity and to as high as 80% in the penumbra (Park *et al.*, 1997, 1998). (Clearly the HT-3 had also worked on the donor stem cells in an autocrine/paracrine fashion.) This experiment enunciated the use of neural stem cells for simultaneous, combined gene therapy *and* cell replacement in the same transplant procedure using the same clone of cells in the same transplant recipient, an intriguing stem cell ability with implications for therapies in other conditions.

A somewhat similar phenomenon of a transgene theoretically working for the benefit of the donor cell may be observed with the unipotent neuroblastic line RN46A, which produces soley serotonergic neurons *in vitro* and *in vivo*. The survival and maturation of these cells toward serotonergic neurons seem to be enhanced by overexpression of BDNF (White *et al.*, 1994; Eaton and Whittemore, 1996). Such cells, after appropriate manipulation, could theoretically be "instructed" to provide a source of replacement serotonergic neurons *in vivo* without the need for engineering the donor cells specifically for serotonin production.

An optimal temporal window of tropism and trophism for neural stem cells by apoptotic neurodegenerative environments appears evident also in the postnatal SC during LMN degeneration induced by neonatal sciatic axotomy, a classic model of specific spinal motoneuron degeneration and SC dysfunction (Flax and Snyder, 1995; Park *et al.*, 1999). LMN are born normally only in the fetal SC. In perliminary studies, C17.2 stem cells, implanted into the L_5 ventral horn 2 to 4 weeks after axotomy, stably engraft for at least 4 months. A subpopulation of these integrated donor progenitors (~5%) differentiate into cells that appear to have "replaced" some of the missing LMNs. If C17.2 cells are transplanted within the first week following axotomy (closer to the peak of active apoptotic degeneration), not only is the number of engrafted cells increased and the proportion of cells resembling LMNs more than tripled (up to 17–20%), but one can even implant the stem cells into the *dorsal* horn (a technically easier, less invasive intervention) and allow them to migrate anteriorly toward the LMN-impoverished ventral horn and as much as 1 mm in the rostral-caudal dimension within the ventral horn. Though these donor-derived neurons may require additional signals to consummate their induction and maturation, these findings further support the hypothesis that developmental signals (which normally are extant only during embryogenesis) may be transiently reexpressed during phases of active degeneration, and that neural stem cells may respond by shifting their differentiation program toward a lineage that could replace the dying cells. Again, it appears that neural stem cells might be attempting to repopulate and reconstitute an area of injury. And again, engrafted progenitors continue to express *lacZ*, suggesting that, as in the asphyxiated brain, implantation of genetically engineered stem cells expressing trophins or other factors might enhance neuronal differentiation, neurite outgrowth, and proper connectivity. Furthermore, it might be possible to treat chronic lesions by expressing "apoptotic signals" (certain intercellular signalling molecules such as cytokines), which emulate the acute phase to which stem cells might then respond.

IX. SOME UNEXPECTED USES OF NEURAL STEM CELL LINES

The ability of neural stem cells to migrate, distribute themselves throughout the brain, and disseminate a foreign gene product has prompted the use of these cells in a number of somewhat unconventional areas of investigation (Table I).

A. Generating Models of Neural Disease

Although one typically thinks of using gene transfer technology for therapeutic purposes, neural stem cells have actually also proven useful for the distribution of "disease-producing" agents throughout normal brains in order to create models of CNS disease where none yet exist. For example, a model of "retrovirus-induced spongiform myeloencephalopathy" (a histopathology that resembles HIV-induced CNS disease and with a motoneuron involvement that resembles amyotrophic lateral sclerosis) was created through the diffuse engraftment of genetically engineered C17.2 stem cells expressing neurovirulent retroviral envelope proteins and replication-competent retroviruses (Lynch *et al.*, 1996). This technique not only created a model of retrovirally mediated neurodegeneration but also permitted a sequential dissection of its underlying pathophysiology by using stem cells variously genetically manipulated to deliver isoforms of molecules and viral strains specific for each stage of the retroviral life cycle. It could be determined, for example, that envelope protein, contrary to a popular hypothesis, is not sufficient for induction of disease nor are early retroviral replication events. Rather, late replication processes, probably within infected host microglia, are actually required. Having generated this model, attempts may now ensue to reverse degeneration using normal or engineered cells. This novel use of neural stem cell lines may help generate and test models of other types of neuropathology via the transfer of putatively toxic molecules or viral particles; it may help to sequentially dissect the role particular factors may play in complex pathologic processes, particularly in instances in which transgenic mice have either been unsuccessful, unhelpful, or too cumbersome to generate (especially when multiple variations of a molecule require testing).

B. Neural Stem Cell-Mediated Approaches to Brain Tumor Therapy

A second benefit of the migratory capacity of neural stem cells may be the emergence of novel approaches to the treatment of some brain tumors. One of the impediments to gene therapy of some tumors (e.g., gliomas) has been the degree to which they expand, migrate widely, and infiltrate normal tissue. In preliminary studies, it appears that neural stem cells (clone C17.2), when implanted into an experimental glioma, will migrate along with and throughout the tumor while continuing to express their foreign reporter gene in juxtaposition to even widely expanded and advancing tumor cells (Aboody-Guterman *et al.*, 1996). Migratory neural progenitors might, therefore, provide a platform for the dissemination of therapeutic genes to tumor cells that previously were inaccessible. These observations may lay the groundwork for a number of potential new gene therapy approaches to previously refractory brain tumors.

C. Stemlike Cells as Packaging Cells for Retroviral Vectors

Up to this point, the extensive distribution of large quantities of viral vector-mediated therapeutic genes to the brain has been disapppointingly limited. It seems quite likely that widely engrafted neural stem cells (clone C17.2), which have been engineered to package replication-defective retroviral vectors (in much the same way as they have been engineered to deliver replication-competent viruses in (Lynch *et al.*, 1996), may serve as platforms for the brain-wide dissemination of these viral vectors. Such a stem cell-based platform for the dissemination of viral vectors may well magnify the efficacy of these vectors. Such an approach would marry viral- and cellular-mediated gene therapy techniques.

X. CONCLUSION

Emerging data suggest that at least a subpopulation of progenitors or stem cells exists in the mammalian CNS whose isolation, propagation, and transplantation will have clinical utility for molecular support therapy and/or cell replacement for some degenerative, developmental, and acquired insults to the CNS. The data summarized in this chapter, affirming the pivotal role stem cells have played in mediating recovery in an array of neurodegeneration animal models, suggest that such strategies are moving beyond the stage of mere speculation. Induction of sprouting; reversal of neuronal atrophy; delivery and replacement of diffusible and nondiffusible factors (including metabolic enzymes, neurotransmitters, neurotrophins, cytotoxic agents, viral vectors, myelin, ECM, etc.); improvement in injury-induced and age-related cognitive deficits; cross-correction of metabolic, neurogenetic, toxic, and perinatal diseases; protection against excitotoxicity and ischemia; repopulation of injured, ischemic, degenerated, diseased, and developmentally impaired neural cells of multiple lineages in multiple regions of brain and spinal cord, all appear today as feasible neuroregenerative strategies.

It is important to note, however, that such progress has been made by systematic analyses of the fundamental biology underlying the *in vitro* and *in vivo* behavior of these stem cell clones in idealized, well-characterized, well-controlled, robust animal models.

It is now time to envision how this biology may be translated to treatment strategies in genuine clinical situations. Most neurologic diseases are characterized by complex pathology, often requiring multiple repair strategies. By virtue of their inherent biologic properties, CNS stem cells seem to possess the multiple therapeutic capabilities demanded. A number of options are possible from the implantation of neural stem cells into the dysfunctional CNS. First, transplanted stem cells may literally differentiate into and replace damaged or degenerated neurons and/or glia. Host cells degenerating not only by cell-intrinsic disease processes (the obvious situation) but also cell-*extrinsic* forces might theoretically be replaced. For example, in some pathologic conditions, host factors may be elaborated within the microenvironment, which are inimical to the survival and/or differentiation of donor stem cells. However, donor cells that are not naturally resistant may be engineered *ex vivo* to *be* resistant to these factors, or to secrete substances that might neutralize those factors, or to express trophic agents that might overcome those factors.

Even those implanted cells that do *not* differentiate into the desired missing cell types may nevertheless serve a therapeutic purpose. Being of CNS origin, donor stem cells may intrinsically provide factors–both diffusible (e.g., trophins) and nondiffusible (e.g., "bridges," cell–cell contact signals, myelin, ECM)–that might enable the injured host to regenerate its own lost cells and/or neural circuitry. In the event such factors are not naturally produced in sufficient quantities, the cells could be genetically engineered before transplantation to become "factories" for sustained local production of substances known to mobilize quiescent *host* progenitor/stem cell pools, to promote regeneration and differentiation of immature nerve cells (either endogenous or, in a paracrine/autocrine fashion, donor), to attract ingrowth of host fibers, to forestall degeneration resulting from insufficiency of a trophin or enzyme in the *milieu*.

Finally, there appears to be a "window" during which signals may be transiently elaborated within the degenerating region to which stem cells may be able to respond (e.g., directed migration; reconstitution of lost cells). During phases of active, apoptotic degeneration, developmental cues may be transiently re-expressed (that normally are extant only during embryonic neurogenesis) to which neural stem cells may react by differentiating toward replacement of the dying cells. In other words, progenitors with the potential for multiple fates might preferentially differentiate toward the repletion of deficient cell types, a possible developmental mechanism with obvious therapeutic value. The treatment of more chronic lesions might theoretically be achieved by "recreating" active degeneration through the delivery of exogenous factors, which emulate the more acute phase to which stem cells are responsive.

Because stem cells may incorporate into host cytoarchitecture in a functional manner, they may prove more than vehicles for "passive" delivery of substances: the regulated release of certain substances through feedback loops may be reconstituted, as might the reformation of essential connections. Although presently available gene transfer vectors usually depend on relaying new genetic information through established neural circuits, which may, in fact, have degenerated, neural stem cells may actually also reconstitute the pathways.

Intriguingly, a single multipotent neural stem cell line may, in certain conditions, theoretically perform many or all of the above-mentioned therapeutic functions, even concurrently in one transplant procedure. Furthermore, because stem cells are sufficiently plastic and may integrate in recipient brains either locally or in a widespread manner depending on method of delivery, host age, and neurodegenerative context, the transfer of molecules and the replacement of cells may be electively delivered in either a site-specific or a more disseminated manner. This adaptability allows the targeting of specific cell groups and the rewiring of specific circuitry in some therapeutic contexts or the treatment of more extensive, multifocal, or global pathologic processes in others, depending on the demands of the disease.

In short, the inherent biologic properties of neural stem cells may elevate them beyond simply being an alternative to fetal tissue in neural transplantation paradigms or simply another gene delivery vehicle in neural gene therapy paradigms. Neural stem cell transplantation, in many ways, has helped to reorient and broaden the paradigmatic scope of neural transplantation as a therapeutic intervention. Although transplantation has traditionally been reserved for neuropathologies, which are localized anatomically (e.g., Parkinson's disease), the use of stem cells, because of their ability to engraft within germinal zones and to migrate and integrate widely, may permit a "cell-based" therapy such as transplantation to tackle conditions that effect the CNS extensively and, through widespread enzyme and/or cell replacement, to address certain therapeutic challenges conventionally consigned solely to pharmacologic or genetic interventions, as well as providing treatments for CNS diseases that heretofore had been refractory to intervention.

Because neural stem cells can populate widely disseminated developing or degenerating CNS regions with cells of multiple lineages, their use as graft material in the brain can be considered analogous to hematopoietic stem-cell-mediated reconstitution and gene transfer in the body. Yet, as mentioned in this chapter, unlike in BMT, this method of delivery does not preclude being able to transport gene products into the cytoarchitecture of circumscribed regions in order to effect selective manipulations and avoid extensive genetic alteration should the clinical context demand it.

XI. FUTURE DIRECTIONS

Future challenges in the isolation, characterization, and utilization of neural stem cells include the following: to determine the parameters that optimize their engraftment and differentiation in various developing and degenerating CNS environments; to discern the triggers that direct the phenotypic fate of stem cells in the brain; to optimize the vectors and methods of gene transfer to and modification of stem cells; to identify the mechanisms that dictate efficiency of foreign gene expression by engrafted cells; to devise efficient systems for regulating the amount of transgene expression in the brain, including providing "safety-switches" for stopping production of a foreign gene product if necessary. These all will require a better understanding of fundamental neural stem cell biology.

In the long run, a better understanding of this biology may allow stem cell lines with a similar repertoire of characteristics to be generated expeditiously from human neural tissue and transplanted without concern for recipient safety. Progress in this regard is already being made by a small group of labs (Sah *et al.*, 1996; Yang *et al.*, 1998; Vescovi *et al.*, 1998). It is interesting in this regard, and perhaps instructive that, although there are a number of methods for growing neural stem cells, propagating them in the manner by which the clones described in this chapter were generated, appears to be among the most efficacious and safest to date, at least in rodent systems. The results attained by these rodent stem cells in the various disease paradigms described above sets a standard, so to speak, to which the product of any stem cell isolating and propagating method must aspire. To achieve less may suggest not truly realizing or unlocking the capabilities of the neural stem cell.

ACKNOWLEDGMENTS

AM-S would like to acknowledge Professor Anders Björklund (Lund, Sweden) for his invaluable contribution and support to some of the work presented in this chapter, and the financial support of the Human Frontier Science Program, and the Swedish Medical Research Council. Some of the work described by EYS was supported by grants from the National Institute of Neurologic Diseases and Stroke (NS33852, NS34247), the American Paralysis Association, the Paralyzed Veterans of America, and by Mental Retardation Research Center grant HD18655. EYS also thanks members of his lab (J. D. Flax, K. I. Park, S. Liu, S. Aurora, C. Yang, K. I. Auguste, S. Nissim, D. Tsai, B. D. Yandava, C. R. Rosario, C. Simonin, A. M. Wills, L. L. Billinghurst, and Karen Aboody-Guterman) and his collaborators (whose names are liberally cited in the text and references) for their valuable contributions to some of the work discussed in this chapter. J. Hayden helped to prepare the computer reconstruction in Fig. 1.

REFERENCES

Aboody-Guterman, K. S., Sena-Esteves, M., Herrlinger, U., Snyder, E. Y., and Breakefield, X. O. (1996). Visualization of brain tumour growth and neural cell migration *in vivo* using retroviral vectors expressing GFP. *Soc Neurosci Abs* 22:370.13.

Aboody-Guterman, K. S., Pechan, P., K. S., Rainov, N. G., Sena-Esteves, M., Synder, E. Y., Wild, P., Schraner, E., Tobler, K., Breakefield, X. O., and Fraefel, C. (1997). Green fluorescent protein as a reporter for retrovirus and helper virus-free HSV-1 amplicon vector-mediated gene transfer into neural cells in culture and *in vivo*. *Neuro Report* 8(17):3801–3808.

Andsberg, G., Kokaia, Z., Björklund, A., Lindvall, O., and Martínez-Serrano, A. (1998). Amelioration of ischemia-induced neuronal death in the rat striatum by NGF-secreting neural stem cells. *Eur J Neurosci* 10:2026–2036.

Anton, R., Kordover, J. H., Maidment, N. T., Manaster, J. S., Kane, D. J., Rabizadeh, S., Scueller, S. B., Yang, J., Rabizadeh, S., Edwards, R. H., and Bredesen, D. E. (1994). Neural-targeted gene therapy for rodent and primate hemiparkinsonism. *Exp Neurol* 127:207–218.

Auguste, K. I., Nakajima, K., Miyata, T., Ogawa, M., Mioshiba, K., and Snyder, E. Y. (1996). Neural progenitor transplantation into newborn reeler cerebellum may rescue certain aspects of mutant cytoarchitecture. *Soc Neurosci Abstr* 22:484.

Aurora, S., Kim, S. U., Lacorraza, H. D., Jendoubi, M., Proia, R. L., and Snyder, E. Y. (1997). Use of murine and human neural progenitors for cross-correction of enzyme replacement in neural cell culture from mouse models of the neurodegenerative gangliosidoses (Tay Sachs Disease), *Soc Neurosci Abstr* 23:1622.

Bartlett, P. F., Reid, H. H., Bailey, K. A., and Bernard, O. (1988). Immortalization of mouse neural precursor cells by the c-myc oncogene. *Proc Natl Acad Sci USA* 85:3255–3259.

Bayer, S. A., and Altman, J. (1998). Development: A summary of neurogenesis and neuronal migration in the rat central nervous system. In: Paxinos, G., ed., *The rat nervous system,* 2nd ed. Orlando: Academic Press.

Bencsics, C., Watchel, S. R., Milstien, S., Hatakeyama, K., Becker, J. B., and Kang, U. J. (1996). Double transduction with GTP cyclohydrolase I and tyrosine hydroxylase is necessary for spontaneous synthesis of L-DOPA by primary fibroblasts. *J Neurosci* 16:4449–4456.

Björklund, A. (1991). Neural transplantation–an experimental tool with clinical possibilities. *Trends Neurosci* 14:319–322.

Björklund, A. (1993). Better cells for brain repair. *Nature* 362:414–415.

Bredesen, D. E. (1995). Neural apoptosis. *Ann Neurol* 38(6):839–851.

Brüstle, O., and McKay, R. D. G. (1996). Neuronal progenitors as tools for cell replacement in the nervous system. *Curr Op Neurobiol* 6:688–695.

Cadusseau, J., and Peschansli, M. (1992). Identifying grafted cells. In: Dunnet, S. B., and Björklund, A., eds., *Neural transplantation, a practical approach.* Oxford, UK: IRL Press, Oxford University Press.

Campell, K., Olsson, M., and Bjorklund, A. (1995). Regional incorporation and site-specific differentiation of strial precursors transplanted to the embryonic forebrain ventricle. *Neuron* 15:1259–1273.

Carpenter, M. K., Winkler, C., Fricker, R., Wong, S. C., Greco, C., Emerich, D., Chen, E.-Y., Chu, Y., Kordower, J., Messing, A., Björklund, A., and Hammang, J. P. (1996). Transplantation of EGF-responsive neural stem cells derived from GFAP-hNGF transgenic mice attenuates excitotoxic striatal lesions. *Soc Neurosci Abs* 22:232.4.

Cattaneo, E., Magrassi, L., Santi, L., Butti, G., McKay, R. D. G., and Pezzota, S. (1993). Transplanting embryonic striatal cell lines into the embryonic rat brain. *Soc Neurosci Abs* 19:107.6.

Cattaneo, E., Magrassi, L., Butti, G., Santi, L., Giavazzi, A., and Pezzotta, S. (1994). A short term analyses of the behavior of conditionally immortalized neuronal progenitors and primary neuroepithelial cells implanted into the fetal rat brain. *Dev Brain Res* 83:197–208.

Chen, K. S., and Gage, F. H. (1995). Somatic gene transfer of NGF to the aged brain: Behavioral and morphological amelioration. *J Neurosci* 15:2819–2825.

Coats, S., Flanagan, W. M., Nourse, J., and Roberts, J. M. (1996). Requirement of p-27^{kip1} for restriction point control of the fibroblast cell cycle. *Science* 272:877–880.

Corti, O., Horellou, P., Colin, P., Cattaneo, E., and Mallet, J. (1996). Intracerebral tetracycline-dependent regulation of gene expression in grafts of neural precursors. *NeuroReport* 7:1655–1659.

Davis, A. A., and Temple, S. (1994). A self-renewing multipotential stem cell in embryonic rat cerebral cortex. *Nature* 372:263–266.

Dunnett, S. B., and Bjorklund, A., eds. (1994). *Functional neural transplantation.* New York: Raven Press.

Eaton, M. J., and Whittemore, S. R. (1996). Autocrine BDNF secretion enhances the survival and serotonergic differentiation of raphe neuronal precursor cells grafted into the adult rat CNS. *Exp Neurol* 140:105–114.

Fischer, W., Wictorin, K., Björklund, A., Williams, L. R., Varon, S., and Gage, F. H. (1987). Amelioration of cholinergic neuron atrophy and spatial memory impairment in aged rats by nerve growth factor. *Nature* 329:65–68.

Fisher, L. J., and Gage, F. H. (1993). Grafting in the mammalian central nervous system. *Physiol Rev* 73:583–616.

Fisher, L. J. (1997). Neural precursor cells: Applications for the study and repair of the central nervous system. *Neurobiology of Disease* 4:1–22.

Flax, J. D., and Snyder, E. Y. (1995). Transplantation of CNS stem-like cells as a possible therapy in a mouse model of spinal cord dysfunction. *Soc Neurosci Abstr* 21:2028.

Frederiksen, K., Jat, P. S., Valtz, N., Levy, D., and McKay, R., *et al.* (1988). Immortalization of precursor cells from the mammalian CNS. *Neuron* 1:439–448.

Frim, D. M., Uhler, T. A., Short, M. P., Ezzedine, Z. D., Klagsbrun, M., Breakefield, X. O., and Isacson, O. (1993a). Effects of biologically delivered NGF, BDNF and bFGF on striatal excitotoxic lesions. *NeuroReport* 4:367–370.

Frim, D. M., Yee, W., and Isacson, O. (1993b). NGF reduces striatal excitotoxic neuronal loss without affecting concurrent neuronal stress. *NeuroReport* 4:655–658.

Gage, F., and Christen, Y., eds. (1997). *Isolation, characterization, and utilization of CNS stem cells*. Berlin Heidelberg: Springer-Verlag.

Gage, F. H., and Fisher, L. J. (1991). Intracerebral grafting: A tool for the neurobiologist. *Neuron* 6:1–12.

Gage, F. H., Kawaja, M. D., and Fisher, L. (1991). Genetically modified cells: Applications for intracerebral grafting. *Trends Neurosci* 14:328–333.

Gage, F. H., Ray, J., and Fisher, L. J. (1995a). Isolation, characterization and use of stem cells from the CNS. *Ann Rev Neurosci* 18:159–192.

Gage, F. H., Coates, P. W., Palmer, T. D., et al. (1995b). Survival and differentiation of adult neuronal progenitor cells transplanted to the adult brain. *Proc Natl Acad Sci USA* 92:11879–11883.

Galaktionov, K., Chen, C., and Beach, D. (1996). Cdc25 cell-cycle phosphatase as a target of c-myc. *Nature* 382:511–517.

Galpern, W. R., Matthews, R. T., Beal, M. F., and Isacson, O. (1996). NGF attenuates 3-nitrotyrosine formation in a 3-NP model of Huntington's disease. *NeuroReport* 7:2639–2642.

Ghosh, A., and Greenberg, M. E. (1995). Distinct roles for bFGF and NT-3 in the regulation of cortical neurogenesis. *Neuron* 15:89–103.

Gussoni, E., Wang, Y., Fraefel, C., Miller, R. G ., Blau, H. M., Geller, A. I., and Kunkel, L. M. (1996). A method to codetect introduced gene and their products in gene therapy protocols. *Nat Biotech* 14:1012–1016.

Gusssoni, E., Blau, H. M., and Kunkel, L. M. (1997). The fate of individual myoblasts after transplantation into muscles of DMD patients. *Nat Med* 3:970–977.

Himes, B. T., Slowska-Baird, J., Boyne, L., Snyder, E. Y., Tessler, A., and Fischer, I. (1995). Grafting of genetically modified cells that produce neurotrophins in order to rescue axotomized neurons in rat spinal cord. *Soc Neurosci Abstr* 21:537.

Iavarone, A., and Massague, J. (1997). Repression of the CDK activator Cdc25A and cell-cycle arrest by cytokine TGF-β in cells lacking the CDK inhibitor p15. *Nature* 387:417–422.

Johe, K. K., Hazel, T. G., Muller, T., Dugich-Djordjevic, M. M., and McKay, R. D. G. (1996). Single factors direct the differentiation of stem cells from the fetal and adult central nervous system. *Genes Dev* 10:3129–3140.

Kang, U. J., Fisher, L. J., Joh, T. H., O'Malley, K. L., and Gage, F. H. (1993). Regulation of dopamine production by genetically modified primary fibroblasts. *J Neurosci* 13:5203–5211.

Kaplit, M. G., and Loewy, A. D., eds. (1995). *Viral vectors: Tools for analysis and genetic manipulation of the nervous system*. San Diego: Academic Press.

Lacorazza, H. D., Flax, J. D., Snyder, E. Y., and Jendoubi, M. (1996). Expression of human β-hexosaminidase alfa-subunit gene (the gene defect of Tay-Sachs disease) in mouse brains upon engraftment of transduced progenitor cells. *Nature Medicine* 2:424–429.

Leavitt, B. R., Canales, M., Snyder, E. Y., and Macklis, J. D. (1996). Multipotent neural precursors transplanted to regions of targeted neuronal apoptosis in adult mouse somatosensory cortex differentiate into neurons & reform callosal projections. *Soc Neurosci Abstr* 22:505.

Leone, G., DeGregori, J., Sears, R., Jakoi, L., and Nevins, J. R. (1997). Myc and Ras collaborate in inducing accumulation of active cyclin E/Cdk2 and E2F. *Nature* 387:422–426.

Lundberg, C., and Björklund, A. (1996). Host regulation of glial markers in intrastriatal grafts of conditionally immortalized neural stem cell lines. *NeuroReport* 7:847–852.

Lundberg, C., Field, P. M., Ajayi, Y. O., Raisman, G., and Björklund, A. (1996a). Conditionally immortalized neural progenitor cell lines integrate and differentiate after grafting to the adult rat striatum: A combined autoradiographic and electron microscopic study. *Brain Res* 737:295–300.

Lundberg, C., Horellou, P., Mallet, J., and Björklund, A. (1996b). Generation of DOPA-producing astrocytes by retroviral transduction of the human tyrosine hydroxylase gene: *In vitro* characterization and *in vivo* effects in the rat Parkinson model. *Exp Neurol* 139:39–53.

Lundberg, C., Winkler, C., Whittemore, S. R., and Björklund, A. (1996c). Conditionally immortalized neural progenitor cells grafted to the striatum exhibit site-specific neuronal differentiation and establish connections with the host globus pallidus. *Neurobiol Disease* 3:33–50.

Lundberg, C., Martínez-Serrano, A., Cattaneo, E., McKay, R. D. G., and Björklund, A. (1997). Survival, integration and differentiation of neural stem cell lines after transplantation to the adult rat striatum. *Exp Neurol*: in press.

Lynch, W. P., Snyder, E. Y., Qualtiere, L., Portis, J. L., and Sharpe, A. H. (1996). Late virus replication events in microglia are required for neurovirulent retrovirus-induced spongiform neurodegeneration: Evidence from neural progenitor-derived chimeric mouse brains. *J Virol* 70:8896–8907.

Martínez-Serrano, A., and Björklund, A. (1996a). Protection of the neostriatum against excitotoxic damage by neurotrophin-producing, genetically modified neural stem cells. *J Neurosci* 16:4604–4616.

Martínez-Serrano, A., and Björklund, A. (1996b). Gene transfer to the mammalian brain using neural stem cells: A focus on trophic factors, neuroregeneration and cholinerergic neuron systems. *Clin Neurosci* 3:301–309.

Martínez-Serrano, A., and Björklund, A. (1998). Ex vivo nerve growth factor gene transfer to the basal forebrain in presymptomatic middle-aged rats prevents the development of cholinergic neuron atrophy and cognitive impairment during aging. *Proc Natl Acad Aci USA* 95:1858–1863.

Martínez-Serrano, A., Lundberg, C., Horellou, P., Fischer, W., Bentlage, C., Campbell, K., McKay, R. D. G., Mallet, J., and Björklund, A. (1995a). CNS-derived neural progenitor cells for gene transfer of nerve growth factor to the adult rat brain: Complete rescue of axotomized cholinergic neurons after implantation into the septum. *J Neurosci* 15:5668–5680.

Martínez-Serrano, A., Fischer, W., and Björklund, A. (1995b). Reversal of age-dependent cognitive impairments and cholinergic neuron atrophy by NGF-secreting neural progenitors grafted to the basal forebrain. *Neuron* 15:473–484.

Martínez-Serrano, A., Fischer, W., Söderström, S., Ebendal, T., and Björklund, A. (1996a). Long-term functional recovery from age-induced spatial memory impairments by nerve growth factor gene transfer to the rat basal forebrain. *Proc Natl Acad Sci USA* 93:6355–6360.

Martínez-Serrano, A., Hantzopoulos, P., and Björklund, A. (1996b). *Ex-vivo* gene transfer of brain-derived neurotrophic factor to the intact rat forebrain: Neurotrophic effects on cholinergic neurons. *Eur J Neurosci* 8:727–735.

Martínez-Serrano, A., Olsson, M., Gates, M., and Björklund, A. (1998). In utero gene transfer reveals survival effects of Nerve Growth Factor on brain cholinergic neurons during development. *Eur J Neurosci* 10:263–271.

McKay, R. D. G. (1997a). Immortal mammalian neuronal stem cells differentiate after implantation into the developing brain. In: Gage, F. H., and Christen, Y., eds., *Gene transfer and therapy in the nervous system*. Berlin-Heidelberg: Springer-Verlag, pp. 76–85

McKay, R. (1997b). Stem cells in the central nervous system. *Science* 276:66–71.

Morrison, S. J., Shah, N. M., and Anderson, D. J. (1997). Regulatory mechanisms in stem cell biology. *Cell* 88:287–298.

Nissim, S., Peel, A. L., Liu, S., Yang, C., Flax, J. D., and Snyder, E. Y. (1997). Neuronally differentiated multipotent progenitor cells may inhibit the neuronal differentiation of surrounding uncommitted, clonally related multipotent neural progenitors via membrane-associated factors. *Soc Neurosci Abstr* 23: in press.

Nourse, J., *et al.* (1994). Interleukin-2-mediated elimination of the $p27^{kip-1}$ cyclin-dependent kinase inhibitor prevented by rapamycin. *Nature* 372:570–573.

Olanow, C. W., Kordower, J. H., and Freeman, T. B. (1996). Fetal nigral transplantation as a therapy for Parkinson's disease. *Trends Neurosci* 19:102–109.

Onifer, S. M., White, L. A., Whittemore, S. R., and Holets, V. R. (1993). *In vitro* labeling strategies for identifying primary neural tissue and a neuronal cell line after transplantation in the CNS. *Cell Transpl* 2:131–149.

Park, K. I., Jensen, F. E., and Snyder, E. Y. (1995). Neural progenitor transplantation for hypoxic-ischemic brain injury in immature mice. *Soc Neurosci Abstr* 21:2027.

Park, K. I., Jensen, F. E., Stieg, P. E., Fischer, I., and Snyder, E. Y. (1997). Transplantation of neural stem-like cells engineered to produce NT-3 may enhance neuronal replacement in hypoxia-ischemia CNS injury. *Soc Neurosci Abstr* 23:346.

Park, K. I., Liu, S., Flax, J. D., Nissim, S., Stieg, P. E., and Snyder, E. Y. (1999). Transplantation of neural progenitor & stem-like cells: Developmental insights may suggest new therapies for spinal cord and other CNS dysfunction. *Journal of Neurotrauma*: in press.

Qian, X., Davis, A. A., Goderie, S. K., and Temple, S. (1997). FGF2 concentration regulated the generation of neurons and glia from multipotent cortical stem cells. *Neuron* 18:81–93.

Renfranz, P. J., Cunningham, M. G., and McKay, R. D. G. (1991). Region-specific differentiation of the hippocampal stem cell line HiB5 upon implantation into the developing mammalian brain. *Cell* 66:713–729.

Rosario, C. M., Yandava, B. D., Kosaras, B., Zurakowski, D., Sidman, R. L., and Snyder, E. Y. (1997). Differentiation of engrafted multipotent neural progenitors towards replacement of missing granule neurons in *meander tail* cerebellum may help determine the locus of mutant gene action. *Development* 124:4213–4224.

Ryder, E. F., Snyder, E. Y., and Cepko, C. L. (1990). Establishment and characterization of multipotent neural cell lines using retrovirus vector-mediated oncogene transfer. *J Neurobiol* 21:356–375.

Sabaate, O., Horellou, P., Vigne, E., *et al.* (1995). Transplantation to the rat brain of human neural progenitors that were genetically modified using adenoviruses. *Nature Genetics* 9:256–260.

Sah, D. W. Y., Ray, J., Richard, N., Leisten, J., and Gage, F. H. (1996). Conditional immortalization of human neuronal, glial, and multi-potent CNS progenitor cells. *Soc Neurosci Abs* 22:21.12.

Schumacher, J. M., Short, M. P., Hyman, B. T., Breakefield, X. O., and Isacson, O. (1991). Intracerebral implantation of nerve growth factor producing fibroblasts protects striatum against neurotoxic levels of excitatory amino acids. *Neuroscience* 45:561–570.

Sheen, V. L., and Macklis, J. D. (1995). Targeted neocortical cell death in adult mice guides migration and differentiation of transplanted embryonic neurons. *J Neurosci* 15(12):8378–8392.

Shihabuddin, L. S., Hertz, J. A., Holets, V. R., and Whittemore, S. R. (1995). The adult CNS retains the potential to direct region-specific differentiation of a transplanted neuronal precursor cell line. *J Neurosci* 15:6666–6678.

Shihabuddin, L. S., Holets, V. R., and Whittemore, S. R. (1996). Selective hippocampal lesions differentially affect the phenotypic fate of transplanted neuronal precursor cells. *Exp Neurol* 139:61–72.

Snyder, E. Y., Yandava, B. D., Pan, Z.-H., Yoon, C., and Macklis, J. D. (1993). Immortalized postnatally derived cerebellar progenitors can engraft & participate in development of multiple structures at multiple stages along mouse neuraxis. *Soc Neurosci Abstr* 19:613.

Snyder, E. Y., and Senut, M.-C. (1997). Use of non-neuronal cells for gene delivery. *Neurobiol Dis* 4:69–102.

Snyder, E. Y., Yoon, C., Flax, J. D., and Macklis, J. D. (1997a). Multipotent neural progenitors can differentiate toward replacement of neurons undergoing targeted apoptotic degeneration in adult mouse neocortex. *Proc Natl Acad Sci USA* 94:11645–11650.

Snyder, E. Y., Park, K. I., Flax, J. D., *et al.* (1997b). The potential of neural "stem-like" cells for gene therapy & repair of the degenerating CNS. *Adv Neurol* 72:121–132.

Snyder, E. Y., Flax, J. D., Yandava, B. D., Park, K. I., Liu, S., Rosario, C. M., and Aurora, S. (1997c). Transplantation & differentiation of neural "stem-like" cells: possible insights into development & therapeutic potential In: Gage, F. H., and Christen, Y., eds. *Research & perspectives in neurosciences: Isolation, characterization, & utilization of CNS stem cells.* Berlin: Springer-Verlag, pp. 173–196.

Snyder, E. Y. (1994). Grafting immortalized neurons to the CNS. *Curr Op Neurobiol* 4:742–751.

Snyder, E. Y., and Wolfe, J. H. (1996). CNS cell transplantation: A novel therapy for storage diseases? *Current Opin in Neurol* 9:126–136.

Snyder, E. Y., and Fisher, L. J. (1996). Gene therapy for neurologic diseases. *Curr Opin in Pediatr* 8:558–568.

Snyder, E. Y. (1995). Retroviral vectors for the study of neuroembryology: Immortalization of neural cells. In: Kaplit, M. G., and Loewy, A. D., eds. *Viral vectors: Tools for analysis and genetic manipulation of the nervous system.* New York: Academic Press, pp. 435–475.

Snyder, E. Y., and Macklis, J. D. (1996). Multipotent neural progenitor or stem-like cells may be uniquely suited for therapy for some neurodegenerative conditions. *Clin Neuroscience* 3:310–316.

Snyder, E. Y. (1998). Neural stem-like cells: Developmental lesions with therapeutic potential. *The Neuroscientist:* in press.

Snyder, E. Y., Deitcher, D. L., Walsh, C., Arnold-Aldea, S., Hartwieg, E. A., and Cepko, C. L. (1992). Multipotent neural cell lines can engraft and participate in development of mouse cerebellum. *Cell* 68:33–51.

Snyder, E. Y., Taylor, R. M., and Wolfe, J. H. (1995). Neural progenitor cell engraftment corrects lysosomal storage throughout the MPS VII mouse brain. *Nature* 374:367–370.

Steiner, P., *et al.* (1995). Identification of a Myc-dependent step during the formation of active G1 cyclin-cdk complexes. *EMBO J* 14:4814–4826.

Stemple, D. L., and Mahantharppa, N. K. (1997). Neural stem cells are blasting off. *Neuron* 18:1–4.

Suhonen, J. O., Peterson, D. A., Ray, J., and Gage, F. H. (1996). Differentiation of adult hippocampus-derived progenitors into olfactory neurons *in vivo*. *Nature* 383:624–627.

Suhr, S., and Gage, F. H. (1993). Gene therapy for neurologic disease. *Arch Neurol* 50:1252–1268.

Svendsen, C. N., Clarke, D. J., Rosser, A. E., and Dunnett, S. B. (1996). Survival and differentiation of rat and human epidermal growth factor-responsive precursor cells following grafting into the lesioned adult central nervous system. *Exp Neurol* 137:376–388.

Taylor, R. M., and Snyder, E. Y. (1997). Widespread engraftment of neural progenitor & stem-like cells throughout the newborn *twitcher* mouse brain. *Transplantation Proceedings* 29:845–847.

Tsai, D., Park, K. I., and Snyder, E. Y. (1999). Some of the dynamics and variables influencing retrovirally-transduced foreign gene expression by engrafted neural stem cells. (submitted).

Vescovi, A. L., Daadi, M., Asham, R., and Reynolds, B. A. (1998). Continuous generation of human catecholaminergic neurons by embryonic CNS stem cells. *Soc Neurosci Abstr* 23:319.

Weiss, S., Reynolds, B. A., Vescovi, A. L., Morshead, C., Craig, C. G., and van der Kooy, D. (1996). Is there a neural stem cell in the mammalian forebrain? *Trends Neurosci* 19:387–393.

White, L., and Whittemore, S. R. (1992). Immortalization of raphe neurons: An approach to neuronal function *in vitro* and *in vivo*. *J Chem Neuroanatomy* 5:327–330.

White, L. A., Eaton, M. J., Castro, M. C., Klose, J., Globus, M. Y.-T., Shaw, G., and Whittemore, S. R. (1994). Distinct regulatory pathways control neurofilament expression and neurotransmitter synthesis in immortalized serotonergic neurons. *J Neurosci* 14:6744–6753.

Whittemore, S. R., and Snyder, E. Y. (1996). Physiological relevance and functional potential of central nervous system-derived cell lines. *Mol Neurobiol* 12:13–38.

Whittemore, S. R., and White, L. (1993). Target regulation of neuronal differentiation in a temperature-sensitive cell line derived from medullary raphe. *Brain Res* 615:27–40.

Winkler, J., Suhr, S. T., Gage, F. H., Thal, L. J., and Fisher, L. J. (1995). Essential role of neocortical acetylcholine in spatial memory. *Nature* 375:484–487.

Yandava, B. D., and Snyder, E. Y. (1995). Widespread engraftment by multipotent neural progenitors as a possible cellular therapy for the *shiverer* mouse brain. *Soc Neurosci Abstr* 21:20.

Yang, C.-H., Flax, J. D., Aurora, S., Simonin, C., Wills, A. M., Billinghurst, L., Jendoubi, M., Sidman, R. L., Kim, S. U., Wolfe, J. H., and Snyder, E. Y. (1998). Engraftable human neural stem cells, propagated by genetic and epigenetic means, can respond to developmental cues, differentiate into deficient neuronal types, and express foreign genes *in vivo*. *Soc Neurosci Abstr* (in press).

CHAPTER 9

Genetic Engineering for CNS Regeneration

XANDRA BREAKEFIELD,* ANDREAS JACOBS,† AND SAM WANG*

*Molecular Neurogenetics Unit, Neurology Department, Massachusetts General Hospital and Harvard Medical School, Boston, Massachusetts, †Max-Planck-Institute for Neurological Research, 50931 Koeln, Germany

Genetic engineering provides a powerful and versatile set of tools for the complex architectural feats needed for nerve repair. A variety of techniques are currently available for genetic engineering of cells, with more being developed. This chapter reviews basic principles of gene delivery and current vectors available for delivery to neural cells, including virus and synthetic vectors. Among the promising vectors for efficient gene delivery to neurons with minimal toxicity are recombinant herpes simplex virus (HSV) with multiple deletions in toxic genes, "gutless" adenovirus, hybrid HSV/AAV amplicons packaged without helper virus, and lentivirus. Modes of delivery *in vivo* include direct injection into the brain and spinal cord parenchyma, or indirect injection into the cerebrospinal fluid or vasculature, with the injury site providing a zone of access across the blood–brain barrier. A variety of cell types can be used for *ex vivo* engineering in culture with subsequent grafting into nervous tissue, including fibroblasts, endothelial cells, Schwann cells, and neuroprogenitor cells. Grafted cells can be designed to be stationary or migratory, or placed in encapsulated biopolymers to provide focal delivery and/or a conduit for neurite regrowth. This plethora of genetic engineering techniques can be used to augment all phases of CNS regeneration, including rescue and sustenance of injured neurons, creation of a pathway to stimulate and facilitate directed neurite extension, and a provision of a conductive environment for recognition and reinnervation of appropriate target tissue.

I. INTRODUCTION—THE CHALLENGE OF CNS REGENERATION

Many schemes have been envisioned for gene therapy of neurologic diseases (Chiocca and Breakefield, 1998; Friedmann, 1994), including CNS regeneration (Senut *et al.*, 1997). In general these have fallen into three categories: (1) replacement of a missing protein or neuroactive molecule, such as enzyme replacement for lysosomal storage disorders (Wolfe *et al.*, 1992; Kaye, 1997) or supplemental synthesis of dopamine for Parkinson's disease (Fisher and Gage, 1993); (2) release of diffusely active growth factors to assist neurons in surviving degenerative insults (Lindvall *et al.*, 1994), for example, to help neurons survive excitotoxic insults associated with brain trauma (Frim *et al.*, 1993c); and (3) alteration of the physiology of neurons by expression of neuroactive proteins, such as efforts to downregulate pain by changing the spectrum of neurotransmitters released by sensory neurons into the spinal cord (Davar, 1997; Wilson *et al.*, 1996). For the most part these can be looked at as changing the *status quo* of neurons to achieve a new stable state. In contrast, CNS nerve regeneration will require a systematic series of genetic and mechanical engineering feats to recreate destroyed nerve pathways over time and space. Although this strategy will incorporate elements of the more static gene therapy strategies, it must coordinate these strategies, as well as more conventional implementation of drugs, protein infusions, and mechanical supports, in a timed series of interventions, and then, ultimately erase the interventional "scaffold" when the structure is completed.

Gene therapy for spinal cord repair must be designed for three basic targets: neurons, extraneuronal environment, and postsynaptic partner; and over three time windows: damage control immediately after injury, a prolonged period of directed neurite extension, and finally recognition and securing of appropriate synaptic contact (Waxman and Kocsis, 1997). In the initial period after nerve injury the most critical task is twofold: to support survival of damaged neurons and to prevent establishment of a hostile environment to neuronal regrowth. Neuronal survival can be enhanced in many of the same ways currently envisioned for neurodegenerative diseases, such as delivery of neurotrophic factors, anti-apoptotic genes and enzymes that decrease free radical formation. Means to decrease the hostility of the environment include inhibition of inflammatory responses and demyelination, and reduction of glial proliferation and scar formation. The means to alter these latter responses are less clear, but examples might be taken from models of gene therapy for inflammatory diseases, such as arthritis (Chernajovsky *et al.*, 1997; Evans and Robbins, 1996), and from reactive cell overgrowth, as in stenosis (Kaneda *et al.*, 1997), or even from strategies to kill proliferating glioma cells (Kramm *et al.*, 1995). Neurite growth could be encouraged by fortifying neurons for extensive process formation, including upregulation of energy metabolism and structural components of axons, such as neurofilaments, actin, and tubulin. The path of regrowth could also be "greased" by release of extracellular matrix proteins, like L1 (Mohajeri *et al.*, 1996), and growth factors (Tuszynski *et al.*, 1996; Horner *et al.*, 1996; Grill *et al.*, 1997) that promote process extension, such as those released from Schwann cells in peripheral nerve regrowth (Xu *et al.*, 1997b). This process of neurite regrowth may take extended periods of time, months to years, depending on the site of spinal cord damage and the tissue being targeted, as even under optimal conditions neurites extend only at a rate of 1 mm per day. The growth of neurites must also be directed along a route leading to the designated synaptic partner. This can be achieved in part by laying down a preferred substratum, but ultimately the neurons must seek out the synaptic partner and communicate with it. During development, neurites are believed to find their targets in part through a chemotrophic gradient achieved by secretion of trophic factors from the target cells (Henderson, 1996). Because following maturation of the adult CNS the target may downregulate secretion of such factors, or physical barriers to diffusion may have arisen, it may be necessary to supply moveable "decoy" targets along the path, as in the "donkey-and-carrot" model. Once the neurites reach their final destination it may then be necessary to remove the decoy and promote contact between the nerve endings and the target cells by synapse-promoting factors, such as agrin for skeletal muscle innervation (Fallon and Hall, 1994).

This chapter will focus primarily on the array of methods currently available to achieve genetic engineering of the adult nervous system. This includes the variety of vectors available for delivery of transgenes to cells in culture and *in vivo*, and their advantages and disadvantages in different scenarios, as well as the different modalities that can be used to introduce vectors or genetically engineered cells into the nervous system. Although use of gene delivery as a therapeutic modality has great potential for repair of CNS damage, as it has for other diseases, researchers should choose experimental models carefully and proceed with great caution in considering human trials. Given the fragility and importance of remaining neuronal pathways in spinal cord injury patients, it becomes especially important to "do no harm." Special concerns must be addressed to problems of toxicity inherent to current gene delivery strategies, including toxicity of vectors and inflammatory immune responses to them, as well as possible pain and involuntary muscle contractions that could result from inappropriate synaptic connections. Still, one must not be overwhelmed by the complexity and enormity of the problem, but be willing to work in a step-by-step, cautious manner toward the larger goal. Given increased understanding by the neuroscientific community of the many molecules that are involved in normal axon growth and synaptic formation, there is an emerging resolve that the instructions for CNS regeneration may be within our grasp.

II. MODES OF GENETIC ENGINEERING

A. Principles of Gene Delivery

The goal of gene therapy methods is to achieve expression of the transgene(s) in target cells *in vivo*. Critical steps needed to achieve this are the entrance of transgenes into cells, their state within cells, and the regulation of transgene expression (Fig. 1). Most current modes of vector entry into cells are not specific to cell type, and rely on the route of delivery to target certain cell populations. Delivery into the nervous system can be achieved by focal, stereotactic injection into the parenchyma, uptake from nerve terminals in the periphery with retrograde transport to neuronal cell bodies within the nervous system, or introduction into the vasculature or intrathecal space.

DNA is usually packaged to neutralize its negative charge and to reduce shear forces on it, thus avoiding degradation and facilitating delivery. In the case of mechanical delivery, DNA can be precipitated with calcium phosphate, bound to gold particles, encapsulated in artificial membranes termed liposomes, and/or complexed with positively charged proteins termed molecular conjugates. Particles can be delivered by electroporation or bombardment, in which case they pass directly through temporary disruptions of the cell membrane. Alternatively, liposomes can fuse to the membrane, thus releasing their contents directly into the cytoplasm; conjugates and cationic lipids can be taken up by endocytosis or pinocytosis. In the latter case they end up within membrane vesicles in the cytoplasm and must exit from those vesicles before they enter the degradative endosomal–lysosomal pathway. Many viruses have fusion proteins that activate at acidic pH in endosomes, thereby releasing the viral genome into cytoplasm (Helenius, 1992; Greber *et al.*, 1993). Mechanical means of gene delivery can be very inefficient, however, with low levels of cellular uptake, degradation within cells, and poor access to the nucleus. Synthetic vectors frequently incorporate specific viral proteins or DNA sequences, sometimes from several different viruses, to promote cell delivery, DNA stability, and transgene expression.

In the case of viral vectors, the transgene is integrated within viral sequences so that it is packaged with viral DNA–RNA in virus capsids and enters cells by the mode characteristic of that particular virus. For virus vectors in current use, the cellular recognition molecules are present on most cells and thus entry is not cell specific, although there may be differences in the relative efficiency of infection among cell types and species for particular viruses. By

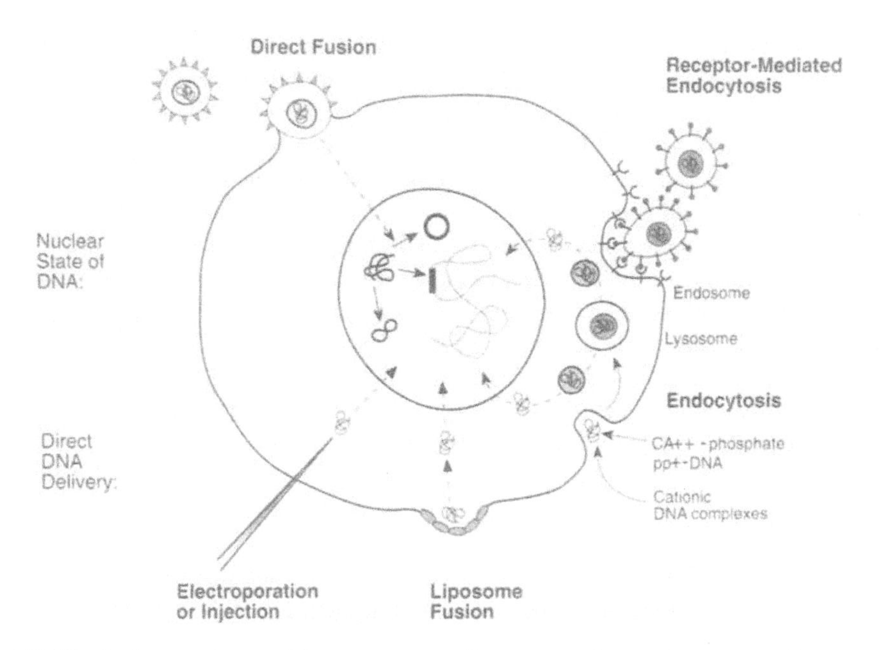

FIGURE 1 Entry and state of foreign DNA in cells. DNA can be introduced into cells in a number of ways. Some viruses, like herpes and retrovirus, have a membrane envelope that fuses to the cell surface, depositing the virions directly into the cytoplasm. Liposomes bearing elements that fuse at neutral pH can also be taken up in this direct manner. Other viruses, like adenovirus and AAV, are taken up by receptor-mediated endocytosis and then use acid-triggered fusigenic proteins to escape from the endosomes and avoid degradation in lysosomes. Receptor-mediated endocytosis also provides the route of entry for Ca^{2+}-phosphate DNA precipitates and molecular conjugates. In addition, DNA can be delivered through the membrane by temporary disruption using electroporation, injection, or particle bombardment. Models of transit through the cytoplasm to the nucleus are less well defined, although some viruses travel on actin and microtubular elements. Some viruses, like HIV, have nuclear localization signals on viral proteins to target the viral genome to the cell nucleus. DNA can pass into the nucleus through nuclear pores, but in some cases, e.g., retrovirus, requires dissolution of the nuclear membrane at mitosis. Within the nucleus DNA, elements can integrate into the cellular genome or exist as replicating or nonreplicating extrachromosomal DNA elements. Extrachromosomal genes can exist in stable (episomal) or free states, and the latter are usually degraded over time.

example, viruses commonly used for gene delivery, herpes virus (usually derived from herpes simplex virus type 1, HSV1) and retrovirus (usually derived from Moloney murine leukemia virus, MoMLV) enter by direct fusion of the viral envelope to the cell membrane so that the capsid is released into the cytoplasm. Adenovirus and adenoassociated virus (AAV) vectors are taken up by receptor-mediated endocytosis.

Most transgenes are designed to be expressed within the cell nucleus and the route from the cytoplasm into the nucleus is not well understood. Transgenes introduced by most vectors can access the nucleus even in nonmitotic cells. In general, however, breakdown of the nuclear membrane facilitates nuclear entry, and host cell DNA replication promotes integration of the foreign DNA and transgene expression. Some virus vectors, like MoMLV retrovirus, can only access the nuclear space when the nuclear membrane breaks down during mitosis; others like AAV and HIV can enter the nucleus and integrate into the genome of dividing as well as nondividing cells. "Free" DNA in the nucleus can be transcriptionally active for some time, but is eventually inactivated, degraded, or lost during mitosis. Some viruses, e.g., HSV and Epstein Barr virus (EBV), can maintain their genome as a stable extrachromosomal (episomal) element in the host cell nucleus. Latent HSV establishes a nucleosomal configuration in some postmitotic cells, e.g., sensory neurons, whereas EBV maintains itself as a replicative element, replicating in-phase with host lymphoid cells. Incorporation of the EBV origin of DNA replication, oriP, and the gene encoding the viral DNA binding protein EBNA1, into an expression plasmid can permit replication, retention, and expression of transgenes over a number of cell generations (Wang and Vos, 1996). Theoretically one could also incorporate mammalian origins of replication into such expression plasmids. Most of the mammalian origins described to date are quite large, >12 kb, and would require large cloning vehicles, but several groups have shown extended retention of plasmids with smaller mammalian origins of DNA replication (Hamlin, 1992). AAV vectors can replicate as episomal elements in the nucleus of the host cell, thus amplifying transgene copy number on site. Within the nucleus transgenes can also integrate into the host cell genome, usually at random sites, but in the case of AAV, preferentially on human chromosome 19q. Genomic integration is a very inefficient process with nonviral DNA sequences, but some viruses have developed means to promote this process, e.g., retrovirus, HIV and AAV, and elements from these viruses can be used to promote transgene integration. Other vectors can express transgenes in the cytoplasm without requiring nuclear import. These include, for example, vectors derived from vaccinia (Weisz and Machamer, 1994), Sindbis (Piper *et al.*, 1994), and bacculovirus (Boyce and Bucher, 1996). Self-promoting DNA transcription cassettes have also been developed to express genes in the cytoplasm without the need for nuclear access by encoding prokaryotic RNA polymerases, which, in turn, can yield expression of transgenes utilizing the appropriate prokaryotic promoter, e.g., T7 (see the following discussion). Messages generated and expressed in the cytoplasm must be able to function without post-transcriptional modifications, which normally occur in the nucleus, including capping the 5′ terminal, splicing of introns, and polyadenylation.

One of the unresolved problems of transgene expression is characterization of appropriate promoters. Expression of transgenes tends to be extinguished over time *in vivo,* even when the DNA is still present in cells. Several factors have been identified that contribute to this problem. In general, viral promoters appear to be recognized as "foreign" in eukaryotic cells and shut down, primarily through DNA methylation. Thus, for example, the cytomegalovirus (CMV) promoter, which can give extremely strong transient transgene expression, is eventually shut off in many mammalian cells (e.g., Doll *et al.,* 1996). In general, if the DNA bearing the transgene is replicating, and thus in an "open" configuration, expression continues. Depending on the site of integration into the host cell nucleus, condensation of the DNA into a nucleosomal configuration can also downregulate gene expression. The identification of moveable locus control regions (LCR), e.g., flanking the globin gene (Talbot *et al.,* 1989), and matrix binding sites (MAR; McKnight *et al.,* 1992), both of which "hold the DNA open" to transcription, has been a promising finding. However, these regions may be cell specific, and none have been identified as yet for neural cells. Mammalian promoters can achieve transgene expression for longer periods *in vivo,* e.g., months, but again are strongly influenced by neighboring DNA elements. The phosphoglycerate kinase (PGK) promoter (a strong housekeeping promoter) has been found to mediate strong transgene expression in a number of cell types *in vivo* (Moullier *et al.,* 1993). In other cases, combinations of enhancer elements and promoters have proven effective in maintaining transgene expression *in vivo,* e.g., the muscle-specific creatinine kinase enhancer element linked to the CMV promoter for differentiated muscle cells (Dai *et al.,* 1992). For neural cells a number of different promoters, many of which have been defined in transgenic mice, may prove useful (see the following discussion).

Important advances are also being made in the areas of self-promoting and inducible promoters. For example, the T7–T7 cytoplasmic expression cassette bears the T7 polymerase under the T7 promoter, and the transgene under the same T7 promoter (Lieber *et al.,* 1993). Messages transcribed via these promoters are rendered "capless" by virtue of an internal ribosomal entry sequence 5′ to the translation initiation codon (Ghattas *et al.,* 1991). By bringing in T7 RNA polymerase or the RNA encoding it, into the cytoplasm along with the transgene, the cassette is primed and thereafter generates more polymerase and transgene on its own. Another self-priming system explored in nuclear vectors is the GAL4 : VP16 transactivating fusion protein in combination with multiple copies of the GAL4-binding site (Sadowski *et al.,* 1988). The GAL4 protein (derived from yeast) binds to the GAL4 DNA site and VP16 (derived from HSV) serves as a potent transactivator of transcription. The fusion protein can drive synthesis of its own message as well as that of a transgene (Glorioso *et al.,* 1995). This system has been made drug-inducible by adding a truncated form of the hormone-binding domain of the progesterone receptor to the GAL4 : VP16 transactivator; this truncated portion no longer binds progesterone, but does bind a steroid analog drug, which can activate the transactivator (Vegeto *et al.*, 1992). Although a number of other inducible promoters have been described (Gossen *et al.*, 1993), the most effective appears to be the prokaryotic-derived tetracycline (tet) regulatable promoter system that incorporates a tet repressor : VP16 transactivator protein and the tet operator linked to the CMV promoter. This comes in both a tet-on (Gossen *et al.*, 1995) and tet-off form (Gossen and Bujard, 1992) and has recently been incorporated in a self-contained retrovirus vector (e.g., Paulus *et al.*, 1996), which is able to tightly control transgene expression even within rodent brain (Yu *et al.*, 1996). Still the ability to reproducibly achieve long-term (>6 month) expression in cells in some tissues has remained elusive. The use of the promoter for the latency-associated transcripts (LATs) in the context of latent HSV vectors in some neurons, e.g., sensory neurons (Glorioso *et al.*, 1995; Dobson *et al.*, 1990), and the use of CMV promoter in lentivirus vectors (Blömer *et al.*, 1997) are the closest researchers have come to the goal of achieving long-term transgene expression.

The field of gene therapy is still very much in an exploratory phase with many techniques being tried, only a few of which have been directly compared to each other. Over the next few years expertise in mechanical and viral gene delivery will evolve into the creation of synthetic vectors that retain viral elements for efficient delivery, at the same time eliminating elements that cause toxicity. This combined with more sophisticated promoter technology should make gene therapy a viable treatment option for neurologic syndromes.

B. Vehicles for Gene Delivery

Gene delivery to cells is usually achieved by introduction of DNA into cells aimed at expression in the nucleus, although some cytoplasmic expression vehicles are also available. In nuclear vectors the DNA itself (or RNA equivalent in the case of retroviruses) contains a promoter-transgene cassette, which may include one or more therapeutic genes, as well as reporter genes, such as *lacZ*, alkaline phosphatase, or (jelly fish) green fluorescent protein (enhanced for expression in mammalian cells), and drug selection genes, e.g., conferring resistance to neomycin, puromycin, or hygromycin. Transgenes can be placed individually under their own promoters or placed in tandem under the same promoter and separated from each other by an internal ribosome entry site (IRES; Ghattas *et al.*, 1991), which allows translational initiation within the message as well as at the beginning of the message. In cytoplasmic expression systems, the IRES can be used to replace the cap function of the message so that it can be translated directly in the cytoplasm without processing in the nucleus. Each virus vector system has its particular size restraints, and synthetic vectors are limited by the size of convenient cloning vehicles and fragility of large DNA fragments.

Expression cassettes can be packaged in a variety of ways to facilitate entry into cells. Although purified DNA can be used for transfection of cells in culture, it does so very inefficiently, because cellular uptake is nonspecific, probably through receptor-mediated endocytosis; transit through the cytoplasm is random, with extensive lysosomal breakdown of DNA; entry into the nucleus is undirected; and the state in the nucleus is undefined, with extensive degradation and, only infrequently, stable integration into the host cell genome. Vectors have been designed to increase the efficiency of each step of this process and the reliability of transgene expression, while at the same time minimizing the toxicity of the delivery process. In the context of the nervous system even the volume of the delivery bolus can be toxic and any inflammatory reaction can have harmful consequences. In general, synthetic vectors are considered to have relatively poor efficiency of gene delivery, but essentially no toxicity, whereas the reverse is true for virus vectors.

Synthetic vectors include DNA projectiles, molecular conjugates, and liposome–DNA complexes. Projectiles consist of plasmid DNA expression cassettes affixed to particles, typically gold particles, which are then shot out of pressure guns into the tissue (Klein and Fitzpatrick-McElligott, 1993). Particles can penetrate a few millimeters into tissues and enter cells without killing them, thereby bypassing the endocytotic route (Carrasco, 1994). This type of delivery has proven able to transitorily effect transgene expression in cells in the brain (Jiao *et al.*, 1993b). Molecular conjugates refer to the ionic association and condensation of DNA (negatively charged) with proteins, typically polylysine or histone (positively charged) to which other proteins or ligands have been covalently bound to aid in cellular uptake and escape from lysosomes (Fritz *et al.*, 1996; Wagner *et al.*, 1991). Thus, for example, the expression constructs can be complexed with transferrin to target cells with transferrin receptors and to facilitate receptor-mediated endocytosis of the complex (Wagner *et al.*, 1991), as well as with the penton protein of the adenovirus virion to mediate exit of DNA from the lysosomal compartment (Blumenthal *et al.*, 1986). Liposomes are either neutral, negatively (anionic) or positively (cationic) charged lipid bilayers that can form a bubble around DNA, condense it, and protect it from degradation (Felgner *et al.*, 1987). By direct fusion of liposomes (typically, neutral and anionic) with the cell membrane, the DNA can be deposited directly into the cytoplasm, thereby reducing degradative processes. Cationic liposomes are taken up by endocytosis and must also escape the endosome–lysosome compartment (Li and Huang, 1997). In some cases, neutral pH fusigenic proteins, such as influenza virus hemagglubinuin subunit HA-2 or

Sendai virus fusion (F) protein, have been incorporated into liposomes to facilitate fusion with the cell membrane (Wagner *et al.*, 1992; Tomita *et al.*, 1996). Other proteins with nuclear localization signals (NLS), such as the high mobility group nuclear protein (HMG1; Kaneda *et al.*, 1989) or the SV40 large T antigen NLS (Fritz *et al.*, 1996), have been bound to the DNA within the liposomes (Wychowski *et al.*, 1986) so that following entry the DNA is directed toward the nucleus. Additional modifications made to liposomes to make them safer and more efficient include incorporation of nontoxic biodegradable lipids and inclusion of pH-sensitive lipid moieties (Liang and Hughes, 1997; Tang and Hughes, 1997).

The types of virus vectors that have been used routinely to introduce genes into the mammalian nervous system are derived from retrovirus, lentivirus, adenovirus, adenoassociated virus (AAV) or herpes simplex virus (Table I; for review see Smith *et al.*, 1995; Ali *et al.*, 1994; Jolly, 1994). All can infect essentially any cell type, but with some variation in efficiency. (The vectors described here are replication-defective, unless indicated otherwise). Each has specific features that define how it can be used most effectively. Traditional retrovirus vectors derived from MoMLV (for review see Sena-Esteves *et al.*, 1996) have the unique feature that they can only integrate their genome into dividing cells (integration is at random sites), apparently needing the breakdown of the nuclear matrix at mitosis to access the cellular genome (Miller *et al.*, 1990). This limits their usefulness for direct injection into the postnatal nervous system, because most cells there are postmitotic, although they could be used in settings of glial cell proliferation or neovascularization, which can accompany CNS injury. Their ability to integrate into the genome allows them to become a functioning gene in infected cells and their progeny. In experimental animal models, retrovirus vectors have been used extensively to deliver genes to cells in culture, which are then grafted into the brain (see *ex vivo* models that follow). Although retrovirus vectors traditionally had the disadvantage of low titers and unstable particles, their usefulness for gene delivery to the nervous system has been expanded by a number of developments. The instability of the virions can be overcome in two ways, by grafting packaging cells, which release the vectors over an extended period into the brain (Short *et al.*, 1990), and by pseudotyping the virion through incorporation of the envelope glycoprotein (G) derived from vesicular stomatitis virus (VSV), which stabilizes the particles and expands their host cell range (Burns *et al.*, 1993). The VSVG-virions can also be concentrated to higher titers. For many years virtually the only packing cell lines used were those derived from immortalized murine cells lines. This has the disadvantage that packaging cells are immune incompatible with many other species and thus rejected quickly before extensive transfer can take place. Further, murine cells are notorious for containing a plethora of endogenous retroviral sequences, which are expressed

TABLE I Current Virus Vectors for Gene Delivery to Nervous System

Virus	Type	Insert (kb)	State in host cell nucleus	Stability	Helper virus	Toxicity
Retrovirus	ssRNA	10	integration[a]	+	−	−
Lentivirus	ssRNA	10	integration	+	−	−/±
Adenovirus	dsDNA	>8	"free"	±	−	±
"Gutless"		30	"free"	±	−	−
AAV	ssDNA	4–5	integration	+	(+)	−
Herpes:	dsDNA					
HSV amplicon		~20	"free"	±	+ or −	±
HSV–AAV amplicon		>20	integration	+	+/−	±/−
Recombinant		~30	episomal[b]	+	−	±/−

[a] Retrovirus only in dividing cells
[b] In some neurons and possibly other cells

at high levels and can recombine with vector sequences, thereby generating wild type virus, or can interfere with packaging of the vectors, thus reducing titers (Coffin, 1996). Several new attempts have been made to generate new packaging modalities, such as co-delivery of vector and packaging functions either through direct transfection (Naviaux *et al.*, 1996) or by infection with adenovirus (Feng *et al.*, 1997) or herpes amplicons (Savard *et al.*, 1997). Traditional retrovirus vectors are packaged in cell lines designed to reduce the chance of generation of wild type virus and are considered essentially nontoxic. These vectors have been used extensively in clinical trials, e.g., for enzyme replacement in adenosine deaminase deficiency (Blaese *et al.*, 1995) and for cancer vaccination strategies (e.g., Dranoff *et al.*, 1997) in the periphery. Murine packaging cell lines releasing a vector with a pro-drug activating gene have also been grafted into patients with brain tumors as a therapeutic strategy (Oldfield *et al.*, 1993).

Probably the most significant advance in retrovirus vectors has been expansion to the use of lentivirus particles (Naldini *et al.*, 1996; Fig. 2). These members of the larger retrovirus family, e.g., HIV, are able to integrate sequences into both dividing and nondividing cells through functions such as nuclear localization signals on the matrix protein and the integrase carried in the virions (Stevenson, 1996). Further integration in nondividing cells can be enhanced by incubation of virions with deoxynucleoside triphosphate (Gao *et al.*, 1993). Although HIV traditionally enters cells through the CD4 receptor that is expressed almost exclusively on T lymphocytes, by pseudotyping with VSVG these virions can infect a wide range of cell types. Recent studies indicate that lentivirus vectors preferentially express transgenes in neurons and can achieve sustained transgene expression under the control of the CMV promoter in rat brain for over six months (Blömer *et al.*, 1997). The newer generation of lentivirus vectors is still considered high risk and has been used to date only for experimental animal work under conditions of high containment (BL3). However, as increased safety features are introduced into these little vectors, containment restrictions will be reduced and it can be assumed that they will be used in clinical trials. Retrovirus and lentivirus vectors have the distinct advantage that once gene transfer is complete, the recipient cells do not express any viral proteins and the transgene becomes, in effect, a cellular gene.

Adenovirus vectors have been the next most popular vehicle to introduce genes into the nervous system (for review see Chiocca and Cotten, 1998). These vectors introduce their DNA apparently in a nonintegrated or "free" state in the cell nucleus. Presumably in this state the virus DNA is gradually degraded by cellular nucleases, leading to eventual clearance from the cells, but in some nondividing cells transgene expression persists for months, e.g., in the brain, albeit with a continuing downward trend (Davidson *et al.*, 1993;

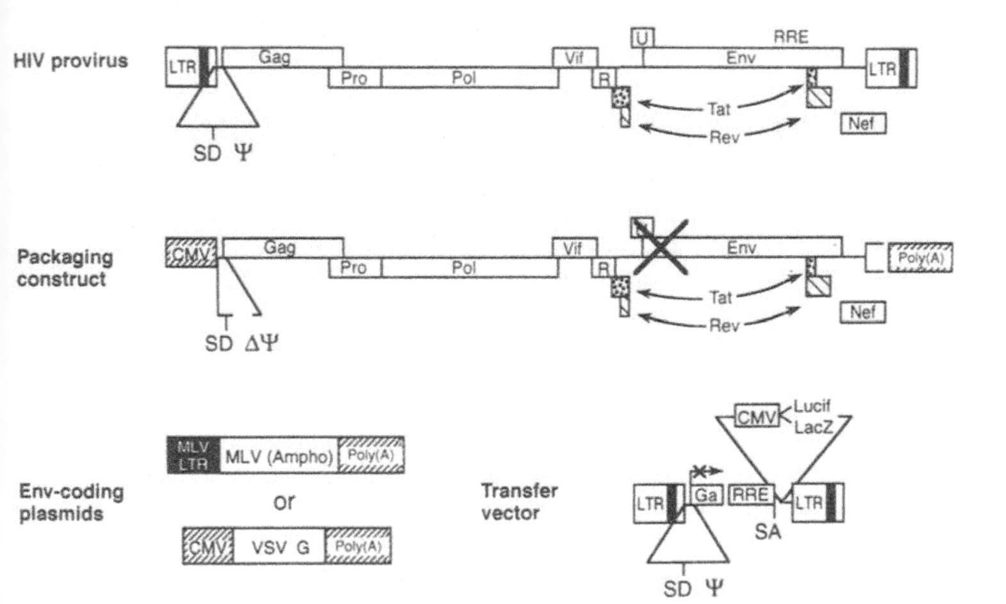

FIGURE 2 Components of the HIV provirus and the three-plasmid expression system used for generating a pseudotyped HIV-based (lentivirus) vector. Only the relevant portion of each plasmid is shown. For the HIV provirus, the coding region of viral proteins, including the accessory proteins, is shown. The splice donor site (SD) and the packaging signal (ψ) are indicated. In the packaging construct the reading frames of Env and U are blocked (X). Two Env-coding plasmids can be used, in one the coding region of the amphotropic MLV envelope is flanked by a MLV LTR and a SV40 poly(A) site, in the other. The VSV G coding region is flanked by the CMV promoter and a poly(A) site. In the transfer vector the *gag* gene is truncated and out of frame (X), and the internal promoter CMV is used to drive expression of either β-galactosidase (*lacZ*) or luciferase cDNA. The Rev responsive element (RRE) and splice acceptor site (SA) are shown. (Reprinted from Naldini *et al.*, 1996.)

Akli *et al.*, 1993). Adenovirus vectors can be grown to very high titer and are relatively stable. They are considered to have low toxicity in the body, and have been used in clinical trials for brain tumors; however, elevated liver functions indicate some pathogenicity (Eck *et al.*, 1996). Their effectiveness has been limited to some extent in humans by a strong immune response to the viral antigens, which makes subsequent treatments difficult due to neutralization of the vector and inflammatory responses (Byrnes *et al.*, 1996). New versions of the vector delete the E2a gene, which reduces this antigenicity (Yang *et al.*, 1994), and retain the E3 gene, which serves to mask expression of virus antigens (Lee *et al.*, 1995), and/or delete open reading frames in E4, which reduces expression of viral proteins (Gao *et al.*, 1996). The major drawback to adenovirus vectors is the expression of virus proteins which can lead to rejection of transfected cells and inflammatory responses (Smith and Eck, 1997), as well as the laboriousness of generating these vectors by recombination into the viral genome. The latter has been simplified by cloning of the adenovirus genome in a cosmid vector to allow direct manipulation (Fu and Deisseroth, 1997). Several groups have recently developed the technology to generate "gutless" or "minichromosome" adenovirus-derived vectors that retain the packaging and replication signals of the viral genome, but eliminate all viral genes (Hardy *et al.*, 1997; Kumar-Singh and Chamberlain, 1996; Fisher *et al.*, 1996; Kochanek *et al.*, 1996). These gutless vectors also have the advantage of an increased transgene capacity to 30 kb. The "helper" adenovirus used in packaging the gutless vectors (Fig. 3) can be separated from the vector by gradient centrifugation or eliminated by Cre-*lox* recombination to remove the packaging signal from the helper genome (Hardy *et al.*, 1997). Extensive gene delivery to the nervous system has been reported with adenovirus vectors (thousands of cells), but some investigators have reported CNS toxicity of these vectors, and this aspect needs further investigation.

Adenoassociated virus (AAV) vectors are derived from a parvovirus that can use either adenovirus or herpes virus for packaging (Muzyczka, 1992; Snyder *et al.*, 1996). Vectors are usually packaged with a helper adenovirus, which is then differentially heat denatured or separated by gradient centrifugation. These small vectors, which hold only about 4 kb of foreign DNA, have the advantage that they can integrate their genome into postmitotic human cells at a specific locus in chromosome 19 (Kotin *et al.*, 1992). The inverted terminal repeats (ITRs) that flank the transgene cassette serve as origins of DNA replication and the transgene is amplified by replication in the host cell nucleus. These ITRs can also assume a "hairpin" configuration that is resistant to exonuclease activity and thus helps to stabilize the transgenes in the host cell nucleus. Further, sequences in these ITRs bind the AAV-encoded Rep isozymes, 68 and 78, which in turn bind ITR-homologous sequences on human chromosome 19 to facilitate integration (Weitzman *et al.*, 1994; Giraud *et al.*, 1994). Transgenes delivered by AAV vectors can also integrate at other sites in the human genome and at a number of locations in the rodent genome. They are considered nontoxic, with the caveat that the residual helper virus may have some infectivity and toxicity. Although these particles are highly stable, it is difficult to prepare them at high titers. rAAV is typically prepared by co-transfection of rAAV vector and helper plasmid containing AAV *rep* and *cap*, followed by superinfection with replication-defective adenovirus in cells that complement this defect. Vector production is limited by low efficiency plasmid transfections and adenovirus toxicity. Current strategies to increase virus titers and maintain consistency between preparations include use of alternative helper virus functions (Conway *et al.*, 1997) or genetically engineered cells that intrinsically provide helper functions (Clark *et al.*, 1996). In

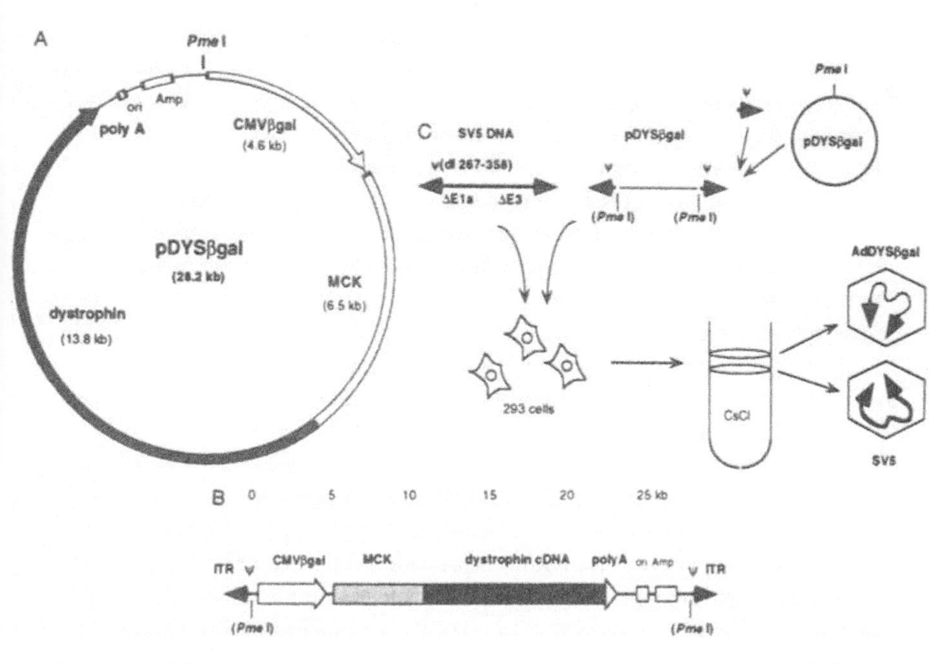

FIGURE 3 Schematic representation of a "gutless" adenovirus vector. (A) Structure of plasmid expression cassette containing the murine MCK promoter, the transgene, in this case the murine 13.8-kb full-length dystrophin cDNA (dystrophin), the 290-bp polyadenylylation signal of the bovine growth hormone gene [poly(A)], and the *E. coli* lacZ gene under the control of the early CMV promoter (CMVβgal). The bacterial origin of replication (ori) and ampicillin resistance gene (Amp) are indicated. This plasmid has a unique *PmeI* restriction endonuclease cleavage site. (B) Structure expression plasmid after linearization with *PmeI* and ligation of the left terminus of adenovirus type 5 containing the viral terminal repeat (ITR) and the full-length packaging signal (ψ) to the ends of the linearized DNA using T4 DNA ligase. The *PmeI* site is destroyed during the ligation step. (C) Schematic outline of the rescue, amplification, and purification of the vector. Adenoviral termini are ligated to both ends of the linearized expression plasmid, which is cotransfected with a DNA–terminal protein complex prepared from an adenovirus mutant that has a partial deficiency for encapsidation and a deletion within both E1a and E3 (ΔE1a, ΔE3). After plaque purification and serial propagation in 293 cells, a cell lysate is subjected to CsCl equilibrium centrifugation to resolve the vector and mutant helper virus. (Reprinted from Kochanek *et al.*, 1996. Copyright 1996 National Academy of Sciences, USA)

one study where a *lacZ*-bearing vector was inoculated into the rodent brain or spinal cord, the transgene was expressed in a moderate number of cells (up to 1000) for a few months (Peel *et al.*, 1997; McCown *et al.*, 1996; Du *et al.*, 1996). It is not clear whether loss of expression reflected a lack of integration or whether it resulted from promoter downregulation.

Two types of vectors derived from herpes simplex virus (HSV) have been generated, termed amplicon and recombinant virus vectors (for review see

Breakefield *et al.*, 1997; Ho, 1994; Johnson and Friedmann, 1994). HSV is a neurotropic virus that preferentially inflects neurons and is taken back up to the cell nucleus by rapid retrograde transport following infection at the nerve terminals. Recombinant HSV vectors are derived by replacement of genes in the virus genome with transgenes. These vectors have the largest capacity for transgenes of the current set of virus vectors, greater than 30 kb. They also have the advantage that the herpes genome can enter a state of latency in some neurons and possibly other nondividing cells. In latency the virus exists as a "benign," stable episomal element in the cell nucleus and is capable of some transcriptional activity (Dobson *et al.*, 1989). Although it has been possible to achieve transgene expression in hundreds of neural cells in the brain with such vectors, expression is usually temporary (Huang *et al.*, 1992). In cases in which the neuronal specific enolase promoter (Andersen *et al.*, 1992) or the neurofilament promoter (Carpenter and Stevens, 1996) has been used to regulate transgene expression, it has been robust for greater than one month, but only in very few neurons in the brain. Redefinition of the LAT promoter has extended the period of transgene expression for up to a year in some types of neurons, including spinal cord neurons (Lachmann and Efstathiou, 1997; Chen *et al.*, 1995a; Lokensgard *et al.*, 1994). Recombinant HSV vectors can be grown to high titers and are stable. Although herpes virus is normally toxic to neural cells, this toxicity can be reduced dramatically through mutations in the genes for the virus transcription factor, ICP4 (the product of the IE3 gene; Johnson *et al.*, 1992); a host cell function shut off gene, UL41 (Kwong and Frenkel, 1989; Krikorian and Read, 1991); and a neurovirulence factor, gamma 34.5 (Chou *et al.*, 1990). Mutant virus deleted for multiple immediate early genes has very low intrinsic toxicity (Wu *et al.*, 1996).

Amplicon vectors are plasmid constructs containing the HSV origin of replication and a virion packaging signal (Spaete and Frenkel, 1982; Geller and Breakefield, 1988; Fraefel *et al.*, 1997). Amplicons are packaged as DNA concatenates into HSV virions in cells expressing HSV helper virus functions. These can be conferred either by infection with a replication defective HSV helper virus in cells that complement the missing function, usually achieving at best a ratio of 1 : 1, amplicon vector : helper virus, or by transfection with a set of cosmids spanning the HSV genome (Cunningham and Davison, 1993) deleted for packaging functions (Fraefel *et al.*, 1996), in which case only vector particles are generated (Fig. 4A). The latter system is much more promising for delivery to the nervous system as the helper virus is invariably toxic due to expression of viral genes. Traditional amplicon vectors remain as "free" elements in the cell nucleus and are presumably degraded with time, although transgene expression and persistence of the vector have been reported for over months using this system (Kaplitt *et al.*, 1991; Starr *et al.*, 1996).

Several new concepts in vector technology are expanding the potential for controlled gene delivery to the nervous system, including the use of hybrid vectors (Jacoby *et al.*, 1997) and the design of targeted vectors (Spear *et al.*, 1997). Hybrid vectors incorporate genetic elements of two or more different viruses. Examples are the HSV–AAV hybrid vector in which a transgene cassette flanked by ITRs is placed in an HSV amplicon plasmid, which also contains the AAV *rep* gene (Johnston *et al.*, 1997; Fig. 4B). The hybrid plasmid is packaged as a concatenate in HSV virions; following infection and delivery to the cell nucleus, the transgene cassette is amplified through replication mediated by the ITRs and Rep and potentially integrated into the host cell genome in both dividing and nondividing cells through the Rep isozymes. A similar type of hybrid vector has been generated by incorporating an ITR flanked transgene cassette into an adenovirus backbone and conjugation of a plasmid bearing the Rep gene to the adenovirus virion (Fisher *et al.*, 1996). HSV amplicons have also been modified to incorporate elements of Epstein Barr virus, the oriP origin of DNA replication and the EBNA1 protein, which allow episomal replication in the host cell nucleus in parallel with replication of the host cell genome (Wang and Vos, 1996). This type of delivery system offers an extended, but not permanent means of retaining transgene cassettes in dividing cells. In the future we can anticipate more of these combinatorial vectors that use isolated, noncoding viral elements to control the fate of the transgene in the cell.

New types of vectors are continually being explored for their utility in different gene delivery paradigms. In circumstances in which the rapid and transient expression of a transgene is sought, for instance in forestaying acute trauma to neurons, cytoplasmic expression systems can be valuable. Two vectors appear suitable for this modality. One is a T7–T7 plasmid expression system (Chen et al., 1995b; see the previous discussion) delivered to the cytoplasm of cells by incorporation into liposomes. This method has been used successfully to express genes in tumor cells in rat brain (Rainov *et al.*, in preparation). Another nontoxic cytoplasmic expression system employs a nonreplicative form of the Sindbis RNA virus that can be directly translated in the cytoplasm and is effective in gene delivery to neurons in the brain (Altmann-Hamamdzic *et al.*, 1997). Another RNA virus that appears especially efficient at gene delivery to neurons is rabies virus (P. Strick, personal communication). Recombinant versions of rabies virus have been used to deliver reporter genes to cells in culture and nonreplicative forms should be suitable for gene delivery *in vivo* (Mebatsion *et al.*, 1996).

Most currently used vectors have broad cell and species tropisms to maximize the efficiency of gene delivery. In some circumstances it may be very important however to target transgene expression to a subset of cells in a tissue. Two strategies have been explored in this regard: modification of the vector surface by incorporation of antibody or receptor ligands, and use of cell specific promoters

A

Modified HSV-1 Cosmid Set (*pac* sequences deleted)

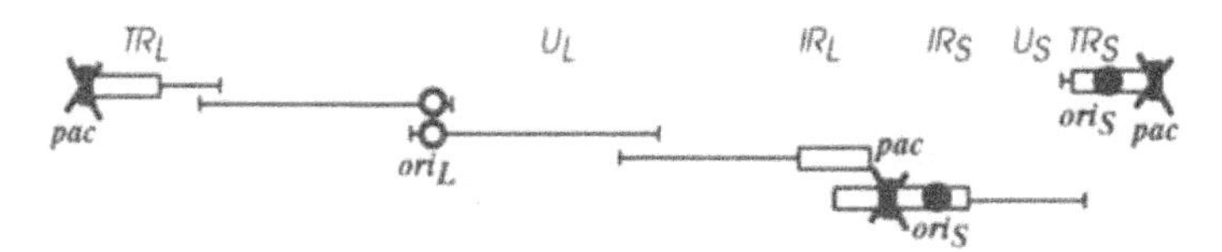

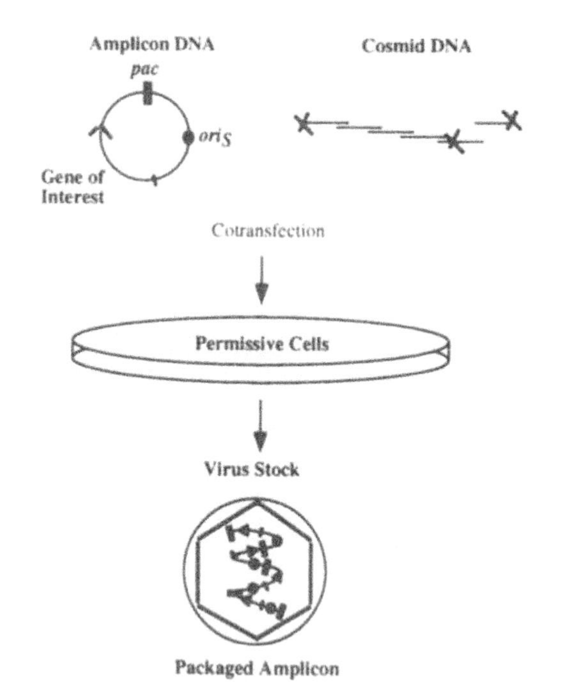

B

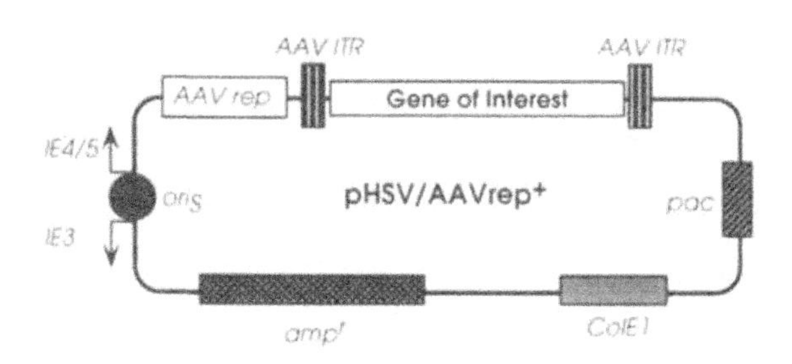

to drive transgene expression (for review see Spear *et al.*, 1997; Michael and Curiel, 1994). Examples would be conjugation of the epidermal growth factor (EGF) ligand to molecular conjugate vectors to effect delivery to cells expressing the EGF receptor (Cristiano and Roth, 1996), expression of the erythropoetin ligand as a recombinant protein with the retrovirus and HSV envelope glycoproteins (Kasahara *et al.*, 1994; Glorioso *et al.*, 1995), and conjugation of peptide ligands to the fiber protein of the adenovirus capsid (Michael *et al.*, 1995; Stevenson *et al.*, 1997). In general, however, with the present state of the art, modification of virions to achieve specific cell targeting generally decreases the overall efficiency of gene delivery (particles needed to transduce a cell) to the targeted cells, as compared to the unmodified virion.

There are a broad range of promoters specific to cells of the nervous system that could be used for cell selective transgene expression in models of CNS regeneration. These promoters include by cell type; the neurofilament light (Charron *et al.*, 1995; Yaworsky *et al.*, 1997) and heavy chains (Schwartz *et al.*, 1994), prion (Baybutt and Manson, 1997), preproenkephalin (Donovan *et al.*, 1992), type II sodium channel (Maue *et al.*, 1990), tyrosine hydroxylase (Min *et al.*, 1994; Wong *et al.*, 1994; Yoon and Chikaraishi, 1992), neuron-specific enolase (Forss-Petter *et al.*, 1990; Twyman and Jones, 1997), platelet-derived growth factor-B chain (Sasahara *et al.*, 1991) and neuron-specific potassium channels (Gan *et al.*, 1996), for neurons; myelin basic protein (Miyao *et al.*, 1993) for oligodendrocytes; glial fibrillary acidic protein (GFAP; Brenner *et al.*, 1994) for astrocytes; and P_0 (Messing *et al.*, 1992) for Schwann cells. Other recently characterized promoters that may be useful for neuron-specific expression include the enhancer–silencer associated with nicotinic receptors (Bessis *et al.*, 1997),

---

FIGURE 4 HSV amplicon vectors. (A) Helper virus-free packaging of HSV-1 amplicon vectors into HSV-1 particles. Cells that are permissive for HSV-1 replication are cotransfected with amplicon plasmid DNA and DNA from a HSV-1 cosmid set that is mutated in the *pac* signals. In the absence of the *pac* signals, the HSV-1 cosmid set cannot generate a packagable HSV-1 genome, but can provide all the transacting functions, required for the replication and packaging of the cotransfected amplicon DNA. Consequently, the resulting amplicon vector stock is free of helper virus. (Reprinted from Fraefel *et al.*, 1997.) (B) Basic structure of the HSV–AAV hybrid amplicon. HSV-1 amplicons typically contain three kinds of genetic elements: prokaryotic sequences for propagation of plasmid DNA in bacteria, including *E. coli* origin of DNA replication (*ColE*1) and an antibiotic resistance gene (e.g., amp^r); sequences from HSV-1, including an ori_s and a *pac* signal to support replication and packaging in mammalian cells in the presence of helper virus functions; and a transcription unit with one or more genes of interest. [The HSV-1 IE3 (ICP4) and IE4/5 (ICP22/47) promoters adjacent to ori_s can be used to direct the expression of the gene(s) of interest as in a standard HSV amplicon.] In the "hybrid" amplicon vector the gene of interest is flanked by adenoassociated virus (AAV) inverted terminal repeats (ITRs) and the AAV *rep* gene is placed outside of this cassette. Reprinted from Fraefel *et al.*, 1997.

microtubule-associated protein 1B (Liu and Fischer, 1997), choline acetyltransferase (Lonnerberg *et al.*, 1996), tau microtubule-associated protein (Sagot *et al.*, 1996), and N-methyl-D-aspartate receptor subunit NR2C (Suchanek *et al.*, 1997). In general, cell specific promoters have the advantage of more stable gene expression as compared to viral promoters in neural cells *in vivo*.

C. Routes of Delivery *in Vivo*

Gene delivery to the nervous system has several confounding factors. First, given the uniqueness and nonreplaceability of neurons, it is critical to minimize damage to these valuable cells. Second, in most cases gene delivery to neurons, and even to most glia, must be on a one-to-one basis, that is, each cell must receive its own transgene. Third, access to the CNS is difficult given the skull and vertebral columns, overlying membrane and blood–brain barrier (BBB). Several schemes have been tested to overcome these difficulties, including *ex vivo* therapy, direct stereotactic inoculation, and access across the BBB or through the CSF.

In *ex vivo* therapy, cells in culture are transfected with a transgene and then grafted back into animals (Fig. 5). Cells can be either continuous cell lines or primary cells. Ideally, appropriate cell lines are syngeneic with the host (although this is not always feasible) nontumorigenic, and free of endogenous pathogens (Isacson and Breakefield, 1997). A number of cell lines have been generated from neural precursor cells that can be passaged in culture, but then differentiate when grafted back into the nervous system. Such lines have been derived from human teratocarcinoma cells (NT2; Younkin *et al.*, 1993), and rodent embryos that have been transfected with immortalizing genes, such as a temperature-sensitive form of SV40 large T antigen (Almazan and McKay, 1992) or the *v-myc* gene (Ryder *et al.*, 1990). The latter strategy has been used to generate lines derived from human and mouse fetal CNS progenitor cells, which can be directed along different lines of neuronal differentiation by exposure to growth factors (Sah *et al.*, 1997; Snyder *et al.*, 1992). These neuronally derived lines do not seem to elicit a strong immune response when grafted into the nervous system of nonsyngeneic experimental animals. However, use of these cells should be considered with caution in regeneration paradigms, as cells passaged in culture can generate additional genetic changes that may predispose them to transformation and potentially for tumor formation. The expression of foreign antigens can also lead to eventual loss of cells through immune rejection, possibly with associated inflammation and release of toxic metabolites. To avoid immune rejection and to contain overgrowth of the graft, cells can be placed in semipermeable capsules, which have a pore size that excludes antibodies, but allows release of neuroactive molecules (e.g., Hoffman *et al.*, 1993; Sagot *et al.*, 1995). Genetic engineering could also be

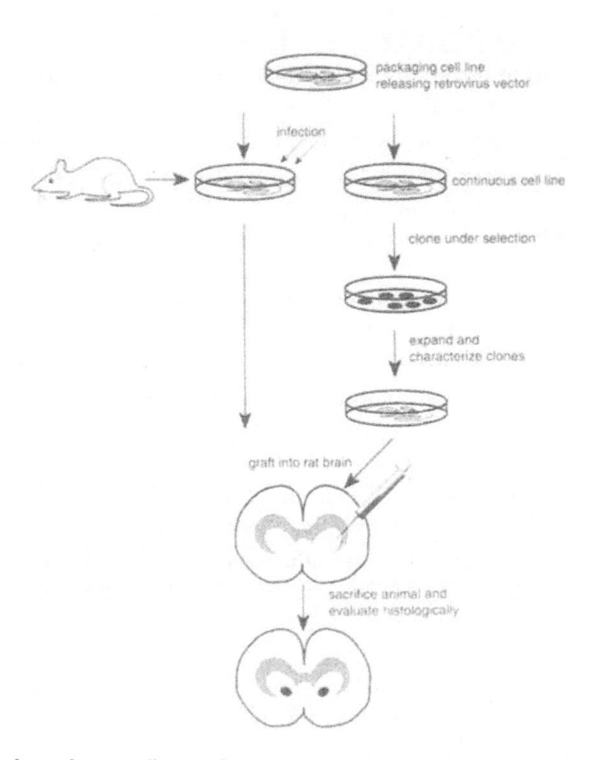

FIGURE 5 Grafting of genetically modified cells into rodents. Conditioned medium from packaging cells is used as a source of retrovirus vector. Transgenes can be delivered to cultured primary cells by serial infection with the vector to maximize the percentage of the population transduced, or by single infection of continuous cell lines and selection of clones representing stable transfected cells. Cells are grafted into the rodent brain by stereotactic surgery, and the effects of these grafts are assessed by histochemical and behavioral analyses. (Modified and reprinted from Fisher and Gage, 1993.)

undertaken to reduce the chances of toxic outcomes. For example, introduction into grafted cells of genes for pro-drug-activating enzymes, such as HSV-TK, which in combination with ganciclovir can kill dividing cells (Moolten, 1986), could provide a means to kill tumor cells if they arise. Further, downregulation of antigen presentation, such as antisense to major histocompatibility (MHC) molecules or expression of viral proteins that block peptide presentation, such as ICP47 (Fruh *et al.*, 1995) or E3 gene product (Lee *et al.*, 1995), could be used to prevent immune rejection.

A number of primary cell types can also be used for *ex vivo* therapy, including neuronal stem cells (for review see Martinez-Serrano and Bjorklund, 1996). Given the availability of vectors that can stably deliver transgenes to nondividing cells, such as AAV, HSV–AAV amplicon vectors and lentivirus, it is not essential that the donor cells are able to divide in culture. One could also obtain cells from transgenic mice that have been engineered to enhance production of particular factors. Neuronal replacement is possible using either embryonic neurons or progenitor cells obtained from the CNS of adult mice (Snyder, 1994). This transplantation paradigm has been used in a number of experimental models of CNS degeneration (Senut *et al.*, 1997) and spinal cord injury (Li and Raisman, 1993), although means to steer neurons toward different neuronal phenotypes is still being investigated. Transplantation of both human and pig fetal mesencephalic tissue, containing embryonic dopaminergic neurons, has been used for neuronal replacement in Parkinson patients with some success (for review see Isacson and Breakefield, 1997; Freeman *et al.*, 1995; Breeze *et al.*, 1995). A number of other cell types that could potentially be obtained from the patient themselves have also been used in CNS transplantation paradigms, including skin fibroblasts (Wolff *et al.*, 1989), astrocytes (Cunningham *et al.*, 1994), myoblasts (Jiao *et al.*, 1993a; Gussoni *et al.*, 1992), Schwann cells (Guenard *et al.*, 1992; Blakemore and Crang, 1985), oligodendrocytes (Lachapelle *et al.*, 1994), endothelial cells (Quinonero *et al.*, 1997; Lal *et al.*, 1994), and testis-derived Sertoli cells (Sanberg *et al.*, 1997). In most paradigms the grafted cells are transfected with genes so that they act as biologic minipumps for release of neuroactive compounds, including a number of neurotrophic factors, e.g., NGF, BDNF, bFGF, or neurotransmitters through expression of the appropriate synthetic enzyme, e.g., tyrosine hydroxylase for DOPA and choline acetyltransferase for acetylcholine (Gage and Fisher, 1991). Some growth factors have been found to have broad effects on many different types of neurons. For example, NGF can protect striatal neurons from injury caused by inhibition of oxidative phosphorylation (Frim *et al.*, 1993b), cholinergic neurons in the basal ganglia from death due to loss of postsynaptic contacts (Rosenberg *et al.*, 1988), hippocampal neurons from ischemic insults (Pechan *et al.*, 1995), and sympathetic neurons from axotomy (Federoff *et al.*, 1992), as well as promote neurite extension following brain and spinal cord lesions (Tuszynski *et al.*, 1996). However, some growth factors show neuron specificity in certain injury paradigms. For example, immortalized fibroblasts transfected with retrovirus vectors mediate biologic delivery of NGF, BDNF, and bFGF (in order of potency) for protection of striatal neurons from excitotoxic lesions (Frim *et al.*, 1993c) and are most effective when placed in the immediate vicinity of the injured neurons (Frim *et al.*, 1993a). BDNF and GDNF appear to be more effective at helping dopaminergic neurons in the substantia nigra survive exposure to toxic conditions (Frim *et al.*, 1994; Choi-

Lundberg *et al.*, 1997; Bilang-Bleuel *et al.*, 1997). A number of recent studies have shown protection of CNS neurons from death following spinal cord injury using this *ex vivo* model of delivery of trophic factors, including the ability of BDNF to rescue retinal neurons from apoptosis following axotomy (Fournier *et al.*, 1997), FGF, BDNF, and NT3 to sustain neurons from spinal cord lesions (Tuszynski *et al.*, 1996; Horner *et al.*, 1996; Grill *et al.*, 1997; Walicke, 1988), and GDNF and CNTF to support motor neurons (Sagot *et al.*, 1995; Sendtner *et al.*, 1992).

Ex vivo gene delivery remains an important strategy in transplantation paradigms for spinal cord injury and CNS degeneration. Three features are especially attractive: (1) the ability to sustain replacement neurons. For example, genes can be delivered to human neuroepithelial stem cells in culture using a virus vector with continued expression of the transgene after grafting in the rat CNS (Sabate *et al.*, 1995). Expression of growth factors or anti-apoptotic proteins, like bcl2 (e.g., Lawrence *et al.*, 1996), may help these neuronal precursors to survive the grafting procedure. This parsimony of neurons becomes especially important when considering transplantation of fetal human neurons in which less than 0.1% survive transplantation and consequently tissue from 10 fetuses is needed for one Parkinson patient; (2) genetically engineered cells can be presented in a variety of forms. Three variations on the context of grafted cells should be considered: (a) allowing cells to fill or line a mechanical conduit, which can direct the route of neurite regeneration (Xu *et al.*, 1997a); and (b) delivering cells that can migrate along with the extending neurite, such as Schwann cells (Lankford *et al.*, 1997) or olfactory ensheathing cells (Goodman *et al.*, 1993); and (c) removable targets, such as encapsulated cells, which could serve as scaffolds during the regeneration process and could then be "collapsed" if necessary once endogenous networks have been restored; (3) biologic delivery of trophic factors has a distinct advantage over infusion of these proteins, because it is possible to achieve steady-state production of biologically significant levels with minimal intervention. Damage can be done by repeated injections of concentrated amounts of the factors, whereas grafting involves a single injection and then the factor is released continuously at constant, lower levels over time. This sustained production seems well suited for processes such as neurite regrowth that occur over an extended period. However, in the early stages of injury, when trophic factors are needed immediately, protein infusion may be the only feasible mode. The apparent safety and efficacy of *ex vivo* delivery, combined with its versatility, make it a likely candidate for human clinical trials in amelioration of CNS injury.

Direct gene delivery involves the inoculation of vectors into the nervous system to transduce endogenous cells. This has typically been done by stereotactic injection into a specific site in the CNS. Only a few μl volume can be injected into the neuronal parenchyma without causing pressure damage and the diluent itself can be toxic. Typically, replication-defective vectors have been used with the extent of gene delivery being determined by the titer of the virus, its diffusibility in the parenchyma, and its efficiency of infecting specific cell types. Titers of recombinant adenovirus vectors are usually the highest of the vectors currently in use, $10^{7-9}/\mu l$. Titers of recombinant HSV are typically around $10^{6-8}/\mu l$, with AAV, gutless adenovirus, HSV lentivirus vectors in the range of $10^{3-6}/\mu l$. Adenovirus, AAV, and retrovirus virions appear to have little selectivity in the endogenous neural cells they infect, although HSV virions seem to preferentially infect neurons (Boviatsis *et al.*, 1994; Kaplitt *et al.*, 1994). The apparent preference of lentivirus vectors for neuronal expression is believed to result from use of the CMV IE1 promoter (Blömer *et al.*, 1997). Adenovirus vectors seem best suited for a "blast" of transgene expression. Vectors that integrate like AAV, lentivirus, and retrovirus have the potential for longer expression, but titers are low and their transgene capacity is relatively small (4–8 kb). HSV vectors, both recombinant and amplicon versions, can give strong transgene expression for extended periods. HSV vectors seem especially well suited for many neurons because they can be taken up efficiently from nerve endings by retrograde transport (Topp *et al.*, 1994), and the LAT promoter can effect long-term transgene expression in a large fraction of sensory neurons (Dobson *et al.*, 1990; Lachmann and Efstathiou, 1997). Several other vectors, e.g., adenovirus, also appear to be transported retrogradely to the nucleus (Ghadge, 1995) and thus provide a means of reaching neuronal cell bodies at some distance.

Three methods have been evaluated to effect disseminated gene delivery in the nervous system using adenovirus and HSV vectors: conditional on-site propagation, intravascular delivery, and intrathecal delivery (Fig. 6). Generation of virus vectors on-site has the advantage that the number and range of vector particles available to infect cells can exceed that achievable with a limited inoculum. However, most virus vectors, e.g., adenovirus, AAV, and HSV, kill the packaging cells and would thus do damage. The only virus vectors that are produced without killing the host cell are retrovirus vectors, including potentially lentivirus vectors. Several groups have tried to develop strategies by which adenovirus (Feng et al., 1997) or HSV amplicon (Savard *et al.*, 1997) vectors can be used to deliver the genes needed to make retrovirus vectors to cells *in vivo,* with subsequent production of retrovirus vectors, but the titers generated are very low in the present configurations. Intravascular delivery has a great deal of promise especially in areas of injury undergoing neovascularization. The new vessels do not have an established blood–brain barrier (BBB) and thus have a relatively high facility to transport molecules, including vectors into the CNS (Muldoon *et al.*, 1995). Many of these studies have been carried out in brain tumors, but would apply to any regions of angiogenesis in the

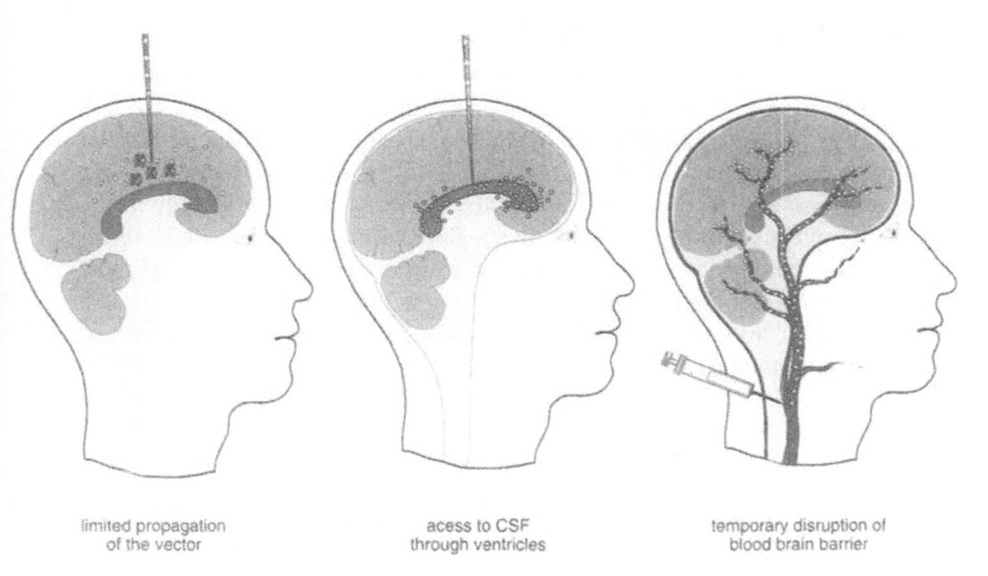

FIGURE 6 Possible modes of disseminated gene delivery to the brain. Stable, high-titer vectors, such as recombinant HSV-1 and adenovirus vectors, can be delivered to brain tumors through three routes (from left to right). (1) Replication-conditional vectors injected into a tumor can spread out to tumor cells by selective on-site propagation in dividing cells and diffusion of the virion within the parenchyma. (2) Vectors injected into the ventricles can spread throughout the CSF and can reach neural cells adjacent to this space. In some regions, including sites of injury or neovascularization, the virus can pass across the brain parenchyma or spinal cord. (3) Virus injected into the carotid artery in combination with osmotic disruption or pharmacologic agents, such as bradykinin or RMP-7, which temporarily open the blood vessels, can reach tumor cells or neural cells in areas of injury selectively because the neovasculature has a weaker blood–tumor barrier, as compared with the blood–brain barrier in normal CNS. (Reprinted from Breakefield *et al.*, 1995.)

CNS. Under conditions of temporary osmotic shock, virus particles can pass across the BBB into normal brain, and to an even greater extent into brain tumors (Neuwelt *et al.*, 1994; Muldoon *et al.*, 1995; Nilaver *et al.*, 1995). Stable, high titer vectors, such as HSV and adenovirus, are needed for this as they are injected directly into the blood stream through the internal carotid artery. Pharmacologic agents, such as bradykinin and RMP-7, can facilitate passage of vectors through neovasculature, but show little-to-no transport to normal brain (Rainov *et al.*, 1995; Barnett *et al.*, submitted). Another interesting feature of neovasculature is that it can attract and incorporate endothelial cells delivered intravascularly (Ojeifo *et al.*, 1995), thus providing a target for *ex vivo* therapies. Because some extent of neovascularization follows any traumatic injury, this may provide a therapeutic window in time and space to the wound area by simple injection of cells or vectors into the blood stream. A third strategy involves injection of vectors into the CSF through the ventricles or intrathecal space. Early studies showed that injection of adenovirus vectors into the CSF yielded transfection of ependymal cells and achieved production of a secreted protein for an extended period of time (Bajocchi *et al.*, 1993). Vectors can also enter into the brain directly from the CSF especially in the choriod plexus and circumventricular regions (Kramm *et al.*, 1995). Again, breakdown of ependymal barriers associated with traumatic injury may allow selective access to the damaged neurons through injection of vectors anywhere in the CSF. Interestingly, Snyder and colleagues (1997) have shown that genetically modified neuronal precursor lines will migrate from the CSF into the brain parenchyma, and within the brain parenchyma will migrate extensive distances to regions of nerve damage. It should also be kept in mind that resident progenitor cells, present even in adults, are positioned on the other side of the ependymal layer and thus directly accessible through the CSF to vectors that could promote their proliferation and differentiation.

III. CONCLUSION

There is a rich array of possible intervention points and times in stimulation of CNS repair using genetic engineering techniques. These can be broken down into three cell categories—neurons, pathways, and innervation targets—and three time intervals—immediately after injury, during neurite regrowth, upon target recognition, and upon reinnervation (Fig. 7). At the time of injury it should be possible to inject vectors and cells directly into the damage site. A number of vectors can be used for direct gene delivery to neurons and glia by injection or through the CSF or blood. The most nontoxic, yet efficient direct gene delivery vectors in the current formulations would be AAV, gutless adenovirus, helper virus-free HSV amplicons, and lentivirus. For most of these vectors, transgene expression would remain on for months, but would eventually be extinguished. For some genes, shut off of expression might be appropriate; for others, in which expression for longer periods is required, subsequent injections might be needed or vectors must be developed that are capable of consistent, extended gene expression. Genes that might be helpful for neurons in the injury period include those coding for anti-apoptotic proteins, like bcl2, or protective proteins for free radicals, like SOD. Glia could be transfected with genes, such as those for growth factors that help to sustain neurons, as well as with anti-inflammatory proteins and inhibitors of glial proliferation. Genes could also be introduced at this time to help increase the regenerative potential of neurons. For neurons this could include the means of elevating metabolic activity, such as the glucose transporter (Ho *et al.*,

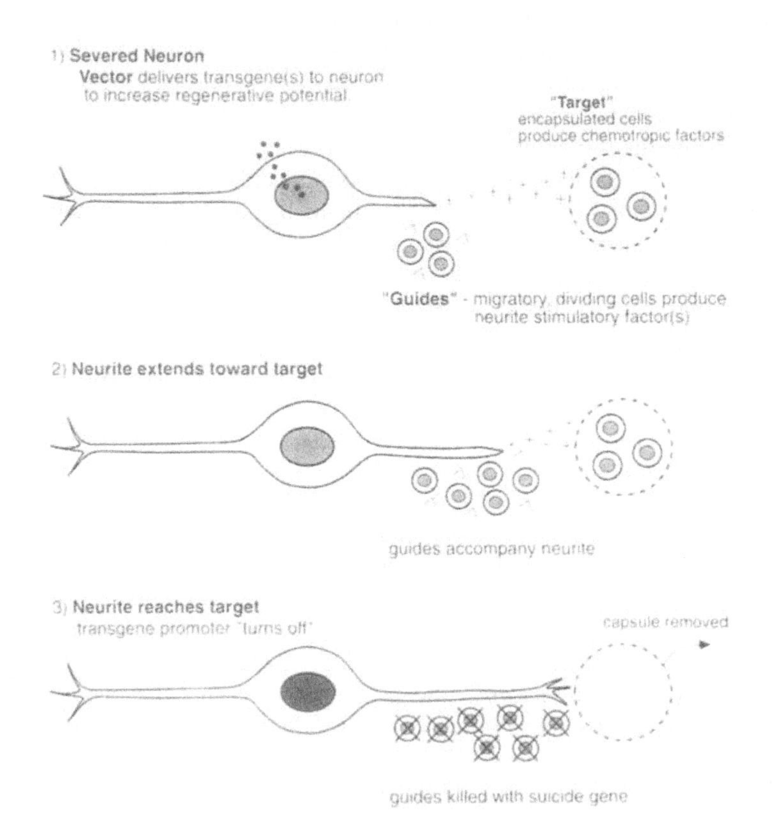

FIGURE 7 Theoretical modes of genetic engineering appropriate for CNS nerve regeneration. Three arenas of gene delivery can be envisioned to promote neurite extension: (a) damaged neurons, (b) path of regrowth, and (c) target of reinnervation. (1) Direct gene delivery to neurons can be used to sustain survival after damage and to fortify the cell for extensive neurite outgrowth. In the pathway, vectors and genetically modified cells can be used to minimize inflammation and glial scar formation and to encourage neurite extension along designated routes. Such "guide" cells can be used in combination with mechanical conduits. A "target" could be set up to provide impetus and direction to neurite outgrowth. This could, for example, be encapsulated, genetically modified cells producing a growth factor attractant that would set up a chemotropic gradient. (2) Once past the area of injury, migratory guide cells could be used that would move along with the growing neurite toward the "target." The target could potentially be moved to maintain a directional concentration gradient as the neurites extended. (3) Ultimately when the neurites reached their normal tissue target, it might be useful to "erase" the scaffold. For neurons, based on current vectors, transgene expression will automatically go off with time. Guide cells could be eliminated by genetically engineering them prior to grafting with a drug-inducible suicide gene. The "target" could be surgically removed encouraging maturation of contact between the neurite and target organ. (Reprinted from Breakefield, 1994.)

1993), and of providing the structural proteins needed for neurite extension, such as tubulin, neurofilament proteins, or GAP43 (Neve *et al.*, 1991); for oligodendrocytes, this could include myelin proteins. It may also prove necessary to graft in neuronal progenitor cells that may be more able than adult neurons to recapitulate the developmental process of tissue innervation.

Once the injury is stabilized the next step seems to be extension of neurites across the injury site. Reduction in glia scar formation, in itself, might enhance this process, but overexpression of other factors along the preferred route might "grease" the route. Glia could be genetically modified on site or grafted to release several classes of proteins: antibody ligands to block the action of inhibitory substances known to be present in the injury site; extracellular matrix proteins, such as L1, to provide a surface neurite extension; and growth factors that attract the neurites. The latter should be delivered in a site-specific manner so that the neurites grow vectorially in the appropriate direction. This could be done by placement of the injection site or by implanting genetically modified glial cells in semipermeable mechanical conduits or capsules. In order to avoid further injury to the spinal cord, these biological devices might be placed outside of it, as in the peripheral nerve segments used for regrowth of CNS neurites from one CNS point to another. Once across the injury impasse, the neurites have to be prepared for a long trek toward their target tissue. Ideally the guide cells would line the route, either through an extended conduit or by migrating along with the nerve growth cone. A number of neural cell types are capable of migration, such as astrocytes, Schwann cells, and olfactory ensheathing cells. These could be genetically engineered in an *ex vivo* paradigm not only to release growth stimulating molecules but also to respond to the same attractant growth factors as the growth cone, so that they would move along with it to the source of growth factor production. The limit of diffusion of the attractant growth factor will dictate how far it can be placed from the extending neurite, and for long distances it may have to be moved along the route over time. Although this would be difficult to do in the CNS *per se* as multiple interventions might be damaging, it might be feasible within a conduit place along the spinal cord proper.

Through these guided routes it might be possible for neurons to extend processes to reach target tissues. Ideally at this juncture one would remove the genetically engineered scaffold cells and allow the neurites to sort out the right target tissue and synapse with it. Problems, such as nonfunctionality, pain, or inappropriate responses, could result if the innervation were inappropriate. Although some growth factors are known that stimulate appropriate synapse formation, like agrin for motor neurons with skeletal muscle, in general the recognition factors for appropriate synapses are largely unknown, and for now one would be relying mostly on the molecular instincts of the neurons and target cells themselves. In fact, the growth cone may "prefer" the growth factor over producing cell "decoy" to the fully matured target cell, in which case the grafted decoy needs to be removed. In the case of encapsulated cells, this could be done surgically. Alternatively, grafted cells could be preprogrammed genetically with a pro-drug-activation suicide gene, so that systemic application of the pro-drug would lead to cell death. An ideal system would be activation of a cell death gene, like ICE, under the control of a tetracycline inducible promoter (e.g., Yu *et al.*, 1996; Gossen and Bujard, 1992).

The field of CNS regeneration is of two minds at the present junction. On the one hand, investigators may feel like primitive humans envisioning a bridge across a wide chasm: knowing that it is possible to achieve, but realizing that the technology to carry out the feat will be complex and is not yet available. On the other hand, there is some hope that neurons through molecular instinct may already "know" how to achieve this feat, and just need some help to overcome some of the obvious stumbling blocks. Fortunately whichever view one holds, the solution is the same: understand the biologic conundrum as thoroughly as possible and try through experimental trial and error to overcome each of the blocks. For no matter how large or small these blocks may seem to be, ultimately the success of the complex process will involve surmounting all of them. In this regard, techniques of genetic engineering provide valuable experimental tools both in understanding the process of nerve repair and in manipulating it. Ultimately some of these tools should prove useful in clinical procedures to augment other therapies for CNS repair, but given concerns about the safety of current modes of gene delivery and the imperative to do no further damage in cases of brain–spinal cord injury, use in clinical trials should be delayed until safety concerns have been allayed through trials in patients with terminal conditions, such as brain tumors.

ACKNOWLEDGMENTS

We are deeply indebted to Ms. Suzanne McDavitt for skilled editorial preparation of this manuscript. We would also like to acknowledge the inspiring leadership of Peter Banyard and John Cavanagh of the International Spinal Research Trust in the field of spinal cord repair. XOB is supported by NINDS grant NS24279, AJ by the Max-Planck-Society, and SW by the Howard Hughes Medical Institute.

REFERENCES

Akli, S., Caillaud, C., Vigne, E., Stratford-Perricaudet, L. D., Poenaru, L., Perricaudet, M., Kahn, A., and Peschanski, M. R. (1993). Transfer of a foreign gene into the brain using adenovirus vectors. *Nature Genetics* 3:224–228.

Ali, M., Lemoine, N. R., and Ring, C. J. A. (1994). The use of DNA virsues as vectors for gene therapy. *Gene Therapy* 1:367–384.

Almazan, G., and McKay, R. (1992). An oligodendrocyte precursor cell line from rat optic nerve. *Brain Res* 579:234–245.

Altman-Hamamdzic, S., Groseclose, C., Ma, J.-X., Hamamdzic, D., Vrindavanam, N. S., Middaugh, I. D., Parratto, N. P., and Sallee, F. R. (1997). Expression of beta-galactosidase in mouse brain: Utilization of a novel nonreplicative Sindbis virus vector as a neuronal gene delivery system. *Gene Therapy* 4:815–822.

Andersen, J. K., Garber, D. A., Meaney, C. A., and Breakfefield, X. O. (1992). Gene transfer into mammalian central nervous system using herpes virus vectors: Extended expression of bacterial lacZ in neurons using the neuron-specific enolase promoter. *Hum Gene Ther* 3:487–499.

Bajocchi, G., Feldman, S. H., Crystal, R. G., and Mastrangeli, A. (1993). Direct *in vivo* gene transfer to ependymal cells in the central nervous system using recombinant adenovirus vectors. *Nature Genetics* 3:229.

Barnett, F. H., Rainov, N., Ikeda, K., Schuback, D. E., Elliot, P., Kramm, C., Chase, M. Qureshí, N., Harsh, G., Chiocca, E. A., and Breakefield, X. O. (1998). Selective delivery of herpes virus vectors to experimental brain tumors using RMP-7. *Cancer Gene Therapy,* in press.

aybutt, H., and Manson, J. (1997). Characterization of two promoters for prion protein (Pr) gene expression in neuronal cells. *Gene* 184:125–131.

Bessis, A., Champtiaux, N., Chatelin, L., and Changeux, J. P. (1997). The neuron-restrictive silencer element: A dual enhancer/silencer crucial for patterned expression of a nicotinic receptor gene in the brain. *Proc Natl Acad Sci USA* 94:5906–5911.

Bilang-Bleuel, A., Revah, F., Colin, P., Locquet, I., Robert, J.-J., Mallet J., and Horellou, P. (1997). Intrastriatal injection of an adenoviral vector expressing glial-cell-line-derived neurotrophic factor prevents dopaminergic neuron degeneration and behavioral impairment in a rat model of Parkinson disease. *Proc Natl Acad Sci USA* 94:8818–8823.

Blaese, R. M., Culver, K. W., Miller, A. D., Carter, C. S., Fleisher, T., Clerici, M., Shearer, G., Chang, L., Chiang, Y., Tolstoshev, P., *et al.* (1995). T lymphocyte-directed gene therapy for ADA-SCID: Initial trial results after 4 years. *Science* 270:475–480.

Blakemore, W. F., and Crang, A. J. (1985). The use of cultured autoogous Schwann cells to remyelinate areas of demyelination in the central nervous system. *J Neurol Sci* 70:207–223.

Blömer, U., Naldini, L., Kafri, T., Trono, D., Verma, I. M., and Gage, F. H. (1997). Highly efficient and sustained gene transfer in adult neurons with lentivirus vector. *J Virol* 71:6641–6649.

Blumenthal, R., Seth, P., Willingham, M. C., and Pasten, I. (1986). pH-dependent lysis of liposomes by adenovirus. *Biochem* 25:2231–2237.

Boviatsis, E. J., Chase, M., Wei, M., Tamiya, T., Hurford, R. K., Kowall, N. W., Tepper, R. I., Breakefield, X. O., and Chiocca, E. A. (1994). Gene transfer into experimental brain tumors mediated by adenovirus, herpes simplex virus (HSV), and retrovirus vectors. *Hum Gene Ther* 5:183–191.

Boyce, F. M., and Bucher, N. L. R. (1996). Baculovirus-mediated gene transfer into mammalian cells. *Proc Natl Acad Sci USA* 93:2348–2352.

Breakefield, X. O. (1994). Gene therapy for spinal cord injury? *ISRT Research Digest* 6:3.

Breakfefield, X. O., Kramm, C. M., Chiocca, E. A., and Pechan, P. A. (1995). Herpes simplex virus vectors for tumor therapy. In: Sobol, R. E., and Scanlon, K. J. (eds.), *The Internet book of gene therapy: Cancer gene therapeutics.* Stamford, Conn.: Appleton and Lange, pp. 41–56.

Breakefield, X. O., Pechan, P., Johnston, K., and Jacoby, D. (1997). Herpes virus vectors. In: Quesenberry, P., Forget, B., and Weissman, S., (eds.), *Stem cell biology and gene therapy.* Wiley-Liss, New York, NY in press.

Breeze, R. E., Wells, T. H. J., and Freed, C. R. (1995). Implantation of fetal tissue for the management of Parkinson's disease: A technical note. *Neurosurg* 36:1044–1047.

Brenner, M., Kisseberth, W. C., Su, Y., Besnard, F., and Messing, A. (1994). GFAP promoter directs astrocyte-specific expression in transgenic mice. *J Neurosci* 14:1030–1037.

Burns, J. C., Friedmann, T., Driever, W., Burrascano, M., and Yee, J.-K. (1993). Vesicular stomatitis virus G glycoprotein pseudotyped retroviral vectors: Concentration to very high titer and efficient gene transfer into mammalian and nonmammalian cells. *Proc Natl Acad Sci USA* 90:8033–8037.

Byrnes, A. P., Wood, M. J. A., and Charlton, H. M. (1996). Role of T cells in inflammation caused by adenovirus vectors in the brain. *Gene Therapy* 3:644–651.

Carpenter, D. E., and Stevens, J. G. (1996). Long-term expression of a foreign gene from a unique position in the latent herpes simplex virus genome. *Human Gene Therapy* 7:1447–1454.

Carrasco, L. (1994). Entry of animal virsues and macromolecules into cells. *FEBS Lett* 350:151–154.

Charron, G., Julien, J.-P., and Bibor-Hardy, V. (1995). Neuron specificity of the neurofilament light promoter in transgenic mice requires the presence of DNA unwinding elements. *J Biol Chem* 270:25739–25745.

Chen, X., Li, Y., Xiong, K., Xie, Y., Aizicovici, S., Snodgrass, R., Wagner, T. E., and Platika, D. (1995b). A novel nonviral cytoplasmic gene expression system and its implications in cancer gene therapy. *Cancer Gene Ther* 2:281–289.

Chen, X., Schmidt, M. C., Goins, W. F., and Glorioso, J. C. (1995a). Two herpes simplex virus type 1 latency-active promoters differ in their contributions to latency-associated transcript expression during lytic and latent infections. *J Virol* 69:7899–7908.

Chernajovsky, Y., Adams, G., Triantaphyllopoulos, K., Ledda, M. F., and Podhajcer, O. L. (1997). Pathogenic lymphoid cells engineered to express TGF beta 1 ameliorate disease in a collagen-induced arthritis model. *Gene Ther* 4:553–559.

Chiocca, E. A., and Breakefield, X. O. (1998). In: Chiocca, E. A., and Breakefield, X. O., eds., *Gene therapy for neurological disorders and brain tumors.* Totowa, N.J.: Humana Press Inc.

Chiocca, S., and Cotten, M. (1998). Characteristics of adenovirus vectors. In: Chiocca, A. E., and Breakefield, X. O. (eds.), *Gene therapy for neurological disorders and brain tumors.* Totowa, N.J.: Humana Press, Inc., pp. 39–52.

Choi-Lundberg, D. L., Lin, Q., Mohajeri, H., Chang, Y. N., Chiang, Y. L., Hay, C. M., Davidson, B. L., and Bohn, M. C. (1997). Dopaminergic neurons protected from degeneration by GDNF gene therapy. *Science* 275:838–841.

Chou, J., Kern, E. R., Whitley, R. J., and Roizman, B. (1990). Mapping of herpes simplex virus-I neurovirulence to gamma 34.5, a gene nonessential for growth in culture. *Science* 250:1262–1266.

Clark, K. R., Voulgaropoulou, F., and Johnson, P. R. (1996). A stable cell line carrying adenovirus-inducible rep and cap genes allows for infectivity titration of adeno-associated virus vectors. *Gene Ther* 3:1124–1132.

Coffin, J. M. (1996). Retroviridae: The viruses and their replication. In: Fields, B. N., and Howley, P. M. (eds.), *Fields virology.* Philadelphia, P.A.: Lippincott-Raven Publishers, pp. 1767–1847.

Conway, J. E., Zolotukhin, S., Muzyczka, N., Hayward, G. S., and Byrne, B. J. (1997). Recombinant adeno-associated virus type 2 replication and packaging is entirely supported by a herpes simplex virus type 1 amplicon expressing rep and cap. *J Virol* 71:8780–8789.

Cristiano, R., and Roth, J. (1996). Epidermal growth factor mediated DNA delivery into lung cancer cells via the epidermal growth factor receptor. *Cancer Gene Therapy* 3:4–10.

Cunningham, C., and Davison, A. J. (1993). A cosmid-based system for constructing mutants of herpes simplex virus type 1. *Virology* 197:116–124.

Cunningham, L., Short, M. P., Breakefield, X. O., and Bohn, M. C. (1994). Nerve growth factor released by transgenic astrocytes enhances the function of adrenal chromaffin cell grafts in a rat model of Parkinson's Disease. *Brain Res.* 658:219–223.

Dai, Y., Roman, M., Naviaux, R. K., and Verma, I. M. (1992). Gene therapy via primary myoblasts: long-term expression of factor IX protein following transplantation *in vivo*. *Proc Natl Acad Sci USA* 89:10892–10895.

Davar, G. (1997). Gene therapy for pain. In: Chiocca EA, and Breakefield, X. O., eds., *Gene therapy for neurological diseases and brain tumors*. Totowa, N.J.: Humana Press, Inc., pp. 419–426.
Davidson, B. L., Allen, E. D., Kozarksy, K. F., Wilson, J. M., and Roessler, B. J. (1993). A model system for *in vivo* gene transfer into the central nervous system using an adenoviral vector. *Nature Genetics* 3:219–223.
Dobson, A. T., Margolis, T. P., Sedarati, F., Stevens, J. G., and Feldman, T. (1990). A latent, nonpathogenic HSV-1-derived vector stably expresses beta-galactosidase in mouse neurons. *Neuron* 5:353–360.
Dobson, A. T., Sederati, F., Devi-Rao, G., Flanagan, W. M., Farrell, M. J., Stevens, J. G., Wagner, E. K., and Feldman, L. T. (1989). Identification of the latency-associated transcript promoter by expression of rabbit beta-globin mRNA in mouse sensory nerve ganglia latently infected with a recombinant herpes simplex virus. *J Virol* 63:3844–3851.
Doll, R. F., Crandall, J. E., Dyer, C. A., Aucoin, J. M., and Smith, F. I. (1996). Comparison of promoter strengths on gene delivery into mammalian brain cells using AAV vectors. *Gene Therapy* 3:437–447.
Donovan, D. M., Takemura, M., O'Hara, B. F., Brannock, M. T., and Uhl, G. R. (1992). Preproenkephalin promoter "cassette" confers brain expression and synaptic regulation in transgenic mice. *Proc Natl Acad Sci USA* 89:2345–2349.
Dranoff, G., Soiffer, R., Lynch, T, Mihm, M., Jung, K., Kolesar, K., Liebster, P., Lam, P., Duda, R., and Mentzer, S. (1997). A phase I study of vaccination with autologous, irradiated melanoma cells engineered to secrete human granulocyte-macrophage colony stimulating factor. *Hum Gene Ther* 7:111–123.
Du, B., Wu, P., Boldt-Houle, D. M., and Terwilliger, E. F. (1996). Efficient transduction of human neurons with an adeno-associated virus vector. *Gene Therapy* 3:254–261.
Eck, S. L., Alavi, J. B., Alavi, A., Tavis, A., Hackney, D., Judy, K., Mollman, J., Phillips, P. C., Wheeldon, E. B., and Wilson, J. M. (1996). Clinical protocol: Treatment of advanced CNS malignancies with the recombinant adenovirus H5.01RSVTK: A phase I trial. *Hum Gene Ther* 7:1465–1482.
Evans, C. H., and Robbins, P. D. (1996). Pathways to gene therapy in rheumatoid arthritis. *Curr Opin Rheumatol* 8:230–234.
Fallon, J. R., and Hall, Z. W. (1994). Building synapses: Agrin and dystroglycan stick together. *Trends Neurosci* 17:469–473.
Federoff, H. J., Geschwind, M. D., Geller, A. I., and Kessler, J. A. (1992). Expression of nerve growth factor *in vivo* from a defective herpes simplex virus 1 vector prevents effects of axotomy on sympathetic ganglia. *Proc Natl Acad Sci USA* 89:1636–1640.
Felgner, P. L., Gadek, T. R., Holm, M., Roman, R., Chan, H. W., Wenz, M., Northrop, J. P., Ringold, G. M., and Danielsen, M. (1987). Lipofection: A highly efficient, lipid-mediated DNA-transfection procedure. *Proc Natl Acad Sci USA* 84:7413–7417.
Feng, M., Jackson, W. H. J., Goldman, C. K., Rancourt, C., Wang, M., Dusing, S. K., Siegal, G., and Curiel, D. T. (1997). Stable *in vivo* gene transduction via a novel adenoviral/retroviral chimeric vector. *Nat Biotechnol* 15:866–870.
Fisher, L. J., and Gage, F. H. (1993). Grafting in the mammalian central nervous system. *Physiol Rev* 73:583–616.
Fisher, K. J., Choi, H., Burda, J., Chen, S., and Wilson, J. M. (1996). Recombinant adenoviruses deleted of all viral genes for gene therapy of cystic fibrosis. *Virol* 217:11–22.
Forss-Petter, S., Danielson, P. E., Catsicas, S., Battenberg, E., Price, J., Nerenberg, M., and Sutcliffe, J. G. (1990). Transgenic mice expression beta-galactosidase in mature neurons under neuron-specific enolase promoter control. *Neuron* 5:187–197.
Fournier, A. E., Beer, J., Arregui, C. O., Essagian, C., Aguayo, A. J., and McKerracher, L. (1997). Brain-derived neurotrophic factor modulates GAP-43 but not T alpha 1 expression in injured retinal ganglion cells of adult rats. *J Neurosci Res* 47:561–572.

Fraefel, C., Breakefield, X. O., and Jacoby, D. (1997). The HSV-1 amplicon. In: Breakefield, X. O., and Chiocca, E. A., eds., *Gene therapy for neurological disorders*. Totowa, N.J.: Humana Press, Inc., pp. 63–82.

Fraefel, C., Song, S., Lim, F., Lang, P., Yu, L., Wang, Y., Wild, P., and Geller, A. (1996). Helper virus-free transfer of herpes simplex virus type 1 plasmid vectors into neural cells. *J Virol* 70:7190–7197.

Freeman, T. B., Olanow, C. W., Hauser, R. A., Nauert, G. M., Smith, D. A., Borlongan, C. V., Sanberg, P. R., Holt, D. A., and Kordower, J. H. (1995). Bilateral fetal nigral transplantation into the postcommissural putament in Parkinson's disease. *Ann Neurol* 38:379–388.

Friedmann, T. (1994). Gene therapy for neurological disorders. *Trends Genet* 10:210–214.

Frim, D. M., Short, M. P., Rosenberg, W. S., Simpson, J., Breakefield, X. O., and Isacson, O. (1993a). Local protective effects of nerve growth factor-secreting fibroblasts against excitotoxic lesions in the rat striatum. *J Neurosurg* 78:267–273.

Frim, D. M., Simpson, J., Uhler, T. A., Short, M. P., Bossi, S. R., Breakefield, X. O., and Isacson, O. (1993b). Striatal degeneration induced by mitochondrial blockade is prevented by biologically delivered NGF. *J Neurosci Res* 35:452–458.

Frim, D. M., Uhler, T. A., Galpern, W., Beal, M. F., Breakefield, X. O., and Isacson, O. (1994). Implanted fibroblasts genetically engineered to produce brain-derived neurotrophic factor prevent 1-methyl-4-phenylpyridinium toxicity to dopaminergic neurons in the rat. *Proc Natl Acad Sci USA* 91:5104–5108.

Frim, D. M., Uhler, T. A., Short, M. P., Ezzedine, Z. D., Klagsbrun, M., Breakefield, X. O., and Isacson, O. (1993c). Effects of biologically delivered NGF, BDNF and bFGF on striatal excitotoxic lesions. *Neuroreport* 4:367–370.

Fritz, J. D., Herweijer, H., Zhang, G., and Wolff, J. A. (1996). Gene transfer into mammalian cells using histone-condensed plasmid DNA. *Hum Gene Ther* 7:1395–1404.

Fruh, K., Ahn, K., Djaballah, H., Sempe, P., Van Endert P. M., Tampe, R., Peterson, P. A., and Yang, Y. (1995). A viral inhibitor of peptide transporters for antigen presentation. *Nature* 375:415–418.

Fu, S., and Deisseroth, A. B. (1997). Use of the cosmid adenoviral vector cloning system for the *in vitro* construction of recombinant adenoviral vectors. *Hum Gene Ther* 8:1321–1330.

Gage, F. H., and Fisher, L. J. (1991). Intracerebral grafting: A tool for the neurobiologist. *Neuron* 6:1–12.

Gan, L., Perney, T. M., Kaczmarek, L. K. (1996). Cloning and characterization of the promoter for a potassium channel expressed in high frequency firing neurons. *J Biol Chem* 271:5859–5865.

Gao, W. Y., Cara, A., Gallo, R. C., and Lor, F. (1993). Low levels of deoxynucleotides in peripheral blood lymphocytes: A strategy to inhibit human immunodeficiency virus type 1 replication. *Proc Natl Acad Sci USA* 90:8925–8928.

Gao, G. P., Yang, T., and Wilson, J. M. (1996). Biology of adenovirus vectors with E1 and E4 deletions for liver-directed gene therapy. *J Virol* 70:8934–8943.

Geller, A. I., and Breakefield, X. O. (1988). A defective HSV-1 vector expresses Escherichia coli beta-galactosidase in cultured peripheral neurons. *Science* 241:1667–1669.

Ghadge, G. D. (1995). CNS gene delivery by retrograde transport of recombination replication-defective adenoviruses. *Gene Therapy* 2:132–137.

Ghattas, I. R., Sanes, J. R., and Majors, J. E. (1991). The encephalomyocarditis virus internal ribosome entry site allows efficient coexpression of two genes from a recombinant provirus in cultured cells and in embryos. *Mol Cell Biol* 11:5848–5859.

Giraud, C., Winocour, E., and Berns, K. I. (1994). Site-specific integration by adeno-associated virus is directed by a cellular DNA sequence. *Proc Natl Acad Sci USA* 91:10039–10043.

Glorioso, J. C., Bender, M. A., Goins, W. F., Fink, D. J., and DeLuca, N. (1995). Herpes simplex virus as a gene-delivery vector for the central nervous system. In: *Viral vectors*. New York: Academic Press, pp. 1–23.

Goodman, M. N., Silver, J., and Jacobberger, J. W. (1993). Establishment and neurite outgrowth properties of neonatal and adult rat olfactory bulb glial cell lines. *Brain Res* 619:199–213.

Gossen, M., and Bujard, H. (1992). Tight control of gene expression in mammalian cells by tetracycline-responsive promoters. *Proc Natl Acad Sci USA* 89:5547–5551.

Gossen, M., Bonin, A. L., and Bujard, H. (1993). Control of gene activity in higher eukaryotic cells by prokaryotic regulatory elements. *Trends Biochem Sci* 18:471–475.

Gossen, M., Freundlieb, S., Bender, G., Muller, G., Hillen, W., and Bujard, H. (1995). Transcriptional activation by tetracyclines in mammalian cells. *Science* 268:1766–1769.

Greber, U. F., Willetts, M., Webster, P., and Helenius, A. (1993). Stepwise dismantling of adenovirus 2 during entry into cells. *Cell* 75:477–486.

Grill, R., Murai, K., Blesch, A., Gage, F. H., and Tuszynski, M. H. (1997). Cellular delivery of neurotrophin-3 promotes corticospinal axonal growth and partial functional recovery after spinal cord injury. *J Neurosci* 17:5560–5572.

Guenard, V., Kleitman, N., Morrissey, T. K., Bune, R. P., and Aebischer, P. (1992). Syngeneic Schwann cells derived from adult nerves seeded in semipermeable guidance channels enhance peripheral nerve regeneration. *J Neurosci* 12:3310–3320.

Gussoni, E., Pavlath, G. K., Lanctot, A. M., Sharma, K. R., Miller, R. G., Steinman, L., and Blau, H. M. (1992). Normal dystrophin transcripts detected in Duchenne muscular dystrophy patients after myoblast transplantation. *Nature* 356:435–438.

Hamlin, J. L. (1992). Mammalian origins of replication. *Bioessays* 14:651–659.

Hardy, S., Kitamura, M., Harris-Stansil, T., Dai, Y., and Phipps, M. L. (1997). Construction of adenovirus vectors through cre-*lox* recombination. *J Virol* 71:1842–1849.

Helenius, A. (1992). Unpacking the incoming influenza virus. *Cell* 69:577–578.

Henderson, C. E. (1996). Role of neurotrophic factors in neuronal development. *Curr. Opinion Neurobiol* 6:64–70.

Ho, D. Y. (1994). Amplicon-based herpes simplex virus vectors. *Meth Cell Biol* 43:191–210.

Ho, D. Y., Mocarski, E. S., and Sapolsky, R. M. (1993). Altering central nervous system physiology with a defective herpes simplex virus vector expressing the glucose transporter gene. *Proc Natl Acad Sci USA* 90:3655–3659.

Hoffman, D., Breakefield, X. O., Short, M. P., and Aebischer, P. (1993). Transplantation of a polymer-encapsulated cell line genetically engineered to release NGF. *Exp Neurol* 122:100–106.

Horner, P. J., McTigue, D. M., Ridet, J.-L., Ray, J., Senut, M.-C. C., Stokes, B. T., and Gage, F. H. (1996). Neurotrophin producing fibroblasts stimulate myelination and supraspinal neuronal sprouting when grafted in the contused rat spinal cord. *J Neurotrauma:* in press.

Huang, Q., Vonsattel, J. P., Schaffer, P. A., Martuza, R. I., Breakefield, X. O., and DiFiglia, M. (1992). Introduction of a foreign gene (Escherichia coli lacZ) into rat neostriatal neurons using herpes simplex virus mutants: A light and electron microscopic study. *Exp Neurol* 115:303–316.

Inamura, T., Nomura, T., Bartus, R. T., and Black, K. L. (1994). Intracarotid infusion of a RMP-7, a bradykinin analog: A method for selective drug delivery to brain tumors. *J Neurosurg* 81:752–758.

Isacson, O., and Breakefield, X. O. (1997). Benefits and risks of hosting animal cells in the human brain. *Nature Med* 3:964–969.

Jacoby, D. R., Fraefel, C., and Breakefield, X. O. (1997). Hybrid vectors: A new generation of virus-based vectors designed to control the cellular fate of delivered genes. *Gene Therapy:* 4:1281–1283.

Jiao, S., Cheng, L., Wolff, J. A., and Yang, N. S. (1993b). Particle bombardment-mediated gene transfer and expression in rat brain tissues. *Biotechnology* 11:497–502.

Jiao, S., Gurevich, W., and Wolff, J. A. (1993a). Long-term correction of rat model of Parkinson's disease by gene therapy. *Nature* 3621:450–453.

Johnson, P. A., and Friedmann, T. (1994). Replication-defective recombinant herpes simplex virus vectors. *Meth Cell Biol* 43:211–230.

Johnston, K. M., Jacoby, D., Pechan, P., Fraefel, C., Borghesani, P., Schuback, D., Dunn, R., Smith, F., and Breakefield, X. O. (1997). HSV/AAV hybrid amplicon vectors extend transgene expression in human glioma cells. *Human Gene Therapy* 8:359–370.

Johnson, P. A., Miyanohara, A., Levine, F., Cahill, T., and Friedman, T. (1992). Cytotoxicity of a replication-defective mutant of herpes simplex virus type 1. *J Virol* 66:2952–2965.

Jolly, D. (1994). Viral vector systems for gene therapy. *Cancer Gene Therapy* 1:51–64.

Kaneda, Y., Morishita, R., and Dzau, V. J. (1997). Prevention of restenosis by gene therapy. *Ann NY Acad Sci* 811:299–308.

Kaneda, Y., Twai, K., and Uchida, T. (1989). Increased expression of DNA cointroduced with nuclear protein in adult rat liver. *Science* 243:375–378.

Kaplitt, M. G., Pfaus, J. G., Kleopoulos, S. P., Hanlon, B. A., Rabkin, S. D., and Pfaff, D. W. (1991). Expression of a functional foreign gene in adult mammalian brain following *in vivo* transfer via a herpes simplex virus type 1 defective viral vector. *Mol Cell Neurosci* 2:320–330.

Kaplitt, M. G., Leone, P., Samulski, R. J., Xiao, X., Pfaff, D. W., O'Malley, K. L., and During, M. J. (1994). Long-term gene expression and phenotypic correction using adeno-associated virus vectors in the mammalian brain. *Nat Genet* 8:148–154.

Kasahara, N., Dozy, A. M., and Kan, Y. W. (1994). Tissue-specific targeting of retroviral vectors through ligand-receptor interactions. *Science* 266:1373–1376.

Kaye, E. (1997). Gene therapy for lysosomal storage diseases. In: Chiocca, E. A., and Breakefield, X. O., eds., *Gene therapy for neurological diseases and brain tumors.* Totowa, N.J.: Humana Press, Inc., pp. 409–418.

Klein, T. M., and Fitzpatrick-McElligott, S. (1993). Particle bombardment: A universal approach for gene transfer to cells and tissues. *Curr Opin Biotechnol* 4:583–590.

Kochanek, S., Clemens, P. R., Mitani, K., Chen, H. H., Chan, S., and Caskey, C. T. (1996). A new adenoviral vector: Replacement of all viral coding sequences with 28 kd of DNA independently expressing both full length dystrophin and beta-galactosidase. *Proc Natl Acad Sci USA* 93:5731–5736.

Kotin, R. M., Linden, R. M., and Berns, K. I. (1992). Characterization of a preferred site on human chromosome 19q for integration of adeno-associated virus DNA by non-homologous recombination. *EMBO J* 11:5071–5078.

Kramm, C. M., Sena-Esteves, M., Barnett, F., Rainov, N., Schuback, D., Yu, J., Peachan, P., Paulus, W., Chiocca, E. A., and Breakefield, X. O. (1995). Gene therapy for brain tumors. *Brain Pathology* 5:345–381.

Krikorian, C. R., and Read, G. S. (1991). *In vitro* mRNA degradation system to study the virion host shutoff function of herpes simplex virus. *J Virol* 65:112–122.

Kumar-Singh, R., and Chamberlain, J. S. (1996). Encapsidated adenovirus minichromosomes allow delivery and expression of a 14 kb dystrophin cDNA to muscle cells. *Hum Mol Genetics* 5:913–921.

Kwong, A. D., and Frenkel, N. (1989). The herpes simplex virus virion host shut off function. *J Virol* 63:4834–4839.

Lachapelle, F., Duhamel-Clerin, E., Gansmuller, A., Baron-Van Evercorren,A., Villarroya, H., and Gumpel, M. (1994). Transplanted transgenically marked oligodendrocytes survive, migrate and myelinate in the normal mouse brain as they do in the shiverer mouse brain. *Eur J Neurosci* 6:814–824.

Lachmann, R. H., and Efstathiou, S. (1997). Utilization of the herpes simplex virus type 1 latency-associated regulatory region to drive stable reporter gene expression in the nervous system. *J Virol* 71:3197–3207.

Lal, B., Indurti, R. R., Couraud, P. O., Goldstein, G. W., and Laterra, J. (1994). Endothelial cell implantation and survival within experimental gliomas. *Proc Natl Acad Sci USA* 91:9695–9699.

Lankford, K. L., Imaizumi, T., and Kocsis, J. D. (1997). Quantitative morphometric analysis of spinal cord remyelination by transplanted glia. *Soc Neurosci:* Abst. #13701.

Lawrence, M. S., Ho, D. Y., Sun, G. H., Steinberg, G. K., and Sapolsky, R. M. (1996). Overexpression of bcl-2 with herpes simplex virus vectors protects CNS neurons against neurological insults *in vitro* and *in vivo*. *J. Neurosci* 16:486–496.

Ledley, F. D. (1996). Pharmaceutical approach to somatic gene therapy. *Pharm Res* 13:1595–1614.

Lee, M. G., Abina, M. A., Haddada, H., and Perricaudet, M. (1995). The constitutive expression of the immunomodulatory gp 19k protein in E1-, E3-adenoviral vectors through reduces the host cytotoxic T-cell response against the vector. *Gene Ther* 2:256–262.

Li, S., and Huang, L. (1997). *In vivo* gene transfer via intravenous administration of cationic lipid-protamine-DNA (LPD) complexes. *Gene Therapy* 4:891–900.

Li, Y., and Raisman, G. (1993). Long axon growth from embryonic neurons transplanted into myelinated tracts of the adult rat spinal cord. *Brain Res* 629:115–127.

Liang, E., and Hughes, J. A. (1998). Characterization of a pH-sensitive surfactant, dodecyl-2-(1′ imidazolyl) propionate (DIP), and preliminary studies in liposome mediated gene transfer. *Biochim Biophys Acta* 1369:39–50.

Lieber, A., Sandig, V., Sommer, W., Bahring, S., and Strauss, M. (1993). Stable high-level gene expression in mammalian cells by T7 phage RNA polymerase. *Meth Enzymol* 217:47–66.

Lindvall, O., Kokaia, Z., Bengzon, J., Elmer, E., and Kokaia, M. (1994). Neurotrophins and brain insults. *Trends Neurosci.* 17:490–496.

Liu, D., and Fischer, I. (1997). Structural analysis of the proximal region of the microtubule-associated protein 1B promooter. *J Neurochem* 69:910–919.

Lokensgard, J. R., Bloom, D C., Dobson, A. T., and Feldman, L. T. (1994). Long-term promoter activity during herpes simplex virus latency. *J Virol* 68:7148–7158.

Lonnerberg, P., Schoenherr, C. J., Anderson, D. J., and Ibanez, C. F. (1996). Cell type-specific regulation of choline acetyltransferase gene expression. Role of the neuron-restrictive silencer element and cholinergic-specific enhancer sequences. *J Biol Chem* 271:33358–33365.

Martinez-Serrano, A., and Bjorklund, A. (1996). Protection of the neostriatum against excitotoxic damage by neurotrophin-producing, genetically modified neural stem cells. *J Neurosci* 16:4604–4616.

Maue, R. A., Kraner, S. D., Goodman, R. H., and Mandel, G. (1990). Neuron-specific expression of the rat brain type II sodium channel gene is directed by upstream regulatory elements. *Neuron* 4:223–231.

McCown, T. J., Xiao, X., Li, J., Breese, G. R., and Samulski, R. J. (1996). Differential and persistent expression patterns of CNS gene transfer by an adeno-associated virus (AAV) vector. *Brain Res* 713:99–107.

McKnight, R. A., Shamay, A., Sankaran, L., Wall, R. J., and Hennighausen, L. (1992). Matrix-attachment regions can impart position-independent regulation of a tissue-specific gene in transgenic mice. *Proc Natl Acad Sci USA* 89:6943–6947.

Mebatsion, T., Schnell, M. J., Cox, J. H., Finke, S., and Conzelmann, K.-K. (1996). Highly stable expression of a foreign gene from rabies virus. *Proc Natl Acad Sci USA* 93:7310–7314.

Messing, A., Behringer, R. R., Hammang, J. P., Palmiter, R. D., Brinster, R. L., and Lemmke, G. (1992). P_0 promoter directs expression of reporter and toxin genes to Schwann cells of transgenic mice. *Neuron* 8:507–520.

Michael, S. I., and Curiel, D. T. (1994). Strategies to achieve targeted gene delivery via the receptor-mediated endocytosis pathway. *Gene Therapy* 1:223–232.

Michael, S. I., Hong, J. S., Curiel, D. T., and Engler, J. A. (1995). Addition of a short peptide ligand to the adenovirus fiber protein. *Gene Therapy* 2:660–668.

Miller, D. G., Adam, M. A., and Miller, A. D. (1990). Gene transfer by retrovirus vectors occurs only in cells that are actively replicating at the time of infection. *Mol Cell Biol* 10:4239–4242.

Min, N., John, T. H., Kim, K. S., Peng, C., Son, I. H. (1994). 5′ upstream DNA sequence of the rat tyrosine hydroxylase gene directs high-level and tissue-specific expression to catecholaminergic neurons in the central nervous system of transgenic mice. *Mol Brain Res* 27:281–289.

Miyao, Y., Shimizu, K., Moriuchi, S., Yamada, M., Nakahira, K., Nakajima, K., Nakao, J., Kuriyama, S., Tsujii, T., and Mikoshiba, K. (1993). Selective expression of foreign genes in glioma cells: Use of the mouse myelin basic protein gene promoter to direct toxic gene expression. *J Neurosci Res* 36:472–479.

Mohajeri, M. H., Bartsch, U., van der Putten, H., Sansig, G., Mucke, L., and Schachner, M. (1996). Neurite outgrowth on non-permissive substrates *in vitro* is enhanced by ectopic expression of the neural adhension molecule L1 by mouse astrocyte. *Eur J Neurosci* 8:1085–1097.

Moolten, F. L. (1986). Tumor chemosensitivity conferred by inserted herpes thymidine kinase genes: A paradigm for a prospective cancer control strategy. *Cancer Res* 56:5276–5281.

Moullier, P., Bohl, D., Heard, J.-M., and Danos, O. (1993). Correction of lysosomal storage in the liver and spleen of MPS VII mice by implantation of genetically modified skin fibroblasts. *Nature Genetics* 4:154–159.

Muldoon, L. L., Nilaver, G., Kroll, R. A., Pagel, M. A., Breakefield, X. O., Chiocca, E. A., Davidson, B. L., Weissleder, R., and Neuwelt, E. A. (1995). Comparison of intracerebral inoculation and osmotic blood brain barrier disruption for delivery of adenovirus, herpesvirus and iron oxide particles to normal rat brain. *Am J Path* 147:1840–1851.

Muzyczka, N. (1992). Use of adeno-associated virus as a general transduction vector for mammalian cells. *Curr Top Microbiol Immunol* 158:97–129.

Naldini, L., Blömer, U., Gallay, P., Ory, D., Mulligan, R., Gage, F. H., Verma, I. M., and Trono, D. (1996). *In vivo* gene delivery and stable transduction of nondividing cells by a lentiviral vector. *Science* 272:263–267.

Naviaux, R. K., Costanzi, E., Haas, M., and Verma, I. (1996). The pCL vector system: Rapid production of helper-free, high-titer, recombinant retroviruses. *J Virol* 70:5701–5705.

Neuwelt, E. A., Weissleder, R., Nilaver, G., Krill, R. A., Roman-Goldstein, S., Szumowski, J., Pagel, M. A., Jones, R. S., Remsen, L. G., and McCormick, C. I. (1994). Delivery of virus-sized iron oxide particles to rodent CNS neurons. *Neurosurg* 34:777–784.

Neve, R. L., Ivins, K. J., Benowitz, L. I., During, M. J., and Geller, A. I. (1991). Molecular analysis of the function of the neuronal growth-associated protein GAP-43 by genetic intervention. *Mol Neurobiol* 5:131–141.

Nilaver, G., Muldoon, U., Kroll, R. A., Pagel, M. A., Breakefield, X. O., Davidson, B. L., and Neuwelt, E. A. (1995). Delivery of herpes virus and adenovirus to nude rat intracerebral tumors following osmotic blood-brain barrier disruption. *Proc Natl Acad Sci USA* 92:9829–9833.

Ojeifo, J. O., Reza, F., Paik, S., Maciag, T., and Zwiebel, J. A. (1995). Angiogenesis-directed implantation of genetically modified endothelial cells in mice. *Cancer Res* 55:2240–2244.

Oldfield, E. H., Ram, Z., Culver, K. W., Blaese, R. M., DeVroom, H. L., and Anderson, W. F. (1993). Gene therapy for the treatment of brain tumors using intra-tumoral transduction with the thymidine kinase gene and intravenous ganciclovir. *Hum Gene Ther* 4:39–69.

Paulus, W., Baur, I., Boyce, F. M., Breakefield, X. O., and Reeves, S. A. (1996). Self-contained, tetracycline-regulated retroviral vector system for gene delivery to mammalian cells. *J Virol* 70:62–67.

Pechan, P. A., Yoshida, T., Panahian, N., Moskowitz, M. A., and Breakefield, X. O. (1995). Genetically modified fibroblasts producing NGF protect hippocampal neurons after ischemia in the rat. *Neuroreport* 6:669–672.

Peel, A. L., Zolotukhin, S., Schrimsher, G. W., Muzyczka, N., and Reier, P. J. (1997). Efficient transduction of green fluorescent protein in spinal cord neurons using adeno-associated virus vectors containing cell type-specific promoters. *Gene Therapy* 4:16–24.

Piper, R. C., Slot, J. W., Li, G., Stahl, P. D., and James, D. E. (1994). Recombinant Sindbis virus as an expression system for cell biology. *Meth Cell Biol* 43:55–78.

Quinonero, J., Tchelingerian, J.-L., Vignais, L., Foignant-Chaverot, N., Colin, C., Horellou, P., Liblau, R., Barbin, G., Strosberg, A. D., Jacque, C., and Couraud, P.-O. (1997). Gene transfer to the central nervous system by transplantation of cerebral and endothelial cells. *Gene Therapy* 4:111–119.

Rainov, N. G., Ikeda, K., Qureshi, N., Grover, S., Herrlinger, U., Pechan P., Chiocca, E. A., Breakefield, X. O., and Barnett, F. H. Intra-arterial delivery of adenovirus vectors and liposome-DNA complexes to experimental brain neoplasms: in preparation.

Rainov, N. G., Zimmer, C., Chase, M., Kramm, C., Chiocca, E. A., Weissleder, R., and Breakefield, X. O. (1995). Selective uptake of viral and monocrystalline particles delivered intra-arterially to experimental brain neoplasms. *Hum Gene Ther* 6:1543–1552.

Rosenberg, M. B., Friedmann, T., Robertson, R. C., Tuszynski, M., Wolff, J. A., Breakefield, X. O., and Gage, F. H. (1988). Grafting genetically modified cells to the damaged brain: Restorative effects of NGF expression. *Science* 242:1575–1578.

Ryder, E. F., Snyder, E. Y., and Cepko, C. L. (1990). Establishment and characterization of multipotent neual cell lines using retrovirus vector-mediated oncogene transfer. *J Neurobiol* 21:356–375.

Sabate, O., Horellou, P., Vigne, E., Colin, P., Perricaudet, M., Buc-Caron, M.-H., Mallet, J. (1995). Transplantation to the rat brain of human neural progenitors that were genetically modified using adenoviruses. *Nature Genetics* 9:256–260.

Sadot, E., Heicklen-Klein, A., Barg, J., Lazarovici, P., and Ginzburg, I. (1996). Identification of a tau promoter region mediating tissue-specific-regulated expression in PC12 cells. *J Mol Biol* 256:805–812.

Sadowski, I., Ma, J., Triczenberg, S., and Ptashne, M. (1988). GAL4-VP16 is an unusually potent transcriptional activator. *Nature* 335:563–564.

Sagot, Y., Tan, S. A., Baetge, E., Schmalbruch, H., Kato, A. C., and Aebischer, P. (1995). Polymer encapsulated cell lines genetically engineered to release ciliary neurotrophic factor can slow down progressive motor neuronopathy in the mouse. *Eur J Neurosci* 7:1313–1322.

Sah, D. W. Y., Ray, J., and Gage, F. H. (1997). Bipotent progenitor cell lines from the human CNS. *Nat Biotech* 15:574–580.

Sanberg, P. R., Borlongan, C. V., Othberg, A. I., Saporta, S., Freeman, T. B., and Cameron, D. F. (1997). Testis-derived Sertoli cells have a trophic effect on dopamine neurons and alleviate hemiparkinsonism in rats. *Nat Med* 3:1129.

Sasahara, M., Fries, J. W., Raines, E. W., Gown, P. M., Westrum, L. E., Frosch, M. P., Bonthron, D. T., Ross, R., and Collins, T. (1991). PDGF B-chain in neurons of the central nervous system, posterior pituitary, and in a transgenic model. *Cell* 64:217–227.

Savard, N., Cosset, F.-L., and Epstein, A. L. (1997). Defective herpes simplex virus type 1 vectors harboring *gag*, *pol*, and *env* genes can be used to rescue defective retrovirus vectors. *J Virol* 71:4111–4117.

Schwartz, M. L., Katagi, C., Bruce, J., and Schlaepfer, W. W. (1994). Brain-specific enhancement of the mouse neurofilament heavy gene promoter *in vitro*. *J Biol Chem* 269:13444–13450.

Sena-Esteves, M., Aghi, M., Pechan, P., Kaye, E., and Breakefield, X. O. (1995). Gene delivery to the nervous system using retroviral vectors. In: Latchman, D., ed., *Genetic manipulation of the nervous system*. Academic Press, pp. 149–180.

Sena-Esteves, M., Aghi, M., Pechan, P., Kaye, E., and Breakefield, X. O. (1997). Gene delivery to the nervous system using retroviral vectors. In: Latchman, D., ed., *Genetic manipulation of the nervous system*. Academic Press, pp. 149–180.

Sendtner, M., Schmalbruch, H., Stockli, K. A., Carroll, P., Kreutzberg, G. W., and Thoenen, H. (1992). Giliary neurotrophic factor prevents degeneration of motor neurons in mouse mutant progressive motor neuronopathy. *Nature* 358:502–504.

Senut, M.-C., Aubert, I., Horner, P. J., and Gage, F. H. (1997). Gene transfer for adult CNS regeneration and aging. In: Chiocca, E. A., and Breakefield, X. O. eds., *Gene therapy for neurological diseases and brain tumors.* Totowa, N.J.: Humana Press Inc., pp. 345–376.

Short, M. P., Choi, B. C., Lee, J. K., Malick, A., Breakefield, X. O., and Martuza, R. L. (1990). Gene delivery to glioma cells in rat brain by grafting of a retrovirus packaging cell line. *J Neurosci Res* 27:427–439.

Smith F., Jacoby, D., and Breakefield, X. O. (1995). Virus vectors for gene delivery to the nervous system. *Rest Neurol Neurosci* 8:21–34.

Smith, J. G., and Eck, S. L. (1997). Immune response to viral vectors. In: Chiocca, E. A., and Breakefield, X. O., eds., *Gene therapy for neurological disorders and brain tumors.* Totowa, N.J.: Humana Press Inc., pp. 147–160.

Snyder, E. Y. (1994). Grafting immortalized neurons to the CNS. *Curr Opinion Neurobiol* 4:742–751.

Snyder, E. Y., Deitcher, D. L., Walsh, C., Arnold-Aldea, S., Hartwieg, E. A., and Cepko, C. L. (1992). Multipotent neural cell lines can engraft and participate in development of mouse cerebellum. *Cell* 68:33–51.

Snyder, E. Y., Park, K. I., Flax, J. D., Liu, S., Rosario, C. M., Yandava, B. D., and Aurora, S. (1997). Potential of neural "stem-like" cells for gene therapy and repair of the degenerating central nervous system. *Adv Neurol* 72:121–132.

Snyder, R. O., Xiao, X., and Samulski, R. J. (1996). Production of recombinant adeno-associated viral vectors. In: Dracopoli, N., Haines, J., Krof, B., Moir, D., Morton, C., Seidman, C., Scidman, J., and Smith, D., eds., *Current protocols in human genetics.* New York: John Wiley and Sons, pp. 1–24.

Spaete, R., and Frenkel, N. (1982). The herpes virus amplicon: A new eucaryotic defective-virus cloning-amplifying vector. *Cell* 30:295–304.

Spear, M., Herrlinger, U., Rainov, N., Pechan, P., Weissleder, R., and Breakefield, X. O. (1997). Targeting gene therapy vectors to CNS malignancies. *J Neurovirol:* in press.

Starr, P. A., Lim, F., Grant, F. D., Trask, L., Lang, P., Yu, L., Geller, A. I. (1996). Long-term persistence of defective HSV-1 vectors in the rat brain is demonstrated by reactivation of vector gene expression. *Gene Therapy* 3:615–623.

Stevenson, M. (1996). Portals of entry: uncovering HIV nuclear transport pathways. *Trends in Cell Biol* 6:9–15.

Stevenson, S. C., Rollence, M., Marshall-Neff, J., and McClelland, A. (1997). Selective targeting of human cells by a chimeric adenovirus vector containing a modified fiber protein. *J Virol* 71:4782–4790.

Suchanek, B., Seeburg, P. H., and Sprengel, R. (1997). Tissue specific control regions of the N-methyl-D-aspartate receptor subunit NR2C promoter. *Biol Chem* 378:929–934.

Talbot, D., Collis, P., Antoniou, M., Vidal, M., Grosveld, F., and Greaves, D. R. (1989). A dominant control region from the human beta-globin locus conferring integration site-independent gene expression. *Nature* 338:352–355.

Tang, F., and Hughes, J. A. (1998). Introduction of a disulfide bond into a cationic lipid enhances transgene expression of plasmid DNA. *Biochem Biophys. Res Commun.* 242:141–145.

Tomita, N., Morishita, R., Higaki, J., Tomita, S., Aoki, M., Ogihara, T., and Kaneda, Y. (1996). *In vivo* gene transfer of insulin gene into neonatal rats by the HVJ-liposome method resulted in sustained transgene expression. *Gene Therapy* 3:477–482.

Topp, L. S., Meade, L. B., and LaVail, J. H. (1994). Microtubule polarity in the peripheral processes of trigeminal ganglion cells; relevance for the retrograde transport of herpes simplex virus. *J Neurosci* 14:318–325.

Tuszynski, M. H., Gabriel, K., Gage, F. H., Suhr, S., Meyer, S., and Rosetti, A. (1996). Nerve growth factor delivery by gene transfer induces differential outgrowth of sensory, motor, and noradrenergic neurites after adult spinal cord injury. *Exp Neurol* 137:157–173.

Twyman, R. M., and Jones, E. A. (1997). Sequences in the proximal 5′ flanking region of the rat neuron-specific enolase (NSE) gene are sufficient for cell type-specific reporter gene expression. *J Mol Neurosci* 8:63–73.

Vegeto, E., Allan, G. F., Schrader, W. T., Tsai, M. J., McDonnell, D. P., O'Malley, B. W. (1992). The mechanism of RU486 antagonism is dependent on the conformation of the carboxy-terminal tail of the human progesterone receptor. *Cell* 69:703–713.

Wagner, E., Cotten, M., Foisner, R., and Birnstiel, M. (1991). Transferrin-polycation-DNA complexes: The effect of polycations on the structure of the complex and DNA DNA delivery to cells. *Proc Natl Acad Sci USA* 88:4255–4259.

Wagner, E. C. P., Zatloukal, K., Cotten, M., and Birnstiel, M. L. (1992). Influenza virus hemagglutinin HA-2 N-terminal fusogenic peptides augment gene transfer by transferrin-polylysine-DNA complexes: Toward a synthetic virus-like gene-transfer vehicle. *Proc Natl Acad Sci* USA 89:7934–7938.

Walicke, P. A. (1988). Basic and acidic fibroblast growth have trophic effects on neurons from multiple CNS regions. *J Neurosci* 8:2618–2627.

Wang, S., and Vos, J.-M. (1996). A hybrid herpes virus infectious vector based on Epstein-Barr virus and herpes simplex virus type 1 for gene transfer into human cells *in vitro* and *in vivo*. *J Virol* 70:8422–8430.

Waxman, S. G., and Kocsis, J. D. (1997). Spinal cord repair: Progress towards a daunting goal. *Neuroscientist* 3:263–269.

Weisz, O. A., and Machamer, C. E. (1994). Use of recombinant vaccinia virus vectors for cell biology. *Meth Cell Biol* 43:137–159.

Weitzman, M. D., Kyostio, S. R. M., Kotin, R. M., and Owens, R. A. (1994). Adeno-associated virus (AAV) Rep proteins mediate complex formation between AAV DNA and its integration site in human DNA. *Proc Natl Acad Sci USA* 91:5808–5812.

Wilson, S. P., Yeomans, D. C., Bender, M. A., and Glorioso, J. (1996). Delivery of enkephalins to mouse sensory neurons by a herpes virus encoding proenkephalin. 26th Ann Meeting Soc Neurosci: Abst. #540.3.

Wolfe, J. H., Deshmane, S. L., and Fraser, N. W. (1992). Herpesvirus vector gene transfer and expression of beta-glucuronidase in the central nervous system of MPS VII mice. *Nature Genetics* 1:379–384.

Wolff, J. A., Fischer, L. J., Xu, L., Jinnah, H. A., Langlais, P. J., Iuvone, P. M., O'Malley, K. L., Rosenberg, M. B., Shimohama S, Friedmann, T., and Gage, F. H. (1989). Grafting fibroblasts genetically modified to produce L-dopa in a rat model of Parkinson disease. *Proc Natl Acad Sci USA* 86:9011–9014.

Wong, S. C., Moffat, M., and O'Malley, K. L. (1994). Sequences distal to the AP1/E box motif are involved in the cell tissue-specific expression of the rat tyrosine hydroxylase gene. *J Neurochem* 62:1691–1697.

Wu, N., Watkins, S. C., Schaffer, P. A., and DeLuca, N. A. (1996). Prolonged gene expression and cell survival after infection by a herpes simplex virus mutant defective in the immediate-early genes encoding ICP4, ICP27, and ICP22. *J. Virol* 70:6358–6369.

Wychowski, C., Benichou, D., and Girard, M. (1986). A domain of SV40 capsid polypeptide VP1 that specifies migration into the cell nucleus. *EMBO J* 5:2569–2576.

Xu, X. M., Bamber, N., Li, H., Zhang, S. X., Lu, X., Aebischer, P., and Oudega, M. (1997a). Extensive axonal regrowth and reentry into host spinal cord of adult rats following the transplantation of Schwann cell containing mino-channels and infusion of two neurotrophins, BDNF, and NT-3, into distal spinal cord. *Soc Neurosci:* Abst. #1644.

Xu, X. M., Chen, A. V. G., Kleitman, N., and Bunge, M. B. (1997b). Bridging Schwann cell transplants promote axonal regeneration from both the rostral and caudal stumps of transected adult rat spinal cord. *J Neurocytol* 26:1–16.

Yang, Y., Nunes, F. A., Berencsi, K., Gonczol, E., Engelhardt, J. F., and Wilson, J. M. (1994). Inactivation of E2a in recombinant adenoviruses improves the prospect for gene therapy in cystic fibrosis. *Nat Genet* 7:362–369.

Yaworsky, P. J., Gardner, D. P., and Kappen, C. (1997). Transgenic analyses reveal developmentally regulated neuron- and muscle-specific elements in the murine neurofilament light chain gene promoter. *J Biol Chem* 272:25112–25120.

Yoon, S. O., and Chikaraishi, D. M. (1992). Tissue-specific transcription of the rat tyrosine hydroxylase gene requires synergy between an AP-1 motif and an overlapping E box-containing dyad. *Neuron* 9:55–67.

Yoshimoto, Y., Lin, Q., Collier, T. J., Frim, D. M., Breakefield, X. O., and Bohn, M. C. (1995). Astrocytes retrovirally transduced with BDNF elicit behavioral improvement in a rat model of Parkinson's disease. *Brain Research* 691:25–36.

Younkin, D. P., Tang, C.-M., Hardy, M., Roddy, U. R., Shi, Q.-Y., and Pleasure, S. J. (1993). Inducible expression of neuronal glutamate receptor channels in the NT2 human cell line. *PNAS* 90:2174–2178.

Yu, J. S., Sena-Esteves, M., Paulus, W., Breakefield, X. O., and Reeves, S. (1996). Retroviral delivery and tetracycline-dependent expression of IL-1β-converting enzyme (ICE) in a rat glioma model provides controlled induction of apoptotic death in tumor cells. *Cancer Res* 56:5423–5427.

PART III

Promoting Recovery in Neurological Disease

CHAPTER 10

Introduction to Parkinson's and Huntington's Disease

JEFFREY H. KORDOWER
Research Center for Brain Repair and Department of Neurological Sciences, Rush Presbyterian St. Luke's Medical Center, Chicago, Illinois

Disorders of the basal ganglia such as Parkinson's disease and Huntington's disease, have received extensive empirical and theoretical attention with respect to experimental therapeutic interventions. Investigations into Parkinson's disease in particular have been particularly intensive, for several reasons. First, the cardinal features of Parkinson's disease, consisting of bradykinesia, resting tremor, cogwheel rigidity, and postural instability (Paulson and Stern, 1997), are linked to the degeneration of a specific and well-understood neural circuit: the dopaminergic nigrostriatal system. Second, excellent rodent and nonhuman primate models of Parkinson's disease exist to test novel therapeutic interventions. Perhaps most importantly, levodopa pharmacotherapy, the mainstay treatment for Parkinson's disease in combination with carbidopa, produces dramatic benefit to patients with this disease (Poewe and Granata, 1997). Basic scientists who study Parkinson's disease should avail themselves of any opportunity to visit a Movement Disorders clinic to watch the extraordinary transformation of a frozen, wheelchair-bound Parkinson's disease patient in the "off state" into a functional and ambulatory individual following administration of levodopa. In addition to the potency of this pharmacological effect, the success of levodopa treatment proves an important biological principle—that

CNS Regeneration

manipulating a single neurotransmitter system (e.g., dopamine) can provide powerful benefits for patients with chronic neurodegenerative disease.

Clearly, the nigrostriatal system is not the only system that undergoes degeneration in PD. As the disease progresses, other symptoms such as dementia or autonomic dysfunction emerge that are not levodopa-responsive. Still, the benefits of this single dopaminergic therapy substantially enhance the quality of patients' lives.

The problem, however, is that the benefits of levodopa therapy diminish five to ten years following its initiation. As the disease progresses, patients require increasing doses of levodopa (Poewe and Granata, 1997). Although patients can still be turned "on" by levodopa, the benefits in treating the cardinal symptoms of Parkinson's disease are invariably offset by the emergence of drug-induced side effects, most notably disabling dyskinesias and dystonia. These side effects are likely due to the unregulated storage and release of dopamine caused by loss of host nigrostriatal neurons. As Parkinson's disease advances, the therapeutic window for clinical benefit in the absence of side effects, such as dyskinesias, can shrink to a point at which real benefit can no longer be achieved. Furthermore, the timing of benefit becomes unpredictable and patients, even after taking levodopa, can turn "off" and become frozen without warning.

Thus although levodopa in many respects can be considered a "miracle" drug, its limitations require that novel therapeutic strategies be established. In this regard, interest in surgical treatments for Parkinson's disease have been rekindled (Olanow *et al.*, 1996; Vitek, 1997). Some of these strategies such as pallidotomy or deep brain stimulation rely on lesioning or disrupting select regions of "downstream" basal ganglia circuitry, which have become overactive as a consequence of nigrostriatal degeneration. These approaches are exciting, and rigorous scientific studies are being performed to validate and characterize their early clinical promise.

In contrast to these destructive approaches, chapters in this section will focus on therapeutic interventions aimed at augmenting the function of the nigrostriatal system. Transplants of a variety of dopaminergic cells have been successful in ameliorating motor deficits in animals with striatal dopamine depletion secondary to experimental lesions or genetic mutations. Brundin and coworkers review the basic science studies that employ fetal nigral transplants in animal models of Parkinson's disease (Chap. 11). These studies demonstrate that fetal grafts can consistently survive, innervate the denervated striatum, and mediate functional recovery in a variety of rodent models of Parkinson's disease. These studies provided the foundation for clinical trials of fetal transplantation for Parkinson's disease, which are still underway and are reviewed in Chapter 14 by Bakay and colleagues. However, practical issues in donor tissue procurement and the enduring ethical debate regarding the use of human donor cells for transplantation has led to a search for alternative donor sources such as cultured stem cells or dopaminergic xenografts. Eberling and colleagues describe their studies using another approach—the transplantation of cells that have been modified to enhance nigrostriatal dopaminergic tone in animal models of Parkinson's disease (Chapter 16).

Although extensive nigrostriatal degeneration has already taken place in Parkinsonian patients at the time of clinical diagnosis, any attempt to preserve residual host circuitry rather than replace lost neurons is an attractive therapeutic approach. Hebert, Hoffer, and coworkers have been leaders in studying the structural and functional consequences of intracerebral administration of the potent dopaminergic neurotrophic factor, the glial-derived neurotrophic factor (GDNF), in animal models of Parkinson's disease. They will review their preclinical studies performed in rodent and nonhuman primate models of Parkinson's disease (Chapter 15), studies that have led to the recent initiation of a multicenter clinical trial testing the efficacy of GDNF in patients with Parkinson's disease.

As these and newer technologies arise, it is hoped that lessons learned in the neural regeneration field will guide future progress. The failure of clinical adrenal cell transplantation studies (Kordower *et al.*, 1998) was foretold by the poor survival and very modest functional improvement seen in animal models of Parkinson's disease. In contrast, the survival, innervation, and functional improvement (albeit incomplete) seen following fetal grafting in rodent and nonhuman primate models of Parkinson's disease have been translated both neuroanatomically and functionally to some patients with Parkinson's disease that have received fetal nigral implants.

Although the etiology of Parkinson's disease remains to be elucidated, the cause of Huntington's disease is clear. Huntington's disease is a fully penetrant, autosomal dominant genetic disorder whose clinical symptoms and underlying neural degeneration result from an expanded region of CAG repeats at the IT 15 locus on chromosome 4 (The Huntington's Disease Collaborative Research Group, 1993). Genetic testing, even *in utero,* allows for the indisputable identification of those individuals at risk who will ultimately suffer from this devastating and invariably fatal disorder. Unlike Parkinson's disease in which symptomatic pharmacotherapy can, for a period of time, provide substantial clinical benefit, little can be done for the relentless psychiatric and motor deficits seen in HD patients.

The exciting discovery of the gene that causes Huntington's disease has furthered research in this area, particularly with regard to novel therapeutic strategies. Pathologically, Huntington's disease is characterized by a loss of medium spiny neurons within the caudate nucleus and putamen (e.g., Ferrante *et al.*, 1985). As a result of the degeneration of these perikarya, which use gamma amino butyric acid (GABA) as well as a variety of neuropeptides as neurotransmitters, there is a loss of inhibitory innervation to the globus pallidus and substantia nigra pars reticulata, effectively disrupting basal ganglia circuitry. Two exciting approaches have been examined preclinically in an attempt to restore the structural and functional integrity of the basal ganglia in rodent and nonhuman primate models of Huntington's disease. One strategy employs replacement of the lost neurons via neural transplantation, a procedure that is now undergoing testing in clinical trials. These studies are reviewed by Bacoud-Lévi and coworkers in Chapter 17. The second strategy employs cellular delivery of neurotrophic factors, not to replace lost neurons but rather to prevent their degeneration in the first place. Clinical trials examining the ability of ciliary neurotrophic factor delivery via genetically modified cells are being planned, and this neuroprotective strategy is reviewed by Emerich and coworkers in Chapter 18.

In general, rigorous systematic studies performed in animal models have been excellent predictors of the structural and functional results seen in clinical trials in Parkinson's disease. If newly emerging treatment strategies for both Parkinson's and Huntington's disease undergo similar testing in rodent and nonhuman primate models, perhaps a new "miracle" treatment similar to levodopa therapy, but with *long-term* benefit, will be discovered for the millions of patients who suffer from these diseases.

REFERENCES

Ferrante, R. J., Kowall, N. W., Beal, M. F., Richardson, E. F., and Martin, J. B. (1985). Selective sparing of a class of striatal neurons in Huntington's disease. *Science* 230:561–563.

The Huntington's Disease Collaborative Research Group. A novel gene containing a trinucleotide repeat that is expanded and unstable on Huntington's disease chromosomes. *Cell* 72:971–983.

Kordower, J. H. Hanbury, R., and Bankiewicz, K. (1998). Neuropathology of dopaminergic transplants in patients with Parkinson's disease. In: Freeman, T. B., and Widner, H., eds., *Clinical neural transplantation*. Academic Press. San Diego CA., 51–77.

Olanow, C. W., Kordower, J. H., and Freeman, T. B. (1996). Fetal nigral transplantation for the treatment of Parkinson's disease. *Trends Neurosci* 19:102–108.

Paulson, H. L., and Stern, M. B. (1997). Clinical manifestations of Parkinson's disease. In: Watts, R. L., and Koller, W. C., eds., *Movement disorders: Neurologic principles and practice*. New York: McGraw-Hill, pp. 183–200.

Poewe, W. and Granata, R. Pharmacological treatment of Parkinson's disease. In: Watts, R. L., and Koller, W. C., eds., *Movement disorders: Neurologic principles and practice*. New York: McGraw-Hill, pp. 201–221.

Vitek, J. K. (1997). Stereotaxic surgery and deep brain stimulation for Parkinson's disease and movement disorders. In: Watts, R. L., and Koller, W. C., eds., *Movement disorders: Neurologic principles and practice*. New York: McGraw-Hill, pp. 237–256.

CHAPTER 11

Grafts of Embryonic Dopamine Neurons in Rodent Models of Parkinson's Disease

PATRIK BRUNDIN, MIA EMGÅRD, AND ULRIKA MUNDT-PETERSEN
Section for Neuronal Survival, Wallenberg Neuroscience Center, Department of Physiology and Neuroscience, Lund University, Lund, Sweden

Studies on embryonic tissue transplants in adult rodent models of Parkinson's disease have made valuable contributions to our understanding of nigrostriatal system physiology and function. Moreover they have generated important preclinical data that has contributed to the design and performance of ongoing human clinical trials. This chapter will review several aspects of this body of work in rodent models, including (1) morphological features of grafts containing dopamine neurons, (2) issues related to the survival of transplanted dopamine neurons and attempts to enhance graft survival, (3) behavioral effects of grafted dopamine neurons, with an emphasis on tests of more complex motor function in which there has been demonstration of little or no functional recovery, and (4) the limited functional effects attained when placing grafts in the nigral as opposed to striatal regions. Finally, this chapter samples the literature for studies regarding the effects of grafted dopamine neurons on the physiology of host brain neurons and speculates on the importance of these effects on achieving functional behavioral recovery.

I. A BRIEF HISTORY AND OBJECTIVES OF THE CURRENT REVIEW

In 1979 two groups independently reported that transplants of embryonic mesencephalic tissue can survive and affect the behavior of an adult recipient

CNS Regeneration

with damage to the nigrostriatal pathway (Björklund and Stenevi, 1979; Perlow *et al.*, 1979). These were the first reports in history that clearly demonstrated that grafted brain neurons can perform functions that they are normally associated with. In retrospect, it was no coincidence that these studies were conducted within the nigrostriatal dopamine system. In the 1970s, the catecholamine histofluorescence method constituted the most powerful tool available for the study of the cell bodies and axons of a select population of neurons. Furthermore, it was possible to selectively lesion catecholaminergic neurons using the neurotoxin 6-hydroxydopamine (6-OHDA). The availability of the robust, unilateral 6-OHDA lesion rat model developed by Ungerstedt and co-workers in the late 1960s (Ungerstedt, 1968), in combination with drugs that had well-defined modes of action on the dopamine system, made the rat dopamine system an ideal target for transplantation studies. By the seventies the strong clinical relevance of grafted dopamine neurons to Parkinson's disease had already added impetus to the research. Today it is clear that grafting studies performed in the dopamine system in many ways have taken a pioneering role within the whole neural transplantation field. Due to the fortunate coincidence of the existence of powerful morphological techniques, well characterized animal models and the clear clinical importance, research on grafted dopamine neurons has provided many important insights that have proven relevant also to other systems.

The objectives of this chapter are, first, to briefly exemplify rodent models of Parkinson's disease that have been used in transplantation studies and to recount basic transplantation methods. Emphasis is placed on studies of grafts in adult, as opposed to neonatal, hosts because of the greater clinical relevance. This chapter also describes morphological features of grafts containing dopamine neurons; the issue of a low survival rate of transplanted dopamine neurons and attempts to circumvent this problem. In a section on behavioral effects of grafted dopamine neurons, emphasis is placed on tests of more complex motor function in which there has been demonstration of little or no functional recovery with nigral implants, and a discussion of why grafts are unable to reverse all types of complex functional deficits. The chapter samples the literature for studies regarding the effects of grafted dopamine neurons on the physiology of the host brain neurons and speculates on the importance of these effects for functional behavioral recovery. The chapter also focuses on the role of nigral grafts that are implanted homotopically in the mesencephalon, for example, near their location in the normal brain. Finally, there is a brief discussion about the future importance of grafting in rodent models of Parkinson's disease.

II. ANIMAL MODELS USED AS RECIPIENTS FOR GRAFTS OF DOPAMINE NEURONS

Because there is more marked degeneration of dopamine neurons in the substantia nigra than in the ventral tegmental area in Parkinson's disease (Agid *et al.*, 1987), this chapter focuses on transplantation studies performed within the mesostriatal as opposed to the mesolimbocortical dopamine system. Therefore, the behavioral deficits in the lesion models described here focus on functions linked to the caudate-putamen proper, that is dorsal aspects of the striatal complex. However, the reader should be aware of that there are also a number of publications on transplants in the mesolimbocortical dopamine system (for review see Reading, 1994).

The neurotoxin 6-OHDA is selectively taken up by catecholaminergic neurons and kills them through a mechanism that is still not fully elucidated, but may well include oxidative stress (Glinka *et al.*, 1997). When injected in adequate amounts into the brain, it can cause extensive and permanent damage to catecholaminergic neurons, such as those giving rise to the nigrostriatal pathway. The relative specificity of 6-OHDA and the possibility to routinely create almost complete lesions, which provide little scope for biochemical or behavioral spontaneous recovery, make 6-OHDA a highly useful toxin.

The unilateral 6-OHDA lesion of the mesostriatal pathway is the gold standard animal model in studies of grafted dopamine neurons. Although rats with bilateral 6-OHDA lesions can be seen as a closer analogy to classical Parkinson's disease, both with regard to a bilateral dopamine depletion and the occurrence of marked akinesia, rats with such lesions are difficult to study because they develop severe aphagia and adipsia (Ungerstedt, 1971a). In contrast, rats with a unilateral lesion both drink and eat and do not suffer from severe akinesia. From a scientific point of view they also have a major advantage in that they present unilateral behavioral deficits. Thus, many tests of motor and sensory function utilize the possibility to directly compare the two sides of the body in the same rat. For example, rats with unilateral lesions exhibit rotational motor asymmetry that is readily quantified and can be amplified by administration of drugs that release dopamine, such as amphetamine (Ungerstedt and Arbuthnott, 1970), or preferentially stimulate supersensitive dopamine receptors directly, such as apomorphine (Ungerstedt, 1971b), in the denervated striatum. Such rats also display easily quantifiable deficits in fine motor skills in the forepaw on the side of the body contralateral to the lesion (Montoya *et al.*, 1991), and have problems in moving the contralateral forepaw when provoked in the so-called "stepping test" (Schallert *et al.*, 1992). The rats also display deficiencies in their ability to integrate sensory stimuli into motor responses, both when the stimuli are simply applied by poking with a sharp object on the side contralateral to the lesion (termed sensory neglect) (Marshall *et al.*, 1980), or when they are presented to rats that are engaged in another activity (eating) in the so-called "disengage" test (Schallert and Hall, 1988).

There have also been a number of studies on grafted dopamine neurons in mouse models of Parkinson's disease. Examples of lesions in the mouse dopamine system that have been utilized are unilateral intrastriatal 6-OHDA lesions (Brundin *et al.*, 1986b); 1-methyl-4-phenyl-1,2,3,6-tetrahydropyridine (MPTP) lesions (Zuddas *et al.*, 1991); and genetically determined degeneration in the Weaver mutant mouse (Triarhou *et al.*, 1990). In this chapter, emphasis is placed on studies in rat models.

III. DONOR TISSUE AND TRANSPLANTATION TECHNIQUES

A. Donor Tissue

Essentially only embryonic, as opposed to mature, dopamine neurons survive transplantation. Furthermore, it is evident that there is an optimal embryonic donor age, that is, beyond a certain developmental stage the survival rate of grafted dopamine neurons drops drastically. If the graft tissue is mechanically dissociated into a "cell suspension," the survival of dopamine neurons is very poor if the donor rat embryos are older than 15 to 16 days (Brundin *et al.*, 1988a), whereas grafts of solid tissue seem to survive well also from slightly older donors (Simonds and Freed, 1990). It is possible to culture dopamine neurons prior to their implantation, either in dissociated monolayer cultures (Brundin *et al.*, 1988a), reaggregates (Strecker *et al.*, 1989), or as solid explants, such as roller tube cultures (Spenger *et al.*, 1996). However, it seems that prolonged maintenance of mesencephalic donor tissue in culture reduces its survival following grafting. Similarly, cryopreservation of mesencephalic donor tissue prior to transplantation is associated with a marked reduction in survival of dopamine neurons (Sauer *et al.*, 1992). Possibly the best pre-graft storage method is also the most simple, because it is now clear that mesencephalic donor tissue can be maintained at 4°C for up to 8 days in a balanced medium supplemented with a compound that reduces lipid peroxidation, without there being an additional loss of graft viability (Grasbon-Frodl *et al.*, 1996).

B. Transplantation Methods

Several different transplantation methods have been employed to deliver embryonic mesencephalic dopamine neurons to the adult rodent striatum. For example, small solid pieces of embryonic ventral mesencephalic tissue can be implanted into the lateral ventricle (Perlow *et al.*, 1979; Simonds and Freed, 1990), adjacent to the striatum, or directly into the brain parenchyma (Strömberg *et al.*, 1989). Alternatively, in many early studies the solid grafts were placed in a premade cavity in the neocortex, in which a vascular bed scar had developed, overlying the caudate-putamen (Björklund and Stenevi, 1979). Perhaps the most widespread technique has been the stereotaxic injection of disrupted mesencephalic tissue into the striatal parenchyma (Björklund *et al.*, 1980; Brundin and Strecker, 1991). The degree of tissue disruption has varied from simply cutting the mesencephalic tissue into small pieces (Strömberg *et al.*, 1989) to subjecting the tissue to enzymatic digestion prior to dissociating it into a single cell suspension (Nikkhah *et al.*, 1993) or a mixture of single cells and aggregates (Brundin *et al.*, 1985). The survival rate of grafted dopamine neurons when using the latter of these three methods has been extensively studied and is only in the range of 5 to 20% (Sauer and Brundin, 1991; Nakao *et al.*, 1994) depending on the precise protocol used. A small number of studies have examined the survival, growth, and function of mesencephalic tissue implanted adjacent to the damaged substantia nigra in adult hosts (discussed later). In these cases, a tissue dissociation protocol has invariably been employed when preparing the donor tissue.

IV. ANATOMICAL AND PHYSIOLOGICAL FEATURES OF NIGRAL GRAFTS

A. Anatomical Features

When nigral tissue is implanted directly into the host striatum, there is a temporary upregulation of various markers for astrogliosis and morphological labels that identify activated microglia (Barker *et al.*, 1996; Duan *et al.*, 1995) around the graft tissue. By around 6 weeks after surgery the glial reaction has progressively resided to almost normal levels. The response is a result of the implantation trauma and not an immunological reaction, because it also occurs following syngeneic (transplants between genetically identical individuals) transplantation.

Dopamine neurons transplanted to the striatum typically arrange themselves around the perimeter of the graft tissue (Fig. 1), effectively innervate the target host striatum (Doucet *et al.*, 1989), and can form synapses with the host striatal neurons (e.g., Freund *et al.*, 1985; Mahalik *et al.*, 1985). When implanted directly into the rat striatum, they extend axons that typically reach 1 to 1.5 mm from the host-graft border, and form a terminal network of about 50% of normal density close to the implant (Doucet *et al.*, 1989). The extent

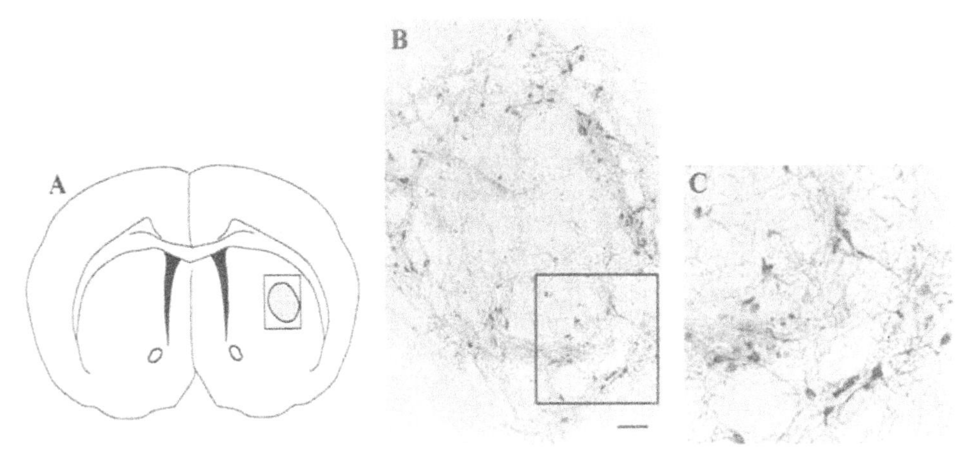

FIGURE 1 (A) Schematic drawing of a coronal section illustrating the placement of a nigral graft (gray shaded area) in the striatum of an adult rat. The box surrounding the graft represents the region visualized in the photograph in panel B. (B) Photomicrograph of the nigral graft shown in panel A. The section is stained with an antibody against tyrosine hydroxylase and reveals several darkly labeled dopamine neurons that are primarily located around the perimeter of the transplant. The host striatum is dopamine-denervated due to prior destruction of the host substantia nigra by an intracerebral injection of 6-hydroxydopamine. Box shows the region displayed in higher magnification in panel C. Scale bar = 100 μm. (C) Several grafted dopamine neuronal cell bodies with processes are visible in this higher magnification of the inset box from panel B. (Data from Emgård-Mattson *et al.*, 1997.)

of innervation of the host striatum is greater if the target is denervated using 6-OHDA prior to grafting (Doucet *et al.*, 1990), and there is also a positive correlation between the number of surviving grafted dopamine neurons and volume of host striatum that is reached by graft-derived fibers (Nakao *et al.*, 1995). The addition of embryonic striatal tissue to embryonic mesencephalic grafts, in so-called cografts, can augment neurite extension from grafted dopamine neurons (Brundin *et al.*, 1986a; Costantini *et al.*, 1994; Emgård-Mattson *et al.*, 1997, Yurek *et al.*, 1990). However, it seems that the observed increase in dopaminergic fiber outgrowth in the presence of embryonic striatal tissue is largely confined to the cografted striatal tissue, rather than extending into the adjacent adult host striatum. Several studies have attempted to increase fiber outgrowth from mesencephalic implants by addition of trophic factors. Basic fibroblast growth factor (bFGF) (Mayer *et al.*, 1993; Takayama *et al.*, 1995; Zeng *et al.*, 1996) and brain-derived neurotrophic factor (BDNF) (Yurek *et al.*, 1995) have both been shown to increase axon growth from mesencephalic transplants. More recently, glial cell line-derived neurotrophic factor (GDNF) has been found to increase the fiber outgrowth from nigral transplants (Rosenblad *et al.*, 1996; Sinclair *et al.*, 1996). As will be described in more detail later, GDNF and bFGF also increase the number of surviving dopamine neurons in the transplants, and to what extent the observed increase in fiber outgrowth is simply secondary to an increased number of surviving dopamine neurons in the graft, rather than a specific neurite outgrowth-promoting effect, is not known.

A few studies have described growth of neuronal connections from the host brain into mesencephalic rat grafts. Anatomical tracing studies indicate that serotonergic axons from the brain stem raphe nuclei innervate intrastriatal transplants, and also a few cortical afferents seem to contact grafted mesencephalic neurons (Chkirate *et al.*, 1993; Doucet *et al.*, 1989). In addition, there is electrophysiological evidence that host cortex, striatum, and brain stem nuclei can innervate transplants of nigral tissue (Arbuthnott *et al.*, 1985; Fisher *et al.*, 1991).

In studies in which the nigral transplants have been placed into the mesencephalon of adult hosts, there is survival of the dopaminergic neuron population, but the fiber outgrowth has typically been relatively sparse (Björklund *et al.*, 1983; Mendez *et al.*, 1996; Nikkhah *et al.*, 1994). The connectivity and functional capacity of such grafts will be dealt with in more detail in the section entitled "Effects of Grafting Homotopically to the Mesencephalon."

B. Physiological Features

The survival of dopamine neurons placed in a 6-OHDA denervated striatum is also reflected by restoration of striatal dopamine levels to around 10 to 30% of normal (Schmidt *et al.*, 1982, 1983; Nikkhah *et al.*, 1993). Examination of the ratio between levels of the dopamine metabolite DOPAC and dopamine reveals ratios that are 50 to 200% higher in grafted striata compared to normal, suggesting that the grafted neurons have an increased transmitter turnover (Schmidt *et al.*, 1982, 1983). Electrophysiological data have demonstrated that there are spontaneously active neurons in grafted mesencephalic tissue and some of them possess characteristics of dopamine neurons (Arbuthnott *et al.*, 1985; Fisher *et al.*, 1991; Wuerthele *et al.*, 1981). Indeed, as assessed using the *in vivo* microdialysis technique, transplanted nigral neurons have a high capacity to spontaneously release dopamine in the host striatum (Brundin *et al.*, 1988b; Strecker *et al.*, 1987; Zetterström *et al.*, 1986) and increase this release in response to amphetamine (Brundin *et al.*, 1988b; Zetterström *et al.*, 1986). The electrophysiological activity and transmitter release are under some degree of autoregulatory control, because the administration of dopamine receptor agonists reduces spontaneous firing (Wuerthele *et al.*, 1981) and levels of dopamine release from the grafts (Strecker *et al.*, 1987).

V. ATTEMPTS TO INCREASE THE SURVIVAL OF GRAFTED DOPAMINERGIC NEURONS

Undoubtedly the number of dopamine neurons that survive in grafts is important for the extent of dopaminergic striatal reinnervation that is achieved, and, as will be elaborated further in a later section, also for the degree and speed of onset of functional recovery. Because the basic survival rate of embryonic dopamine neurons unfortunately is as low as 5 to 20% in grafts of dissociated tissue, there seems to be great scope for improvement. The precise reason why the dopamine neurons in the graft tissue succumb is not known, although there is recent evidence that at least some of them die through apoptosis during the first 2 to 3 weeks after transplantation surgery (Mahalik *et al.*, 1994). Furthermore, studies that have assessed the number of surviving neurons at different time points after implantation suggest that the vast majority of the dopamine neurons that die in the grafts do so within one week following surgery (Duan *et al.*, 1995; Barker *et al.*, 1996).

Throughout the 1980s there were several attempts to increase the yield of dopamine neurons, albeit with little success. The emergence of several neurotrophic factors that were active on dopamine neurons cultured *in vitro* provided new hope. As mentioned earlier, there have been several attempts to improve transplant outcome by the addition of trophic factors to the graft tissue. A fibroblast cell line genetically engineered to produce bFGF that was cografted with embryonic mesencephalic dopamine neurons was found to increase the survival rate of the dopamine neurons severalfold in the rat striatum (Takayama *et al.*, 1995). Intracerebral infusions or direct supplementation of the nigral graft tissue with bFGF also increase graft survival (Mayer *et al.*, 1993; Zeng *et al.*, 1996). Similarly, recent studies have shown that GDNF not only enhances fiber outgrowth from intracerebrally transplanted dopamine neurons, but can also markedly improve their survival rate (Rosenblad *et al.*, 1996; Sinclair *et al.*, 1996). However, grafted dopamine neurons do not respond to neurotrophic factors in a nonselective fashion, because several neurotrophic factors, some of which are effective on dopamine neurons *in vitro*, have been found not to improve the survival of grafted dopamine neurons. For example, nerve growth factor (Sauer *et al.*, 1993; Zeng *et al.*, 1996), BDNF (Sauer *et al.*, 1993; Yurek *et al.*, 1996), neurotrophin-3 (Haque *et al.*, 1996a), and kFGF (Haque *et al.*, 1996a) do not increase dopamine neuron survival in different grafting studies.

In parallel to these studies on effects of neurotrophic factors, another recent series of experiments have identified oxidative stress as a cause of neuronal death in the nigral implants. Thus, grafts of transgenic mesencephalic tissue overexpressing the scavenger enzyme superoxide dismutase survive four times better than transplants prepared from wildtype littermate embryos (Nakao *et al.*, 1995). Also, the use of lazaroids (a class of compounds designed to reduce lipid peroxidation, one consequence of oxidative stress) can improve the survival of both cultured (Frodl *et al.*, 1994; Othberg *et al.*, 1997) and grafted (Nakao *et al.*, 1994) dopamine neurons two- to three-fold. These latter experiments have inspired the use of lazaroids in ongoing clinical trials with nigral grafts in parkinsonian patients in Lund, Sweden.

VI. EFFECTS OF GRAFTED DOPAMINE NEURONS ON BEHAVIOR

There are a large number of reports illustrating the capacity of grafted dopamine neurons to affect behavior in rodents with a unilateral 6-OHDA lesion of the nigrostriatal pathway. The vast majority of these studies have concerned drug-induced rotation for which nigral grafts are very effective. Thus, amphetamine-induced rotational asymmetry can be reversed by 50% by as few as 400 to 500 surviving grafted tyrosine hydroxylase immunopositive (dopaminergic) neurons (Sauer and Brundin, 1991), a number that represents only around 2 to 3% of the normal number of dopaminergic neurons innervating the striatum. There is a clear correlation between survival of the grafted dopamine neurons and the extent of recovery in the rotation test (Brundin *et al.*, 1985, 1988a; Sauer and Brundin, 1991; Nakao *et al.*, 1994, 1995). Why so few neurons are necessary to completely reverse the lesion induced under amphetamine influence is still not fully understood. Among several possible explanations, the following concepts are attractive: the grafted dopamine neurons lack the normal host regulatory input, are metabolically hyperactive, and release increased amounts of dopamine per neuron following amphetamine stimulation; the host striatal neurons are supersensitive and possess increased numbers of dopamine receptors; rotational behavior is governed from only a small part of the striatum and reinnervation restricted to only this crucial region is sufficient to elicit the behavioral response. Nigral transplants also ameliorate rotation in response to dopamine receptor agonists such as apomorphine (for detailed review see Brundin *et al.*, 1994). Typically there is not a complete reversal of the motor asymmetry in tests using directly acting dopamine receptor agonists, which is in stark contrast to grafted rats subjected to amphetamine-induced rotation, which often even are "overcompensated" (i.e., they rotate more contralaterally than ipsilaterally to the lesion; Brundin *et al.*, 1994).

Spontaneous (not under the influence of a drug) behavioral deficits can also be reversed in rats with unilateral 6-OHDA lesions. For example, sensory neglect, which appears contralateral to the lesion, is significantly ameliorated by mesencephalic implants that innervate the ventrolateral caudate-putamen

(Fig. 2) (Dunnett *et al.*, 1981a, 1983a; Mandel *et al.*, 1990). Interestingly, this location of graft innervation is relatively ineffective at reversing deficits in rotational behavior described earlier, which requires that the transplants recon-

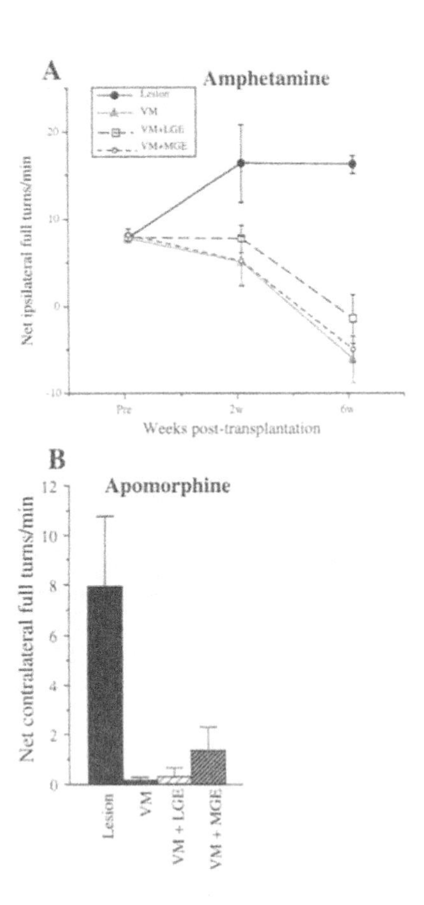

FIGURE 2 (A) Changes in amphetamine-induced (2.5 mg/kg, ip) motor asymmetry in rats with unilateral 6-OHDA lesions grafted with nigral tissue. One group of rats received pure ventral mesencephalic tissue (VM), whereas the two other groups were implanted with "cografts" that contained both ventral mesencephalic tissue and tissue from the striatal anlagen in the lateral ganglionic eminence (VM+LGE) or from the adjacent medial ganglionic eminence (VM+MGE). One group of rats with unilateral 6-OHDA lesions was left as control (Lesion). By 6 weeks after graft surgery, rats in all three grafted groups display complete reversal of the ipsilateral rotational asymmetry induced by the lesion. Indeed the mean rotation scores are in fact lower than zero, which indicates that several rats are overcompensated, i.e., display more turns contralateral than ipsilateral to lesion. (B) The same groups of rats as in panel A are tested for apomorphine-induced (0.05 mg/kg, s.c.) rotation at 13 to 15 weeks posttransplantation. Again, there is virtually complete amelioration of the 6-OHDA-induced motor response, this time in response to a directly acting dopamine receptor agonist. (Data from Emgård-Mattson *et al.*, 1997.)

stitute the dopamine innervation of more dorsal and central regions of the host striatum (Dunnett *et al.*, 1981a, 1983a; Mandel *et al.*, 1990). Conversely, grafts in such dorsal–central areas do not ameliorate deficits in sensorimotor neglect, which clearly illustrates that the rat striatum is functionally heterogeneous. Although the experiments on sensory neglect illustrate that nigral grafts have the capacity to affect spontaneous behaviors in rats, there are several examples of tests in which such grafts are ineffective. For instance, "disengage behavior" (Schallert and Hall, 1988; described in an earlier section) may not be restored, even in grafted rats that display recovery in the sensory neglect test (Mandel *et al.*, 1990), although one study reports partial amelioration of this deficit (Nikkhah *et al.*, 1993). Similarly it seems difficult to achieve transplant-induced recovery in contralateral forepaw function in rats with unilateral 6-OHDA lesions (Abrous *et al.*, 1993; Emgård-Mattson *et al.*, 1997; Montoya *et al.*, 1990; Olsson *et al.*, 1995) in the so-called staircase test (Fig. 3). In this test of fine motor function, food-deprived rats retrieve, with either the left or right paw, food pellets from a series of platforms that are increasingly difficult to reach in a stepwise fashion. One study purports that a subset of transplanted rats can improve forepaw function contralateral to the lesion also in this test when multiple, small intrastriatal implants of nigral tissue are made (Nikkhah *et al.*, 1993), but in a follow-up study the same investigators did not detect any beneficial effect of similar grafts in the staircase test (Olsson *et al.*, 1995). An interesting recent addition to the repertoire of behavioral tests

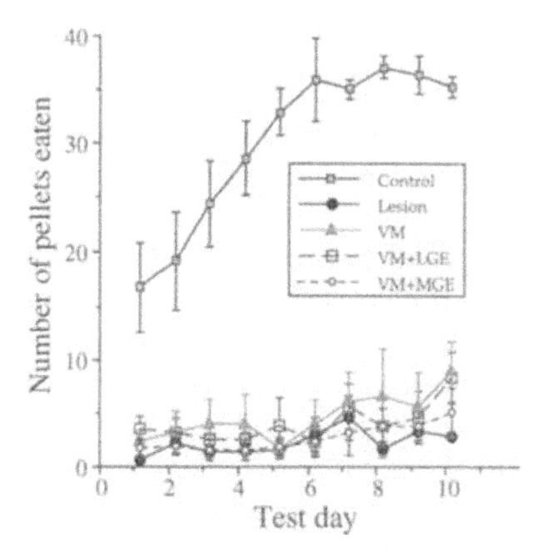

FIGURE 3 The same grafted and lesion control groups as described in Fig. 2 are tested over 10 days for fine motor function in the forepaw contralateral to the lesion using the so-called staircase test, at 9 weeks after transplantation surgery. In addition, one group of normal control rats were studied. Interestingly, although the grafted groups all displayed beavioral recovery in the two tests of drug-induced rotation described in Fig. 2, there was no amelioration of the deficit in forepaw function in the more complex staircase task. (Data from Emgård-Mattson *et al.*, 1997.)

in this model is the "stepping test" (Schallert *et al.*, 1992). In this examination of spontaneous motor function of the forepaws, there is indeed clear evidence of beneficial effects of intrastriatal nigral transplants (Olsson *et al.*, 1995).

Grafted rats with bilateral 6-OHDA lesions have also been the subject of extensive behavioral investigation. As indicated earlier, these animals exhibit regulatory impairments of aphagia and adipsia (Ungerstedt, 1971a). These spontaneous behaviors have not been possible to restore when grafting into adult recipients (Dunnett *et al.*, 1981b, 1983b), but there are reports that nigral grafts placed into neonates can counteract the appearance of such symptoms when the hosts sustain bilateral dopamine depletion as adults (Schwarz and Freed, 1987). This may reflect an increased degree of anatomical graft-host integration, which is likely to occur in the more plastic neonatal brain environment. Even though regulatory impairments are not amenable to transplant-induced recovery, it is clear that nigral grafts placed in adult rats with bilateral 6-OHDA lesions are not functionally ineffective in the absence of activating drugs and are able to, for example, increase spontaneous locomotor behavior (Dunnett *et al.*, 1983b).

VII. EFFECTS OF GRAFTED DOPAMINE NEURONS ON HOST NEURONS

Following 6-OHDA lesion of the nigrostriatal pathway there is an increase in the spontaneous firing activity of striatal neurons. Nigral grafts can reduce this activation and striatal neurons located within graft-innervated areas display firing rates that are only about one-half to one-third of those recorded from areas outside this zone (Di Loreto *et al.*, 1996; Strömberg *et al.*, 1985).

More insight into the profound effects of nigral grafts on the physiology of host striatal neurons has been gained from the study of striatal neurotransmitters and receptors. Disruption of dopamine neurotransmission in the striatum by a nigral lesion causes an increased activity in GABA–enkephalinergic cells that project to the pallidum (the rodent analogue to the external pallidal segment), and a decreased activity in GABA–dynorphin substance P-ergic neurons that project to the substantia nigra and the entopenduncular nucleus (the rodent analogue to the internal pallidal segment) (for review see Albin *et al.*, 1989). These changes are accompanied by pronounced changes in the expression of striatal neuropeptides (e.g., enkephalin) and increased glutamic acid decarboxylase (GAD) activity, indicating increased GABA production and release. There is also supersensitivity of postsynaptic dopamine receptors in response to a 6-OHDA lesion, which is particularly consistent for the D2 dopamine receptor subtype. Mesencephalic transplants effectively counteract several of the secondary effects observed in striatal neurons after disruption of dopaminergic neurotransmission, such as increases in GAD mRNA expression (Segovia *et al.*, 1991; Cenci *et al.*, 1997), preproenkephalin mRNA expression (Bal *et al.*, 1993; Cadet *et al.*, 1991; Cenci *et al.*, 1993; Mendez *et al.*, 1993; Sirinathsinghji and Dunnett, 1991), and enkephalin immunoreactivity (Manier *et al.*, 1991). In addition, several studies (see Blunt *et al.*, 1992; Chritin *et al.*, 1992; Savasta *et al.*, 1992) have examined dopamine D2 receptor binding in the 6-OHDA denervated and grafted striatum and have found that the receptor binding increases significantly in response to denervation; the increase is counteracted by nigral grafts and displays a tendency toward being overcompensated by the transplants. Interestingly, some of the studies mentioned here describe that the graft-induced effects on striatal neurons extend further beyond the area reached by dopaminergic innervation from the grafted cells (Chritin *et al.*, 1992; Manier *et al.*, 1991; Sirinathsinghji and Dunnett, 1991). Possibly this is mediated via striatal interneurons rather than diffusion of dopamine over long distances, because it is likely to be metabolized rapidly in the extracellular space (for discussion see Brundin *et al.*, 1987). A similar pattern of widespread striatal changes is seen for the expression of the proto-oncogene *c-fos*. Thus, administration of direct or indirect dopaminergic agonists, which induce supranormal expression of *c-fos* in the 6-OHDA denervated striatum, elicits only a normal expression of Fos immunoreactivity throughout the striatum in rats with nigral transplants (Abrous *et al.*, 1992; Cenci *et al.*, 1992b). One study (Cenci *et al.*, 1992a) that combined Fos immunohistochemistry with neuroanatomical tracing indicates that Fos is preferentially expressed in the substance P-ergic *striatonigral* cell population, rather than the enkephalinergic *striatopallidal* neurons. Consequently, the expression of mRNA for preprotachykinin, the precursor molecule for substance P that is downregulated in striatonigral neurons following dopamine denervation, is also normalized following implantation of embryonic dopamine neurons (Cenci *et al.*, 1993; Mendez *et al.*, 1993). However, in this case the normalization is restricted to the portion of the caudate-putamen that is reinnervated by the transplant (Cenci *et al.*, 1993).

In summary, there is ample evidence that transplanted dopamine neurons have extensive effects on neurotransmission in striatal neurons in the recipient brain. In some cases, these effects extend into striatal regions that lie beyond the zone actually innervated by graft-derived dopaminergic fibers and may therefore be mediated indirectly, via local striatal interneurons. On the whole, the findings suggest that the dopaminergic innervation provided by nigral grafts has profound, chronic effects on host striatal neurons. A long-standing spontaneous dopamine release at specialized synaptic sites is likely to underlie these changes and almost certainly plays a crucial role in mediating the recovery in spontaneous behaviors. Although it has not yet been addressed in specific experiments, it is likely that grafts that do not grow a sufficiently dense dopaminergic terminal plexus to change gene expression in host striatal neurons are unlikely to ameliorate deficits in spontaneous motor functions.

VIII. EFFECTS OF GRAFTING HOMOTOPICALLY TO THE MESENCEPHALON

Most studies with nigral transplants have been performed by implanting the tissue ectopically into the striatum. Of course, this reduces the chances of the transplant receiving appropriate anatomical, regulatory inputs from the host, which has been considered one contributory factor to why ectopic grafts of dopamine neurons are unable to reverse certain spontaneous lesion-induced behavioral deficits. Therefore, early studies addressed this problem (Björklund *et al.*, 1983) and implanted mesencephalic tissue in the host substantia nigra and at sites along the course of the nigrostriatal pathway. They observed only a limited fiber outgrowth locally, and saw no functional effects in the hosts. Since then, several methods have been applied to increase fiber outgrowth from nigral grafts placed in the mesencephalon and to guide the dopaminergic fibers to their target. One such early attempt involved the use of a peripheral nerve graft. A sciatic nerve was placed outside the skull and had one end placed adjacent to the solid neural graft on the dorsal mesencephalon, with the other stump inserted into the dorsal aspect of the caudate-putamen (Gage *et al.*, 1985). Some dopaminergic neurons extended fibers along the full length of the peripheral nerve bridge, but they rarely gave rise to a functional terminal network in the 6-OHDA denervated striatum. Dunnett and colleagues (1989) grafted dissociated nigral tissue into mesencephalon and proceeded to implant various types of tissue or substrates along an oblique intracerebral injection tract connecting the graft with the striatum. Implants of embryonic olfactory bulb, cultured astrocytes, or laminin-coated microspheres were all ineffective environments for axonal growth from the grafted nigral neurons, whereas tracts of embryonic striatal tissue supported some growth that in a few rats may have exerted small effects on the amphetamine-induced rotation test (Dunnett *et al.*, 1989). More recently, the same team has employed fibroblast growth factor-4-transfected RN-22 schwannoma cells to construct the "bridge" from the mesencephalic graft site to the striatum and has seen several dopaminergic axons innervating the host striatum, causing a reduction in amphetamine-induced rotation (Brecknell *et al.*, 1996). Another recent study by Zhou and associates (1996) has demonstrated that the scar tissue formed after an excitotoxic lesion can also form a hospitable environment for the growth of grafted neurons. Thus, they grafted nigral tissue into the mesencephalon of rats with unilateral 6-OHDA lesions and injected the excitotoxin kainic acid along an oblique pathway connecting the intramesencephalic graft and the striatum.

They found evidence for a graft dopaminergic innervation of the striatum, growth of nondopaminergic grafted neurons into the host striatum and, finally, projections from host striatal neurons back into the graft. *In vivo* voltammetry, which can monitor changes in extracellular levels of dopamine and its metabolites, suggested that there was recovery of striatal dopamine release in the grafted rats and in some cases there was also evidence of functional effects in the amphetamine-induced rotation test (Zhou *et al.*, 1996). Although this experiment illustrates the validity of the approach of grafting nigral tissue homotopically, additional experiments are needed to fully assess the functional capacity of the grafts; further, the use of excitotoxin gives the technique a limited clinical value.

Results from recent studies that have revisited the original paradigm of grafting nigral tissue homotopically to the mesencephalon without providing an artificial "bridge" are of particular interest. It seems that notable fiber outgrowth into the host striatum from homotopic nigral grafts of rat donor origin can only be obtained if the host is neonatal (Nikkhah *et al.*, 1995), whereas in adult hosts the grafted neurons rarely grow rostrally toward the striatum (Mendez *et al.*, 1996; Nikkhah *et al.*, 1994). Interestingly, such intranigral grafts in adult hosts that only provide the local mesencephalic environment (e.g., the substantia nigra pars reticulata) with a dopaminergic innervation can partially reverse contralateral rotational asymmetry induced by directly acting dopamine receptor agonists (Nikkhah *et al.*, 1994; Olsson *et al.*, 1995), but are totally ineffective on the ipsilateral motor asymmetry induced by amphetamine administration (Mendez *et al.*, 1996; Nikkhah *et al.*, 1994; Olsson *et al.*, 1995). In tests of spontaneous behavior, the effects of the homotopic nigral grafts are also minor. There is only partial amelioration, less marked than with intrastriatal nigral transplant, of forelimb akinesia in the "stepping test," and there are no improvement in complex forepaw function in the staircase test following intranigral implantation (Olsson *et al.*, 1995). In contrast to transplants of rat donor origin, human dopamine neurons xenografted to the rat mesencephalon can extend axons all the way to the host striatum (Wictorin *et al.*, 1990). However, the behavioral capacity of this type of graft has not been evaluated.

In summary, it seems that there are some functional effects of restoring dopaminergic neurotransmission in the substantia pars reticulata, best illustrated by reductions in dopamine agonist-induced rotation. Nevertheless, the inability of axons from homotopically placed nigral rat grafts to reach the host striatum limits their utility in adult transplant recipients. Recent studies on various strategies to provide a hospitable bridge on which embryonic nigral neurons can grow to reinnervate the striatum show some promise. However, it remains to be seen whether the repair of nigrostriatal circuitry with the "bridge" approach has anything to offer in functional terms that cannot already be achieved with intrastriatal implants of mesencephalic tissue.

IX. FUTURE DIRECTIONS

Transplantation experiments in the nigrostriatal system have played a key role in the development of this research field. What are the reasons to believe that this position will be maintained also in the future? For investigators studying neural transplantation from a basic science point of view, the rat with damage to the substantia nigra has two main assets. First, as mentioned in the introduction, there are several biological and technical experimental features of the nigrostriatal dopamine system that make it a very attractive model for the study of brain repair. These features facilitate the study and understanding of, for example, basic mechanisms that promote neuronal survival and stimulate/inhibit axonal outgrowth in the adult brain; the correlation between brain integrity and behavioral function; and the possibility to use gene transfer technology to promote repair or genetically engineered nonneuronal cells as donor tissue.

Second, basic research with grafts of embryonic dopamine neurons in rats are likely to continue to play a crucial role in the development of ongoing clinical neural grafting trials. Today the variability in graft survival and clinical outcome between patients, shortage of suitable human embryonic donor material, and incomplete symptomatic relief in even successful cases, all constitute problems (Lindvall, 1997). Grafting experiments in rodents can contribute to improving the survival of grafted human dopamine neurons (e.g., by use of antioxidants or trophic factors) and to enhancing the volume of host brain that they can innervate (e.g., by use of neurotrophic factors). Also the scientific development of alternative sources of donor tissue, such as porcine neurons, requires the use of rodent models. However, experiments geared at understanding why certain parkinsonian symptoms (e.g., tremor, gait disturbance) respond less to graft therapy in patients may be more difficult to perform in rodents. Although some insight may be gained from studying more complex motor functions in rodents and alternative grafting techniques, there is a large difference between rodent and human motor functions. Experiments conducted in nonhuman parkinsonian primates, and the clinical trials themselves, are more likely to provide valuable information in this case.

ACKNOWLEDGMENTS

The authors' own work described in this chapter was supported by the Swedish Medical Research Council, The Medical Faculty at Lund University and the Kock, Crafoord and Segerfalk Foundations.

REFERENCES

Abrous, D. N., Torres, E. M., Annett, L. E., Reading, P. J., and Dunnett S. B. (1992). Intrastriatal dopamine-rich grafts induce a hyperexpression of Fos protein when challenged with amphetamine. *Exp Brain Res* 91:181–190.

Abrous, D. N., Shaltot, A. R., Torres, E. M., and Dunnett, S. B. (1993). Dopamine-rich grafts in the neostriatum and/or nucleus accumbens: Effects on drug-induced behaviours and skilled paw-reaching. *Neuroscience* 53:187–197.

Agid, Y., Javoy-Agid, F., and Ruberg, M. (1987). Biochemistry of neurotransmitter in Parkinson's disease. In: Marsden, C. D., Fahn, S., Eds)., *Movement disorder,* Vol. 2. London: Butterworths, pp. 166–230.

Albin, R. L., Young, A. B., and Penney, J. B. (1989). The functional anatomy of basal ganglia disorders. *Trends Neurosci* 12:366–375.

Arbuthnott, G., Dunnett, S. B., and MacLeod, N. (1985). Electrophysiological properties of single units in mesencephalic transplants in rat brain. *Neurosci Lett* 57:205–210.

Bal, A., Savasta, M., Chritin, M., Mennicken, F., Abrous, D. N., Le Moal, M., Feuerstein, C., and Herman, J. P. (1993). Transplantation of fetal nigral cells reverses the increase of preproenkephalin mRNA levels in the rat striatum caused by 6-OHDA lesion of the dopaminergic nigrostriatal pathway: A quantitative *in situ* hybridization study. *Mol Brain Res* 18:221–227.

Barker, R. A., Dunnett, S, B., Faissner, A., and Fawcett, J. W. (1996). The time course of loss of dopamine neurons and the gliotic reaction surrounding grafts of embryonic mesencephalon to the striatum. *Exp Neurol* 141:79–93.

Björklund, A., and Stenevi, U. (1979). Reconstruction of the nigrostriatal pathway by intracerebral nigral transplants. *Brain Res* 177:555–560.

Björklund, A., Schmidt, R. H., and Stenevi, U. (1980). Functional reinnervation of the neostriatum in the adult rat by use of intraparenchymal grafting of dissociated cell suspensions from the substantia nigra. *Cell Tissue Res* 212:39–45.

Björklund, A., Stenevi, U., Schmidt, R. H., Dunnett, S. B., and Gage, F. H. (1983). Intracerebral grafting of neuronal cell suspensions. II. Survival and growth of nigral cell suspensions implanted in different brain sites. *Acta Physiol Scand* 522(Suppl.):9–18.

Blunt, S. B., Jenner, P., and Marsden, C. D. (1992). Autoradiographic study of striatal D_1 and D_2 dopamine receptors in 6-OHDA-lesioned rats receiving foetal ventral mesencephalic grafts and chronic treatment with L-DOPA and carbidopa. *Dev Brain Res* 582:299–311.

Brecknell, J. E., Haque, N. S., Du, J. S., Muir, E. M., Fidler, P. S., Hlavin, M. L., Fawcett, J. W., and Dunnett, S. B. (1996). Functional and anatomical reconstruction of the 6-hydroxydopamine lesioned nigrostriatal system of the adult rat. *Neuroscience* 71:913–925.

Brundin, P., Barbin, G., Strecker, R.E., Isacson, O., Prochiantz. A., and Björklund A. (1988a). Survival and function of dissociated dopamine neurons grafted at different developmental stages or after being cultured *in vivo*. *Dev Brain Res* 39:233–243.

Brundin, P., Duan, W.-M., and Sauer, H. (1994). Functional effects of mesencephalic dopamine neurons and adrenal chromaffin cells grafted to the striatum. In: Dunnett, S. B., and Björklund, A., eds., *Functional neural transplantation.* Raven Press, New York, pp. 9–46.

Brundin, P., Isacson, O., and Björklund, A. (1985). Monitoring of cell viability in suspensions of embryonic CNS tissue and its use as a criterion for intracerebral graft survival. *Brain Res* 331:251–259.

Brundin, P., Isacson, O., Gage, F. H., and Björklund, A. (1986a). Intrastriatal grafting of dopamine-containing neuronal cell suspensions: Effects of mixing with target or non-target cells. *Dev Brain Res* 24:77–84.

Brundin, P., Isacson, O., Gage, F. H., Prochiantz, A., and Björklund, A. (1986b). The rotating 6-hydroxydopamine lesioned mouse as a model for assessing functional effects of neuronal grafting. *Brain Res* 366:346–349.

Brundin, P., and Strecker, R.E. (1991). Preparation and intracerebral grafting of dissociated fetal brain tissue in rats. In: Conn, P.M., ed., *Methods in neurosciences,* Vol. 7, *Lesions and transplantation.* Academic Press, San Diego, pp. 305–326.

Brundin, P., Strecker, R. E., Londos, E., and Björklund, A. (1987). Dopamine neurons grafted unilaterally to the nucleus accumbens affect drug-induced circling and locomotion. *Exp Brain Res* 69:183–194.

Brundin, P., Strecker, R. E., Widner, H., Clarke, D. J., Nilsson, O. G., Åstedt, B., Lindvall, O., and Björklund, A. (1988b). Human fetal dopamine neurons grafted in a rat model of Parkinson's disease: Immunological aspects, spontaneous and drug-induced behaviour, and dopamine release. *Exp Brain Res* 70:192–208.

Cadet, J. L., Zhu, S. M., and Angulo, J. A. (1991). Intrastriatal implants of fetal mesencephalic cells attenuate the increases in striatal proenkephalin mRNA observed after unilateral 6-hydroxydopamine-induced lesions of the striatum. *Brain Res Bull* 27:707–711.

Cenci, M. A., Campbell, K., and Björklund, A. (1993). Neuropeptide-mRNA expression in the 6-hydroxydopamine-lesioned striatum reinnervated by fetal dopaminergic transplants: Differential effects of the grafts on preproenkephalin-, preprotachykinin-, and prodynorphin mRNA levels. *Neuroscience* 57:275–296.

Cenci, M. A., Campbell, K., and Björklund, A. (1997). Glutamic acid decarboxylase gene expression in the dopamine-denervated striatum: Effects of intrastriatal fetal nigral transplants or chronic apomorphine treatment. *Mol Brain Res* 48:149–155.

Cenci, M. A., Campbell, K., Wictorin, K., and Björklund, A. (1992a). Striatal c-fos induction by cocaine or apomorphine occurs preferentially in output neurons projecting to the substantia nigra in the rat. *Eur J Neurosci* 4:376–380.

Cenci, M. A., Kalén, P., Mandel, R. J., Wictorin, K., and Björklund, A. (1992b). Dopaminergic transplants normalize amphetamine- and apomorphine-induced Fos expression in the 6-hydroxydopamine lesioned striatum. *Neuroscience* 46:943–957.

Chkirate, M., Vallée, A., and Doucet, G. (1993). Host striatal projections into fetal ventral mesencephalic tissue grafted to the striatum of immature or adult rat. *Exp Brain Res* 94:357–362.

Chritin, M., Savasta, M, Mennicken, F., Bal, A., Abrous, D. N., Le Moal, M., Feuerstein, C., and Herman, J. P. (1992). Intrastriatal dopamine-rich implants reverse the increase of dopamine D_2 receptor mRNA levels caused by lesion of the nigrostriatal pathway: A quantitative *in situ* hybridization study. *Eur J Neurosci* 4:663–672.

Costantini, L. C., Vozza, B. M., and Snyder-Keller, A. M. (1994). Enhanced efficacy of nigral-striatal cotransplants in bilaterally dopamine-depleted rats. *Exp Neurol* 127:219–231.

Di Loreto, S., Florio, T., Capozzo, A., Napolitano, A., Aorno, D., and Scarnati, E. (1996). Transplantation of mesencephalic cell suspension in dopamine-denervated striatum of the rat. *Exp Neurol* 138:318–326.

Doucet, G., Murata Y., Brundin, P., Bosler O., Mons N., Geffard M., Ouimet, C. C., and Björklund, A. (1989) Host afferents into intrastriatal transplants of fetal ventral mesencephalon. *Exp Neurol* 106:1–19.

Doucet, G., Brundin, P., Descarries, L., and Björklund, A. (1990). Effect of prior dopamine denervation on survival and fiber outgrowth from intrastriatal fetal mesencephalic grafts. *Eur J Neurosci* 2:279–290.

Duan, W.-M., Widner, H., and Brundin, P. (1995). Temporal pattern of host responses against intrastriatal grafts of syngenic, allogenic or xenogenic embryonic neuronal tissue in rats. *Exp Brain Res* 104:227–242.

Dunnett, S. B., Rogers, D. C., and Richards, S. J. (1989). Nigrostriatal reconstruction after 6-OHDA lesions in rats: Combination of dopamine-rich nigral grafts and nigrostriatal "bridge" grafts. *Exp Brain Res* 75:523–535.

Dunnett, S. B., Björklund, A., Schmidt, R. H., Stenevi, U., and Iversen, S. D. (1983a). Intracerebral grafting of neuronal cell suspensions. IV. Behavioural recovery in rats with unilateral 6-OHDA lesions following implantation of nigral cell suspensions in different brain sites. *Acta Physiol Scand* 522(Suppl.):29–37.

Dunnett, S. B., Björklund, A., Schmidt, R. H., Stenevi, U., and Iversen, S. D. (1983b). Intracerebral grafting of neuronal cell suspensions. V. Behavioural recovery in rats with bilateral 6-OHDA lesions following implantation of nigral cell suspensions. *Acta Physiol Scand* 522(Suppl.):39–47.

Dunnett, S. B., Björklund, A., Stenevi, U., and Iversen, S. D. (1981a). Grafts of embryonic substantia nigra reinnervating the ventrolateral striatum ameliorate sensorimotor impairments and akinesia in rats with 6-OHDA lesions of the nigrostriatal pathway. *Brain Res* 229:209–217.

Dunnett, S. B., Björklund, A., Stenevi, U., and Iversen, S. D. (1981b). Behavioural recovery following transplantation of substantia nigra in rats subjected to 6-OHDA lesions of the nigrostriatal pathway. II. Bilateral lesions. *Brain Res* 229:457–470.

Emgård-Mattson, M., Karlsson, J., Nakao, N., and Brundin, P. (1997). Addition of lateral ganglionic eminence to rat mesencephalic grafts affects fiber outgrowth but does not enhance function. *Cell Transpl* 6:277–286.

Fisher, L. J., Young, S. J., Tepper, J. M., Groves, P. M., and Gage, F. H. (1991). Electrophysiological characteristics of cells within mesencephalon suspension grafts. *Neuroscience* 40:109–122.

Frodl, E. M., Nakao, N., and Brundin, P. (1994). Lazaroids improve the survival of cultured rat embryonic mesencephalic neurons. *NeuroReport* 5:2393–2396.

Freund, T. F., Bolam, J. P., Björklund, A., Stenevi, U., Dunnett, S. B., Powell, J. F., and Smith, A. D. (1985). Efferent synaptic connections of grafted dopaminergic neurons reinnervating the host neostriatum: A tyrosine hydroxylase immunocytochemical study. *J Neurosci* 5:603–616.

Gage, F. H., Stenevi, U., Carlstedt, T., Foster, G., Björklund, A., and Aguayo, A. J. (1985). Anatomical and functional consequences of grafting mesencephalic neurons into a peripheral nerve "bridge" connected to the denervated striatum. *Exp Brain Res* 60:584–589.

Glinka, Y., Gassen, M., and Youdim, M. B. H. (1997). Mechanism of 6-hydroxydopamine neurotoxicity. *J Neural Transm* 50(Suppl.):55–66.

Grasbon-Frodl, E. M., Nakao, N., and Brundin, P. (1996). The lazaroid U-83836E improves the survival of rat embryonic mesencephalic tissue stored at 4°C and subsequently used for cultures or intracerebral transplantation. *Brain Res Bull* 39:341–348.

Haque, N. S., Hlavin, M. L., Du, J. S., Fawcett, J. W., and Dunnett, S. B. (1996a). *In vivo* effects of kFGF on embryonic nigral grafts in a rat model of Parkinson's disease. *NeuroReport* 6:2177–2181.

Haque, N. S., Hlavin, M. L., Fawcett, J. W., and Dunnett, S. B. (1996b). The neurotrophin NT4/5, but not NT3, enhances the efficacy of nigral grafts in a rat model of Parkinson's disease. *Brain Res* 712:45–52.

Lindvall, O. (1997). Neural transplantation: A hope for patients with Parkinson's disease. *NeuroReport* 8:iii–x.

Mahalik, T. J., Finger, T. E., Strömberg, I., and Olson, L. (1985). Substantia nigra transplants into denervated striatum of the rat: Ultrastructure of graft and host interconnections. *J Comp Neurol* 240:60–70.

Mahalik, T. J., Hahn, W. E., Clayton, G. H., and Owens, G. P. (1994). Programmed cell death in developing grafts of fetal substantia nigra. *Exp Neurol* 129:27–36.

Mandel, R. J., Brundin, P., and Björklund, A. (1990). The importance of graft placement and task complexity for transplant-induced recovery of simple and complex sensorimotor deficits in dopamine devervated rats. *Eur J Neurosci* 2:888–894.

Manier, M., Abrous, D. N., Feuerstein, C., Le Moal, M., and Herman, J. P. (1991). Increase of striatal methionin enkephalin content following lesion of the nigrostriatal dopaminergic pathway in

adult rats and reversal following the implantation of embryonic dopaminergic neurons: A quantitative immunohistochemical analysis. *Neuroscience* 42:427–439.
Marshall, J. F., Berrios, N., and Sawyer, S. (1980). Neostriatal dopamine and sensory inattention. *J Comp Physiol Psychol* 94:833–846.
Mayer, E., Dunnett, S. B., and Fawcett, J. W. (1993). Basic fibroblast growth factor promotes the survival of embryonic ventral mesencephalic dopaminergic neurons. II. Effects on nigral transplants *in vivo*. *Neuroscience* 56:389–398.
Mendez, I. M., Naus, C. C. G., Elisevich, K., and Flumerfelt, B. A. (1993). Normalization of striatal proenkephalin and preprotachykinin mRNA expression by fetal substantia nigra grafts. *Exp Neurol* 119:1–10.
Mendez, I., Sadi, D., and Hong, M. (1996). Reconstruction of the nigrostriatal pathway by simultaneous intrastriatal and intranigral dopaminergic transplants. *J Neurosci* 16:7216–7227.
Montoya, C. P., Astell, S., and Dunnett, S. B. (1990). Effects of nigral and striatal grafts on skilled forelimb use in the rat. *Prog Brain Res* 82:459–466.
Montoya, C. P., Campbell-Hope, L. J., Pemberton, K. D., and Dunnett, S. B.(1991). The "staircase" test: A measure of independent forelimb reaching and grasping abilities in rats. *J Neurosci Meth* 36:219–228.
Nakao, N., Frodl, E. M., Duan, W.-M., Widner, H., and Brundin, P. (1994). Lazaroids improve the survival of grafted rat embryonic dopamine neurons. *Proc Natl Acad Sci USA* 91:12408–12412.
Nakao, N., Frodl, E. M., Widner, H., Carlson, E., Eggerding, F. A., Epstein, C. J. and Brundin, P. (1995). Overexpressing Cu/Zn-superoxide dismutase enhances survival of transplanted neurons in a rat model of Parkinson's disease. *Nature Med* 1:226–231.
Nikkhah, G., Duan, W.-M., Knappe, U., and Björklund, A. (1993). Restoration of complex sensorimotor behavior and skilled forelimb use by a modified nigral cell suspension transplantation approach in the rat Parkinson model. *Neuroscience* 56:33–43.
Nikkhah, G., Bentlage, C., Cunningham, M. G., and Björklund, A. (1994). Intranigral fetal dopamine grafts induce behavioral compensation in the rat Parkinson model. *J Neurosci* 14:3449–3461.
Nikkhah, G., Cunningham, M. G., Cenci, M. A. McKay, R. D., and Björklund, A. (1995). Dopaminergic microtransplants into the substantia nigra of neonatal rats with bilateral 6-OHDA lesions: Evidence for anatomical reconstruction of the nigrostriatal pathway. *J Neurosci* 15:3548–3561.
Olsson, M., Nikkhah, G., Bentlage, C., and Björklund, A. (1995). Forelimb akinesia in the rat Parkinson model: Differential effects of dopamine agonists and nigral transplants as assessed by a new stepping test. *J Neurosci* 15:3863–3875.
Othberg, A., Keep, M., Brundin, P., and Lindvall, O. (1997). Tirilazad mesylate improves survival of rat and human embryonic dopamine neurons *in vitro*. *Exp Neurol* 147:498–502.
Perlow, M. J., Freed, W. J., Hoffer, B. J., Seiger, Å., Olson, L., and Wyatt, R. J. (1979). Brain grafts reduce motor abnormalities produced by destruction of nigrostriatal dopamine system. *Science* 204:643–647.
Reading, P. J. (1994). Neural transplantation in the ventral striatum. In: (Dunnett, S. B., and Björklund, A., eds.,) *Functional neural transplantation*. Raven Press, New York, pp. 197–216.
Rosenblad, C., Martinez-Serrano, A., and Björklund, A. (1996). Glial cell line-derived neurotrophic factor increases survival, growth and function of intrastriatal fetal nigral dopaminergic grafts. *Neuroscience* 75:979–985.
Sauer, H., and Brundin, P. (1991). Effects of cool storage on survival and function of intrastriatal ventral mesencephalic grafts. *Rest Neurol Neurosci* 2:123–135.
Sauer, H., Frodl, E. M., Kupsch, A., ten Bruggencate, G., and Oertel, W. H. (1992). Cryopreservation, survival and function of intrastriatal fetal mesencephalic grafts. *Exp Brain Res* 90:54–62.
Sauer, H., Fisher, W., Nikkhah, G., Wiegand, S. J., Brundin, P., Lindsay, R., and Björklund, A. (1993). BDNF enhances function rather than survival of intrastriatal dopamine cell-rich grafts. *Brain Res* 626:37–44.

Savasta, M., Mennicken, F., Chritin, M., Abrous, D. N., Feuerstein, C., Le Moal, M., and Herman, J. P. (1992). Intrastriatal dopamine-rich implants reverse the changes in dopamine-D2 receptor densities caused by 6-hydroxydopamine lesion of the nigrostriatal pathway in rats: An autoradiographic study. *Neuroscience* 46:729–738.

Schallert, T., and Hall, S. (1988). "Disengage" sensorimotor deficit following apparent recovery from unilateral dopamine depletion. *Behav Brain Res* 30:15–24.

Schallert, T., Norton, D., and Jones, T.A. (1992). A clinically relevant unilateral rat model of Parkinson's disease. *J Neural Transpl Plast* 3:332–333.

Schmidt, R. H., Ingvar, M., Lindvall, O., Stenevi, U., and Björklund, A. (1982). Functional activity of substantia nigra grafts reinnervating the striatum: Neurotransmitter metabolism and [^{14}C] 2- deoxy-D-glucose autoradiography. *J Neurochem* 38:737–748.

Schmidt, R. H., Björklund, A., Stenevi, U., Dunnett, S. B., and Gage, F. H. (1983). Intracerebral grafting of neuronal cell suspensions. III. Activity of intrastriatal nigral suspension implants as assessed by measurements of dopamine synthesis and metabolism. *Acta Physiol Scand* 522(Suppl.):19–28.

Schwartz, S., and Freed, W. J. (1987). Brain tissue transplantation in neonatal rats prevents a lesion-induced syndrome of adipsia, aphagia, and akinesia. *Exp Brain Res* 65:449–454.

Segovia, J., Castro, R., Notario, V., and Gale, K. (1991). Transplants of fetal substantia nigral regulate glutamic acid decarboxylase gene expression in host striatal neurons. *Mol Brain Res* 10:359–362.

Simonds, G. R., and Freed, W. J. (1990). Effects of intraventricular substantia nigra allografts as a function of donor age. *Brain Res* 530:12–19.

Sinclair, S. R., Svendsen, C. N., Torres, E. M., Martin, D., Fawcett, J. F., and Dunnett, S. B. (1996). GDNF enhances dopaminergic cell survival and fiber outgrowth in embryonic nigral grafts. *Neuro Report* 7:2547–2552.

Sirinathsinghji, D. J. S., and Dunnett, S. B. (1991). Increased proenkephalin mRNA levels in the rat neostriatum following lesion of the ipsilateral nigrostriatal dopamine pathway with 1-methyl-4-phenylpyridinium ion (MPP^+): Reversal by embryonic nigral dopamine grafts. *Mol Brain Res* 9:263–269.

Spenger, C., Haque, N. S. K., Studer, L., Evtouchenko, L., Wagner, B., Bühler, B., Lendahl, U., Dunnett, S. B., and Seiler, R. W. (1996). Fetal ventral mesencephalon of human and rat origin maintained *in vitro* and transplanted to 6-hydroxydopamine-lesioned rats gives rise to grafts rich in dopaminergic neurons. *Exp Brain Res* 112:47–57.

Strecker, R. E., Sharp, T., Brundin, P., Zetterström, T., Ungerstedt, U., and Björklund, A. (1987). Autoregulation of dopamine release and metabolism by intrastriatal nigral grafts as revealed by intracerebral microdialysis. *Neuroscience* 22:169–178.

Strecker, R. E., Miao, R., and Loring, J. F. (1989). Survival and function of aggregate cultures of rat fetal dopamine neurons grafted in a rat model of Parkinson's disease. *Exp Brain Res* 76:315–322.

Strömberg, I., Johnson, S., Hoffer, B., and Olson, L. (1985). Reinnervation of the dopamine-denervated striatum by substantia nigra transplants: Immunohistochemical and electrophysiological correlates. *Neuroscience* 14:981–990.

Strömberg, I., Almquist, P., Bygdeman, M., Finger, T. E., Gerhardt, G., Granholm, A.-C., Mahalik, T. J., Seiger, Å., Olson, L., and Hoffer, B. (1989). Human fetal mesencephalic tissue grafted to dopamine-denervated striatum of athymic rats: Light- and electron-microscopical histochemistry and *in vivo* chronoamperometric studies. *J Neurosci* 9:614–624.

Takayama, H., Ray, J., Raymon, H. K., Baird, A., Hogg, J., Fisher, L. J., and Gage, F. H. (1995). Basic fibroblast growth factor increases dopaminergic graft survival and function in a rat model of Parkinson's disease. *Nature Med* 1:53–58.

Triarhou, L. C., Low, W., and Ghetti, B. (1990). Dopamine neuron grafting to the weaver mouse striatum. *Prog Brain Res* 82:187–195.

Ungerstedt, U. (1968). 6-hydroxydopamine induced degeneration of central monoamine neurons. *Eur J Pharmacol* 5:107–110.

Ungerstedt, U. (1971a). Aphagia and adipsia after 6-hydroxydopamine induced degeneration of the nigro-striatal dopamine system. *Acta Physiol Scand* 367(Suppl.):95–122.

Ungerstedt, U. (1971b). Post-synaptic supersensitivity after 6-hydroxydopamine induced degeneration of the nigro-striatal dopamine system. *Acta Physiol Scand* 367(Suppl):49–68.

Ungerstedt, U., and Arbuthnott, G. W. (1970). Quantitative recording of rotational behavior in rats after 6-hydroxy-dopamine lesions of the nigrostriatal dopamine system. *Brain Res* 24:485–493.

Wictorin, K., Brundin, P., Sauer, H., Lindvall, O., and Björklund, A. (1992). Long distance axonal growth from human dopaminergic mesencephalic neuroblasts implicated along the nigral pathway in 6-hydroxydopamine lesioned adult rats. *J Comp Neurol* 323:475–494.

Wuerthele, S. M., Freed, W. J., Olson, L., Morihisa, J., Spoor, L., Wyatt, R. J., and Hoffer, B. J. (1981). Effect of dopamine agonists and antagonists on the electrical activity of substantia nigra neurons transplanted into the lateral ventricle of the rat. *Exp Brain Res* 44:1–10.

Yurek, D. M., Collier, T. J., and Sladek, J. R., Jr. (1990). Embryonic mesencephalic and striatal co-grafts: Development of grafted dopamine neurons and functional recovery. *Exp Neurol* 109:191–199.

Yurek, D. M., Lu, W., Hipkens, S., and Wiegand, S. J. (1996). BDNF enhances the functional reinnervation of the striatum by grafted fetal dopamine neurons. *Exp Neurol* 137:105–118.

Zeng, B. Y., Jenner, P., and Marsden, C. D. (1996). Altered motor function and graft survival produced by basic fibroblast growth factor in rats with 6-OHDA lesions and fetal ventral mesencephalic grafts are associated with glial proliferation. *Exp Neurol* 139:214–226.

Zetterström, T., Brundin, P., Gage, F. H., Sharp, T., Isacson, O., Dunnett, S. B., Ungerstedt, U., and Björklund A. (1986). *In vivo* measurement of spontaneous release and metabolism of dopamine from intrastriatal nigral grafts using intracerebral microdialysis. *Brain Res* 362:344–349.

Zhou, F. C., Chiang, Y. H., and Wang, Y. (1996). Constructing a new nigrostriatal pathway in the parkinsonian model with bridged neural transplantation in substantia nigra. *J Neurosci* 16:6965–6974.

Zuddas, A., Corsini, G. U., Barker, J. L., Kopin, I. J., and di Porzio, U. (1991). Specific reinnervation of lesioned mouse striatum by grafted mesencephalic dopaminergic neurons. *Eur J Neurosci* 3:72–85.

CHAPTER 12

Fetal Grafts in Parkinson's Disease

Primate Models

JOHN R. SLADEK, JR.,* TIMOTHY J. COLLIER,† JOHN D. ELSWORTH,** ROBERT H. ROTH,** JANE R. TAYLOR** AND D. EUGENE REDMOND, JR.**

*Department of Neuroscience, The Chicago Medical School, N. Chicago, Illinois, †Department of Neurological Sciences, Rush Medical School, Chicago, Illinois, **Departments of Psychiatry and Pharmacology, Yale Medical School, New Haven, Connecticut

Improvements in parkinsonism have been reported in patients after embryonic ventral mesencephalic (VM) tissue implantation into either the caudate nucleus, or into the putamen, or into both regions (e.g., Freed *et al.*, 1992; Henderson *et al.*, 1991; Lindvall *et al.*, 1993; Peschanski *et al.*, 1994; Spencer *et al.*, 1992; Widner *et al.*, 1992). Initial results from clinical trials have been encouraging; however, there is no consensus about the optimal site(s) for clinical benefit, or age of the donor tissue. Because reinnervation of the entire striatum is unlikely with tissue from a single donor, the distribution of the implanted tissue within a site(s) also may be of critical functional importance. Given that these issues cannot be easily addressed in clinical studies, experimental models of parkinsonism are essential for optimizing transplantation techniques. Specifically, primate studies are essential to link important rodent studies with human applications. For example, the most effective placement of the grafted tissue within the functionally and anatomically heterogeneous large primate brain is unlikely to be identified using simpler dopamine-depletion models in the rat brain. Their complex behavioral repertoire and their manifestation of classic parkinsonism should provide a useful and predictive method of investigating the extent and nature of the functional changes resulting from fetal tissue grafts.

CNS Regeneration

I. INTRODUCTION

Parkinson's disease is a neurodegenerative disorder that is characterized by a progressive loss of dopaminergic neurons of the substantia nigra that project to the neostriatum (Barbeau, 1969; Birkmayer and Hornykiewicz, 1976). Cardinal signs of the disease include a slowing and decrease in movement, resting tremor of the limbs, and postural imbalance. Although the administration of pharmacotherapy such as levodopa or carbidopa can result in substantial symptomatic relief during the early stages of the disease, drug effectiveness can decline with prolonged use as more cells are lost, leading to severe side effects (Fahn, 1992). Research in rodents and nonhuman primates since the late 1970s has demonstrated that replacement of lost dopamine is possible through the intracerebral grafting of embryonic dopaminergic neurons from the ventral mesencephalon. These cells are capable of surviving the transplant procedure and resupplying the neostriatum with sufficient dopamine to restore motor disturbances that were altered in lesion-induced animal models of parkinsonism (Bjorklund and Stenevi, 1979; Freed *et al.*, 1980; Dunnett *et al.*, 1981). The results are so profound in the nonhuman primate that resting tremor, postural instability, bradykinesia, and other symptoms can be reversed as a result of the survival of grafted dopaminergic neurons of the substantia nigra (Taylor *et al.*, 1991). Clinical experiments have been performed utilizing information gained from these studies with encouraging results that parallel the cell survival seen in nonhuman primates as described as follows.

II. THERAPEUTIC APPROACHES TO PARKINSON'S DISEASE

Prior to 1961, therapy for the symptoms of Parkinson's disease included the use of neurosurgical lesions to various subcortical structures associated with the control and modulation of movement as well as the administration of some drugs that attempted to balance the motor problems (Meyers, 1942). The discovery that parkinsonian symptoms were produced by a loss of dopamine from the substantia nigra suggested that patients might benefit from administration of levodopa, the naturally occurring amino acid precursor of dopamine. By 1967, Cotzias and colleagues reported that high doses of orally administered levodopa produced improvement in the symptoms of Parkinson's disease (Cotzias *et al.*, 1967). Subsequently, co-administration of carbidopa, a drug that selectively blocks the conversion of levodopa to dopamine in peripheral tissues, but not in brain, permitted the reduction in the oral dose of levodopa administered and helped to prevent unwanted peripheral side effects such as changes in blood pressure. Although the combination of levodopa and carbidopa advanced the treatment of Parkinson's disease, it did not represent a cure. Moreover, long-term use led to the emergence of drug-related side effects such as on-off periods, for example, daily, alternating periods during which drugs were maximally effective and patients enjoyed relatively normal movement ("on" periods), versus periods of diminished drug effect that left patients immobile ("off" periods). Also, as the disease progresses levodopa doses can be elevated that produce an overactivity of the dopamine system that is characterized by unwanted bursts of involuntary movement, termed dyskinesias. Balancing the dosage to relieve symptoms while avoiding side effects becomes increasingly problematic as the disease progresses. Eventually, any significant therapeutic effect of levodopa may be lost. Thus, considerable effort has been devoted to the discovery of alternate therapeutic strategies such as neural transplantation. The use of the nonhuman primate as a test of feasibility is described next.

A. An Animal Model of Parkinson's Disease

The organic compound MPTP (1-methyl-4-phenyl-1,2,3,6-tetrahydropyridine) selectively destroys dopamine neurons of the substantial nigra in monkeys (Fig. 1) and produces many of the classic symptoms of Parkinson's disease. MPTP-lesioned monkeys develop a parkinsonian syndrome that mimics the disease in its behavioral features, neuronal cell loss, depletion of the neurotransmitter dopamine, and responsiveness to the standard therapy of levodopa or carbidopa (Burns *et al.*, 1983). These findings are important because all of these symptoms are observed in a species closely related to humans. The study of these MPTP-treated parkinsonian monkeys has provided insights into potential risk factors and causes of Parkinson's disease with resultant new therapies that have been brought to clinical trials. Early studies of the metabolism of MPTP indicated that this molecule was a preferred substrate of the enzyme monoamine oxidase-B (MAO-B) with an affinity equal to that of its natural substrate dopamine. MAO-B acted on MPTP to generate the metabolite MPP^+. The conversion of MPTP to MPP^+, and the accumulation of MPP^+ in substantia nigra dopamine neurons led to dopamine neuron death. A next logical step was to determine whether blockade of the action of MAO-B would prevent the toxicity of MPTP. Two groups of researchers independently demonstrated that giving monkeys a relatively nonspecific MAO inhibitor drug (pargyline) prior to MPTP administration prevented the development of parkinsonian behavior, dopamine depletion, and substantia nigra cell loss that typically is produced by MPTP exposure. One of these studies additionally showed that a specific MAO-B inhibitor called deprenyl also prevented MPTP toxicity (Langston *et al.*, 1984). Although it was not clear whether the chemical reaction that lent MPTP its dopamine cell toxicity was related to the events that produce

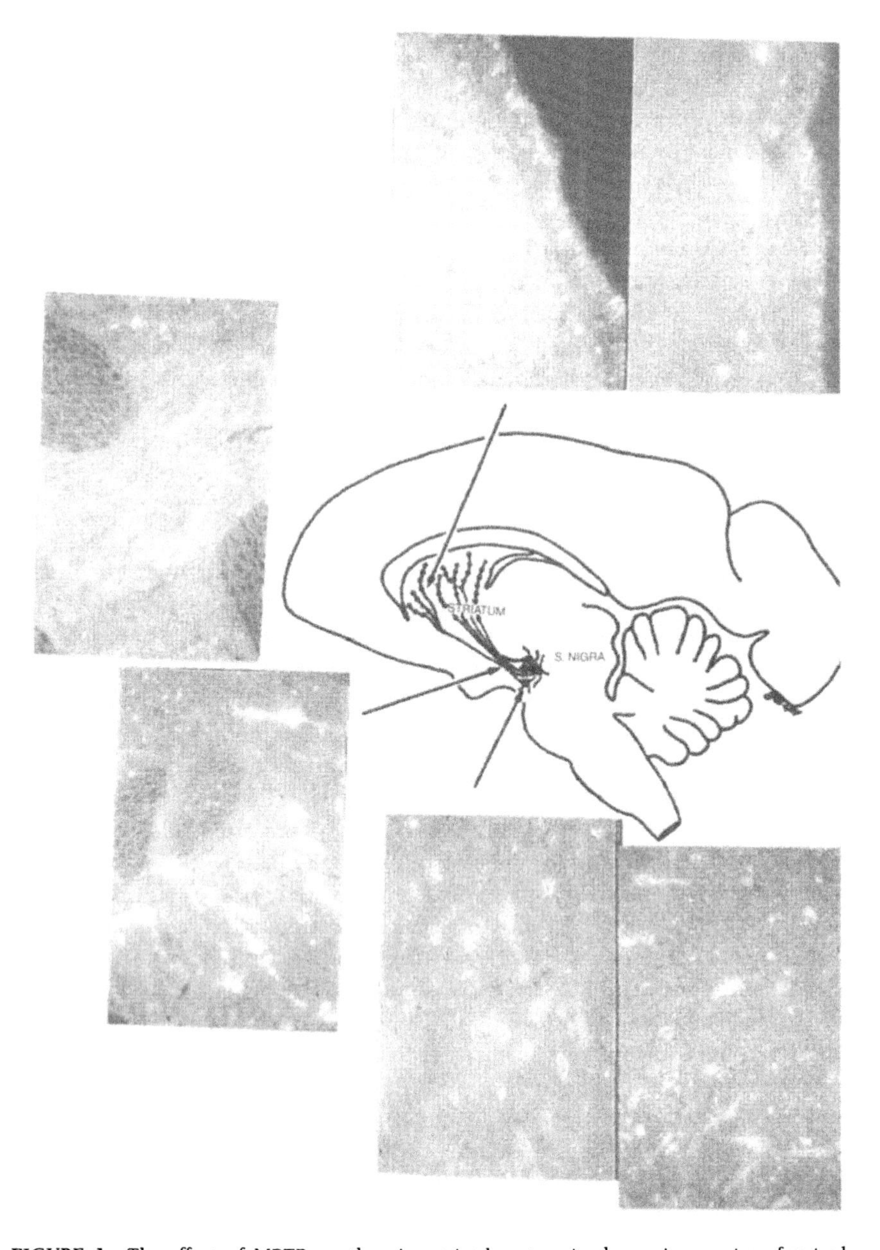

FIGURE 1 The effect of MPTP on the nigrostriatal system is shown in a series of paired histofluorescence images. In the upper right, a loss of fluoresence in dopamine terminals is seen 30 days after injection of the neurotoxin. The images to the left represent regions of preterminal axons that are normal in the upper image, but are enlarged in the lower image after MPTP, which is typical of degenerating processes. The two images in the lower right represent the normal substantia nigra (left) and the lesioned nucleus (right). The latter has fewer dopaminergic neurons and an increase in degenerating neurons as marked by the presence of lipofuscin.

Parkinson's disease, it suggested that if an endogenous toxin or environmental toxin contributed to Parkinson's disease, it might operate via a chemical mechanism similar to that of MPTP. Other lines of evidence suggested that interfering with the action of MAO-B might be beneficial for Parkinson's disease patients. First, MAO-B normally was active in breaking down dopamine in the brain. Therefore, slowing this process might allow dopamine to exist for longer periods of time in the brain, and provide for greater activity from the remaining dopamine. Second, products of dopamine breakdown produced by MAO-B include hydrogen peroxide and reactive oxygen radicals that, in high enough concentration, are toxic to dopamine neurons. Taken together, these observations suggested that drugs that diminish the activity of MAO-B and inactivate oxygen radicals may be useful as therapeutic agents for Parkinson's disease.

B. Surgical Replacement of Dopamine Neurons: Neural Grafting of Embryonic Tissue

Neural grafting is now a straightforward approach among therapies for animal models of Parkinson's disease. Because abnormalities of the motor system in Parkinson's disease can be attributed to the death of dopamine neurons, one way to treat the disease is to attempt to replace dopamine neurons. Drawing upon a concept that had been developed near the turn of the nineteenth century (Thompson, 1890; Ranson, 1914; Dunn, 1917) and upon emerging knowledge concerning the growth properties of the developing nervous system, scientists in Sweden reported the successful transplantation of embryonic neural tissue to the anterior eye chamber in rats (Olson and Malmfors, 1970). They found that if embryonic brain cells were removed at a time prior to completing their connections with other brain regions, they could be transplanted into adult animals, survive this grafting, and establish connections with the host brain probably because they retained their capacity for further growth and development. Subsequently, Freed and colleagues reported that this strategy could be used with embryonic dopamine neurons and that grafts into rats with experimentally induced damage to the substantia nigra survived transplantation and replaced a supply of dopamine to the striatum, improving the symptoms of this rat model of Parkinson's disease (Freed *et al.*, 1980; see also Chapter 11).

Following years of important tests in rats, brain cell transplantation was attempted in a species more closely related to human. In 1986, Sladek and associates published an initial report of successful embryonic dopamine neuron transplantation in MPTP-treated parkinsonian monkeys. During the next 10 years of studies in monkeys our work and that of others (reviewed next) have demonstrated that embryonic dopamine neurons derived from appropriately aged donors exhibit excellent survival, growth, and integration with the dopamine-depleted host striatum (Fig. 2). These grafted cells provide a long-term source of replacement dopamine that is effective in diminishing the behavioral signs of MPTP-induced parkinsonism in these monkeys. Although certain issues unique to the use of neural grafts in human patients remain unresolved, including whether Parkinson's disease will afflict the replacement cells and what effect continued use of levodopa may have on grafted cells, transplanted dopamine neurons theoretically could provide a cellular source of dopamine throughout the remaining life span of a patient.

C. Overview of Transplants in Nonhuman Primates

Much of the early work on grafting embryonic dopamine neurons in nonhuman primate models of Parkinson's disease dealt with reexamination of issues pre-

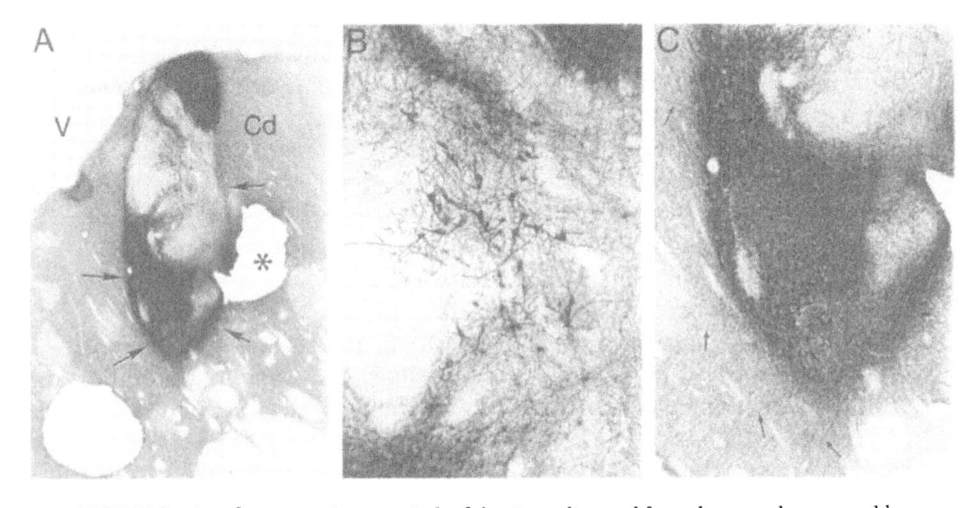

FIGURE 2 A graft representing one-sixth of the tissue dissected from the ventral mesencephlon occupies a position deep within the caudate nucleus left of the midline. The graft appears as an elongated mass of tissue, filled with dopamine neurons stained for tyrosine hydroxlase. A circular blank area seen to the left of the graft represents tissue that was removed by a micropunch dissection for the measurement of dopamine and its metabolites. Another micropunch (*) closer to the graft contained significantly elevated dopamine attributed to cells from the graft. In B, the central area of the graft is enlarged to reveal the dense cluster of dopamine neurons and the extensive fiber outgrowth. In C, neurites (arrows) are evident as they exit the graft to supply the surrounding neuropil of the host brain.

viously studied in rodent models (see Chapters 6 and 11). These included determination of the optimal age of donor tissue for transplantation, methods for preparation and implantation of tissue for grafting, the extent that dopamine could be replaced in the host brain, and the functional efficacy of dopamine replacement via implants. Although the outcome of rodent studies made definite predictions concerning all of these variables, it remained unclear whether such findings would translate directly to the nonhuman primate models. Indeed, these early basic studies were important in establishing a firm foundation for subsequent clinical trials in human patients.

1. Optimal Gestational Age of Donor Tissue for Grafting

The accumulated experience of studies implanting ventral mesencephalic tissue of varying gestational age in nonhuman primates confirms the prediction of rodent studies that earlier gestation tissue survives and grows better than late gestation tissue. Implantation of mesencephalic tissue from the first 35 to 50 days of gestation (approximately 165 days total gestation for rhesus and African green monkeys studied) was found to yield good to excellent survival of grafted cells and, in most cases, extensive neurite outgrowth from grafted neurons (Bakay *et al.*, 1985, 1987, 1988; Bankiewicz *et al.*, 1990a; Collier *et al.*, 1994; Sladek *et al.*, 1986, 1993a). In contrast, experiments grafting ventral mesencephalic tissue from mid- to late-gestation embryos revealed modest to poor cell survival and limited reinnervation of the host target tissue (Annett *et al.*, 1997; Collier *et al.*, 1994; Morihisa *et al.*, 1984; Sladek *et al.*, 1986). There exist a few reports that are not entirely consistent with this generalization. Implantation of mesencephalic tissue from a 40-day gestation *Macaca fascicularis* embryo was reported to yield poor survival of grafted dopamine neurons (Dubach *et al.*, 1988), and three studies in marmosets suggested that mid-gestation tissue, 60 to 80 days of gestation in a 145-day full gestation species, could yield good survival of grafted cells (Fine *et al.*, 1988; Annett *et al.*, 1990, 1997). However, the authors of this latter study acknowledged that within this gestational range, younger tissue survived transplantation better. Taken together, grafting studies in monkeys have supported the prediction made by studies in rodents that relatively early gestation embryonic donor tissue, around the time of initial differentiation and prior to extensive neurite innervation of the striatal targets of the dopamine neurons, provides the optimal tissue source for transplantation.

2. Techniques for Preparation and Implantation of Tissue

Grafting studies in nonhuman primates have explored three main variables with regard to tissue preparation and implantation. First, the physical nature of the implanted tissue has been studied, comparing implants of small pieces of embryonic tissue to enzymatically dissociated cell suspensions. Both grafts consisting of tissue pieces (Bankiewicz *et al.*, 1990a; Collier *et al.*, 1994; Redmond *et al.*, 1986a; Sladek *et al.*, 1987a,b, 1988a,b, 1993a,b) and cell suspensions (Bakay *et al.*, 1985, 1987, 1988; Fine *et al.*, 1988; Annett *et al.*, 1990) have been reported to yield good survival and neurite outgrowth. A single study simultaneously compared the two preparations with the same gestational age donor material in the same experimental model (Fine *et al.*, 1988). Observations based on one pair of monkeys indicated that cell suspension grafts yielded better reinnervation of the surrounding host striatum than grafted tissue fragments. A more detailed comparison of graft survival and integration across a range of gestational age donors for solid versus suspension grafts in primates remains to be conducted.

Second, the method of tissue implantation has been varied in studies examining optimal conditions for fetal graft survival. Cavitation or some other injury to the host striatum prior to tissue grafting has been examined as a means of improving graft survival, and is quite distinct from the minimal tissue damage associated with stereotaxic implantation. Studies in rodents provide the rationale for using pre-grafting tissue damage as a preparatory step to cell implantation. This work indicated that damage induced a proliferation of blood vessels at the implantation site, providing a nutritive environment for grafted tissue (Bjorklund and Stenevi, 1979). In addition, prior damage was accompanied by accumulation of chemical factors that may be neurotrophic for the graft (Nieto-Sampedro *et al.*, 1982). These approaches have only rarely been directly compared within a given experiment, however. Both surgical cavitation of the host striatum prior to transplantation (Bankiewicz *et al.*, 1990a) and stereotaxic methodology (Sladek *et al.*, 1993a) have been used with solid mesencephalic tissue grafts, and both yielded good survival of grafted tissue. However, differences in graft-host morphology were apparent: with prior cavitation, neurites of grafted dopamine neurons were reported to remain confined to the graft, with no evidence of growth into the surrounding host striatum, whereas stereotaxic implantation yielded grafts with extensive neurite extension into the surrounding host brain (Figs. 2, 5, and 6). In addition, cavities were associated with apparent growth of dopaminergic neurites from surviving host neurons directed toward the area of damage, whereas stereotaxic implantation did not appear to induce this host response. One study examined a more modest form of prior tissue damage, insertion of a stylet similar in size to the cannula used to stereotaxically implant tissue, one week prior to grafting, as compared to no prior damage (Bakay *et al.*, 1988). Morphological observations in one monkey suggested that prior damage increased survival of grafted dopamine neurons by 240%. Although the interaction between injury responses of the host brain and the viability of embryonic mesencephalic grafts have not been examined in detail, the existing evidence suggests that the same significant damage to the host striatum that is required to invoke a robust compensatory response by the lesioned mesostriatal dopamine system may also yield glial and inflammatory responses that are not beneficial to the grafted tissue. The best survival and integration of grafted dopamine neurons appear to be associated with minimizing damage to the host brain through stereotaxic implantation of small amounts of grafted tissue as described later.

A third factor that may influence graft viability in the host is the time interval between collection of donor tissue and implantation into the brain. In particular, it has been of interest to determine whether cryopreservation and storage of tissue for extended periods of time, with its attendant practical advantages, can yield graft survival equivalent to implantation of tissue that is freshly harvested. The benefits of being able to store, test for safety, and readily transport tissue intended for transplantation have been described thoroughly (Houle and Das, 1980; Jensen *et al.*, 1987; Robbins *et al.*, 1990). However, study of cryopreserved tissue grafts in monkeys to date has focused upon demonstrating the feasibility of cryopreservation (Collier *et al.*, 1987) rather than on a detailed study of morphology and function of fresh versus frozen tissue. Embryonic monkey mesencephalic tissue cryopreserved and stored for up to 25 days prior to transplantation has been demonstrated to yield surviving dopamine neurons, albeit in grafts judged to be smaller than those seen following grafting of fresh tissue of similar gestation. It is noteworthy, however, that these studies utilized mid- to late-gestation tissue, and similar work should be done involving earlier gestation tissue comparable to that likely to be employed in clinical grafting situations. To date, the relative utility of cryopreserved tissue has not been resolved in experiments of grafting in monkeys.

3. Do Mesencephalic Grafts Replace Dopamine?

The capacity of grafted embryonic dopamine neurons to produce and release dopamine in monkey hosts has been assessed by three types of measures. First, cerebrospinal fluid levels of levodopa (Bakay *et al.*, 1985, 1987) or homovanillic acid (HVA) (Redmond *et al.*, 1986a; Sladek *et al.*, 1987a,b; Taylor *et al.*, 1990) have been measured in the cerebrospinal fluid of dopamine-depleted parkinsonian monkeys and compared to levels measured in monkeys receiving embryonic tissue grafts. Reductions of these markers were detected following MPTP lesions of the dopamine system, and recovery to near baseline values was observed over 1 to 2 months following grafting. However, at least one study indicated that cerebrospinal fluid HVA did not correlate with graft-induced behavioral recovery (Bankiewicz *et al.*, 1990b), and the variability of these values in cerebrospinal fluid make it unclear whether these measures are a useful reflection of striatal dopamine activity.

A second means of assessing production of dopamine and its metabolites after grafting has been the performance of neurochemical assays on tissue punches collected adjacent to and distant from ventral mesencephalic grafts placed in the striatum (Sladek *et al.*, 1988a,b, 1993b; Elsworth *et al.*, 1994). In monkeys with optimal gestation grafts, dopamine levels measured in striatal tissue immediately adjacent to grafts exhibited 4 to 8-fold increases over tissue levels distant from the grafts, attaining dopamine levels approximately 50% of normal, and routinely in the range of dopamine levels measured in behaviorally asymptomatic MPTP-treated monkeys.

A third means of assessing dopamine production from grafted neurons has been the determination of the pattern of binding of a radiolabeled cocaine analog as a marker for the dopamine transporter. This has been used as a means of visualizing neurite extension from grafted neurons into the surrounding host striatum (Elsworth *et al.*, 1994). Comparisons of binding in one untreated monkey, one monkey treated with MPTP only, and one monkey treated with MPTP, and subsequently grafted with early gestation embryonic mesencephalic tissue into the caudate nucleus showed that in contrast to the near-total loss of specific binding in the MPTP-treated nontransplanted monkey, the grafted animal exhibited increased binding surrounding the grafts. Approximately 60% of the caudate nucleus exhibited increased binding, as compared to the MPTP-only monkey, with 4 to 8-fold increases in dopamine content measured from micropunched tissue adjacent to the grafts. Most proximal to the grafts, binding density was not different from that measured in the untreated control monkey.

4. Behavioral Recovery in Monkeys Receiving Mesencephalic Grafts

Measures of spontaneous activity, often including scored parkinsonian-like behavior, (Bakay *et al.*, 1985, 1987, 1988; Bankiewicz *et al.*, 1990; Fine *et al.*, 1988; Redmond *et al.*, 1986a; Sladek *et al.*, 1988a,b; Taylor *et al.*, 1990), spontaneous and drug-induced rotational behavior in monkeys with unilateral lesions of the mesencephalic dopamine system (Bankiewicz *et al.*, 1990; Fine *et al.*, 1988; Dubach *et al.*, 1988; Annett *et al.*, 1990, 1994), and various types of volitional or skilled behavior, often involving reaching tasks (Bankiewicz *et al.*, 1990; Taylor *et al.*, 1990; Annett *et al.*, 1990), all have been reported to show some degree of improvement following successful embryonic dopamine neuron grafts. Two studies raise issues of particular interest. First, assessment of changes in motor and cognitive behavior in MPTP-treated monkeys receiving dopamine grafts into the caudate nucleus as compared to lesioned monkeys either receiving inappropriate grafts into the caudate comprised of embryonic cerebellar tissue, or misplaced dopamine grafts located in the cingulate cortex, indicated that only grafts of ventral mesencephalic dopamine-rich tissue placed into the striatum yielded behavioral improvement (Taylor *et al.*, 1990, 1991).

No behavioral effects were detected associated with nonspecific surgical injury or any general characteristic of embryonic tissue per se. Second, improvements in behavior following grafting may be variable across the types of behavior monitored (Annett *et al.*, 1990, 1994, 1995). Although improvements in spontaneous and drug-induced rotational behaviors suggested that embryonic dopaminergic grafts clearly were functional, little or no improvement was observed in tests of hand preference, skilled hand use, and sensory neglect. Thus, even in the presence of surviving grafts, the amount of dopamine produced by grafted tissue and the specific neural location of this dopamine replacement may provide for variable effects on behavior.

Taken together, evidence provided by grafting studies in dopamine-depleted monkeys indicate that embryonic dopamine neurons of the optimal developmental stage can survive transplantation in numbers ranging from hundreds to thousands of dopamine neurons per graft, extend neurites into the surrounding host striatum, produce significant increases in the dopamine content of striatal tissue adjacent to grafts, and provide significant amelioration of some behavioral deficits over a period of several months. These early studies did not resolve issues associated with the optimal physical form of the grafted tissue, the number and location of implants for best therapeutic outcome, and the relative efficacy of cryopreserved tissue in nonhuman primates.

D. Toward Human Applications

In 1989, a report by scientists in Sweden described their initial attempt at grafting human embryonic dopamine neurons into patients with Parkinson's disease (Lindvall *et al.*, 1989). During the next 5 years, a limited number of studies found variable, modest improvement in the symptoms of Parkinson's disease following neural grafting. Interestingly, the one report of marked improvement following dopamine neuron transplantation was in two of the young adults that had ingested MPTP and had been assessed at Stanford University almost 10 years previously (Widner *et al.*, 1992). The marked success following grafting in these MPTP parkinsonian humans, in contrast to initial results from grafting in older individuals suffering from idiopathic Parkinson's disease, raised a variety of questions about whether grafts could be effective in elderly individuals, and whether the disease was less treatable than MPTP poisoning. Aside from the patients involved, the other critical difference in the successful grafting study was the amount of embryonic tissue grafted: 2 to 8 times more tissue was placed into the MPTP patients than had been utilized in other human studies. This increase in the number of dopamine neurons grafted was important, and in 1995 a report was published by Kordower and colleagues in which dramatic improvement of the symptoms of Parkinson's disease was achieved following grafting of similarly large amounts of embryonic dopamine neurons (Kordower *et al.*, 1995). To date, two of the Parkinson's disease patients that had received embryonic tissue grafts have come to autopsy. The first patient died 18 months after transplant surgery, due to causes unrelated to tissue grafting. Study of his brain revealed robust survival of grafted dopamine neurons and clear signs of behaviorally significant dopamine replacement (Fig. 11c). Although the long-term efficacy of neural grafts remains to be determined, and the philosophical and practical issues associated with the use of embryonic tissue remain controversial, it is possible that neural grafting could provide enduring therapy for Parkinson's disease.

III. RESULTS IN A NONHUMAN PRIMATE MODEL OF PARKINSONISM

The information to follow attempts to summarize key findings from our ongoing study of the ability of grafts of embryonic mesencephalic tissue to reverse MPTP-induced parkinsonism in a nonhuman primate model. These studies are a useful prerequisite to the consideration of human clinical trials, and currently provide critical adjunct information regarding a number of important factors critical to cell survival and functional improvement.

A. Optimal Embryo Age Is Younger Than Expected

We have examined nearly 200 animals that have received transplants of either embryonic cells or control procedures such as sham surgery or the implantation of non-dopaminergic cells to test the effects of grafted dopaminergic neurons on motor disability produced by the degeneration of dopaminergic neurons following the administration of MPTP. These grafts can increase dopamine content as measured in tissue punches taken near the location of well-developed grafts. These increases in dopamine content averaged 500% near grafts (Elsworth *et al.*, 1996a) and were nonexistent at sites that were not targets for grafts. Immunohistochemical staining for tyrosine hydroxylase, the rate limiting enzyme in the biosynthesis of dopamine and norepinephrine, revealed animals with over 28,000 dopaminergic neurons in multiple grafts derived from a single embryo. However, variability exists in the percentage of dopamine neuroblasts that survive grafting, which could be related to the age of the donor tissue. For example, developing neurons that have extended axons by the time of dissection are subject to degeneration induced by axotomy. Alternately, cells derived from developmental ages prior to neurogenesis of the substantia nigra might lack appropriate cues for cellular differentiation, including proper axon guidance to an appropriate target.

In an earlier study, survival of dopaminergic neurons was studied in 14 primate graft recipients. Grafts were selected from pregnant monkeys by ultrasonography, which permitted the selection of donor tissue samples from a range of embryonic days, such as E30–E47 (Sladek *et al.*, 1993b). The entire ventral mesencephalon was dissected from each donor and divided into six equal-sized blocks of solid (i.e., nondissociated) tissue. These samples were implanted into six separate sites bilaterally within the caudate nucleus of each animal. Thus, each site received one-sixth of the donor tissue that contained potential dopaminergic neurons. Cell counts then could be made of the total number of surviving dopaminergic neurons from each animal donor. Brains were collected at 3 to 6 months after surgery. The number of tyrosine hydroxylase positive neurons found within grafted neurons ranged from 1,300 to 28,400 per host; 9 of 14 animals contained between 8,200 and 28,400 dopamine neurons. The numbers of dopamine neurons in grafts seen at each age examined were E30(3,500), E38(13,780), E39(9,288), E41(21,296), E42(n = 4; 1,300, 18,000, 11,700, 12,136), E44(n = 4; 3104, 4,100, 15,064, 28,400), E47(n = 2; 8,200, 13,344).

Cell survival may be influenced by the state of differentiation of the developing neuron. Neurogenesis of the substantia nigra in an old world macaque, the Rhesus monkey, is believed to occur between days E36 and E43 (Levitt and Rabic, 1982). Comparable neurogenesis data does not exist in the species used in our studies, the African green monkey, but because both animals are old world macques with similar gestational periods, it is possible that neurogenesis occurs in a similar time sequence. Assuming this to be true, the average dopamine neuronal survival of donors (n = 11) taken during putative neurogenesis (E36–E43 ± 24 hr) of the substantia nigra was 12,560 ± 2456. The adult African green monkey has about 80,000 DA neurons in both substantia nigrae. Thus, an average survival of 12,560 neurons represents about 16% of the adult population; the best survival obtained (i.e., 28,400) is approximately 36%, which is substantially higher than results obtained from studies using xenografted human cells in rodents (Brundin *et al.*, 1988). These findings suggest that graft survival is improved significantly when donor tissue is taken during the developmental period that includes neurogenesis. The timing of these events in the African green monkey initially was estimated from data available for neurogenesis in the rhesus monkey. However, our preliminary work (Fig. 3) demonstrated that many dopaminergic neurons already were present in the ventral mesencephalon by the first day of putative neurogenesis (i.e., by E35), suggesting that neurogenesis may occur somewhat earlier in the African green monkey (Bundock *et al.*, 1997, 1998). In addition, the demonstration of early axonal outgrowth and the formation of a prominent

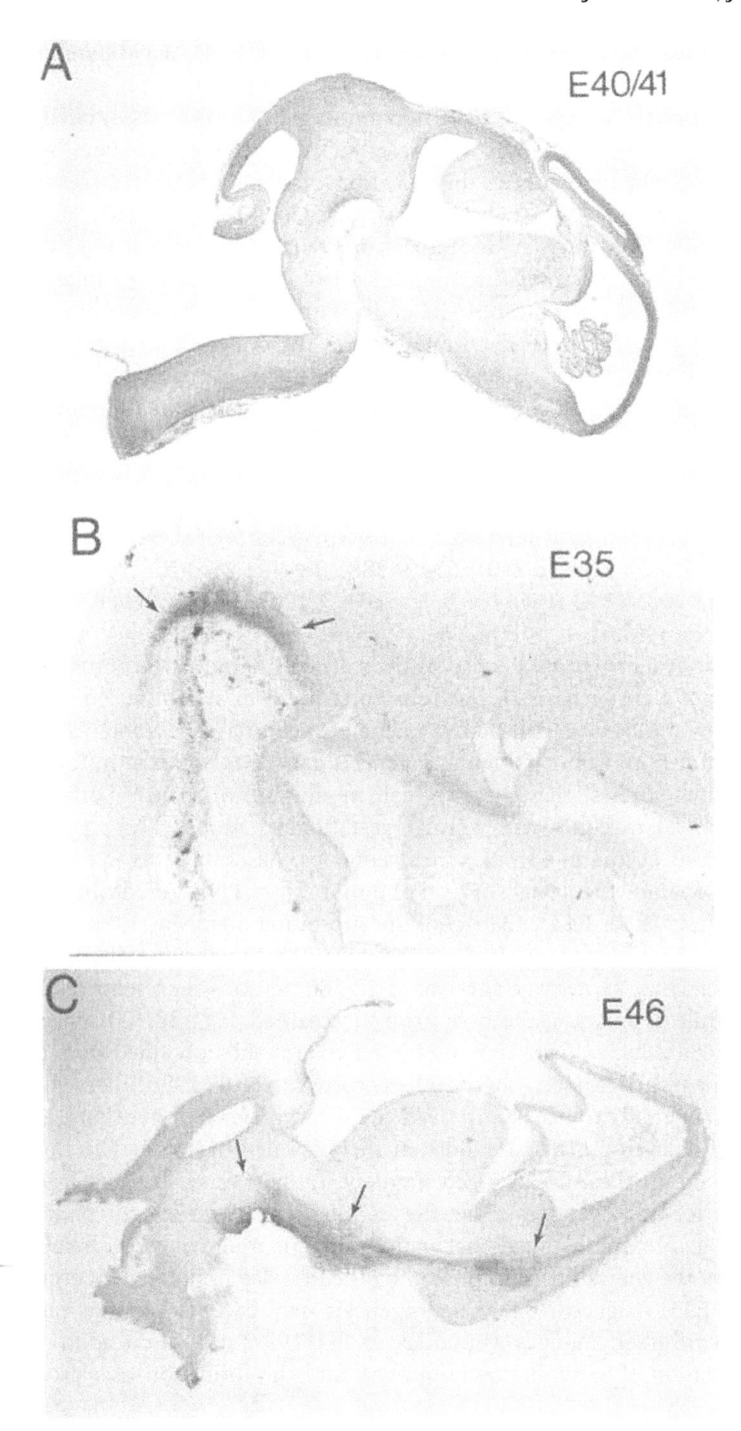
A
E40/41
B
E35
C
E46

ascending bundle of fibers by E40 suggested that younger donor tissue may be more desirable for grafting. Of interest, grafts from an early donor age (E30), which is younger than the estimated onset of neurogenesis, contained over 3000 surviving DA neurons. In order to understand how the stage of neuronal development at the time of transplantation may be affecting graft survival, it will be necessary to determine the precise timing of neurogenesis, migration, and differentiation using "birth dating" techniques such as the incorporation of 5-bromo-2′-deoxyuridine (BrdU) into DNA during the synthesis phase of the cell cycle (Bundock *et al.*, 1998). Preliminary analysis in our laboratory in the African green monkey over a wide range of developmental ages indicates that BrdU and tyrosine hydroxylase can be colocalized successfully using immunocytochemical techniques. BrdU-labeled cells were seen at their sites of origin, in migratory routes, and in the fully developed substantia nigra. Studies in progress of brains labeled over a range of developmental ages will permit the determination of (1) the precise timing of neurogenesis of dopamine neurons, (2) the spatial distribution of developing dopamine neurons in the mesencephalon, and (3) the donor age and dissection technique needed to maximize grafting of young neurons. Preliminary observations suggest that neurogenesis of the mesencephalic dopamine neurons may occur substantially sooner than anticipated in the African green monkey, which could move the transplant window up to even earlier time points than used currently.

B. Transmitter and Nucleus Phenotype Is Maintained in Grafts

It was not known if primate neurons taken from their developmentally correct locations (ventral mesencephalon) would develop normally because of potential differences in the trophic or tropic environment of the implant site (striatum). Although tropic factors for axon guidance and synapse formation would be expected to exist in the striatum because it does represent the target of nigrostriatal dopamine axons, one would not expect that trophic factors for

---

FIGURE 3 A demonstrates a parasagittal view of E40–41 embryo stained for tyrosine hydroxylase. The magnification is too low to be able to distinguish the cells that are seen to advantage in B and C. In B the dopamine neuroblasts have migrated to the floor of the mesencephlon by day E35. (arrows). Neurites rapidly grow rostrally from the mesencephalic groups of dopamine neurons to course through the forebrain in route to the striatum (arrows) as seen in C for an E46 embryo. At this stage, the neurons would be less likely to survive grafting due to the extensive axonal outgrowth.

neuronal survival would be present (except by coincidence) because the cell bodies of the dopamine neurons migrate from the germinal epithelium of the ventricle to the ventral mesencephalon instead of the striatum. Hence, there was the potential for poor survival of dopamine neurons in the primate if the milieu of essential developmental factors and cues was missing in the new target in the adult brain. Instead, it was found that grafts of ventral mesencephalic tissue into the adult striatum reproduced the organotypic appearance of the dopamine cell groups (i.e., A 8–10) as illustrated in Figure 4.

This finding extends to other central nuclear groups as well. Neurons of a specific nucleus of the hypothalamus, for example, may develop into pyramidal neurons if implanted into the cerebral cortex at some optimal time and with favorable growth factors present in the host environment. It is well known that neurons of peripheral sympathetic ganglia can be influenced by a specific growth factor to shift their transmitter phenotypic (Landis, 1980). Indeed, one of the first successful studies of neuronal plasticity demonstrated a shift in perikarya size, shape, and the number of processes to mimic a new environment when spinal ganglion cells were grafted into the cerebral cortex (Ranson, 1914). In nonhuman primates, grafted neurons have maintained both their geometric phenotype for a particular central nucleus as well as their readily identifiable transmitter-specific identity. Dorsal pontine grafts resulted in the appearance of two types of neurons reminiscent of the locus and subcoeruleal nuclei. They maintained their strong expression for tyrosine hydroxylase as demonstrated immunocytochemically. The same was true for hypothalamic dopaminergic neurons of the tuberoinfundiblar nucleus and the dopaminergic neurons of the mesencephalon, even when these cells were grown in the cerebral cortex. Clearly the signal that controls the transmitter phenotype was not altered by the relocation of neurons to less-appropriate sites with respect to their "normal" targets, but it is of interest that their neurites were less extensive then when the same cells were grafted into an appropriate target. Specifically, dopaminergic neurons of the substantia nigra grafted into the cerebral cortex did not show robust outgrowth into the cortex. Instead, the fibers remained within the grafts, suggesting a lack of directional signal to boost the ultimate success of the outgrowth in reaching appropriate targets.

C. Grafts Can Occupy and Supply Large Regions of the Target Area

Despite encouraging findings that grafted dopamine neurons can survive and release dopamine in the parkinsonian monkey brain, the animals may not show complete behavioral recovery. One factor that may prove critical to the degree of behavioral recovery seen after neural grafting is the extent to which

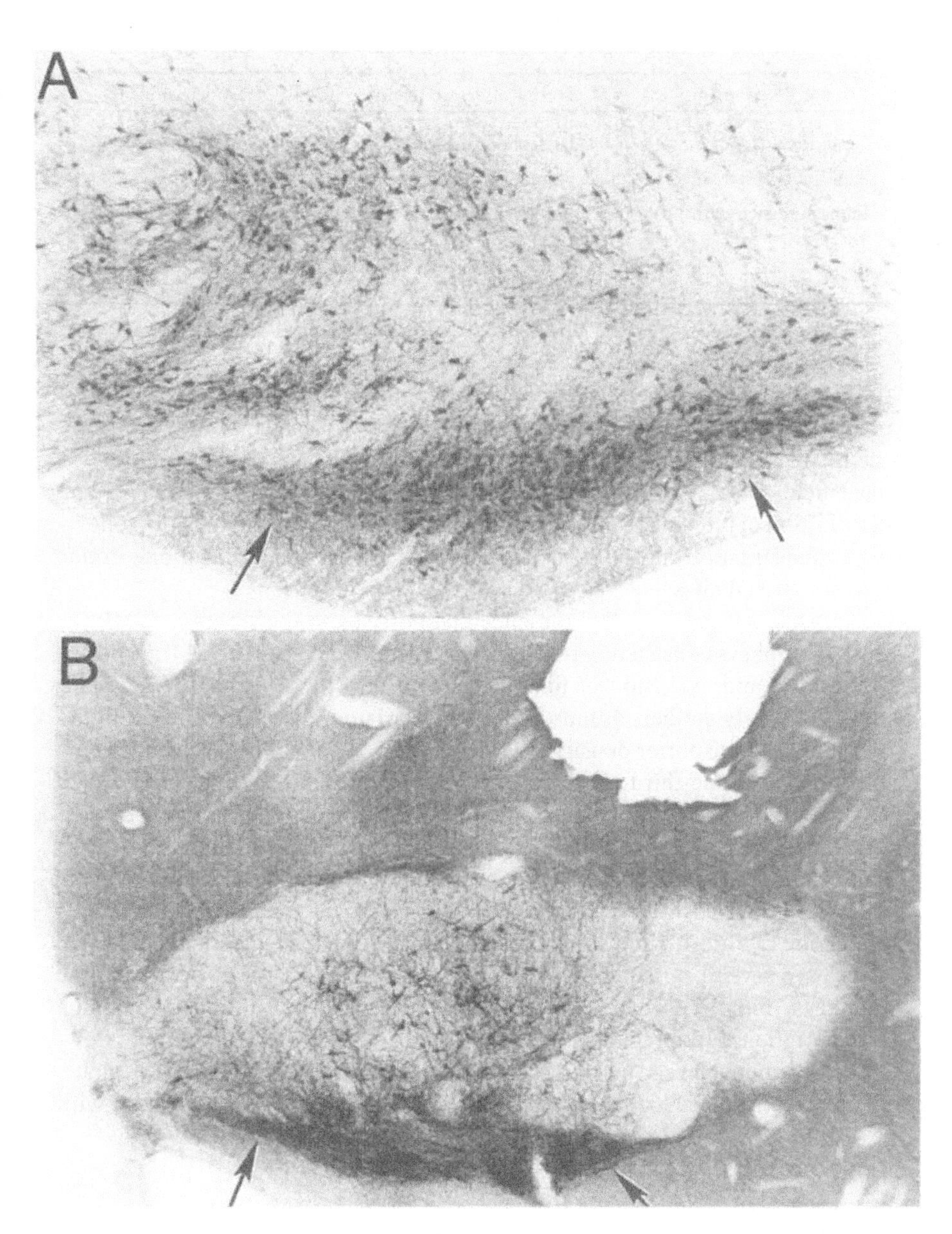

FIGURE 4 A and B compare the normal substantia nigra (A) in the African green monkey to a graft (B) seen after 6 months of survival in a host caudate nucleus. The position of the graft has been rotated 90 degrees counterclockwise so that the dense cluster of zona compacta neurons (arrows) appears to the bottom of each illustration. Here a dense packing of dopamine neurons is reminiscent of the normal substantia nigra and more widely distributed, large, multipolar neurons are seen dorsal to the compact zone. This cytoarchitectural arrangement reflects organotypic arrangments that are characteristic of the mesencephlic dopamine cell groups following transplantation.

TABLE I Factors Affecting Degree of Reinnervation Following Neural Grafting

Donor determinants	Technical determinants	Host determinants
1 Gestational Age	1 Dissection of fetal tissue	1 Extent of lesion
2 Supply of growth factors	2 Storage of fetal tissue	2 Immune status
3 Immune match with host	3 Implantation technique	3 Supply of growth factors
	4 Number of sites grafted	4 Time allowed between grafting and examination
	5 Amount of tissue per site	

the host striatum receives graft-derived dopamine. There are several determinants of dopaminergic reinnervation after grafting (Table I). In order to study the degree of reinnervation that is achieved in our severe-MPTP model, we have held several factors as constant as possible, and then measured the distribution of tyrosine hydroxylase and dopamine transporter densities following transplantation of undissociated ventral mesencephalon from one donor into six striatal sites.

Transplantation of ventral mesencephalon tissue to the striatum of MPTP-treated monkeys consistently resulted in an increased density of DA transporter in striatum and extension of fiber outgrowth into the host brain (Figures 2 and 5). In early studies, 4 mm tissue slabs containing a grafted site exhibited elevated DA transporter density (Elsworth *et al.*, 1994, 1996b), accounting for approximately one-third of the volume of the grafted striatal nucleus (caudate or putamen). Preliminary data from studies grafting greater amounts of fetal tissue to each site, or grafting a larger number of sites, indicate that these strategies substantially increase the proportion of the striatum that is reinnervated (Elsworth *et al.*, 1998). Even with this increased replacement of dopaminergic tone to the striatum, some parkinsonian motor abnormalities can persist. There are several possible reasons for the remaining parkinsonism in these monkeys. First, it is conceivable that the degree of elevation in dopamine concentration in the regions reached by graft fibers remains subnormal; however, dopamine levels in micropunches taken in the vicinity of the grafts (Elsworth *et al.*, 1996a) and dopamine transporter densities associated with

FIGURE 5 A single graft placed within the caudate nucleus can have a far-reaching influence because of the extensive outgrowth of dopamine fibers. This low-power view illustrates not only the graft, but the density of the surrounding caudate as well as outflow (arrows) through the internal capsule (IC) to the adjacent putamen. The graft is seen to advantage in B and is the same graft that was illustrated in Figure 12.4B following rotation to mimic the position of the substantia nigra *in situ*. This phenomenon of outflow is seen again in Figure 12.6.

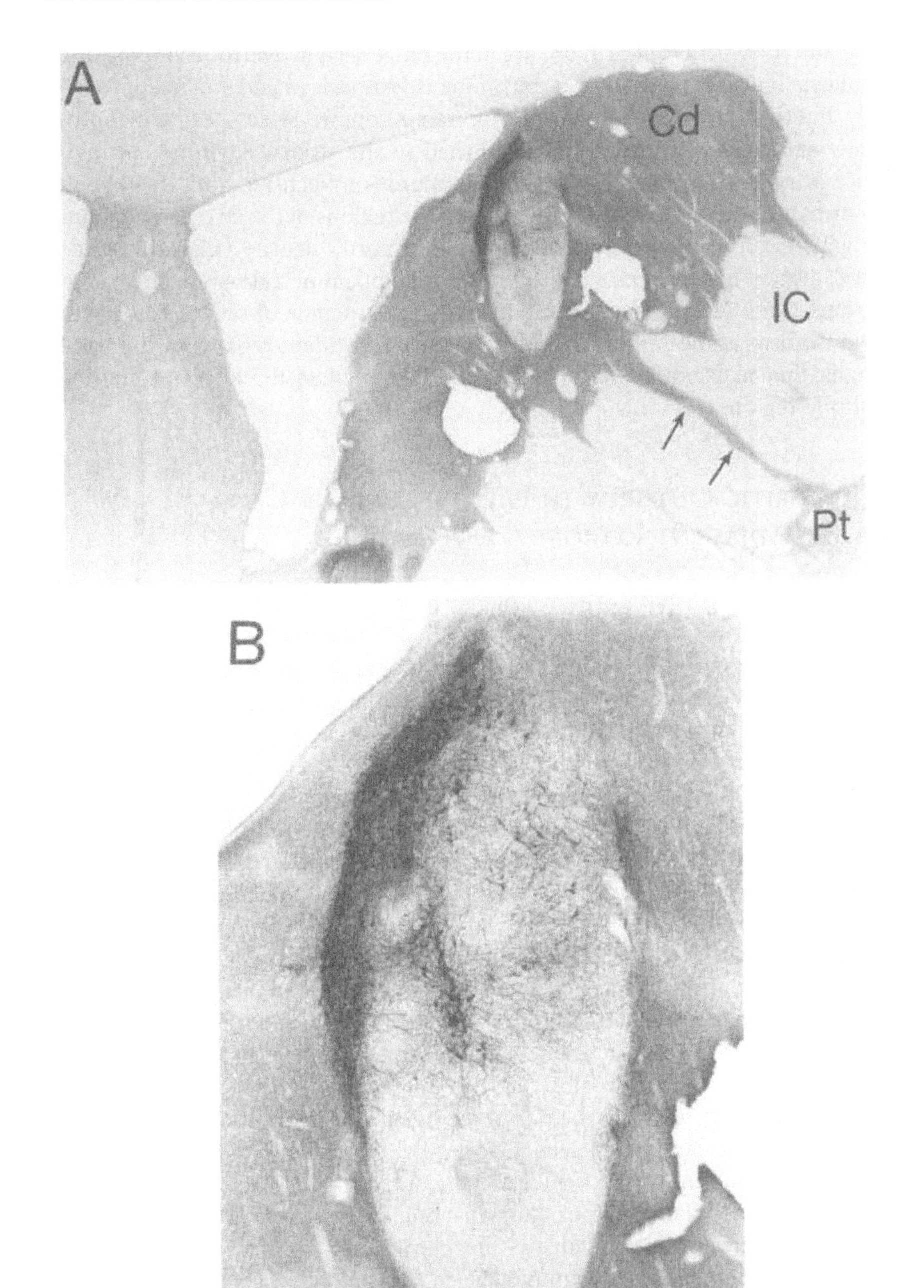
A
Cd
IC
Pt
B

the grafts (Elsworth *et al.*, 1996b) are in the range seen in overtly asymptomatic monkeys. It also is possible that dopamine is synthesized and stored appropriately, but that release of the neurotransmitter is compromised because inappropriate synaptic connections are established in the striatal environment. Evidence of synapse formation with grafted neurons (Wechsler *et al.*, 1997) and of normalization of D2 receptor binding in the regions of the striatum containing graft-induced increases in dopamine transporter density (Elsworth *et al.*, 1998) argues for the existence of sufficient dopamine release from grafted neurons. Another important component in the outcome of neural grafting in parkinsonism is the importance of dopamine-dependent regions of the basal ganglia that may not be fully reinnervated, as well as the effect of grafting multiple regions.

D. Neuritic Outgrowth Is Extensive and Can Invade Adjacent Regions

The extent to which grafts of dopaminergic neurons can resupply *both* a targeted brain site and host structures *beyond* a grafted region has been examined. For example, in some clinical studies of fetal dopaminergic neuronal grafting, patients have received neural grafts in either the caudate, the putamen, or both; in some instances, grafts into only one of these two striatal targets resulted in general improvement raising questions regarding mechanisms of graft-induced recovery. We tested the effect of graft placement into caudate and putaminal regions that were located in proximity to the internal capsule, because the capsule has interfascicular gray regions of striatal origin termed "pontis grisei caudato lenticulares" (Riley, 1943) that could serve as potential channels for axonal growth into the adjacent nucleus (Fig. 6). Ten monkeys were grafted with solid pieces of ventral mesencephalon bilaterally into either the putamen or the caudate, and were examined for neuroanatomical and biochemical markers of viable grafts between 3 and 6 months after surgery. Survival of grafted DA neurons was comparable to that reported earlier, with as many as 6000 DA neurons observed in a single graft. Dense tyrosine hydroxylase labeling was present in the dorsal putamen in animals that received grafts solely within the caudate nucleus. Conversely, tyrosine hydroxylase immunolabeling was present in the ventrolateral caudate when grafts were placed in the underlying putamen (Fig. 6). Dopamine content was elevated in these sites of "crossover" fibers, but was not elevated in regions lacking tyrosine hydroxylase reinnervation. Immunolabeled fibers could be traced from grafted neurons into the surrounding neuropil of the striatum, and appeared continuous with fibers that coursed through the internal capsule en route to an adjacent

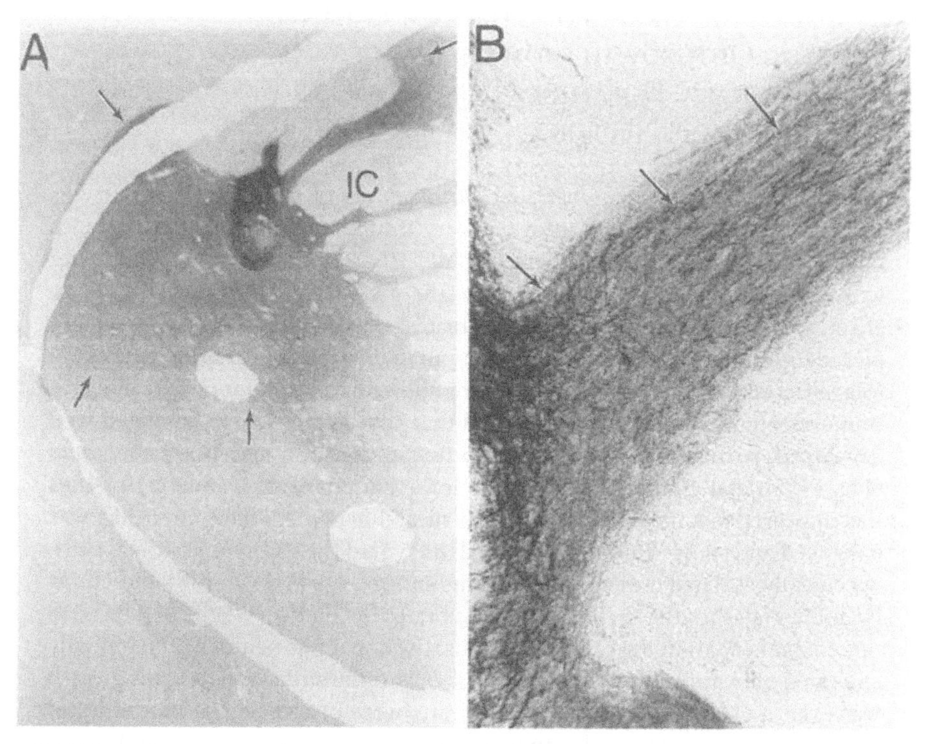

FIGURE 6 In A, a reasonably small, but densely packed graft shows an area of dopamine resupply that extends both ventral and dorsal to the graft (arrows). The fibers that extend dorsally are seen to advantage in B. They course between the fascicles of the external capsule through an area designated the pontis grisea caudato lenticulares. The importance of this finding may be that solitary grafts can influence both the caudate and putamen and, if so, could reduce the need for the total number of grafts to produce behavorial improvement in humans.

nucleus. This phenomenon was not observed in sham-operated animals. These fibers did not appear to invade the caudate from the underlying nucleus accumbens, and there was no density gradient in the putamen that would favor the interpretation that sprouting occurred from fibers of nigrostriatal origin; the latter phenomenon would exhibit increased fiber growth in the caudal putamen, with a density gradient decrease into the rostral striatum.

Thus, the ability of dopaminergic neurons to supply both the caudate and putamen nuclei from a single graft suggests a mechanism whereby a graft placed into one striatal nucleus can influence the adjacent nucleus. This finding could affect the design of transplant protocols with respect to the location and number of grafts needed for attaining full reinnervation of the striatum.

E. Embryonic Grafts of Striatum Enhance Neuritic Outgrowth from Grafts of Mesencephalic Dopamine Neurons and from Host Dopamine Neurons

Yurek and colleagues demonstrated that cografts of ventral mesencephalon and striatum achieved behavioral recovery from a dopamine deficit rat more rapidly than sole nigral grafts (Yurek *et al.*, 1990). This appeared to correlate with enhanced neurite outgrowth from the grafted DA neurons. This tropic effect was confirmed in nonhuman primates: nigral and striatal cografts were placed into either the caudate nucleus or putamen of 10 MPTP-treated monkeys (Sladek *et al.*, 1993a). Solid grafts were allowed to grow for 3 to 6 months. Immunocytochemical analysis revealed that striatal grafts were large and well developed, particularly when placed in juxtaposition to mesencephalic grafts (Fig. 7). Striatal grafts also received a dense pattern of dopamine varicosities that appeared maximal in apposition to striatal neurons. Fiber outgrowth was enhanced and showed a directionality that favored the embryonic striatal cells. Specifically, isolated mesencephalic grafts showed fiber growth in all directions from the graft into the surrounding host brain. In contrast, when cografts were placed 2 mm apart in the caudate nucleus, the dopamine neurites preferentially extended outward from the nigral grafts toward the striatal grafts. This growth was even more dramatic when the cografts were placed on opposite sides of the brain. In this instance, dopamine neurites grew across the midline through the corpus callosum, toward the striatal graft in the contralateral caudate nucleus. Sprouting of host dopaminergic fibers into the isolated striatal grafts also was seen in the lateral portion of the caudate nucleus. These findings suggest the presence of striatally derived tropic influences from the adult host on grafted as well as host dopaminergic neurons. The presence of tyrosine hydroxylase-immunolabeled fibers of host origin in the striatal grafts suggests a mechanism that might be useful in the treatment of early stage parkinsonism through the stimulation of the residual population of dopaminergic neurons of the substantia nigra by a developmentally derived trophic or tropic factor(s).

FIGURE 7 This series of illustrations of co-grafts depict in A, a low-power view of bilateral grafts which are seen to advantage in B–D. In B, the graft on the left side of the brain reveals a dorsal portion that is densely supplied with dopamine and a ventral portion that contains dopamine neurons. This is a co-graft of embryonic striatum (St) that was placed dorsal to embryonic substantia nigra (Sn) as indicated by the horizontal dashed line. Cells in the nigral portion, as seen in C and D, preferentially supplied fibers dorsally into the embryonic striatum. In D, individual fibers are seen to penetrate the cluster of striatal neurons.

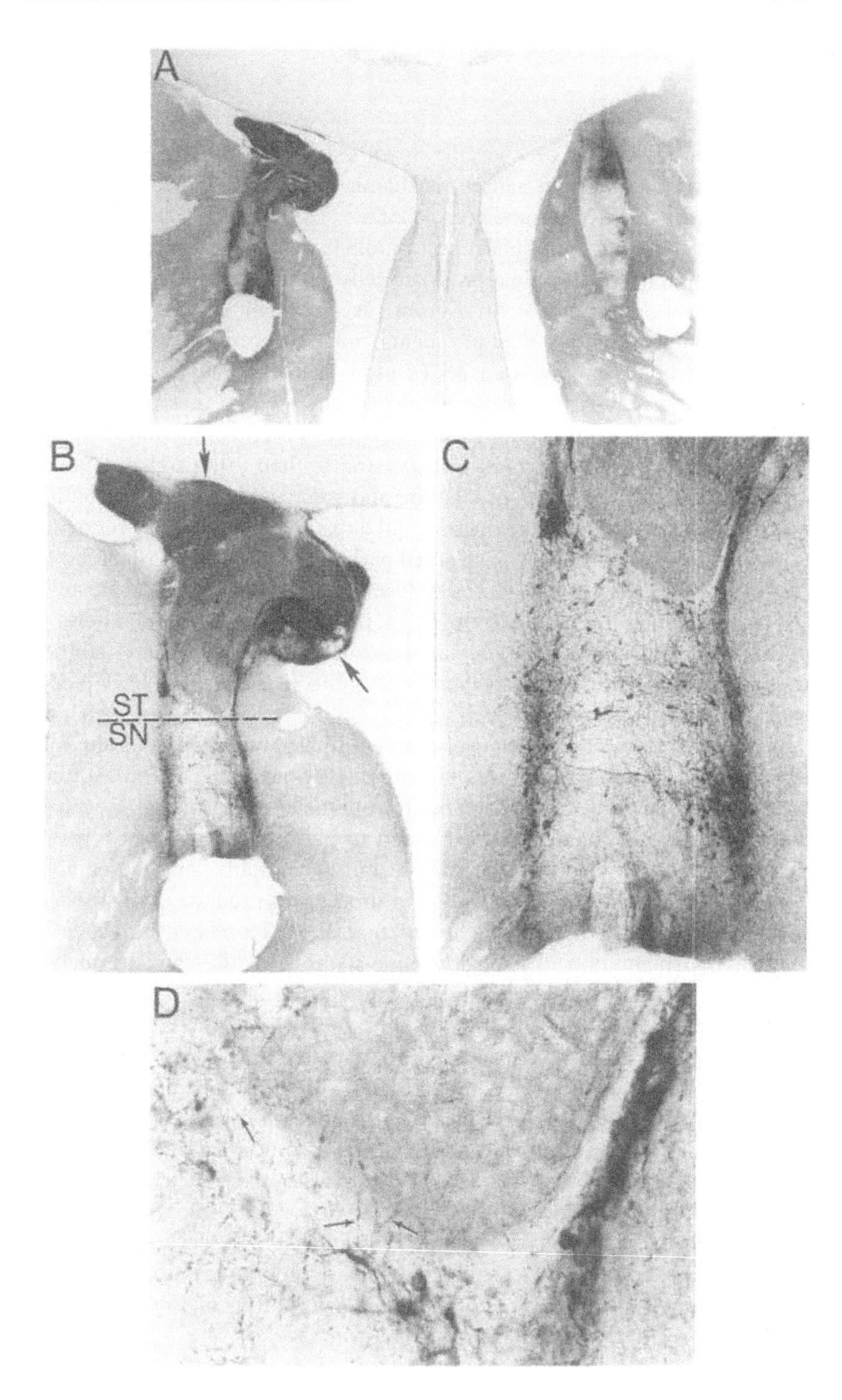
A
B
C
D
ST
SN

F. Sham Surgery Does Not Produce Host Sprouting

Depending on the nature of a specific hypothesis that is undergoing testing, control groups in studies on neural transplantation include animals that have been rendered parkinsonian and receive either (1) no fetal cell implantation, (2) introduction of an empty injection cannula into the target region ("sham" graft), or (3) implantation of inappropriate cells to the target region. However, the inclusion of appropriate control groups in transplant studies in primates has received more attention than in rodents, notably because monkeys appear to have varying susceptibility to MPTP, and primates that fail to become severely parkinsonian tend to recover behaviorally without therapeutic intervention. This problem in variability has sometimes been compounded by the tendency to use fewer subjects in primate studies than using rodent studies. Furthermore, there has been speculation, and some evidence, that sprouting or regeneration of residual host mesostriatal dopamine neurons may represent the basis of recovery in some transplanted parkinsonian monkeys (Bankiewicz *et al.*, 1990a; Fiandaca *et al.*, 1988). Because understanding the mechanism of graft-induced amelioration of parkinsonism in primates is critical to efforts to improve the technique for possible application to humans, we have studied the effects of sham surgery and inappropriate grafts in the severe-MPTP model in the nonhuman primate.

Because dopamine innervation of the ventral striatum is less sensitive to MPTP toxicity than that of the dorsal striatum (Elsworth *et al.*, 1996b), sprouting, if present, would be expected to occur from the ventral striatum. However, in our severe-MPTP model we have failed to observe sprouting of tyrosine hydroxylase positive neurons from this region. Quantitative autoradiography supports the view that sham implantation procedures produce no increase in dopamine transporter density in the striatum of MPTP-treated monkeys. In contrast, transplantation of fetal dopamine neurons to the caudate nucleus or putamen of MPTP-treated monkeys resulted in a significant elevation of dopamine transporter density (Elsworth *et al.*, 1996b). Measurements of dopamine concentration in the striatum revealed that grafts of late-stage fetal ventral mesencephalon (beyond E50), fetal cerebellum, or sham implantation did not increase dopamine concentration near the graft site, whereas transplantation of early (E38–E49) gestational age fetal ventral mesencephalon (Sladek *et al.*, 1993a) was associated with marked elevation of caudate nucleus dopamine concentration in the vicinity of the graft (Elsworth *et al.*, 1996a). These morphological and biochemical data strongly indicate that in the severely parkinsonian MPTP-treated African green monkey, sham surgery or inappropriate tissue transplantation does not result in regeneration or sprouting of dopamine terminals. So how are these data reconciled with reports of sprouting of dopamine fibers following sham surgery or inappropriate grafts in the MPTP-treated primate? The most likely explanation is that differences in the model or transplantation technique are responsible for the observed differences in sprouting. We have speculated previously (Elsworth *et al.*, 1996b) that the level of MPTP-induced striatal dopamine depletion is a major factor in determining the behavioral outcome of neural grafting. One extreme example is a chronic, severely lesioned striatum, in which there is little or no capacity for regeneration. Here, dopamine-based behavioral recovery would be due entirely to graft-derived dopamine. At the other extreme would be a recent partially lesioned striatum in which regeneration could be induced by tissue-derived or host-derived growth factors. Here, the amount of dopamine that is derived from the graft that is necessary for behavioral recovery would be minimal. In fact, in MPTP-treated monkeys that were not in the severe category, we have observed some sprouting and occasionally the expression of small intrinsic tyrosine hydroxylase positive neurons in the striatum. This is not to infer that one situation is better than the other, and in fact the extreme examples given above may both be useful in different situtations. The more severe lesion may be akin to later-stage parkinsonism, whereas the lesser lesion may be a model of mild parkinsonism. However, as alluded to earlier, the use of a monkey with a substantial reserve of host nigrostriatal dopamine fibers is more complicated and attributing behavioral recovery to one particular factor may not be possible.

G. Grafts from Multiple Donors Can Result in More Extensive Recovery of Dopamine

To determine the capability of the host brain to accept tissue from more than one donor and to enhance the total amount of DA recovery in the caudate nucleus, experiments have been performed in which twice the amount of fetal tissue has been grafted in our standard primate paradigms (Sladek *et al.*, 1998). Solid ventral mesencephalic tissue was collected from fetuses at embryonic days 40 to 42 and grafted into each of eight MPTP-treated monkeys as shown in Fig. 8. Each host adult received tissue from two donors. Six penetrations were made per brain, each carrying tissue from one-third of a donor ventral mesencephalon. Upon neuroanatomical analysis, grafts were comparable to those seen in prior experiments and contained dense clusters of dopaminergic neurons (Fig. 8). Neuritic outgrowth was extensive from most grafts, and extended into the host caudate nucleus. Although some clusters of dopamine neurons appeared near the periphery of grafts, many were distributed uniformly throughout the grafts. Several grafts were larger than those seen in our prior

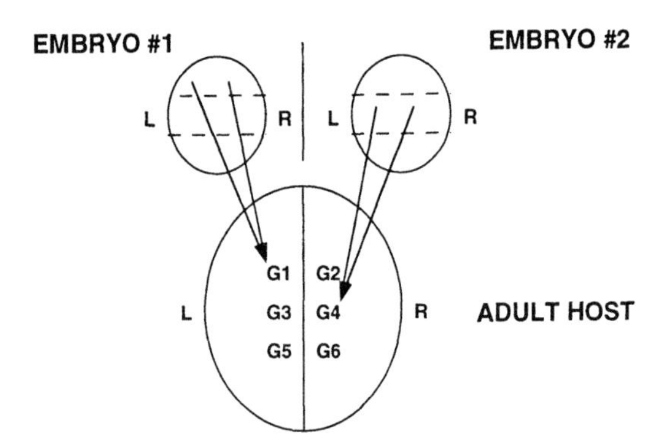

FIGURE 8 This schematic demonstrates the distribution of tissue from two embryos grafted into a single adult. All tissue from one embryo was grafted to one side of the adult brain. Thus, twice as much tissue in this "maxi-load" experiment was grafted into each of our six standard implantation sites.

studies and extended the entire distance from the lateral ventricle to the anterior limb of the internal capsule.

H. Appropriate Synapses Are Formed between Host and Graft

Synaptophysin immunolabeling has been used to determine whether synaptic remodeling occurs after fetal dopaminergic neuronal grafting in primates. This integral membrane protein in the presynaptic vesicle has been shown to be a reliable marker of terminal differentiation. Sections from several animals were examined that contained robust grafts double-labeled for synaptophysin and tyrosine hydroxylase in an attempt to assess the relationship of synaptophysin immunoreactivity to grafted dopamine neurons (Sortwell *et al.*, 1998). Results showed a dense pericellular array of synaptophysin varicosities, in juxtaposition to dopamine cell bodies in the grafts. This synaptophysin labeling appeared more robust than that normally seen within the host striatum. It also was seen in the form of linear profiles along nerve fibers, a finding that is reminiscent of histochemical patterns of monoaminergic neuron staining. Cografts of mesencephalic and striatal tissue to the striatum were also examined and contained prominent synaptophysin-labeled fibers within the grafts as well as fibers that extended outside the grafts. This suggests that the monoaminergic fibers not only have marked axonal and dendritic growth from their cell bodies, but that this growth is capable of extending into the host brain and of forming synapses. These findings suggest that functional synaptic reorganization occurs in the primate brain following grafting of mesencephalon as well as striatum, and taken together with ultrastructural analyses provides evidence for the formation of functional synapses (Wechsler *et al.*, 1997).

The fine morphology and transmitter phenotypes of grafts to the striatum, and particularly of mesencephalic plus striatal cografts, have been examined using several different immunolabels including DARPP-32, substance-P, glutamic acid decarboxylase (GAD), and metenkephalin (Wechsler *et al.*, 1995). DARPP-32 immunoreactive fibers within co-grafts appear to be organized in a manner similar to those of the intact striatonigral pathways. The mesencephalic component of striatal and nigral cografts contain abundant dopamine neurons. Many of these neurons are contained within a network of small diameter DARPP-32-positive fibers that resemble the organization of the substantia nigra *in situ,* and exhibit a greater density than those seen in solitary mesencephalic grafts. Some of the DARPP-32 and substance-P-positive fibers appeared juxtaposed to dopamine neurons, thus displaying an arrangement similar to that seen *in situ.* These enlargements are suggestive of potential points of synaptic contact. The source of this potential input to the dopaminergic neurons in the center of the grafts most likely is the cografted embryonic striatal tissue, but that can only be determined unequivocally with the use of host-specific or graft-specific markers. Correlative light microscopy reveals that substance-P immunoreactive fibers cross the interface between host and graft. In some instances, these fibers appear to originate in the striatal patch compartments of the host caudate nucleus. Substance-P immunolabeling was most concentrated in those graft regions that also contained TH-immunoreactive neurons, and some substance-P-positive enlargements were found in juxtaposition to TH-immunolabeled neurons and processes. At the electron microscopic level, many of these enlargements appeared to resemble synaptic contacts (Wechsler *et al.*, 1997).

I. Cryopreservation Can Result in Substantial Cell Survival

Prior studies utilizing either embryonic rat or human cryopreserved neurons demonstrated survival of DA neurons, a finding that was encouraging from a qualitative perspective (Collier *et al.*, 1987). Specifically, cryopreserved dopamine neurons were observed in well-circumscribed grafts and compared favorably in size and shape to those neurons seen in fresh tissue grafts from the ventral mesencephalon. Cell numbers, on the order of a hundred neurons per graft, were comparable to the best-surviving fresh grafts that were obtained at the time (1985–1988). Subsequent studies, however, have demonstrated that nonhuman primate neurons harvested during the period of neurogenesis of

DA neurons show significantly greater graft survival, as described earlier. Cryopreservation of cells taken during neurogenesis was examined in a subsequent experiment that also utilized a computer-assisted freezing apparatus in an attempt to stabilize certain technical variables as well as to provide a consistent freezing protocol.

Six adult male African green monkeys were grafted with E44 ventral mesencephalic tissue that was dissected in such a manner that precisely one-half of the mesencephalic tissue was implanted as fresh, solid grafts while the other half was cryopreserved, stored briefly, thawed, and implanted into the contralateral side of the same recipient brain. Thus, three fresh grafts were placed into the right caudate nucleus while three equal-sized grafts of cryopreserved tissue were implanted into the left caudate nucleus. Frozen cell grafts were implanted within 3 to 4 hours of placement of the fresh grafts. Because the embryonic mesencephalon was bisected into equilateral halves, the same potential number of dopamine neuroblasts was implanted into each hemisphere. Grafts were allowed to grow for 3 months. Brains were removed for immunohistochemical analysis of dopamine cell and fiber pattern analysis.

Grafts of cryopreserved dopamine neurons were found in all six animals. Two of six animals showed substantial grafts and one had grafts that were comparable in size and DA neuron number to fresh grafts (Fig. 9). Grafts were characterized by the presence of DA neurons that conformed to the known cytoarchitectonic features of ventral mesencephalic DA neurons. The average

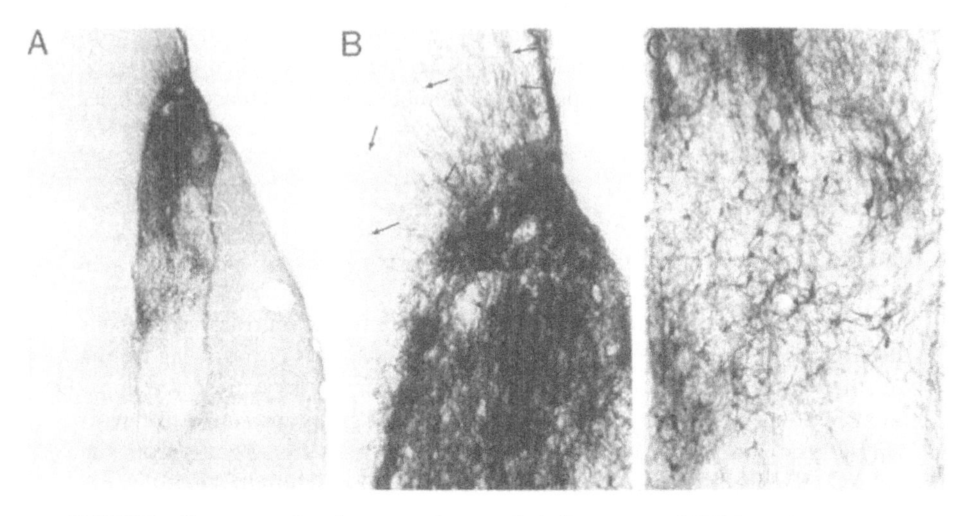

FIGURE 9 Cryopreserved grafts can produce survival of as many as 1,000 dopamine neurons in a single graft as seen in A. In B, fiber outgrowth (arrows) is comparable to that seen in fresh-tissue grafts, and cell survival depicted in C likewise is comparable to that seen in unfrozen tissue.

number of DA neurons seen on the side of the brain that contained cryopreserved neurons for the six animals was 532, in comparison to grafts of fresh tissue that averaged 6020. Even though some counts were incomplete due to minor tissue loss during histological processing, fewer DA neurons were seen in grafts of cryopreserved tissue. Nevertheless, some cryopreserved grafts were comparable to fresh grafts, both qualitatively and quantitatively. For example, two of three frozen cell grafts in the best animal each contained over 1000 DA neurons, which probably are underestimates because the cell packing density was so great that it was impossible to visualize all dopamine positive neurons in these grafts. Fiber outgrowth from these two cryopreserved grafts was comparable to that seen in fresh grafts (Fig. 8). These grafts were positioned deep within the head of the caudate nucleus and were generally morphologically indistinguishable from fresh grafts.

These preliminary observations support the view that cryopreservation of primate fetal neural tissue may be refined to obtain significant survival after transplantation. However, the substantial variability in dopamine neuron survival impacts functional improvement following transplantation. Thus, it will be necessary to develop a more consistent and effective cryopreservation procedure for preserving fetal dopamine neuroblasts to enhance viability and function after transplantation.

J. Micrografts Appear To Optimize Cell Survival and Resupply of the Striatal Target

Our most recent experiments suggest that optimal grafts can be achieved when smaller pieces of solid tissue are implanted. This experiment was designed to test the survival characteristics of solid tissue grafts of smaller dimensions than those used routinely, following a suggestion that "micrografts" provided significant advantages in rodent transplant studies (Nikkah *et al.*, 1993, 1994). A potential advantage is that such grafts may be more rapidly vascularized because the depth of tissue through which vessels must grow is reduced and penetration of nutrients through the interstitial space would be enhanced with thinner grafts. Also, the reduced size of the tissue might favor a greater number of placements and result in less local disruption at the implantation site. Eight adult monkeys were implanted according to the protocol illustrated in Figure 9. Tissue grafts were 50% smaller than those of prior experiments and were placed into twice as many locations within the caudate nucleus.

Excellent survival was seen in the micrografts (Fig. 10). Considerable outgrowth of TH-immunolabeled neurites occured throughout the host striatum and the reinnervation was so complete that a patch-matrix phenomenon was apparent. The placement of two grafts (instead of one) at each anteroposterior

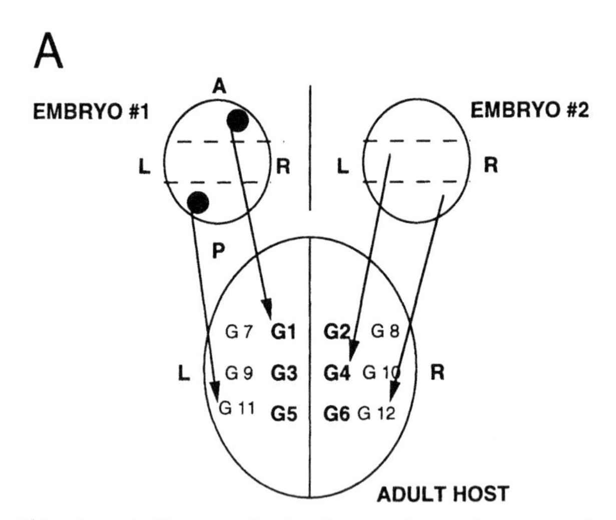

FIGURE 10 This schematic illustrates the distribution of tissue from two embryos into twice as many sites, which was done in an attempt to take advantage of presumptive availability of trophic factors in the adult host striatum. A successful graft is illustrated in B and is characterized by thousands of dopamine neurons and extensive outgrowth into the host brain.

level of the striatum resulted in more complete supply patterns of dopamine-rich fibers throughout the striatal neuropil that seen previously. Moreover, grafts of TH-positive neurons were found both medially and laterally in the caudate, discounting the notion that graft survival is enhanced by proximity to the lateral ventricle; no substantive difference could be seen qualitatively in these grafts.

K. Behavioral Improvement in Primates

Several studies have investigated the *functional* effects of ventral mesencephalon allografts into various striatal sites. Typical sites for grafting include the caudate nucleus and the putamen in MPTP-treated monkeys, and findings have been compared to functional restoration observed in control-grafted (i.e., cerebellar) and sham surgical techniques. Because significant variation has been found in the response of individual animals to MPTP, and evidence exists for "spontaneous" recovery in parkinsonian MPTP-treated monkeys, it has been essential to evaluate the level of lesion severity and to compare findings to control subjects in any study using MPTP as a model over time (Taylor *et al.*, 1994, 1997). The stability of deficits is greater when the clinical signs include those unique to Parkinson's disease in more severely affected subjects, including severe immobility, tremor, bradykinesia, and akinesia. However, seriously im-

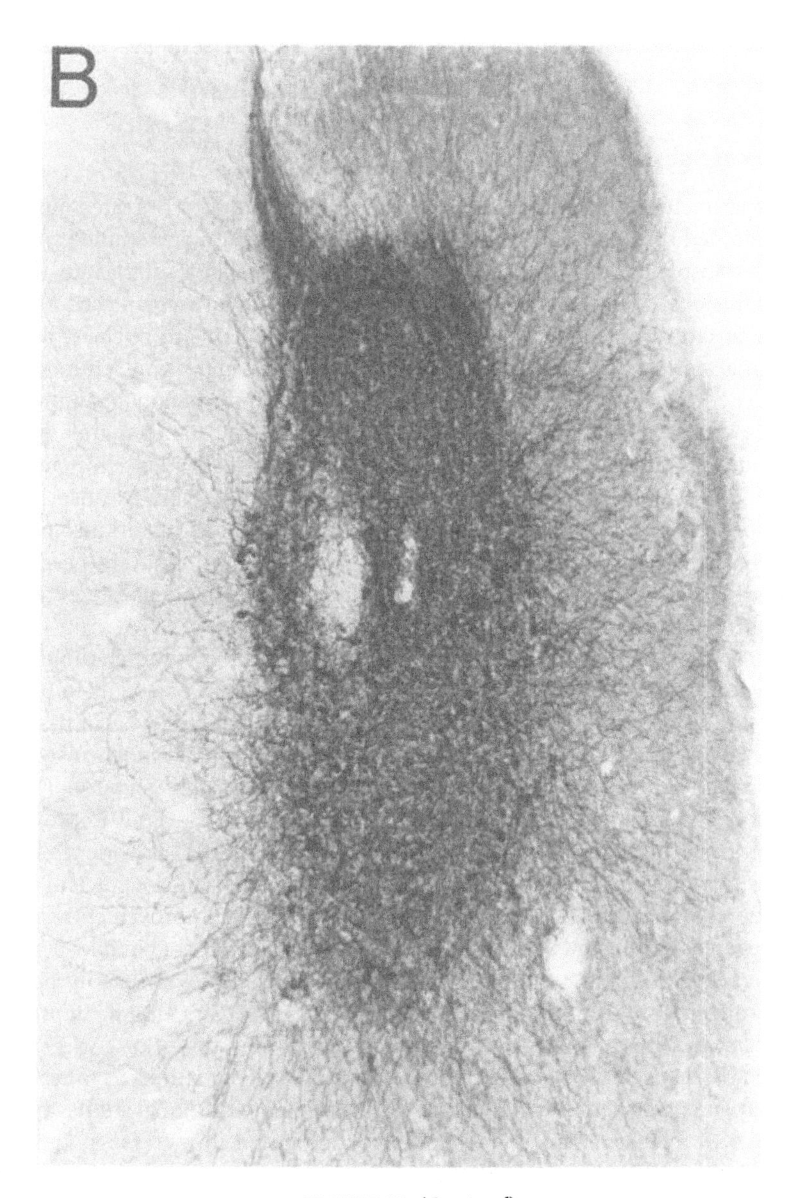

FIGURE 10 (*Continued*)

paired, bilateral MPTP-treated parkinsonian animals require additional medical and nursing care, which is justified if the predictive power of the model is greater than that of other models. Thus, more recent behavioral studies have examined the functional effects of grafts exclusively in severely parkinsonian monkeys.

1. Grafts of Substantia Nigra into the Caudate

Early experiments with moderately to severely debilitated MPTP-treated monkeys examined the behavioral effects of three groups of monkeys studied after bilateral transplantation: (1) animals grafted with substantia nigra into the caudate nucleus; (2) animals grafted with substantia nigra tissue placed outside the striatum or cerebellar tissue placed into the caudate nucleus; and (3) normal control monkeys that received substantia nigra grafts into the caudate (Redmond *et al.*, 1986a,b, Taylor *et al.*, 1991). Grafts of nondopaminergic tissue appeared behaviorally ineffective compared to substantia nigra tissue grafted into the caudate nucleus. These data provided some evidence for graft-mediated recovery in severely affected MPTP-treated subjects, because the inappropriately grafted subjects did not show improvements in parkinsonian signs. Nevertheless, additional subjects and control conditions were needed to replicate and extend initially positive findings and to evaluate the extent to which recovery requires graft-derived dopamine release.

Later studies confirmed that early mesencephalic tissue-grafted subjects with severe parkinsonism showed significant functional improvements in parkinsonism (Taylor *et al.*, 1995) as well as increased dopamine levels in tissue samples taken close to the graft compared with samples from a similar rostral-caudal punch taken away from the graft 8 to 9 months after surgery. This difference in dopamine levels significantly correlated with overall improvements in parkinsonian summary scores and provided additional evidence for the role for mesencephalic tissue implantation into the caudate nucleus in restoration of function. However, the significant functional improvements seen in the severely parkinsonian subjects after nigral grafts were not complete (see Taylor *et al.*, 1995). Although function gradually improved over time in these severely affected subjects, a complete reversal of parkinsonism, despite impressive grafts and increased striatal dopamine levels near the grafts, was not observed. Perhaps the time course (8 to 9 months) was not sufficient to result in a complete reversal of MPTP-induced parkinsonism. Additional studies that attempted to further enhance graft-host integration and to increase amounts of tissue within the striatum have been pursued in recent studies.

2. Grafts of Substantia Nigra into the Putamen

The ability of nigral grafts placed into the caudate or putamen to improve functional outcomes has also been investigated in primates after bilateral MPTP lesions. In initial studies, putaminal grafts did not result in significant functional improvement (Taylor *et al.*, 1995). However, these conclusions were limited by surgical bleeding complications that contributed to group variability. More recent studies have indicated some behavioral improvements in MPTP-treated monkeys with putaminal grafts compared to sham-grafted subjects (Taylor *et al.*, 1998, submitted). However, the consistency and degree of functional improvement was inferior to that observed if grafts were placed into the caudate nucleus ($n = 8$). In both groups, significant increases in dopamine levels, neuronal growth, and dopamine neuron numbers were detected. Overall, these studies conclude that in the MPTP-treated monkey, implantation of cells into the caudate nucleus results in more consistent and extensive improvement in parkinsonism signs than grafts placed into the putamen.

3. Age of Donor Tissue

Grafts from younger fetal tissue (<4 cm crown-rump length) showed greater quantitative neuronal survival than grafts from older fetuses (>14 cm crown-rump length), as first documented by Sladek and associates (1993a). Behavioral and functional analyses of these groups of subjects confirmed that behavioral improvement occured in the groups with the greatest surviving nigral tissue placed into the caudate. Recent data support a relationship between stage of neurogenesis, graft integration, and host connectivity. Tissue from E38–44 (±2 days) exhibited excellent survival in the host brain, but survival significantly dropped 1 week after completion of neurogenesis. Moreover, neither sham surgery nor cerebellar tissue implantation in severely parkinsonian monkeys resulted in significant improvement in parkinsonian signs. The lack of clinical effects in recipients of these control grafts was similar to observations in severe parkinsonian monkeys that received no surgery. These data confirmed the hypothesis that grafts of "early" nigral tissue enhance functional improvement due to better graft survival, integration, and outgrowth, with resulting increases in dopamine production in the striatum.

4. Effects of Implants of Non-Dopaminergic Tissue or Sham Surgery: Can Lesions Alone Induce Functional Recovery in MPTP-Induced Parkinsonism?

Some controversy has existed with regard to whether simply lesioning the striatum in the course of a surgical grafting procedure is sufficient to induce some functional recovery in MPTP-induced parkinsonian monkeys. Lesion-induced sprouting, lesion-induced production of neurotrophic factors, and "mini-pallidotomy" are potential mechanisms of recovery that could result from simply introducing needles or grafting cells into the brain. However, previously it was reported that primate controls with "inappropriate" tissue-type grafts or inappropriate graft placement (Redmond *et al.*, 1986) failed to improve. Cerebellar grafts into the caudate in a small number of primates also failed to ameliorate MPTP-induced parkinsonism (Taylor *et al.*, 1991). More recently, these earlier reports were confirmed and extended to include the effects of sham surgery (injection of lactated Ringer's solution without graft tissue) in a group of primates (Taylor *et al.*, 1995). "Inappropriate" non-dopaminergic tissue implantation or sham surgery produced little functional benefit compared to subjects that received fetal mesencephalic grafts. Although some have suggested that host recovery is entirely responsible for behavioral improvement, leading to proposals to carry out sham surgical implantations in patients, the study by Taylor and colleagues (1995) found no recovery in subjects with cerebellar tissue, sham surgery or no surgical intervention. There appeared to be only minor changes in parkinsonian scores in these control groups compared with subjects grafted with developmentally "early" substantia nigra. It is possible that the extensive nursing and supportive care given to these debilitated monkeys may have also resulted in small functional improvements in control groups. The fact that increased dopamine concentrations are detectable within substantia nigra grafts (Elsworth *et al.*, 1995, 1996a) supports the likelihood that recovery seen in recipients of mesencephalic fetal tissue grafts surpass the magnitude of other possible restorative factors in control grafts.

5. Effects of Location of Graft Placement on Functional Recovery

Different profiles of behavioral recovery have been observed after various paradigms of ventral mesencephalic graft placement into the striatum in rodents (Nikkah *et al.*, 1993, 1994). These results suggest that an increased distribution of grafted tissue ("micrografts") in the striatum could significantly increase the outgrowth and integration of the tissue and, critically, the functional effects of the graft. Indeed, it was suggested that increased distribution of tissue may be necessary to restore more "complex" or widespread function in rodents. Improved sensorimotor behavior, as measured by disengage behavior and skilled forelimb use, as well as improved graft-derived dopamine fiber growth, was found only after "micro" grafts in comparison with tissue implanted as "macro" grafts. These studies suggest that previous failures to improve function may have been due to technical difficulties rather than with the reliability of the model.

In recent studies, effects of grafts from two donors implanted into the caudate have been examined together with the ability of this procedure to enhance functional recovery ("maxi-*load*" $n = 8$). Subjects were compared with those receiving tissue from one donor, but distributed in 12 sites ("maxi-*spread*" $n = 8$) rather than a "standard" single donor implantation, but with six implantation sites. Surgery was performed 4 months after MPTP to induce a severe and stable parkinsonian syndrome. Subjects were examined for up to 6 months after grafting. As a group, signs of parkinsonism were reduced in the maxi-spread group to a greater extent than the maxi-load group, although both groups of severely parkinsonian subjects showed evidence of functional improvement (unpublished data). These data suggest that the distribution of the grafted fetal tissue may be a critical factor in optimizing functional recovery.

IV. CONCLUSION

These studies provide important information in support of the hypothesis that recovery of function after fetal nigral grafting in primate models is dependent on such factors as the type of tissue implanted, the site(s) of the graft placement within the brain, and the distribution of the graft within the brain. These data argue against the contention that sham-surgery or non-dopaminergic tissue can have significant functional effects. Moreover, these data suggest that it may not be the absolute amount ("maxi-load") of the tissue implanted but rather the distribution ("maxi-spread") of the tissue implanted that maximally supports functional recovery. Ongoing experiments are aimed at addressing whether improvements in recovery may be augmented by combining these techniques (i.e., tissue from two donors distributed into 12 sites, "maxi-load plus spread"). Other methods for restoring the intrinsic circuitry, such as bridge grafts from the substantia nigra to the striatum and/or implantation of nigral tissue within substantia nigra, rather than ectopic graft placement in the striatum alone, also may be necessary to restore function fully in subjects with severe and stable parkinsonism. We also have found that co-grafts of embryonic striatal tissue and nigral tissue can enhance growth (Sladek *et al.*, 1993a; Collier *et al.*, 1994); the functional effects of such manipulations are being assessed in symptomatic MPTP-treated monkeys.

V. SUMMARY

Improvements in parkinsonism have been reported in patients after embryonic ventral mesencephalic tissue implantation into either the caudate nucleus, into the putamen, or into both regions (e.g., Freed *et al.*, 1992; Henderson *et al.*, 1991; Lindvall *et al.*, 1990; Peschanski *et al.*, 1994; Spencer *et al.*, 1992; Widner *et al.*, 1992). Initial results from clinical trials have been encouraging; however, there is no consensus regarding the optimal choice of grafting site(s) to achieve clinical benefit, nor of optimal age of the donor tissue. Because reinnervation of the entire striatum is unlikely with tissue from a single donor, the distribution of the implanted tissue within a site(s) also may be of critical functional importance. Given that these issues cannot be easily addressed in clinical studies, experimental models of parkinsonism are essential for optimizing transplantation techniques. Specifically, primate studies are essential to link important rodent studies with human applications. For example, the most effective placement of grafted tissue within the functionally and anatomically heterogeneous large primate brain is unlikely to be identified using simpler dopamine-depletion models in the rat brain. The complex primate behavioral repertoire and the faithful elicitation of classic parkinsonian symptoms in primates provide a useful and reliable method for investigating the extent and nature of functional changes resulting from fetal tissue grafts.

For the past 14 years, we and others have examined the potential for intracerebral grafts of early-stage developing dopamine neurons and/or neuroblasts to restore lesion-induced motor imbalance in the nonhuman primate. The MPTP-induced primate model of Parkinson's disease offers perhaps the best animal model of a human neurodegenerative syndrome developed to date (Collier *et al.*, 1988). That embryonic neurons can restore motor balance with relatively modest graft survival is remarkable, and has raised important questions about possible clinical uses in Parkinson's patients (e.g., Sladek and Shoulson, 1988; Bakay, 1993; Bakay and Sladek 1993; Bjorklund, 1993; Goetz *et al.*, 1993; Kopin *et al.*, 1993; Sladek, 1996). Even though the future application of this technology is discussed elsewhere in this volume, it is important to emphasize the key role played by the nonhuman primate model in testing the feasibility of the fetal-grafting approach. The scaling-up process is substantial and is illustrated in part in Fig. 11, which compares simple size differences between rat, monkey, and human brains stained immunocytochemically for tyrosine hydroxylase. The volume of the striatum increases by an order of magnitude from one species to the next, helping to explain that a single graft in a rat can effectively restore rotational asymmetry following a lesion, whereas six or more grafts are required in monkeys and 18 or more grafts are thought to be required in humans to influence functional recovery. Other factors that support the importance of studying the monkey brain as a step toward human applications include the close neuroanatomical, behavioral, and neurochemical similarities between human and monkey species.

The findings of Kordower and colleagues demonstrating robust survival of grafted embryonic dopamine neurons in a patient with Parkinson's disease who improved following surgery dramatically illustrates the importance of primate research for designing human clinical trials. Grafts in this autopsy patient compare remarkably well to observations consistently made in monkey studies (Fig. 11). Graft shape and cytoarchitectural organization is indistinguishable in the two species, and clearly fetal grafts in both species are developing and interacting in response to similar cues from the host environment. Thus, the provocative findings of Kordower and colleagues support the contin-

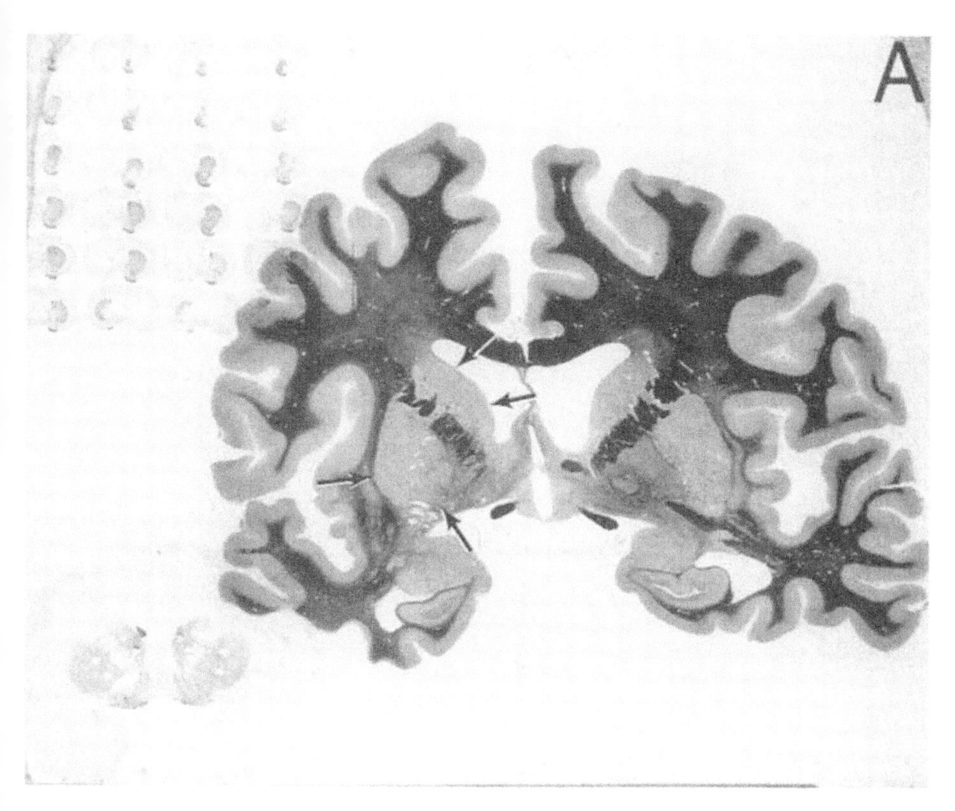

FIGURE 11 The use of nonhuman primate model of parkinsonism has proven to have a predictive value for the implantation of embryonic cells in human brains. In A, the relative sizes of the rat, monkey, and human brain are illustrated on three superimposed microscope slides. The upper-left-hand corner shows 24 sections through a rat brain stained for tyrosine hydroxylase following a unilateral lesion of the nigrostriatal pathway. Consequently, the dopamine field innervation appears dark. By comparison, in the lower left corner there is a single monkey brain stained similarly. It reveals the presence of two transplants and in fact is the same brain illustrated in Figure 12.7A. A human brain stained for myelin is seen in the right half of the image. The relative size differences are profound and account for the need to implant more advanced species with greater numbers of grafts to fully supply the striatum. B and C compare successful grafts from our studies in nonhuman primates with the first successful survival of massive numbers of dopamine neurons as demonstrated in the report by Kordower and associates (courtesy J. Kordower). A graft of human tissue is seen in C and mimics the arrangement, shape, and cell survival, as illustrated in B for monkeys.

uation of human clinical trials: well-differentiated grafts were found and were filled with dopamine neurons that extended neurites into the dopamine-compromised host brain, and the patient exhibited improvements in "on-off

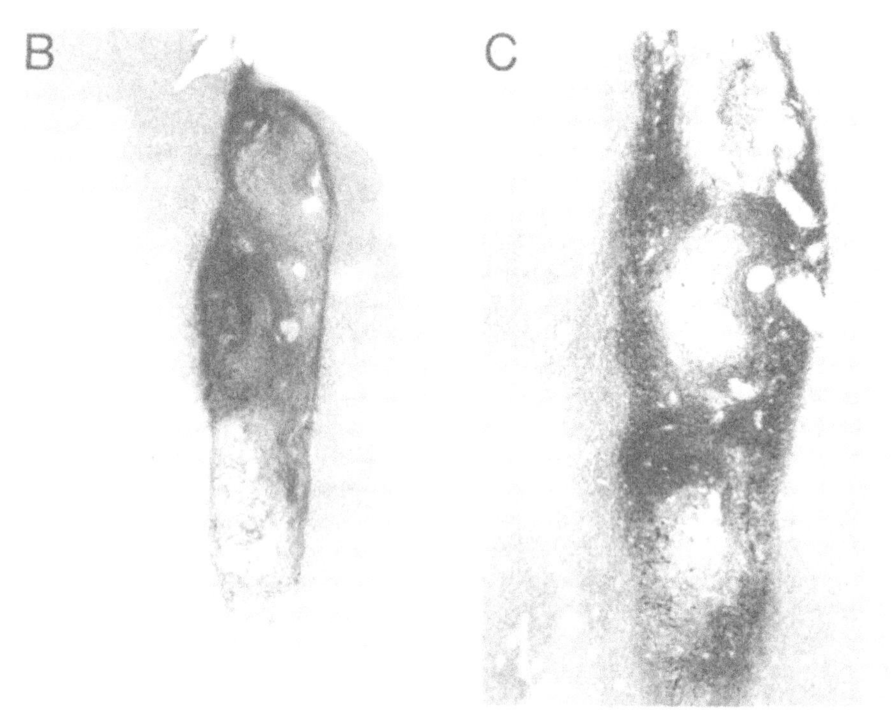

FIGURE 11 (*Continued*)

time," reduced dyskinesia, and a lack of adverse effects. Moreover, substantial clinical improvement occurred despite the presence of a progressive neurodegenerative disease in a relatively old patient (in contrast to the younger subjects used in primate studies). Although several questions remain to be resolved, including when to intervene surgically in the course of a progressive neurological disorder and whether a source of cells can be identified that eliminates the requirement for using human embryonic tissue, it is clear that methodology has been developed and tested to a degree that can potentially support the safe and effective use of this neurosurgical intervention in additional Parkinson's disease patients in the future.

ACKNOWLEDGMENTS

The research upon which this review is based was supported by PO 1 NS24302-13 and RSA MH 00643 (DER) from the National Institutes of Health, the Axion Research Foundation, and the St. Kitts Biomedical Research Foundation. The authors are indebted to Mrs. Barbara Blanchard for skilled technical assistance and preparation of photographic images, to Mrs. Terrie Nierenberger and Ms. Deidre Erlenborn for help with preparation of the final text, to Carol Sortwell, Ph.D., and Ms. Bridgett Gavin for their contributions to the test material, and to Robert Wechsler, Ph.D., and Elizabeth Bundock for the inclusion of selected findings from their doctoral studies.

REFERENCES

Annett, L. E., Dunnett, S. B., Martel, F. L., Rogers, D. C., Ridley, R. M., Baker, H. F., and Marsden, C. D. (1990). A functional assessment of embryonic dopaminergic grafts in the marmoset. *Prog Brain Res* 82:535–542.

Annett, L. E., Martel, F. L., Rogers, D. C., Ridley, R. M., Baker, H. F., and Dunnett, S. B. (1994). Behavioral assessment of the effects of embryonic nigral grafts in marmosets with unilateral 6-OHDA lesions of the nigrostriatal pathway. *Exper Neurol* 125:228–246.

Annett, L. E., Torres, E. M., Ridley, R. M., Baker, H. F., and Dunnett, S. B. (1995). A comparison of the behavioral effects of embryonic nigral graft in the caudate nucleus and in the putamen of marmosets with unilateral 6-OHDA. *Exp Brain Res* 103:355–371.

Annett, L. E., Torres, E. M., Clarke, D. J., Ishida, Y., Barker, R. A., Ridley, R. M., Baker, H. F., and Dunnett, S. B. (1997). Survival of nigral grafts within the striatum of marmosets with 6-OHDA lesions depends critically on donor embryo age. *Cell Transplant* 6:557–569.

Bakay, R. A. (1993). Neurotransplantation: A clinical update. *Acta Neurochir Suppl* (Wien) 58:8–16.

Bakay, R. A. E., Barrow, D. L., Fiandaca, M. S., Iuvone, P. M., Schiff, A., and Collins, D. C. (1987). Biochemical and behavioral correction of MPTP Parkinson-like syndrome by fetal cell transplantation. *Annal NY Acad Sci* 495:623–640.

Bakay, R. A. E., Fiandaca, M. S., Barrow, D. L., Schiff, A., and Collins, D. C. (1985). Preliminary report on the use of fetal tissue transplantation to correct MPTP-induced Parkinson-like syndrome in primates. *Appl Neurophysiol* 48:358–361.

Bakay, R. A. E., Fiandaca, M. S., Sweeney, K. M., Colbassani, H. J., Jr., and Collins, D. C. (1988). Delayed stereotactic transplantation technique in non-human primates. *Prog Brain Res* 78:463–471.

Bakay, R. A., and Sladek, J. R., Jr. (1993). Fetal tissue grafting into the central nervous system: Yesterday, today, and tomorrow [see comments]. *Neurosurgery* 33:645–647.

Bankiewicz, K. S., Plunkett, R. J., Jacobowitz, D. M., Porrino, L., Di Porzio, U., London, W. T., Kopin, I. J., and Oldfield, E. H. (1990a). The effect of fetal mesencephalon implants on primate MPTP-induced parkinsonism. *J Neurosurg* 72:231–244.

Bankiewicz K. S., Plunkett, R. J., Mefford, I., Kopin, I. J., and Oldfield, E. H. (1990b). Behavioral recovery from MPTP-induced parkinsonism in monkeys after intracerebral tissue implants is not related to CSF concentrations of dopamine metabolites. *Prog Brain Res* 82:561–571.

Barbeau, A. (1969). Parkinson's disease as a systemic disorder. Third Symposium of Parkinson's Disease. Gillingham, F. J., and Donaldson, I. M. L., eds. Edinburgh, 66–73.

Birkmayer, W., Hornykiewicz, O., eds. (1976). Advances in parkinsonism: Biochemistry, physiology, treatment. Fifth International Symposium on Parkinson's Disease, Vienna. Basel: Roche.

Bjorklund, A (1993). Neurobiology. Better cells for brain repair [news; comment]. *Nature* 362:414–415.

Bjorklund, A., and Stenevi, U. (1979). Reconstruction of the nigrostriatal dopamine pathway by intracerebral nigral transplants. *Brain Res* 177:555–560.

Brundin, P., Barbin, G., Strecker, R. E., Isacson, O., Prochiantz, A., and Bjorklund, A. (1988). Survival and function of dissociated dopamine neurons grafted at different developmental stages or after being cultured *in vitro*. *Dev Brain Res* 39:233–243.

Bundock, E. A., Blanchard, B., Evans, L., Elsworth, J. D., Roth, R. H., Redmond, D. E., Jr., and Sladek, J. R., Jr. (1997). Development of the nigrostriatal tract in the non-human primate. *Soc. for Neurosci Abstr* 23:1426.

Bundock, E. A., Blanchard, B., Evans, L., Elsworth, J. D., Roth, R. H., Redmond, D. E., Jr., and Sladek, J. R., Jr. (1998). Neurogenesis of the substantia nigra in the non-human primate. *Exper Neurol* (in press).

Burns, R. S., Chiueh, C. C., Markey, S. P., Ebert, M. H., Jacobowitz, D. M., and Kopin, I. J. (1983). A primate model of parkinsonism: selective destruction of dopaminergic neurons in the pars compacta of the substantia nigra by N-methyl-4-phenyl-1,2,3,6-tetrahydropyridine. *Proc Natn Acad Sci* 80:4546–4550.

Collier, T. J., and Sladek, J. R., Jr. (1988). Neural transplantation in animal models of neurodegenerative disease. *NIPS* 3:204–206.

Collier, T. J., Elsworth, J. D., Taylor, J. R., Sladek, J. R., Jr., Roth, R. H., and Redmond, D. E., Jr. (1994). Peripheral nerve-dopamine neuron co-grafts in MPTP-treated monkeys: Augmentation of tyrosine hydroxylase-positive fiber staining and dopamine content in host systems. *Neurosci* 61:875–889.

Collier, T. J., Redmond, D. E., Roth, R. H., Elsworth, J. D., and Sladek, J. R., Jr. (1988). Reversal of experimental parkinsonism in African green monkeys following fetal dopamine neuron transplantation. *Progress in Parkinson's Research:* 211–216.

Collier, T. J., Redmond, D. E., Jr., Roth, R. H., Elsworth, J. D., Taylor, J. R., and Sladek, J. R., Jr. (1997). Metabolic energy capacity of dopaminergic grafts and the implanted striatum in parkinsonian nonhuman primates as visualized with cytochrome oxidase histochemistry. *Cell Transplant* 6:135–140.

Collier, T. J., Redmond, D. E., Jr., Sladek, C. D., Gallagher, M. J., Roth, R. H., and Slade, J. R., Jr. (1987). Intracerebral grafting and culture of cryopreserved primate dopamine neurons. *Brain Res* 436:363–366.

Collier, T. J., Sladek, C. D., Gallagher, M. J., Blanchard, B. C., Daley, B. F., Foster, P. N., Redmond, D. E., Jr., Roth, R. H., and Sladek, J. R., Jr. (1988). Cryopreservation of fetal rat and non-human primate mesencephalic neurons: Viability in culture and neural transplantation. *Prog Brain Res* 78:631–636.

Cotzias, G. C., van Woert, M. H., and Schiffer, L. M. (1967). Aromatic amino acids and modification of parkinsonism. *N Eng J Med* 276:374–379.

Dubach, M., Schmidt, R. H., Martin, R., German, D. C., and Bowden, D. M. (1988). Transplant improves hemiparkinsonian syndrome in nonhuman primate: Intracerebral injection, rotometry, tyrosine hydroxylase immunohistochemistry. *Prog Brain Res* 78:491–496.

Dunn, E. H. (1917). Primary and secondary findings in a series of attempts to transplant cerebral cortex in the albino rat. *J Comp Neurol* 27:565–582.

Dunnett, S. B., and Annett, L. E. (1991). Nigral transplants in primate models of parkinsonism. In: Lindvall, O., Bjorklund, A., and Widner, H., eds. *Neural Transplantation of a Practical Approach.* IRL Press Oxford: pp. 1–19.

Dunnett, S. B., Bjorklund, A., Steveni, U., and Iverson, S. D. (1981). Behavioral recovery following transplantation of substantia nigra in rats subjected to 6-OHDA lesions of the nigrostriatal pathway I. unilateral lesions. *Brain Res* 215:147–161.

Ellis, J. E., Byrd, L. D., and Bakay, R. A. E. (1992). A method for quantitating motor deficits in non-human primates following MPTP-induced hemiparkinsonism and co-grafting. *Exp. Neurol* 115:376–387.

Elsworth, J. D., Al-Tikriti, M. S., Sladek, J. R., Jr., Taylor, J. R., Innis, R. B., Redmond, D. E., Jr., and Roth, R. H. (1994). Novel radioligands for the dopamine transporter demonstrate the presence of intrastriatal nigral grafts in the MPTP-treated monkey: Correlation with improved behavioral function. *Exper Neurol* 126:300–304.

Elsworth, J. D., Sladek, J. R., Jr., Taylor, J. R., Collier, T. J., Redmond, D. E., Jr., and Roth, R. H. (1995). Early gestational mesencephalon grafts, but not later gestational mesencephalon, cerebellum, or sham grafts, increase dopamine in caudate nucleus of MPTP-treated monkeys. *Neuroscience* 72:477–484.

Elsworth, J. D., Brittan, M. S., Taylor, J. R., Sladek, J. R., Jr., al-Tikriti, M. S., Zea-Ponce, Y., Innis, R. B., Redmond, D. E., Jr., and Roth, R. H. (1996b). Restoration of dopamine transporter density in the striatum of fetal ventral mesencephalon-grafted, but not sham-grafted, MPTP-treated parkinsonian monkeys. *Cell Transplant* 5:315–325.

Elsworth, J. D., Taylor, J. R., Sladek, J. R., Jr., Brittan, M. S., Redmond, D. E., Jr., and Roth, R. H. (1998). Effect of fetal mesencephalic graft volume and distribution on dopaminergic reinnervation of the parkinsonian primate striatum. *Exper Neurol* (in press).

Fahn, S. (1992). Adverse effects of levodopa. In: Olanow, C. W., and Lieberman, A., eds. *Scientific Basis for Therapy in Parkinson's Disease.* United Kingdom: Parthenon Press, p. 89.

Fiandaca, M. S., Bakay, R. A., Sweeney, K. M., and Chan, W. C. (1988). Immunologic response to intracerebral fetal neural allografts in the rhesus monkey. *Prog Brain Res* 78:287–296.

Fine, A., Hunt, S. P., Oertel, W. H., Nomoto, M., Chong, P. N., Bond, A., Waters, C., Temlett, J. A., Annett, L., Dunnett, S., Jenner, P., and Marsden, C. D. (1988). Transplantation of embryonic marmoset dopaminergic neurons to the corpus striatum of marmosets rendered parkinsonian by 1-methyl-4-phenyl-1,2,3,6-tetrahydropyridine. *Prog Brain Res* 78:479–489.

Freed, C. R., Breeze, R. E., Rosenberg, N. l., Schneck, S. A., Kriek, E., Qi, J. X., Lone, T., Zhang, Y. B., Snyder, J. A., Wells, T. H., Ramig, L. O., Thompson, L., Mazziotta, J. C., Huang, S. C., Grafton, S. T., Brooks, D., Sawle, G., Schroter, G., and Ansari, A. A. (1992). Survival of implanted fetal dopamine cells and neurologic improvement 12 to 46 months after transplantation for Parkinson's disease. *N Engl J of Med* 327(22):1549–1555.

Freed, W. J., Perlow, M. J., Karoum, F., Seiger, A., Olson, L., Hoffer, B., and Wyatt, B. J. (1980). Restoration of dopamine function by grafting of fetal substantia nigra to the caudate nucleus: Long term behavioral, biochemical and histological studies. *Ann Neurol* 8:510–522.

Goetz, C. G., De Long, M. R., Penn, R. D., and Bakay, R. A. (1993). Neurosurgical horizons in Parkinson's disease [see comments]. *Neurology* 43:1–7.

Henderson, B. T. H., Clough, C. G., Hughes, R. C., Hitchcock, E. R., and Kenny, B. G. (1991). Implantation of human fetal ventral mesencephalon to the right caudate nucleus in advanced Parkinson's disease. *Arch Neurol* 48:822–827.

Hopkins-Dunn, E. (1917). Primary and secondary findings in a series of attempts to transplant cerebral cortex in the albino rat. *J Comp Neurol* 27:565–582.

Houle, J. D., and Das, G. D. (1980). Freezing of embryonic neural tissue and its transplantation in the rat brain. *Brain Res* 192:570–574.

Jensen, S., Sorensen, R., and Zimmer, J. (1987). Cryopreservation of fetal rat brain tissue later used for intracerebral transplantation. *Cryobiol* 24:120–134.

Kopin, I. J., Bankiewicz, K. S., Plunkett, R. J., Jacobowitz, D. M., and Oldfield, E. H. (1993). Tissue implants in treatment of parkinsonian syndromes in animals and implications for use of tissue implants in humans. *Adv Neurol* 60:707–714.

Kordower, J. H., Freeman, T. B., Snow, B. J., Vingerhoets, F. J. G., Mufson, E. J., Sanberg, P. R., Hauser, R. A., Smith, D. A., Nauert, G. M., Perl, D. P., and Olanow, C. W. (1995). Neuropathological evidence of graft survival and striatal reinnervation after the transplantation of fetal mesencephalic tissue in a patient with Parkinson's disease. *N Engl J Med* 332:1118–1124.

Langston, J. W., Forno, L. S., Robert, C. S., and Irwin, I. (1984). Selective nigral toxicity after systemic administration of 1-methyl-4-phenyl-1,2,5,6-tetrehydropyridine (MPTP) in the squirrel monkey. *Brain Res* 292:390–394.

Landis, S. C. (1980). Developmental changes in the neurotransmitter properties of dissociated sympathetic neurons: A cytochemical study of the effects of medium. *Dev Biol* 77:349–361.

Levit, P., and Rakic, P. (1982). The time of genesis, embryonic origin and differentiation of the brain stem monoamine neurons in the rhesus monkey. *Dev Brain Res* 4:35–57.

Lindvall, O., Brundin, P., and Widner, H. (1990). Grafts of fetal dopamine neurons survive and improve motor function in Parkinson's disease. *Science* 247:574–577.

Lindvall, O., Rechncrona, S., Brundin, P., Gustavii, B., Astedt, B., Widner, H., Lindholm, T., Bjorklund, A., Leenders, K. L., and Rothwell, J. C. (1989). Human fetal dopamine neurons grafted into the striatum in two patients with Parkinson's disease: A detailed account of methodology and 16 month follow-up. *Arch Neurol* 46:615–631.

Meyers, R. (1942). The modification of alternating tremors, rigidity and festination by surgery of the basal ganglia. *Res Publ Assn Nerv Ment Dis* 21:602–665.

Morihisa, J. M., Nakamura, R. K., Freed, W. J., Mishkin, M., and Wyatt, R. J. (1984). Adrenal medulla grafts survive and exhibit catecholamine-specific fluorescence in the primate brain. *Exper Neurol* 84:643–653.

Nieto-Sampedro, M., Lewis, E. R., Cotman, C. W., et al. (1982). Brain injury causes a time-dependent increase in neurotrophic activity at the lesion site. *Science* 217:860–861.

Nikkhah, G., Cunningham, M. G., Jodicke, A., Knappe, U., and Björklund, A. (1994). Improved graft survival and striatal reinnervation by microtransplantation of fetal nigral cell suspensions in the rat Parkinson model. *Brain Res* 633:133–143.

Nikkhah, G., Duan, W.-M., Knappe, U., Jodicke, A., and Björklund, A. (1993). Restoration of complex sensorimotor behavior and skilled forelimb use by a modified nigral cell suspension transplantation approach in the rat Parkinson model. *Neuroscience* 56:33–43.

Olson, L., and Malmfors, T. (1970). Growth characteristics of adrenergic nerves in the adult rat: Fluorescence histochemical and 3H-noradrenaline uptake studies using tissue transplantation to the anterior chamber of the eye. *Acta Physiol Scand Suppl* 348:1–112.

Peschanski, M., Defer, G., N'Guyen, J. P., Monofort, J. C., Remy, P., Geny, C., Samson, Y., Hantraye, P., and Jeny, R. (1994). Bilateral motor improvement and alteration of L-dopa effect in two patients with Parkinson's disease following intrastriatal transplantation of foetal ventral mesencephalon. *Brain* 117:487–499.

Plunkett, R. J., Saris, S. C., Bankiewicz, K. S., Ikejiri, B., and Weber, R. J. (1989). Implantation of dispersed cells into primate brain. *J Neurosurg* 70:441–445.

Ranson, S. W. (1914). Transplantation of the spinal ganglion, with observations of the significance of the complex types of spinal ganglion cells. *J Comp Neurol* 24:547–558.

Redmond, D. E., Jr., Sladek, J. R., Jr., Roth, R. H., Collier, T. J., Elsworth, J. D., Deutch, A. Y., and Haber, S. (1986a). Fetal neuronal grafts in monkeys given methylphenyltetrahydropyridine. *Lancet* May 17:1125–1127.

Redmond, D. E., Jr., Sladek, J. R., Jr., Roth, R. H., Collier, T. J., Elsworth, J. D., Deutch, A. Y., and Haber, S. (1986b). Transplants of primate neurons [letter]. *Lancet* 2:1046.

Riley, H. A. (1943). *An Atlas of the Basal Ganglia, Brain Stem and Spinal Cord* [book]. Baltimore: The Williams & Wilkins Co.

Robbins, R. J., Torres-Aleman, I., Leranth, C., Bradberry, C. W., Deutch, A. Y., Welsh, S., Roth, R. H., Spencer, D., Redmond, D. E. Jr., and Naftolin, F. (1990). Cryopreservation of human brain tissue. *Exper Neurol* 107:208–213.

Sladek, J. R., Jr., Collier, T. J., Haber, S. N., Roth, R. H., and Redmond, D. E., Jr. (1986). Survival and growth of fetal catecholamine neurons transplanted into primate brain. *Brain Res Bull* 17:809–818.

Sladek, J. R., Jr., Collier, T. J., Haber, S. N., Deutch, A. Y., Elsworth, J. D., Roth, R. H., and Redmond, D. E., Jr. (1987a). Reversal of parkinsonism by fetal nerve cell transplants in primate brain. *Ann NY Acad Sci* 495:641–657.

Sladek, J. R., Jr., Redmond, D. E., Jr., Collier, T. J., Haber, S. N., Elsworth, J. D., Deutch, A. Y., and Roth, R. H. (1987b). Transplantation of fetal dopamine neurons in primate brain reverses MPTP induced parkinsonism. *Prog Brain Res* 71:309–323.

Sladek, J. R., Jr., and Shoulson, I. (1988). Neural transplantation: A call for patience rather than patients. *Science* 240:1386–1388.

Sladek, J. R., Jr., Redmond, D. E., Jr., and Roth, R. H. (1988a). Transplantation of fetal neurons in primates. *Clin Res* 36:200–204.

Sladek, J. R., Jr., Redmond, D. E., Jr., Collier, T. J., Blount, J. P., Elsworth, J. D., Taylor, J. R., and Roth, R. H. (1988b). Fetal dopamine neural grafts: Extended reversal of methylphenyltetrahydropyridine-induced parkinsonism in monkeys. *Prog Brain Res* 78:497–506.

Sladek, J. R., Jr., Redmond, D. E., Jr., Collier, T. J., Elsworth, J. D., and Roth, R. H. (1989). Transplantation advances in Parkinson's disease. *Mov Disord* 4(Suppl) 1:120–125.

Sladek, J. R., Jr., Collier T. J., Elsworth, J. D., Taylor, J. R., Roth, R. H., and Redmond, D. E., Jr. (1993a). Can graft-derived neurotrophic activity be used to direct axonal outgrowth of grafted dopamine neurons for circuit reconstruction in primates? *Exp Neurol* 124:134–139.

Sladek, J. R., Jr., Elsworth, J. D., Roth, R. H., Evans, L. E., Collier, T. J., Cooper, S. J., Taylor, J. R., and Redmond, D. E., Jr. (1993b). Fetal dopamine cell survival after transplantation is dramatically improved at a critical donor gestational age in nonhuman primates. *Exper Neurol* 122:16–27.

Sladek, J. R., Jr. (1996). Neural transplants work. *News Physiol Sci* 11:298–299.

Sladek, J. R., Jr., Collier, T. J., Elsworth, J. D., Roth, R. H., Taylor, J. R., and Redmond, D. E., Jr. (1998). Intrastriatal grafts from multiple donors do not result in a proportional increase in survival of dopamine neurons in nonhuman primates. *Cell Trans* 7:87–96.

Sortwell, C. E., Blanchard, B. C., Collier, T. J., Elsworth, J. D., Taylor, J. R., Roth, R. H., Redmond, D. E., Jr., and Sladek, J.R., Jr. (1998). Pattern of synaptophysin immunoreactivity within mesencephalic grafts following transplantation in a parkinsonian primate model. *Brain Res* (in press).

Spencer, D. D., Robbins, R. J., Naftolin, F., Marek, K. L., Vollmer, T., Leranth, C., Roth, R. H., Price, L. H., Gjedde, A., Bunney, B. S., Sass, K. J., Elsworth, J. D., Kier, L. E., Makuch, R., Hoffer, P. B., and Redmond, D. E., Jr. (1992). Unilateral transplantation of human fetal mesencephalic tissue into the caudate nucleus of parkinsonian patients: Functional effects for 18 months. *New England J of Medicine* 327(22):1541–1548.

Taylor, J. R., Elsworth, J. D., Roth, R. H., Collier, T. J., Sladek, J. R., Jr., and Redmond, D. E., Jr. (1990). Improvements in MPTP-induced object retrieval deficits and behavioral deficits after fetal nigral grafting in monkeys. *Prog Brain Res* 82:543–559.

Taylor, J. R., Elsworth, J. D., Roth, R. H., Sladek, J. R., Jr., Collier, T. J., and Redmond, D. E., Jr. (1991). Grafting of fetal substantia nigra to striatum reverses behavioral deficits induced by MPTP in primates: A comparison with other types of grafts as controls. *Exper Brain Res* 85:335–348.

Taylor, J. R., Elsworth, J. D., Roth, R. H., Sladek, J. R., Jr., and Redmond, D. E., Jr. (1994). Behavioral effects of MPTP administration in the vervet monkey: A primate model of Parkinson's disease. In: Woodruff, M. L., and Nonneman, A. J., eds. *Toxin-Induced Models of Neurological Disorders* New York: Plenum Press, pp. 139–174.

Taylor, J. R., Elsworth, J. D., Sladek, J. R., Jr., Collier, T. J., Roth, R. H., and Redmond, D. E., Jr. (1995). Sham surgery does not ameliorate MPTP-induced behavioral deficits in monkeys. *Cell Transplantation* 4:13–26.

Taylor, J. R., Elsworth, J. D., Roth, R. H., Sladek, Jr., J. R., and Redmond, D. E., Jr. (1997). Severe long-term MPTP-induced parkinsonism in the vervet monkey (*Cercopithecus aethiops sabaeus*). *Neuroscience* 81:475–755.

Taylor, J. R., Elsworth, J. D., Roth, R. H., Sladek, Jr., J. R., and Redmond, D. E. Jr. (1998-submitted). Behavioral effects of embryonic nigral grafts in the putamen of monkeys with severe MPTP-induced parkinsonism.

Thompson, W. G. (1890). Successful brain grafting. *NY Med J* 51:701–702.

Wechsler, R. T., Blanchard, B. C., Collier, T. J., Elsworth, J. D., Taylor, J. R., Roth, R. H., Redmond, D. E., Jr., and Sladek, J. R., Jr. (1995). Interactions of fetal mesencephalic and fetal striatal grafts with host brain in a parkinsonian primate model. *Exper Neurol* 135:172.

Wechsler, R. T., Leranth, C., Collier, T. J., Elsworth, J. D., Redmond, D. E., Jr., Roth, R. H., Taylor, J. R., and Sladek, J. R., Jr. (1997). Electron microscopic examination of input to grafts of fetal ventral mesencephalon in a parkinsonian primate model. *Exper Neurol* 143:336.

Widner, H., Tetrud, J. W., and Rehncrona, S., et al. (1992). Bilateral fetal mesencephalic grafting in two patients with parkinsonism induced by 1-methyl-4-phenyl-1,2,3,6-tetrahydropyridine (MPTP). *N Engl J Med* 327:1556–1563.

Yurek, D. M., Collier, T. J., and Sladek, J. R., Jr. (1990). Embryonic mesencephalic and striatal co-grafts: Development of grafted dopamine neurons and functional recovery. *Exper Neurol* 109:191–199.

CHAPTER 13

Immunobiology and Neuroscience of Xenotransplantation in Neurological Disease

OLE ISACSON,[*,**,††] TERRENCE DEACON,[*] AND JAMES SCHUMACHER[*,†]

[*]Neuroregeneration Laboratory, McLean Hospital, Belmont, Massachusetts, [†]Neurological Associates, Sarasota, Florida, [**]Program in Neuroscience, Harvard Medical School, Boston, Massachusetts, [††]Department of Neurology, Massachusetts General Hospital, Boston, Massachusetts

The understanding of complex immune responses to potentially useful clinical treatments such as xenotransplantation is of great importance. The relatively mild immunological reactions in the CNS to allografts or xenografts have provided an empirical basis for clinical xenotransplantation of neural tissue. Our results demonstrate that porcine neural xenotransplantation presents fewer obstacles for clinical transplantation than other forms of xenotransplantation, including heart, islet cells, kidneys, and muscle grafting. Nonetheless, transplant immunobiology of the brain indicates that stimuli presented by the implanted donor tissue to the host lead to an immunization process in the host. Following graft acceptance, the xenografts are rejected upon withdrawal of the immunosuppression (such as cyclosporine). The current understanding of the rejection process of xenografts in the brain indicates that, although antibody-dependent and complement-mediated cytotoxicity plays an early role, the most prominent factor in long-term CNS transplant rejection is T-cell-mediated toxicity.

CNS Regeneration

I. RATIONALE FOR RESEARCH ON NEURAL XENOTRANSPLANTATION IN PARKINSON'S DISEASE

Previous fetal allogeneic neural transplants have been repeatedly shown to cause functional recovery in animal models of neurodegenerative diseases (Perlow *et al.*, 1979; Björklund *et al.*, 1981; Low *et al.*, 1982; Gage *et al.*, 1984; Isacson *et al*, 1984, 1986; Dunnett and Annett, 1991; Hantraye *et al.*, 1992) and to improve motor dysfunction in several patients with Parkinson's disease (Freed *et al.*, 1992; Spencer *et al.*, 1992; Widner *et al.*, 1992; Lindvall *et al.*, 1994). To obtain human fetal tissue for neural transplantation, complex methodological and ethical guidelines have been devised (Greenley *et al.*, 1989; Vawter *et al.*, 1990; Turner and Kearney, 1993). Although evidence is accumulating in favor of the therapeutic potential of allogeneic neural transplantation in Parkinson's (and Huntington's) disease, the limited availability of human fetal neural tissue may pose serious limitations on clinical application of this potential treatment modality (Sanberg and Isacson, 1994). Neural xenotransplantation, defined here as transplantation of fetal neuroblasts derived from homologous neural structures of a different (mammalian) species into the adult brain, may circumvent many of the limitations associated with the use of human fetuses, although basic questions remain for neurobiological, immunological, and safety aspects. Neuroblasts at precisely defined embryonic ages can be prepared in large quantities and in sterile fashion from the fetuses of especially bred pathogen-free xenogeneic donors for clinical use.

II. IMMUNOLOGY OF NEURAL XENOGRAFTS

The understanding of complex immune responses to potentially useful clinical treatments such as xenotransplantation is of great importance. The mechanisms involved in immunological rejection of neural tissue transplanted to the central nervous system (CNS) are not fully understood (Pakzaban and Isacson, 1994; Wood *et al.*, 1996). On the other hand, the mild immunological reactions in the CNS to allografts or xenografts relative to immunological responses seen elsewhere in the body has provided a rationale and empirical basis for clinical xenotransplantation of neural tissue (Mason *et al.*, 1986; Finsen *et al.*, 1988a; Pakzaban and Isacson, 1994; Pedersen *et al.*, 1995). Recent results indeed demonstrate that functional neural xenotransplantation may present less obstacles for clinical transplantation than other forms of xenotransplantation, including heart, islet cells, kidneys, and muscle grafting (Pakzaban and Isacson, 1994). Transplant immunobiology of the brain involves knowledge of potential stimuli presented by the implanted donor tissue to the host, and also the host repertoire of inflammatory responses, and finally the immunization processes that may occur in the host. Although the most optimal clinical situation would be a treatment that would make the graft acceptable and nonrejectable, currently all studies suggest that it is only with cyclosporine and combinations of systemic immune suppression by which sufficient immunosuppressive effects are achieved. Moreover, the xenografts are rejected upon withdrawal of the immunosuppression. The current understanding of the rejection process of xenografts in the brain indicates a prominent role of T-cell-mediated toxicity different than antibody-dependent and complement-mediated cytotoxicity seen in hyperacute xenotransplant rejection (Pakzaban and Isacson, 1994; Galpern *et al.*, 1996; Deacon *et al.*, 1997; Duan, 1997).

III. FACTORS INFLUENCING NEURAL XENOGRAFT SURVIVAL

The parameters that may influence neural xenograft survival can be divided into two categories: those specific to transplantation across species lines and those that apply to all neural transplantation. Factors that are specifically relevant to xenotransplantation include the choice of the donor and host species (Freed *et al.*, 1988; Dymecki and Freed, 1989), use of immunosuppression (Brundin *et al.*, 1985; Inoue *et al.*, 1985), duration of immunosuppression (Brundin *et al.*, 1989), and immune competence of the host (Mason *et al.*, 1986). Factors that are generally important to survival of neural grafts include regional differences in resilience of neurons to *ex vivo* manipulation (Detta and Hitchcock, 1990), the developmental stage of the donor tissue (Schmidt *et al.*, 1981; Brundin *et al.*, 1986; Zimmer *et al.*, 1988; Simonds and Freed, 1990; Freeman *et al.*, 1991; Sladek *et al.*, 1993), the site of the transplant (intraparenchymal vs intraventricular) (Nishino *et al.*, 1986; Poltorak *et al.*, 1988; Heim *et al.*, 1993), the presence of a preceding lesion (Isacson *et al.*, 1987a; Niijima *et al.*, 1990; Sladek *et al.*, 1993), and the method of transplantation (solid tissue vs cell suspension) (Heim *et al.*, 1993). In addition, the effect of time on xenograft survival (Inoue *et al.*, 1985; Finsen *et al.*, 1988b; Nakashima *et al.*, 1988; Marion *et al.*, 1990), and the specificity of the histological method used for identification of donor tissue must be kept in mind when comparing graft survival data from different studies. In the absence of immunosuppression, survival of neural xenografts in the rat brain was significantly affected by the choice of donor species (Pakzaban and Isacson, 1994). When the donor species are ordered according to phylogenetic distance from the rat host (Walker, 1964; Goodman *et al.*, 1971; Goodman, 1972), xenograft survival decreases statistically (Pakzaban and Isacson, 1994). In the presence of cyclosporine, the decrease in graft survival as a function of phylogenetic distance was not observed (Pakzaban and Isacson, 1994).

Analyzing the various factors that contribute to the immunologically privileged status of the brain, perhaps the most important of these is the relative paucity of MHC class I and class II antigens in the adult (host) and the developing (donor) brain (Lampson, 1987). The fact that the immunological privilege is not absolute can in part be attributed to induction of MHC antigens in neural tissue under certain conditions, including transplantation (Lampson and Fisher, 1984; Wong *et al.*, 1984; Mason *et al.*, 1986; Nicholas *et al.*, 1987; Finsen *et al.*, 1991). Induction of MHC class II antigens in the xenograft would trigger the afferent arm of the immune response, mediated by class-II-restricted helper T cells, whereas expression of class I antigens would render the graft a target for the efferent limb of this response, mediated by class-I-restricted cytotoxic T cells.

In the normal adult brain, expression of MHC class I antigens is limited to the endothelial cells, whereas MHC class II expression is virtually absent except for rare expression on microglia (Darr *et al.*, 1984; Lampson and Hickey, 1986; Lampson, 1987; Widner *et al.*, 1988). MHC class I expression can be induced on all components of neural tissue (and class II on astrocytes and microglia) by exposure to γ-interferon (γ-IFN), tumor necrosis factor-a, or various inflammatory conditions (Lampson and Fisher, 1984; Wong *et al.*, 1984, 1985; Fierz *et al.*, 1985; Traugott *et al.*, 1985; Massa *et al.*, 1986; Lavi *et al.*, 1987). Similarly, whereas no MHC expression has been found in the embryonic brain (Hofman *et al.*, 1984; Lampson, 1987; Lampson and Siegel, 1988), induction of both class I and class II antigens has been documented on donor cells after transplantation of fetal neural tissue into the adult brain (Mason *et al.*, 1986; Nicholas *et al.*, 1987; Finsen *et al.*, 1991; Isacson *et al.*, 1995). Elaboration of cytokines during the inflammatory response that follows the surgical trauma of transplantation is postulated to underlie the induction of MHC antigens in fetal neural grafts (Widner *et al.*, 1988; Widner, 1993). Finsen and associates have compared the time course of MHC expression on various donor and host cells after neural allo- and xenotransplantation (Finsen *et al.*, 1991). They find a more prominent and persistent astroglial and microglial reaction in neural tissue-block xenografts, accompanied by induction of both classes of MHC antigens on graft and host cells. These changes are concurrent with heavy infiltration of the graft by bone-marrow-derived inflammatory cells and cytoarchitectural destruction of the graft.

The extent to which an MHC-restricted T cell response actually contributes to neural xenograft rejection has been questioned. It has been proposed that MHC-restricted mechanisms play a more important role in neural allograft rejection, although direct or antibody-mediated complement activation and antibody-dependent cellular cytotoxicity constitute the major mechanisms of xenograft rejection (Widner, 1993). However, the success of cyclosporin and other strategies that interfere with T cell function in abrogating neural xenograft rejection attests to the importance of MHC-restricted T cell immunity in neural xenotransplantation (Brundin *et al.*, 1985; Inoue *et al.*, 1985; Finsen *et al.*, 1988b; Geist *et al.*, 1991; Honey *et al.*, 1991a, 1991b; Pakzaban *et al.*, 1995). Xenogeneic MHC antigens may be directly recognized by host T cells across certain species barriers (Bluestone *et al.*, 1987; Kievits *et al.*, 1989; Bontrop *et al.*, 1990), or may be processed by the host's antigen-processing cells and presented to the T cells in the context of self MHC (Lindahl and Bach, 1976; Sloan *et al.*, 1991). Either mechanism would support a prominent role for T cells in xenograft rejection. The inflammatory response observed in the brain may also be involved in rejection processes, particularly with the observed release of cytokines that promote inflammations such as IFN-γ, tumor necrosis factor α (TNF-α) and interleukin 6 (IL-6). However, although many of these factors are known to be active in inflammation, their role in xenograft rejection is not clear. If IFN-γ can change the vascular and cell adhesive environment of the brain, as well as activate glial and microglial cells, this factor may also have a role in stimulating or inducing the rejection of xenografts (Sethna and Lampson, 1991; Subramanian *et al.*, 1995). Moreover, IFN-γ has been shown to activate both MHC-I and MHC-II expression, which otherwise is generally absent in the adult brain.

Recently, examining T-cell and macrophage-mediated rejections of xenografts has resulted in a better understanding of the response to xenogeneic cells (Pakzaban and Isacson, 1994; Deacon *et al.*, 1997; Duan, 1997). In fetal neural tissue transplantation, data from our and other laboratories confirm that T-cell activation is inhibited after xenotransplantation if immune suppression such as cyclosporine is present. The most prominent feature of the immune response seems to be a progressive activation of interferon gamma (IFN-γ) peaking between 9 and 15 days after neural xenotransplantation, which coincides with the most active period of rejection. Moreover, other cytokines, such as IL-2, IL-4, and IL-10, also seem to be induced around day 7 and in most cases continue for at least 30 days after transplantation (Sethna and Lampson, 1991; Pakzaban and Isacson, 1994; Duan, 1997). Nevertheless, a set of results that favor a view of selective clonal T-cell expansion and cytotoxicity directed at specific xenogeneic antigens in neural transplantation is the fact that CD4 and CD8 T cells, macrophages, and microglial cells will have a higher local cell density and accumulation in conditions of simple lesion paradigms (without neural transplants). We and others have demonstrated larger lymphocytic–macrophage infiltrates and glial reactions following direct excitotoxic lesions than after neural grafting (Isacson *et al.*, 1987b; Coffey *et al.*, 1990; Pakzaban and Isacson, 1994; Duan, 1997). Thus, some of the inflammatory responses seen in association with grafting could actually indicate that cell death or survival may be closely regulated by cytokines. For example, an increased expression of IL-10, which has been shown to be anti-inflammatory, could indicate that the grafts of fetal tissue have a capacity to downregulte the immune reactions of the host. It has also been reported that IL-10 can downregulate TNF-α production (Benveniste *et al.*, 1994). The local release of factors or stimuli from fetal transplants could thus reduce the effects of immunologically active host systems. Several studies have also evaluated how the graft's endothelium and glial cells may actively participate, or conversely inhibit, the activation of T-cell-mediated graft rejections. By examining the various inflammatory cell adhesive and immunological activations caused by these grafts over time, the exact sequence of events involved in neural transplant rejection may be determined. In our studies, we have seen relatively little inflammation in association with neural xenografts (Pakzaban *et al.*, 1995; Galpern *et al.*, 1996; Deacon *et al.*, 1997). In cases of transplantation to humans, we have seen a few T cells at the graft–host interface (Deacon *et al.*, 1997). However, only when there has been apparent vascular damage at the injection site have an increased number of T cells been clearly identified, thereby not proving any mounted host response to the tissue presence of xeno-antigens. Finally, cell death seen in fetal xenografts is not necessarily immunologically derived because it could also be due to lack of trophic factors and inappropriate growth substrates for fetal neural cells. Cytokine induction may also be mechanistically linked to growth factor responses. Thus, in development of transplanted fetal tissues and late control of cell-growth and/or cell-death, there may be an interplay and synchronicity between the immune system and trophic factor regulation (Pakzaban and Isacson, 1994; Ludlam *et al.*, 1995; Lynn and Wong, 1995; Rothwell and Hopkins, 1995).

IV. IMMUNOPROTECTION STRATEGIES

The comparative ease by which mammalian neural tissue can be isolated gives neural xenotransplantation a relative advantage over methodologies requiring production of cells *de novo* such as genetically engineered cells. Although this is encouraging for the field of xenotransplantation, the limited knowledge about rejection mechanisms and the limited knowledge about effective immunosuppressive methods and safety issues stand as obstacles to making neural xenotransplantation a routine procedure for patients with neurodegenerative disease. Moreover, the current need for systemic cyclosporine (CsA) immunosuppression for extended xenotransplant survival highlights the need for more research into alternative immunological strategies to prevent neural transplant rejection. Novel findings toward alternative immunosuppressive or immunomodulatory treatments are of medical interest, because systemic immunosuppression by cyclosporine is a potentially toxic procedure that may render the patient both unable to fend off common viral infections, fungal and even cancerous formations as well as increasing the likelihood of neurotoxicity (McDonald *et al.*, 1996). The primary mechanism by which cyclosporine provides immune suppression appears to be that of reducing T-cell activation through binding to a cyclosporine-binding protein through a cofactor, subsequently binding to calcineurin, which in turn modifies intracellular calcium levels and IL-2 production of T-helper cells (Liu *et al.*, 1991). In regular organ transplantation, cyclosporine in combination with other factors has provided a sufficient immune suppression. The serum levels of cyclosporine needed in clinical neural xenotransplantation, for which our neuroscience laboratory has provided some basic data for clinical application, suggests that cyclosporine cytotoxicity can be controlled. However, side effects of cyclosporine such as damage to the kidneys, as well as infections and some cancers have been reported in other clinical scenarios. Lately, focus has also been placed on the potential neurotoxicity of cyclosporine-like compounds (McDonald *et al.*, 1996). The potential role of MHC-I in graft rejection processes or tolerance induction has been investigated (Pakzaban *et al.*, 1995). The $F(ab')_2$ MHC-I masked enhancement of neural xenograft survival, or delay of rejection, appears to be an active process (like tolerance) because our recent work (Fine *et al.*, unpublished observations) shows that combinations of $F(ab')_2$ treatment and cyclosporine are counterproductive. It would be useful to know how tolerance induction may correlate with activation of specific cytokines and presence of cell types participating in inflammatory responses. Given the importance of MHC-restricted T-cell-mediated mechanisms in neural (xeno-)graft rejection, novel immunoprotection strategies have been devised that target specific steps in this process. Injection of monoclonal antibodies directed against helper T-cell surface glycoprotein CD4 has been shown to protect neural allo- and xenografts against rejection (Nicholas *et al.*, 1990; Honey *et al.*, 1991a; Wood *et al.*, 1996). Inhibition of T-cell clonal expansion by administration of antibodies directed against the interleukin-2 receptor promotes survival of neural allografts and xenografts, and induces specific tolerance to subsequent neural allografts (Honey *et al.*, 1991b; Wood *et al.*, 1993). Transplantation of immunoselected neuroepithelial cells that do not express MHC class I after exposure to IFN-γ has been shown to result in improved allograft survival (Bartlett *et al.*, 1990). Masking of xenogeneic MHC class I antigens on pancreatic islet cells by $F(ab')_2$ monoclonal antibody fragments prevents xenograft rejection and induces tolerance to subsequent unmasked xenografts (Faustman and Coe, 1991). We have observed that similar masking of MHC class I antigens on porcine neuroepithelial cells improves porcine neural xenograft survival in the rat host (Pakzaban *et al.*, 1995). The resulting grafts are, however, smaller and have fewer neuronal elements than those in cyclosporin-treated animals (Pakzaban *et al.*, 1995). Although the degree of graft protection may fall short of that obtained by cyclosporin immunosuppression, such targeted immunoprotective strategies may further enhance xenograft survival. Further manipulation of the immunological determinants of neural xenograft rejection may eventually unlock the potential of neural xenotransplantation as a treatment modality. Tolerance induction and related mechanisms (Ridge *et al.*, 1996) are likely to be involved in the $F(ab')_2$-mediated effects.

V. EXPERIMENTAL STUDIES OF PORCINE NEURAL TRANSPLANTS FOR NEURODEGENERATIVE DISEASE

A. Transplantation of Porcine Xenogeneic Dopaminergic Cells to Animal Models of Parkinson's Disease: Morphology and Organization of Fetal Porcine Ventral Mesencephalic Grafts

In situ hybridization for pig-repeat element (PRE) DNA in porcine ventral mesencephalon (VM) grafts (containing developing dopamine neurons) has demonstrated localization of the majority of porcine-derived nuclei within the confines of VM grafts (Isacson *et al.*, 1995; Galpern *et al.*, 1996; Deacon *et al.*, 1997). Donor-derived cells were uniformly distributed whereas the donor tyrosine hydroxylase positive (TH+) dopaminergic neurons formed clusters within the graft. PRE DNA labeling revealed a distinct graft–host boundary. Porcine cells were not found outside the graft in the host striatal gray matter yet were observed in host white matter tracts, particularly the internal capsule and corpus callosum. Because no TH+ or neurofilament positive (NF70+) neuronal cell bodies were observed outside the graft, these cells most likely correspond to the cluster of differentiation antigen 44 labeled (CD44+) glia that have migrated into the host white matter (Fig. 1). In large grafts, the donor-derived TH+ fibers reinnervated the entire extent of the striatum. Axons from the graft grew into the gray matter of the surrounding host striatum, avoiding the penetrating white matter tracts. This pattern of reinnervation was comparable to the normal striatal innervation by TH+ axons originating from the intact contralateral SN, and the density of the TH+ fibers in the grafted striatum approached that observed in the intact striatum. In all of these animals, TH+ neuron staining was totally absent in SN ipsilateral to the 6-OHDA lesion, indicating that the dense striatal innervation was entirely donor derived. No TH+ axons were present in the striatum of rats without surviving grafts

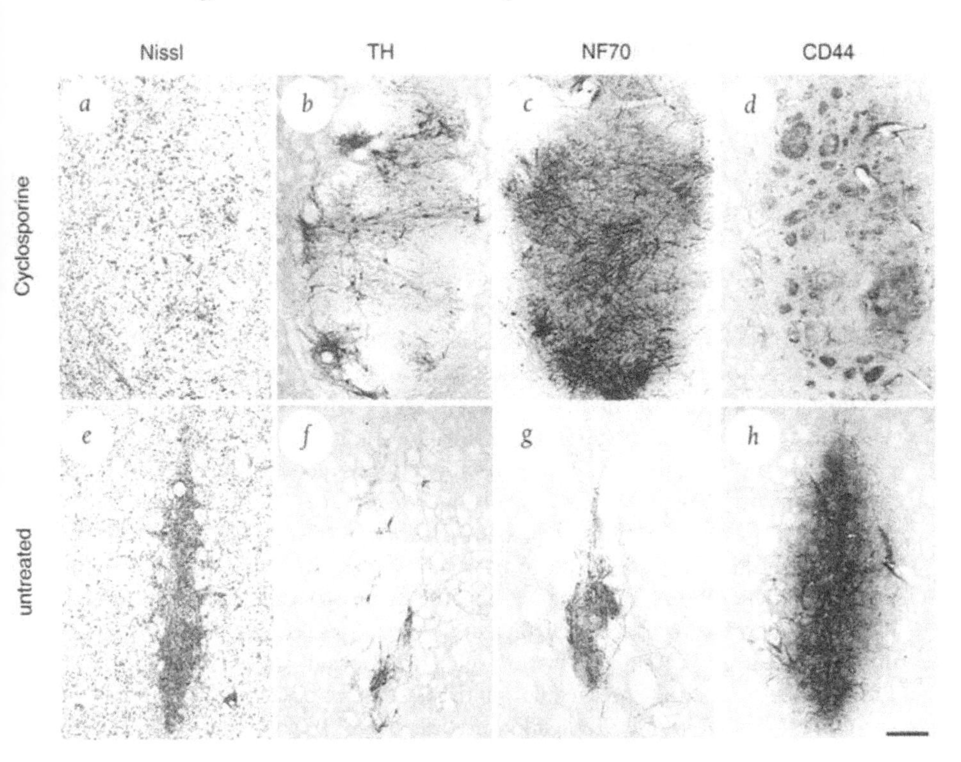

FIGURE 1 Fetal neural xenotransplant morphology, integration and rejection process: Structure of a normal and partially rejected porcine fetal cell suspension xenografts in an animal model of Parkinson's disease. Adjacent coronal sections through a representative porcine ventral mesencephalon (VM) xenograft of a cyclosporine A-treated rat (a–d) and an untreated rat (e–h) stained for Nissl a & e, tyrosine hydroxylase (TH) b & f, pig-specific neurofilament (NF70) c & g, and pig-specific glial (CD44) d & h. The partially rejected graft (below) shows evidence of cellular accumulation (e), loss of neuronal content (f & g), and concentration of glia (h). Scale bar = 200 μm.

(Isacson *et al.*, 1995; Galpern *et al.*, 1996) (Fig. 1). In contrast to the CsA-treated rats, grafts from non-CsA rats were characteristically small and condensed, with few surviving neurons as demonstrated by the limited number of TH- and NF70-immunostained cells. Surviving grafts from this group contained varying degrees of small-cell infiltration. There was greater survival of the glial components of the graft relative to the neuronal components in this group. In contrast to the CsA-treated rats, CD44+ immunostaining was usually concentrated in the graft core and fewer fibers were seen in the white matter tracts. Of particular interest, no significant host tissue damage was apparent beyond the transplant site in rats that did not have surviving grafts at sacrifice

(Isacson *et al.*, 1995; Pakzaban *et al.*, 1995; Galpern *et al.*, 1996; Deacon *et al.*, 1997).

B. Expression of Cell and Immune Markers in Neural Porcine Xenografts

The neuronal and glial organization of ventral mesencephalic xenografts in the CsA-treated and non-CsA-treated groups was assessed by Nissl staining as well as TH, NF70, and CD44 immunostaining. Nissl-stained sections of CsA-treated rats showed large, neuron-rich grafts that were well integrated with the host tissue. The transplants contained large numbers of TH+ cell bodies with the characteristic morphology of substantia nigra DA neurons. Immunostaining of the porcine VM xenografts for TH revealed organotypic features of the SN, with large clusters of TH+ neurons at the graft perimeter and a dense network of TH+ neuronal fibers within the transplant and extending from the graft to innervate the surrounding host striatum. NF70 immunostaining further demonstrated that the TH-negative regions of the grafts were densely filled with porcine-derived neurons and fibers. CD44 immunostaining of donor-derived glia revealed a distinct glial component within the graft with the most dense labeling observed within the host white matter tracts penetrating the graft (Isacson *et al.*, 1995; Galpern *et al.*, 1996; Deacon *et al.*, 1997).

To determine if the internal organization of striatal xenografts was altered by $F(ab')_2$ treatment, the cellular composition of $F(ab')_2$-treated grafts was compared to that of CsA-treated grafts. On Nissl-stained sections, grafts in both groups were composed of multiple clusters of large neuron-like cells surrounded by bands of smaller cells. The neuronal (and striatal) phenotype of the large-cell clusters was confirmed by the close correspondence between these regions and the TH-positive regions on adjacent sections. Comparison of NF70 and CD44 immunostaining on adjacent sections revealed a consistent segregation of the NF70-immunoreactive (NF70-IR) neuronal and CD44-IR glial components of the grafts in both $F(ab')_2$- and CsA-treated animals. In both groups, the NF70-IR regions corresponded to the neuronal clusters, whereas the CD44-IR glia were distributed along the bands that surrounded and separated the neuronal clusters. Although the cytoarchitectonic organization of the grafts was similar in both groups, the relative proportion of NF70-IR elements appeared to be greater in CsA-treated grafts (Pakzaban *et al.*, 1995).

Immunostaining for porcine MHC-I in selected grafts revealed persistent expression of donor MHC-I in the periphery of the surviving xenografts. Importantly, the pattern of expression of donor MHC-I corresponded to the distribution of donor-derived CD44-IR glia in both treatment groups (Pakzaban *et al.*, 1995). The NF70-IR neuronal clusters were devoid of MHC-I expression on adjacent sections. In some brains (in all three groups), necrotic graft remnants were detected on Nissl-stained sections. These regions, which were interspersed among *intact* host striatal white matter tracts, were infiltrated with small cells on Nissl stain. Immunostaining of adjacent sections with OX1 (labeling lymphocytes and activated microglia (Finsen *et al.*, 1991) and OX42 (labeling phagocytes and microglia (Finsen *et al.*, 1991) revealed an overlapping pattern of accumulation of host inflammatory cells and/or activated microglia in these regions. The inflammatory infiltrate was fairly localized and spared the striatal white matter tracts and most of the remaining neuron-rich portions of the grafts (Figs. 2 and 3).

C. Effect of Immunosuppression on Porcine Neural Xenografts

An important finding is the preferential expression of MHC-I in graft regions populated by donor glia. Using porcine CD44 and NF70 immunoreactivity as markers of donor-derived glia and neurons, respectively, we have consistently observed a segregation of donor glial and neuronal populations in maturing porcine neural xenografts (Deacon *et al.*, 1994) (Fig. 4). Although the mechanism for this segregation is unknown, it facilitates specific immunohistochemical characterization of graft neuronal and glial elements (e.g., with regard to MHC-I expression). Our findings confirm the suggestions that graft neurons may express lower levels of MHC than the nonneuronal elements in the graft parenchyma (Nicholas *et al.*, 1987). In addition, the greater expression of MHC-I on graft glia is consistent with the observation that expression of MHC-I occurs more readily in glial than neuronal cell lines in culture (Lampson and Fisher, 1984; Lampson, 1987). Importantly, the preferential expression of MHC-I on graft glia does not necessarily imply a greater susceptibility of these nonneuronal elements to immune rejection, as has been suggested previously (Nicholas *et al.*, 1987). Once a specific MHC-restricted T cell response is mounted, all graft cells may come under attack by secreted effector molecules or by secondarily recruited non-MHC-restricted effector cells (Lampson, 1987). In such settings, the proliferative capacity of graft glia, not shared by graft neurons, may in fact confer a selective survival advantage to the former. Consistent with this view, we have observed a preferential preservation of donor-derived glia in the smaller $F(ab')_2$ and CsA-treated grafts (Fig. 5).

Our observation that MHC-I expression is fairly restricted to graft glia may have important implications for alternative means of protecting neural xenografts against immune rejection. Induction of maximal MHC-I expression on donor cells with IFN-γ *in vitro* (Wong *et al.*, 1984, 1985), followed by immunoselection and transplantation of cells lacking MHC-I may not only

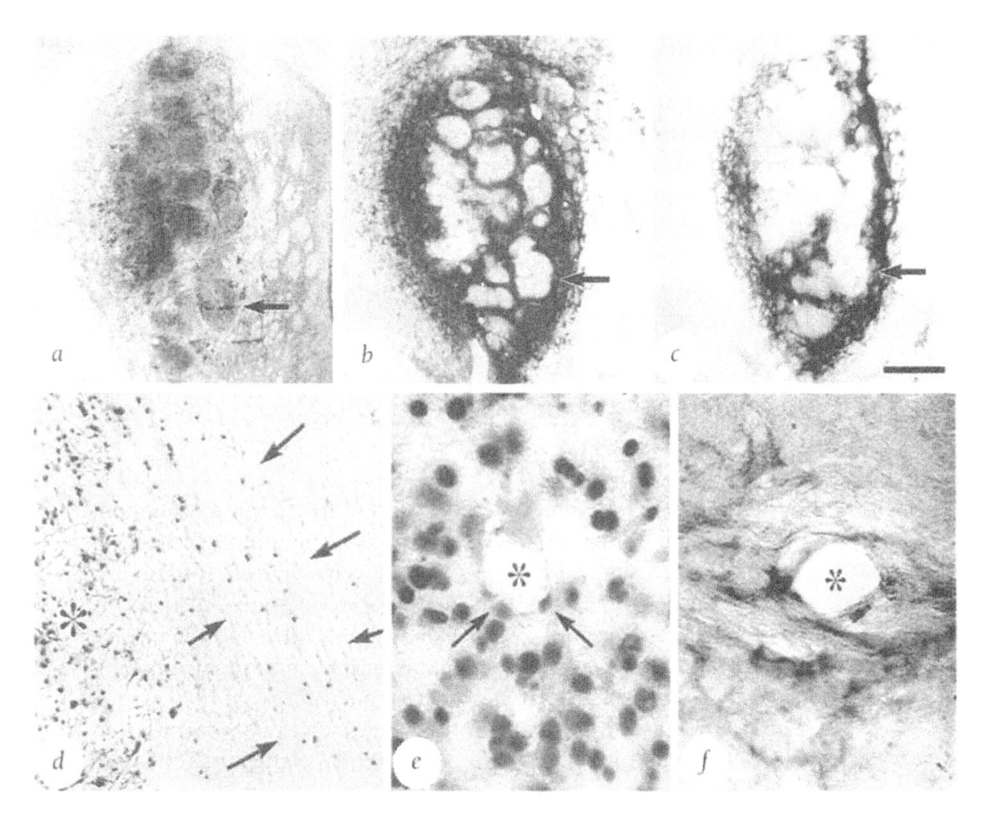

FIGURE 2 Cell and immune markers for porcine xenografts: Porcine neural xenografts in a rodent brain demonstrated with donor-specific markers. Top: MHC-I expressing cells (c) in a striatal xenograft appear to be largely confined to the glial component by 4 months post-transplantation as demonstrated by comparison with pig-specific neuronal (a) and glial (b) cell markers in the same graft. Bottom: Pig-specific glial marker (CD44) and *in situ* hybridization using a probe that recognizes a pig-specific DNA repeat element (PRE): donor cell migration out of the graft (d) demonstrated by the presence of PRE+ nuclei (black dots) within the corpus callosum (between the arrows). The section was double labeled with TH and counterstained with nuclear fast red. The majority of PRE+ cells are located within the confines of the graft (asterisk). A small fraction of PRE+ nuclei are observed (arrows) within the endothelium of blood vessels (asterisk) penetrating the graft (e) and pig glial cells and fibers, immunoreactive for CD44, are also identified within the endothelium (f).

protect the xenograft against MHC-I restricted immune attack, but may also enrich the graft in neuronal elements. Such a strategy has been reported to enhance neural allograft survival (Bartlett *et al.*, 1990). Alternatively, immunoselection can be performed on the basis of donor glial markers such as CD44, obviating the need for *ex vivo* induction of MHC-I. However, it remains to be determined whether depletion of the immunogenic glia from the donor cell

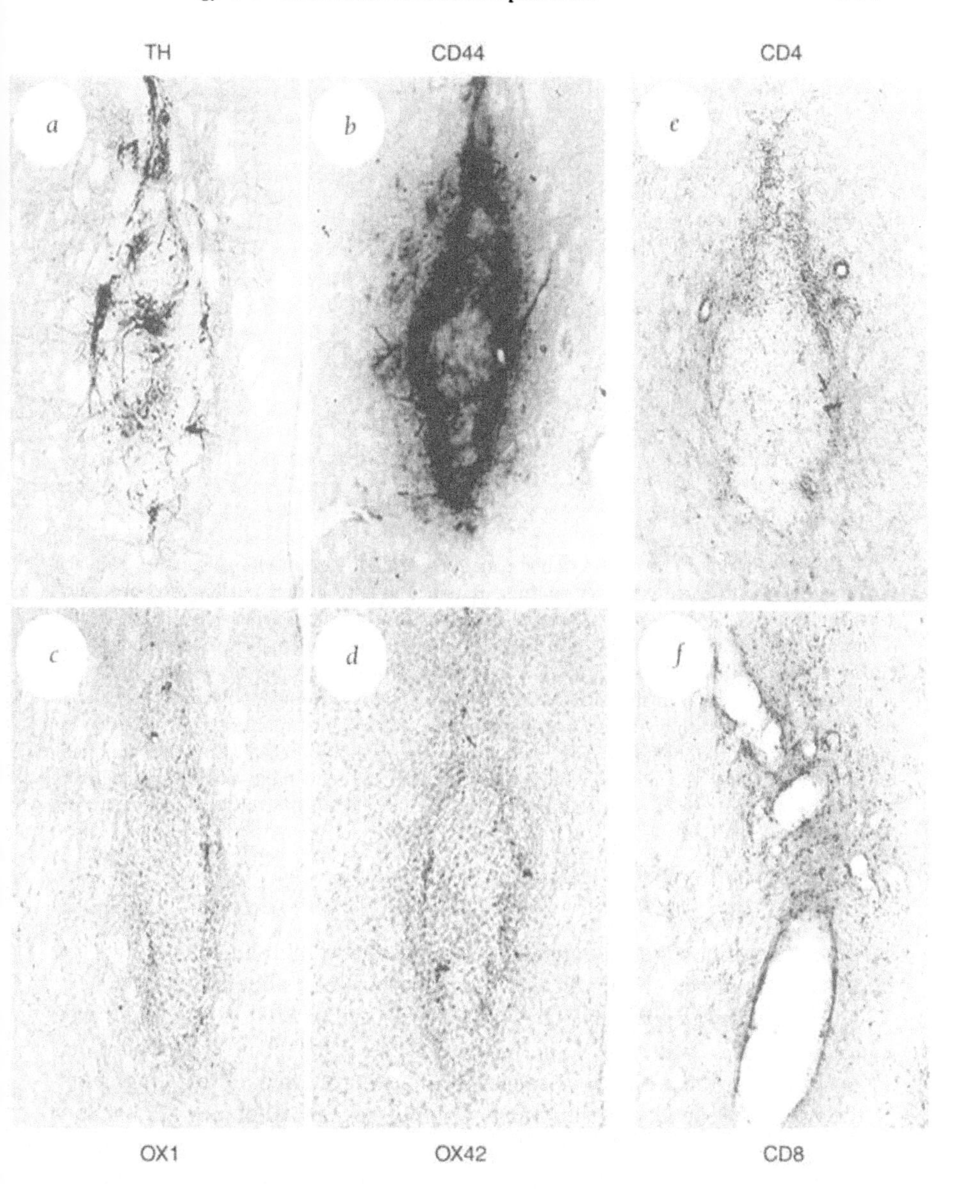

FIGURE 3 Immunosuppression and immunomodulation of porcine fetal VM cell suspension xenografts: Rat host immunological response to a VM xenograft demonstrated by markers for rat immune cell response, including T-helper cells (CD4), cytotoxic T cells (CD8), all leucocytes (OX1), and macrophages and dendritic cells (OX42), in a pig xenograft immunomodulated by $F(ab')_2$ masking of pig MHC class-1 prior to transplantation. Adjacent sections stained for TH+ neurons (a) and CD44+ pig glia (b) are shown for comparison to OX1 (c) and OX42 (d) immunostaining. Xenograft exhibiting signs of immune activation of CD4+ cells (e) and a rejected necrotic graft (f) showing accumulation of CD8+ cells.

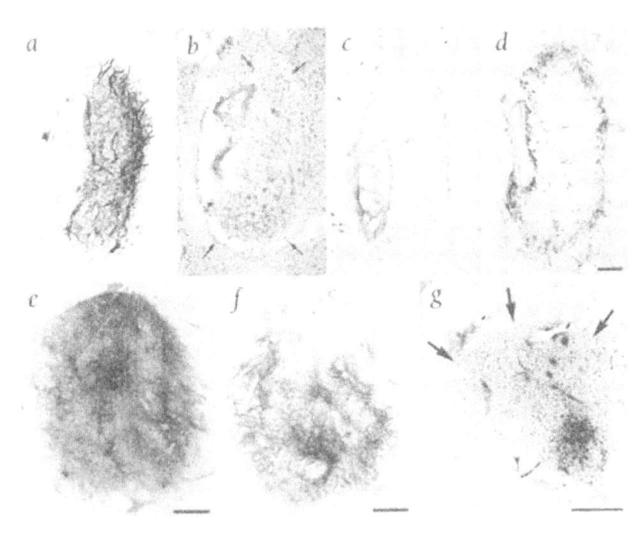

FIGURE 4 Preliminary characterization of patient response to neural cell implantation: Human immune response to porcine VM xenografting examined in post mortem putamen sections from a patient in a clinical safety trial of porcine xenotransplantation for Parkinson's disease. Death occurred after a pulmonary embolism 7.5 months after transplantation. Top: Small region from one of 3 TH+ and NF70+ (a) a graft that demonstrated some activation of host immune cells as indicated by hematoxylin and eosin staining (b), and immunostaining for human CD3 (c), and human MHC-II (d). Bottom: Another graft site from the same patient demonstrating distribution of pig CD44+ glia within the graft site (e) and GFAP+ glia (not pig-specific) also mostly confined to the graft (f). Donor cell migration into the host putamen for up to 4mm from the implantation site (asterisk) is demonstrated by the distribution PRE+ nuclei (g) (arrows indicate extent of distribution). Scale bars: a-d 50μm; e-f 200 μm; g 1mm.

suspension would adversely affect subsequent maturation and integration of neural xenografts by eliminating putative glial trophic influences.

The mechanism by which $F(ab')_2$ masking of donor MHC-I interferes with allograft (Osorio *et al.*, 1994) and xenograft rejection (Faustman and Coe, 1991) long after the probable dissociation and destruction of $F(ab')_2$ fragments is thought to depend on induction of donor-specific tolerance (Faustman and Coe, 1991). Establishment of tolerance has been documented in mice transplanted with rat insulinoma cells pretreated with $F(ab')_2$ fragments to donor MHC-I (Faustman and Coe, 1991). The recipients did not mount an immune response against subsequent *untreated* rat xenografts. The mechanism of tolerance in this setting is not understood and may involve interference of $F(ab')_2$ fragments with high-affinity binding of MHC-I to the T-cell receptor (TCR), perhaps by impeding appropriate secondary aggregation of MHC-I molecules into tetrameric structures (Krishna *et al.*, 1992), which in turn

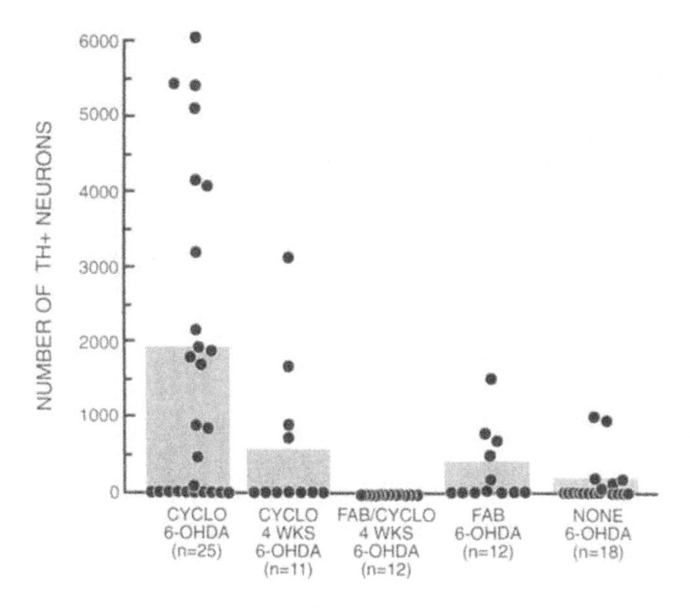

FIGURE 5 Survival of dopamine neurons in porcine xenografts under 5 different immune modulation regimens in the rat as assessed by numbers of surviving (TH+) cells 16 weeks post-implantation in a 6-OHDA lesioned Parkinson rat model: The conditions included: CYCLO (continuous cyclosporine administration); CYCLO-4 WKS (cyclosporine administration for the first 4 weeks only); FAB/CYCLO ($F(ab')_2$ masking plus 4 weeks cyclosporine administration); FAB ($F(ab')_2$ masking); NONE (no immune suppression). Numbers of cases as indicated. Each condition produced some grafts that contained no surviving TH+ cells. However, all conditions, but the FAB–CYCLO condition, produced some grafts with surviving TH+ cells. In this series of pilot experiments, statistical analysis only supported the conclusion that long-term survival was significantly better with continuous cyclosporine treatment, and significantly worse with a combination of cyclosporine and MHC-I masking $F(ab')_2$ (ANOVA, $p < 0.05$, and hypothesis testing using various nonparametric tests). Cell counts are estimated using the Abercrombie correction.

impedes T-cell activation and clonal expansion. In absence of a stimulus for T-cell proliferation, TCR occupancy can result in induction of clonal anergy (DeSilva *et al.*, 1991). A similar state of anergy and graft acceptance is observed after administration of monoclonal antibodies to intercellular adhesion molecule-1 (ICAM-1) and leukocyte function associated antigen-1 (LFA-1) (Isobe *et al.*, 1992). The interaction of LFA-1 on the T-cell surface with its ligand ICAM-1 on the antigen-presenting cell provides the necessary co-stimulatory signals for T-cell activation and cytolysis (Springer, 1990). The prolonged state of donor-specific tolerance induced by antibody-mediated interference with LFA-1–ICAM-1 binding may resemble that induced by $F(ab')_2$-mediated interference with TCR–MHC-I binding. Alternatively, binding of $F(ab')_2$ fragments to MHC-I can alter the configuration of the alpha-3 locus of MHC-I (Spertini *et al.*, 1992). The alpha-3 locus is the binding site for the CD8 co-receptor (Salter *et al.*, 1990), which increases the avidity of TCR for the target molecule (Norment *et al.*, 1988). Interference of $F(ab')_2$ fragments with the proper binding and function of the CD8 co-receptor may be important to the mechanism by which tolerance is induced in this context.

VI. SAFETY ISSUES THAT MUST BE CONSIDERED FOR NEURAL XENOTRANSPLANTATION

The single greatest risk factor to the patients in neural cell xenotransplantation is rejection of the implanted cells (Fig. 4). The FDA-guided clinical safety trials that have been conducted by several specialist transplant teams in United States have not resulted in any obvious rejections or fatalities to date due to surgical procedures. The second major risk factor is side effects from immunosuppressive treatments, such as cyclosporine. The low cyclosporine dosage used in the FDA-guided safety study of neural xenotransplantation for Parkinson's and Huntington's disease in United States (Galpern *et al.*, 1996; Deacon *et al.*, 1997) and absence of accompanying immunosuppressive regimens (commonly used in human organ transplantation) may have contributed to the lack of side effects observed.

Concerns of other risk factors are difficult to assess on the basis of clinical data. Recently, the propagation of viral particles from one species to another has received widespread attention. The idea that endogenous viral sequences (it has been estimated that about 1% of DNA sequences in mammals contain some reminiscence to viral sequences) could (1) get activated, (2) form active viral particles, (3) form pathogenic particles, or (4) form transmissible pathogenic particles has been forwarded on the basis of *in vitro* experiments with kidney cancer cells known to produce retroviruses (Patience *et al.*, 1997). Although an apparent viral safety risk of the sort suggested cannot be discounted on theoretical grounds, the likelihood that the four events necessary for pathological viral transmission would happen in sequence in one patient transplanted with normal pathogen-free fetal neurons appears miniscle. On the other hand, it is of outstanding importance to determine all risk factors associated with neural xenotransplantation very carefully, despite the fact that pig retroviruses generally are not transmissible to other species (Woods *et al.*, 1973; Lieber *et al.*, 1975). For this reason we propose examination for any transmission of viral particles from the xenogeneic donor by screening for retroviruses based on standard assays. Peripheral blood lymphocytes can be prepared and then RNA extracted and assessed for pig DNA sequences using primers for known retrovirus sequences (Woods *et al.*, 1973; Lieber *et al.*, 1975; Wolf *et al.*, 1988; Patience *et al.*, 1997). The gels of such PCRs can then reveal if any pig retroviral material is present. Likewise, peripheral rat or nonhuman primate organs can be analyzed using the same assays in transplanted animals (Fig. 5). A chromosomal insertion of a pig retroviral sequence, while unlikely, could theoretically occur, and therefore purification of DNA from blood and tissues followed by PCRs for amplification of any porcine cDNA sequence could determine if such an event can happen. The possible false positives that may occur with such sensitive assays can be controlled for by assaying animals transplanted with the vehicle solution the cells are prepared in, or non-pig cells.

In terms of relative risk assessment for neural xenotransplantation for neurodegenerative diseases, current gene therapy trials using attenuated or replication-deficient viruses for gene delivery (such as adenoviruses) probably pose a low but relatively higher risk for pathology or modified viral transmission than would the presence of endogenous retroviral DNA sequences in porcine cells, because porcine DNA sequences have yet to be demonstrated to be able to produce active viruses in nontransformed pig cells, with subsequent transmission causing pathology to hosts. Furthermore, screened and pathogen-free porcine tissue appears to be an unlikely source of infectious agents, compared to other xenogeneic sources. In a recent document to the FDA, 44 emminent US and European microbiologists and immunologists proposed that porcine donor tissue be the only type used on a limited scale until all risk factors can be assessed with other types of xenotransplant donor tissue to man (Aldrete *et al.*, 1996).

ACKNOWLEDGMENTS

This work was supported in part by a PHS grant NIH NS30064 and funds from McLean Hospital and Massachusetts General Hospital. We appreciate the editorial assistance of Ms. S. Pohlman.

REFERENCES

Aldrete, J., Allan, J. S., Blomberg, J., Burger, H., Chanh, T., Clements, J., Coffin, J., Desrosiers, R., Essex, M., Fultz, P. N., Gauntt, C. J., *et al.*, (1996). Response to draft public health service guidelines on infectious disease issues in xenotransplantation. Southwest Foundation for Biomedical Research.

Bartlett, P. F., Rosenfeld, J., Bailey, K. A., Cheesman, H., Harvey, A. R., and Kerr, R. S. C. (1990). Allograft rejection overcome by immunoselection of neural precursor cells. *Prog Brain Res* 82:153–160.

Benveniste, E. N., Kwon, J., Chung, W. J., Sampson, J., Pandya, K., and Tang, L. P. (1994). Differential modulation of astrocyte cytokine gene expression by TGF-beta. *J Immunol* 153:5210–5221.

Björklund, A., Stenevi, U., Dunnett, S. B., and Iversen, S. D. (1981). Functional reactivation of the deafferented neostriatum by nigral transplants. *Nature* 289:497–499.

Bluestone, J. A., Pescovitz, M. D., Frels, W. I., Singer, D. S., and Hodes, R. J. (1987). Cytotoxic T lymphocyte recognition of a xenogeneic major histocompatibility complex antigen expressed in transgenic mice. *Eur J Immunol* 17:1035–1041.

Bontrop, R. E., Elferink, D. G., Otting, N., Jonker, M., and de Vries, R. R. P. (1990). Major histocompatibility complex class II-restricted antigen presentation across a species barrier: Conservation of restriction determinants in evolution. *J Exp Med* 172:53–59.

Brundin, P., Nilsson, O. G., Gage, F. H., and Björklund, A. (1985). Cyclosporin A increases survival of cross-species intrastriatal grafts of embryonic dopamine-containing neurons. *Exp Brain Res* 60:204–208.

Brundin, P., Nilsson, O. G., Strecker, R. E., Lindvall, O., Åstedt, B., and Björklund, A. (1986). Behavioral effects of human fetal dopamine neurons grafted in a rat model of Parkinson's disease. *Exp Brain Res* 65:235–240.

Brundin, P., Widner, H., Nilsson, O. G., Strecker, R. E., and Björklund, A. (1989). Intracerebral xenografts of dopamine neurons: The role of immunosuppression and the blood–brain barrier. *Exp Brain Res* 75:195–207.

Coffey, P. J., Perry, V. H., and Rawlins, J. N. (1990). An investigation into the early stages of the inflammatory response following ibotenic acid-induced neuronal degeneration. *Neuroscience* 35:121–132.

Darr, A. S., Fuggle, S. V., and Fabre, J. W. (1984). The detailed distribution of HLA-A, B, C antigens in normal human organs. *Transplantation* 38:287–292.

Deacon, T., Schumacher, J., Dinsmore, J., Thomas, C., Palmer, P., Kott, S., Edge, A., Penney, D., Kassissieh, S., Dempsey, P., and Isacson, O. (1997). Histological evidence of fetal pig neural cell survival after transplantation into a patient with Parkinson's disease. *Nature Med* 3:350–353.

Deacon, T. W., Pakzaban, P., Burns, L. H., Dinsmore, J., and Isacson, O. (1994). Cytoarchitectonic development, axon-glia relationships and long distance axon growth of porcine striatal xenografts in rats. *Exp Neurol* 130:151–167.

DeSilva, D. R., Urdahl, K. B., and Jenkins, M. K. (1991). Clonal anergy is induced *in vitro* by T cell receptor occupancy in the absence of proliferation. *J Immunol* 147:3261–3267.

Detta, A., and Hitchcock, E. (1990). The selective viability of human foetal brain cells. *Brain Res* 520:277–283.

Duan, W.-M. (1997). Immunological and inflammatory responses against intrastriatal neural grafts in the rat. *Thesis, Lund University.*

Dunnett, S., and Annett, L. (1991). In: Lindvall, O., Björklund, A., and Widner, H., ed., *Intracerebral transplantation in movement disorders.* Amsterdam: Elsevier, pp. 27–50.

Dymecki, J., and Freed, W. J. (1989). Results of cross-species transplantation of substantia nigra to the lateral ventricle of the brain of rats with experimentally-induced Parkinson disease. *Neuropatol Pol* 27:557–571.

Faustman, D., and Coe, C. (1991). Prevention of xenograft rejection by masking donor HLA class I antigens. *Science* 252:1700–1702.

Fierz, W., Endler, B., Reske, K., Wekerle, H., and Fontana, A. (1985). Astrocytes as antigen presenting cells. I. Induction of Ia antigen expression on astrocytes by T cells via immune interferon and its effect on antigen presentation. *J Immunol* 134:3785–3793.

Finsen, B., Oteruleo, F., and Zimmer, J. (1988a). Immunocytochemical characterisation of the cellular immune response to intracerebral xenografts of brain tissue. *Prog Brain Res* 78:261–270.

Finsen, B., Poulsen, P. H., and Zimmer, J. (1988b). Xenografting of fetal mouse hippocampal tissue to the brain of adult rats. Effect of cyclosporin A treatment. *Exp Brain Res* 70:117–133.

Finsen, B. R., Sorensen, T., Castellano, B., Pedersen, E. B., and Zimmer, J. (1991). Leukocyte infiltration and glial reactions in xenografts of mouse brain tissue undergoing rejection in the aduly rat brain. A light and electron microscopical immunocytochemical study. *J Neuroimmunol* 32:159–183.

Freed, C., Breeze, R., Rosenberg, N., Schneck, S., Kriek, E., Qi, J.-X., Lone, T., Zhang, Y.-B., Snyder, J., Wells, T., Ramig, L., Thompson, L., Mazziotta, J., Huang, S., ST, G., Brooks, D., Sawle, G., Schroter, G., and Ansari, A. (1992). Survival of implanted fetal dopamine cells and neurologic improvement 12 to 24 months after transplantation for Parkinson's disease. *N Engl J Med* 327:1549–1555.

Freed, W. J., Dymecki, J., Poltorak, M., and Rodgers, C. R. (1988). Intraventricular brain allografts and xenografts: Studies of survival and rejection with and without systemic sensitization. *Prog Brain Res* 78:233–241.

Freeman, T. B., Spence, M. S., Boss, B. D., Spector, D. H., Strecker, R. E., Olanow, C. W., and Kordower, J. H. (1991). Development of dopaminergic neurons in the human substantia nigra. *Exp Neurol* 113:344–353.

Gage, F., Björklund, A., Stenevi, U., Dunnett, S., and Kelly, P. (1984). Intrahippocampal septal grafts ameliorate learning impairments in aged rats. *Science* 225:533–536.

Galpern, W. G., Burns, L. H., Deacon, T. W., Dinsmore, J., and Isacson, O. (1996). Xenotransplantation of porcine fetal ventral mesencephalon in a rat model of Parkinson's disease: Functional recovery and graft morphology. *Exp Neurol* 140:1–13.

Geist, M. J., Maris, D. O., and Grady, M. S. (1991). Blood–brain barrier permeability is not altered by allograft or xenograft fetal neural cell suspension grafts. *Exp Neurol* 111:166–174.

Goodman, J. M. (1972). *Phylogenetic development of vertebrate immunity*. New York: MSS Information Corp.

Goodman, M., Barnabas, J., Matsuda, G., and Moore, G. W. (1971). Molecular evolution in the descent of man. *Nature* 233:604–613.

Greenley, H. T., Hamm, T., Johnson, R., Price, C. R., Weingarten, R., and Raffin, T. (1989). The ethical use of human fetal tissue in medicine. *N Engl J Med* 320:1093–1096.

Hantraye, P., Riche, D., Maziere, M., and Isacson, O. (1992). Intrastriatal transplantation of cross-species fetal striatal cells reduces abnormal movements in a primate model of Huntington disease. *Proc Natl Acad Sci USA* 89:4187–4191.

Heim, R. C., Willingham, G., and Freed, W. J. (1993). Comparison of solid intraventricular and dissociated intraparenchymal fetal substantia nigra grafts in a rat model of Parkinson's disease: Impaired graft survival is associated with high baseline rotational behavior. *Exp Neurol* 122:5–15.

Hofman, F. M., Danilovs, J. A., and Taylor, C. R. (1984). HLA-DR (Ia)-positive dendritic-like cells in human fetal non-lymphoid tissues. *Transplantation* 37:590–594.

Honey, C. R., Charlton, H. M., and Wood, K. J. (1991a). Rat brain xenografts reverse hypogonadism in mice immunosuppressed with anti-CD4 monoclonal antibody. *Exp Brain Res* 85:149–152.

Honey, C. R., Clarke, D. J., Dallman, M. J., and Charlton, H. M. (1991b). Human neural graft function in rats treated with anti-interleukin II receptor antibody. *Neuroreport* 1:247–249.

Inoue, H., Kohsaka, S., Yoshida, K., Ohtani, M., Toya, S., and Tsukada, Y. (1985). Cyclosporin A enhances the survivability of mouse cerebral cortex grafted into the third ventricle of rat brain. *Neurosci Lett* 54:85–90.

Isacson, O., Brundin, P., Kelly, P. A. T., Gage, F. H., and Björklund, A. (1984). Functional neuronal replacement by grafted neurons in the ibotenic acid-lesioned striatum. *Nature* 311:458–460.

Isacson, O., Dawbarn, D., Brundin, P., Gage, F. H., Emson, P. C., and Björklund, A. (1987a). Neural grafting in a rat model of Huntington's disease: Striosomal-like organization of striatal grafts as revealed by acetylcholinesterase histochemistry, immunocytochemistry and receptor autoradiography. *Neuroscience* 22:481–497.

Isacson, O., Deacon, T. W., Pakzaban, P., Galpern, W. R., Dinsmore, J., and Burns, L. H. (1995). Transplanted xenogeneic neural cells in neurodegenerative disease models exhibit remarkable axonal target specificity and distinct growth patterns of glial and axonal fibres. *Nature Med* 1:1189–1194.

Isacson, O., Dunnett, S. B., and Björklund, A. (1986). Graft-induced behavioral recovery in an animal model of Huntington's disease. *Proc Natl Acad Sci USA* 83:2728–2732.

Isacson, O., Fischer, W., Wictorin, K., Dawbarn, D., and Björklund, A. (1987b). Astroglial response in the excitotoxically lesioned neostriatum and its projection areas in the rat. *Neuroscience* 20:1043–1056.

Isobe, M., Yagita, H., Okumura, K., and Ihara, A. (1992). Specific acceptance of cardiac allograft after treatment with antibodies to ICAM-1 and LFA-1. *Science* 255:1125–1127.

Kievits, F., Wijfels, J., Lokhorst, W., and Ivanyi, P. (1989). Recognition of xeno (HLA, SLA) major histocompatibility complex antigens by mouse cytotoxic T cells is not H-2 restricted: A study with transgenic mice. *Proc Natl Acad Sci USA* 86:617–620.

Krishna, S., Benaroch, P., and Pillai, S. (1992). Tetrameric cell-surface MHC class I molecules. *Nature* 357:164–167.

Lampson, L. A. (1987). Molecular bases of the immune response to neural antigens. *Trends Neurosci* 10:211–216.

Lampson, L. A., and Fisher, C. A. (1984). Weak HLA and b2-microglobulin expression of neuronal cell lines can be modulated by interferon. *Proc Natl Acad Sci USA* 81:6476–6480.

Lampson, L. A., and Hickey, W. F. (1986). Monoclonal antibody analysis of MHC expression in human brain biopsies: Tissue ranging from "histologically normal" to that showing different levels of glial tumor involvement. *J Immunol* 136:4054–4062.

Lampson, L. A., and Siegel, G. (1988). Defining the mechanisms that govern immune acceptance or rejection of neural tissue. *Prog Brain Res* 78:243–247.

Lavi, E., Suzumura, A., Zoltick, P. W., Murasko, D. M., Silberberg, D. H., and Weiss, S. R. (1987). Tumor necrosis factor induces MHC class I antigen expression on mouse astrocytes. *J Neuroimmunol* 16:102.

Lieber, M. M., Sherr, C. J., Benveniste, R. E., and Todaro, G. J. (1975). Biologic and immunologic properties of porcine type C viruses. *Virology* 66:616–619.

Lindahl, K. F., and Bach, F. H. (1976). Genetic and cellular aspects of xenogeneic mixed leukocyte culture reaction. *J Exp Med* 144:305–318.

Lindvall, O., Sawle, G., Widner, H., Rothwell, J. C., Björklund, A., Brooks, A., Brundin, P., Frackowiak, R., Marsden, C. D., Odin, P., and Rehncrona, S. (1994). Evidence for long-term survival and function of dopaminergic grafts in progressive Parkinson's disease. *Ann Neurol* 35:172–180.

Liu, J., Farmer, J. D., Lane, W. S., Friedman, J., Weissman, I., and Schreiber, S. L. (1991). Calcineurin is a common target of cyclophilin-cyclosporin A and FKBP-FK506 complexes. *Cell* 66:807–815.

Low, W., Lewis, P., Bunch, S., Dunnett, S., Thomas, S., Iversen, S., Björklund, A., and Stenevi, U. (1982). Functional recovery following neural transplantation of embryonic septal nuclei in adult rats with septohippocampal lesions. *Nature* 300:260–262.

Ludlam, W. H., Chandross, K. J., and Kessler, J. A. (1995). LIF- and IL-1 beta-mediated increases in substance P receptor mRNA in axotomized, explanted or dissociated sympathetic ganglia. *Brain Res* 685:12–20.

Lynn, W. S., and Wong, P. K. (1995). Neuroimmunodegeneration: do neurons and T cells use common pathways for cell death? *FASEB J* 9:1147–1156.

Marion, D. W., Pollack, I. F., and Lund, R. D. (1990). Patterns of immune rejection of mouse neocortex transplanted into neonatal rat brain, and effects of host immunosuppression. *Brain Res* 519:133–143.

Mason, D. W., Charlton, H. M., Jones, A. J., Lavy, C. B., Puklavec, M., and Simmonds, S. J. (1986). The fate of allogeneic and xenogeneic neuronal tissue transplanted into the third ventricle of rodents. *Neuroscience* 19:685–694.

Massa, P. T., Dörries, R., and ter Meulen, V. (1986). Viral particles induce Ia antigen expression on astrocytes. *Nature* 320:543–546.

McDonald, J. W., Goldberg, M. P., Gwag, B. J., Chi, S.-I., and Choi, D. W. (1996). Cyclosporine induces neuronal apoptosis and selective oligodendrocyte death in cortical cultures. *Ann Neurol* 40:750–758.

Nakashima, H., Kawamura, K., and Date, I. (1988). Immunological reaction and blood-brain barrier in mouse-to-rat cross-species neural graft. *Brain Res* 475:232–243.

Nicholas, M. K., Antel, J. P., Stefansson, K., and Arnason, B. G. W. (1987). Rejection of fetal neocortical neural transplants by H-2 incompatible mice. *J Immunol* 139:2275–2283.

Nicholas, M. K., Chenelle, A. G., Brown, M. M., Stefansson, K., and Arnason, B. G. W. (1990). Prevention of neural allograft rejection in the mouse following *in vivo* depletion of L3T4+ but not LYT-2+ T-lymphocytes. *Prog Brain Res* 82:161–167.

Niijima, K., Araki, M., Ogawa, M., Nagatsu, I., Sato, F., Kimura, H., and Yoshida, M. (1990). Enhanced survival of cultured dopamine neurons by treatment with soluble extracts from chemically deafferentiated striatum of adult rat brain. *Brain Res* 528:151–154.

Nishino, H., Ono, T., Takahashi, J., Kimura, M., Shiosaka, S., and Tohyama, M. (1986). Transplants in the peri- and intraventricular regions grow better than those in the central parenchyma of the caudate. *Neurosci Lett* 64:184–190.

Norment, A. M., Salter, R. D., Parham, P., Engelhard, V. H., and Littman, D. R. (1988). Cell-cell adhesion mediated by CD8 and MHC class I molecules. *Nature* 336:79–81.

Osorio, R. W., Ascher, N. L., and Stock, P. G. (1994). Prolongation of *in vivo* mouse islet allograft survival by modulation of MHC class I antigen. *Transplantation* 57:783–788.

Pakzaban, P., Deacon, T. W., Burns, L. H., Dinsmore, J., and Isacson, O. (1995). A novel mode of immunoprotection of neural xenotransplants: Masking of donor major histocompatibility complex class I enhances transplant survival in the CNS. *Neuroscience* 65:983–996.

Pakzaban, P., and Isacson, O. (1994). Neural xenotransplantation: reconstruction of neuronal circuitry across species barriers. *Neuroscience* 62:989–1001.

Patience, C., Takeuchi, Y., and Weiss, R. A. (1997). Infection of human cells by an endogenous retrovirus of pigs. *Nature Med* 3:282–286.

Pedersen, E. B., Poulson, F. R., Zimmer, J., and Finsen, B. (1995). Prevention of mouse-rat xenograft rejection by a combination therapy of cyclosporin A, prednisolone and azathioprine. *Exp Brain Res* 106:181–186.

Perlow, M. J., Freed, W. J., Hoffer, B. J., Seiger, A., Olson, L., and Wyatt, R. J. (1979). Brain grafts reduce motor abnormalities produced by destruction of the nigrostriatal dopamine system. *Science* 204:643–647.

Poltorak, M., Freed, W. J., Sternberger, L. A., and Sternberger, N. H. (1988). A comparison of intraventricular and intraparenchymal cerebellar allografts in rat brain: Evidence for normal phosphorylation of neurofilaments. *J Neuroimmunol* 20:63–72.

Ridge, J. P., Fuchs, E. J., and Matzinger, P. (1996). Neonatal tolerance revisited: Turning on newborn T cells with dendritic cells. *Science* 271:1723–1726.

Rothwell, N. J., and Hopkins, S. J. (1995). Cytokines and the nervous system II: Actions and mechanisms of action. *TINS* 18:130–136.

Salter, R. D., Benjamin, R. J., Wesley, P. K., Buxton, S. E., Garrett, T. P., Clayberger, C., Krensky, A. M., Norment, A. M., Littman, D. R., Parham, P., Spertini, F., Chatila, T., and Geha, R. S.

(1990). A binding site for the T-cell co-receptor CD8 on the alpha 3 domain of HLA-A2. *Nature* 345:41–46.

Sanberg, P. R., and Isacson, O. (1994). In: *Cell transplantation for Huntington's disease.* Austin: R. G. Landes Co., pp. 73–82.

Schmidt, R. H., Björklund, A., and Stenevi, U. (1981). Intracerebral grafting of dissociated CNS tissue suspensions: A new approach for neural transplantation to deep brain sites. *Brain Res* 218:347–356.

Sethna, M. P., and Lampson, L. A. (1991). Immune modulation within the brain: Recruitment of inflammatory cells and increased major histocompatibility antigen expression following intracerebral injection of IFN-gamma. *J Neuroimmunol* 34:121–132.

Simonds, G. R., and Freed, W. J. (1990). Effect of intraventricular substantia nigra allografts as a function of donor age. *Brain Res* 530:12–19.

Sladek, J. R., Elsworth, J. D., Roth, R. H., Evans, L. E., Collier, T. J., Cooper, S. J., Taylor, J. R., and Redmond, D. E. (1993). Fetal dopamine cell survival after transplantation is dramatically improved at a critical donor gestational age in nonhuman primates. *Exp Neurol* 122:16–27.

Sloan, D. J., Wood, M. J., and Charlton, H. M. (1991). The immune response to intracerebral neural grafts. *Trends Neurosci* 14:341–346.

Spencer, D., Robbins, R., Naftolin, F., Marek, K., Vollmer, T., Leranth, C., Roth, R., Price, L., Gjedde, A., Bunney, B., Sass, K., Elsworth, J., Kier, E., Makuch, R., Hoffer, P., and Redmond, D. (1992). Unilateral transplantation of human fetal mesencephalic tissue into the caudate nucleus of patients with Parkinson's disease. *N Engl J Med* 327:1541–1548.

Spertini, F., Chatila, T., and Geha, R. S. (1992). Engagement of MHC class I molecules induces cell adhesion via both LFA-1-dependent and LFA-1-independent pathways. *J Immunol* 148:2045–2049.

Springer, T. A. (1990). Adhesion receptors of the immune system. *Nature* 346:425–434.

Subramanian, T., Pollack, L. F., and Lund, R. D. (1995). Rejection of mesencephalic retinal xenografts in the rat induced by systemic administration of recombinant interferon-γ. *Exp Neurol* 131:157–162.

Traugott, U., Scheinberg, L., and Raine, C. (1985). On the presence of Ia-positive endothelial cells and astrocytes in multiple sclerosis lesions and its relevance to antigen presentation. *J Neuroimmunol* 8:1–14.

Turner, D. A., and Kearney, W. (1993). Scientific and ethical concerns in neural fetal tissue transplantation. *Neurosurgery* 33:1031–1037.

Vawter, D. E., Kearney, W., Gervais, K. G., Caplan, A. L., Garry, D., and Tauer, C. (1990). The use of human fetal tissue: Scientific, ethical and policy concerns. Center for Biomedical Ethics Report. Minneapolis: University of Minnesota.

Walker, E. P. (1964). *Mammals of the world.* Baltimore: The Johns Hopkins Press.

Widner, H. (1993). In: Lindvall, O, ed., *Basic and clinical aspects of neuroscience,* 5. Heidelberg: Springer-Verlag, pp. 63–74.

Widner, H., Brundin, P., Björklund, A., and Möller, E. (1988). Immunological aspects of neural grafting in the mammalian central nervous system. *Prog Brain Res* 78:303–307.

Widner, H., Tetrud, J., Rehncrona, S., Snow, B., Brundin, P., Gustavii, B., Björklund, A., Lindvall, O., and Langston, J. W. (1992). Bilateral fetal mesencephalic grafting in two patients with Parkinsonism induced by 1-methyl-4-phenyl-1,2,3,6-tetrahydropyridine (MPTP). *N Engl J Med* 327:1556–1563.

Wolf, D., Richter-Landsberg, C., Short, M. P., Cepko, C., and Breakefield, X. O. (1988). Retrovirus-mediated gene transfer of beta-nerve growth factor into mouse pituitary line AtT-20. *Mol Biol Med* 5:43–59.

Wong, G. H. W., Barlett, P. F., Clark-Lewis, I., Battye, F., and Schrader, J. W. (1984). Inducible expression of H-2 and Ia antigens on brain cells. *Nature* 310:688–691.

Wong, G. H. W., Barlett, P. F., Clark-Lewis, I., Mckimm-Breschkin, J. L., and Schrader, J. W. (1985). Interferon-γ induces the expression of H-2 and Ia antigens on brain cells. *J Neuroimmunol* 7:255–278.

Wood, M. J. A., Sloan, D. J., Dallman, M. J., and Charlton, H. M. (1993). Specific tolerance to neural allografts induced with an antibody to the interleukin 2 receptor. *J Exp Med* 177:597–603.

Wood, M. J. A., Sloan, D. J., Wood, K. J., and Charlton, H. M. (1996). Indefinite survival of neural xenografts induced with anti-CD4 monoclonal antibodies. *Neuroscience* 70:775–789.

Woods, W. A., Papas, T. S., Hirumi, H., and Chirigos, M. A. (1973). Antigenic and biochemical characterization of the C-type particle of the stable porcine kidney cell line PK-15. *J Virol* 12:1184–1186.

Zimmer, J., Finsen, B., Sorensen, T., and Poulsen, P. H. (1988). Xenografts of mouse hippocampal tissue. Exchange of laminar and neuropeptide specific nerve connections with the host rat brain. *Brain Res Bull* 20:369–379.

CHAPTER 14

Restorative Surgical Therapies for Parkinson's Disease

ROY A. E. BAKAY,* JEFFREY H. KORDOWER,† AND PHILIP A. STARR,*

*Department of Neurological Surgery, Emory University School of Medicine, Atlanta, Georgia,
†Research Center for Brain Repair and Department of Neurological Sciences, Rush Presbyterian St. Luke's Medical Center, Chicago, Illinois

Parkinson's disease is a progressive degenerative disease without a cure. Medical therapy provides symptomatic relief for a limited time before the development of motor fluctuations and adverse drug responses limits their effectiveness. Central nervous system transplantation provides the possibilities of long-term symptomatic relief and ultimately the hope of a cure through repair and restoration of neurological function. The first clinical investigations grafted adrenal medullary tissue into the striatum. Analysis of these studies point to the need to use preclinical data to plan surgery, the need for standard outcome measures and controls, and the value of autopsy data. After a false start with adrenal medullary grafting, fetal mesencephalic grafts have impressive failures and successes. The examination of both gives insight into patient selection, surgical techniques, and biological factors to optimize outcomes. The application of neurotrophic factors to host reparative processes and grafting may further outcome results. The proof of principle has been made. Careful clinical investigations are still needed to prove safety and efficacy before transplantation can be included in the therapeutic armamentarium.

Recently, surgical treatment for Parkinson's disease (PD) has gained renewed interest. Ablative surgery and deep brain stimulation procedures attempt to

CNS Regeneration

alleviate parkinsonian signs by creating a compensatory blockage in neural circuits that are abnormally active in PD. Although highly successful for relieving most symptoms, these procedures do not cure. There is something conceptually wrong with making lesions in an already damaged brain. Another set of surgical therapies, referred to here as "restorative," seeks to directly replace the missing circuitry in PD. The pathology of PD arises from the degenerating dopaminergic neuron in the substantia nigra pars compacta (SNc) and the subsequent loss of dopaminergic projections to the basal ganglia. Restorative therapies fall into two categories—intracerebral transplantation of dopamine-secreting tissues, and intracerebral delivery of neurotrophic factors that may arrest and reverse SNc degeneration.

I. CELL TRANSPLANTATION

PD is the neurodegenerative disorder most amenable to a transplantation paradigm because (1) it is a focal disease involving predominantly the loss of one cell type, dopaminergic neurons of the SNc; (2) there are limited and specific projection areas from these neurons; and (3) dopaminergic modulation is tonic and nonselective in that there is not a need for specific one-to-one synaptic contact. Because the major projections from SNc innervate the striatum, the strategy of all clinical transplantation efforts thus far has been to place the dopamine-secreting tissue within the striatum.

Preclinical transplantation strategies were first tested in the 6-hydroxydopamine (6-OHDA) rat model of PD (Bakay, 1990; Bjorklund and Steveni, 1979; Freed, 1983; Perlow *et al.*, 1979). Injection of the neurotoxic dopamine analog 6-OHDA into the nigrostriatal pathway of the rat results in death of dopaminergic neurons within the SNc, and subsequent loss of dopamine in the SNc projection areas. When tested with dopamine receptor agonists (apomorphine) or uptake blockers (amphetamine), a unilaterally 6-OHDA lesioned rat rotates at a fairly consistent, reproducible, and quantifiable rate (Ungerstedt, 1971). The ability to diminish drug-induced rotational behavior in this model has been used as the predominant screening test for many new potential PD treatments. However, even though drug-induced rotation is an easily quantified behavioral manifestation of dopamine depletion, it may not be the best physiological correlate of human parkinsonism (Annett *et al.*, 1995).

The availability of a primate model of PD has augmented our understanding of PD and greatly facilitated transplantation research. The primate parkinsonian model was developed in the 1980s following the discovery that the intravenous injection of the 1-methyl-4-phenyl-1,2,3,6-tetrahydropyridine (MPTP), a neurotoxic by-product of illicit drug synthesis causes parkinsonism in man (Davis *et al.*, 1979). When administered systematically to primates, MPTP produces a syndrome of tremor, rigidity, and bradykinesia (Burns *et al.*, 1983; Langston *et al.*, 1982). MPTP primate model is responsive to L-dopa, and resembles Parkinson's disease in every way except that it is nonprogressive.

A. Transplantation of Adrenal Medulla into Striatum

Transplantation of nonneuronal dopamine-secreting tissues could prove effective for treatment of PD, provided that local nonsynaptic dopamine release in the host striatum is sufficient for behavioral recovery. The best-studied paraneuronal tissue is the adrenal medulla. Although adrenal chromaffin cells *in situ* produce much more epinephrine than dopamine, isolated adrenal chromaffin cells grafted into brain tissue increase dopamine expression (Freed *et al.*, 1983) and correct drug-induced rotation in 6-OHDA rats (Freed, 1983; Freed *et al.*, 1981). Even in these early studies there was a warning suggesting that clinical trails might fail. Although grafting of chromaffin cells into the striatum of the 6-OHDA rat model ameliorate motor abnormalities in young rats, the symptomatic improvement was not evident in adult rats (Freed, 1983). Nevertheless, adrenal medulla autografting (grafting self to self) entered clinical trials rapidly in the early 1980s, prior to fetal tissue allografting (grafting to different individuals of the same species), because the technique avoids the immunological and ethical issues associated with transplantation of fetal allografts (Vawter *et al.*, 1992). While initial reports failed to demonstrate benefit (Backlund *et al.*, 1985), others claimed significant benefit from this procedure (Jiao *et al.*, 1989; Madrazo *et al.*, 1987). Carefully performed and evaluated studies showed only very modest improvement along with significant surgical morbidity (Allen *et al.*, 1989; Apuzzo *et al.*, 1990; Bakay, 1991; Bakay *et al.*, in press, 1990; Boyer and Bakay, 1995; Goetz *et al.*, 1989, 1991; Lindvall *et al.*, 1987; Olanow *et al.*, 1990).

Although many benefits of adrenal medullary grafting were reported, the most consistent improvements were in increased "on" time and improvement in motor function. These effects were lost after 1 to 2 years. Transient complications were frequent, as might be anticipated in the performance of two major operations on elderly patients. Reported major complications ranged from 0 to 100% (American Association of Neurological Surgeons General Registry for Adrenal and Fetal Transplantation—GRAFT Project). In an evaluation of over 126 patients operated on in the United States, an average 19% had medical complications, 9% had abdominal complications, and 13% had intracranial complications (Bakay, 1991).

Autopsy studies of these transplanted patients revealed that few, if any, of the grafts survived (Forno and Langston, 1991; Hirsh *et al.*, 1990; Hurtig *et al.*, 1989; Jankovic, 1989; Kordower *et al.*, 1991; Peterson *et al.*, 1989; Waters *et al.*, 1990). Of the seven published autopsy reports, only one documented the presence of viable tyrosine hydroxylase immunoreactive (TH-ir) cells indicating survival of dopaminergic neurons but only a few TH-ir cells were observed (Kordower *et al.*, 1991). Adrenal cortical cells were present in many cases suggesting contamination of the graft and possibly better survival of these cells compared to chromaffin cells (Hurtig *et al.*, 1989; Peterson *et al.*, 1989; Waters *et al.*, 1990). All cases demonstrated necrotic graft tissue infiltrated with macrophages and lymphocytes. Autopsy reports also documented surgical complications with damage to the septum, corpus callosum, and thalamus (Forno and Langston, 1991). With little long-term benefit, significant morbidity, and documented failure of graft survival, all adrenal medullary grafting in the United States ended in 1990 (AANS GRAFT Project).

Although adrenal chromaffin cells failed to survive grafting, some patients demonstrated marked symptomatic improvement following grafting. Clinical benefit in the absence of surviving chromaffin cells would appear to be a paradox. Whereas the improvement could be attributed to placebo effect, there are four lines of evidence that indicate that this is not the only explanation. First, the improvement was observed not immediately after surgery but usually several months later. Second, in a well-studied cohort, benefit from the grafts was observed for at least four years (Diamond *et al.*, 1994). Third, the degree of improvement appears to correlate not with quantity or quality of the adrenal tissue but with the degree of residual host striatal TH-ir innervation (Bakay *et al.*, 1990). Thus, the character of the host tissue and not the graft was a critical factor in those patients who improved. Fourth, at least two postmortem examinations demonstrated that, despite poor chromaffin cell survival, there was sprouting of the host residual mesostriatal fibers (Hirsh *et al.*, 1990; Kordower *et al.*, 1991). Primate studies have indicated that adrenal grafting, or simply making an inflammatory lesion in the striatum, can induce sprouting of host dopaminergic fibers (Bankiewicz *et al.*, 1991, 1988; Fiandaca *et al.*, 1988; Hansen *et al.*, 1988). The sprouting is believed to be the result of neurotrophic factors released by the graft and/or host and has led to new understanding regarding the host-grafting relationship in transplantation. This neurotrophic effect is possibly the mechanism underlying the modest degrees of improvement observed in human adrenal medullary graft trials.

Adrenal chromaffin graft survival and differentiation into a neuronal phenotype are enhanced by treating the grafts with nerve growth factor (NGF) after implantation (Stromberg *et al.*, 1985). Based upon this, one patient has been reported to have an intrastriatal placement of an adrenal autograft, followed by 23 days intraparenchymal infusion of NGF around the graft (Olson *et al.*, 1991). Clinical improvements in this patient were longer-lasting than in the same group's original effort with adrenal autografting (Backlund *et al.*, 1985), but were nevertheless modest. Long-term follow-up has not been reported.

Sural nerve–adrenal medulla co-grafts have been shown to augment the survival of adrenal chromaffin cells in MPTP-lesioned monkeys relative to grafts of adrenal medulla alone (Kordower *et al.*, 1990; Watts *et al.*, 1995; Watts and Bakay, 1991). These grafts have been shown to potentiate recovery of motor function in parkinsonian monkeys (Watts *et al.*, 1995; Watts and Bakay, 1991). Based upon these studies, clinical trials were initiated in which patients have undergone co-grafting of adrenal medulla tissue with peripheral nerve as a source of NGF (Date *et al.*, 1995, 1996; Kopyov *et al.*, 1996; Lopez-Lozano *et al.*, 1993; Watts *et al.*, 1997; Zhang *et al.*, 1994). As might be anticipated from preclinical studies, long-term improvements were observed (Date *et al.*, 1995, 1996; Watts *et al.*, 1997). Postmortem analysis from a bilateral caudate sural nerve–adrenal medullary co-graft demonstrated moderate TH-ir chromaffin cell survival and sprouting in a patient who had moderate and persistent improvement (Date *et al.*, 1996). Laboratory work continues in an effort to find clinically applicable means of enhancing the survival of adrenal chromaffin grafts (Chalmers *et al.*, 1992; Cunningham *et al.*, 1994; Lopez-Lozano *et al.*, 1997). This work does not appear ready for clinical trial.

Both the poor survival of adrenal chromaffin cells and the sprouting of host dopaminergic mesostriatal systems could have been predicted from experimental studies. Poor adrenal graft survival is generally seen in MPTP-lesioned parkinsonian monkeys (Bankiewicz *et al.*, 1988; Fiandaca *et al.*, 1988; Hansen *et al.*, 1988; Morihisa *et al.*, 1984). It appears that the presence of nonchromaffin cell constituents (fibroblasts and endothelial cells) are responsible for the poor adrenal medulla graft survival because isolation of the bovine chromaffin cells results in excellent graft survival (Schueler *et al.*, 1993, 1995). The sprouting of host dopaminergic fibers following adrenal chromaffin grafts was first observed in animal models of PD. Bohn and associates (1987) demonstrated a robust increase in TH-ir within the striatum of MPTP-treated mice despite poor adrenal medullary graft survival. Both Bankiewicz and colleagues (1988) and Fiandaca and associates (1988) found that similar sprouting of host mesostrial fibers occurs in MPTP-lesioned monkeys following adrenal medullary transplantation. When the grafts are placed into the head of the caudate nucleus, TH-ir fibers can be traced from the nucleus accumbens to the graft site (Beck *et al.*, 1995). The nucleus accumbens receives dopaminergic inputs from the ventral tegmental area, which is relatively spared following MPTP lesioning (Burns *et al.*, 1983; Langston *et al.*, 1984). The TH-ir axonal sprouting is not specific for the adrenal tissue itself because it occurs following nondopaminergic grafting (Bankiewicz *et al.*, 1991; Fiandaca *et al.*, 1988) or following interleukin-1 injection (Wang *et al.*, 1994). Clearly, the clinical investigations were premature and based on inadequate outcome measure in a disease that is highly fluctuant and easily subject to marked placebo effects.

The lessons to be learned from these early clinical transplantation studies are (1) preclinical studies, especially with the MPTP-induced parkinsonian primate model, are highly predictive for clinical trials; (2) meticulous attention must be applied to surgical techniques that are based on the preclinical studies; (3) without a proper control cohort and sophisticated outcome measures, performed by movement disorder experts, therapeutic effects in PD can be easily misinterpreted; and, (4) postmortem studies are critical to understanding the mechanism of the transplantational therapeutic effect or lack of it.

II. TRANSPLANTATION OF FETAL MESENCEPHALON INTO STRIATUM

In the developing mammalian fetus, the ventral mesencephalon (VM) contains the dopaminergic cell bodies that are the precursors of the mature SNc. In both rodent (Arbuthnott *et al.*, 1985; Bjorklund and Steveni, 1979; Doucet *et al.*, 1989; Hattori *et al.*, 1992; Kondoh and Low, 1994; Meyer *et al.*, 1995; Perlow *et al.*, 1979) and primate (Annett *et al.*, 1995; Bakay *et al.*, 1987; Bankiewicz *et al.*, 1990; Beck *et al.*, 1995; Kordower *et al.*, 1992; Taylor *et al.*, 1991) models of PD, grafts of fetal VM tissue, placed heterotopically (not in the normal location) into the striatum, have been shown to survive, reinnervate portions of the striatum, and correct abnormal motor behaviors. The mechanism of fetal VM graft function has been extensively studied in rats (Bjorklund and Steveni, 1979). Grafts of fetal VM show spontaneous electrical activity and are capable of releasing physiologically relevant levels of dopamine in response to depolarization (Stromberg *et al.*, 1989; van Horne *et al.*, 1990). Synaptically mediated graft–host interactions are probably important for graft function. Fetal VM can form ultrastructurally normal synapses on to denervated neurons of the host striatum (Meyer *et al.*, 1995; Stromberg *et al.*, 1989). In addition, host cortical and striatal neurons appear to innervate the graft (Doucet *et al.*, 1989; Meyer *et al.*, 1995). Dopamine release from intrastriatal grafts can be increased by electrical stimulation of host cortex (Arbuthnott *et al.*, 1985) and by pharmacological blockade of glutamate reuptake (Kondoh and Low, 1994) as is true of normal striatal tissue in normal rats. Motor activities that increase striatal dopamine in normal rats, such as running on a treadmill, also increase striatal dopamine in grafted animals from which host dopaminergic innervation has been completely depleted (Hattori *et al.*, 1992). Thus, both anatomical and physiological studies (Lin *et al.*, 1993; Stromberg *et al.*, 1989) of intrastriatal fetal VM grafts suggest that synaptically mediated host–graft interactions occur and result in a more complex level of function than would be expected from implantation of an unregulated dopamine pump. Human fetal VM grafted into the rat shows the same electrophysiological (Stromberg *et al.*, 1989; Van Horne *et al.*, 1990), anatomical (Brundin *et al.*, 1988; Clarke *et al.*, 1988; Stromberg *et al.*, 1989), and pharmacological responses (Brundin *et al.*, 1988).

A. Clinical Trials of Human Fetal Tissue Transplants

1. Patient Characteristics and Clinical Outcomes

As of 1996, at least 14 groups have reported at least 200 patients have undergone transplantation of human fetal VM cells to the striatum for PD. (Bakay, 1990; Boyer and Bakay, 1995; Defer *et al.*, 1996; Folkerth and Durso, 1996; Freed *et al.*, 1992; Freeman *et al.*, 1995; Hitchcock, 1995; Hitchcock *et al.*, 1994; Hoffer *et al.*, 1992; Huang *et al.*, 1989; Iacono *et al.*, 1992; Kopyov *et al.*, 1996; Lindvall, 1997; Lindvall *et al.*, 1990, 1994; Lopez-Lozano *et al.*, 1995, 1997; Madrazo *et al.*, 1990; Meyer *et al.*, 1995; Molina *et al.*, 1994; Peschanski *et al.*, 1994; Redmond *et al.*, 1990; Remy *et al.*, 1995; Spencer *et al.*, 1992; Subrt *et al.*, 1991; Widner *et al.*, 1992; Wenning *et al.*, 1992; Yurek *et al.*, 1990; Zhang *et al.*, 1994). Unreported cases may be many times this number. These reports are summarized in Table I. All transplanted patients were diagnosed as idiopathic PD, except for two patients transplanted by the Lund group, who suffered from MPTP-induced parkinsonism (Widner and Brundin, 1988). One of the four patients in the Yale group was subsequently found at autopsy to have striatonigral degeneration (Redmond *et al.*, 1990; Spencer *et al.*, 1992). All patients had relatively advanced disease, with Hoehn and Yahr (H&Y) stages of about 3 when "on" (time of optimal medication effect) and 4 to 5 when "off" (time without medication effect). Most patients were 40 to 60 years old and had suffered from PD for 5 to 20 years prior to surgery. Patients were primarily symptomatic from bradykinesia and rigidity rather than from tremor.

Outcomes have been highly variable. Observations common to many studies were that improvements were rarely immediate but took several months. There was a worsening in many patients' motor symptoms during the first 4 to 6 weeks after surgery and sometimes transiently increased dyskinesias. There were several transient psychiatric complications including hallucinations (Fisher *et al.*, 1991; Peschanski *et al.*, 1994), panic attacks (Spencer *et al.*, 1992), and obsessive–compulsive disorders (Peschanski *et al.*, 1994). Major complications reported include a brain abscess (Lindvall *et al.*, 1994), seizures (Freeman *et al.*, 1995; Spencer *et al.*, 1992), and cerebral hemorrhage (Freeman

TABLE Ia Current Experience: Fetal Transplantation for Parkinson's Disease in Humans–U.S.A. Experience

		No. Patients Treated		No. Implants		Fetal Tissue					
Investigator	Techniques	Unilateral	Bilateral	Caudate	Putamen	No. donor/ patient	Post-gestational age (weeks)	Use of immuno-suppression	Length follow-up (months)	Function ratings scales	Reported outcome
Freed *et al.*, 1992 Colorado	Stereotactic	2 —	— 5	4 —	6 12 to 16	1 1 to 2	5 to 6 5 to 6	Every other patient	12 to 46	UPDRS HY	Mild to Moderate motor Improvement ↓ L-dopa need ↓ Off time
Spencer *et al.*, 1992 Connecticut	Stereotactic	4	—	2 to 4	—	1	5 to 9	Yes	4 to 18	UPDRS HY	Mild motor improvement ↓ L-dopa need
Iacono *et al.*, 1992 Arizona	Delayed stereotactic	—	5	—	—	2 to 3	—	Yes	2 to 22	HY	Improved motor ↓ L-dopa need ↓ on-off
Kopyov *et al.*, 1996 California	Stereotactic	9	13	Mixed Patterns	—	1 to 3	6 to 10	Yes	6 to 24	UPDRS CAPIT	Mild to Moderate motor improvement ↓ L-dopa need improvement improved UPDRS
Freeman *et al.*, 1995 Florida	Stereotactic	—	4	—	6 to 8	6 to 8	6.5 to 9	Yes	6	UPDRS CAPIT	Moderate motor improvement Improved UPDRS ↓ off time

UPDRS = Unified Parkinson's Disease Rating Scale; HY = Hoehn Yahr Staging System; SEADL = Schwab-England Activities of Daily Living Scale; WRS = Webster Rating Scale; NUDS = Northwestern University Disability Scale; CAPIT = Core Assessment Program for Intracerebral Transplantation.
Modified with permission from Boyer and Bakay.

TABLE Ib Current Experience: Fetal Transplantation for Parkinson's Disease in Humans–Non USA Stereotactic

		No. Patients Treated		No. Implants		Fetal Tissue					
Investigator	Techniques	Unilateral	Bilateral	Caudate	Putamen	No. donor/ patient	Post-gestational age (weeks)	Use of immuno-suppression	Length follow-up (months)	Function ratings scales	Reported outcome
Subrt *et al.*, 1991 Prague	Stereotactic	3	—	1	—	1	7 to 8	No	6 to 12	H & Y	Mild motor improvement ↓ L-dopa need
Wu *et al.*, 1994* China	Stereotactic	4	—	1 to 2	—	1	12 to 14	Yes	14 to 23	UPDRS	Moderate motor improvement
		1	—	many	—	1		Yes	9	SEADL WEBSTER	
Defer *et al.*, 1996 France	Stereotactic	4	—	1	3	1 to 3	6 to 9	Yes	12 to 24	UPDRS	Improved UPDRS
		1	—	—	3					CAPIT	Moderate motor improvement ↓ L-dopa need
Hitchcock *et al.*, 1994, 1995 UK	Stereotactic	24	—	1	—	1	12 to 19	No	>12	WRS	Mild to moderate motor improvement
		12	—	—	1	1	12 to 19	No		NUDS	
		—	12	1	—		12 to 19	No			No change L-dopa
Lindvall *et al.*, 1990,	Stereotactic	—	2**	0 to 1	3	6 to 8	7 to 9	Yes	12 to 24	UPDRS	Marked
1994		2	—	1	2	4	8 to 10	Yes	15		Modest
Widner *et al.*, 1992 Sweden		2	—	—	3	4	6 to 8	Yes		SEADL	Moderate motor improvement

* 1 Patient grafted with fetal adrenal and FM tissue; all had Vim Thalamotomy.

** MPTP Patients

TABLE Ic Current Experience: Fetal Transplantation for Parkinson's Disease in Humans–Non USA Craniotomy

Investigator	Techniques	No. Patients Treated		No. Implants		Fetal Tissue		Use of immuno-suppression	Length follow-up (months)	Function ratings scales	Reported outcome
		Unilateral	Bilateral	Caudate	Putamen	No. donor/ patient	Post-gestational age (weeks)				
Lopaz-Lorenzo *et al.*, 1995, 1997 Spain	Craniotomy	10	—	1	—	1	6 to 8	Yes	60	UPDRS NUDS	Motor improvement improved UPDRS ↓ L-dopa need
Madrazo *et al.*, 1990 Mexico	Craniotomy microsurgery	7	—	1	—	1	12 to 14	Yes	6 to 19	HY UPDRS SEADL	Motor improvement ↓ L-dopa need
Molina *et al.*, 1994 Cuba	Stereotactic craniotomy microsurgery	7 30	9 —	1 1	2 —	1 1	8 to 13 6 to 12	Yes Yes	3 to 18 39 to 60	UPDRS CAPIT	Moderate motor improvement ↓ L-dopa ↓ Off phase
Zhang *et al.*, 1994 Poland	Craniotomy microsurgery	3	—	1	—	1	11 to 12	Yes	30	HY	Motor improvement No change L-dopa ↓ off

et al., 1995). There were several complications of immunosuppressive therapy requiring its cessation (Lopez-Lozano *et al.*, 1995, 1997). In most studies there was modest benefit in most but not all patients, particularly in reduction of the amount of time spent "off," reduction in drug-induced dyskinesias (which may reflect lowering the medications), and some improvement in fine motor tasks. The majority of patients did not have a significant change in the H&Y functional status, and very few were able to discontinue medication.

2. Transplantation Protocols

All groups used human fetal tissue dissected from the ventral mesencephalon. With rare exception (Madrazo *et al.*, 1990; Yurek *et al.*, 1990), most centers used tissue from elective abortions rather than spontaneous abortions. Treatment protocols differed widely with respect to several important variables: exact location of transplant (head of caudate, putamen, or both), surgical technique used (closed stereotactic injection or open craniotomy), unilateral versus bilateral implantation, age of fetal tissue used, number of fetuses used to provide donor tissue, interval between harvest and implantation, and use of immunosuppressive therapy (Table I). Even combining grafting with lesioning procedures has been attempted (Yurek *et al.*, 1990).

3. Immunological Concerns about Allografts

Degree of immune response to allographic donor tissue has not been well characterized in clinical studies, and preclinical studies are needed to aid in defining the role of immunosuppression for allografting. Immunologic response to fetal CNS allografts in primates may be restricted to a local inflammatory reaction (Fiandaca *et al.*, 1988; Freed *et al.*, 1996; Mahalik *et al.*, 1985; Poltorak and Freed, 1989). Essential to the initiation of graft rejection is the expression of major histocompatibility complex (MHC) antigens by the graft and the presentation of antigens to the immunological competent cells of the host (Bakay *et al.*, in press; Freed *et al.*, 1996; Lawrence *et al.*, 1990; Poltorak and Freed, 1989; Widner and Brundin, 1989). MHC expression was detected in associated connective tissues but not in neural cells of the developing brain and spinal cord (Grabowska and Lampson, 1992; Hitchcock, 1995; Skoskiewicz *et al.*, 1985). Under normal conditions, brain tissue does not express class I MHC antigens, but in the presence of inflammation, the expression is upregulated. It has not been shown that neural tissue can express class II MHC antigens which are critical to systemic transplantation rejection (Skoskiewicz *et al.*, 1985). Microglia (Poltorak and Freed, 1989) and astrocytes (Vidovic *et al.*, 1990) can express MHC class I and II antigens and may be responsible for direct sensitization after grafting if they are allowed to present their antigens to the systemic immune surveillance. However, microglia may also stimulate the arrest of immune responses in the CNS (Hirshberg and Schwartz, 1995).

Evidence from primate models suggests that allografts to the brain do not evoke donor-specific lymphocytotoxic sensitization (Bakay *et al.*, in press; Fiandaca *et al.*, 1988). Several human CNS-grafted patients have been studied and also lack humoral or cellular systemic sensitization (Freed *et al.*, 1996). Autopsy data also fail to show rejection (Kordower *et al.*, 1997), but both microglia and peripheral immune cells were observed in the grafts (Fig. 1, see color insert). Because the grafted neurons appeared healthy, (Kordower *et al.*, 1995, in press, 1996, in press); It is unclear if these would have eventually destroyed the grafts. It should not be assumed that these cells would eventually destroy the grafts. A similar response is observed in both FM-grafted MPTP primates without evidence of host sensitization and sham-operated monkeys in response to CNS trauma (Bakay *et al.*, in press). The use of multiple fetuses and staggered grafting times may result in higher chances for rejection. The use of immunosuppressive therapy for fetal CNS grafting remains controversial (Freed *et al.*, 1996; Lindvall, 1997). Although most investigators begin immunosuppression preoperatively, Freed and colleagues (1992) used immunosuppression on every other patient, and graft rejection was not observed. This group no longer uses immunosuppression.

4. Graft Survival by PET Scanning

Severity of dopaminergic cell loss in PD can be assessed noninvasively using positron emission tomography (PET) (Vingerhoets *et al.*, 1994). A tracer of dopaminergic metabolic activity, such as 6-[^{18}F] L-DOPA (FD), is used as the positron-emitting source. The concentration of the FD positron source in the brain following IV administration reflects the presence of uptake by dopaminergic cells. Several groups have used PET to follow survival of fetal transplants in the striatum (Freed *et al.*, 1992; Freeman *et al.*, 1995; Lindvall *et al.*, 1994; Peschanski *et al.*, 1994; Remy *et al.*, 1995; Widner *et al.*, 1992; Wenning *et al.*, 1997). PET FD uptake increases in successful grafting (Lindvall *et al.*, 1994) and has been observed in autopsy-proven cases with TH-ir cell survival (Kordower *et al.*, 1995, in press, 1996, in press). Remy and colleagues (1995) correlated increased FD uptake with clinical improvement and Wenning and colleagues (1997) observed decreased FD uptake in the contralateral nongrafted side where the disease continued to progress. In the patients followed for the longest time, there is PET evidence for graft survival at six years (Lindvall *et al.*, 1994; Wenning *et al.*, 1997). Tracer uptake by PET scanning cannot distinguish between survival of transplanted tissue versus sprouting of host dopaminergic tissue, which is known to occur with lesioning of the striatum (Bankiewicz *et al.*, 1988; Fiandaca *et al.*, 1988). However, short of autopsy studies, PET is currently the best method to follow graft survival.

5. Graft Survival by Autopsy Studies

There have been several reports of autopsies on patients who have received fetal cell transplants for PD. Hitchcock and colleagues (in press) used second trimester fetal tissue with poor TH-ir cell survival. At Yale, a patient receiving fetal tissue of 11 weeks gestational age demonstrated no TH-ir cell survival (Redmond *et al.*, 1990). Preclinical studies would predict poor graft survival for fetal tissue older than 9.5 weeks (Freeman *et al.*, 1995). The use of cryopreserved fetal tissue and misdiagnosis may also be associated with poor graft survival (Kordower *et al.*, in press; Redmond *et al.*, 1990).

Two patients transplanted at the University of South Florida died 18 months after grafting from causes unrelated to the transplant (Kordower *et al.*, 1995, in press, 1996, in press). Both patients had received bilateral grafts into the putamen of tissue from 6 to 7 fetuses, age 6.4 to 9 weeks. The patients had been immunosuppressed for 6 months after transplantation. Many TH-ir cells were observed in the grafts (up to 49,336 per graft). Clusters of TH-ir cells within the grafts displayed classic organotypic organization, thus appearing morphological identical to the SNc (Fig. 2). There was dense TH-ir innervation within the graft and many fibers that crossed the graft–host interface. These fibers reinnervated the adjacent striatum in a patch-matrix pattern and extended neuronal processes up to 7 mm into the surrounding normal brain (Fig. 3). The grafted-derived TH-ir fibers formed synaptic contacts with the host neurons (Kordower *et al.*, in press). There was no evidence of host-derived TH-ir sprouting, and in nongrafted regions TH-ir innervation was negligible. These studies demonstrated that human fetal neural allografts can survive and robustly reinnervate host tissue, and suggested that long-term immunosuppression (greater than 6 months posttransplant) may not be needed for survival of the allografts, even if multiple donors are used.

6. Which Transplant Protocols Were Most Effective?

Unfortunately, few of the clinical studies used standardized outcome measures that allow direct comparisons. A small number of very well-documented patients have had dramatic clinical benefit with good evidence of graft survival, and it is worth considering them individually. The fourth patient from the Swedish series of idiopathic PD patients has been followed for 3 years, and is the only reported patient off of all dopaminergic drugs (Lindvall *et al.*, 1994).

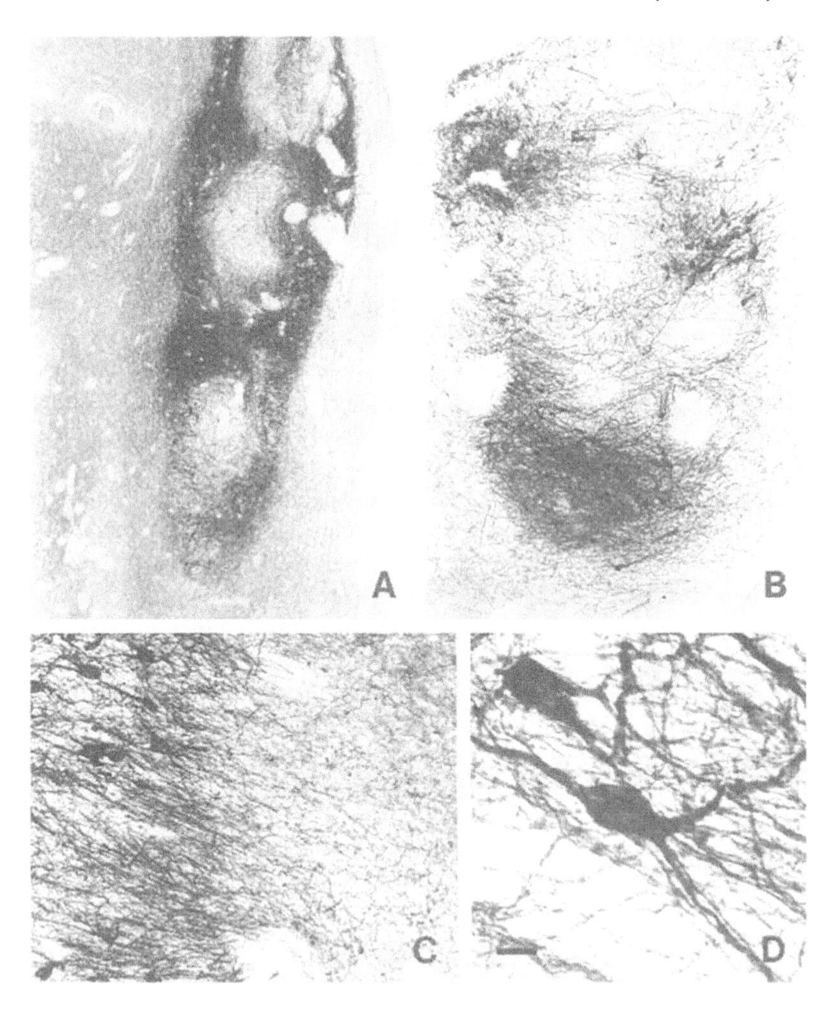

FIGURE 2 TH-ir-stained sections through the post-commissural putamen of a patient receiving fetal nigral allografts. (A) Low-power view of a large graft with TH-ir neurons. (B) These neurons were clustered in a classic organotypic cytoarchitectonic arrangement. (C) Fibers could be seen crossing the graft-host interface. (D) Individual neurons displayed morphological profiles similar to nigral neurons *in situ*. Scale bar in D represents the following magnifications: A = 3D1000 = B5m; B = 3D250 = B5m; C = 3D100 = B5μm; D = 3D16.5 = B5μm.

The patient's "off" periods have disappeared. Putaminal uptake of FD has normalized on the operated side. Symptoms continued to improve during the second year after surgery, then remained stable. Two patients with MPTP-induced parkinsonism, also operated on by the Lund group, have also had great benefit, with 50-point decreases in Unified Parkinson's Disease Rating

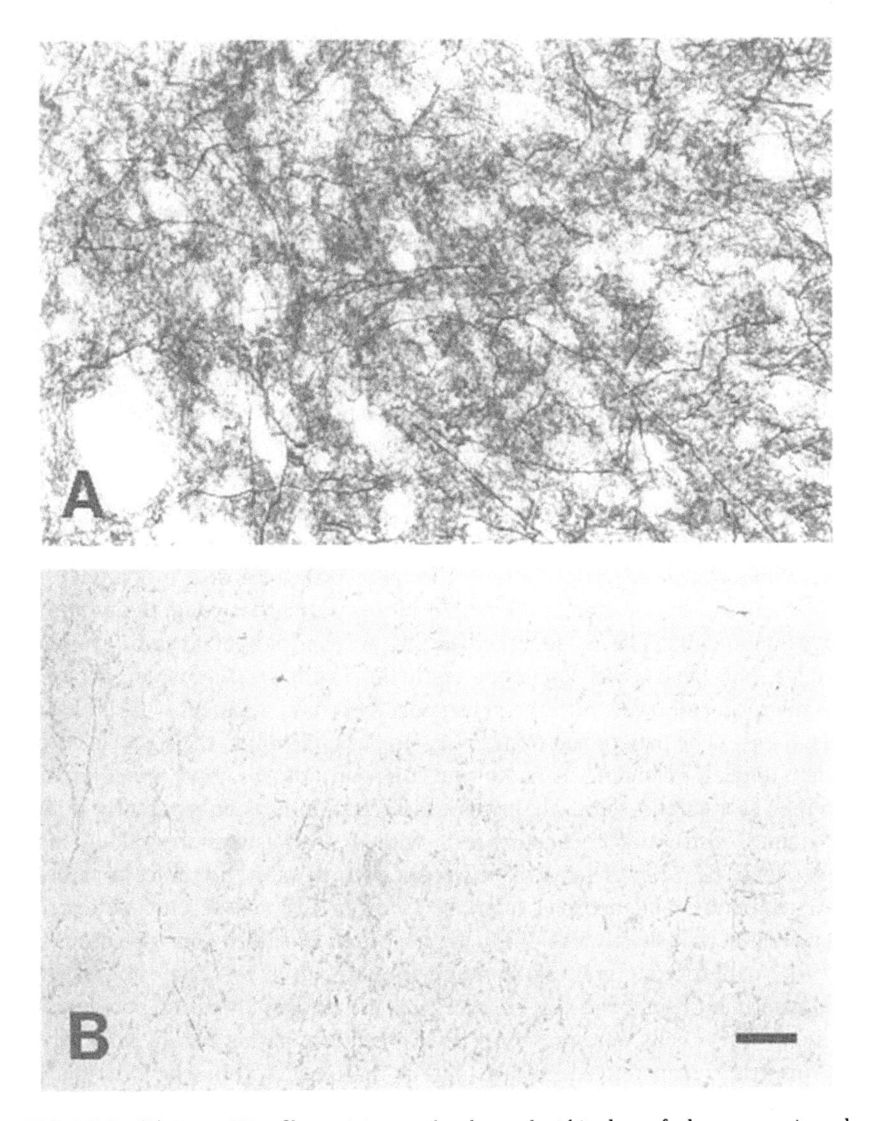

FIGURE 3 (A) Dense TH-ir fiber staining can be observed within the grafted post-commissural putamen of a PD patient. (B) In contrast, sparse TH-ir staining is found in the ungrafted anterior putamen. Scale bar represents 100 = B5μm for both panels A and B.

Scale (UPDRS) scores (both "on" and "off") 2 years posttransplant, reductions in medications, elimination of drug-induced dyskinesias, and continued decrease in rigidity, bradykinesia, and fluctuation in motor performance during the second year (Widner *et al.*, 1992). The patients in the University of South

Florida group (two of whom were the subject of autopsy investigations described above) also have improvement with a 22-point decrease in total UPDRS when "off," reduction in percent "off" times from 30 to 12%, and reduction in the percent time "on" with dyskinesia from 44 to 3.8% (Freeman *et al.*, 1995; Kordower *et al.*, 1995).

Features common to the best-documented patients showing best results are use of younger (6 to 9 weeks gestation) tissue, implantation of tissue from multiple fetal sources at multiple sites, implantations into the putamen or putamen and caudate rather than just the caudate alone, and surgery using stereotactically guided injection rather than open craniotomy. There is data to question this conclusion (Lopez-Lozano *et al.*, 1997). Although long-term immunosuppression may not be necessary for graft survival, most patients with the best clinical results had at least short-term (6 months postsurgery) immunosuppression (Kordower *et al.*, 1995; Lindvall, 1997). In order to better assess the benefit of fetal transplantation for PD two randomized, double-blinded studies have been sponsored by the National Institutes of Health and are in progress.

The importance of understanding the basic principles of transplantation prior to clinical application is illustrated in a case that has come to autopsy. Folkerth and Durso (1996) documented the presence of bone, mature hyaline cartilage, hair shafts, and squamous epithelium with keratinous debris after placement of fetal tissue into the ventricular system of a patient with Parkinson's disease. The patient had tissue from fetuses grafted into the right caudate nucleus and left putamen. Then a suspension from a third embryo was injected into the ventricular system. The patient died 23 months later apparently from ventricular obstruction secondary to growth of the suspension graft in the fourth ventricle. Scientifically, there were serious flaws in the choice of transplant technique: (1) incorrect fetal age (too old); (2) tissue obtained from wrong region (not strictly VM); (3) poor dissection (included non-CNS tissue); and (4) inappropriate grafting technique (intraventricular suspension grafts are dangerous) (Kondower *et al.*, in press). At the autopsy 23-month postoperatively, no TH-ir cells were present. The clinical data on this patient cannot be interpreted. Postoperative clinical status is discussed without baseline data. The data is based on patient and family reports, and did not include standardized UPDRS rating scales. The observations were confounded by manipulations of antiparkinsonian medications. The PET scan is uninterpretable due to the lack of a preoperative scan. This report documents what can occur when neural transplants are performed in an unscientific fashion. The key point is the same as discussed under adrenal medullary grafting, i.e., failure to take advantage of preclinical information, to perform precise transplantation technique based on established preclinical data, and failure to conduct rigorous clinical studies will result in a poor outcome as documented by the autopsy.

B. Unresolved Issues in Clinical Transplantation

Many problems must be solved before cell transplantation becomes widely applicable in the treatment of PD. First, a source of tissue other than human fetuses must be found, and second, the transplant procedure must be refined so as to produce more complete and consistent resolution of motor abnormalities in PD patients. In the most successful human fetal transplant protocols, less than 10% of implanted dopaminergic cells survive, and reinnervation of host structures occurs only over a few millimeters, necessitating the use of a large number of fetal cells and many passes of an injection needle. In the future, the efficiency of the procedure could be improved by providing transplanted tissue with a source of neurotrophic support and/or chemotactic guidance. In animal models, this has been achieved by treating grafts *in situ* with neurotrophins, which promote survival and sprouting of dopaminergic neurons, or by co-transplanting fetal mesencephalon with fetal striatum (Zhang *et al.*, 1994). One small group of patients has undergone co-grafting of fetal mesencephalon with fetal striatum, but more patients and longer follow-up will be necessary to determine if survival of fetal mesencephalon or clinical outcome is enhanced by this co-grafting paradigm (Meyer *et al.*, 1995).

C. Need for Alternative Tissue Sources

The most successful clinical transplant protocols have also been those implanting the most tissue, often requiring 6 to 8 fetuses per procedure (Bakay, 1990; Defer *et al.*, 1996; Freeman *et al.*, 1995; Hoffer *et al.*, 1994; Kordower *et al.*, 1995; Lindvall *et al.*, 1990, 1994; Peschanski *et al.*, 1994; Widner *et al.*, 1992). Each fetus must be obtained from an elective abortion taking place 6 to 8 weeks after conception (Freeman *et al.*, 1995). To harvest enough appropriately aged tissue for a therapeutic PD transplantation program is logistically difficult on a local basis and impossible on a national basis. There are simply not enough fetuses available to meet the needs of a large-scale PD treatment program, and with increasing use of the abortion pill RU384, there will be even fewer fetuses available. Ethical concerns will present problems for many physicians and patients (Vawter *et al.*, 1992). Therefore, a great variety of alternative tissue sources are being explored. Xenografts (grafts between different species) and genetically engineered cells are the two most actively researched areas for transplantation.

1. Dopamine-Secreting Xenografts

The use of porcine xenografts for neuronal cell transplantation is actively investigated in the laboratory, and in a phase I clinical trial with PD patients, as an alternative to human fetal allografts (Deacon *et al.*, 1995; Edge and Dinsmore, 1997). In rats, systemically immunosuppressed with cyclosporine, porcine fetal VM cells can survive, extend processes, and integrate into surrounding brain tissue (Deacon *et al.*, 1994; Isacson *et al.*, 1995). One autopsy study has been published on a PD patient who died (of a pulmonary embolus) 7 months after intrastriatal transplantation of porcine fetal VM (Deacon *et al.*, 1995). The patient had been immunosuppressed with cyclosporine. There was evidence of graft survival and extension of processes into the host brain, although cell survival and growth were not nearly as robust as in the best human fetal VM allograft transplantation (Kordower *et al.*, 1995). This autopsy suggests the need for more cells to be implanted in more sites (proposed phase II trial).

Immune-mediated rejection of xenografts is, in general, even more vigorous than that of allografts (Edge and Dinsmore, 1997). This could limit the clinical utility of xenografts. Transplant rejection is mediated mainly by MHC antigens. One strategy, in order to circumvent rejection, is to use an antibody fragment directed against MHC antigens to induce tolerance of the foreign tissue in the host environment. Treating pancreatic and hepatic xenografts just prior to transplantation with the variable region fragment $F(ab')_2$ of a monoclonal antibody against MHC antigens induces tolerance in the host (Faustman and Coe, 1991). This tolerance is long lasting even though the transplant is treated only once. By using antibody fragments from which the constant region F(c) has been removed, activation of the complement cascade- and antibody-mediated destruction is avoided while cellular immunity is modified. This MHC masking technique has been applied to porcine neural xenografts. When pretreated with $F(ab')_2$ fragments of a monoclonal antibody against MHC-I antigens, intrastriata xenografts of porcine fetal VM survive in nonimmunosuppressed host rats as well as untreated xenografts in immunosuppressed (cyclosporine-treated) rats (Pakzaban *et al.*, 1995). It is unknown if this technique is effective in humans.

Another strategy to shelter xenografts from the host immune system is to encapsulate xenograft tissue within a synthetic, biologically compatible polymer coating that allows diffusion of xenograft-produced molecules but prevents immune attack of the xenograft (Aebischer *et al.*, 1991, 1994). In hemiparkinsonian MPTP monkeys, implantation of polymer-encapsulated PC12 cells, a dopamine-secreting cell line derived from a rat pheochromocytoma, is effective in ameliorating parkinsonian symptoms (Aebischer *et al.*, 1994). A disadvantage of implanting encapsulated tissue is that graft–host synaptic interactions, which may be an important mechanism of behavioral recovery in fetal cell transplantation, are prevented.

2. Genetically Engineered Autologous Tissue

To completely circumvent the immunologic complexity of xeno- or allotransplantation, a patient's own tissues could be removed, genetically engineered to adopt a useful neuronal phenotype, then reimplanted into the host brain. Primary fibroblasts have been engineered to express tyrosine hydroxylase using retrovirus-mediated transfection; these cells then secrete L-dopa (Wu *et al.*, 1994). Implantation of autologous L-dopa secreting fibroblasts in the 6-OHDA model of PD results in behavioral improvement for at least 8 weeks postimplantation, but the behavioral improvements decline after 2 weeks and the implanted fibroblasts show some signs of neoplasia (Fisher *et al.*, 1991). The use of plasmid-transfected primary cultured muscle cells, in the same experimental paradigm, has resulted in more stable behavioral amelioration, with evidence of continued L-dopa secretion, up to 6 months posttransplant (Jiao *et al.*, 1993). As with polymer-encapsulated tissue, a disadvantage of using genetically engineered nonneuronal tissues is that it is not possible for such tissues to form functional synapses with host tissue. A promising technique is to use retroviral vectors to genetically modify fetal neurons so that in culture (30°C) they continue to divide and are thus immortalized, but at higher temperatures (38–39°), are rendered permanently amitotic (Anton *et al.*, 1994). These cells are then ideal for transplantation, and when transfected to produce TH, have been shown in both rat and monkey hemiparkinsonian models to reverse behavioral deficits without evidence of tumor formation (Bredesen *et al.*, 1991). In the future, neuronal progenitor cells combined with gene therapy may be a powerful technique for repair and restoration of neurological function. However, long-term expression of foreign genes in transfected cells is difficult to achieve with present gene transfer methods.

D. Optimal Target for Transplantation

A basic assumption in previous clinical work on transplantation in PD is that depletion of striatal dopamine is fundamental to the pathophysiology of disordered movement, and thus any transplantation strategy should seek to restore striatal dopamine levels. However, there are now a number of human transplant recipients documented to have robust dopaminergic reinnervation of the striatum (by PET or autopsy studies), but who have experienced only modest recovery from the motor abnormalities of PD (Defer *et al.*, 1996; Freed *et al.*, 1992; Freeman *et al.*, 1995; Hoffer *et al.*, 1992; Kopyov *et al.*, 1996; Peschanski *et al.*, 1994). This suggests that restoration of striatal dopaminergic innervation alone may not be sufficient for complete recovery from PD.

In addition to the long-distance nigrostriatal pathway, there are other dopaminergic fibers arising from SNc, which innervate different nuclei. These extrastriatal dopaminergic pathways are also affected in PD. The dendrodendritic dopaminergic innervation of substantia nigra pars reticulata (SNr) is one such pathway (Bjorklund and Lindvall, 1975). SNr is a nucleus that neighbors the SNc, and functions (along with the globus pallidus internus) as the major motor output nuclei of the basal ganglia (Alexander *et al.*, 1990). Restoration of dopamine in SNr, independent of dopamine in the striatum, appears to play an important role in the correction of parkinsonism in animal models (Johnson and Becker, 1997; Nikkah *et al.*, 1994). Future human transplant protocols might test intranigral or combined intranigral and intrastriatal transplantation of dopaminergic tissue rather than rely solely on heterotopic placement of fetal substantia nigra tissue into striatum.

III. INTRACEREBRAL DELIVERY OF NEUROTROPHINS

Although intracerebral transplantation of dopamine-secreting cells can correct dopamine depletion in host tissue, the transplantation paradigm does not necessarily reverse the fundamental pathology of PD, which is progressive loss of dopaminergic cells of the SNc. There has been interest in identifying a neurotrophic factor that could halt or reverse the loss of SNc neurons. Such a factor could provide a more direct and effective treatment for PD and could be particularly useful for early PD, prior to the extreme loss of SNc cells. Although postmortem studies suggest this is not the case (Kordower *et al.*, in press, 1996) it is possible that transplanted cells are subject to the same degenerative process as host dopaminergic cells. These same neurotrophic factors could prove useful for transplantation studies (Lindsay, 1995; Lindvall, 1997; Zabek *et al.*, 1994).

Several neurotrophins, including brain-derived neurotrophic factor (BDNF), neurotrophin 4/5, and glial cell line-derived neurotrophic factor (GDNF) enhance the survival of mesencephalic dopaminergic neurons in tissue culture, and rescue dopaminergic neurons from the effects of axotomy *in vivo* (Lindsay, 1995). Of these factors, GDNF has been shown to be the most potent, and is the only one shown to protect dopaminergic neurons from toxic insults *in vivo*. GDNF is a member of the transforming growth factor β superfamily, isolated in 1993 on the basis of its ability to support the survival and differentiation of embryonic midbrain dopaminergic neurons in culture (Lin *et al.*, 1993). Intracerebral injection of GDNF has been shown to protect adult rodent midbrain dopaminergic neurons from death following axotomy-induced deprivation of target-derived trophic support (Beck *et al.*, 1995). GDNF enhances survival of SNc dopaminergic neurons, and ameliorates motor abnormalities in rodent models of Parkinson's disease (Hoffer *et al.*, 1994; Kearns and Gash, 1995; Tomac *et al.*, 1995). In the MPTP-treated primate model of PD, intracerebral administration of GDNF results in improvements in tremor, rigidity, and bradykinesia, as well as increases in the number of dopaminergic neurons within the lesioned SNc (Gash *et al.*, 1996). Interestingly, striatal dopamine remained depleted after GDNF treatment, supporting the possibility that amelioration of parkinsonian motor signs may not depend strictly on restoration of striatal dopamine. A clinical trial of intraventricular administration of GDNF in PD patients began in 1996 (Lindsay, 1995).

IV. CONCLUSION

Clearly, fetal nigral transplantation is still an experimental strategy. The team needed to perform a clinical transplant trial at a minimum includes gynecologists, neural transplant biologists, stereotactic neurosurgeons, neuroradiologists, infectious disease specialists, and neurologists with movement disorder expertise. The single most important consideration is patient selection. Each case begins and ends with the neurological evaluation. The procurement of fetal tissue requires special ethical and technical considerations. The single most difficult aspect of performing a fetal graft is the procurement and accurate dissection of the fetal tissue. Dissecting the ventral mesencephalon, or any other fetal brain region requires experience and expertise. Clinical fetal transplants should not be attempted without demonstrating the ability to dissect and reproducibly transplant human fetal cells into an animal model. This is not a procedure that uses whatever equipment is available but requires specific instrumentation. Each aspect of the procedure must be carefully planned and executed. The proof of the principle has been made (Kordower *et al.*, 1995; Lindvall, 1997).

Numerous improvements in the fetal transplantation procedure are on the horizon, which may dramatically impact upon the level of clinical recovery seen in graft recipients. The use of co-grafts, neurotrophic factors, lazaroids, anti-apoptotic agents, and many other techniques may significantly impact upon the survival of grafted neurons and the volume of tissue reinnervated. But even if less fetal donor tissue is needed and the reinnervation pattern can be extended across a wider area of the basal ganglia, another source of tissue needs to be developed. The key to future knowledge is understanding the repair processes in the CNS. Future neurosurgeons will not only make use of both CNS transplantation and transfection to restore function, but ultimately prevent the loss of function in PD and other neurodegenerative diseases.

REFERENCES

Aebischer, P., Goddard, M., Signore, A. P., and Timpson, R. L. (1994). Functional recovery in hemiparkinsonian primates transplanted with polymer-encapsulated PC12 cells. *Exp Neurol* 126:151–158.

Aebischer, P., Tresco, P. A., Winn, S. R., Greene, L. A., and Jaeger, C. B. (1991). Long-term cross-species brain transplantation of a polymer encapsulated dopamine-secreting cell line. *Exp Neurol* 111:269–275.

Alexander, G. E., Crutcher, M. D., and DeLong, M. R. (1990). Basal ganglia-thalamocortical circuits: Parallel substrates for motor, oculomotor, "prefrontal" and "limbic" functions. *Prog Brain Res* 85:119–1476.

Allen, G. S., Burns, R. S., Tulipan, N. B., and Parker, R. A. (1989). Adrenal medullary transplantation to the caudate nucleus in Parkinson's disease. *Arch Neurol* 46:487–491.

Annett, L. E., Torres, E. M., Ridley, R. M., Baker, H. F., and Dunnett, S. B. (1995). A comparison of the behavioral effects of embryonic nigral grafts in the caudate nucleus and in the putamen of marmosets with unilateral 6-OHDA lesions. *Exp Brain Res* 103:355–371.

Anton, R., Kordower, J. H., Manaster, J. S., *et al.* (1994). Neural-targeted gene therapy for rodent and primate hemiparkinsonism. *Exp Neurol* 127:199–206.

Apuzzo, M. L. J., Neal, J. H., Waters, C. H., Appley, A. J., Boyd, S. D., Couldwell, W. T., Wheelock, V. H., and Weiner, L. P. (1990). Utilization of unilateral and bilateral stereotactically placed adrenomedullary-striatal autografts in parkinsonian humans: Rationale, techniques, and observations. *Neurosurgery* 26:746–757.

Arbuthnott, G., Dunnett, S., and MacLeod, N. (1985). Electrophysiological properties of single units in dopamine-rich mesencephalic transplants in rat brain. *Neurosci Lett* 57:205–210.

Backlund, E. O., Granberg, P. O., and Hamberger, B. (1985). Transplantation of adrenal medullary tissue to striatum in parkinsonism. First clinical trials. *J Neurosurg* 62:169–173.

Bakay, R. A. E. (1990). Transplantation in the central nervous system: A therapy of the future. *Neurosurgery Clinics of North America—Vol. 1*, pp. 881–895.

Bakay, R. A. E. (1991). Selection criteria for CNS grafting into Parkinson's disease patients. In: Lindvall, O., Bjorklund, A., and Widner, H., eds., *Intracerebral transplantation in movement disorders*. Elsevier Science Publishers, B. V., pp. 137–148.

Bakay, R. A. E., Barrow, D. L., Fiandaca, M. S., Iuvone, P. M., Schiff, A., and Collins, D. C. (1987). Biochemical and behavioral correction of MPTP Parkinson-like syndrome by fetal cell transplantation. *Ann NY Acad Sci* 495:623–640.

Bakay, R. A. E., Boyer, K. L., Ansari, A., and Freed, C. R. (in press). Immunological responses to injury and grafting in the central nervous system of non-human primates. *Cell Transplantation*.

Bakay, R. A. E., Watts, R. L., Freeman, A., Iuvone, P. M., Watts, N., and Graham, S. D. (1990). Preliminary report on adrenal-brain transplantation for parkinsonism in man. *Stereotact Funct Neurosurg* 54 + 55:312–323.

Bankiewicz, K. S., Plunkett, R. J., Jacobowitz, D. M., Porrino, L., di Porzio, U., London, W. T., Kopin, I. J., and Oldfield, E. H. (1990). The effect of fetal mesencephalon implantation primate MPTP-induced parkinsonism. Histochemical and behavioral studies. *J Neurosurg* 7:231–244.

Bankiewicz, K. S., Plunkett, R. J., and Jacobowitz, D. M. (1991). Fetal nondopaminergic neural implants in parkinsonian primates. Histochemical and behavioral studies. *J Neurosurg* 74:97–104.

Bankiewicz, K. S., Plunkett, R. J., Kopin, I. J., London, W. T., and Oldfield, E. H. (1988). Transient behavioral recovery in hemiparkinsonian primates after adrenal medullary autografts. In: Gash, D. M., and Sladek, J., Jr., eds., *Transplantation into the mammalian CNS; Prog Brain Res* 78:543–550.

Beck, K. D., Valverde, J., Alexi, T., Poulsen, K., Moffat, B., Vandlen, R. A., Rosenthal, A., and Hefti, F. (1995). Mesencephalic dopaminergic neurons protected by GDNF from axotomy-induced degeneratonin the adult brain. *Nature* 373:339–341.

Bjorklund, A., and Lindvall, O. (1975). Dopamine in dendrites of the substantia nigra neurons: Suggestions for a role in dendritic terminals. *Brain Res* 83:531–537.

Bjorklund, A., and Steveni, U. (1979). Reconstruction of the nigrostriatal pathway by intrastriatal nigral transplants. *Brain Res* 177:555–560.

Bohn, M. C., Cupit, L., Marciano, F., and Gash, D. M. (1987). Adrenal medulla grafts enhance recovery of striatal dopaminergic fibers. *Science* 237:913–916.

Boyer, K. L., and Bakay, R. A. E. (1995). The history, theory, and present status of brain transplantation. *Neurosurgery Clinics in N Am* 6(1):113–125.

Bredesen, D. E., Manaster, J. S., Rayner, S., *et al.* (1991). Functional improvement in parkinsonism following transplantation of temperature-sensitive immortalized neural cells. *Neurology* 41:325, 1991.

Brundin, P., Strecker, R. E., Widner, H., *et al.* (1988). Human fetal dopamine neurons grafted in a rat model of Parkinson's disease: Immunological aspects, spontaneous and drug-induced behavior, and dopamine release. *Exp Brain Res* 70:192–208.

Burns, R. S., Chiueh, C. C., Markey, S. P., Ebert, M. N., Jacobowitz, D. M., and Kopin, I. J. (1983). A primate model of parkinsonism: Selective destruction of DA neurons in the pars compacta of the substantia nigra by n-methyl-1,2,3,6-tetrahydropyridine. *Proc Natl Acad Sci USA* 80:4546–4550.

Chalmers, G. R., Niijima, K., Patterson, P. H., Peterson, D. A., Fisher, L. J., and Gage, F. H. (1992). Chromaffin cells cografted with NGF-producing fibroblasts exhibit neuronal features. *Soc Neurosci Abstr* 18:782.

Clarke, D. J., Brundin, P., Strecker, R. E., Nilsson, O. G., Bjorklund, A., and Lindvall, O. (1988). Human fetal dopamine neurons grafted in a rat model of Parkinson's disease: Ultrastructural evidence of synapse formation using tyrosine hydroxylase immunocytochemistry. *Exp Brain Res* 73:115–126.

Cunningham, L. A., Short, M. P., Breakefield, O. X., and Bohn, M. C. (1994). Nerve growth factor released by transgenic astrocytes enhances the function of adrenal chromaffin cell grafts in a rat model of Parkinson's disease. *Brain Res* 658:219–231.

Date, I., Asari, S., and Ohmoto, T. (1995). Two-year follow-up study of a patient with Parkinson's disease and severe motor fluctuations treated by co-grafts of adrenal medulla and peripheral nerve into bilateral caudate nuclei: Case report. *Neurosurgery* 37:515–518.

Date, I., Imaoka, T., Miyoshi, Y., Ono, T., Asari, S., and Ohmoto, T. (1996). Chromaffin cell survival and host dopaminergic fiber recovery in a patient with Parkinson's disease treated by cografts of adrenal medulla and pretransected peripheral nerve. *J Neurosurg* 84:685–689.

Davis, C. G., Williams, A. C., Markey, S. P., Ebert, M. H., Caine, E. D., Reichert, C. M., and Kopin, I. J. (1979). Chronic parkinsonism secondary to intravenous injections of meperidine analogues. *Psych Res* 1:249–254.

Deacon, T. W., Pakzaban, P., Burns, L. H., Dinsmore, J., and Isacson, O. (1994). Cytoarchitectonic development, axon-glia relationships, and long distance axon growth of porcine striatal xenografts in rats. *Exp Neurol* 130:151–167.

Deacon, T., Shumacher, J., Dinsmore, J., Thomas, C., Palmer, P., Kott, S., Edge, A., Penney, D., Kassissieh, S., Dempsey, P., and Isacson, O. (1995). Histological evidence of fetal pig neural cell survival after transplantation into a patient with Parkinson's disease. *Nature Med* 3:350–353.

Defer, G. L., Geny, C., Ricolfi, F., Fenelon, G., Monfort, J. C., Remy, P., Villafane, G., Jeny, R., Samson, Y., Keravel, Y., Gaston, A., Degos, J. D., Peschanski, M., Cesaro, P., and Nguyen, J. P. (1996). Long-term outcome of unilaterally transplanted parkinsonian patients. I. Clinical approach. *Brain* 119:41–50.

Diamond, S. G., Markham, C. H., Rand, R. W., Becker, D. P., and Treciokas, L. J. (1994). Four-year follow-up of adrenal-to-brain transplants in Parkinson's disease. *Arch Neurol* 51:559–563.

Doucet, G., Murata, Y., Brundin, P., Bosler, O., Mons, N., Geffard, M., Ouimet, C. C., and Bjorklund, A. (1989). Host afferent into intrastriatal transplants of fetal ventral mesencephalon. *Exp Neurol* 106:1–19.

Edge, A. S. B., and Dinsmore, J. (1997). Xenotransplantation in the central nervous system. *Xeno* 5(2):23–25.

Faustman, D., and Coe, C. (1991). Prevention of xenograft rejection by masking donor HLA class I antigens. *Science* 252:1700–1702.

Fiandaca, M. S., Bakay, R. A. E., Sweeney, K. M., *et al.* (1988). Immunological response to intracerebral fetal neural allografts in the Rhesus monkey. *Prog Brain Res* 78:287.

Fiandaca, M. S., Kordower, J. H., Hansen, J. T., Jiao, S. S., and Gash, D. M. (1988). Adrenal medullary autografts into the basal ganglia of Cebus monkeys. Injury-induced regeneration. *Exp Neurol* 102:76–91.

Fisher, L. J., Jinnah, H. A., Kale, L. C., Higgins, G. A., and Gage, F. H. (1991). Survival and function of intrastriatally grafted primary fibroblasts genetically modified to produce L-dopa. *Neuron* 6:371–380.

Folkerth, R. D., and Durso, R. (1996). Survival and proliferation of nonneural tissues, with obstruction of cerebral ventricles, in a parkinsonian patient treated with fetal allografts. *Neurology* 46:1219–1224.

Forno, L. S., and Langston, J. W. (1991). Unfavorable outcome of adrenal medullary transplant for Parkinson's disease. *Acta Neuropathol* 81:691–694.

Freed, C. R., Breeze, R. E., Rosenberg, N. L., Schneck, S. A., Kriek, E., Qi, J., Lone, T., Zhang, Y., Snyder, J. A., Wells, T. H., Ramig, L. O., Thompson, L., Mazziotta, J. C., Huant, S. C., Grafton, S. T., Brooks, D., Sawyle, G., Schroter, G., and Ansari, A. (1992). Survival of implanted fetal dopamine cells and neurological improvement 12 to 46 months after transplantation for Parkinson's disease. *N Engl J Med* 22:1549–1555.

Freed, C. R., Breeze, R. E., Schneck, S. A., Bakay, R. A. E., and Ansari, A. A. (1996). Fetal neural transplantation for Parkinson disease. In: Rich, R. R., ed., *Clinical immunology: Principles and practice.* New York: Mosby, pp. 1677–1687.

Freed, W. J. (1983). Functional brain tissue transplantation: Reversal of lesion-induced rotation by intraventricular substantial nigra and adrenal medulla grafts, with a note on intracranial retinal grafts. *Biol Psychiatry* 18:1205.

Freed, W. J., Morihisa, J. M., Spoor, E., Hoffer, B. J., Olson, L., Seiger, A., and Wyatt, R. J. (1981). Transplanted adrenal chromaffin cells in rat brain reduce lesion-induced rotational behavior. *Nature* 292:351–352.

Freed, W. J., Farouk, K., Spoor, H. E., Morihisa, J. M., Olson, L., Wyatt, R. J. (1983). Catecholamine content of intracerebral adrenal medulla grafts. *Brain Res* 269:184–189.

Freeman, T. B., Olanow, C. W., Hauser, R. A., Nauert, G. M., Smith, D. A., Borlongan, C. V., Sanberg, P. R., Holt, D. A., Kordower, J. H., Vingerhoets, F. J. G., Snow, B. J., Caine, D., and Gauger, L. L. (1995). Bilateral fetal nigral transplantation into the postcommissural putamen in Parkinson's disease. *Ann Neurol* 38:379–388.

Freeman, T. B., Sanberg, P. R., Nauert, G. M., Boss, B. D., Spector, D., and Olanow, C. W. (1995). The influence of donor age on the survival of solid and suspension intraparenchymal human embryonic nigral grafts. *Cell Transplant* 4:141–154.

Gash, D. M., Zhang, Z., Ovadia, A., Cass, W. A., Yi, A., Simmerman, L., Russell, D., Martin, D., Lapchak, P., Collins, F., Hoffer, B. J., and Gerhardt, G. A. (1996). Functional recovery in parkinsonian monkeys treated with GDNF. *Nature* 380:252–255.

Goetz, C. G., Olanow, C. W., Koller, W. C., Penn, R. D., Cahill, D., Morantz, R., Stebbins, G., Tanner, C. M., Klawans, H. L., Shannon, K. M., Comella, C. L., Witt, T., Cox, C., Waxman,

M., and Gauger, L. (1989). Multicenter study of autologous adrenal medullary transplantation to the corpus striatum in patients with advanced Parkinson's disease. *N Engl J Med* 320:337–341.

Goetz, C. G., Stebbins, G. T., Klawans, H. L., Koller, W. C., Grossman, R. G., Bakay, R. A., and Penn, R. D. (1991). United Parkinson Foundation Neurotransplantation Registry on adrenal medullary transplants: Presurgical, and 1- and 2-year followup. *Neurology* 41:1719–1722.

Grabowska, A., and Lampson, L. A. (1992). Expression of class I and II major histocompatibility (MHC) antigens in the developing CNS. *J Neural Transplantation Placticity* 3:204–205.

Hansen, J. T., Kordower, J. H., Fiandaca, M. S., Jiao, S. S., and Gash, D. M. (1988). Adrenal medullary autografts into the basal ganglia of Cebus monkeys: Graft viability and fine structure. *Exp Neurol* 102:65–75.

Hattori, S., Li, Q. M., Matsui, N., Hashitani, T., and Nishino, H. (1992). Treadmill running combined with microdialysis evaluates motor deficits and improvement following dopaminergic grafts in 6-OHDA lesioned rats. *Restor Neurol Neurosci* 4:165.

Hirsh, E. C., Duyckaerts, C., Javoy-Agid, F., Hauw, J. J., and Agid, Y. (1990). Does adrenal graft enhance recovery of dopaminergic neurons in Parkinson's disease. *Ann Neurol* 27:676–682.

Hirshberg, D. L., and Schwartz, M. (1995). Macrophage recruitment to acutely injured central nervous system is inhibited by a resident factor: A basis for an immune-brain barrier. *J Neuroimmunol* 61:89–96.

Hitchcock, E. (1995). Current trends in neural transplantation. *Neurol Res* 17:33–37.

Hitchcock, E. H., Whitwell, H. L., Sofroniew, M. V., and Bankiewicz, K. S. (1994). Survial of TH-positive and neuromelanin-containing cells in patients with Parkinson's disease after intrastriatal grafting of fetal ventral mesencephalon. *Exp Neurol* 129:3 (abstract).

Hoffer, B. J., Leenders, K. L., Young, D., Gerhardt, G., Zerbe, G. O., Bygdeman, M., Seiger, H., Olson, L., Stromberg, I., and Freedman, R. (1992). Eighteen-month course of two patients with grafts of fetal dopamine neurons for severe Parkinson's disease. *Exp Neurol* 188:243–252.

Hoffer, B. J., Hoffman, A., Bowenkamp, K., Huettl, P., Hudson, J., Martin, D., Lin, L. F., and Gerhardt, G. A. (1994). Glial cell line-derived neuronotrophic factor reverses toxin-induced injury to midbrain dopaminergic neurons *in vivo*. *Neurosci Lett* 182:107–111.

Huang, S., Pei, G., Kang, F. A., *et al.* (1989). Transplant operation of human fetal substantia nigra tissue to caudate nucleus in Parkinson's disease: First clinical trials. *Clin J Neurosurg* 5(3):210–213.

Hurtig, H., Joyce, J., Sladek, J. R., and Trojanowski, J. Q. (1989). Postmorten analysis of adrenal-medulla-to-caudate autograft in a patient with Parkinson's disease. *Ann Neurol* 25:607–614.

Iacono, R. P., Tang, Z. S., Mazziotta, J. C., Grafton, S., and Hoehn, M. (1992). Bilateral fetal grafts for Parkinson's disease: 22 months' results. *Stereotactic Funct Neurosurg* 58:84–87.

Isacson, O., Deacon, T. W., Pakzaban, P., Galpern, W. R., Dinsmore, J., and Burns, L. H. (1995). Transplanted xenogeneic neural cells in neurodegenerative disease models exhibit remarkable axonal target specificity and distinct growth patterns of glial and axonal fibres. *Nature Med* 1:1189–1194.

Jankovic, J. (1989). Adrenal medullary autografts in patients with Parkinson's disease. *N Engl J Med* 32:325–326 (letter).

Jiao, S., Ding, Y., Zhang, W., Cao, J., Zhang, G., Zhang, Z., Ding, M., Zhang, Z., and Meng, J. (1989). Adrenal medullary autografts in patients with Parkinson's disease. *N Engl J Med* 321:324–325 (letter).

Jiao, S., Gurevich, V., and Wolff, J. A. (1993). Long-term correction of a rat model of Parkinson's disease by gene therapy. *Nature* 362:450–453.

Johnson, R. E., and Becker, J. B. (1997). Intranigral grafts of fetal ventral mesencephalic tissue in adult 6-hydroxydopamine-lesioned rats can induce behavioral recovery. *Cell Transplant* 6:267–276.

Kearns, C. M., and Gash, D. M. (1995). GDNF protects nigral dopamine neurons against 6-hydroxydopamine *in vivo. Brain Res* 672:104–111.

Kondoh, T., and Low, W. C. (1994). Glutamate uptake blockade induces striatal dopamine release in 6-hydroxydopamine rats with intrastriatal grafts: Evidence for host modulation of transplanted dopamine neurons. *Exp Neurol* 127:191–198.

Kopyov, O. V., Jacques, D., Lieberman, A., Duma, C. M., and Rogers, R. L. (1996). Clinical study of fetal mesencephalic intracerebral transplants for the treatment of Parkinson's disease. *Cell Transplant* 5:327–337.

Kordower, J. H., Cochran, E., Penn, R. D., and Goetz, C. G. (1991). Putative chromaffin cell survival and enhanced host-derived TH-fiber innervation following a functional adrenal medulla autograft for Parkinson's disease. *Ann Neurol* 29:405–412.

Kordower, J. H., Felten, D. L., and Gash, D. M. (1992). Dopaminergic transplants for Parkinson's disease. In: Olanow, C. W., and Lieberman, A. eds., *Scientific basis for therapy in Parkinson's disease.* New York Parthenon Press, Lancs, pp. 175–224.

Kordower, J. H., Fiandaca, M. S., Notter, M. F. D., Hansen, J. T., and Gash, D. M. (1990). NGF-like trophic support from peripheral nerve for grafted adrenal chromaffin cells. *J Neurosurg* 73:418–428.

Kordower, J. H., Freeman, T. B., Snow, B. J., Vingerhoets, F. J. G., Mufson, E. J., Sanberg, P. R., Hauser, R. A., Smith, D. A., Nauert, G. M., Perl, D. P., and Olanow, C. W. (1995). Neuropathological evidence of graft survival and striatal reinnervation after the transplantation of fetal mesencephalic tissue in a patient with Parkinson's disease. *N Engl J Med* 332:1118–1124.

Kordower, J. H., Hanbury, R., and Bankiewicz, K. S. (In press). *Neuropathology of dopaminergic transplants in patients with Parkinson's disease.*

Kordower, J. H., Rosenstein, J. M., Collier, T. J., Burke, M. A., Chen, E. Y., Li, J. M., Martel, L., Levey, A. E., Mufson, E. J., Freeman, T. B., and Olanow, C. W. (1996). Functional fetal nigral grafts in a patient with Parkinson's disease: Chemoanatomic, ultrastructural, and metabolic studies. *J Comp Neurol* 370:203–230.

Kordower, J. H., Rosenstein, J. M., Collier, T. J., Chen, E. Y., Li, J. L., Mufson, E. J., Sanberg, P., Freeman, T. B., Olanow, C. W. (In press). Functional fetal nigral implants in a second patient with Parkinson's disease. In: *Movement disorders.*

Kordower, J. H., Styren, S., Clarke, M., DeKosky, S. T., Olanow, C. W., and Freeman, T. B. (1997). Fetal grafting for Parkinson's disease: Expression of immune markers in two patients with functional fetal nigral implants. *Cell Transpl* 6:213–219.

Langston, J. W., Forno, L. S., Rebert, C. S., and Irwin, I. (1984). Selective nigral toxicity after systemic administration of 1-methyl-4-phenyl-1,2,5,6-tetrahydropyrine (MPTP) in the squirrel monkey. *Brain Res* 292:390.

Lawrence, J. M., Morris, R. J., Wilson, D. J., *et al:* (1990). Mechanisms of allograft rejection in the rat brain. *Neuroscience* 37:431–462.

Lin, L. F., Doherty, D. H., Lile, J. D., Bextesh, S., and Collins, F. (1993). GDNF: A glial cell-line derived neurotrophic factor for midbrain dopaminergic neurons. *Science* 260:1130–1132.

Lindsay, R. M. (1995). Neuron saving schemes. *Nature* 373:289–290 (news and comments).

Lindvall, O. (1997). Neural transplantation: A hope for patients with Parkinson's disease. *NeuroReport* 8(14):iii–x.

Lindvall, O., Backlund, E. O., Farde, L., Sedvall, G., Freedman, R., Hoffer, B., Nobin, A., Seiger, A., and Olson, L. (1987). Transplantation in Parkinson's disease: Two cases of adrenal medullary grafts to the putamen. *Ann Neurol* 22:457–468.

Lindvall, O., Brundin, P., Widner, H., Rehncrona, S., Gustavii, G., Frackowiak, R., Leenders, K. L., Sawle, G., Rothwell, J. C., Marsden, C. D., and Bjorklund, A. (1990). Grafts of fetal dopamine neurons survive and improve motor function in Parkinson's disease. *Science* 247:574–577.

Lindvall, O., Sawle, G., Widner, H., Rothwell, J. C., Bjorklund, A., Brooks, D., Brundin, P., Frackowiak, R., Marsden, C. D., Odin, P., Rehncrona, S. (1994). Evidence for long-term survival and function of dopaminergic grafts in progressive Parkinson's disease. *Ann Neurol* 35:172–180.

Lopez-Lozano, J. J., Bravo, G., Abascal, J., Brera, B., and Santos, H. (1993). Co-transplantation of peripheral nerve and adrenal medulla in Parkinson's disease. *Lancet* 339:430 (letter).

Lopez-Lozano, J. J., Bravo, B., Brera, B., Dargallo, J., Salmean, J., Uria, J., Insausti, J., and Millan, I. (1995). Long-term follow-up in 10 Parkinson's disease patients subjected to fetal brain grafting in to a cavity in the caudate nucleus: The Clinica Puerta de Hierro experience. *Transplantation Proc* 27:1395–1400.

Lopez-Lozano, J. J., Bravo, G., Brera, B., Millan, I., Dargallo, J., Salmean, J., and Inausti, J. (1997). Long-term improvement in patients with severe Parkinson's disease after implantation of fetal ventral messencephalic tissue in a cavity of the caudate nucleus: 5-year follow-up in 10 patients. *J Neurosurg* 86:931–942.

Madrazo, I., Drucker-Colin, R., Diaz, V., Martinez-Mata, J., Torres, C., and Becerril, J. J. (1987). Open microsurgical autograft of adrenal medulla to the right caudate nucleus in two patients with intractable Parkinson's disease. *N Engl J Med* 316:831–834.

Madrazo, I., Franco-Bourland, R., Ostrosky-Solid, F., Aguilera, M., Cuevas, C., Zamorano, C., Morelos, A., Magallon, E., and Guizar-Sahagun, G. (1990). Fetal homotransplants (ventral mesencephalon and adrenal tissue) to the striatum of Parkinson subjects. *Arch Neurol* 47:1281–1285.

Mahalik, T. J., Finger, T. E., Stromberg, I., and Olson, L. (1985). Substantia nigra transplants into denervated striatum of the rat: Ultrastructure of graft and host interconnections. *J Comp Neurol* 240:60–70.

Meyer, C. H., Detta, A., and Kudoh, C. (1995). Hitchcock's experimental series of focal implants for Parkinson's disease: Cografting of ventral mesencephalon and striatum. *Acta Neurochir Suppl* 64:1–4.

Molina, H., Quinones-Molina, R., Munoz, J., Alvarez, L., Alamino, A., Ortega, I., Ohye, C., Macias, R., Piedra, J., and Gonzalez, C. (1994). Neurotransplantation in Parkinson's disease: From open microsurgery to bilateral stereotactic approach: First clinical trial using microelectrode recording. *Stereotact Funct Neurosurg* 62:204–208.

Morihisa, J. M., Nakamura, R. K., Freed, W. J., Mishkin, M., and Wyatt, R. J. (1984). Adrenal medulla grafts survive and exhibit catecholamine-specific fluorescence in the primate brain. *Exp Neurol* 84:643–653.

Nikkah, G., Bentlage, C., Cunningham, M. G., and Bjorklund, A. (1994). Intranigral fetal dopamine grafts induce behavioral compensation in the rat Parkinson mode. *J Neurosci* 14:3449–3461.

Olanow, C. W., Koller, W., Goetz, C. G., Stebbins, G. T., Cahill, D. W., Gauger, L. L., Morantz, R., Penn, R. D., Tanner, C. M., Klawans, H. L., Snannon, K. M., Comells, C. L., and Witt, T. (1990). Autologous transplantation of adrenal medulla in Parkinson's disease: 18 month results. *Arch Neurol* 47:1286–1289.

Olson, L. O., Backlund, E. O., Ebendal, T., Freedman, R., Hamberger, B., Hansson, P., Hoffer, B., Lindblom, U., Meyerson, B., Stromberg, I., Sydow, O., and Sieger, A. (1991). Intraputaminal infusion of nerve growth factor to support adrenal medullary autografts in Parkinson's disease. One year follow-up of first clinical trial. *Arch Neurol* 48:373–381.

Pakzaban, P., Deacon, T. W., Burns, L. H., Dinsmore, J., and Isacson, O. (1995). A novel mode of immunoprotection of neural xenotransplants: Masking of donor major histocompatibility complex class I enhances transplant survival in the central nervous system. *Neuroscience* 65:983–996.

Perlow, M. J., Freed, W. J., Hoffer, B. J., Seiger, A., Olson, L., and Wyatt, R. J. (1979). Brain grafts reduce motor abnormalities produced by destruction of nigrostriatal dopamine system. *Science* 204:643–647.

Peschanski, M., Defer, G., N'Guyen, J. P., Ricolfi, F., Monfort, J. C., Remy, P., Geny, C., Samson, Y., Hantraye, P., Jeny, R., Gaston, A., Keravel, Y., Degos, J. D., and Cesaro, P. (1994). Bilateral motor improvement and alteration of L-dopa effect in two patients with Parkinson's disease following intrastriatal transplantation of foetal ventral mesencephalon. *Brain* 117:487–499.

Peterson, D. I., Price, M. L., and Small, C. S. (1989). Autopsy findings in a patient who had an adrenal-to-brain transplant for Parkinson's disease. *Neurology* 39:235–238.

Poltorak, M., and Freed, W. J. (1989). Immunological reactions induced by intracerebral transplantations: Evidence that host microglia but not astroglia are the antigen-presenting cells. *Exp Neurol* 103:222–233.

Redmond, D. E., Leranth, C., Spencer, D. D., Robbins, R., Vollmer, T., Kim, J. H., Roth, R. H., Dwork, A., and Naftolin, F. (1990). Fetal neural graft survival. *Lancet* 336:820–822 (letter).

Remy, P., Samson, Y., Hantraye, P., Fontaine, A., Defer, G., Mangin, J. F., Fenelon, G., Geny, C., Ricolfi, F., Frouin, V., N'Guyen, J., Jeny, R., Degos, J. D., Peschanski, M., and Cesaro, P. (1995). Clinical correlates of [18-F] fluorodopa uptake in five grafted parkinsonian patients. *Ann Neurol* 38:580–588.

Schueler, S. B., Ortega, J. D., Sagen, J., and Kordower, J. H. (1993). Robust survival of isolated a novel hypothesis of adrenal graft viability. *J Neurosci* 13:4496–4510.

Schueler, S. B., Sagen, J., Pappas, G. D., and Kordower, J. H. (1995). Long-term viability of isolated bovine adrenal medullary chromaffin cells following intrastriatal transplantation. *Cell Transplantation* 4(1):55–64.

Skoskiewicz, M. J., Colvin, R. C., Scheenberger, E. E., *et al.* (1985). Widespread and elective distribution of major histocompatibility complex-determined antigens *in vivo* by interferon. *J Exp Med* 152:1645–1664.

Snyder, E. Y. (October 1994). Grafting immortalized neurons to the CNS. *Curr Opin Neurobiol* 4(5):742–751.

Spencer, D. D., Robbins, R. J., Naftolin, F., Marek, K. L., Vollmer, T., Leranth, C., Roth, R. H., Price, L. H., Gjedde, A., Bunney, B. S., Sass, K. J., Elsworth, J. D., Kier, E. L., Makuch, R., Hoffer, P. B., and Redmond, D. E. (1992). Unilateral transplantation of human fetal mesencephalic tissue into the caudate nucleus of patients with Parkinson's disease. *N Engl J Med* 22:1542–1548.

Stromberg, I., Almqvist, P., Bygdeman, M., Finger, T. E., Gerhardt, G., Granholm, A., Mahalik, T. J., Seiger, A., Olson, L., and Hoffer, B. (1989). Human fetal mesencephalic tissue grafted to dopamine-denervated striatum of athymic rats: Light- and electron-microscopical histochemistry and *in vivo* chronoamperometric studies. *J Neurosci* 9:614–624.

Stromberg, I., Herrera-Marschitz, M., Ungerstedt, U., Ebendahl, T., and Olson, L. (1985). Chronic implants of chromaffin tissue into the dopamine-denervated striatum: Effects of NGF on graft survival, fiber growth, and rotational behavior. *Exp Brain Res* 60:335–349.

Subrt, O., Tichy, M., Vladyka, V., and Hurt, K. (1991). Grafting of fetal dopamine neurons in Parkinson's disease: The Czech experience with severe akinetic patients. *Acta Neurochir Suppl* 52:51–53.

Taylor, J. R., Elsworth, J. D., Roth, R. H., Sladek, J. R., Collier, T. J., and Redmond, D. E. (1991). Grafting of fetal substantia nigra to striatum reverses behavioral deficits induced by MPTP in primates: A comparison with other types of grafts as controls. *Exp Brain Res* 85:335–348.

Tomac, A., Lindqvist, E., Lin, L. F. H., Ogren, S. O., Young, D., Hoffer, B. J., and Olsen, L. (1995). Protection and repair of the nigrostriatal dopaminergic system by GDNF *in vivo*. *Nature* 373:335–339.

Ungerstedt, U. (1971). Striatal dopamine release after amphetamine or nerve degeneration revealed by rotational behavior. *Acta Physiol Scan Suppl* 367:49–68.

van Horne, C. G., Mahalik, T., Hoffer, B., Bygdeman, M., Almqvist, P., Stieg, P., Seiger, A., Olson, L., and Stromber, I. (1990). Behavioral and lectrophysiological correlates of human mesencephalic dopaminergic xenograft function in the rat striatum. *Brain Res Bull* 25:325–334.

Vawter, D., Gervais, K., and Caplan, A. (1992). Risks of fetal tissue donation to women. *J Neural Transplantation Plasticity* 3:322.

Vidovic, M., Sparacio, S. M., Elovitz, M., and Beneviste, E. N. (1990). Induction and regulation of class II major histocompatability complex mRNA expression in astrocytes by interferon-gamma and tumor necrosis factor-alpha. *J Neuroimmunol* 30:89–200.

Vingerhoets, F. J. G., Schulzer, M., and Snow, B. J. (1994). Reproducibility of the fluorodopa PET indices in Parkinson's disease. *Mov Disord* 9:119.

Wang, J., Bankiewicz, K., Plunkett, R., and Oldfield, E. (1994). Intrastriatal implantation of interleukin-1. Reduction of parkinsonism in rats by enhancing neuronal sprouting from residual dopaminergic neurons in the ventral tegmental area of the midbrain. *J Neurosurg* 80:484–490.

Waters, C., Itabashi, H. H., Apuzzo, M. I. J., and Weiner, L. P. (1990). Adrenal to caudate transplantation-postmortem study. *Mov Disord* 5:3–5.

Watts, R. L., Mandir, A. S., and Bakay, R. A. (1995). Intrastriatal cografts of autologous adrenal medulla and sural nerve in MPTP-induced parkinsonian macaques: Behavioral and anatomical assessment. *Cell Transplant* 4:27–38.

Watts, R. L., and Bakay, R. A. E. (1991). Autologous adrenal medulla-to-caudate grafting for parkinsonism in humans and in nonhuman primates. *Frontiers in Neuroendocrinology* 12(4):357–378.

Watts, R. L., Thyagarajan, S., Freeman, A., Goetz, C. G., Penn, R. D., Stebbins, G. T., Kordower, J. H., and Bakay, R. A. E. (1997). Effect of stereotaxic intrastriatal cografts of autologous adrenal medulla and peripheral nerve in parkinson's disease: Two-year follow-up study. *Experimental Neurology* 147:510–517.

Widner, H., and Brundin, P. (1988). Immunological aspects of grafting in the mammalian central nervous system: A review and speculative synthesis. *Brain Res Rev* 13:287–324.

Widner, H., Tetrud, J., Rehncrona, S., Snow, B., Brundin, P., Gustavii, B., Bjorklund, A., Lindvall, O., and Langston, J. W. (1992). Bilateral fetal mesencephalic grafting in two patients with parkinsonism induced by 1-methyl-4-phenyl-1,2,3,6- tetrahydropyridine (MPTP). *N Engl J Med* 22:1556–1563.

Wenning, G. K., Odin, P., Morrish, P., *et al.* (1997). Short-and long-term survival and function of unilateral intrastriatal dopaminergic grafts in Parkinson's disease. *Ann Neurol* 42:95–107.

Wolff, J. A., Fisher, L. J., Xu, L., Jinnah, H. A., Langlais, P. J., Iuvone, P. M., O'Malley, K. L., Rosenberg, M. B., Shimohamai, S., Friedmann, T., and Gage, F. H. (1989). Grafting fibroblasts genetically modified to produce L-dopa in rat model of Parkinson's disease. *Proc Natl Acad Sci* 86:9011–9014.

Wu, C. Y., De Zhou, M., Bao, X.-F., Zhang, Q.-L., Sun, W., Li, F.-Z., and Zhao, J. J. (1994). The combined method of transplantation of foetal substantia nigra and stereotactic thalamotomy for Parkinson's disease. *Br J Neurosurg* 8:709–716.

Yurek, D. M., Collier, T. J., and Sladek, J. R. (1990). Embryonic mesencephalic and striatal co-grafts: Development of grafted dopamine neurons and functional recovery. *Exp Neurol* 109:191–199.

Yurek, D. M., Lu, W., Hipkens, S., and Wiegand, S. J. (1996). BDNF enhances the functional re-innervation of the striatum by grafted fetal dopamine neurons. *Exp Neurol* 137:105–118.

Zabek, M., Mazurowski, W., Dymecki, J., *et al.* (1994). A long term follow-up of foetal dopaminergic neuronal transplantation into the brains of three parkinsonian patients. *Res Neurol Neurosci* 6:97–106.

Zhang, W. C., Ding, Y. J., Cao, J. K., Du, J. X., Zhang, G. F., and Liu, Y. J. (1994). Intracerebral co-grafting of Schwann's cells and fetal adrenal medulla in the treatment of Parkinson's disease. *Chin Med J* 107:583–588.

CHAPTER 15

Functional Effects of GDNF in Normal and Parkinsonian Rats and Monkeys

MELEIK A. HEBERT,* BARRY J. HOFFER,* ZHIMING ZHANG,† WAYNE A. CASS,† ALEXANDER F. HOFFMAN,* DON M. GASH,† AND GREG A. GERHARDT*

**Neuroscience Training Program, Departments of Pharmacology and Psychiatry, and the Rocky Mountain Center for Sensor Technology, University of Colorado Health Sciences Center, Denver, Colorado and †Department of Anatomy and Neurobiology, University of Kentucky, Lexington, Kentucky*

This chapter summarizes animal experiments, in which administration of glial cell line-derived neurotrophic factor (GDNF) was shown to produce marked and long-lasting changes in dopaminergic neuronal indices. The robust and persistent effects resulting from GDNF treatment *in vivo* support the hypothesis that this factor may be efficacious in the treatment of Parkinson's disease (PD). A combination of behavioral, neurochemical, and immunocytochemical methods were used to study the effects of this novel peptide in normal and hemiparkinsonian rats and rhesus monkeys. In all studies, it was found that effects from a single GDNF administration can be observed for weeks to months. A single injection of GDNF into the substantia nigra of normal rats elicited small, but significant increases in locomotion, lasting 7 to 10 days. The behavioral effects over a 3-week observation period following GDNF administration in normal rhesus monkeys were small, with blinded observers unable to distinguish between GDNF- and vehicle-treated animals. *In vivo* electrochemical recordings performed in the ipsilateral striatum of normal rats and monkey showed that GDNF treatment resulted in a two-fold increase in stimulus-evoked dopamine (DA) overflow. HPLC results indicated that the increases in DA release within the rat and monkey striatum were not due to alterations in the whole tissue levels of DA in these regions. However, DA levels were significantly increased in the substantia nigra of GDNF recip-

CNS Regeneration

ients in both species. In normal rats, GDNF produced a dose-dependent sprouting of tyrosine hydroxylase (TH)-positive neurites toward the injection site and increased TH-immunoreactivity in the ipsilateral striatum. Histological assays performed in normal monkeys revealed that GDNF administration led to significant increases in dopamine-neuron perikaryal size and the numbers of TH-positive axons and dendrites. In 6-OHDA-lesioned rats and MPTP-lesioned monkeys, drug-induced turning behavior and parkinsonian motor deficits were attenuated in GDNF recipients, respectively. Whole tissue levels of DA were significantly increased in the ipsilateral substantia nigra of the lesioned rats and monkeys following GDNF treatment, with DA content in this region returning to normal levels in hemiparkinsonian rats. A single injection of this peptide also led to an increase in TH-positive cell size in MPTP-lesioned rhesus monkeys. Taken together, the data from our laboratories suggest that GDNF exerts long-lasting trophic effects on DA neurons *in vivo* and may be useful as a potential therapeutic agent in humans suffering from PD.

I. GDNF: A TROPHIC FACTOR FOR DA NEURONS—A POTENTIAL THERAPY FOR PD?

Parkinson's disease (PD) is one of the most common neurological disorders of the elderly. It is characterized by a progressive degeneration of midbrain dopaminergic (DAergic) neurons with a subsequent loss of dopamine (DA) input to the striatum (Bernheimer *et al.*, 1973; Hornykiewicz and Kish, 1987). This loss leads to the cardinal symptoms of bradykinesia, rigidity, resting tremor, and akinesia seen in PD (Hornykiewicz and Kish, 1987). Common treatment strategies have focused on the replacement of the lost dopaminergic neurotransmission either pharmacologically using the dopamine precursor levodopa and other drugs (Rinne and Sonninen, 1968; Kopin, 1993) or by grafting catecholamine-producing cells to the basal ganglia (Bjorklund and Stenevi, 1979; Perlow *et al.*, 1979). These therapies relieve many of the symptoms in the early to middle stages of the disease but do not arrest or attenuate the progressive degeneration of nigral DA neurons. Consequently, interventions that slow or inhibit the progression of neuronal degeneration are currently being investigated. One such approach involves trophic factor administration.

Although there is little evidence that deficiencies of trophic factors are associated with the etiology of PD (Hornykiewicz, 1993), several of these factors have been shown to produce significant beneficial effects on DAergic neurons and have been postulated to be effective in the treatment of PD (Lindsay *et al.*, 1993; Magal *et al.*, 1993; Hyman *et al.*, 1994; Olson, 1994; Olson *et al.*, 1994). Brain-derived neurotrophic factor (BDNF) and neurotrophin-4/5 have been shown to support the survival of DA neurons in culture (Spina *et al.*, 1992; Beck *et al.*, 1993; Hynes *et al.*, 1994). Ciliary neurotrophic factor (CNTF) has been shown to prevent the degeneration of axotomized substantia nigra afferents *in vivo* (Hagg and Varon, 1993). However, the most striking effects of a neurotrophic factor on DA-containing neurons have resulted from the administration of a newly described peptide, glial cell line-derived neurotrophic factor (GDNF; Lin *et al.*, 1993).

Perhaps surprisingly, GDNF was not identified and characterized by modern molecular techniques, but in a conditioned media assay aimed at finding a neurotrophic factor for substantia nigra DA neurons (Lin *et al.*, 1993, 1994). Initial studies demonstrated that the supernatant from the rat B49 glial cell line exerted potent and relatively specific trophic effects on embryonic midbrain DAergic cells in culture. These effects included maintenance in DA cell number and increases in cell size, neurite length, and DA uptake. GDNF was purified as the protein responsible for the trophic actions.

The specificity of this trophin for DA was determined in several assays. GDNF was shown to have an EC_{50} of 1.2 pM (36 pg/ml) for DA uptake, whereas high-affinity uptake of γ-aminobutyric acid (GABA) or serotonin was not affected by GDNF given at a concentration 30,000 times higher than the EC_{50} for DA uptake. Moreover, GDNF does not appear to influence the density of astrocytes in DAergic cell cultures, nor does it alter the content of glial fibrillary acidic protein (GFAP) in these cells (Lin *et al.*, 1993). More recently, GDNF has also been shown to promote the survival and neurite extension of cultured DA neurons damaged by the neurotoxin 1-methyl-4-phenylpyridinium (Hou *et al.*, 1996).

Cloning of GDNF suggests that this factor is a distant member of the transforming growth factor beta (TGFβ) superfamily. GDNF acts as a disulfide-bonded dimer, each portion of the mature protein consists of 134 amino acid residues, with 93% identity between the human and rat sequences. The naturally occurring dimer has a molecular weight of 30kDa and is glycosylated. The mature human GDNF expressed in *Escherichia coli* is not glycosylated but exhibits the same biological potency *in vitro* as the GDNF isolated from a natural source (Lapchak *et al.*, 1996). For the experiments discussed in this chapter, recombinant human GDNF (rhGDNF) was used.

GDNF has been shown to be a considerably more potent DA neuron survival factor, both *in vitro* and *in vivo,* than any other peptide studied to date (for review see Lindsay, 1995). Although GDNF was originally characterized on the basis of its DAergic selectivity (Lin *et al.*, 1993, 1994), it is important to note that this trophic factor has been shown to enhance both dopaminergic and nondopaminergic systems *in vivo*. Recent studies have reported alterations in noradrenergic, cholinergic, and GABAergic systems with GDNF administration (Arenas *et al.*, 1995; Price *et al.*, 1996). It has yet to be determined whether the observed changes in nondopaminergic systems are distinct or whether they are intimately linked to alterations seen in DAergic systems.

II. EFFECTS OF GDNF ON MIDBRAIN DOPAMINERGIC NEURONS IN VIVO

A. Behavioral, Functional, and Neurochemical Effects of GDNF in Normal Rats

Following the discovery of GDNF as a "dopaminotrophic" factor for cultured midbrain neurons, it was essential to establish its potency as a trophic factor for dopaminergic neurons of the nigrostriatal system *in vivo*. Two studies from our laboratory have yielded behavioral, immunocytochemical, and neurochemical data that suggest that a single intracranial injection of GDNF adjacent to the substantia nigra elicits profound and long-lasting upregulation of the midbrain dopaminergic system in normal rats (Hudson *et al.*, 1995; Hebert *et al.*, 1996).

The long-term behavioral effects of a single intranigral GDNF administration were investigated by monitoring animals for changes in locomotor behaviors that have been linked to dopaminergic pathways (Ungerstedt, 1971b; Zigmond and Striker, 1989). For these measurements both total distance traveled and average movement speed were recorded one and three weeks post administration using an automated activity monitor. As compared to vehicle-injected animals, GDNF-treated rats manifested small, but significant increases in total distance traveled, which lasted 7 to 10 days after single intracranial injections, and these effects of GDNF eventually returned to control levels (Hudson *et al.*, 1995; Hebert *et al.*, 1996). At three weeks, highly significant increases in average movement speed were observed. Thus, the infusions of GDNF were seen to produce significant increases in spontaneous locomotor activity one week following treatment and average movement speed three weeks post injection. These behaviors are consistent with upregulation of DA systems using indirect dopamine agonists such as amphetamine or methylphenidate (Goodman and Gilman, 1990).

To examine changes in the functional capacity of nigrostriatal neurons, *in vivo* electrochemical recordings were performed. The dynamics of potassium (K+)-evoked DA overflow in different regions of the rat striatum were measured one and three weeks post GDNF administration (Hebert *et al.*, 1996). This technique allows selective, rapid measurements of stimulus-evoked release of DA and the DA reuptake process (Gerhardt *et al.*, 1986, 1991; Gerhardt, 1995). No significant effects of GDNF on K+-evoked overflow of DA were observed in the striatum of rats at the one-week time point. However, at three weeks post injection, the overall amplitudes of K+-evoked signals obtained from all recordings sites were two-fold greater than control. The high degree of spatial resolution provided by *in vivo* electrochemical recordings allowed us to address the possibility of regional differences in the effects of GDNF on signal amplitude. GDNF-induced increases in release were uniform along a dorsoventral axis, whereas the anterior regions of the striatum appeared to be more affected by GDNF than the posterior striatum. This may be due to GDNF affecting the higher levels of DA, which have been described in the anterior striatum (Szostak *et al.*, 1989). By analyzing the temporal dynamics of the K+-evoked responses recorded at the three-week time point, we found that GDNF treatment resulted in K+-evoked DA overflow signals that were increased in amplitude but were cleared more rapidly so that the signal duration remained unchanged. This suggests that although GDNF enhances the amount of DA that can be released, the dopamine transporters appear to be capable of controlling this increase in DA output.

In vivo microdialysis measurements were used to study basal levels of DA and DA metabolites, as well as K+- and d-amphetamine-induced DA release (Hebert *et al.*, 1996). Although basal levels of DA were unaffected, small but significant increases in both DOPAC and HVA were found in the GDNF-treated group. Both K+-depolarization of the DA terminals and d-amphetamine-induced displacement of DA resulted in highly significant increases in DA release in the striatum ipsilateral to GDNF treatment. These results further confirm and extend the effects of GDNF observed with the electrochemical measures of potassium-evoked overflow of DA. Additionally, the observed increases in stimulus-induced overflow of DA support the hypothesis that GDNF may be affecting the amounts of DA that are available to be released.

To investigate the effects of GDNF on TH-immunoreactivity in DA neurons, immunocytochemical studies were performed three weeks post GDNF injections (Hudson *et al.*, 1995). Marked increases in the numbers of dopamine fibers were seen in the mesencephalon on the injected side, suggesting that GDNF can induce neuritogenesis in adult midbrain dopamine neurons. None of the brains injected with vehicle showed any increase of TH staining surrounding the injection sites or in the striatum. Upregulation of TH, the rate-limiting enzyme in DA biosynthesis, is consistent with the increased dopaminergic activity (Roth *et al.*, 1975). However, it must be noted that TH-immunohistochemistry does not distinguish between DAergic and noradrenergic cells, and therefore results must be interpreted accordingly.

The effects of GDNF on the synthesis and storage of DA, norephinephrine (NE) and serotonin (5-HT), and their metabolites, were analyzed in two separate studies (Hudson *et al.*, 1995; Hebert *et al.*, 1996) using high-pressure liquid chromatography coupled with electrochemical detection (HPLC-EC). The earlier study reported that significant increases in DA levels within the ipsilateral substantia nigra and ventral tegmental area occurred at both one and three weeks following GDNF injections. Conversely, Hebert and coworkers (1996) found a large increase in nigral DA one week following GDNF treatment, whereas at three weeks after GDNF administration DA levels within the substantia nigra approached normal. The metabolites 3, 4 dihydroxyphenylacetic acid (DOPAC) and homovanillic acid (HVA) were slightly increased (Fig. 1). In both studies, a small but significant decrease in DA levels was observed in the striatum ipsilateral to the injection, with DOPAC and HVA levels remaining unchanged. It should be noted that a difference in injection procedures may be responsible for the disparity of the two studies (see Hebert *et al.*, 1996). Despite the disagreement in nigral DA levels between the two studies, the neurochemical measurements suggest that the increases in stimulated release of DA observed with *in vivo* electrochemistry and microdialysis cannot be explained by an overall increase in DA content within the terminal fields (Hebert *et al.*, 1996). Taken together, the experiments in normal rats strongly supports the hypothesis that GDNF enhances DA signaling through a presynaptic release mechanism(s).

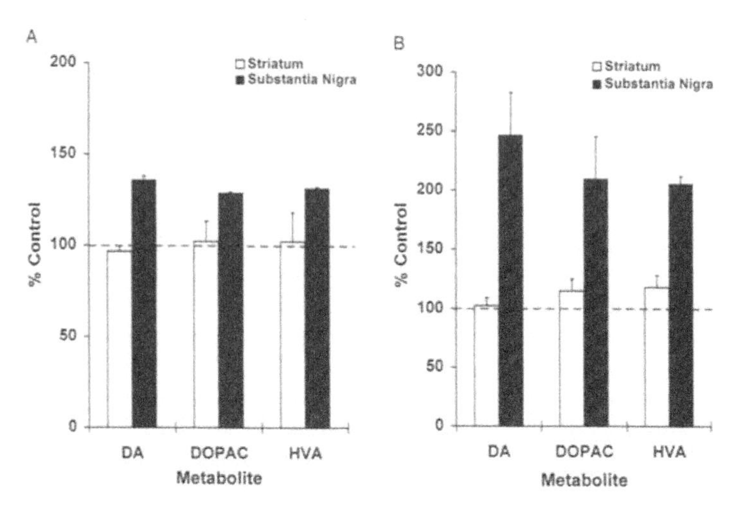

FIGURE 1 Comparative values of DA, DOPAC, and HVA whole tissue content within the ipsilateral cell body and terminal regions of GDNF-treated animals as a percent of the ipsilateral content in vehicle-treated animals. (A) Percent control levels in normal rats three weeks following GDNF injections. (B) Percent control levels in normal rhesus monkeys three to four weeks following GDNF administration.

B. Morphological and Functional Effects of Intranigrally Administered GDNF in Normal Rhesus Monkeys

In another study, the effects of a single intranigral injection of 150μg rhGDNF into the rhesus monkey were investigated (Gash *et al.*, 1995). Although crucial and informative, the initial studies involving GDNF treatment in rats are limited in their relevance to the human. Rodents have a much smaller nervous system that differs significantly in numerous neuroanatomical and neurochemical parameters from the human. In contrast, nonhuman primates possess a central nervous system and behavioral repertoire much closer to the human than the rodent. In these experiments, animals were evaluated prior to and then for three to four weeks following a single injection of 150μg rhGDNF into the right substantia nigra using MRI-guided stereotactic procedures (Gash *et al.*, 1995). In addition to monitoring home cage activity and behavior, *in vivo* electrochemical measurements of K+-evoked DA release were recorded in the caudate nucleus and putamen, referred to as the striatum, three weeks post GDNF administration to assess functional changes in the nigrostriatal DAergic system. Following euthanasia, multiple discrete tissue punches were taken from the caudate nucleus, putamen, nucleus accumbens, globus pallidus, ventral tegmental area, and substantia nigra for neurochemical analyses. Brain tissue from vehicle- and GDNF-treated animals was also processed for histological analysis of DA-containing neurons.

The behavioral effects over the three-week observation period following GDNF treatment were small, with blinded observers unable to distinguish between GDNF- and vehicle-treated animals (Gash *et al.*, 1995). The most consistent behavioral response to GDNF treatment was weight loss, which averaged about 7% of total body weight over the three-week period after trophic factor administration; conversely, vehicle-treated animals averaged 3% weight gain over the same period. None of the other behavioral responses that might be expected from stimulating central DAergic pathways, such as nausea, dyskinesia, or stereotypic behavior, were observed in the monkeys. Even though the GDNF injections were into the right nigral region, no unilateral effects on motor behavior were observed.

Following behavioral measures, *in vivo* electrochemical measurements of K+-evoked DA overflow were recorded in the ipsilateral striatum 3 weeks post GDNF administration. The amplitudes of the potassium-evoked responses recorded in both regions were significantly increased after the single 150μg intranigral administration of GDNF (Gash *et al.*, 1995). The GDNF-induced increase in K+-stimulated DA overflow parallels that observed in rats (Hebert *et al.*, 1996). The signals recorded from the striatum of GDNF-treated monkeys were almost two-fold greater than those measured in vehicle-treated primates (Gash *et al.*, 1995). These data suggest that 150μg of GDNF, a dose somewhat analogous to 10μg in the rats, is sufficient to produce long-lasting changes in the function of nigrostriatal DAergic neurons.

Interestingly, the GDNF-enhanced DA release cannot be related to alterations in the storage or synthesis of DA within the striatum. Whole-tissue neurochemical measurements performed with HPLC-EC have shown that GDNF affects DA levels within the cell bodies of nigrostriatal neurons, but not levels within the terminal fields (Gash *et al.*, 1995). In fact, DA levels within the ipsilateral substantia nigra and ventral tegmental area of rhesus monkeys treated with GDNF were approximately two-fold greater than those in the contralateral region and in control animals. Homovanillic acid (HVA), the major DA metabolite in the rhesus monkey, was found to be significantly elevated in the substantia nigra, ventral tegmental area, and striatum, with a trend toward increased HVA levels in the putamen. The HVA–DA turnover ratios were not significantly different between vehicle and GDNF recipients. Levels of serotonin and its metabolite 5-HIAA tended to be higher in the substantia nigra and basal ganglia on both sides in GDNF-treated monkeys, but these did not reach statistical significance. Figure 1B illustrates the modest changes in levels of DA and metabolites observed within the striatum of the rhesus monkey and the much larger increases measured in the substantia nigra.

Tyrosine hydroxylase-immunoreactivity suggested that GDNF treatment leads to increases in DA cell size and number in the ipsilateral mesencephalon, similar to those observed in culture (Lin *et al.*, 1993) and adult rodent DA neurons *in vivo* (Hoffer *et al.*, 1994; Bowenkamp *et al.*, 1995; Hudson *et al.*, 1995). Unique to the nonhuman primate studies was the observation that GDNF can sufficiently diffuse to stimulate TH-positive axons and dendrites over at least a 2-mm rostral-caudal gradient (Gash *et al.*, 1995). Histological data suggested that there was some tissue damage resulting from the intranigral injections of 150μg of GDNF, but DA neurons appeared relatively unaffected by the injections, probably due to the neuroprotective and neurorestorative effects of GDNF reported for rodent midbrain dopamine neurons (Beck *et al.*, 1995; Bowenkamp *et al.*, 1995; Kearns and Gash, 1995; Tomac *et al.*, 1995). The results indicate that, in the adult rhesus monkey, a single intranigral GDNF injection induces a significant upregulation of mesencephalic DA neurons lasting for weeks. Thus, the nonhuman primate studies are in good agreement with results seen from single injections of GDNF into the rat substantia nigra.

C. GDNF Attenuates Symptoms in Animal Models of Parkinson's Disease

1. Regenerative Effects of GDNF in 6-OHDA Lesioned Rats

A second series of experiments involved GDNF administration in hemiparkinsonian animals. PD models provide an *in vivo* evaluation of the efficacy of treatment strategies in restoring DAergic function. A rat model that reproduces the neurochemical deficits seen in PD involves a unilateral injection of a selective catecholamine neurotoxin, 6-hydroxydopamine (6-OHDA), into the medial forebrain bundle (Ungerstedt and Arbuthnott, 1970; Ungerstedt, 1971a, b). This widely used animal model manifests pronounced behavioral asymmetries, including hemispatial neglect and pharmacologically induced rotational behavior, reminiscent of hemiparkinsonism (Marshall and Ungerstedt, 1977a,b). Rats that have been unilaterally lesioned in this manner rotate contralaterally in response to systemic administration of low doses of apomorphine, a DA agonist. The magnitude of rotations is considered to accurately reflect the degree of DAergic degeneration (Ungerstedt and Arbuthnott, 1970; Marshall and Ungerstedt, 1977a,b). All lesioned animals used in the study exhibited a stable rotation pattern and turned >300 times contralateral to the lesion after a low dose (0.05 mg/kg) of apomorphine. We have previously shown that such animals have a loss of striatal DA $\geq$ 95% and a DA depletion in the substantia nigra of about 70 to 75% (Hudson *et al.*, 1993).

Lesioned rats meeting the behavioral criteria received various doses (0.1–100.0μg) of GDNF or vehicle intranigrally (Hoffer *et al.*, 1994). Following the 100μg GDNF administration, there was a rapid and long-lasting profound decrease in rotational behavior. The diminution in rotations was evident one week following GDNF treatment and remained at a significantly reduced level for five weeks.

Analogous to studies in normal animals, neurotransmitters and metabolites were quantified in the striatum and the substantia nigra using HPLC-EC methods. In rats that received vehicle injections into the lesioned substantia nigra, there was a marked nigral DA depletion, similar to that reported after 6-OHDA alone. In contrast, five weeks after animals received 100μg of GDNF, DA and DOPAC levels within the nigra were restored to normal levels. GDNF treatment (100μg) had no significant effects on 5-HT and 5-HIAA levels within the striatum or substantia nigra. No changes in striatal DA, which was considerably reduced on the lesioned side ($\geq$99%), were produced by any dose of GDNF at the five-week time point (Hoffer *et al.*, 1994). These data demonstrate that intranigral injection of 100μg of GDNF elicits marked and long-lasting behavioral and neurochemical changes suggesting a reversal of 6-OHDA-induced DA depletion.

An intriguing observation in this experiment using hemiparkinsonian rats was that GDNF produced an increase in DA and DA metabolite content within the substantia nigra without any apparent changes in these levels within the lesioned striatum (Fig. 2). Therefore, it appears as though normalization of DA levels within the substantia nigra correlates with the decrease in rotations observed at five weeks. These neurochemical results, combined with the behavioral measurements, support the hypothesis that DA levels in the substantia nigra may play a major role in apomorphine-induced rotational behavior (Robertson and Robertson, 1988, 1989; Hudson *et al.*, 1993).

2. Functional Recovery in Parkinsonian Monkeys Treated with GDNF

The effects of GDNF administered to rhesus monkeys with hemiparkinsonian features induced by the infusion of MPTP through the right cartoid artery (Smith *et al.*, 1993) were also investigated (Gash *et al.*, 1996). In humans and nonhuman primates, MPTP induces neurochemical, neuropathological, and behavioral features with numerous similarities to those found in idiopathic Parkinson's disease (Brooks *et al.*, 1987; Langston *et al.*, 1983, 1984). MPTP-treated nonhuman primates display the cardinal symptomology of PD: bradykinesia, rigidity, and balance and gait abnormalities (Bankiewicz *et al.*, 1986;

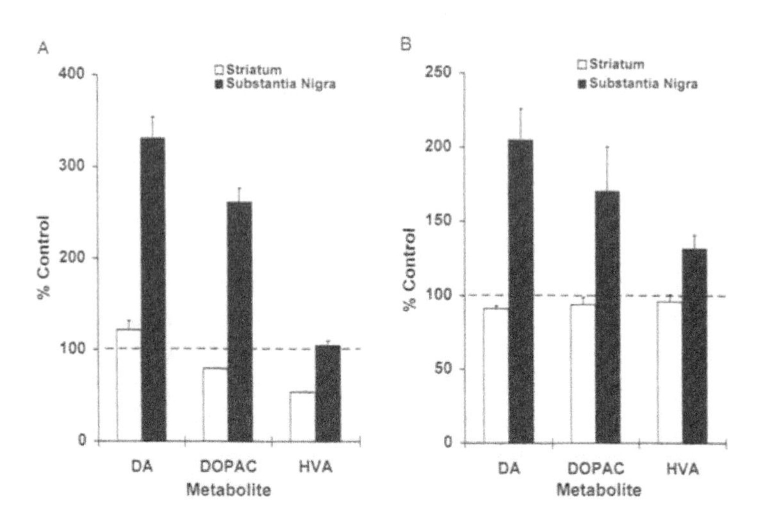

FIGURE 2 Comparative values of DA, DOPAC, and HVA whole tissue content within the ipsilateral cell body and terminal regions of GDNF recipients as a percent of the ipsilateral content of these neurochemicals in lesioned animals treated with vehicle. (A) Percent control levels in 6-OHDA-lesioned rats five weeks following GDNF administration. (B) Percent control levels in MPTP-lesioned monkeys three weeks following GDNF treatment.

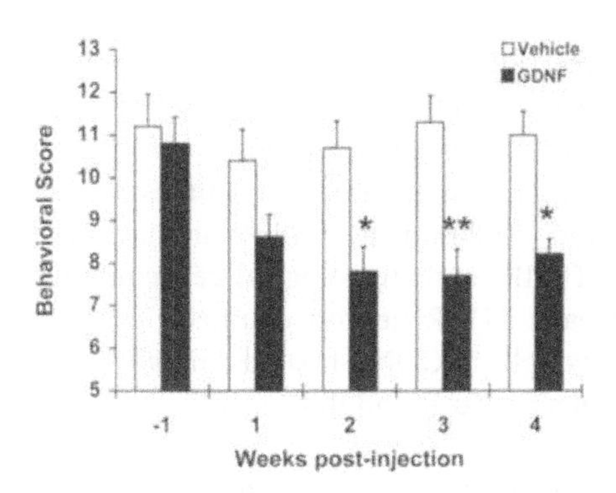

FIGURE 3 MPTP-lesioned monkeys with continuously expressed hemiparkinsonian features, rated between 8 and 14 on our nonhuman primate rating scale (Ovadia *et al.*, 1995) and were randomly assigned to treatment groups prior to GDNF treatment. Parkinsonian features such as bradykinesia, rigidity (arm and leg), tremor (arm and leg), fine motor movement, posture, and balance were scored blindly by at least two raters from videotapes of biweekly standardized testing periods prior to and for four weeks following GDNF administration. The data are shown as the mean ± s.e.m. As the rating scale was nonlinear, the nonparametic Mann-Whitney *U*-test was used to analyze differences between groups, ($^{**}p < 0.01$, $^{*}p < 0.05$). (See Gash *et al.*, 1996 for more detailed information.)

Kurlan *et al.*, 1991a, 1991b; Smith *et al.*, 1993). Histological and neurochemical alterations in the brain induced by MPTP administration also resemble those found in PD. Symptomatic monkeys have significant reductions in TH-immunoreactivity within the midbrain and decreased DA levels in the striatum and midbrain (German *et al.*, 1988; Kastner *et al.*, 1994).

Three routes of GDNF delivery were tested in these experiments: intranigral, intracaudate, and intracerebroventricular (ICV). GDNF was not administered until the animals were at least three months post MPTP treatment and displayed stable parkinsonian symptoms. Six hemiparkinsonian rhesus monkeys were treated with GDNF: two received intranigral injections of 150μg, two 450μg intracaudate injections, and two 450μg ICV infusions. Seven animals served as controls. Prior to and following GDNF administration the animals were analyzed for changes in behavior, using a nonhuman primate hemiparkinsonian rating scale (Ovadia *et al.*, 1995). Although no significant behavioral changes were observed in the control group, the GDNF recipients showed functional improvements that lasted throughout the four-week test period.

The ICV-treated animals were also assessed for the ability to respond to repeated dosing of GDNF. The sample size was increased by adding one additional animal to the 450μg treatment group and three animals that received 100μg of GDNF ICV. In both ICV GDNF treatment groups, the effects of trophic factor administration were very striking by the third week following initial infusion. Three cardinal symptoms of PD, bradykinesia, rigidity, and postural instability, were significantly improved by ICV infusions of GDNF in MPTP-lesioned monkeys (Gash *et al.*, 1996).

The intracaudate and intranigral injected animals were sacrificed four weeks after the first injection. Midbrains from these animals were processed for TH-immunoreactivity (TH-IR) and stereological cell counting (Fig. 4). Cell size was measured for each TH-IR neuron counted. There was a trend toward increased TH-positive cells in GDNF-treated animals, but the increase was not significant. Neuronal size was found to be significantly larger in the GDNF recipients.

Multiple tissue punches were taken from the basal ganglia and midbrain of ICV-treated monkeys for HPLC-EC determinations of DA and its metabolites (Gash *et al.*, 1996). Significant increases in DA were measured in the substantia nigra, ventral tegmental area, and globus pallidus, but DA, DOPAC, and HVA

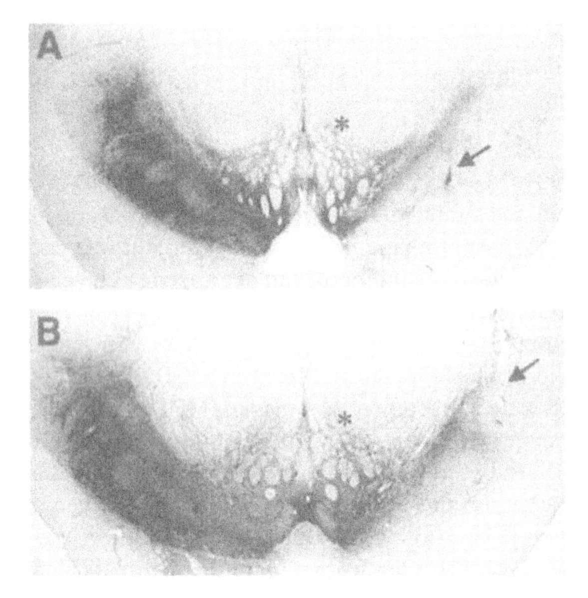

FIGURE 4 (A) In this coronal section through the midbrain region of a MPTP treated rhesus monkey, the loss of TH immunocytochemical staining in the right substantia nigra is demonstrated. Reduced TH fiber density was also evident in the less lesioned left nigral area. The needle tract (arrow) for delivery of vehicle into the right nigral region of this animal was also stained darkly because of reactive macrophages to the local tissue injury. The ventral tegmental area (*) was identified by the oculomotor nerve rootlets. (B) Significantly increased TH staining was seen in both the severely lesioned right nigra as well as the less injured left nigra of animals receiving intranigral injections of 150μg GDNF. The needle tract (arrow) was still evident three weeks after the injection.

levels within the striatum were not significantly affected (Fig. 2B). Taken together, the findings from MPTP-lesioned monkeys demonstrate that intracerebrally administered GDNF partially restores DA levels in the midbrain, stimulates the growth of surviving DA neurons, and significantly improves motor functions.

III. CONCLUSION

The rapidly increasing knowledge about the role of neurotrophic factors for neuronal development, nerve fiber formation, and maintenance of adult neurons has led to a search for factors that might exert such trophic influences on midbrain DA neurons, and potentially be of therapeutic value in the treatment of PD. Recent studies using positron emission tomography (PET) with [^{18}F]-fluorodopa and [^{11}C]-nomifensine have shown that the degree of dopamine denervation, reported from postmortem brain samples, may not reflect the capacity of DA neurons to take up levodopa in the earlier stages of the disease (Leenders *et al.*, 1990). In fact, despite the significant reduction in striatal DA levels in PD patients, both nomifensine-sensitive high-affinity DA transporters and levodopa uptake are present (Leenders *et al.*, 1990; Lloyd *et al.*, 1975). Furthermore, the degree of nigrostriatal cell loss observed postmortem may be far less than the extent of DA depletion seen in the putamen of patients with PD (Leenders *et al.*, 1988) due to the fact that loss of DA neurons may not coincide with the 80 to 90% loss of DA content within the substantia nigra (German *et al.*, 1992). Altogether these results indicate that PD may have an intermediate phase in which DA neurons could be stimulated to function normally and/or to manufacture dopamine at increased levels if given appropriate trophic support. The long-term stimulatory effects of GDNF on nigral DA neurons in normal and hemiparkinsonian rats and rhesus monkeys suggests that this trophic factor may be efficacious in the treatment of PD.

The most striking feature of GDNF is the observed pronounced and long-lived changes on DAergic content and presynaptic function that occur with a single administration of this factor. The observed *in vivo* effects resulting from GDNF administration are in stark contrast to reported effects from other factors considered to have trophic actions on DA neurons. Treatments with BDNF and NT-4/5 have been shown to enhance the survival of cultured nigral neurons, but both have failed to protect against the decline of striatal DAergic function following medial forebrain transection (Hyman *et al.*, 1994; Knusel *et al.*, 1992; Lapchak *et al.*, 1993; Spina *et al.*, 1992; Hynes *et al.*, 1994). Repeated intranigral infusion of high concentrations of CNTF has been shown to prevent degeneration of axotomized adult rat substantia nigra DA neurons *in vivo;* however, the rescued neurons failed to express TH-immunoreactivity (Hagg and Varon,

1993). Of all factors investigated to date, GDNF is the only trophic factor that has been shown to enhance several dopaminergic functional indices *in vivo*. Whatever GDNF's mode of action(s) may be, it is clear that the long-term stimulatory effects of GDNF on nigral DA neurons in normal and hemiparkinsonian animals are quite remarkable and provide evidence that this trophic factor may be efficacious in treating Parkinson's disease.

REFERENCES

Arenas, E., Trupp, M., Akerud, P., and Ibanez, C. F. (1995). GDNF prevents degeneration and promotes the phenotype of brain noradrenergic neurons *in vivo*. *Neuron* 15:1465–1473.

Bankiewicz, K. S., Oldfield, E. H., Chiueh, C. C., Doppman, J. L., Jacobowitz, D. M., and Kopin, I. J. (1986). Hemiparkinsonism in monkeys after unilateral internal cartoid artery infusion of 1-methyl-4-phenyl-1,2,3,6-tetrahydropyridine (MPTP). *Life Sci* 39:7–16.

Beck, K., Knusel, B., and Hefti, F. (1993). The nature of the trophic action of brain-derived neurotrophic factor, des(1-3)-insulin-like growth factor-1, and basic fibroblast growth factor on mesencephalic dopaminergic neurons developing in culture. *Neuroscience* 52:855–866.

Beck, K. D., Valverde, J., Alexi, T., Poulsen, K., Moffat, B., Vandlen, R. A., Rosenthal, A., and Hefti, F. (1995). Mesencephalic dopaminergic neurons protected by GDNF from axotomy-induced degeneration in the adult brain. *Nature* 373:339–341.

Bernheimer, H., Birkmayer, W., Hornykiewicz, O., Jellinger, K., and Seitelberger, F. (1973). Brain dopamine and the syndromes of Parkinson and Huntington. Clinical, morphological and neurochemical correlations. *J Neurol Sci* 20:415–455.

Björklund, A. and Stenevi, U. (1979). Reconstruction of the nigrostriatal dopamine pathway by intracerebral nigral transplants. *Brain Res* 177:555–560.

Bowenkamp, K. E., Hoffman, A. F., Gerhardt, G. A., Henry, M. A., Biddle, P. T., Hoffer, B. J., and Granholm, A.-C. (1995). Glial cell line-derived neurotrophic factor supports survival of injured midbrain dopaminergic neurons. *J Comp Neurol* 355:479–489.

Brooks, B. A., Eidelberg, E., and Morgan, W. W. (1987). Behavioral and biochemical studies in monkeys made hemiparkinsonian by MPTP. *Brain Res* 419:329–332.

Gash, D. M., Zhang, Z., Cass, W. A., Ovadia, A., Simmerman, L., Martin, D., Russell, D., Collins, F., Hoffer, B. J., and Gerhardt, G. A. (1995). Morphological and functional effects of intranigrally administered GDNF in normal rhesus monkeys. *J Comp Neurol* 363:345–358.

Gash, D. M., Zhang, Z., Ovadia, A., Cass, W. A., Yi, A., Simmerman, L., Russell, D., Martin, D., Lapchak, P. A., Collins, F., Hoffer, B. J., and Gerhardt, G. A. (1996). Functional recovery in parkinsonian monkeys treated with GDNF. *Nature* 380:252–255.

Gerhardt, G. A., Rose, G. M., and Hoffer, B. J. (1986). Release of monoamines from striatum of rat and mouse by local application of potassium: Evaluation of a new *in vivo* electrochemical technique. *J Neurochem* 46:842–850.

Gerhardt, G. A., Palmer, M. R., and Granholm, A.-C. (1991). Age-induced changes in single locus coeruleus brain transplants grown in oculo: An *in vivo* electrochemical study. *Neurobiol Aging* 12:487–494.

Gerhardt, G. A. (1995). Rapid chronocoulometric measurements of norepinephrine overflow and clearance in CNS tissues. In: (Boulton, A., Baker, G., and Adams, R. N., eds), Neuromethods: Voltammetric methods in brain systems, Vol. 27. Humana Clifton, N.J.: Humana Press, pp. 117–151.

German, D. C., Dubach, M. Askari, S., Speciale, S. G., and Bowden, D. M. (1988). 1-methyl-4-phenyl-1,2,3,6-tetrahydropyridine-induced parkinsonian syndrome in Macaca fasicularis: Which midbrain dopaminergic neurons are lost? *Neuroscience* 24:161–174.

German, D. C., Manaye, K. F., Sonsalla, P. K., and Brooks, B. A. (1992). Midbrain dopaminergic cell loss in Parkinson's disease and MPTP-induced parkinsonism: Sparing of calbindin-D28k-containing cells. *Ann NY Acad Sci* 648:42–62.

Goodman and Gilman's *The pharmacological basis of therapeutics* (1990). New York: Pergamon Press.

Hagg, T., and Varon, S. (1993). Ciliary neurotrophic factor prevents degeneration of adult rat substantia nigra dopaminergic neurons. *Proc Natl Acad Sci USA* 90:6315–6331.

Hebert, M. A., van Horne, C. G., Hoffer, B. J., and Gerhardt, G. A. (1996). Functional effects of GDNF in normal rat striatum: Presynaptic studies using *in vivo* electrochemistry and microdialysis. *J Pharmacol Exper Ther* 279(3):1181–1190.

Hoffer, B. J., Hoffman, A., Bowenkamp, K., Huettl, P., Hudson, J., Martin, D., Lin, L. F. H., and Gerhardt, G. (1994). Glial cell line-derived neurotrophic factor reverses toxin-induced injury to midbrain dopaminergic neurons *in vivo*. *Neurosci Lett* 182:107–111.

Hornykiewicz, O., and Kish, S. (1987). Biochemical pathology of Parkinson's disease. *Adv Neurol* 45:19–34.

Hornykiewicz, O. (1993). Parkinson's disease and the adaptive capacity of the nigrostriatal dopamine system: Possible neurochemical mechanisms. *Adv Neurol* 60:140–147.

Hou, J. G. G., Lin, L. F. H., and Mytilineou, C. (1996). Glial cell line-derived neurotrophic factor exerts neurotrophic effects on dopaminergic neurons *in vitro* and promotes their survival and regrowth after damage by 1-methyl-4-phenylpyridinium. *J Neurochem* 66:74–82.

Hudson, J. L., van Horne, C. G., Stromberg, I., Brock, S., Clayton, J., Masserano, J., Hoffer, B. J., and Gerhardt, G. A. (1993). Correlation of apomorphine- and amphetamine-induced turning with nigrostriatal dopamine content in unilateral 6-hydroxydopamine lesioned rats. *Brain Res* 626:167–174.

Hudson, J., Granholm, A.-C., Gerhardt, G. A., Henry, M. A., Hoffman, A., Biddle, P., Leela, N. S., Mackerlova, L., Lile, J. D., Collins, F., and Hoffer, B. J. (1995). Glial cell line-derived neurotrophic factor augments midbrain dopaminergic circuits *in vivo*. *Brain Res Bull* 36:425–432.

Hyman, C., Juhasz, M., Jackson, C., Wright, P., Ip, N. Y., and Lindsay, R. M. (1994). Overlapping and distinct actions of the neurotrophins BDNF, NT-3 and NT-4/5 on cultured dopaminergic and GABAergic neurons of the ventral mesencephalon. *J Neurosci* 4(1):335–347.

Hynes, M. A., Poulsen, K., Armanini, M., Birkenmeier, L., Phillips, H., and Rosenthal, A. (1994). Neurotrophin-4/5 is a survival factor for embryonic midbrain dopaminergic neurons in enriched cultures. *J Neurosci Res* 37:144–154.

Kastner, A., Herrero, M. T., Hirsch, E. C., Guillem, J., Luquin, M. R., Javoy-Agid, F., Obeso, J. A., and Agid, Y. (1994). Decreased tyrosine hydroxylase content in the dopaminergic neurons of MPTP-intoxicated monkeys: Effect of levodopa and GM1 ganglioside therapy. *Ann Neurol* 36:206–214.

Kearns, C. M., and Gash, D. M. (1995). GDNF protects nigral dopamine neurons against 6-hydroxydopamine *in vivo*. *Brain Res* 672:104–111.

Knusel, B., Beck, K. D., Winslow, J. W., Rosenthal, A., Burton, L. E., Widmer, H. R., Nikolics, K., and Hefti, F. (1992). Brain-derived neurotrophic factor factor administration protects basal forebrain cholinergic but not nigral dopaminergic neurons from degenerative changes after axotomy in the adult rat brain. *J Neurosci* 12:4391–4402.

Kopin, I. J. (1993). The pharmacology of Parkinson's disease therapy: An update: *Annu Rev Pharmacol Toxicol* 33:467–495.

Kurlan, R., Kim, M. H., and Gash, D. M. (1991a). The time course and magnitude of spontaneous recovery of parkinsonism produced by intracarotid administration of 1-methyl-4-phenyl-1,2,3,6-tetrahydropyridine to monkeys. *Ann Neurol* 29:677–679.

Kurlan, R., Kim, M. H., and Gash, D. M. (1991b). Oral levodopa dose-response study in MPTP-induced hemiparkinsonian monkeys: Assessment with a new rating scale for monkey parkinsonism. *Movement Disorders* 6:111–118.

Lapchak, P. A., Beck, K. D., Araujo, D. M., Irwin, I., Langston, J. W., and Hefti, F. (1993). Chronic intranigral administration of brain-derived neurotrophic factor produces striatal dopaminergic hypofunction in unlesioned adult rats and fails to attenuate the decline of striatal dopaminergic function following medial forebrain bundle transection *Neuroscience* 53:639–650.

Lapchak, P. A., Jiao, S., Miller, P. J., Williams, L. R., Cummins, V., Inouye, G., Matheson, C. R., and Yan, Q. (1996). Pharmacological characterization of glial cell line-derived neurotrophic factor (GDNF): Implications for GDNF as a therapeutic molecule for treating neurodegenerative diseases. *Cell Tissue Res* 286:179–189.

Langston, J. W., Ballard, P., Tetrud, J. W., and Irwin, I. (1983). Chronic parkinsonism in humans due to a product of meperidine-analog synthesis. *Science* 219:979–980.

Langston, J. W., Langston, E. B., and Irwin, I. (1984). MPTP-induced parkinsonism in human and non-human promotes—clinical and experimental aspects. *Acta Neurol Scand* 100:49–54.

Leenders, K., Aquilonius, S.-M., Bergstrom, K., Bjurling, P., Crossman, A., Eckernas, S.-A., Gee, A., Hartvig, P., Lundqvist, H., Langstrom, B., Rimland, A., and Tedroff, J. (1988). Unilateral MPTP lesion in a rhesus monkey, effects on the striatal dopaminergic system measured *in vivo* with PET using various novel tracers. *Brain Res* 445:61–67.

Leenders, K., Salmon, E., Tyrrell, P., Perani, D., Brooks, D., Sager, H., Jones, T., Marsden, C., and Frackowiak, R. (1990). The nigrostriatal dopaminergic system assessed *in vivo* by positron emission tomography in healthy volunteer subjects and patients with Parkinson's disease. *Arch Neurol* 471:290–298.

Lin, L.-F. H., Doherty, D. H., Lile, J. D., Bektesh, S., and Collins, F. (1993). GDNF: A glial cell line-derived neurotrophic factor for midbrain dopaminergic neurons. *Science* 260:1130–1132.

Lin, L.-F. H., Zhang, T. J., Collins, F., and Armes, L. G. (1994). Purification and initial characterization of rat B49 glial cell line-derived neurotrophic factor. *J Neurochem* 63:758–768.

Lindsay, R. M., Altar, C. A., Cedarbaum, J. M., Hyman, C., and Wiegand, S. J. (1993). The therapeutic potential of neurotrophic factors in the treatment of Parkinson's disease. *Exp Neurol* 124:103–118.

Lindsay, R. M. (1995). Neuron saving schemes. *Nature* 373:289–290.

Lloyd, K., Davidson, L., and Hornykiewicz, O. (1975). The neurochemistry of Parkinson's disease: Effect of L-dopa therapy. *J Pharmacol Exp Ther* 195:453–464.

Magal, E., Burnham, P., Varon, S., and Louis, J.-C. (1993). Convergent regulation by ciliary neurotrophic factor and dopamine of tyrosine hydroxylase expression in cultures of rat substantia nigra. *Neuroscience* 52:867–881.

Marshall, J. F., and Ungerstedt, U. (1977a). Supersensitivity to apomorphine following destruction of the ascending dopamine neurons: Quantification using the rotation model. *Eur J Pharmacol* 41:351–367.

Marshall, J. F., and Ungerstedt, U. (1977b). Striatal efferent fibers play a role in maintaining rotational behavior in the rat. *Science* 198:62–64.

Olson, L. (1994). Neurotrophins in neurodegenerative disease: Theoretical issues and clinical trials. *Neurochem Intl* 25(1):1–3.

Olson, L., Backman, L., Ebendal, T., Eriksdotter-Jonhagen, M., Hoffer, B., Humpel, C., Freedman, R., Giacobini, M., Meyerson, B., and Nordberg, A. (1994). Role of growth factors in degeneration and regeneration in the CNS: Clinical experiences with NGF in Parkinson's and Alzheimer's diseases. *J Neurol* 242 (1 Suppl. 1):S12–15.

Ovadia, A., Zhang, Z., and Gash, D. M. (1995). Increased susceptibility to MPTP toxicity in middle-aged rhesus monkeys. *Neurobiol Aging* 16:931–937.

Perlow, M., Freed, W., Hoffer, B., Seiger, A., Olson, L., and Wyatt, R. (1979). Brain grafts reduce motor abnormalities produced by CNS damage. *Science* 204:643–647.

Price, M. L., Hoffer, B. J., and Granholm, A. C. (1996). Effects of GDNF on fetal septal forebrain transplants *in oculo*. *Exp Neurol* 141:181–189.

Rinne, U. K., and Sonninen, V. (1968). A double blind study of l-dopa treatment in Parkinson's disease. *Eur Neurol* 1:180–191.

Robertson, G. and Robertson, H. (1988). Evidence that the substantia nigra is a site of action for L-DOPA. *Neurosci Lett* 89:204–208.

Robertson, G., and Robertson, H. (1989). Evidence that L-DOPA-induced rotational behavior is dependent on both striatal and nigral mechanisms. *J Neurosci* 9:3326–3331.

Roth, R. H., Walters, J. R., Murrin, L. C., and Morgenroth, V. H., III (1975). Dopaminergic neurons: Role of impulse flow and presynaptic receptors in the regulation of tyrosine hydroxylase. In: Usdin, E., and Bunney, W. E., eds., *Pre- and postsynaptic receptors*. New York: Marcel Dekker, pp. 5–58.

Smith, R. D., Zhang, Z., Kurlan, R., McDermott, M., and Gash, D. M. (1993). Developing a stable bilateral model of parkinsonism in rhesus monkeys. *Neuroscience* 52:7–16.

Spina, M., Squinto, S., Miller, J., Lindsay, R., and Hyman, C. (1992). Brain-derived neurotrophic factor protects dopamine neurons against 6-hydroxydopamine and N-methyl-4-phenylpyridinium ion toxicity: Involvement of the glutathione system. *J Neurochem* 59:99–106.

Szostak, C., Jakubovic, A., Phillips, A. G., and Fibiger, H. C. (1989). Neurochemical correlates of conditioned circling within localized regions of the striatum. *Exp Brain Res* 75:430–440.

Tomac, A., Lindquist, E., Lin, L.-F. H., Orgen, S. O., Young, D., Hoffer, B. J., and Olson, L. (1995). Protection and repair of the nigrostriatal dopaminergic system by GDNF *in vivo*. *Nature* 373:335–339.

Ungerstedt, U., and Arbuthnott, G. W. (1970). Quantitative recording of rotational behavior in rats after 6-hydroxydopamine lesions of the nigrostriatal dopamine system. *Brain Res* 24:485–493.

Ungerstedt, U. (1971a). Adipsia and aphagia after 6-hydroxydopamine induced degeneration of the nigrostriatal dopamine system. *Acta Physiol Scand* 82:96–122.

Ungerstedt, U. (1971b). Postsynaptic supersensitivity after 6-hydroxydopamine induced degeneration of the nigrostriatal dopamine system. *Acta Physiol Scand* 82:69–92.

Zigmond, M. J., and Stricker, E. M. (1989). Animal models of parkinsonism using selective neurotoxins: Clinical and basic implications. *Int Rev Neurobiol* 31:1–79.

CHAPTER 16

Genetic Engineering for Parkinson's Disease

JAMIE L. EBERLING,* EUGENE O. MAJOR,† DEA NAGY,*
AND KRZYSTOF S. BANKIEWICZ†

**Center for Functional Imaging, Lawrence Berkeley National Laboratory, University of California, Berkeley, California, †Laboratory of Molecular Medicine and Neuroscience, NINDS, NIH, Bethesda, Maryland*

Gene therapy offers new alternatives for the treatment of Parkinson's disease and other CNS disorders. Current gene therapy approaches include the transplantation of genetically modified autologous cells into target CNS sites, or the direct transfer of genetic material into CNS cells using viral or synthetic vectors to produce and deliver therapeutic agents to specific brain sites. Here we discuss two major therapeutic strategies: (1) dopamine replacement, aimed at increasing dopamine levels by the transplantation of dopamine producing cells or by introducing genes that express one or more of the enzymes or precursors required for dopamine synthesis, and (2) neuroprotection and neuroregeneration, aimed at protecting intact dopaminergic cells from degeneration or restoring the function of degenerating cells. Although more preclinical work is needed in appropriate animal models addressing such issues as optimal gene delivery and the regulation of gene expression, both *ex vivo* and *in vivo* gene therapy approaches show great promise for the future treatment of Parkinson's disease.

I. INTRODUCTION

Parkinson's disease (PD) is a progressive disorder that results in the degeneration of the nigrostriatal pathway and a deficiency in tyrosine hydroxylase (TH) that is necessary for the synthesis of the neurotransmitter, dopamine (DA) (Agid *et al.*, 1987). Although compensatory processes work early in the disease to upregulate the amount of TH available for the production of DA, as the disease progresses the system is no longer able to produce the required levels (Bernheimer *et al.*, 1973; Kebabian and Calne, 1979; Lee *et al.*, 1978). Clinical symptoms begin to appear at a point when DA levels in the striatum have been reduced by about 80%, and 60% of the cells in the substantia nigra have degenerated (Agid *et al.*, 1987; Bernheimer *et al.*, 1973).

Current treatments for PD are aimed at compensating for loss of activity in dopaminergic neurons. Oral L-dopa administration continues to be the mainstay of current therapy. L-dopa is transported into DA cells where it is converted to DA by aromatic amino acid decarboxylase (AADC), thus circumventing the rate-limiting TH deficiency. However, L-dopa treatment is limited in several respects: (1) the inability to achieve site-specific delivery results in unwanted side effects and limits the amount of the drug that can be given, (2) unpredictable "on-off" effects occur due to the inability to maintain sustained drug levels in the CNS, and (3) remaining DA cells continue to degenerate. Gene therapy is a rapidly advancing field that has produced potential alternatives for the treatment of neurodegenerative disorders. The application of gene therapy to the treatment of PD is focused on increasing brain DA levels either through the transplantation of genetically modified autologous cells, or through the direct transfer of genetic material into CNS cells using viral or synthetic vectors to produce and deliver therapeutic agents to specific brain sites. Both rodent and nonhuman primate models of PD have been used to evaluate the safety and efficacy of new therapies. Here we discuss two major gene therapy approaches: (1) DA replacement by the introduction of DA or key DA precursors and enzymes, and (2) neuroprotection and neuroregeneration using growth factors.

II. PRIMATE MODELS OF PARKINSON'S DISEASE

Although rodent models are extensively used to evaluate novel treatments for PD, there are considerable neurochemical and neuroanatomical differences between the human and rodent nervous systems. By contrast, the nonhuman primate has a similar nervous system and behavioral characteristics to humans, and the 1-methyl-4-phenyl-1,2,3,6-tetrahydropyridine (MPTP) lesioned monkey is widely regarded as the best preclinical model of PD. In addition, surgical complications related to the intracranial introduction of cellular implants or the direct delivery of vectors that may be manifested by neurological deficits are more likely to be observed in monkeys rather than in smaller animal species (Bringas *et al.*, 1996). The issue of targeting large areas of the brain can also be more adequately addressed in nonhuman primates because the rhesus monkey brain is closer in volume and anatomy to the human brain than is the rodent brain. In addition, repeated biochemical analyses such as *in vivo* microdialysis and cerebrospinal fluid (CSF) sampling can be easily performed in the monkeys, together with postmortem analyses to yield biochemical and histological information (Bankiewicz, 1991; Bankiewicz *et al.*, 1986; Elsworth *et al.*, 1987).

The neurotoxin MPTP is selective for dopaminergic cells and produces a clinical syndrome, biochemical alterations, and pathological changes that closely resemble PD (Burns *et al.*, 1983; Davis *et al.*, 1979; Langston *et al.*, 1983). Like PD, the main sites of MPTP neurotoxicity are the DA terminals in the striatum and the DA cell bodies in the substantia nigra pars compacta (SNc). Although the time course of neurodegenerative changes in the nigrostriatal pathway is accelerated in the MPTP model, compensatory and physiologic changes are similar to those that occur in idiopathic PD. Thus, functional and clinically relevant behavioral outcomes of gene therapy may be tested in monkey models of PD (Bankiewicz, 1991; Bankiewicz *et al.*, 1986; Burns *et al.*, 1983; Henderson *et al.*, 1991).

A. Hemiparkinsonian Model

The selection of the appropriate MPTP primate model is an important factor in evaluating various gene therapy strategies. The systemic administration of MPTP produces bilateral damage to nigrostriatal dopaminergic neurons and a severe bilateral parkinsonian syndrome requiring careful treatment to assure survival. By contrast, the infusion of MPTP into one carotid artery produces damage to the dopaminergic cells on the side of infusion with little damage to the cells in the opposite hemisphere, resulting in a hemiparkinsonian syndrome on the side of the body opposite to infusion. DA levels in the lesioned hemisphere are reduced by up to 98% resulting in a condition simulating end-stage PD (Bankiewicz *et al.*, 1986). The hemiparkinsonian model is particularly useful because supplemental L-dopa treatment is not required for survival, and the intact hemisphere provides a control for studies of biochemical, morphological, and behavioral measures related to the lesion and evaluations of therapeutic interventions. This model is particularly well suited for evaluating DA replacement therapies, such as the neurotransplantation of DA-producing cells.

B. Overlesioned Hemiparkinsonian Model

The overlesioned hemiparkinsonian model was developed in an effort to mimic different severities of PD to examine the relationship between the clinical stage of PD and the underlying neurochemical and pathological changes. This model also allows evaluation of various therapeutic interventions, the value of which may differ according to the stage of the disease. Unilateral intracarotid MPTP administration, followed by systemic administration of MPTP, produces a near-complete lesion and a condition analogous to end-stage PD on the intracarotid infused side, and a partial lesion and a moderate to severe parkinsonian condition on the contralateral side (Bankiewicz *et al.*, 1997a, 1997b). As with the hemiparkinsonian model, the end-stage condition may serve as a target for grafting DA-producing cells (DA replacement), whereas the partially lesioned side may be useful for evaluating therapies aimed at protecting the remaining cells and/or rescuing dying cells. The creation of a partial lesion is especially important for evaluating neuroregenerative therapies, because sprouting and regeneration may no longer be possible in the end-stage condition.

III. DA REPLACEMENT

PD is particularly suited to gene therapy because damage is largely restricted to a circumscribed population of neurons, the DA neurons in the nigrostriatal pathway. As shown in Fig. 1, DA replacement approaches target dopaminergic receptors in the striatum either by transplanting DA-producing cells, or by the direct transfer of genes expressing one or more components of the biosynthetic pathway to drive local synthesis of DA in remaining cells. The local delivery of therapeutic agents to targeted dopaminergic cells avoids some of the problems associated with drug treatments that have more extensive effects on other cell

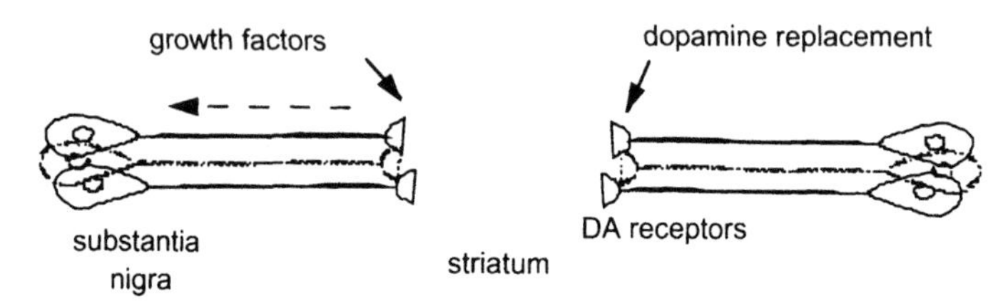

FIGURE 1 In Parkinson's disease (PD), the progressive death of dopaminergic cells in the substantia nigra is responsible for decreased levels of dopamine (DA) in the striatum resulting in upregulation of dopaminergic receptors. It might be possible to slow and/or even reverse this process by providing dopaminergic growth factors in this region using the gene therapy approach. Such growth factors would be released locally for action at the terminal (striatum) and/or transported to the cell body (substantia nigra). For example, using an *in vivo* approach, a growth factor gene could be introduced into the dopaminergic neuron providing *in situ* production of the required growth factor and preventing degeneration. Another approach would target dopaminergic receptors by introducing genetically modified cells or infecting resident striatal cells that would release DA or its precursor, L-dopa.

types and are typically accompanied by numerous side effects. Fluctuations in plasma levels of L-dopa are common following oral administration, resulting in "off" time during which motor function is poor. Although the continuous intravenous infusion of L-dopa may reduce these fluctuations, this is not a practical method of administration and may have toxic effects due to the low solubility of L-dopa and poor stability in nonacidic solutions (Chase *et al.*, 1987). Gene-therapy, on the other hand, can enable the continuous delivery of DA to targeted cells assuming there is a sufficient supply of all components of the biosynthetic pathway. For example, a continuous supply of TH will have limited benefits if amino acid decarboxylase (AADC) levels are greatly reduced. In addition, DA production is regulated by cofactors that are required for TH and AADC activity, as shown in Fig. 2 (also see Leff *et al.*, in press). This is an inherent limitation of the DA replacement approach and limits the usefulness of gene therapy because the introduction, expression, and regulation of multiple genes are required for the system to function properly.

IV. NEUROPROTECTION AND NEUROREGENERATION

Both neuroprotection and neuroregeneration are important when considering potential treatments for PD. Neuroprotection–neuroregeneration strategies target the presynaptic DA cells in the substantia nigra (see Fig. 1) in order to protect intact cells from degeneration, or to rescue cells that have already begun to degenerate. If started early enough in the disease, neuroprotective agents may stop the progression of neurodegeneration and essentially halt the disease process, whereas neuroregenerative treatments may improve or reverse parkinsonian symptomatology in more advanced cases of PD and restore nor-

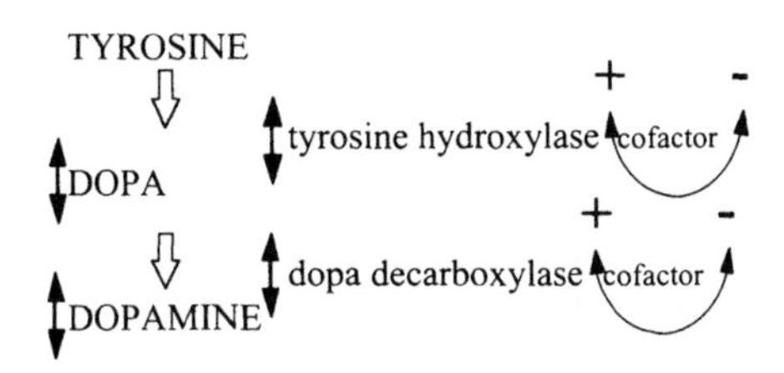

FIGURE 2 Dopamine synthesis is dependent on multiple factors. Both tyrosine hydroxylase (TH) and dopa decarboxylase (DDC) enzymes are required for dopamine production and both of these enzymes are regulated by co-factors. Thus, therapeutic levels of dopamine depends on multiple factors that are depleted to different degrees in Parkinson's disease. Successful restoration of dopamine would require the introduction, expression, and regulation of multiple genes. This is possible only if stable gene expression is achieved.

mal DA function in damaged nigrostriatal neurons. Several growth factors have been shown to have both neuroprotective and neuroregenerative effects. The treatment of human diseases with growth factors is in early stages, but human clinical trials have begun for several neurological disorders (Yuen and Mobley, 1996). Several known growth factors, such as basic fibroblast growth factor (bFGF), brain-derived neurotrophic factor (BDNF) (Beck *et al.*, 1993; Hyman *et al.*, 1991; Spina *et al.*, 1992), truncated insulin-like growth factor-1 (tIGF1) (Beck *et al.*, 1993), and glial cell line-derived neurotrophic factor (GDNF) have been shown to exert variable effects on dopaminergic activity, with GDNF acting most selectively on dopaminergic neurons (Hudson *et al.*, 1995; Lin *et al.*, 1993). GDNF has been shown to increase dopaminergic activity and improve clinical signs of parkinsonism in rodent and primate models of PD (Beck *et al.*, 1995; Bowenkamp *et al.*, 1996, 1995; Gash *et al.*, 1996; Hoffer *et al.*, 1994; Kearns and Gash, 1995; Sauer *et al.*, 1995; Tomac *et al.*, 1995; Tseng *et al.*, 1997; Winkler *et al.*, 1996). The recently discovered growth factors neurturin and persephin are structurally related to GDNF and have neurotrophic effects on DA neurons as well (Horger *et al.*, 1998; Milbrandt *et al.*, 1998). Although neurotrophic effects in animal models of PD do not necessarily predict neuroprotection and neuroregeneration in idiopathic PD, studies showing that growth factors are effective in several animal models using different neurotoxins indicate that they may be effective regardless of the mechanism of neurodegeneration. It is premature to speculate on the efficacy of growth factors for preventing cell death or increasing DA function in PD, but the results of animal studies indicate that this approach has great promise for future clinical application.

V. *EX VIVO* GENE THERAPY USING AUTOLOGOUS CELLS

One method to deliver therapeutic agents locally in the CNS is to transplant cells that synthesize and secrete L-dopa or one of the components of the DA biosynthetic pathway after *ex vivo* genetic modification. For example, a PD patient's own cells could be genetically modified to produce TH. Most augmentation strategies have focused on TH because, as mentioned before, it is the rate-limiting enzyme for the DA synthesis. Depending on the cell type being used as a vehicle, it may be necessary to express a second enzyme, GTP cyclohydrolase I, to produce the required cofactor for TH activity, tetrahydrobiopterin (BH_4) (Leff *et al.*, in press; Levine *et al.*, 1981). These cells would then be grafted into appropriate sites in the striatum to drive the local supply of L-dopa at sites in the brain normally innervated by dopaminergic neurons. The L-dopa secreted by these cells may then be taken up by remaining neuronal and nonneuronal cells, converted to DA, and released either in a normally regulated or unregulated fashion to ameliorate symptoms in these patients.

The choice of the cell used as the delivery vehicle must meet a number of criteria. The cell (1) should not elicit an inflammatory response from the host, (2) should not migrate in the host brain, and, importantly, (3) should not form tumors after *in vivo* placement (Tuszynski and Gage, 1996). Although autologous cells from the patient meet these criteria, it is difficult to prepare such cells with rigorous quality control. For example, evaluation of the autologous cells for chromosomal alterations during propagation, detection of viruses or other microbial contaminants introduced during establishment of the cells and transduction procedures, and phenotypic changes of the cells during the multiple mitotic events needed to accomplish stable transgene expression require large stocks of cells with homogeneous characteristics. This is not easily accomplished if biopsy tissues from the patient are used for the propagation of cells and their genetic engineering. It is also not likely that such candidate cells would be derived from the nervous system, but rather from an anatomically accessible source. Skin fibroblasts have been the primary choice for experimental studies and result in good survival in the striata of MPTP-treated primates. In addition, TH-positive autologous fibroblasts were detected by immunoreactivity and *in situ* hybridization four months following implantation (Bankiewicz *et al.*, 1995, 1997c). However, genetically engineered fibroblasts expressing TH do not exhibit long-term transgene expression in the rat. One reason that may account for the gradual decline in gene expression has come from cell culture studies suggesting that fibroblast transgene expression is downregulated by the inflammatory cytokines, TNF-a, IL-1b, and TGF-b1 (Schinstine *et al.*, 1997). These cytokines are released from macrophages and microglial cells that can invade grafted fibroblasts.

An alternative to using autologous cells would be human nervous system-derived cell lines that could meet the above criteria. Several human fetal CNS cell lines have been developed that may be candidates for gene delivery into the CNS to either repopulate neurons in the degenerating brain or deliver replacement neurotransmitters or survival-promoting growth factors. Recently, Sah and associates (1997) reported the immortalization of a human fetal brain progenitor cell. The v-myc oncogene expressed by a tetracycline-regulated promoter was used to immortalize the cells. One clone expressed a neuronal phenotype. Interestingly, the cells responded to extrinsic factors in culture resulting in differentiation into either neurons or astrocytes (Sah *et al.*, 1997), suggesting that environmental or extrinsic factors can redirect the differentiation pathway. Another example of a human brain cell with potential for *ex vivo* gene therapy is the astrocyte. Recently, human fetal astrocytes were transduced with a retroviral vector to express mouse beta nerve growth factor (NGF). Although not a cell line, these fetal astrocytes released biologically active NGF for several weeks in culture (Lin *et al.*, 1997). Because neuronal survival in human fetal brain tissue grafts is estimated to be only in the range of 5 to 6%, the delivery of growth factors may be necessary in order to enhance the viability of the grafted tissue as well as the remaining dopamineric neurons. Human fetal astrocytes genetically engineered to produce such factors may provide a source of growth factors.

A human fetal brain-derived cell line has been described that constitutively expresses NGF and can be readily engineered to produce transgenes (Major *et al.*, 1985; Tornatore *et al.*, 1996). The cell line, SVG, was immortalized using an SV40 replication origin-defective mutant that synthesizes the viral T protein. This protein is responsible for the immortalization of the cells most likely interfering with cell cycle-regulated proteins such as p53 and Rb (retinoblastoma protein). The cells have many phenotypic characteristics of astrocytes including morphology as shown in Fig. 3a, as well as biological properties of cells more typically described as progenitor cells. Clones of SVG cells have been made expressing TH, so-called SVG-TH cells. Grafts of these cells into 6-hydroxydopamine-lesioned rats ameliorated the abnormal rotational behavior characteristic of the chemical lesioning (Tornatore *et al.*, 1996). The SVG cells also have been implanted into the caudate and putamen of normal rhesus monkeys with little signs of immune reactivity at the site of the graft, no tumor formation after nine months, no evidence of cell migration, and sustained viability of the cells (Tornatore *et al.*, 1996, 1993). In addition, SVG cells can be readily transduced through DNA transfer techniques or even microinjection as shown in Fig. 3b. The cells can carry multiple copies of a transgene in a viral vector that contains a functional SV40 origin of DNA replication sequence in an episomal state, even with cell division (see Fig. 3c). Viral DNA transfected into these cells can replicate to a high copy number and can be segregated into dividing nuclei as the cell undergoes division. Transgenes can also become integrated into the cell chromosome if selection is made using a selectable marker, such as geniticin (G418). Grafting of these cells into MPTP-lesioned primates will determine their potential use as a tool for *ex vivo* gene therapy.

FIGURE 3 Immortalized cell line from human fetal brain, SVG. (A) Subconfluent culture of cells on minimal essential medium and fetal bovine serum with morphological characteristics of neuroglial cells. (B) Microinjection of a viral DNA vector which can replicate in the nucleus of the cell and express its gene product. (C) Dividing nuclei of cells with replicating vector shown in cell division using in situ DNA hybridization. Viral vector system is derived from replication origin sequences of the polyomavirus JCV.

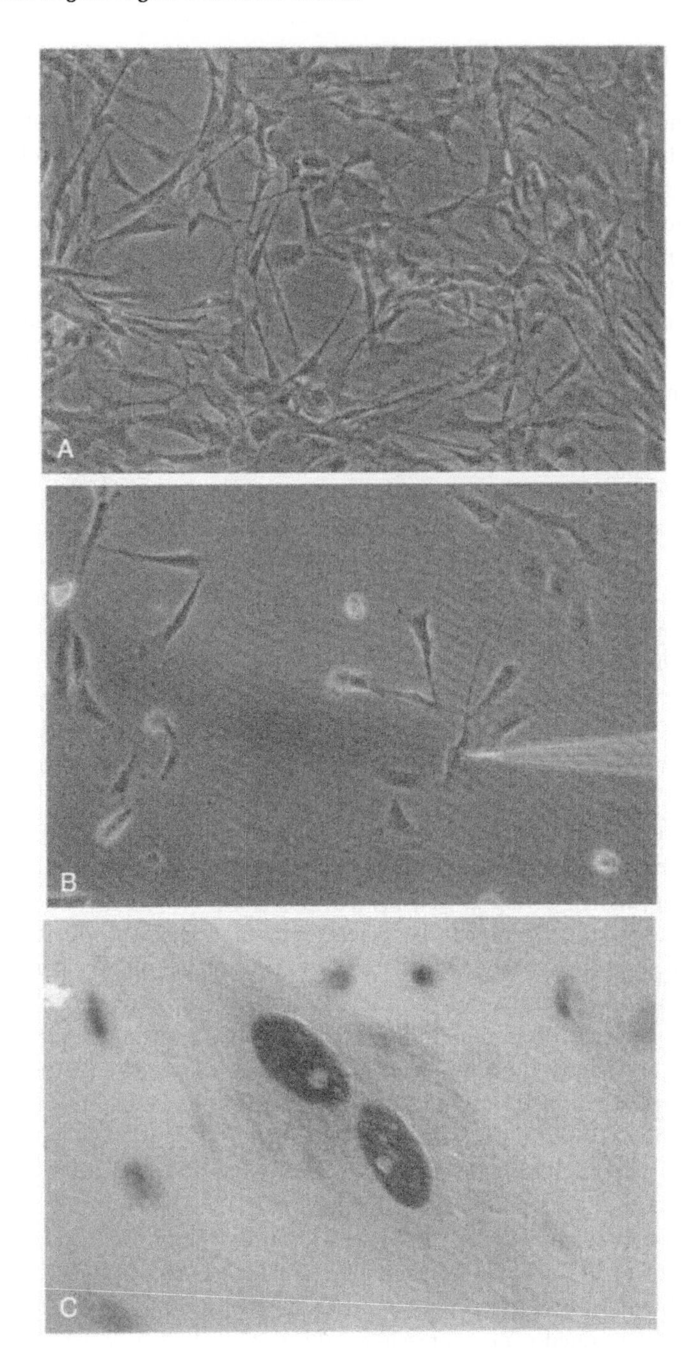
A
B
C

VI. *IN VIVO* GENE TRANSFER

Significant advances have recently been made in the introduction of genetic material into nonreplicating mammalian cells *in vivo*. Several vehicles have been used for *in vivo* transfer of cDNA sequences (see Chapter 9). Currently it is difficult to predict which of these *in vivo* delivery systems will prove to be optimal and clinically applicable. Some of the systems now in development include herpes simplex viral vectors (Federoff *et al.*, 1992; Geller and Breakefield, 1988; Geller and Freese, 1990), adenoviral vectors (Caillaud *et al.*, 1993; Le Gal La Salle *et al.*, 1993), adeno-associated viral vectors (Muzyczka, 1992; Samulski *et al.*, 1989), lentivirus vectors (Naldini *et al.*, 1996; Zufferey *et al.*, 1997), and direct plasmid DNA transfer (Acsadi *et al.*, 1991; Jiao *et al.*, 1992; Wolff *et al.*, 1990).

The interest in herpes viruses as vectors stems from their natural neurotropic characteristics and the size of the viral genome for insertion of nucleotide sequences. However, technical problems have arisen using such a large and complex viral genome making it difficult to produce virus vectors free of contaminating helper, essentially wild-type virions. There have been many generations of herpes virus vectors tested with continual loss of viral genes not essential for transgene expression and vector replication. A herpes vector amplicon has been developed that is a plasmid-based system with HSV cis acting elements for DNA replication and packaging. The vector also has transcriptional control elements for expression of the genes of interest. The advantage of the HSV amplicon seems to be the relatively large region suitable for gene insertion. Although the possible limit of the amplicon may be as high as 140 kb, in actual testing it may be ten-fold less (Lu and Federoff, 1995). Multiple copies of a gene can be placed into the amplicon with the use of inducible promoters, i.e., glucocorticoid, to turn "on and off" the gene expression. Amplicon vectors can also be constructed as bicistronic expressing clones using the internal ribosome-binding site, or IRES, sequences. Although the amplicon HSV-derived vectors have an advantage of transducing postmitotic neurons, gene expression from this system seems to be limited to a few weeks after grafting (Federoff *et al.*, 1997).

The development of AAV vectors has been slowed by the necessity to use adenovirus as a helper to package the vector DNA. The relatively small size of the gene that can be inserted is also a limiting feature. However, for genes of sizes up to 5 kb, the AAV vector system has some advantages. Like the lentiviral and HSV amplicon systems, AAV can transduce postmitotic cells. Reports of sustained gene expression indicate that AAV may persist for over one year (Xiao *et al.*, 1997), an advantage over other viral vectors. In addition, the small size of AAV, approximately 20 nm in diameter, is an advantage for the distribution in the brain. One of the major technical hurdles in preparing AAV vectors has been the separation of AAV from the helper adenovirus. This obstacle may now have been overcome by supplying the packaging functions for virion assembly in an adeno DNA plasmid. The proteins that make up the icosahedral structure that encapsidates the AAV DNA is made from nonreplicating adeno genomic DNA (Ferrari *et al.*, 1997). Consequently, the production of AAV does not require packaging cell lines because the sequences for *trans* function can be co-transfected into the same cell that replicates the AAV sequences.

The potential for site-specific integration of AAV DNA has been documented more recently targeting chromosome 19 near the location for a replication protein gene (Xiao *et al.*, 1997). In the absence of the *rep* gene products, integration does not preferentially use the site on chromosome 19. Although the mechanism for this observation is not known, it does open the possibility to use AAV DNA in a plasmid form for gene transfer and expression to bypass the use of a virion particle (Ferrari *et al.*, 1997). There is also evidence that AAV vectors do not need to integrate for functional gene expression but instead can exist as a double-stranded, episomal DNA. This observation is consistent with other human viruses such as Epstein Barr and the human polyomavirus, (JCV), which can either integrate or remain episomal depending on the cell target for infection.

AAV seems able to infect different cell types in the CNS, including neurons, glia, and endothelial cells. When AAV has been injected directly into the rat brain, infection and gene expression has occurred in the hippocampus, caudate, olfactory tubercle, and cerebellum. The cell type most frequently transduced is the multipolar neuron (Bankiewicz *et al.*, 1996; Xiao *et al.*, 1997). However, there are differences in levels and duration of expression in different regions of the brain and in different cells. Most probably these differences reflect the use of promoter sequences derived from cytomegalovirus (CMV). Future experimental design must take into account the gene expression system as well as the delivery vehicle in the CNS.

The experimental development of efficient biological delivery vehicles, either viral vectors and their DNA or cell systems, must be coordinated with the choice of transcriptional promoter sequences. The choice of the delivery vehicle for a therapeutic molecule is only part of the challenge. For long-term, effective expression of the transgene much more experimental emphasis needs to be placed on the molecular regulation of the gene of interest. The cell targeting of these potential therapeutic molecules requires not only a way to deliver them to target sites, but also to insure their long-term or conditional expression.

VII. LOCAL DELIVERY OF GENE THERAPY VECTORS INTO THE CNS

Delivery of macromolecular therapeutic agents into the CNS is a challenging endeavor. Nonlipophilic or highly charged systematically administered proteins must be modified in some way to bypass the BBB. For example, some success has come from enhancing the lipid solubility of proteins (Gregoriadis, 1976), increasing brain capillary permeability by chemical modification (Pardridge, 1991), and using transferrin receptor monoclonal antibodies as transport vectors across the BBB (Saltzman *et al.*, 1991). However, there are problems associated with such methods that may reduce the therapeutic efficacy of the agent or result in systemic toxicity.

Direct infusion of proteins into the cerebral ventricles is a delivery method that can result in high protein concentration in tissues near the ventricular system, but penetration into other paranchymal regions is quite limited (Blasberg *et al.*, 1975; Ferguson *et al.*, 1991) and clearance into the peripheral circulation is rapid. Direct *intraparenchymal* injection of proteins into the brain at a target site, either by bulk flow or bolus injection, can achieve high drug concentrations at specific sites (Hargraves and Freed, 1986; Lieberman *et al.*, 1995). Bulk flow (e.g., convection-enhanced delivery) has several advantages over bolus injections. Bulk flow delivery (Fig. 4, see color insert), which results from a pressure gradient, distributes large molecules over an area up to several centimeters in relatively homogeneous tissue concentrations. This method is useful for intrastriatal delivery of large molecules and viral vectors (Bankiewicz *et al.*, 1997c, 1996). Delivery by bolus injection relies on diffusion alone, limiting the distance of drug distribution in the brain and producing high drug concentrations at the site of delivery with potentially toxic effects. When viral vectors are injected into the CNS an overproduction of gene product in a limited area may result in local toxicity, which can give rise to unwanted side effects. In addition, because of the limited tissue distribution, injections at several brain sites are typically required in order to cover the target region. By contrast, when an AAV-TH vector was infused into the striatum of monkeys using the bulk flow method, the number of TH-immunoreactive cells and the volume of gene transfer was three to five-fold greater than when the same volume was administered by the bolus injection technique (Bankiewicz *et al.*, 1997d). Bulk flow delivery allows for an even distribution of viral vectors and a uniform area of gene transfer.

VIII. REGULATION OF DA SYNTHESIS AND RELEASE

The regulation of gene expression is important for achieving therapeutic levels of DA in models of Parkinson's disease. This is true for both DA replacement and neuroprotection–neuroregeneration strategies. For example, the administration of TH or AADC, either by direct *in vivo* gene delivery or by *ex vivo* gene therapy, does not assure that therapeutic levels of DA will be achieved. The synthesis of DA from tyrosine requires TH, the co-factor tetrahydrobiopterin (BH_4) and AADC (see Fig. 2). Appropriate levels of each component of the biosynthetic pathway must be maintained for the production of DA. One way of achieving this is by the systemic administration of a pro-drug. For example, the systemic administration of the co-factor BH_4 may augment DA synthesis following the introduction of TH-expressing genes. The ability to regulate gene expression, and thus dopaminergic activity, is important in order to maintain therapeutic levels of DA.

IX. *IN VIVO* DETECTION OF GENE EXPRESSION

The assessment of gene transfer and expression is difficult *in vivo*. An *in vivo* method of detecting gene expression, other than behavioral measures, would be important for evaluating the efficacy of the procedure and for assessing DA production in order to regulate DA levels. Although DA metabolites can be measured in plasma, CSF, and urine, these are not good indicators of gene expression. Neurochemical imaging, using positron emission tomography (PET), provides an *in vivo* method for monitoring the efficacy of gene transfer techniques for increasing DA production. PET enables longitudinal studies in order to evaluate the long-term therapeutic efficacy of gene therapy, and provides physiological measures that can be correlated with behavioral measures. In addition, unlike postmortem studies, PET studies enable real-time measures of gene expression. Figure 5 (see color insert) shows PET images for a hemiparkinsonian monkey before and after infusion with an AAV vector encoding the AADC gene. The PET images show increased striatal uptake of the AADC tracer, 6-[^{18}F]fluoro-L-*m*-tyrosine (FMT) following AAV infusion, indicating increased AADC activity. PET should be useful for preclinical studies in nonhuman primates, as well as for humans during clinical trials in order to monitor changes in DA function over time.

X. CONCLUSION

Since the introduction of L-dopa in the late 1960's, the treatment of PD has been largely symptomatic (Cotzias *et al.*, 1967). Although symptomatic treatments are fairly effective early in the disease, they lose their effectiveness as the disease progresses and are accompanied by adverse side effects. Pharmacological treatments have improved, but side effects continue to be a problem and although an effort has been made to develop new drugs aimed at neuroprotection, these treatments have met with limited success (Parkinson, 1989, 1996a, 1996b). The goal of gene therapy is to introduce genetic material into the brain to increase the production of DA, or to protect and/or regenerate DA neurons. Although gene therapy shows great promise for the treatment of PD, preclinical work in animal models is needed in order to address critical issues such as the method and site of gene delivery, length of gene expression, and therapeutic efficacy of this approach before human gene therapy can be performed safely and responsibly.

REFERENCES

Acsadi, G., Jiao, S. S., Jani, A., Duke, D., Williams, P., Chong, W., and Wolff, J. A. (1991). Direct gene transfer and expression into rat heart *in vivo*. *New Biol* 3:71–81.

Agid, Y., Javoy-Agid, F., and Ruberg, M. (1987). Biochemistry of neurotransmitters in Parkinson's disease. In: Marsden, C. D., and Fahn, S., ed.), *Movement disorders 2*. Stoneham, Mass.: Butterworth and Company Publishers, pp. 166–230.

Bankiewicz, K. S. (1991). MPTP-induced Parkinsonism in non-human primates. *Methods Neurosci* 7:168–182.

Bankiewicz, K. S., Nagy, D., Conway, J., Emborg, M., McLaughlin, W., Leff, S., and Forno, L. (1995). Histological evaluation of graft survival and gene expression in parkinsonian non-human primates at 4 months after striatal grafting of fibroblasts genetically modified with tyrosine hydroxylase cDNA. *Soc Neurosci Abstr* 21:316.

Bankiewicz, K. S., Emborg, M. E., McLaughlin, W. W., Pivirotto, P. J., Bringas, J. R., Hundal, R. S., Yang, B. Y., Nagy, D., Snyder, R., Zhou, S. Z., Eberling, J. L., and Jagust, W. J. (1997a). Utilization of a non-human model of Parkinson's disease for therapeutic applications of gene therapy. *Exp Neurol* 145:S15.

Bankiewicz, K. S., Hundal, R. S., Pivirotto, P. J., McLaughlin, W. W., Bringas, J. R., Yang, B. Y., Emborg, M. E., Wu, F. F., and Irwin, I. (1997b). Behavioral and biochemical changes in MPTP-treated monkeys after L-dopa administration. *Exp Neurol* 145:S14.

Bankiewicz, K. S., Leff, S. E., Nagy, D., Jungles, S., Rokovich, J., Spratt, K., Cohen, L., Libonati, M., Snyder, R. O., and Mandel, R. J. (1997c). Practical aspects of the development of *ex vivo* and *in vivo* gene therapy for Parkinson's disease. *Exp Neurol* 144:147–156.

Bankiewicz, K. S., Nagy, D., Bringas, J. R., Conway, J., Morton, M., Pivirotto, P., McLaughlin, W. W., Hundal, R., Leff, S., Kaspar, B., and Snyder, R. O. (1997d). AAV-mediated gene expression in the striatum of hemiparkinsonian monkeys. Abstract, 5th International Neurotransplantation Meeting, San Diego.

Bankiewicz, K. S., Oldield, E. H., Chiueh, C. C., Doppman, D. M., Jacobowitz, D. M., and Kopin, I. J. (1986). Hemiparkinsonism in monkeys after unilateral internal carotid artery infusion of 1-methyl-4-phenyl-1,2,3,6-tetrahydropyridine (MPTP). *Life Sci* 39:7–16.

Bankiewicz, K. S., Snyder, R., Zhou, S. Z., Morton, M., Conway, J., and Nagy, D. (1996). Adeno-associated (AAV) viral vector-mediated gene delivery in non-human primates. *Soc Neurosci Abstr* 22:768.

Beck, K. D., Knusel, B., and Hefti, F. (1993). The nature of the trophic action of brain-derived neurotrophic factor, des(1-3)-insulin-like growth factor-1, and basic fibroblast growth factor on mesencephalic dopaminergic neurons developing in culture. *Neuroscience* 52:855–866.

Beck, K. D., Valverde, J., Alexi, T., Poulsen, K., Moffat, B., Vandlen, R. A., Rosenthal, A., and Hefti, F. (1995). Mesencephalic dopaminergic neurons protected by GDNF from axotomy-induced degeneration in the adult brain. *Nature* 373:339–341.

Bernheimer, H., Birkmayer, W., Hornykiewicz, O., Jellinger, K., and Seitelberger, F. (1973). Brain dopamine and the syndromes of Parkinson and Huntington. *J Neurol Sci* 20:415–455.

Blasberg, R., Patlak, C., and Fenstermacher, J. (1975). In chemotherapy: Brain tissue profiles after ventriculociste fusion. *J Pharmacol Exp Ther* 2:73–83.

Bowenkamp, K. E., David, D., Lapchak, P. L., Henry, M. A., Granholm, A.-C., Hoffer, B. J., and Mahalik, T. J. (1996). 6-hydroxydopamine induces the loss of the dopaminergic phenotype in substantia nigra neurons of the rat: A possible mechanism for restoration of the nigrostriatal circuit mediated by glial cell line-derived neurotrophic factor. *Exp Brain Res* 111:1–7.

Bowenkamp, K. E., Hoffman, A. F., Gerhardt, G. A., Henry, M. A., Biddle, P. T., Hoffer, B. J., and Granholm, A.-C. E. (1995). Glial cell line-derived neurotrophic factor supports survival of injured midbrain dopaminergic neurons. *J Comp Neurol* 355:479–489.

Bringas, J. R., Kutzscher, E. M., Yang, B. M., Hundal, R. S., Nagy, D., Mclaughlin, W. W., Leff, S., Emborg, M. E., Jungels, S., and Bankiewicz, K. S. (1996). Implantation parameters for reliable delivery of cells into the striatum of monkeys using an automated delivery system. *Soc Neurosci Abst* 22:319.

Burns, R. S., Chiueh, C. C., Markey, S. P., Ebert, M. H., Jacobowitz, D. M., and Kopin, I. J. (1983). A primate model of parkinsonism: Selective destruction of dopaminergic neurons in the pars compacta of the substantia nigra by N-methyl-4-phenyl-1,2,3,6-tetrahydropyridine. *Proc Natl Acad Sci USA* 80:4546–4550.

Caillaud, C., Akli, S., Vigne, E., Koulakoff, A., Perricaudet, M., Poenaru, L., Kahn, A., and Berwaldnetter, Y. (1993). Adenoviral vector as a gene delivery system into cultured rat neuronal and glial cells. *Eur J Neurosci* 5:1287–1291.

Chase, T. N., Juncos, J., Serrati, C., Fabbrini, G., and Bruno, G. (1987). Fluctuation in response to chronic levodopa therapy: Pathogenetic and therapeutic considerations. *Adv Neurol* 45:477–480.

Cotzias, G. C., van Woert, M. H., and Schiffer, L. M. (1967). Aromatic amino acids and modification of parkinsonism. *N Eng J Med* 276:374–379.

Davis, G. C., Williams, A. C., Markey, S. P., Ebert, M. H., Caine, E. D., Reichert, C. M., and Kopin, I. J. (1979). Chronic parkinsonism secondary to intravenous injection of meperidine analogues. *Psychiat Res* 1:249–254.

Elsworth, J. D., Deutch, A. Y., Redmond, D. E., Jr., Sladek, J. R., Jr., and Roth, R. H. (1987). Effects of 1-methyl-4-phenyl-1,2,3,6-tetrahydropyridine (MPTP) on catecholamines and metabolites in primate brain and CSF. *Brain Res* 415:293–299.

Federoff, H. J., Brooks, A., Muhkerjee, B., and Corden, T. (1997). Somatic gene transfer approaches to manipulage neural networks. *J Neurosci Methods* 71:133–142.

Federoff, H. J., Geschwind, M. D., Geller, A. I., and Kessler, J. A. (1992). Expression of nerve growth factor *in vivo* from a defective herpes simplex virus 1 vector prevents effects of axotomy on sympathetic ganglia. *Proc Natl Acad Sci U S A* 89:1636–1640.

Ferguson, I. A., Schweitzer, J. B., Bartlett, P. F., and Johnson, E. M., Jr. (1991). Receptor-mediated retrograde transport in CNS neurons after intraventricular administration of NGF and growth factors. *J Comp Neurol* 313:680–692.

Ferrari, F. K., Xiao, X., McCarty, D., and Samulski, R. J. (1997). New developments in the generation of ad-free, high-titer rAAV gene therapy vectors. *Nature Med* 3:1295–1297.

Gash, D. M., Zhang, Z., Ovadia, A., Cass, W. A., Yi, A., Simmerman, L., Russell, D., Martin, D., Lapchak, P. A., Collins, F. C., Hoffer, B. J., and Gerhardt, G. A. (1996). Functional recovery in GDNF-treated parkinsonian monkeys. *Nature* 380:252–255.

Geller, A. I., and Breakefield, X. O. (1988). A defective HSV-1 vector expresses Escherichia coli beta-galactosidase in cultured peripheral neurons. *Science* 241:1667–1669.

Geller, A. I., and Freese, A. (1990). Transfection of cultured central nervous system neurons with defective HSV-1 vector which expresses b-galactosidase. *Proc Natl Acad Sci USA* 87:1149–1153.

Gregoriadis, G. (1976). The carrier potential of liposomes in biology and medicine. *N Engl J Med* 295:704–710.

Hargraves, R., and Freed, W. J. (1986). Chronic intrastriatal dopamine infusions in rats with unilateral lesions of the substantia nigra. *Life Sci* 40:959–966.

Henderson, B. T., Clough, C. G., Hughes, R. C., Hitchcock, E. R., and Kenny, B. G. (1991). Implantation of human ventral mesencephalon to the right caudate nucleus in advance Parkinson's disease. *Arch Neurol* 48:822–827.

Hoffer, B. J., Hoffman, A., Bowenkamp, K., Huettl, P., Hudson, J. D. M., Lin, L. F., and Gerhardt, G. A. (1994). Glial cell line-derived neurotrophic factor reverses toxin-induced injury to midbrain dopaminergic neurons *in vivo*. *Neurosci Lett* 182:107–111.

Horger, B., Nishimura, M., Armanini, M., Wang, L.-C., Poulsen, K., Rosenblad, C., Kirik, D., Moffat, B., Simmons, L., Johnson, E., Milbrandt, J., Rosenthal, A., Bjorklund, A., Vandlen, R., Hynes, M., and Phillips, H. (1998). Neurturin exerts potent actions on survival and function of midbrain dopaminergic neurons. *J Neurosci* 18:4929–4937.

Hudson, J., Granholm, A. C., Gerhardt, G. A., Henry, M. A., Hoffman, A., Biddle, P., Leela, N. S., Mackerlova, L., Lile, J. D., Collins, F., and Hoffer, B. J. (1995). Glial cell line-derived neurotrophic factor augments midbrain dopaminergic circuits *in vivo*. *Brain Res Bull* 36:425–432.

Hyman, C., Hofer, M., Barde, Y. A., Juhasz, M., Yancopoulos, G. D., Squinto, S. P., and Lindsay, R. M. (1991). BDNF is a neurotrophic factor for dopaminergic neurons of the substantia nigra. *Nature* 350:230–232.

Jiao, S., Williams, P., Berg, R. K., Hodgeman, B. A., Liu, L., Repetto, G., and Wolff, J. A. (1992). Direct gene transfer into nonhuman primate myofibers *in vivo*. *Hum Gene Ther* 3:21–33.

Kearns, C. M., and Gash, D. M. (1995). GDNF protects nigral dopamine neurons against 6-hydroxydopamine *in vivo*. *Brain Res* 672:104–111.

Kebabian, J. W., and Calne, D. B. (1979). Multiple receptors for dopamine. *Nature* 277:93–96.

Langston, J. W., Ballard, P. A., Tetrud, J. W., and Irwin, I. (1983). Chronic parkinsonism in humans due to a product of meperidine-analog synthesis. *Science* 219:979–980.

Le Gal La Salle, G., Robert, J. J., Berrard, S., Ridoux, V., Stratford-Perricaudet, L. D., Perricaudet, M., and Mallet, J. (1993). An adenovirus vector for gene transfer into neurons and glia in the brain. *Science* 259:988–990.

Lee, T., Seeman, P., Rajput, A., Farley, I. J., and Hornykiewicz, O. (1978). Receptor basis for dopaminergic supersensitivity in Parkinson's disease. *Nature* 273:59–61.

Leff, S. E., Rendahl, K. G., Spratt, S. K., Kang, U. J., and Mandel, R. J. (In press). *In vivo* L-dopa production by genetically modified primary rat fibroblast or 9L gliosarcoma cell grafts via co-expression of GTP cyclohydrolase I with tyrosine hydroxylase. *Exp Neurol.*

Levine, R. A., Miller, L. P., and Lovenberg, W. (1981). Tetrahydrobiopterin in striatum: Localization in dopamine nerve terminals and role in catecholamine synthesis. *Science* 214:919–921.

Lieberman, D. M., Laske, D. W., Morrison, P. R., Bankiewicz, K. S., and Oldfield, E. H. (1995). Convection-enhanced distribution of large molecules in gray matter during interstitial drug infusion. *J Neurosurg* 82:1021–1029.

Lin, L. F., Doherty, D. H., Lile, J. D., Bektesh, S., and Collins, F. (1993). GDNF: A glial cell line-derived neurotrophic factor for midbrain dopaminergic neurons. *Science* 260:1130–1132.

Lin, Q., Cunningham, L. A., Epstein, L. G., Pechan, P. A., Short, M. P., Fleet, C., and Bohn, M. C. (1997). Human fetal astrocytes as an *ex vivo* gene therapy vehicle for delivering biologically active nerve growth factor. *Human Gene Therapy* 8:331–339.

Lu, B., and Federoff, H. J. (1995). Herpes simplex virus type 1 amplicon vectors with glucocorticoid-inducible gene expression. *Human Gene Therapy* 6:419–428.

Major, E. O., Miller, A. E., Mourrain, P., Traub, R. G., de Widt, E., and Sever, J. (1985). Establishment of a line of human fetal glial cells which supports JC virus multiplication. *Proc Natl Acad Sci* 82:1257–1261.

Milbrandt, J., deSauvage, F. J., Fahrner, T. J., Baloh, R. H., *et al.* (1998). Persephin, a novel neurotrophic factor related to GDNF and neurturin. *Neuron* 20:245–253.

Muzyczka, N. (1992). Use of adeno-associated virus as a general transduction vector for mammalian cells. *Curr Top Microbiol Immunol* 158:97–129.

Naldini, L., Blomer, U., Gallay, P., Ory, D., Mulligan, R., Gage, F., Verma, I., and Tronto, D. (1996). *In vivo* gene delivery and stable transduction of nondividing cells by lentiviral vector. *Science* 727:263–267.

Pardridge, W. M. (1991). *Peptide drug delivery to the brain*. New York: Raven Press.

Parkinson Study Group. (1989). Effects of tocopherol and deprenyl treatment on the progression of disability in early Parkinson's disease. *N Eng J Med* 321:574–578.

Parkinson Study Group. (1996a). Impact of deprenyl and tocopherol treatment on Parkinson's disease in DATATOP subjects not requiring levodopa. *Ann Neurol* 39:29–36.

Parkinson Study Group. (1996b). Impact of deprenyl and tocopherol treatment on Parkinson's disease in DATATOP patients requiring levodopa. *Ann Neurol* 39:37–45.

Sah, D. W. Y., Ray, J., and Gage, F. H. (1997). Bipotent progenitor cell lines from the human CNS. *Nature Biochem* 15:574–580.

Saltzman, W. M., Radomsky, M. L., Whaley, K. J., and Cone, R. A. (1991). Drugs released from polymers: Diffusion and elimination in brain tissue. *Chem Eng Sci* 46:2429–2444.

Samulski, R. J., Chang, L. S., and Shenk, T. (1989). Helper-free stocks of recombinant adeno-associated viruses: Normal integration does not require viral gene expression. *J Virol* 63:3822–3828.

Sauer, H., Rosenblad, C., and Bjorklund, A. (1995). Glial cell line-derived neurotrophic factor but not transforming growth factor beta 3 prevents delayed degeneration of nigral dopaminergic neurons following striatal 6-hydroxydopamine lesion. *Proc Natl Acad Sci USA* 92:8935–8939.

Schinstine, M., Jasodhara, R., and Gage, F. H. (1997). Potential effects of cytokines on transgene expression in primary fibroblasts implanted into rat brain. *Mol Brain Res* 47:195–201.

Spina, M. B., Squinto, S. P., Miller, J., Lindsay, R. M., and Hyman, C. (1992). Brain-derived neurotrophic factor protects dopamine neurons against 6-hydroxydopamine and N-methyl-4-phenylpyridinium ion toxicity: Involvement of the glutathione system. *J Neurochem* 59:99–106.

Tomac, A., Lindqvist, E., Lin, L.-F. H., Ogren, S. O., Young, D., Hoffer, B. J., and Olson, L. (1995). Protection and repair of the nigrostriatal dopaminergic system by GDNF *in vivo*. *Nature* 373:335–339.

Tornatore, C., Baker-Carins, B., Hamilton, R., Meyers, K., Yadid, G., Cheng, S., Tanner, V., Cummins, A., Atwood, W., and Major, E. O. (1996). Expression of tyrosine hydroxylase in the human fetal glial cell line SVG: *In vitro* characterization and transplantation into the rodent striatum. *Cell Transpl* 5:145–163.

Tornatore, C., Bankiewicz, K., Lieberman, D., and Major, E. O. (1993). Implantiation and survival of a human fetal brain cerived cell line in the basal ganglia of the non-human primate, rhesus monkey. *J Cell Biochem* 17E:227.

Tseng, J. L., Baetge, E., Zurn, A. D., and Aebischer, P. (1997). GDNF reduces drug-induced rotational behavior after medial forebrain bundle transection by a mechanism not involving striatal dopamine. *J Neurosci* 17:325–333.

Tuszynski, M. H., and Gage, F. H. (1996). Somatic gene therapy for nervous system disease. In: *Growth factors as drugs for neurological and sensory disorders*. Chichester (Ciba Foundation Symposium 196), Wiley, pp. 85–97.

Winkler, C., Sauer, H., Lee, C. S., and Bjorklund, A. (1996). Short-term GDNF treatment provides long-term rescue of lesioned nigral dopaminergic neurons in a rat model of Parkinson's disease. *J Neurosci* 16:7206–7215.

Wolff, J. A., Malone, R. W., Williams, P., Chong, W., Acsadi, G., Jani, A., and Felgner, P. L. (1990). Direct gene transfer into mouse muscle *in vivo*. *Science* 247:1465–1468.

Xiao, X., Li, J., McCown, T. J., and Samulski, R. J. (1997). Gene transfer by adeno-associated virus vectors into the central nervous system. *Exp Neurol* 144:113–124.

Yuen, E. C., and Mobley, W. C. (1996). Therapeutic potential of neurotrophic factors for neurological disorders. *Ann Neurol* 40:346–354.

Zufferey, R., Nagy, D., Mandel, R. J., Naldini, L., and Trono, D. (1997). Multiply attenuated lentiviral vector achieves efficient gene delivery *in vivo*. *Nature Biotech* 15:871–875.

CHAPTER 17

Fetal Neural Grafts for the Treatment of Huntington's Disease

Anne-Catherine Bachoud-Lévi,*,† Pierre Brugières,**
Jean-Paul Nguyen,* Bassam Haddad,†† Hamed Chaker,***
Philippe Hantraye,††† Pierre Cesaro,*,† and Marc Peschanski*

**INSERM U421, IM3, Faculté de Médecine, †Service de Neurologie, Hôpital Henri-Mondor, **Service de Neuroradiologie, Hôpital Henri-Mondor, ††Centre Hospitalier Intercommunal de Crétil, ***Hôpital Esquirol, †††CEA-CNRS URA 2210, Service Hospitalier Frédéric Joliot, SHFJ, DRM, DSV*

Fetal neural grafts of striatal tissue are currently under study for the treatment of Huntington's disease (HD) in at least three different groups, including ours.

The validity of this therapeutic approach, from a general viewpoint, is based upon three main types of data. First, there is now wide agreement on the fact that HD is primarily characterized by selective vulnerability of a specific neuronal population ("medium-size spiny" neurons) in the striatum. Second, a very large set of experimental data, obtained in rats and nonhuman primates, has demonstrated that transplanted fetal striatal cells are able to substitute, anatomically and functionally, for previously injured host striatal neurons. Third, clinical neural transplantation has proven safe and effective, in the hands of skilled researchers, for the treatment of patients with another neurodegenerative disease that affects the basal ganglia, namely Parkinson's disease.

Setting up of the clinical program for HD in Créteil has required two specific steps to be taken: (1) performing a number of complementary experiments within the framework of an explicit "pre-clinical protocol" aimed at answering questions directly related to the clinical application, and (2) creating clinical scales for the follow-up of the patients, because these scales were essentially missing.

In the absence of any validated clinical results in grafted HD patients at this time, this chapter has been devoted to the presentation of the experimental bases of our clinical trials with human fetal neural grafts, and to the design of our clinical trials.

CNS Regeneration

Huntington's disease (HD) is a genetic, autosomal dominant disease with complete penetrance that specifically affects the brain. It inexorably provokes a profound cognitive impairment, motor abnormalities, and severe psychiatric disorders. The disease can start clinically at any time in life but begins, in general, during the fourth decade. This means that it affects young adults who are working and, in many cases, have young children. They rapidly lose their autonomy and need constant care, first by the family then in a hospital. They die after 10 to 20 years (17 years as a mean) after the initiation of symptoms, showing profound dementia and abnormal motor behavior (see Bird and Coyle, 1986 for review). The gene responsible for the disease has been localized on the short arm of chromosome 4 (Gusella *et al.*, 1983) and the molecular defect has recently been identified (Huntington's disease coll. res. group, 1993) as an abnormal repeat of CAG triplets in a gene (IT15) that encodes a protein (Huntingtin) with no known function. An intense search for the cell pathology attached to this molecular defect is currently ongoing, but these efforts have not yet been successful (see Sharp and Ross, 1996 for review).

Clinically identified HD is not frequent because its prevalence, for clinically diagnosed cases, amounts to 1 in 10,000 in most western European countries. Gene carriers who will, unavoidably, develop the disease but do not yet present clinical symptoms are 2 in 10,000 because clinical signs appear, in average, around age 35 and progress for an average of 17 years before death, during which patients clinically affected are counted. The total prevalence of HD is, therefore, around 3 in 10,000 in Western countries, which makes HD more common than most other genetic affectations.

The toll of the disease is, furthermore, much heavier than what these numbers tell, because the disease strongly affects the patient's family as well as the patients themselves. Most probably as a consequence of the repeated loss of one major adult member of the families at each generation, the socioeconomical status of patients, and their family, is often low. There is, in addition, a very high incidence of secondary effects of the disease such as suicides, which are much more frequent among patients in the early years of the disease than in the rest of the population. Spouses are prone to psychological and even psychiatric disorders, among which anxiety and depression can be of particular importance, and require specific medical care. The existence of genetic testing has added a further dimension to this tragic picture by creating three types of new potential sources of psychological disorders. The potential gene carriers are split into three categories: (1) the "condemned" ones who are among the only human beings who know the terrible conditions in which they are going to die. These individuals are aware of their limited life expectancy, and indeed almost know the date of their death. They have to live with this knowledge. (2) "Escaped" patients are continuously under the pressure of the pathological environment created by the disease they have miraculously avoided. Like people escaping a plane crash or another catastrophe, they may develop guilt and even pathological depression. (3) "Refuzniks," the large majority of patients, do not want to know their risk of developing the disease and do not take the test. They cannot ignore, nevertheless, that they may give a deadly disease to their children. Further, they suffer the guilt provoked from imposing on their relatives the risk of a brutal familial disruption.

In the absence of a precise knowledge of the *cause* of the disease, and of the function of the Huntingtin protein, one has to rely on general features of HD to try to design therapeutic strategies. HD is mainly characterized by a loss of neurons in a localized brain area, the striatum, and this loss is progressive over several years, with little, if any, "threshold effect" (i.e., the clinical signs appear as soon as a small portion of the neuronal population degenerates or becomes dysfunctional). These two basic features allow two types of therapeutic strategies to be designed that are, in fact, complementing each other. The first type is a "substitution therapy," in which lost neurons are replaced. In the case of HD, the only tool for such a substitution is the transplantation of fetal striatal neurons for the goal of rewiring the neuronal pathways disrupted by the loss of the host neurons (and not, as in Parkinson's disease, a replenishment of the area with a specific neurotransmitter). The present chapter is devoted to this type of potential therapy. The second type is a "protective therapy" that aims at preserving the patient's neurons that have not yet died. This can be achieved, theoretically, by a large number of means that will be analyzed in Chapter 18. Although the presentations of the two types of strategies are different, one should keep in mind that they could, eventually, be used concomitantly, because neuronal substitution is needed to regain lost functions while protection has to be promoted to avoid a continuous loss of the host neurons.

I. EXPERIMENTAL BASES OF FETAL STRIATAL NEURAL GRAFTS IN HD PATIENTS

Fetal neural grafting for the substitution of striatal neurons in patients with HD is essentially based upon three different experimental data sets. First, neuropathological, clinical, and imaging results indicate that HD is a localized vulnerability of a specific neuronal population ("medium-spiny" neurons) in the striatum; therefore, the substitution of striatal neurons may be sufficient to restore major functions in HD patients. Second, transplanted fetal striatal cells have been demonstrated to integrate into the host brain and to substitute anatomically and functionally for previously injured host striatal neurons in rats and nonhuman primates. Third, clinical neural transplantation has proven safe and effective for restoring lost functions in patients with another type of neurodegenerative disease affecting the basal ganglia, Parkinson's disease.

A. HD Affects Primarily the Striatum

A large body of pathological observations strongly suggest that HD is primarily a striatal disease. A long-lasting debate exists, however, over the specificity of this major striatal lesion in HD patients because other degenerating areas have also been recognized. In the striatum, both caudate nucleus and putamen are severely atrophied and up to 60% of the striatum can be lost in common forms (Lange *et al.*, 1976). The Golgi type II (medium-size spiny) neurons, that is, the neurons that project outside of the striatum, are the most directly affected. The neurodegeneration evolves along a caudal to rostral (Dunlap, 1927; McCaughey, 1961) and a dorsomedial to ventrolateral gradient (Vonsattel *et al.*, 1985). It is accompanied by an astrogliosis (Bruyn, 1968; Lange *et al.*, 1976; Myers *et al.*, 1991). Defined stages of striatal atrophy and clinical evolution are well correlated (Vonsattel *et al.*, 1985). One interesting feature of the extrastriatal territories that are affected is their synaptic and/or functional link with the caudate nucleus and putamen. Thus, the *globus pallidus* loses up to 40% of neurons, the atrophy being even more massive because of the additive loss of striatopallidal axons. A similar picture is seen in the *substantia nigra (pars reticulata),* in the subthalamic nucleus (Waters *et al.*, 1988; Oyanagi *et al.*, 1989), and in some thalamic regions (Dom *et al.*, 1976). Upstream from the striatum, neuronal degeneration is observed in the cerebral cortex (De la Monte *et al.*, 1988). This loss is not diffuse, but rather is localized in layers III, V, and VI (Bruyn, 1968; Cudkowicz and Kowall, 1990; Sotrel *et al.*, 1991), which contain corticostriatal neurons (Oka, 1980). All these data suggest that extrastriatal cell losses may be a consequence of striatal degeneration, with secondary degeneration of neurons upstream and downstream from the primary area of lesion.

The extent of cell loss has been addressed experimentally in animals. Target neurons indeed undergo transsynaptic degeneration after lesions of a main afferent system (see Eidelberg *et al.*, 1989). Afferent neurons to an area of neurodegeneration also undergo complex anatomo-functional alterations: first, shrinkage of terminal axonal branches and synaptic boutons, then, several months later, somato-dendritic atrophy, and finally, neuronal degeneration (Peschanski and Besson, 1987; Marty and Peschanski, 1994, 1995; Marty *et al.*, 1994; Emerich *et al.*, 1997).

Massive striatal lesions in HD may, therefore, be the primary cause of neuronal degeneration in extra-striatal areas, although this cannot be definitely ascertained at present. This result is of critical importance for a neural transplantation approach, because one can only hope to substitute for a limited area of the host brain, and thus to provide, indirectly, ways for other neurons to survive.

Functional imaging studies have, further, demonstrated that neuronal dysfunction in the striatum can be seen before clinical symptoms occur and before striatal atrophy can be detected by CT scan (Kuhl *et al.*, 1982; Hayden *et al.*, 1986; Mazziotta *et al.*, 1987; Grafton *et al.*, 1990). A progressive bilateral hypometabolism of the caudate nuclei (37 to 91% decrease) parallels the evolution of the disease (Kuhl *et al.*, 1982; Hayden *et al.*, 1986; Leenders *et al.*, 1986; Mazziotta *et al.*, 1987; Berent *et al.*, 1988; Reid *et al.*, 1988; Brooks and Frackowiak, 1989; Kuwert *et al.*, 1990; Otsuka *et al.*, 1993). The relationship between striatal hypometabolism and clinical symptoms, not only extrapyramidal symptoms and hyperkinesia, but also higher brain functions, has been established (Berent *et al.*, 1988; Weinberger *et al.*, 1988) pointing also to a striatal origin of the cognitive deficits in HD.

Comparative neuropsychological testing of patients with HD, Parkinson's disease with dementia, or Alzheimer's disease also favors a link between the subcortical lesion and dementia in HD (see Podoll *et al.*, 1988 for review, Beatty *et al.*, 1988; Brandt *et al.*, 1988; Granholm and Butters, 1988; Heindel *et al.*, 1989, 1990; Illes, 1989; Salmon *et al.*, 1989; Tröster *et al.*, 1989; Hodges *et al.*, 1990, 1991; Pillon *et al.*, 1991; Randolph, 1991). It is proposed that a progressive lesion of the striatum may "disconnect" the prefrontal cortex and interrupt the flow of information emerging from the cortex, thus mimicking a frontal syndrome in the absence of a primary lesion of the frontal cortex, at least initially.

B. Fetal Neural Grafts in Animal Models of HD

The use of intrastriatal transplantation of fetal striatal tissue for HD is based upon a very large number of studies carried out over the past 12 years in animal models of HD. The issues raised by clinical transplantation in patients with Huntington's disease are partly different from those raised by neural transplantation in Parkinson's disease. For Parkinson's disease, the main goal to be achieved is the release of a neuroactive substance (dopamine) in a localized region of the brain (striatum). Fetal dopaminergic neurons have "only" to reinnervate the host striatum (see discussion in Björklund *et al.*, 1987). With HD, the problems to be solved are somewhat different. Disappearance of striatal neurons interrupts the flow of information originating upstream of the striatum in the cerebral cortex, and circulating toward striatal target zones such as the pallidum and the pars reticulata of the substantia nigra. To be fully functional, the flow must be reconstructed by the transplanted fetal neurons. For this aim to be reached, specific graft-to-host reinnervation must occur, but so too must host-to-graft axonal regeneration. Results obtained in experimental studies show that this is possible, at least partially.

The maturation of grafted medium-size spiny neurons that form the vast majority of the striatal neuronal population is similar to that observed *in situ* (McAllister *et al.*, 1985; Clarke *et al.*, 1988; Helm *et al.*, 1990; Labandeira-Garcia *et al.*, 1991; Defontaines *et al.*, 1992). Fetal striatal grafts express various neuropeptides such as substance P, metenkephalin, somatostatin, neuropeptide Y and neurotransmitters such as γ-aminobutyric acid or acetylcholine (Isacson *et al.*, 1985, 1987; Walker *et al.*, 1987; Roberts and DiFiglia, 1988; Graybiel *et al.*, 1989; Wictorin *et al.*, 1989a; Zhou and Buchwald, 1989; Clarke and Dunnett, 1990; Giordano *et al.*, 1990; Liu *et al.*, 1990; Sirinathsinghji *et al.*, 1990). D1 and D2 receptors, as well as muscarinic receptors, are also present in striatal grafts (Isacson *et al.*, 1987; Deckel *et al.*, 1988a, 1988b; Graybiel *et al.*, 1989; Liu *et al.*, 1990; Mayer *et al.*, 1990; Helm *et al.*, 1991). Administration of dopamine agonists in rats bearing striatal transplants induces cell responses comparable to those observed in the normal striatum (Sirinathsinghji *et al.*, 1988; Dragunow *et al.*, 1990; Liu *et al.*, 1991).

Regeneration of host afferent axons into the graft has been a major issue of experimental studies of transplants for HD. Afferent axons have been shown to persist for a long period of time in an area of neurodegeneration (Peschanski and Besson, 1987; Nothias *et al.*, 1988; Marty *et al.*, 1994) and axonal regeneration has been demonstrated after grafting, though to different degrees. Monoaminergic systems, and in particular the massive dopaminergic innervation, regenerate rapidly and penetrate deeply into the transplants (Pritzel *et al.*, 1986; Clarke *et al.*, 1988; Wictorin *et al.*, 1988, 1989b; Defontaines *et al.*, 1992), whereas projections originating from the cortex or the thalamus only partially regenerate (about one-third of cortical and thalamic neurons grow axons into the graft (Pritzel *et al.*, 1986; Wictorin *et al.*, 1988). Reinnervation is most dense in the peripheral areas of the transplants (Wictorin *et al.*, 1988; Wictorin and Björklund, 1989). Regenerated connections are functional, as demonstrated *in vivo* by stimulating cortical afferent fibers and *in vitro* brain slices (Rutherford *et al.*, 1987).

Graft-host reconnections have also been demonstrated, in particular to the *globus pallidus* (Isacson *et al.*, 1986; Pritzel *et al.*, 1986; Sirinathsinghji *et al.*, 1988; Wictorin and Björklund, 1989; Campbell *et al.*, 1995). Fine structural analysis has confirmed the establishment of appropriate synaptic contacts (Wictorin *et al.*, 1990). In contrast, projections toward more caudal target areas (e.g., the *pars reticulata* of the *substantia nigra*) have not been observed in rat allografts, although they did exist in a case of human xenograft to the rat brain.

Behavioral studies have confirmed the reconstruction of a functional system in animals with fetal grafts. Excitotoxic striatal lesions provoke major motor disorders, and intrastriatal transplantation of fetal striatal neurons (but not other cell types) leads to an improvement in those behaviors. Rat allografts reduce the hyperactivity induced by a bilateral excitotoxic striatal lesion (Deckel *et al.*, 1983; Isacson *et al.*, 1984) and decreases pharmacologically induced rotation (Sanberg *et al.*, 1986, 1987; Dunnett *et al.*, 1988; Norman *et al.*, 1988, 1989). Rat xenografts noticeably reduce the incidence of apomorphine-induced choreiform movements in baboons with unilateral excitotoxic striatal lesions (Hantraye *et al.*, 1992). Positive results for the paw-reaching task (that requires fine motor control) have also been obtained in rat allografts (Dunnett *et al.*, 1988; Montoya *et al.*, 1989; Valouskova *et al.*, 1990).

Higher brain functions are difficult to study in rodents, and up to now only preliminary data in nonhuman primate models of HD are available. In the rat, a modest improvement has been noted in delayed alternation tasks (Deckel *et al.*, 1986; Isacson *et al.*, 1986; Mayer *et al.*, 1991). The development of a model of cognitive testing in the nonhuman primate (Palfi *et al.*, 1996) has allowed us to test potential therapeutic effects of intrastriatal neural grafts. Preliminary results indicate a complete recovery of grafted primates in the object retrieval detour task, whereas lesioned controls do not improve over time (Palfi *et al.*, in press).

C. Clinical Trials of Neural Grafting in Parkinson's Disease Have Proven the Technique to Be Safe and Effective

In the past few years, more than 150 patients with Parkinson's disease have received intrastriatal transplantation of fetal dopaminergic neurons. The expertise gained from these surgeries is useful in terms of designing and performing reliable neural transplantation protocols in HD patients. Intracerebral cell transplantation is a feasible and essentially safe surgical procedure. Transplantation in patients with HD would rely upon the same techniques that have been developed for patients with Parkinson's disease, with the exception of the selection of the fetal tissue to be grafted. Results from fetal grafting studies in patients with Parkinson's disease published by half a dozen groups in the world are, therefore, of major interest (Freed *et al.*, 1990, 1992; Lindvall *et al.*, 1990, 1992, 1994; Henderson *et al.*, 1991; Spencer *et al.*, 1992; Widner *et al.*, 1992; Peschanski *et al.*, 1994; Freeman *et al.*, 1995; Defer *et al.*, 1996). Although details are beyond the scope of this chapter, it is worth mentioning that most clinical results point to a benefit for the patients with Parkinson's disease (see review in Lindvall, 1994; Olanow *et al.*, 1996).

Converging lines of evidence strongly suggest that grafted fetal cells survive and develop in the host brain. There is, in particular, a strong correlation between clinical results and 18F-fluorodopa striatal uptake, as visualized by positron emission tomography (PET, Lindvall *et al.*, 1990, 1992; Sawle *et al.*, 1992; Widner *et al.*, 1992; Rémy *et al.*, 1995; Kordower *et al.*, 1995; Freeman *et al.*, 1995). Direct confirmation of the survival of grafted dopaminergic neurons has recently been obtained from the autopsy of two patients who died, from a non-related cause, 18 months after grafting (Kordower *et al.*, 1995, 1996). Around 100,000 dopaminergic neurons have been evidenced by tyrosine hydroxylase-immunostaining in the transplanted striatum of these patients. These neurons provide a large reinnervation of the implanted striatum, validating the extrapolation previously made from animal experiments. In parallel, one might therefore expect fetal human striatal neurons transplanted into the striatum of HD patients to integrate and colonize the grafted nucleus in a manner comparable to that demonstrated in animal models. The results presented by Kordower and associates have also solved, at least partly, the issue of immunosuppression because the autopsied patients had received immunosuppressive therapy with cyclosporine for 6 months only, and transplanted cells had survived for 12 additional months in the absence of immunosuppression.

II. SETUP FOR NEURAL TRANSPLANTATION IN HD PATIENTS

The actual preparation of the clinical application of neural transplants to patients with HD has required two additional sets of studies aimed at (1) defining the location of the appropriate site in the fetus and the gestational age at which the dissection should be performed, and (2) defining precise clinical and paraclinical tests to evaluate the efficacy of the treatment.

A. Definition of the Tissue to Be Grafted

Until 1993 most studies analyzing rat allografts used striatal primordium taken from the entire ganglionic eminence (Nieuwenhuys, 1977; Smart and Sturrock, 1979; Fentress *et al.*, 1981; Marchand and Lajoie, 1986). This eminence is, however, divided into a medial and a lateral region. Isacson and his colleagues (Pakzaban *et al.*, 1993; Deacon *et al.*, 1994) have provided evidence that cells identified with the usual markers of medium-size spiny neurons are restricted to the lateral region. Cultures from lateral and medial regions confirmed these data (Nakao *et al.*, 1994). Selective transplantation of this lateral region has led to large grafts that contained identified striatal neurons in larger numbers than previously found in rat allografts (so-called "P-zones") (Pakzaban *et al.*, 1993; Deacon *et al.*, 1994; Olsson *et al.*, 1995). The issue has been raised, however, as to whether such a selective dissection might actually be detrimental by excluding interneuronal populations (e.g. cholinergic neurons) that might secondarily migrate into the developing striatum. Replication of many studies that had demonstrated the anatomo-functional integration and benefit of non-selectively dissected grafts will probably be required before one can safely use the selective dissection protocol in HD patients. This is an important issue because a correlation between functional results and the proportion of P-zones in striatal transplants has been reported in the rat (Nakao *et al.*, 1996). Because the amount of tissue grafted cannot be indefinitely expanded, increasing the proportion of striatal (i.e., "useful") neurons in the tissue by carefully selecting the region to be dissected could prove to be important for clinical trials (Brundin *et al.*, 1996).

Analysis of human fetal development using cytological techniques led to a similar localization of the striatal primordium in ganglionic eminences (Johnston, 1923; Cooper, 1946; Hewitt, 1958, 1961). Using DARPP-32 antibodies to selectively identify medium spiny neurons, striatal neurons have been shown to accumulate in the ventral half of the lateral ganglionic eminence in fetuses between 7 and 9 weeks of gestation (Naimi *et al.*, 1996). It is not as easy to define the germinative zone from which striatal neurons are generated, but it is likely that the region is also located in the lateral eminence. Thus it seems reasonable to dissect the human lateral ganglionic eminence based upon the results obtained in rodents by Pakzaban and associates in 1993, including its most ventral part and the periventricular germinal zone. Indeed, intrastriatal transplantation of this region in immunosuppressed rats has led to the survival of striatal neurons (Grasbon-Frodl *et al.*, 1996; Naimi *et al.*, 1996), although the proportion of P-zones in these xenografts may not be as large as in rat allografts.

Fetal age is another major element of the tissue procurement procedure. Neurons are better retrieved just after neurogenesis when the region to be dissected is easier to localize, but the dissection should be done before morphological differentiation is too far advanced (Brundin *et al.*, 1988). For striatal neurons, neurogenesis starts around 7 weeks of gestation (Naimi *et al.*, 1996) and continues beyond the tenth week postconception, the limit for legal abortion in France and several other countries. It is, therefore, possible to use tissues dissected out of 7- to 9.5-week-old fetuses obtained from elective abortions to carry out intracerebral transplantation in patients with HD.

B. Clinical Evaluation

Except for some early pharmacological trials, neural transplantation is one of the first attempts to find a cure for Huntington's disease in which the goal is to restore, not only to preserve, function. This means that few attempts had been made before neural transplantation to analyze specifically the natural history of the disease and to create tools that would be efficient to reveal symptomatic improvements associated with treatment. However, a large number of studies had already demonstrated and quantified alterations in cognitive, motor, or psychiatric function, but longitudinal follow-up was not a major design characteristic in these studies. In the early 1990's, when clinical neural transplantation was envisioned for HD, only one long-term follow-up study, based upon scaling of dependence and ability, had been performed (Shoulson *et al.*, 1989; Myers *et al.*, 1991; Bamford *et al.*, 1995). This study had already indicated, however, a major variability in the evolution rate in HD patients. In a scale going from 100 to 0 by increments of 10, some patients lost only 10 points over 5 years, whereas others could lose up to 40. The symptomatic evolution could also follow either a linear or a steplike course. These results confirmed the view of most clinicians that a major effort had to be made to design more precise clinical scales specifically addressing each aspect of the disease. These scales should also provide indices that could reveal disease progression in addition to the presence and staging of the disease. This, for instance, precluded the use of symptoms like the disappearance of the long-latency reflex that are mostly all-or-none phenomena.

Two different, though partially overlapping, working groups have worked extensively toward these goals: the HSG (Huntington Study Group) in the United States, and the CAPIT-HD (Core Assessment Program for Intracerebral Transplantation in Huntington's Disease) committee formed by European and American experts. The specific aims of these two teams were somewhat different, but complementary. The HSG designed a scale (Huntington Study Group, 1996) following the format of the Unified Parkinson's Disease Rating Scale (called the Unified Huntington's Disease Rating Scale, UHDRS) to "prospectively evaluate all patients with HD and individuals at risk for HD using a single instrument that combined many of the important elements of (previous) scales," i.e., an instrument that was intended to be used by a very large number of clinicians, "for repeated administration during clinical research studies" in a large number of patients. Altogether, the completion of one UHDRS takes about 30 minutes. The CAPIT-HD committee designed a scale (CAPIT-HD committee, 1996) following the format of the CAPIT-PD, the protocol setup to evaluate the effects of neural transplantation in Parkinson's disease (Langston *et al.*, 1992), to address the need for "a thorough and wide-ranging protocol," i.e., an instrument that was intended to be used by a few large teams, comprising many clinicians with different expertise, in a small- or very small- number of patients. Altogether, the completion of one CAPIT-HD session, with full neuropsychological and psychiatric batteries, PET and MRI imaging, neurological examination and, possibly, neurophysiology, takes many hours (in fact,

24 hours in our center). UHDRS is integrated, as one scale among many, in the CAPIT-HD. Depending upon the type of study, and the type of trial, one should, therefore, use one or the other scale. As far as neural transplantation or cell and gene therapy is concerned, the CAPIT-HD is clearly mandatory.

The comprehensive description and analysis of the CAPIT-HD (Table I) is clearly beyond the scope of this chapter and the reader is referred to the original publication as well as to the papers that will present the first results obtained using it. CAPIT-HD is, indeed, an experimental tool that will need refinement and improvement. In contrast to the situation that led to the emergence of the CAPIT (for Parkinson's disease), the work of the committee could not be based on validated longitudinal follow-up scales. Choices have, therefore, been made more on the basis of the experts' evaluation of the envisioned potential of scales and tests rather than of their demonstrated value.

TABLE I Neuropsychological Scales Used in the CAPIT-HD

Tasks	CAPIT-HD
Global cognitive decline	
WAIS shortened form	±
NART	±
MMSE	+
MDRS	±
Attention	
Stroop	+
Trail A	+
Planning	
Trail B	+
Wisconsin	+
Categorical fluency	+
Letter fluency	+
Language	
Syllabic repetition	±
Picture naming	+
Token Test	+
Memory	
Digit span	+
Corsi blocks	+
Digit span backward	+
Lists	+
Rivermead stories	+
Visuo spatial	
VOSP	+

+: mandatory, ± optional

The goal was to design a common instrument and, in the absence of a very strong basis for one or another test, the committee has included as many as seemed both useful and feasible. This does not restrict the value of the scale that has been designed. Quite the opposite, this indicates that the scale should be mandatory for work in cell and gene therapy in HD, because it encompasses as many possibilities to evaluate the patients as seemed possible. In the future, modifications of the scale will proceed naturally, through new meetings of the experts of the field, as a consequence of its use in clinical trials. A recent meeting of the CAPIT-HD committee (London, March 1997) has indeed suggested a number of modifications to increase the practically, scientific utility, and patient acceptability of the CAPIT-HD protocol. These suggestions were based upon the experience gained by different groups, including ours. It has, in particular, been determined that, for neuropsychology, the full battery should be used on a yearly basis.

In addition to the patients, we feel that family need should be very carefully considered by clinical teams involved in neural transplantation (as well as for any other surgical approach of HD). Families with HD are particularly stressed and unstable. This relates not only to the fact that the entire family has to face the continuous deterioration of one of its key adult members, but it also has to deal with secondary psychological problems. The spouse has the main burden of taking responsibility for the whole family, and he or she often feels guilty for the children who may eventually develop the disease. We have experienced the need to offer psychological support to the spouses of the patients involved in our protocols. Children are, as described by all HD clinicians, under a particular stress as soon as they become aware of the risk of developing the disease. Again, this necessitated specific care in our studies. In addition to these issues directly related to HD, the participation in protocols that have potential therapeutic outcomes introduces another potentially damaging stress to these families for several reasons. First, hope can be overwhelming. The Huntington Study Group has recently published a statement detailing the need for a very cautious and comprehensive explanation to the patients and relatives that current protocols are phase I trials to evaluate safety and tolerance rather than efficacy. This is clearly an essential requirement because the patients and relatives will tend to expect beneficial outcomes. Serious consequences, including suicide and depression, may result from insufficient or indiscrete preparation of families involved in these fetal transplantation protocols. Second, families who are involved in our protocols have, over the first years of the disease, tried to reorganize to ensure some stability. By involving the patient and most often the spouse together into a very time-consuming and stressful protocol that includes periods of time in the hospital, surgeries, etc., we are actually destabilizing delicate family balance. Particular care should be taken to minimize this as much as possible. The consequences of what could be called "iatrogenic familial instability" have to be considered, particularly in the way it may affect children.

In conclusion, protocols using neural transplants in an attempt to obtain recovery of functions in HD patients are ongoing. They are based upon a wealth of experimental and clinical data that allow biologists and clinicians to entertain the possibility of a therapeutic effect. There are, however, many uncertainties, including clear differences between the experimental models that have been used and the disease itself. Nevertheless, clinical trials appear to be based upon valid suppositions. It is absolutely indispensable that the patients and relatives know very clearly that these are only therapeutic trials based upon reliable and suggestive preclinical data, and that they are participating in an objectively designed experimental protocol. Very carefully organized trials, on the basis of the specifically designed CAPIT-HD, will be helpful, even if the clinical outcome is not as good as hoped. Badly run trials will be detrimental.

ACKNOWLEDGMENTS

The studies of our group reported in this paper have been supported by grants from INSERM, CNAMTS, Association Huntington de France and Association Française contre les Myopathies. Other members of the Cíeteil/Orsay Huntington Research group for Experimental therapy and Assessment ("COHREA") are: Bartolomeo, P, Boissé, NF, Bourdet, C, Dallabarba, GF, Degos, JD, Ergis, AN, Grandmougin, T, Jémy R, Lefaucheur, JP, Lisovoski, F, Payous E, Rémy, P. The authors wish to thank all the members of the European Network NEST-HD who have continuously allowed us to benefit from their expertise and enlightened discussions. NEST-HD was supported by a grant from the European Community Biomed 2 program. The authors are indebted to Drs. Jeff Kordower and Mark Tuszynski for revising this paper. The skillful secretarial assistance of Mrs. Véronique Ribeil is gratefully acknowledged.

REFERENCES

Bamford, K. A., Caine, E. D., Kido, D. K., Cox, C., and Shoulson, I. (1995). A prospective evaluation of cognitive decline in early Huntington's disease: Functional and radiographic correlates. *Neurology* 45:1867–1873.

Beatty, W. W., Salmon, D. P., Butters, N., Heindel, W. C., and Granholm, E. L. (1988). Retrograde amnesia in patients with Alzheimer's disease or Huntington's disease. *Neurobiology of Aging* 9:181–186.

Berent, S., Giordani, B., Lehtinen, S., Markel, D., Penney, J. B., Buchtel, H. A., Starosta-Rubinstein, S., Hichwa, R., and Young, A. B. (1988). Positron emission tomographic scan investigations of Huntington's disease: Cerebral metabolic correlates of cognitive function. *Brain* 23:541–546.

Bird, E. D., and Coyle, J. T. (1986). Huntington's disease. In: *Clinical neurochemistry*. London: Academic Press Inc., pp. 1–57.

Björklund, A., Lindvall, O., Isacson, O., Brundin, P., Strecker, R. E., and Dunnett, S. B. (1987). Mechanisms of action of intracerebral neural implants. *TINS* 10:509–516.

Brandt J., Folstein, S. E., and Folstein, M. F. (1988). Differential cognitive impairment in Alzheimer's disease and Huntington's disease. *Ann Neurol* 23:555–561.

Brooks, D. J., and Frackowiak, R. S. J. (1989). PET and movement disorders. *J Neurol Neurosurg Psy* (Suppl.):68–77.

Brundin, P., Fricker, R. A., and Nakao, N. (1996). Paucity of P-zones in striatal grafts prohibit commencement of clinical trials in Huntington's disease. *Neuroscience* 71:895–897.

Brundin, P., Fricker, R. A., and Nakao, N. (1996). Paucity of P-zones in striatal grafts prohibit commencement of clinical trials in Huntington's disease. *Neuroscience* 71:895–897.

Brundin, P., Strecker, R. E., Widner, H., Clarke, D. J., Nilsson, O. G., Astedt, B., Lindvall, O., and Björklund, A. (1988). Human fetal dopamine neurons grafted in a rat model of Parkinson's disease: Immunological aspects, spontaneous and drug-induced behaviour, and dopamine release. *Exp Brain Res* 70:192–208.

Bruyn, G. W. (1968). Huntington's chorea: Historical, clinical and laboratory synopsis. In: Vinken, P. J., and Bruyn, G. W., eds., *Handbook of clinical neurology*. Amsterdam: North-Holland Publ., pp. 298–378.

Campbell, K., Wictorin, K., and Björklund, A. (1995). Neurotransmitter-related gene expression in intrastriatal striatal transplants. II. Characterization of efferent projecting graft neurons. *Neuroscience* 64:35–47.

CAPIT-HD committee, Quinn, N., Brown, R., Craufurd, D., Goldman, S., Hodges, J., Kieburtz, K., Lindvall, O., MacMillan, J., and Roos, R. (1996). Core assessment program for intracerebral transplantation in Huntington's disease (CAPIT-HD). *Movement Disorders* 11:143–150.

Clarke, D. J., and Dunnett, S. B. (1990). Ultrastructural organization within intrastriatal striatal grafts. In: Dunnett, S. B., and Richards, S. J., eds., *Neural transplantation. From molecular basis to clinical applications*. Amsterdam: Elsevier, pp. 407–415.

Clarke, D. J., Dunnett, S. B., Isacson, O., Sirinathsinghji, D. J. S., and Björklund, A. (1988). Striatal grafts in rats with unilateral neostriatal lesions. I. Ultrastructural evidence of afferent synaptic inputs from the host nigrostriatal pathway. *Neuroscience* 24:791–801.

Cooper, E. R. A. (1946). The development of the human red nucleus and corpus striatum. *Brain* 69:34–44.

Cudkowicz, M., and Kowall, N. W. (1990). Degeneration of pyramidal projection neurons in Huntington's disease cortex. *Ann Neurol* 27:200–204.

De la Monte, S. M., Vonsattel, J. P., and Richardson, E. P. (1988). Morphometric demonstration of atrophic changes in the cerebral cortex, white matter, and neostriatum in Huntington's disease. *J Neuropathol Exp Neurol* 47:516–525.

Deacon, T. W., Pakzaban, P., and Isacson, O. (1994). The lateral ganglionic eminence is the origin of cells committed to striatal phenotypes: neural transplantation and developmental evidence. *Brain Res* 668:211–219.

Deckel, A. W., Moran, T. H., and Robinson, G. (1988a). Receptor characteristics and recovery of function following kainic acid lesions and fetal transplants of the striatum. II. Dopaminergic systems. *Brain Res* 474:39–47.

Deckel, A. W., Moran, T. H., and Robinson, G. (1988b). Receptor characteristics and recovery of function following kainic acid lesions and fetal transplants of the striatum. I. Cholinergic systems. *Brain Res* 474:27–38.

Deckel, A. W., Moran, T. H., and Robinson, G. (1986). Behavioral recovery following kainic acid lesions and fetal implants of the striatum occurs independent of dopaminergic mechanisms. *Brain Res* 363:383–385.

Deckel, A. W., Robinson, R. G., Coyle, J. T., and Sanberg, P. R. (1983). Reversal of long-term locomotor abnormalities in the kainic acid model of Huntington's disease by day 18 fetal striatal implants. *Eur J Pharmacol* 93:287–288.

Defer, G., Geny, C., Ricolfi, F., Fenelon, G., Monfort, J. C., Remy, P., Villafane, G., Jeny, R., Samson, Y., Keravel, Y., Gaston, A., Degos, J. D., Peschanski, M., Cesaro, P., and Nguyen, J. P. (1996). Long-term outcome of unilaterally transplanted parkinsonian patients: i: clinical approach. *Brain* 119:41–50.

Defontaines, B., Peschanski, M., and Onteniente, B. (1992). Host dopaminergic afferents affect the development of embryonic striatal neurons. *Neuroscience* 48:857–869.

Dom, R., Malfroid, M., and Baro, F. (1976). Neuropathology of Huntington's chorea: Studies of the ventrobasal complex of the thalamus. *Neurology* 26:64–68.

Dragunow, M., Williams, M., and Faull, R. L. M. (1990). Haloperidol induces Fos and related molecules in intra-striatal grafts derived from fetal striatal primordia. *Brain Res* 530:309–311.

Dunlap, C.B. (1927). Pathologic changes in Huntington's chorea with special reference to the corpus striatum. *Arch Neurol Psych* 18:867–943.

Dunnett, S. B., Isacson, O., Sirinathsinghji, D. J. S., Clarke, D. J., and Björklund, A. (1988). Striatal grafts in rats with unilateral neostriatal lesions. III. Recovery from dopamine-dependent motor asymmetry and deficits in skilled paw reaching. *Neuroscience* 24:813–820.

Eidelberg, E., Nguyen, L. H., Polich, R., and Walden, J. G. (1989). Transsynaptic degeneration of motoneurones caudal to spinal cord lesions. *Brain Res Bull* 22:39–45.

Emerich, D. F., Winn, S. R., Hantraye, P., Peschanski, M., Chen, E. Y., Mcdermott, P., Baetge, E. E., and Kordower, J. H. (1997). Encapsulated CNTF-producing cells protect monkeys in a model of Huntington's disease. *Nature* 386:395–398.

Fentress, J. C., Stanfield, B. B., and Cowan, W. M. (1981). Observations on the development of the striatum in mice and rats. *Anat Embryol* 163:275–298.

Freed, C. R., Breeze, R. E., Rosenberg, N. L., Schneck, S. A., Kriek, E., Qi, J. X., Lone, T. M., Zhang, Y. B., Snyder, J. A., Wells, T. H., Ramig, L. O., Thompson, L., Mazziotta, J. C., Huang, S. C., Grafton, S. T., Brooks, D., Sawle, G., Schroter, G., and Ansari, A. A. (1992). Survival of implanted fetal dopamine cells and neurologic improvement 12 to 46 months after transplantation for Parkinson's disease. *N Engl J Med* 327:1549–1555.

Freed, C. R., Breeze, R. E., Rosenberg, N. L., Schneck, S. A., Wells, T. H., Barrett, J. N., Grafton, S. T., Huang, S. C., and Eidelberg, D. (1990). Transplantation of human fetal dopamine cells for Parkinson's disease: Results at 1 year. *Arch Neurol* 47:505–512.

Freeman, T. B., Olanow, C. W., Hauser, R. A., Nauert, G. M., Smith, D. A., Borlongan, C. V., Sanberg, P. R., Holt, D. A., Kordower, J. H., Vingerhoets, F. J., *et al.* (1995). Bilateral fetal nigral transplantation into the postcommissural putamen in Parkinson's disease. *Ann Neurol* 38:379–388.

Giordano, M., Ford, L. M., Shipley, M. T., and Sanberg, P. R. (1990). Neural grafts and pharmacological intervention in a model of Huntington's disease. *Brain Res Bull* 25:453–465.

Grafton, S. T., Mazziotta, J. C., Pahl, J. J., St. George-Hyslop, P., Haines, J. L., Gusella, J., Hoffman, J. M., Baxter, L. R. and Phelps, M. E. (1990). A comparison of neurological, metabolic, structural, and genetic evaluations in persons at risk for Huntington's disease. *Ann Neurol* 28:614–621.

Granholm, E., and Butters, N. (1988). Associative encoding and retrieval in Alzheimer's and Huntington's disease. *Brain and Cognition* 7:335–347.

Grasbon-Frodl, E. M., Nakao, N., Lindvall, O., and Brundin, P. (1996). Phenotypic development of the human embryonic striatal primordium: A study of cultured and grafted neurons from the lateral and medial ganglionic eminences. *Neuroscience* 73:171–183.

Graybiel, A. M., Liu, F. C., and Dunnett, S. B. (1989). Intrastriatal grafts derived from fetal striatal primordia. I. Phenotopy and modular organization. *J Neurosci* 9:3250–3271.

Gusella, J. F., Wexler, N. S., Conneally, P. M., Naylor, S. L., Anderson, M. A., Tanzi, R. E., Watkins, P. C., Ottina, K., Wallace, M. R., and Sakaguchi, A. Y. (1983). A polymorphic DNA marker genetically linked to Huntington's disease. *Nature* 306:234–238.

Hantraye, P., Riche, D., Mazière, M., and Isacson, O. (1992). Intrastriatal transplantation of cross-species fetal striatal cells reduces abnormal movements in a primate model of Huntington's disease. *Proc Natl Acad Sci USA* 89:4187–4191.

Hayden, M. R., Martin, W. R. W., Stoess, A. J., Clark, C., Hollenberg, S., Adam, M. J., Ammann, W., Harrop, R., Rogers, J., Ruth, T., Sayre, C., and Pate, B. D. (1986). Positron emission tomography in the early diagnosis of Huntington's disease. *Neurology* 36:888–894.

Heindel, W. C., Salmon, D. P., and Butters N. (1990). Pictorial priming and cued recall in Alzheimer's and Huntington's disease. *Brain and Cognition* 13:282–295.

Heindel, W. C., Salmon, D. P., Shults, C. W., Walicke, P. A., and Butters, N. (1989). Neuropsychological evidence for multiple implicit memory systems: A comparison of Alzheimer's, Huntington's, and Parkinson's disease patients. *J Neurosci* 9:582–587.

Helm, G. A., Palmer, P. E., and Bennett, J. P. (1990). Fetal neostriatal transplants in the rat: A light and electron microscopic Golgi study. *Neuroscience* 37:735–756.

Helm, G. A., Robertson, M. W., Jallo, G. I., Simons, N., and Bennett, J. P. (1991). Development of dopamine D1 and D2 receptors and associated second messenger systems in fetal striatal transplants. *Exp Neurol* 111:181–189.

Henderson, B. T. H., Clough, C. G., Hughes, R. C., Hitchcock, E. R., and Kenny, B. G. (1991). Implantation of human fetal ventral mesencephalon to the right caudate nucleus in advanced Parkinson's disease. *Arch Neurol* 48:822–827.

Hewitt, W. (1958). The development of the human caudate and amygdaloid nuclei. *J Anat* 92:377–382.

Hewitt, W. (1961). The development of the human internal capsule and lentiform nucleus. *J Anat* 95:191–199.

Hodges, J. R., Salmon, D. P., and Butters, N. (1990). Differential impairment of semantic and episodic memory in Alzheimer's and Huntington's diseases: A controlled prospective study. *J Neurol Neurosurg Psy* 53:1089–1095.

Hodges, J. R., Salmon, D. P., and Butters, N. (1991). The nature of the naming deficit in Alzheimer's and Huntington's disease. *Brain* 114:1547–1558.

Huntington's disease coll. res. group (1993). A novel gene containing a trinucleotide repeat that is expanded and unstable in Huntington's disease chromosome. *Cell* 72:1–18.

Huntington Study Group (1996). Unified Huntington's disease rating scale: Reliability and consistency. *Movement Disorders* 11:136–142.

Illes, J. (1989). Neurolinguistic features of spontaneous language production dissociate three forms of neurodegenerative diseases: Alzheimer's, Huntington's and Parkinson's. *Brain and Language* 37:628–642.

Isacson, O., Brundin, P., Gage, F. H., and Björklund, A. (1985). Neural grafting in a rat model of Huntington's disease. Progressive neurochemical changes after neostriatal ibotenate lesions and striatal tissue grafting. *Neuroscience* 16:799–817.

Isacson, O., Brundin, P., Kelly, P. A. T., Gage, F. H., and Björklund, A. (1984). A functional neuronal replacement by grafted neurons in the ibotenic acid-lesioned striatum. *Nature* 11:458–460.

Isacson, O., Dawbarn, D., Brundin, P., Gage, F. H., Emson, P. C., and Björklund, A. (1987). Neural grafting in a rat model of Huntington's disease: Striosomal-like organization of striatal grafts as revealed by immunocytochemistry and receptor autoradiography. *Neuroscience* 22:481–497.

Isacson, O., Dunnett, S. B., and Björklund, A. (1986). Graft-induced behavioural recovery in an animal model of Huntington's disease. *Proc Natl Acad Sci USA* 83:2728–2732.

Johnston, J. B. (1923). Further contributions to the study of the evolution of the forebrain. *J Comp Neurol* 35:337–412.

Kordower, J. H., Freeman, T. B., Snow, B. J., Vingerhoets, F. J. G., Mufson, E. J., Sanberg, P. R., *et al.* (1995). Neuropathological evidence of graft survival and striatal reinnervation after the transplantation of fetal mesencephalic tissue in a patient with Parkinson's disease. *New Engl J Med* 332:1118–1124.

Kordower, J. H., Rosenstein, J. M., Collier, T. J., Burke, M. A., Chen, E. Y., Li, J. M., Martel, L., Levey, A. E., Mufson, E. J., Freeman, T. B., and Olanow, C. W. (1996). Functional fetal nigral grafts in a patient with Parkinson's disease: Chemanatomic, ultrastructural, and metabolic studies. *J Comp Neurol* 370:203–230.

Kuhl, D. E., Phelps, M. E., Markham, C. H., Metter, E. J., Riege, W. H., and Winter, J. (1982). Cerebral metabolism and atrophy in Huntington's disease determined by 18FDG and computed tomographic scan. *Ann Neurol* 12:425–434.

Kuhl, D. E., Phelps, M. E., Markham, C. H., Metter, E. J., Riege, W. H., and Winter, J. (1982). Cerebral metabolism and atrophy in Huntington's disease determined by 18FDG and computed tomographic scan. *Ann Neurol* 12:425–434.

Kuwert, T., Lange, H. W., Langen, K. J., Herzog, H., Aulich, A., and Feinendegen, L. E. (1990). Cortical and subcortical glucose consumption measured by PET in patients with Huntington's disease. *Brain* 113:1405–1423.

Labandeira-Garcia, J. L., Wictorin, K., Cunningham, E. T., and Björklund, A. (1991). Development of intrastriatal striatal grafts and their afferent innervation from the host. *Neuroscience* 42:407–426.

Lange, H., Thorner, G., Hopf, A., and Schroder, K. F. (1976). Morphometric studies of the neuropathological changes in choreatic diseases. *J Neurol Sci* 28:401–425.

Langston, W. J., Widner, H., Goetz, C. G., Brooks, D., Fahn, S., Freeman, T., and Watts, R. (1992). Core assessment program for intracerebral transplantations (CAPIT). *Mov Disord* 7:2–13.

Leenders, K. L., Frackowiak, R. J. S., Quinn, N., and Marsden, C. D. (1986). Brain energy metabolism and dopaminergic function in Huntington's disease measured *in vivo* using positron emission tomography. *Mov Disord* 1:69–77.

Lindvall, O. (1994). Neural transplantation in Parkinson's disease. In: Dunnett, S. B., and Björklund, A., eds., *Functional neural transplantation*. New York: Raven Press Ltd, pp. 103–137.

Lindvall, O., Brundin, P., Widner, H., Rehncrona, S., Gustavii, B., Frackowiak, R., Leenders, K. L., Sawle, G., Rothwell, J. C., Marsden, C. D., and Björklund, A. (1990). Grafts of fetal dopamine neurons survive and improve motor function in Parkinson's disease. *Science* 247:574–577.

Lindvall, O., Sawle, G., Widner, H., Rothwell, J. C., Björklund, A., Brooks, D., Brundin, P., Frackowiak, R., Marsden, C. D., Odin, P., and Rehncrona, S. (1994). Evidence for long-term survival and function of dopaminergic grafts in progressive Parkinson's disease. *Ann Neurol* 35:172–180.

Lindvall, O., Widner, H., Rehncrona, S., Brundin, P., Odin, P., Gustavii, B., Frackowiak, R., Leenders, K. L., Sawle, G., Rothwell, J. C., Björklund, A., and Marsden, C. D. (1992). Transplantation of fetal dopamine neurons in Parkinson's disease—one-year clinical and neurophysiological observations in two patients with putaminal implants. *Ann Neurol* 31:155–165.

Liu, F. C., Graybiel A. M., Dunnett, S. B., Baughman, R. W. (1990). Intrastriatal grafts derived from fetal striatal primordia: II. Reconstitution of cholinergic and dopaminergic systems. *J Comp Neurol* 295:1–14.

Liu, F. C., Dunnett, S. B., Robertson, H. A., and Graybiel, A. M. (1991). Intrastriatal grafts derived from fetal striatal primordia. III. Induction of modular patterns of Fos-like immunoreactivity by cocaine. *Exp Brain Res* 85:501–506.

Marchand, R., and Lajoie, L. (1986). Histogenesis of the striopallidal system in the rat: Neurogenesis of its neurons. *Neuroscience* 17:573–590.

Marty, S., and Peschanski, M. (1994). Effects of target deprivation on the morphology and survival of adult dorsal column nuclei neurons. *J Comp Neurol* 356:523–536.

Marty, S., and Peschanski, M. (1994). Fine structural alteration in target-deprived axonal terminals in the rat thalamus. *Neuroscience* 62:1121–1132.

Marty, S., Weinitz, J. M., and Peschanski, M. (1994). Target dependence of adult neurons: Pattern of terminal arborizations. *J Neurosci* 14:5257–5266.

Mayer, E., Brown, V. J., Dunnett, S. B., and Robbins, T. W. (1991). Striatal graft-associated recovery of a lesion-induced performance deficit in the rat requires learning to use the transplant. *Eur J Neurosci* 4:119–126.

Mayer, E., Heavens, R. P., and Sirinathsinghji, D. J. S. (1990). Autoradiographic localization of D1 and D2 dopamine receptors in primordial striatal tissue grafts in rats. *Neurosci Lett* 109:271–276.

Mazziotta, J. C., Phelps, M. E., Pahl, J. J., Huang, S. C., Baxter, L. R., Riege, W. H., Hoffman, J. M., Kuhl, D. E., Lanto, A. B., Wapenski, J. A., and Markham, C. H. (1987). Reduced cerebral glucose metabolism in asymptomatic subjects at risk for Huntington's disease. *N Engl J Med* 316:357–362.

McAllister, J. P., Walker, P. D., Zemanick, M. C., Weber, A. B., Kaplan, L. I., and Reynolds, M. A. (1985). Morphology of embryonic neostriatal cell suspensions transplanted into adult neostriata. *Dev Brain Res* 23:282–286.

McCaughey, W. T. E. (1961). The pathologic spectrum of Huntington's chorea. *J Nerv Ment Dis* 133:91–103.

Montoya, C. P., Astell, S., and Dunnett, S. B. (1989). Effects of nigral and striatal grafts on skilled forelimb use in the rat. In: Dunnett, S. B., and Richards, S. J., eds., *Neural transplantation. From molecular basis to clinical applications. Prog in brain res.* Amsterdam: Elsevier, pp. 459–466.

Myers, R. H., Sax, D. S., Koroshetz, W. J., Mastromauro, C., Cupples, L. A., Kiely, D. K., Pettengill, F. K., and Bird, E. D. (1991). Factors associated with slow progression in Huntington's disease. *Arch Neurol* 48:800–804.

Naimi, S., Jeny, R., Hantraye, P., Peschanski, M., Riche, D. (1996). Ontogeny of human striatal DARPP-32 neurons in fetuses and following xenografting to the adult rat brain. *Exp Neurol* 137:15–25.

Nakao, N., Frodl, E. M., Duan, W-M., Widner, H., and Brundin, P. (1994). Lazaroids improve the survival of grafted embryonic dopamine neurons. *Proc Natl Acad Sci USA* 91:12408–12412.

Nakao, N., Grasbon-Frodl, E. M., Widner, H., and Brundin, P. (1996). DARPP-32-rich zones in grafts of lateral ganglionic eminence govern the extent of functional recovery in skilled paw reaching in an animal model of Huntington's disease. *Neuroscience* 74:959–970.

Nieuwenhuys, R. (1977) Aspects of the morphology of the striatum. In: Cools, A. R., Lohman, A. H. M., and Van Den Bercken, J. H. L., eds., *Psychobiology of the striatum.* Amsterdam: Elsevier, pp. 1–19.

Norman, A. B., Calderon, S. F., Giordano, M., and Sanberg, P. R. (1988). Striatal tissue transplants attenuate apomorphine-induced rotational behavior in rats with unilateral kainic acid lesions. *Neuropharmacology* 27:333–336.

Norman, A. B., Giordano, M., and Sanberg, P. R. (1989). Fetal striatal tissue grafts into excitotoxin-lesioned striatum: Pharmacological and behavioral aspects. *Pharmacol Biochem Behav* 34: 139–147.

Nothias, F., Isacson, O., Wictorin, K., Peschanski, M., and Björklund, A. (1988). Morphological alteration of the thalamic afferents in the excitotoxically lesioned striatum. *Brain Res* 461:349–354.

Oka, H. (1980). Organization of the cortico-caudate projections. *Exp Brain Res* 40:203–208.

Olanow, C. W., Kordower, J. H., Freeman, T. B. (1996). Fetal nigral transplantation as a therapy for Parkinson's disease. *Trends Neurosci* 19:102–109.

Olsson, M., Campbell, K., Wictorin, K., and Björklund, A. (1995). Projection neurons in fetal striatal transplants are predominantly derived from the lateral ganglionic eminence. *Neuroscience* 69:1169–1182.

Otsuka, M., Ichiya, Y., Kuwabara, Y., Hosokawa, S., Sasaki, M., Fukumura, T., Masuda, K., Goto, I., and Kato, M. (1993). Cerebral glucose metabolism and striatal 18F-dopa uptake by PET in cases of chorea with or without dementia. *J Neurol Sci* 115:153–157.

Oyanagi, K., Takeda, S., Takahashi, H., Ohama, E., and Ikuta, F. (1989). A quantitative investigation of the substantia nigra in Huntington's disease. *Ann Neurol* 26:13–19.

Pakzaban, P., Deacon, T. W., Burns, L. H., and Isacson, O. (1993). Increased proportion of acetylcholinesterase-rich zones and improved morphological integration in host striatum of fetal grafts derived from the lateral but not the medial ganglionic eminence. *Exp Brain Res* 97:13–22.

Palfi, S., Coudé, F., Riche, D., Brouillet, E., Dautry, C., Nittoux, V., Chibois A., Peschauski, N., and Hautroye, P. Fetal striatal allografts reverse cognitive deficits in a primate model of Huntington's disease. *Nature Ned,* in press.

Palfi, S., Ferrante, R. J., Brouillet, E., Beal, M. F., Dolan, R., Guyot, M. C., Peschanski, M., and Hantraye, P. (1996). Chronic 3-nitropropionic acid treatment in baboons replicates the cognitive and motor deficits of Huntington's disease. *J Neurosci* 16:3019–3025.

Peschanski, M., and Besson, J. M. (1987). Structural alteration and possible growth of afferents after kainate lesion in the adult rat thalamus. *J Comp Neurol* 258:185–203.

Peschanski, M., Defer, G., NGuyen, J. P., Ricolfi, F., Monfort, J. C., Remy, P., Geny, C., Samson, Y., Hantraye, P., Jeny, R., Gaston, A., Keravel, Y., Degos, J. D., and Cesaro, P. (1994). Bilateral motor improvement and alteration of L-DOPA effect in two patients with Parkinson's disease following intrastriatal transplantation of foetal ventral mesencephalon. *Brain* 117:487–499.

Pillon, B., Dubois, B., Ploska, A., and Agid, Y. (1991). Severity and specificity of cognitive impairment in Alzheimer's, Huntington's, and Parkinson's diseases and progressive supranuclear palsy. *Neurology* 41:634–643.

Podoll, K., Caspary, P., Lange, H. W., and Noth, J. (1988). Language functions in Huntington's disease. *Brain* 111:1475–1503.

Pritzel, M., Isacson, O., Brundin, P., Wiklund, L., and Björklund, A. (1986). Afferent and efferent connections of striatal grafts implanted into the ibotenic acid lesioned neostriatum in adult rats. *Exp Brain Res* 65:112–126.

Randolph, C. (1991). Implicit, explicit, and semantic memory functions in Alzheimer's disease and Huntington's disease. *J Clin Exp Neuropsy* 13:479–494.

Reid, I. C., Besson, J. A., Best, P. V., Sharp, P. F., Gemmell, H. G., and Smith, F. W. (1988). Imaging of cerebral blood flow markers in Huntington's disease using single photon emission computed tomography. *J Neurol Neurosurg Psychiatry* 51:1264–1268.

Rémy, P., Samson, Y., Hantraye, P., Fontaine, A., Defer, G., Mangin, J. F., Fénelon, G., Geny, C., Ricolfi, F., Frouin, V., Nguyen, J. P., Degos, J. D., Peschanski, M., and Cesaro, P. (1995). Neural grafting in five parkinsonian patients: Correlations between PET and clinical evolution. *Ann Neurol* 38:580–588.

Roberts, R. C., and DiFiglia, M. (1988). Localization of immunoreactive GABA and enkephalin and NADPH-diaphorase-positive neurons in fetal striatal grafts in the quinolinic-acid-lesioned rat neostriatum. *J Comp Neurol* 274:406–421.

Roberts, R. C., and DiFiglia, M. (1990). Long-term survival of GABA-, enkephalin, NADPH-diaphorase and calbindin-D28k-containing neurons in fetal striatal grafts. *Brain Res* 532:151–159.

Rutherford, A., Garcia-Munoz, M., Dunnett, S. B., and Arbuthnott, G. W. (1987). Electrophysiological demonstration of host cortical inputs to striatal grafts. *Neurosci Lett* 83:275–281.

Salmon, D. P., Kwo-on-Yuen, P. F., Heindel, W. C., Butters, N., and Thal, L.J. (1989). Differentiation of Alzheimer's disease and Huntington's disease with the dementia rating scale. *Arch Neurol* 46:1204–1208.

Sanberg, P. R., Henault, M. A., and Deckel, A. W. (1986). Locomotor hyperactivity: Effects of multiple striatal transplants in an animal model of Huntington's disease. *Pharmacol Biochem Behav* 25:297–300.

Sanberg, P. R., Henault, M. A., Hagenmeyer-Houser, S. H., Giordano, M., and Russel, K. H. (1987). Multiple transplants of fetal striatal tissue in the kainic acid model of Huntington's disease. Behavioral recovery may not be related to acetylcholine esterase. *Ann NY Acad Sci* 495:781–785.

Sawle, G. V., Bloomfield, P. M., Björklund, A., Brooks, D. J., Brundin, P., Leenders, K. L., Lindvall, O., Marsden, C. D., Rehncrona, S., Widner, H., and Frackowiak, R. S. J. (1992). Transplantation of fetal dopamine neurons in parkinson's disease - PET <F-18>6-L-Fluorodopa Studies in two patients with putaminal implants. *Ann Neurol* 31:166–173.

Sharp, A. H., and Ross, C. A. (1996). Neurobiology of Huntington's disease. *Neurobio Disease* 3:3–15.

Shoulson, I., Odoroff, C., Oakes, D., Behr, J., Goldblatt, D., Caine, E., Kennedy, J., Miller, C., Bamford, K., Rubin, A., *et al.* (1989). A controlled clinical trial of baclofen as protective therapy in early Huntington's disease. *Ann Neurol* 25:252–259.

Sirinathsinghji, D. J. S., Dunnett, S. B., Isacson, O., Clarke, D. J., Kendrick, K., and Björklund, A. (1988). Striatal grafts in rats with unilateral neostriatal lesions. II. *In vivo* monitoring of GABA release in globus pallidus and subtantia nigra. *Neuroscience* 24:803–811.

Sirinathsinghji, D. J. S., Morris, B. J., Wisden, W., Northrop, A., Hunt, S. P., and Dunnett, S. B. (1990). Gene expression in striatal grafts. I. Cellular localization of neurotransmitter mRNAs. *Neuroscience* 34:675–686.

Smart, I. H. M., and Sturrock, R. R. (1979). Ontogeny of the neostriatum. In: Divac, I., and Oberg, R. G. E., eds., *The neostriatum*. Oxford: Pergamon, pp. 127–146.

Sotrel, A., Paskevich, P. A., Kiely, D. K., Bird, E. D., Williams, R. S., and Myers, R. H. (1991). Morphometric analysis of the prefrontal cortex in Huntington's disease. *Neurology* 41:1117–1123.

Spencer, D. D., Robbins, R. J., Naftolin, F., Marek, K. L., Vollmer, T., Leranth, C., Roth, R. H., Proce, L. H., Gjedde, A., Bunney, B. S., Sass, K. J, Elsworth, J. D., Kier, E. L., Makuch, R., Hoffer, P. B., and Redmond, E. (1992). Unilateral transplantation of human fetal mesencephalic tissue into the caudate nucleus of patients with Parkinson's disease. *N Engl J Med* 327:1541–1548.

Tröster, A. I., Salmon, D. P., McCullough, D., and Butters, N. (1989). A comparison of the category fluency deficits associated with Alzheimer's and Huntington's disease. *Brain and Language* 37:500–513.

Valouskova, V., Bracha, V., Bures, J., Hernandez-Mesa, N., Marcias-Gonzales, R., Muzurova, Y., and Nemecek, S. (1990). Unilateral striatal grafts induce behavioral and electrophysiological asymmetry in rats with bilateral kainate lesions of the caudate nucleus. *Behav Neurosci* 104:671–680.

Vonsattel, J. P., Myers, R. H., Stevens, T. J., Ferrante, R. J., Bird, E. D., and Richardson, E. P. (1985). Neuropathological classification of Huntington's disease. *J Neuropath Exp Neurol* 44, 559–577.

Walker, P. D., Chovanes, G. I., and McAllister, J. L. (1987) Identification of acetylcholinesterase-reactive neurons and neuropil in neostriatal transplants. *J Comp Neurol* 259:1–12.

Waters, C. M., Peck, R., Rossor, M., Reynolds, G. P., and Hunt, S. P. (1988). Immunocytochemical studies on the basal ganglia and substantia nigra in Parkinson's disease and Huntington's chorea. *Neuroscience* 25:419–438.

Weinberger, D. R., Berman, K. F., Iadarola, M., Driesen, N., and Zec, R. F. (1988). Prefrontal cortical blood flow and cognitive function in Huntington's disease. *J Neurol Neurosurg Psy* 51:94–104.

Wictorin, K., and Björklund, A. (1989). Connectivity of striatal grafts implanted into the ibotenic acid lesioned striatum. II. Cortical afferents. *Neuroscience* 30:297–311.

Wictorin, K., Clarke, D. J., Bolam, J. P., and Björklund, A. (1990). Fetal striatal neurons grafted into the ibotenate lesioned striatum: Efferent projections and synaptic contacts in the host globus pallidus. *Neuroscience* 37:301–315.

Wictorin, K., Isacson, O., Fischer, W., Nothias, F., Peschanski, M., and Björklund, A. (1988). Connectivity of striatal grafts implanted into the ibotenic acid-lesioned striatum. I. Subcortical afferents. *Neuroscience* 27:547–562.

Wictorin, K., Ouimet, C. C., and Björklund, A. (1989a). Intrinsic organization and connectivity of intrastriatal striatal transplants in rats as revealed by DARPP-32 immunohistochemistry: Specific connections with the lesioned host brain. *Eur J Neurosci* 1:690–701.

Wictorin, K., Simerly, R. B., Isacson, O., Swanson, L. W., and Björklund, A. (1989b). Connectivity of striatal grafts implanted into the ibotenic acid lesioned striatum. III. Efferent projecting graft neurons and their relation to host afferents within the grafts. *Neuroscience* 30:313–330.

Widner, H., Tetrud, J., Rehncrona, S., Snow, B., Brundin, P., Gustavii, B., Björklund, A., Lindvall, O., and Langston, J (1992). Bilateral fetal mesencephalic grafting in two patients with parkinsonism induced by 1-methyl-4-phenyl-1,2,3,6-tetrahydropyridine (MPTP). *N Engl J Med* 327:1556–1563.

Zhou, F.C., and Buchwald, N. (1989). Connectivities of the striatal grafts in adult rat brain: A rich afference and scant striatonigral efference. *Brain Res* 504:15–30.

CHAPTER 18

Cellular Delivery of Neurotrophic Factors as a Potential Treatment for Huntington's Disease

DWAINE F. EMERICH,* JEFFREY H. KORDOWER,† OLE ISACSON**

*Department of Neuroscience, Alkermes, Inc., Cambridge MA, †Research Center for Brain Repair and Department of Neurological Sciences, Rush Presbyterian St. Luke's Medical Center, Chicago, Illinois, **Neuroregeneration Laboratory, Harvard Medical School and McLean Hospital, Belmont, Massachusetts

Huntington's Disease (HD) is a devastating neurodegenerative disorder with no effective treatments for the behavioral symptoms or the associated neural degeneration. In recent years, the advent of appropriate animal models has permitted the evaluation of multiple therapeutic strategies. One of the more promising approaches currently under preclinical investigation and clinical consideration is the use of neurotrophic factors that might slow the progression or even prevent the onset of the behavioral and pathological consequences of HD. This chapter highlights the use of transplanted genetically modified cells as a means of delivering neurotrophic factors. Studies in both rodent and nonhuman primate models of HD suggest that trophic factors such as NGF and CNTF exert marked neuroprotective effects upon the vulnerable striatal neurons that degenerate in HD. Although the mechanisms by which these factors exert their beneficial effects remain unclear, their clearcut potency in both rodent and primate models of HD provide the hope that a means of preventing or slowing the relentlessly progressive motor and cognitive declines in HD may be forthcoming. Furthermore, excitotoxicity has been implicated in a variety of pathological conditions including ischemia, and neurodegenerative diseases such as Huntington's, Parkinson's, and Alzheimer's. Accordingly, biologically delivered neurotrophic factors may provide one means of preventing the cell loss and associated behavioral abnormalities of these and possibly other human disorders.

I. INTRODUCTION

Huntington's Disease (HD) is an inherited neurodegenerative disease characterized by a relentlessly progressive movement disorder with psychiatric and cognitive deterioration. The prevalence of HD is between 5 and 10 per 100,000 (Conneally, 1984) and is invariably fatal. On average, patients suffer 17 years of symptomatic illness. HD is found in all regions of the world, even in remote locations. Clinically, this disease is characterized by an involuntary choreiform (dancelike) movement disorder, psychiatric and behavioral changes, and dementia (Greenamyre and Shoulson, 1994). The age of onset is usually between the mid 30's to late fifties, although juvenile (<20 years of age) and late onset (>65 years of age) HD occurs (Conneally, 1984; Farrer and Conneally, 1985). Although medications may reduce the severity of chorea or diminish behavioral symptoms, there are no effective treatments (Penney *et al.*, 1990) because current approaches do not increase survival or substantially improve quality of life as it relates to cognitive state, gait disorder, or dysphagia (Shoulson, 1981).

Pathogenetically, HD is a disorder characterized by a programmed premature death of cells, predominantly in the caudate nucleus and the putamen. Initially, the disease affects medium-sized spiny neurons that contain gamma amino butyric acid (GABA). These cells receive afferent input from cortical glutamatergic and nigral dopaminergic neurons and provide efferent projections to the globus pallidus. Large aspiny interneurons and medium aspiny projection neurons are less affected and degenerate later in the disease process (Ferrante *et al.*, 1985, 1987; Graveland *et al.*, 1985; Kowell *et al.*, 1987; Vonsattle *et al*, 1985). Other subcortical and cortical brain regions are involved, but the degree of degeneration varies and does not correlate with the severity of the disease (Kowell *et al.*, 1987; Vonsattel *et al.*, 1985; Greenamyre and Shoulson, 1994) and is dwarfed by the striatal changes.

In 1993, it was discovered that an unstable expansion of a CAG trinucleotide repeat in the IT15 gene located near the telomere of the short arm of chromosome 4 produces the disease (The Huntington's Disease Collaboration Research Group, 1993; Ashizawa *et al.*, 1994). Although it is believed that the HD mutation causes an increase in excitatory neurotransmission or produces a bioenergetic cell defect, which confers increased sensitivity to ambient levels of excitation, the precise mechanism of neuronal degeneration remains unknown. Establishing a mechanism of cell death may offer hope for specific medical therapies to prevent or slow the degenerative process. This progress, while encouraging, does not obviate the need to develop novel interventions for patients in whom the neurodegenerative process is clinically manifest or in whom degenerative changes are inevitable.

II. EXPERIMENTAL STRATEGIES FOR TREATING HD

Currently, therapy for HD is limited and does not favorably influence the progression of the disease. However, recent advances in our understanding of the genetic and pathogenetic events that cause neural degeneration provide optimism that novel experimental therapeutic strategies may be devised for the treatment of HD. The ability to devise novel therapeutic strategies in HD is tightly linked to the characterization of highly relevant animal models. The initial development of models of HD concentrated on either nonspecific mechanical/electrolytic lesions of the striatum or the pharmacological induction of dyskinesia-like behaviors in rodents (Emerich and Sanberg, 1992 for a review). The difficulties of establishing causal or even suggestive relationships between drug-induced neural changes and specific behavioral alterations together with the inherent lack of anatomical and behavioral specificity resulting from global striatal lesions quickly made these models obsolete. More recently, it has been suggested that the neuronal death occurring in HD is related to an underlying endogenous excitotoxic process (Albin and Greenamyre, 1992; Beal, 1992, 1995; Greenamyre and Obrien, 1986; Olney, 1989). Two *in vivo* systems model alterations in excitotoxic processes and have been used extensively in laboratory HD experiments. One such model is the intrastriatal administration of excitatory amino acids. Although initial studies employed kainic acid (Coyle and Schwarcz, 1976), the excitotoxic model has recently been refined when it was shown that quinolinic acid (QA) more accurately modeled the pathology of HD (Beal *et al.*, 1986, 1989, 1991). In particular, QA lesions of the striatum spare both medium spiny neurons that stain for NADPH-diaphorase and large aspiny cholinergic neurons; both of which are spared in HD. Excitotoxic lesions of the striatum in experimental animals also mimic motor and cognitive changes observed in HD (Block *et al.*, 1993; Sanberg *et al.*, 1989). A second model of striatal neurodegeneration uses injections of the mitochondrial toxin 3-nitropropionic acid (3-NP) to create HD-like lesions within the striatum (Beal *et al.*, 1993).

The advent of appropriate animal models has permitted the evaluation of multiple therapeutic strategies. These strategies have generally fallen into one of two categories. The first encompasses replacement strategies including pharmacological manipulation of altered transmitter levels and replacement of degenerative neural systems via tissue/cell transplantation. The practical appeal of pharmacotherapy is offset by the anatomical and behavioral complexity of HD, together with the difficulty of titrating effective treatment regimens in the face of the ongoing degeneration of multiple neural systems. The logic of replacing lost populations of neurons is equally, if not more, appealing. Animal studies have clearly demonstrated that grafted tissue can integrate within the excitotoxic lesioned brain and promote recovery of motor and cognitive function (Isacson *et al.*, 1984, 1987, 1989; Wictorin and Bjorklund, 1989; Wictorin *et al.*, 1989; Hantraye *et al.*, 1990; Bjorklund *et al.*, 1994). Given the positive results acquired in the animal models to date, clinical trials evaluating fetal tissue grafting in HD patients have recently been initiated. Although replacement strategies hold hope for the treatment of HD, they do not affect the continued and insidious degeneration of the striatum and related structures. A treatment that stopped or slowed these changes might be more therapeutic than neurochemical or cell replacements. Toward this end, it has been suggested that trophic factors may slow the progression of cellular dysfunction and ultimately prevent the degeneration of populations of striatal neurons that are vulnerable in HD (Tatter *et al.*, 1995 for a review). Infusions of trophic factors or grafts of trophic factor-secreting cells prevent degeneration of striatal neurons destined to die from excitotoxic insult or mitochondrial dysfunction (Emerich *et al.*, 1994a, 1996, 1997a,b; Kordower *et al.*, 1996; Frim *et al.*, 1993a–d, 1994; Schumacher *et al.*, 1991). The use of trophic factors in a neural protection strategy may be particularly relevant for the treatment of HD. Unlike other neurodegenerative diseases, genetic screening can identify virtually all individuals at risk who will ultimately suffer from HD. This provides a unique opportunity to design treatment strategies that can intervene prior to the onset of striatal degeneration. Thus, instead of replacing neuronal systems that have already undergone extensive neuronal death, trophic factor strategies can be designed to support host systems destined to die at a later time in the organism's life.

III. NEUROTROPHIC FACTOR THERAPY IN CNS DISEASES

A. Use of Genetically Modified Cells for Treating CNS Disorders

If trophic factors prove to be a worthwhile therapeutic strategy, the method of delivery may be critical. To date, long-term administration of growth factors has been limited to intraventricular infusions using cannulae or pumps. These routes of administration require repeated injections or refilling of pump reservoirs to maintain specific drug levels and avoid the degradation of the therapeutic agent in solution. Additionally, chronic low-dosage infusion of compounds is difficult to sustain using current pump technology. Current pump technology is also suitable for ventricular but not parenchymal delivery. An alternative method is the implantation of cells that have been genetically modified to produce a therapeutic molecule (Gage *et al.*, 1987; Rosenberg *et al.*, 1988; Breakefield, 1989; Kawaja *et al.*, 1991, 1992; Levivier *et al.*, 1995). This avoids the problem of degradation and repeated refilling while allowing a localized distribution within the cerebrospinal fluid or parenchyma. The use of immortalized cell lines for delivery of trophic molecules avoids many of these concerns by providing a continuous *de novo* cellular source of the desired molecule, the dose of which can theoretically be adjusted with specific promoters.

One means of using genetically modified cells to deliver neurotrophic factors to the CNS is to encapsulate them within a semipermeable polymer membrane (Emerich *et al.*, 1992, 1994a–c). Single cells or small clusters of cells can be enclosed within a selective, semipermeable membrane barrier that admits oxygen and required nutrients and releases bioactive cell secretions, but restricts passage of larger cytotoxic agents from the host immune defense system. Use of a selective membrane both eliminates the need for chronic immunosuppression of the host and allows the implanted cells to be obtained from nonhuman sources, thus avoiding the cell-sourcing constraints that have limited the clinical application of generally successful investigative trials of unencapsulated cell transplantation. Polymer encapsulated cell implants have the advantage of being retrievable if the transplant should produce undesired effects, or if the cells should need to be replaced. Cross-species immunoisolated cell therapy has been validated in small and large animal models of Parkinson's disease (Emerich *et al.*, 1992; Kordower *et al.*, 1995), HD (Emerich *et al.*, 1994a, 1996, 1997a,b), amyotropic lateral sclerosis (Sagot *et al.*, 1995), and Alzheimer's disease (Emerich *et al.*, 1994b; Kordower *et al.*, 1994c).

B. Initial Studies of Neurotrophic Factors Using Genetically Engineered Cells in Animal Models of Huntington's Disease

Our working model that predicts whether a neuron will degenerate in the face of trauma, age, or disease focuses on a balance between a cells intrinsic neuronal resilience and the time-dependent nature of the insult (Isacson, 1993). Prior to the insult, genes are expressed within the cell that, at different times during the life of the cell, make it more vulnerable or resistant to irreversible damage. This model of neuronal homeostasis suggests that there are physiological conditions in which a particular neuron or brain region is somehow protected. Conversely, a hypothetical resilience curve for an anatomically defined sick population of neurons in chronic neurodegenerative disease borders close to the threshold of neuronal death. This model also suggests that therapeutic interventions, such as the delivery of neurotrophic factors, might reduce the probability for degeneration in those neuronal populations that approach their threshold for degeneration (Isacson, 1993).

In attempting to increase the resilience (or decrease the degenerative threshold) of nerve cells destined to die in HD, we delivered neurotrophic factors locally into the striatum via the implantation of genetically engineered fibroblasts. In this regard, we have explored the consequences of implantation of immortalized fibroblasts genetically engineered to produce nerve growth factor (NGF), basic fibroblast growth factor (bFGF), and brain derived neurotrophic factor (BDNF) in models of striatal neuronal degeneration (Schumacher *et al.*, 1991; Frim *et al.*, 1993a–d; Galpern *et al.*, 1996). We found that NGF- and bFGF-secreting implants were most effective in reducing excitotoxicity in the neostriatum. In contrast, BDNF-secreting cells were ineffective in preventing striatal neural degeneration although these grafts had significant protective effects for dopaminergic substantia nigra neurons (Frim *et al.*, 1994c).

In preparing cell lines capable of locally secreting biologically active neuroprotective factor for initial tests of neuroprotection, an immortalized Rat I cell line was transfected with retrovirus vectors containing (1) a mouse cDNA (N.8) encoding a near full-length preproNGF precursor (Short *et al.*, 1990), (2) a full-length human cDNA for preproBDNF, or (3) a hybrid mini-gene consisting of a 19 amino acid peptide from the amino terminal of IgG protein preceding a near full-length bovine bFGF cDNA (FXS) (Rogelj *et al.*, 1988). Animals in the NGF experimental groups were grafted (Schumacher *et al.*, 1991; Frim *et al.*, 1993c, 1993d, 1994b) with either control Rat I fibroblast cells producing 6.7 pg NGF/10^5 cells/hr) or Rat I-N.8 2 cells producing 177.6 pg NGF/10^5 cells/hr), respectively. The BDNF groups were grafted with either Rat I cells (conditioned media that did not produce BDNF or support dorsal root ganglion (DRG) explant survival) or Rat 1-BDNF #7 cells (conditioned media that produced BDNF as assessed by ELISA methods). The bFGF groups were grafted with either control Rat I cells or with Rat I-FXS-4a bFGF[+] cells, which had mitogenic activity in an NIH-3T3 cell transformation bioassay equivalent to a bFGF production of approximately 5.4 ng/10^5 cells/hr. For implantation, cells were harvested with trypsin and suspended in 0.1 M phosphate buffered saline, pH 7.4, with 1.0 mg/l $CaCl_2$, 1.0 mg/l $MgCl_2$, and 0.1% glucose. On day 8 of the procedure, animals in the excitotoxic paradigm received an infusion of 120 nmol of QA, an NMDA-receptor agonist previously used for experimental striatal excitotoxic lesioning paradigms (Schumacher *et al.*, 1991; Frim *et al.*, 1993c, 1994b). Animals were sacrificed approximately 2 weeks after graft implantation in the NGF, BDNF, and bFGF experimental groups.

Striatal lesions were readily seen by characteristic neuronal loss around the QA infusion tract after cresyl violet staining in control grafted rats. In contrast, NGF- and bFGF-secreting grafts significantly protected against neuronal loss, whereas the BDNF-secreting graft used showed no apparent neuroprotective effects against excitotoxicity (Frim *et al.*, 1993c). Because the vast majority of striatal neurons are GABA-ergic, the observed sparing of Nissl-stained cells indicated a neuroprotection of this critical population of cells which are vulnerable in HD. Furthermore, NGF- and bFGF- secreting fibroblasts spared striatal cholinergic neurons within the lesioned areas of the neostriatum (Schumacher *et al.*, 1991; Frim *et al.*, 1993d), a neuronal population that degenerates later in the HD disease process.

C. Intrastriatal Transplants of Encapsulated NGF Secreting Cells into Intact Rats: Effects upon Cholinergic and Noncholinergic Striatal Neurons

It appeared from initial studies that NGF delivered from genetically modified cells could prevent the loss of both cholinergic and noncholinergic neurons in a rodent model of HD. To further characterize this neuroprotective effect, we examined the ability of polymer-encapsulated NGF-producing fibroblasts to influence normal intact populations of striatal neurons (Kordower *et al.*, 1996). Using an immunoisolatory polymeric device, encapsulated baby hamster kidney cells (BHK), which were previously transfected with a DHFR-based expression vector containing the human NGF gene (BHK-NGF), were transplanted unilaterally into the striatum of normal rats. Control rats received identical transplants of encapsulated BHK cells that were not transfected with the NGF construct (BHK-Control). Rats were sacrificed 1, 2, or 4 weeks postimplantation and cell size and optical density for immunohistochemical staining was performed. At each time point, ChAT-immunoreactive (ChAT-ir) neurons hypertrophied in response to the BHK-NGF secreting grafts but not in response to control grafts. Furthermore, ChAT-ir striatal neurons displayed an enhanced optical density reflecting a putative increase in ChAT production in rats receiving BHK-NGF. In addition to changes in ChAT-ir, encapsulated BHK-NGF secreting grafts induced a significant hypertrophy of noncholinergic neuropeptide Y (NPY)-ir neurons. This hypertrophy was observed 1, 2, and 4 weeks postimplantation. In contrast to ChAT-ir, the BHK-NGF grafts did not influence the optical density of NPY-ir neurons.

To determine whether the hypertrophy of cholinergic and noncholinergic striatal neurons by encapsulated BHK-NGF grafts was dependent upon the continued presence of the implant, rats received BHK-NGF grafts for one week. The grafts were then retrieved and the animals allowed to survive for an additional 3 weeks. The size of ChAT-ir and NPY-ir striatal neurons in these animals was compared to rats receiving BHK-NGF grafts for 4 weeks. The hypertrophy of both ChAT-ir and NPY-ir induced by the transplant dissipated upon removal of the graft indicating that chronic NGF is required for these effects to occur.

NGF delivered from encapsulated cells clearly influenced both cholinergic and noncholinergic neurons in a reversible manner. However, no data existed concerning the extent of diffusion of neurotrophic factors such as NGF from encapsulated cells. An understanding of the spread of NGF, or any therapeutic molecule, is an essential piece of information that would ultimately determine the feasibility of its delivery into discrete CNS sites. To assess the spread of NGF from the capsules, additional rats received grafts of either encapsulated BHK-NGF cells or BHK-Control cells and were sacrificed 1 week postimplantation using a specially designed fixation procedure and staining protocol to visualize NGF. The spread of NGF secreted from the capsule was determined to be 1 to 2 mm (Kordower *et al.*, 1996). Interestingly, we were able to visualize neurons within the striatum that bound, internalized, and transported NGF secreted from the BHK-NGF implant to the striatal neuron perikarya. In BHK-Control grafted animals, staining was limited to the host basal forebrain system and no staining was observed within the striatum. These data provide *de facto* evidence that NGF transplants can influence both cholinergic and noncholinergic neurons, that NGF diffuses a considerable distance in the striatum, and that NGF transfer may provide the basis for further evaluating this approach in animal models of HD.

D. Transplants of Encapsulated Cells Genetically Modified to Secrete NGF: Rescue of Cholinergic and Noncholinergic Neurons Following Striatal Lesions

Having determined that encapsulated NGF-producing fibroblasts could effect cholinergic as well as noncholinergic striatal neurons, the next series of studies evaluated the neuroprotective effects of BHK-NGF cells in a rodent model of HD. Rats received unilateral NGF-producing or control BHK cell implants into the lateral ventricle (Emerich *et al.*, 1994a). Three days later, all animals received unilateral injections of QA (225 nmols) or saline vehicle into the ipsilateral striatum. Rats were sacrificed 4 weeks later. Nissl-stained sections revealed a comprehensive shrinkage of the size of the lesion within the striatum following NGF treatment with a concomitant diminution of GFAP-ir astrocytosis. Qualitatively, rats receiving BHK-NGF grafts displayed significantly more ChAT-ir and NADPH-positive neurons within the striatum ipsilateral to the lesion relative to BHK-Control-grafted animals. As measured by ELISA,

NGF was released by the encapsulated BHK-NGF cells prior to implantation and following removal. Morphology of retrieved capsules revealed numerous viable and mitotically active BHK cells. These results supported the initial studies by Frim and associates (1993a–d) and suggested that implantation of polymer-encapsulated NGF-secreting cells can be used to protect cholinergic and noncholinergic neurons from excitotoxic damage in a rodent model of HD.

In order to determine if the anatomical protection afforded by NGF treatment was manifest as a functional protection, QA-lesioned animals were tested on a battery of behavioral tests. Approximately two weeks following surgery, animals were tested for apomorphine-induced rotation behavior. Animals receiving QA lesions together with BHK-Control cells showed a pronounced rotation response to apomorphine that increased over repeated test sessions. In contrast, animals that received BHK-NGF cells rotated significantly less than those animals receiving BHK-Control cells or QA alone (Fig. 1). To further characterize the extent of behavioral recovery produced by NGF in QA-lesioned rats, a series of animals were bilaterally implanted with BHK-NGF or BHK-Control cells. One week later, these animals received bilateral intrastriatal injections of QA (150 nmol) and were tested for changes in locomo-

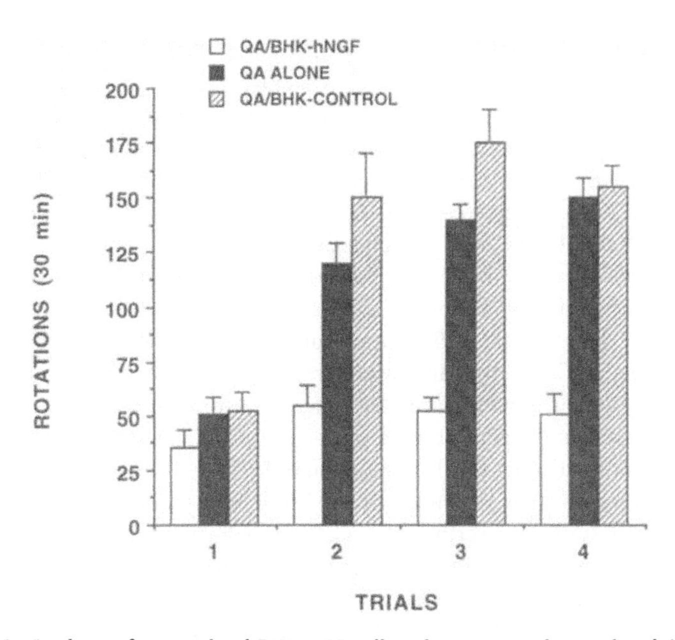

FIGURE 1 Implants of encapsulated BHK-NGF cells reduce apomorphine-induced (1 mg/kg) rotations in rats after unilateral intrastriatal injections of QA. This figure shows the mean + SEM number of rotations in a 30-minute period. Animals were tested over 4 trials with each separated by 3 to 4 days.

tor activity and responsiveness to haloperidol in a catalepsy test. Animals receiving QA lesions without BHK cell implants or implants of BHK-Control cells showed a pronounced hyperactivity when tested in automated activity chambers as well as a diminished cataleptic response to haloperidol. In contrast, animals receiving BHK cells producing NGF showed a significant attenuation of the hyperactivity produced by QA. These same animals showed a normal cataleptic response to haloperidol further indicating the anatomical protection afforded by NGF in this model was paralleled by a robust behavioral protection.

E. Transplants of Encapsulated Cells Genetically Modified to Secrete CNTF: Rescue of Cholinergic and GABAergic Neurons Following Striatal Lesions

When delivered via genetically modified cells, NGF has a clear and potent effect upon both normal and damaged cholinergic and noncholinergic striatal neurons. It remains important, however, to determine whether these effects are limited to NGF and if not to compare the potency of NGF to other neurotrophic factors in these model systems. Ciliary neurotrophic factor (CNTF) is a member of the alpha-helical cytokine superfamily, which has well-documented functions in the peripheral nervous system (Lin *et al.*, 1989; Stockli *et al.*, 1989, 1991; Arakawa *et al.*, 1990; Sendtner *et al.*, 1990, 1992; Apfel *et al.*, 1993; Forger *et al.*, 1993; Masu *et al.*, 1993). Recently it has become clear that CNTF also influences a wide range of CNS neurons. CNTF administration prevents the loss of cholinergic, dopaminergic, and GABAergic neurons in different CNS lesion paradigms (Clatterbuck *et al.*, 1993; Hagg and Varon, 1993; Hagg *et al.*, 1993) Importantly, an initial study demonstrated that infusions of CNTF into the lateral ventricle prevented the loss of Nissl-positive striatal neurons following QA administration. To confirm and expand upon these studies we conducted a series of studies examining the ability of CNTF to protect against QA lesions (Emerich *et al.*, 1996). In these studies, animals received intraventricular implants of CNTF producing BHK cells followed by QA lesions as described above. Animals were again tested for rotation behavior, but were also examined for their ability to retrieve food pellets using a staircase apparatus. Rats receiving BHK-CNTF cells rotated significantly less than animals receiving BHK-Control cells. No behavioral effects of CNTF were observed on the staircase test. An analysis of Nissl-stained sections demonstrated that the size of the lesion was significantly reduced in those animals receiving BHK-CNTF cells (1.44 ± 0.34 mm^2) compared with those animals receiving control implants (2.81 ± 0.25 mm^2) (Fig. 2b,c). Quantitative analysis of striatal neurons further demonstrated that both ChAT and glutamic acid decarboxylase-(GAD) ir neurons were protected by BHK-CNTF implants. The loss of ChAT-ir neurons in those animals receiving CNTF implants was 12% compared to 81% in those animals receiving control cell implants (Fig. 2d–f). Similarly, the loss of GAD-ir neurons was attenuated in animals receiving CNTF-producing cells (20%) compared to those animals receiving control cell implants (72%) (Fig. 2g–i). In contrast, a similar loss of NADPH-diaphorase-positive cells was observed in the striatum of both implant groups (65–78%). Analysis of retrieved capsules revealed numerous viable and mitotically active BHK cells that continued to secrete CNTF (Fig. 2a)

A separate series of animals were tested on a battery of behavioral tests to determine the extent or behavioral protection produced by CNTF (Emerich *et al.*, 1997a). Bilateral infusions of high doses of QA produced a significant loss of body weight and mortality that was prevented by prior implantation with CNTF-secreting cells. Moreover, QA produced a marked hyperactivity, an inability to use the forelimbs to retrieve food pellets in a staircase test, increased the latency of the rats to remove adhesive stimuli from their paws, and decreased the number of steps taken in a bracing test that assessed motor rigidity. Finally, the QA-infused animals were impaired in tests of cognitive function—the Morris water maze spatial learning task and a delayed nonmatching to position operant test of working memory. Prior implantation with CNTF-secreting cells prevented the onset of all the above deficits so that implanted animals were nondistinguishable from shamlesioned controls (see Table I; Emerich *et al.*, 1997a). At the conclusion of behavioral testing, 19 days following QA, the animals were sacrificed for neurochemical determination of striatal ChAT and GAD levels. This analysis revealed that QA decreased striatal ChAT levels by 35% and striatal GAD levels by 45%. In contrast, CNTF-treated animals did not exhibit any decrease in ChAT levels and only a 10% decrease in GAD levels (see Table II).

These results extended previous findings and began to define the extent of both the quantitative and qualitative aspects of CNTF's behaviorally protective effects. The observation that CNTF may prevent the occurrence of both motor and nonmotor deficits following QA has particular relevance for the treatment of HD, which is characterized by a wide range of behavioral alterations. Huntington's chorea, as the name implies, is a disorder most often associated with pronounced motor changes. However, neurological and psychiatric changes frequently occur as much as a decade prior to onset of the motor symptoms. Indeed, given the severity and persistence of the cognitive changes in HD, any viable neuroprotective strategy would necessarily need to exert beneficial effects on cognitive and psychiatric symptoms.

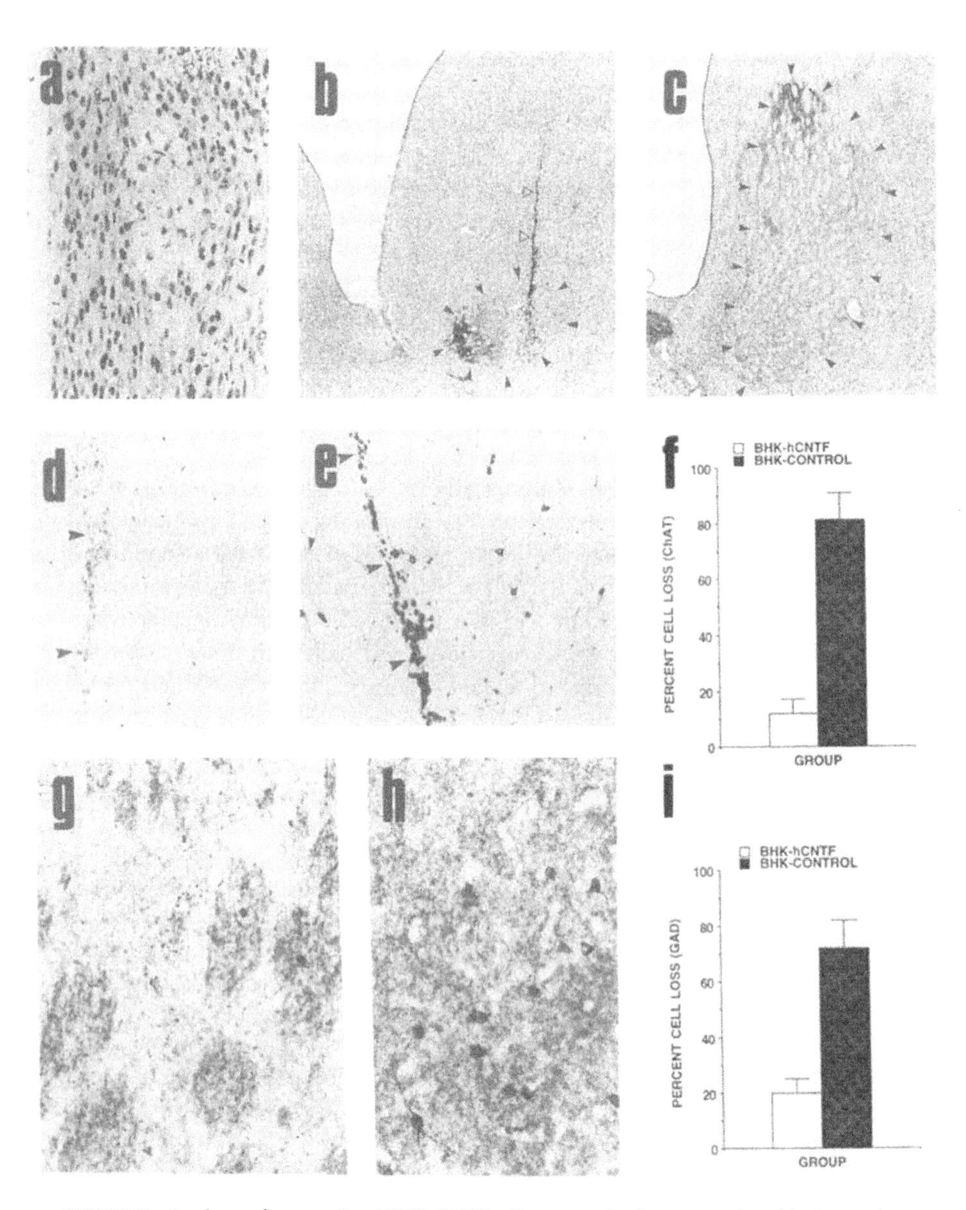

FIGURE 2 Implants of encapsulated BHK-CNTF cells prevent the damage produced by intrastriatal injections of QA. Panel a shows a photomicrograph of hemotoxylin and eosin stained BHK cells in a capsule that was retrieved from a QA-lesioned rat 70 days following implantation. Note the numerous viable BHK cells packed at a moderate to high density within the capsule. Nissl-stained sections revealed that QA produced large spherical lesions within the striatum (b) that were largely prevented by CNTF treatment (c). In addition to the protection evident in Nissl-stained sections, CNTF prevented the loss of ChAT and GAD-positive striatal neurons. QA produced large (70–80%) losses of ChAT-(d) and GAD-positive neurons (g). In contrast, CNTF cells minimized this loss resulting in only a 12% loss of ChAT-positive neurons (e and f) and a 20% loss of GAD-positive neurons (h and i).

TABLE I Behavioral Protection Produced by NGF and CNTF in Rodent Models of Huntington's Disease

Behavioral/functional measure	Neurotrophic factor	
	NGF	CNTF
General		
Body weight	+	+
Mortality	+	+
Motor Tests		
Apomorphine-induced rotation	+	+
Skilled paw use (staircase test)	−	+/−
Bracing	NE	+
Tactile adhesive removal	NE	+
Balance (rotarod)	NE	+
Locomotor activity	+	+
Haloperidol-induced catelepsy	+	NE
Cognitive Tests		
Morris water maze	NE	+
Delayed non-match to position operant task	NE	+

NE = not examined, + = positive outcome, − = negative outcome.

F. Transplants of Encapsuled Cells Genetically Modified to Secrete CNTF: Preservation of Basal Ganglia Circuitry in Nonhuman Primates

An essential prerequisite before the initiation of clinical trials for HD is the demonstration that trophic factors can provide neuroprotection in a nonhuman primate model of HD. Toward this end, three cynomolgus monkeys received unilateral intrastriatal implants of polymer-encapsulated BHK-CNTF cells

TABLE II Striatal ChAT and GAD Neurochemistry

Neurochemical measure	Group		
	Lesion + CNTF	Control	Lesion only
ChAT	267.0 (11.2)	266.3 (13.3)	172.6 (38.7)*
GAD	26.4 (1.4)	29.5 (1.13)	15.8 (5.2)*

ChAT and GAD levels are expressed as mean (± SEM) nmol ACh or GABA/mg protein. *$p < 0.05$ lesion + CNTF vs lesion only.

(Emerich *et al.*, 1997b). The remaining three monkeys served as controls and received BHK-Control cells. One week later, all animals received injections of QA into the ipsilateral caudate and putamen. All monkeys were sacrificed 3 weeks later for histological analysis.

Within the host striatum, QA induced a characteristic lesion of intrinsic neurons together with a substantial atrophy of the striatum. In BHK-Control-implanted monkeys, Nissl-stained sections revealed extensive lesions in the caudate nucleus and putamen that were elliptical in shape. The lesion area in the three BHK-Control animals averaged 317.72 mm^3 ($\pm$26.01) in the caudate and 560.56 mm^3 ($\pm$83.58) in the putamen. Many of the neurons that could be identified were shrunken and displayed a dystrophic morphology. In contrast, the size of the lesion was significantly reduced in BHK-CNTF-implanted monkeys. The lesion area in the three BHK-CNTF monkeys averaged 83.72 mm^3 ($\pm$24.07) in the caudate and 182.54 mm^3 ($\pm$38.45) in the putamen. Numerous healthy-appearing Nissl-stained neurons were observed within the striatum of BHK-CNTF-implanted rats following the QA lesion even in regions proximal to the needle tract.

The neuroprotective effects of CNTF were examined in further detail by quantifying the loss of specific cell types within the striatum (Table 3). Lesioned monkeys receiving BHK-Control implants displayed a significant loss (caudate = 89.0%; putamen = 94.1%) of GAD-ir neurons within the striatum ipsilateral to the transplant. Many remaining GAD-ir striatal neurons appeared atrophic relative to neurons on the contralateral side. The excitotoxic degeneration of GAD-ir neurons in both the caudate nucleus and putamen was significantly attenuated in monkeys receiving implants of polymer-encapsulated BHK-CNTF cells as these monkeys displayed only a 64.0% and 64.1% reduction in GAD-ir neurons in the caudate and putamen, respectively, relative to the contralateral side. In addition to protecting the GABAergic cell bodies, CNTF implants sustained the striatal GABAergic efferent pathways. Enhanced DARPP-32 immunoreactivity (a marker for GABAergic terminals) was seen within the globus pallidus and pars reticulata of the substantia nigra. Optical density measurements revealed a significant reduction (49.0 $\pm$ 5.2%) in DARPP-32 immunoreactivity within the globus pallidus in BHK-Control animals ipsilateral to the lesion. This reduction was significantly attenuated 12.0 $\pm$ 4.3%) in BHK-CNTF implanted animals. Likewise, the reduction in the optical density of DARPP-32-immunoreactivity in the pars reticulata of animals receiving BHK-Control implants (17 $\pm$ 1.8%) was significantly attenuated (4.3 $\pm$ 0.4%) by CNTF implants.

ChAT-ir and diaphorase-positive neurons were also protected by CNTF administration. Monkeys receiving BHK-Control implants displayed significant reductions in ChAT-positive neurons with the caudate (80.3%) and putamen (87.9%) compared to only a 41.2% (caudate) and 54.1% (putaman) in monkeys

TABLE III Neuronal Cell Counts in QA-Lesioned Monkeys

	Caudate			Striatal region	Putamen	
	Intact side	Lesioned/implanted side	% loss	Intact side	Lesioned/implanted side	% loss
GAD-positive Neurons						
CNTF	18088 (437)	6516 (1233)	64.0*	17731 (604)	6374 (851)	64.1*
Control	19298 (1691)	2133 (784)	89.0	16674 (867)	978 (545)	94.1
ChAT-positive Neurons						
CNTF	2259 (188)	1328 (147)	41.2*	2327 (209)	1069 (36)	54.1*
Control	2239 (164)	441 (189)	80.3	2933 (172)	354 (253)	87.9
NADPH-d-positive Neurons						
CNTF	3339 (553)	1733 (489)	48.1*	3508 (824)	1647 (389)	53.1*
Control	3405 (428)	548 (318)	83.9	3393 (106)	458 (347)	86.5

*$p < 0.05$, CNTF vs Control.

receiving BHK-CNTF implants (both $p < 0.05$). Similarly, the loss of NADPH-d-positive neurons in BHK-Control-implanted monkeys (caudate = 83.9%; putamen = 86.6%) was significantly attenuated (caudate = 48.1%; putamen = 53.1%) in monkeys receiving encapsulated CNTF implants.

These data clearly demonstrate that intrastriatal grafts of CNTF-producing fibroblasts protect GABAergic, cholinergic, and NADPH-d-containing neurons. However, the use of trophic factors, or any novel therapeutic strategy for HD has been impeded by our inability to provide a rationale for how these approaches would influence critical nonstriatal regions such as the cerebral cortex that also degenerate in this disorder. Accordingly, we quantified the size of neurons from layer V of monkey motor cortex from a series of Nissl-stained sections in each of the control and CNTF-treated monkeys. QA produced a marked retrograde degeneration of cortical neurons in this region known to project to the striatum. Although the neuron number was unaffected, monkeys receiving BHK-Control implants displayed a significant atrophy (27%) of neurons in layer V of the motor cortex ipsilateral to the lesion. This atrophy was significantly attenuated (6%) in BHK-CNTF-implanted animals. Further analysis demonstrated that the atrophy of cortical neurons was not due to general volumetric changes but rather that it occurred preferentially in the medium-sized (300–400 μm and 400–500 μm cross-sectional area) neurons of the motor cortex that project to the striatum.

Together these data provide the first demonstration that a therapeutic intervention can influence the degeneration of striatal neurons and disruption of basal ganglia circuitry in a primate model of HD. Not only are GABAergic neurons viable in CNTF-treated monkeys, but DARPP-32-ir reveals that the two critical GABAergic efferent projections from striatum to globus pallidus and the pars reticulata of the substantia nigra are sustained in these animals. Moreover, CNTF implants exerted a robust neuroprotective effect on the cortical neurons innervating the striatum. These data indicate that a major component of the basal ganglia loop circuitry, the cortical → striatal → globus pallidus/substantia nigra outflow circuitry is sustained by cellular delivery of CNTF.

IV. MECHANISM OF NEUROTROPHIC FACTOR EFFECTS IN MODELS OF HUNTINGTON'S DISEASE

Although the neuroprotective effects of NGF and CNTF transplants upon striatal neurons in models of HD are striking, the mechanisms underlying these effects remain to be fully elucidated. Under most circumstances, trophic factors function through receptor-coupled mechanisms. Signal transduction for NGF occurs via binding to its high-affinity tropomycin-related kinsase (trk A) receptor. CNTF actions occur following binding to the alpha subunit of the CNTF receptor (CNTFRα). In the rodent and primate striatum, trkA receptors are localized exclusively within cholinergic perikarya (Steininger *et al.*, 1993; Sobreviela *et al.*, 1994; Kordower *et al.*, 1994b). Thus it is unclear by what mechanism grafts of NGF-producing cells provide trophic support to the noncholinergic striatal neurons. Similarly, CNTFRα-mRNA and protein expression is low within the rodent and primate striatum (MacLennan *et al.*, 1994, 1996; Kordower *et al.*, 1997), suggesting that the trophic influences provided by CNTF-secreting transplants are not mediated through receptor-related events.

Interestingly, there appears to be a dichotomy with regards to the neuronal populations protected by NGF depending upon the method of neurotrophin delivery (see Table IV). Infusions of recombinant NGF spare only cholinergic striatal interneurons destined to degenerate following excitotoxic lesions of the striatum (Davies and Beardsall, 1992; Kordower *et al.*, 1994b, Venero *et al.*, 1994). Similarly, intrastriatal infusion of NGF specifically induces hypertrophy of cholinergic neurons (Gage *et al.*, 1989; Bartus *et al.*, 1996) and specifically increases ChAT mRNA (Venero *et al.*, 1994) in intact animals. These findings are not surprising because immunohistochemical studies have demonstrated that the trk A receptor that transduces the NGF signal is located exclusively within cholinergic interneurons in the striatum (Steinenger *et al.*, 1993; Sobreviela *et al.*, 1994; Kordower *et al.*, 1994b). Because cholinergic degeneration is not a central pathology in HD, sparing these neurons would likely not be a useful treatment strategy. However, a consensus has emerged demonstrating that cellular delivery of NGF clearly provides trophic influences to noncholinergic, trkA-negative striatal neurons. As detailed earlier, a number

TABLE IV The Effects of NGF on CNS Neurons Following Infusion or Cellular Delivery

Reference	Method	Neurons affected
Beardsall and Davies, 1992	Infusion	cholinergic
Venero *et al.*, 1994	Infusion	cholinergic
Kordower *et al.*, 1994c	Infusion	cholinergic
Schumacher *et al.*, 1991	NGF-Fibroblast Graft	cholinergic and noncholinergic
Frim *et al.*, 1993a-d	NGF-Fibroblast Graft	cholinergic and noncholinergic
Emerich *et al.*, 1994a	NGF-Fibroblast Graft	cholinergic and noncholinergic
Chen *et al.*, 1994	NGF-Fibroblast Graft	cholinergic and noncholinergic
Kordower *et al.*, 1996	NGF-Fibroblast Graft	cholinergic and noncholinergic
Martinez-Serrano and Björklund 1996	NGF-Stem Cell Graft	cholinergic and noncholinergic
Kordower *et al.*, 1997	NGF-Stem Cell Graft	cholinergic and noncholinergic

of studies have demonstrated that transplants of NGF-secreting cells rescue cholinergic and noncholinergic striatal neurons from excitotoxic degeneration (Schumacher *et al.*, 1991; Emerich et al., 1994a; Frim *et al.*, 1993b–d, 1994). Moreover, NGF-secreting cells induce hypertrophy in both cholinergic and noncholinergic striatal neurons in intact animals (Kordower *et al.*, 1996). Common to all these studies was the use of a fibroblast as a cellular vehicle to deliver the NGF. These findings led us to hypothesize that cellular delivery of NGF plus an additional factor secreted by fibroblasts combined to provide a novel sphere of trophism that included both cholinergic and noncholinergic striatal neurons (Kordower *et al.*, 1996). However, we (Kordower *et al.*, in press) and others (Martinez-Serrano and Björklund, 1996) have recently demonstrated that grafts of stem cells that have been genetically modified to secrete NGF also prevent the degeneration of GABA-ergic and NADPH-expressing striatal neurons, in addition to cholinergic cells. It remains likely that the protection of ChAT-positive neurons in this and previous studies employing cellular delivery of NGF results from the neurotrophin binding to trkA receptors located upon cholinergic neurons. It is unlikely, however, that an additional factor common to both stem cells and fibroblasts synergizes with cellularly delivered NGF to provide comprehensive neuroprotection to cholinergic and noncholinergic striatal neurons. As such, the mechanism by which NGF-secreting grafts provide trophic support for noncholinergic, non-trkA, expressing cells remains elusive.

The type of neuronal cell death in HD suggests that prevention of cellular energy compromise or excitotoxicity may be of therapeutic value. Previous cell culture studies have shown a survival value of selected neurotrophic factors for striatal neurons. BDNF, NT-3, and NT-4/5, but not NGF, have enhanced the survival and differentiation of fetal striatal GABAergic neurons *in vitro* (Mizuno *et al.*, 1994; Widmer and Hefti, 1994; Ventimiglia *et al.*, 1995). These factors are most effective in protecting neurons containing calcium-binding proteins, which could mean a common mechanism of action. PDGF (Nakao *et al.*, 1994) and BDNF (Nakao *et al.*, 1995; Ventimiglia *et al.*, 1995) have also been shown to promote neurite outgrowth and increase cell body size of striatal neurons *in vitro*. These effects are consistent with a developmental and regulatory role of these neurotrophic factors on striatal neurons *in vivo*.

Of interest to HD paradigms, bFGF (Freese *et al.*, 1992) and BDNF (Nakao *et al.*, 1995) can protect against glutamate receptor-mediated toxicity in cultured striatal neurons. The mechanism behind this protection may be improved calcium buffering (Nakao *et al.*, 1995) because BDNF increases expression of calcium-binding proteins in caudate-putamen neurons (Mizuno *et al.*, 1994; Widmer and Hefti, 1994; Ventimiglia *et al.*, 1995). *In vivo*, NGF (Schumacher *et al.*, 1991; Frim *et al.*, 1993d; Emerich *et al.*, 1994) and bFGF (Frim *et al.*, 1993c) produced by genetically engineered fibroblasts reduce striatal neuronal loss induced by neurotoxic glutamate receptor agonists. When NGF is coadministered in time with quinolinate, only TrkA-expressing cholinergic neurons are spared (Venero *et al.*, 1994). Despite *in vitro* studies showing that BDNF is trophic for striatal GABAergic neurons (Mizuno *et al.*, 1994; Widmer and Hefti, 1994; Nakao *et al.*, 1995; Ventimiglia *et al.*, 1995) and protective against NMDA toxicity (Nakao *et al.*, 1995), BDNF has yet to be shown to protect against excitotoxic lesions *in vivo* (Frim *et al.*, 1993d).

NGF-producing fibroblast grafts can also reduce lesions caused by mitochondrial impairment following administration of 3-NP (Frim *et al.*, 1993b). Catalase expression in and around the NGF grafts indicates that this antioxidative enzyme may play a role in the neuroprotection (Frim *et al.*, 1994b). Other mechanisms involved in NGF-mediated neuroprotection include regulation of the sodium/potassium ATPase (Varon and Skaper, 1983) as well as alterations in other antioxidant enzymes such as glutathione peroxidase and superoxide dismutase (Nisticò *et al.*, 1992). Inhibition of neuronal NOS by 7-nitroindazole protects against 3-NP induced lesions and attenuates hydroxyl radical and peroxynitrite formation (Schulz *et al.*, 1995). NGF neuroprotection against 3-NP may be associated with inhibition of neuronal NOS or increased scavenging of superoxide radicals as NGF attenuates 3-NP induced formation of 3-nitrotyrosine (Galpern *et al.*, 1996). Although the precise cellular events that cause neuronal death are unknown, increased intracellular calcium is known to activate numerous deleterious enzyme cascades resulting in the production of superoxide radicals and nitric oxide. Peroxynitrite, formed by the reaction of superoxide and nitric oxide, is highly reactive (Beckman *et al.*, 1990) and may be involved in the neurodegenerative processes of HD (Galpern *et al.*, 1996).

Systemic administration of the succinate dehydrogenase inhibitor 3-NP produces bilateral striatal lesions and behavioral abnormalities characteristic of HD (Beal *et al.*, 1993; Simpson and Isacson, 1993). 3-NP results in increased generation of hydroxyl radicals and 3-nitrotyrosine (Schulz *et al.*, 1995; Galpern *et al.*, 1996), the product of peroxynitrite-mediated nitration of tyrosine (Ischiropoulos *et al.*, 1992). The elevations in hydroxyl radicals and 3-nitrotyrosine formation as well striatal lesion size are attenuated in copper/zinc superoxide dismutase transgenic mice, suggesting a decrease in 3-NP-induced peroxynitrite formation due to increased superoxide scavenging. Furthermore, because inhibition of neuronal nitric oxide synthase (nNOS) by 7-nitroindazole (7-NI) protects against striatal lesions and reduces the increase in 3-nitrotyrosine associated with 3-NP administration, involvement of nitric oxide in 3-NP neurotoxicity is likely Schulz *et al.*, 1995). 3-nitrotyrosine is formed by peroxynitrite-mediated tyrosine nitration, and the reduced production of this marker in NGF[+] grafted rats (Galpern *et al.*, 1996) suggests that the neuroprotective effect of NGF on striatal neurons may be mediated in part by reduced peroxynitrite formation. Furthermore, as the reaction of superoxide and nitric oxide generates peroxynitrite, NGF may increase the scavenging of superoxide radicals and/or may decrease the production of nitric oxide, possibly by inhibiting nNOS. We have shown by the histological analysis of Nissl- and DARPP-32-stained sections that there is no visible neuronal damage at the time of neurochemical analysis (2.5 hours after 3-NP dosing) (Galpern *et al.*, 1996). However, at 7 days after systemic administration of 3-NP, there is bilateral degeneration of the lateral striatum (Beal *et al.*, 1993; Frim *et al.*, 1993b). Because 3-NP-induced striatal lesions are reduced by the prior removal of cortical-striatal glutamatergic input via decortication, it is possible that the cell death is a result of excitotoxicity secondary to impaired energy metabolism (Beal *et al.*, 1993; Simpson and Isacson, 1993). Moreover, the findings that the hydroxyl radical production and 3-nitrotyrosine formation associated with 3-NP administration are decreased in mice overexpressing SOD and in rats treated with the nNOS inhibitor 7-NI (Schulz *et al.*, 1995) indicate a role for both superoxide and nitric oxide as well as peroxynitrite in 3-NP-mediated neuronal death. The findings of reduced levels of 3-nitrotyrosine in NGF[+] rats as compared to the NGF[−] rats suggests that the neuroprotective effects of NGF may involve a reduction in peroxynitrite formation (Galpern *et al.*, 1996). Future studies need to further clarify the mechanism of NGFs and other trophic factors effects on striatal neurons. It will be particularly important to determine if CNTF and other factors exert their effects via a mechanism similar to that of NGF.

V. CONCLUSION

HD is a devastating neurodegenerative disorder with no effective treatment for the behavioral symptoms or the associated neural degeneration. In recent years, the advent of appropriate animal models has permitted the evaluation of multiple therapeutic strategies. This chapter has highlighted one of the more exciting therapeutic possibilities, which suggests that cellular delivery of neurotrophic factors may be one means of treating the neuropathological and behavioral consequences of excitotoxicity. Although the mechanisms by which these factors exert their beneficial effects remain unclear, their clearcut potency in both rodent and primate models of HD provide the hope that a means of preventing or slowing the relentlessly progressive motor and cognitive declines in HD may be forthcoming. Furthermore, excitotoxicity has been implicated in a variety of pathological conditions including ischemia and neurodegenerative diseases such as Huntington's, Parkinson's, and Alzheimer's (Olney, 1989). Accordingly, biologically delivered neurotrophic factors may provide one means of preventing the cell loss and associated behavioral abnormalities of these and possibly other human disorders.

ACKNOWLEDGMENTS

We thank Dr. Wendy R. Galpern for discussion of these issues and the help by Ms. Sandra Pohlman in preparation of this manuscript.

REFERENCES

Albin, R. L., and Greenamyre, J. T. (1992). Alternative excitotoxic hypothesis. *Neurology* 42:733–738.

Anderson, K. D., Panayotatos, N., Cordoran, T., Lindsay, R. M., and Wiegand, S. J. (1996). Ciliary neurotrophic factor protects striatal output neurons in an animal model of Huntington's disease. *Proc Natl Acad Sci* 93:7346–7351.

Apfel, S. C., Arezzo, J. C., Moran, M., and Kessler, J. A. (1993). Effects of administration of ciliary neurotrophic factor on normal motor and sensory peripheral nerves in vivo. *Brain Res* 604:1–6.

Arakawa, Y., Sendtner, M., and Thoenen, H. (1990). Survival effect of ciliary neurotrophic factor (CNTF) on chick embryonic motoneurons in culture: Comparison with other neurotrophic factors and cytokines. *J Neurosci* 10:3507–3515.

Ashizawa, T., Wong, L-J. C., Richards, C. S., Caskey, C. T., and Jankovic, J. (1994). CAG repeat size and clinical presentation in Huntington's disease. *Neurology* 44:1137–1143.

Bartus, R. T., Dean, R. L., Abelleira, S., Vinod, C., and Kordower, J. H. (1996). Dissociation of p75 receptors and nerve growth factor neurotrophic effects:Lack of p75 immunoreactivity in striatum following physical trauma, excitotoxicity and NGF administration. *Restor Neurol Neurosci* 10:49–59.

Beal, M. F. (1992). Does impairment of energy metabolism result in excitotoxic neuronal death in neurodegenerative illnesses? *Ann Neurol* 31:119–130.

Beal, M. F. (1995). Aging, energy, and oxidative stress in neurodegenerative diseases. *Ann Neurol* 38:357–366.

Beal, M. F., Brouillet, E., Jenkins, B. G., Ferrante, R. J., Kowall, N. W., Miller, J. M., Storey, E., Srivastava, R., Rosen, B. R., and Hyman, B. T. (1993). Neurochemical and histologic characterization of striatal excitotoxic lesions produced by the mitochondrial toxin 3-nitropropionic acid. *J Neurosci* 13:4181–4192.

Beal, M. F., Ferrante, R. J., Swartz, K. J., and Kowall, N. W. (1991). Chronic quinolinic acid lesions in rats closely resemble Huntington's disease. *J Neurosci* 11:1649–1659.

Beal, M. F., Kowall, N. W., Ellison, D. W., Mazurek, M. F., Swartz, K. J., and Martin, J. B. (1986). Replication of the neurochemical characteristics Huntington's disease by quinolinic acid. *Nature* 321:168–171.

Beal, M. F., Kowall, N. W., Swartz, K. J., Ferrante, R. J., and Martin, J. B. (1989). Differential sparing of somatostatin-neuropeptide Y and cholinergic neurons following striatal excitotoxin lesions. *Synapse* 3:38–47.

Beckman, J. S., Beckman, T. W., Chen, J., Marshall, P. A., and Freeman, B. A. (1990). Apparent hydroxyl radical production by peroxynitrite: Implications for endothelial injury from nitric oxide and superoxide. *Proc Natl Acad Sci USA* 87:1620–1624.

Bjorklund, A., Campbell, K., Sirinathsinghji, D. J. Fricker, R. A., and Dunnett, S. B. (1994). Functional capacity of striatal transplants in the rat Huntington model. In: Dunnett, S. B., and Bjorklund, A., eds., *Functional neural transplantation*. New York: Raven Press, pp. 157–195.

Block, F., Kunkel, M., and Schwarz, M. (1993). Quinolinic acid lesion of the striatum induces impairment in spatial learning and motor performance in rats. *Neurosci Lett* 149:126–128.

Breakefield, X. O. (1989). Combining CNS transplantation and gene transfer. *Neurobiol Aging* 10:647–648.

Clatterbuck, R. E., Price, D. L., and Koliatsos, V. E. (1993). Ciliary neurotrophic factor prevents retrograde neuronal death in the adult central nervous system. *Proc Natl Acad Sci* 90:2222–2226.

Conneally, P. M. (1984). Huntington's disease: Genetics and epidemiology. *Am J Hum Genet* 36:506–526.

Coyle, J. T., and Schwarcz, R. (1976). Lesion of striatal neurons with kinaic acid provides a model for Huntington's chorea. *Nature* 263:244–246.

Davies, S. W., and Beardsall, K. (1992). Nerve growth factor selectively prevents excitotoxin induced degeneration of striatal cholinergic neurons. *Neurosci Lett* 140:161–164.

Emerich, D. F., and Sanberg, P. R. (1992). Animal models in Huntington's disease. In: Boulton, A. A., Baker, G. B., and Butterworth, R. F., eds., *Neuromethods,* Vol. 17, *Animal models of neurological disease.* N.J.: Humana Press, pp. 65–134.

Emerich, D. F., Cain, C. K., Greco, C., Saydoff, J. A., Hu, Z.-Y., Liu, H., and Lindner, M. D. (1997a). Cellular delivery of human CNTF prevents motor and cognitive dysfunction in a rodent model of Huntington's disease. *Cell Transpl* 6:249–266.

Emerich, D. F., Hammang, J. P., Baetge, E. E., and Winn, S. R. (1994a). Implantation of polymerencapsulated human nerve growth factor-secreting fibroblasts attenuates the behavioral and neuropathological consequences of quinolinic acid injections into rodent striatum. *Exper Neurol* 130:141–150.

Emerich, D. F., Winn, S. R., Chen, E.-Y., Chu, Y., McDermott, P., Baetge, E., and Kordower, J. H. (1997b). Protection of basal ganglia circuitry by encapsulated CNTF-producing cells in a primate model of Huntington's disease. *Nature* 386:395–399.

Emerich, D. F., Winn, S. R., Christenson, L., Palmatier, M., Gentile, F. T., and Sanberg, P. R. (1992). A novel approach to neural transplantation in Parkinson's disease: Use of polymerencapsulated cell therapy. *Neurosci Biobehav Rev* 16:437–447.

Emerich, D. F., Winn, S. R., Harper, J., Hammang, J. P., Baetge, E. E., and Kordower, J. H. (1994b). Transplantation of polymer-encapsulated cells genetically modified to secrete human nerve growth factor prevents the loss of degenerating cholinergic neurons in nonhuman primates. *J Comp Neurol* 349:148–164.

Emerich, D. F., Winn, S. R., Lindner, M. D., Frydel, B. R., and Kordower, J. H. (1996). Implants of encapsulated human CNTF-producing fibroblasts prevent behavioral deficits and striatal degeneration in a rodent model of Huntington's disease. *J Neurosci* 16:5168–5181.

Farrer, L. A., and Conneally, P. M. (1985). A genetic model for age at onset in Huntington's disease *Am J Hum Genet* 37:350–357.

Ferrante, R. J., Beal, M. F., Kowall, N. W., Richardson, E. P., and Martin, J. B. (1987). Sparing of acetylcholinesterase-containing striatal neurons in Huntington's disease. *Brain Res* 415:178–182.

Ferrante, R. J., Kowall, N. W., and Beal, M. F. (1985). Selective sparing of a class of striatal neurons in Huntington's disease. *Science* 230:561–563.

Forger, N. G., Roberts, S. L., Wong, V., and Breedlove, S. M. (1993). Ciliary neurotrophic factor maintains motoneurons and their target miscles in developing rats. *J Neurosci* 13:4720–4726.

Freese, A., Finklestein, S. P., and DiFiglia, M. (1992). Basic fibroblast growth factor protects striatal neurons *in vitro* from NMDA-receptor mediated excitotoxicity. *Brain Res* 575:351–355.

Frim, D. M., Short, M. P., Rosenberg, W. S., Simpson, J., Breakefield, X. O., and Isacson, O. (1993a). Local protective effects of nerve growth factor-secreting fibroblasts against excitotoxic lesions in the rat striatum. *J Neurosurg* 78:267–273.

Frim, D. M., Simpson, J., Uhler, T. A., Short, M. P., Bossi, S. R., Breakefield, X. O., and Isacson, O. (1993b). Striatal degeneration induced by mitochondrial blockade is prevented by biologically delivered NGF. *J Neurosci Res* 35:452–458.

Frim, D. M., Uhler, T. A., Short, M. P., Exxedine, Z. D., Klagsbrun, M., Breakefield, X. O., and Isacson O. (1993c). Effects of biologically delivered NGF, BDNF, and bFGF on striatal excitotoxic lesions. *NeuroReport* 4:367–370.

Frim, D. M., Uhler, T. A., Galpern, W. R., Beal, M. F., Breakefield, X. O., and Isacson, O. (1994a). Implanted fibroblasts genetically engineered to produce brain-derived neurotrophic factor prevent 1-methyl-4-phenylpyridinium toxicity to dopaminergic neurons in the rat. *Proc Natl Acad Sci USA* 91:5104–5108.

Frim, D. M., Wullner, U., Bear, M. R., and Isacson, O. (1994b). Implanted NGF-producing fibroblasts induce catalase and ATP levels but do not affect glutamate receptor binding or NMDA receptor expression in the rat striatum. *Exp Neurol* 128:172–180.

Frim, D. M., Yee, W. M., and Isacson, O. (1993d). NGF reduces striatal excitotoxic neuronal loss without affecting concurrent neuronal stress. *NeuroReport* 4:655–658.

Gage, F. H., Batchelor, P., Chen, K. S., Higgins, G. A., Koh, S., Deputy, S., Rosenberg, M. B., Fisher, W., and Bjorklund, A. (1989). NGF receptor reexpression and NGF-mediated cholinergic neuronal hypertrophy in the damaged adult neostriatum. *Neuron* 2:1177–1184.

Gage, F. H., Wolf, J. A., Rosenberg, M. B., Xu, L., and Yee, J. K. (1987). Grafting genetically modified cells to the brain: Possibilities for the future. *Neuroscience* 23:795–807.

Galpern, W. R., Matthews, R. T., Beal, M. F., and Isacson, O. (1996). NGF attenuates 3-nitrotyrosine formation in a 3-NP model of Huntington's disease. *NeuroReport* 7:2639–2642.

Graveland, G. A., Williams, R. S., and DiFiglia, M. (1985). Evidence of degenerative and regenerative changes in neostriatal spiny neurons in Huntington's disease. *Science* 227:770–773.

Greenamyre, J. T., and O'Brien, C. F. (1986). The role of glutamate in neurotransmission and in neurologic disease. *Arch Neurol* 43:1058–1063.

Greenamyre, J. T., and Shoulson, I. (1994). Huntington's disease. In: Calne, D., ed., *Neurodegenerative diseases*. Philadelphia, Saunders Press, Inc., pp. 685–704.

Hagg, T., and Varon, S. (1993). Ciliary neurotrophic factor prevents degeneration of adult rat substantia nigra dopaminergic neurons *in vivo*. *Proc Natl Acad Sci* 90:6315–6319.

Hagg, T., Quon, D., Higaki, J., and Varon, S. (1993). Ciliary neurotrophic factor prevents neuronal degeneration and promotes low affinity NGF receptor expression in the adult rat CNS. *Neuron* 8:145–158.

Hantraye, P., Riche, D., Maziere, M., and Isacson, O. (1990). A primate model of Huntington's disease: Behavioral and anatomical studies of unilateral excitotoxic lesions of the caudate-putamen in the baboon. *Exp Neurol* 108:91–104.

Isacson, O. (1993). On neuronal health. *Trends Neurosci* 16:306–308.

Isacson, O., Brundin, P., Kelly, P. A. T., Gage, F. H., and Bjorklund, A. (1984). Functional neuronal replacement by grafted striatal neurons in the Ibotenic acid lesioned rat striatum. *Nature* 311:458–35.

Isacson, O., Dawbarn, D., Brundin, P., Gage, F. H., Emson, P. C., and Bjorklund, A. (1987). Neural grafting in a rat model of Huntington's disease: Striosomal-like organization of striatal grafts as revealed by immunocytochemistry and receptor autoradiography. *Neuroscience* 22:401–497.

Isacson, O., Riche, D., Hantraye, P., Sofroniew, M. V., and Maziere, M. (1989). A primate model of Huntington's disease: Cross-species implantation of striatal precursor cells to the excitotoxically lesioned baboon caudate-putamen. *Exp Brain Res* 75:213–220.

Ischiropoulos, H., Zhu, L., Chen, J., Tsai, M., Martin, J. C., Smith, C. D., and Beckman, J. S. (1992). Peroxynitrite-mediated tyrosine nitration catalyzed by superoxide dismutase. *Arch Biochem Biophys* 298:431–437.

Kawaja, M. D., Fagan, A. M., Firestein, B. L., and Gage, F. H. (1991). Intracerebral grafting of cultured autologous skin fibroblasts into the rat striatum:an assessment of graft size and ultrastructure. *J Comp Neurol* 307:695–706.

Kawaja, M. D., Rosenberg, M. B., Yoshida, K., and Gage, F. H. (1992). Somatic gene transfer of nerve growth factor promotes the survival of axotomized septal neurons and the regeneration of their axons in adult rats. *J Neurosci* 12:2849–2864.

Kordower, J. H., Charles, V., Baryer, R., Bartus, R. T., Putney, S., Walus, L. R., and Friden, P. M. (1994a). Intravenous administration of a transferrin receptor antibody-nerve growth factor conjugate prevents the degeneration of cholinergic striatal neurons in a model of Huntington's disease. *Proc Natl Acad Sci* 91:9077–9080.

Kordower, J. H., Chen, E.-Y., Winkler, C., Fricker, R., Charles, V., Messing, A., Mufson, E. J., Wong, S. C., Rosenstein, J. M., Bjorklund, A., Emerich D. F., Hammang, J., and Carpenter, M. K. (1998). Grafts of stem cells genetically modified to secrete NGF: Neuronal rescue, neurite sprouting, and preservation of the blood-brain barrier in a rodent model of Huntington's disease. *J Comp Neurology* 387:96–113.

Kordower, J. H., Chen, E.-Y., Mufson, E. J., Winn, S. R., and Emerich, D. F. (1996). Intrastriatal implants of polymer-encapsulated cells genetically modified to secrete human NGF: Tropic effects upon cholinergic and noncholinergic neurons. *Neuroscience* 72:63–77.

Kordower, J. H., Chen, E.-Y., Sladek, J. R., and Mufson, E. J. (1994b). Trk immunoreactivity in the monkey central nervous system I: Forebrain. *J Comp Neurol* 349:20–35.

Kordower, J. H., Chu, Y.-P., and MacLennan, A. J. (1997). Ciliary neurotrophic factor receptor alpha immunoreactivity in the monkey central nervous system. *J Comp Neurol* 377:365–380.

Kordower, J. H., Liu, Y.-T., Winn, S. R., and Emerich, D. F. (1995). Encapsulated PC12 cell transplants into hemiparkinsonian monkeys: A behavioral, neuroanatomical and neurochemical analysis. *Cell Transplantation* 4:155–171.

Kordower, J. H., Winn, S. R., Liu, Y.-T., Mufson, E. J., Sladek, J. R., Jr., Baetge, E. E., Hammang, J. P., and Emerich, D. F. (1994c). The aged monkey basal forebrain:Rescue and sprouting of axotomized basal forebrain neurons after grafts of encapsulated cells secreting human nerve growth factor. *Proc Natl Acad Sci* 91:10898–10902.

Kowall, N. W., Ferrante, R. J., and Martin, J. B. (1987). Pattern of cell loss in Huntington's disease. *Trends Neurosci* 10:24–29.

Levivier, M., Przedborski, S., Bencsics, C., and Kang, U.-J. (1995). Intrastriatal implantation of fibroblasts genetically engineered to produce brain-derived neurotrophic factor prevents degeneration of dopaminergic neurons in a rat model of Parkinson's disease. *J Neurosci* 15:7810–7820.

Lin, L.-F.H., Mismer, D., Lile, J. D., Armes, L. G., Butler, E. T., Vannice, J. L., and Collins, F. (1989). Purification, cloning, and expression of ciliary neurotrophic factor (CNTF). *Science* 246:1023–1025.

Martinez-Serrano, A., and Bjorklund, A. (1996). Protection of the neostriatum against excitotoxic damage by neurotrophin-producing genetically modified neural stem cells. *J Neurosci* 16:4604–4616.

Masu, Y., Wolf, E., Holtmann, B., Sendtner, M., Brem, G., and Thoenen H. (1993). Disruption of the CNTF gene results in motor neuron degeneration. *Nature* 365:27–32.

MacLennan, J. A., Gaskin, A. A., and Lado, D. C. (1994). CNTFR receptor alpha mRNA expression in rodent and developing rat. *Mol Brain Res* 25:251–256.

MacLennan, J. A., Vinson, E. N., Marks, L., McLaurain, D. L., Pfiefer, M., and Lee, M. (1996). Immunohistochemical localization of ciliary neurotrophic factor receptor alpha expression in the rat nervous system. *J Neurosci* 16:621–630.

Mizuno, K., Carnahan, J., and Nawa, H. (1994). Brain-derived neurotrophic factor promotes differentiation of striatal GABAergic neurons. *Dev Biol* 165:243–256.

Nakao, N., Brundin, P., Funa, K., Lindvall, O., and Odin, P. (1994). Platelet-derived growth factor exerts trophic effects on rat striatal DARPP-32-containing neurons in culture. *Exp Br Res* 101:291–296.

Nakao, N., Brundin, P., Funa, K., Lindvall, O., and Odin, P. (1995). Trophic and protective effects of brain-derived neurotrophic factor on striatal DARPP-32-containing neurons *in vitro*. *Dev Brain Res* 90:92–101.

Nisticò, G., Ciriolo, M. R., Fiskin, K., Iannone, M. De Martino A., and Rotilio, G. (1992). NGF restores decrease in catalase activity and increases superoxide dismutase and glutathione peroxidase activity in the brain of aged rats. *Free Radic Biol Med* 12:177–181.

Olney, J. W. (1989). Excitatory amino acids and neuropsychiatric disorders. *Biol Psychiatr* 26:505–525.

Penney, J. B., Young, A. B., and Shoulson, I. (1990). Huntington's disease in Venezuela: 7 years of follow-up on symptomatic and asymptomatic individuals. *Movement Disorders* 5:93–99.

Rogelj, S., Weinberg, R. A., Fanning, P., and Klagsbrun, M. (1988). Basic fibroblast growth factor fused to a signal peptide transforms cells. *Nature* 331:173–175.

Rosenberg, M. B., Friedman, T., Robertson, R. C., Tuszynski, M., Wolff, J. A., Breakefield, X. O., and Gage, F. H. (1988). Grafting genetically modified cells to the damaged brain: Restorative effects of NGF expression. *Science* 242:1575–1578.

Sagot, Y., Tan, S. A., Baetge, E., Schmalbruch, H., Kato, A. C., and Aebischer, P. (1995). Polymer encapsulated cell lines genetically engineered to release ciliary neurotrophic factor can slow down progressive motor neuronopathy in the mouse. *Eur J Neurosci* 7:1313–1322.

Sanberg, P. R., Calderon, S. F., Giordano, M., Tew, J. M., and Norman, A. B. (1989). The quinolinic acid model of Huntington's disease: Locomotor abnormalities. *Exper Neurol* 105:45–53.

Schulz, J. B., Matthews, R. T., Jenkins, B., Ferrante, R. J., Siwek, D., Henshaw, D. R., Cipolloni, P. B., Mecocci, P., Kowall, N. W., Rosen, B. R., and Beal, M. F. (1995). Blockade of neuronal nitric oxide synthase protects against excitotoxicity *in vivo*. *J Neurosci* 15:8419–8429.

Schumacher, J. M., Short, M. P., Hyman, B. T., Breakefield, X. O., and Isacson, O. (1991). Intracerebral implantation of nerve growth factor-producing fibroblasts protects striatum against neurotoxic levels of excitatory amino acids. *J Neurosci* 45:561–570.

Sendtner, M., Kreutzberg, G. W., and Thoenen, H. (1990). Ciliary neurotrophic factor prevents the degeneration of motor neurons after axotomy. *Nature* 345:440–341.

Sendtner, M., Schmalbruch, H., Stockli, K. A., Carroll, P., Kreutzberg, G. W., and Thoenen, H. (1992). Ciliary neurotrophic factor prevents degeneration of motor neurons in mouse mutant progressive motor neuronpathy. *Nature* 358:502–504.

Shoulson, I. (1981). Functional capacities in patients treated with neuroleptic and antidepressant drugs. *Neurology* 31:1333–1335.

Short, M. P., Rosenberg, M. B., Ezzedine, E. D., Gage, F. H., Friedmann, T., and Breakefield, X. O. (1990). Autocrine differentiation of PC12 cells mediated by retroviral vectors. *Devl Neurosci* 12:34–45.

Simpson, J. R., and Isacson, O. (1993). Mitochondrial impairment reduces the threshold for *in vivo* NMDA-mediated neuronal death in the striatum. *Exp Neurol* 121:57–64.

Sobreviela, T., Clary, D. O., Reichardt, L. F., Brandabur, M. M., Kordower, J. H., and Mufson, E. J. (1994). TrkA immunoreactive profiles in the central nervous system: Colocalization with neurons containing p75 nerve growth factor receptor, choline acetyltransferase, and serotonin. *J Comp Neurol* 350:587–611.

Steininger, T. L., Wainer, B. H., Klein, R., Barbacid, M., and Palfrey, H. C. (1993). High-affinity nerve growth factor receptor (Trk) immunoreactivity is localized in cholinergic neurons of the basal forebrain and striatum in the adult rat brain. *Brain Res* 61:330–335.

Stockli, K. A., Lillien, L. E., Naher-Noe, M., Breitfeld, G., Hughes, R. A., Raff, M. C., Thoenen, H., and Sendtner, M. (1991). Regional distribution, developmental changes and cellular localization of CNTF-mRNA and protein in the rat brain. *J Cell Biol* 115:447–459.

Stockli, K. A., Lottspeich, F., Sendtner, M., Masiakowski, P., Carrol, P., Gotz, R., Lindholm, D., and Thoenen, H. (1989). Molecular cloning, expression, and regional distribution of rat ciliary neurotrophic factor. *Nature* 342:920–923.

Tatter, S. B., Galpern, W. R., and Isacson, O. (1995). Neurotrophic factor protection against excitotoxic neuronal death. *The Neuroscientist* 5:286–297.

The Huntington's Disease Collaborative Research Group. (1993). A novel gene containing a trinucleotide repeat that is expanded and unstable on Huntington's disease chromosomes. *Cell* 72:971–978.

Varon, S., and Skaper, S. D. (1983). The Na+, K+ pump may mediate the control of nerve cells by nerve growth factor. *Trends Biol Sci*:22–25.

Venero, J. L., Beck, K. D., and Hefti, F. (1994). Intrastriatal infusion of nerve growth factor after quinolinic acid prevents reduction of cellular expression of choline acetyltransferase messenger RNA and trkA messenger RNA, but not glutamate decarboxylase messenger RNA. *Neuroscience* 61:257–268.

Ventimiglia, R., Mather, P. E., Jones, B. E., and Lindsay, R. M. (1995). The neurotrophins BDNF, NT-3, and NT-4/5 promote survival and morphological and biochemical differentiation of striatal neurons *in vitro*. *Eur J Neurosci* 7:213–222.

Vonsattel, J. P., Ferrante, R. J., and Stevens, T. J. (1985). Neuropathologic classification of Huntington's disease. *J Neuropathol Exp Neurol* 44:559–577.

Wictorin, K., and Bjorklund, A. (1989). Connectivity of striatal grafts implanted into the ibotenic acid lesioned striatum: II. Cortical afferents. *Neuroscience* 30:297–311.

Wictorin, K, Simerly, R. B., Isacson, O., Wanson, L. W., and Bjorklund, A. (1989). Connectivity of striatal grafts implanted into the ibotemic acid lesioned striatum. III. Efferent projecting neurons and their relationships to host afferents within the grafts. *Neuroscience* 30:313–330.

Widmer, H. R., and Hefti, F. (1994). Neurotrophin-4/5 promotes survival and differentiation of rat striatal neurons developing in culture. *Eur J Neurosci* 6:1669–1679.

CHAPTER 19

Regenerative Strategies for Alzheimer's Disease

MARK H. TUSZYNSKI,*,† AND JEFFREY H. KORDOWER**

*Department of Neurosciences, University of California, San Diego, La Jolla, California, and †Veterans Affairs Medical Center, San Diego, California, **Research Center for Brain Repair and Department of Neurological Sciences, Rush Presbyterian St. Luke's Medical Center, Chicago, Illinois

Alzheimer's disease (AD) is the most common neurodegenerative disorder, currently afflicting an estimated 4 to 5 million Americans and over 8 million people worldwide. Half of individuals over the age of 85 develop AD. The incidence of AD will double in the next 25 years as the proportion of the elderly population increases, expanding a human tragedy that already extracts an enormous toll from its victims and their caregivers. The financial cost to society is also great, with current direct costs of caring for AD victims of 20 to 40 billion dollars per year in the United States (Ernst and Hay, 1994).

Treatment for AD is currently limited to pharmacological agents that enhance the action of acetylcholine in the brain, including acetylcholinesterase inhibitors such as tacrine. However, the clinical benefits of anticholinesterases are modest and efficacy is found in only a minority of patients (Davis *et al.*, 1992). Further, adverse effects are common. Other pharmacological agents that block acetylcholinesterase function have recently been developed that have an improved adverse effect profile compared to tacrine, yet the overall efficacy of this compensatory strategy for AD therapy remains modest.

More recently, neuroprotective strategies have been examined. AD patients who received the antioxidant vitamins E or the monoamine inhibitor selegeline display an attenuated rate of disease progression, although the effect was modest (Sano *et al.*, 1997). The neuroprotective effects of anti-inflammatory agents and of estrogen-like compounds are also being investigated. These approaches in large part aim to either compensate for neuronal loss once it has already occurred, or to reduce neuronal degeneration using putative cellular-protective agents.

In the following three chapters, new approaches for promoting neuronal rescue and replacement that specifically target brain systems affected by AD will be discussed. In Chapters 20 and 21, a cholinergic *neuroprotective* strategy will be discussed that is based upon delivering specific and potent growth factors to the brain. These growth factors, nerve growth factor (NGF) and fibroblast growth factor-2 (FGF-2), specifically prevent the degeneration of cholinergic neurons and entorhinal cortex neurons, respectively, in animal models. NGF and novel methods for its delivery to the brain will be discussed in Chapter 20. Protection of the entorhinal cortex in AD, and a discussion of relevant issues with regard to modulating neuronal plasticity in the brain, will be the subject of Chapter 21. A cholinergic *replacement* strategy will be discussed in Chapter 22 in which cholinergic basal forebrain (CBF) neurons are transplanted into the brain. Together, these methods represent a unique approach to the neurobiology of nervous system disease that may substantially improve AD treatment in the coming decade.

REFERENCES

Davis, K. L., Thal, L. J., Gamzu, E. R., Davis, C. S., Woolson, R. F., Gracon, S. I., Drachman, D. A., Schneider, L. S., Whitehouse, P. J., and Hoover, T. M. (1992). A double-blind, placebo-controlled multicenter study of tacrine for Alzheimer's disease. The Tacrine Collaborative Study Group. *New Engl J Med* 327:1253–1259.

Ernst, R. L., and Hay, J. W. (1994). The US economic and social costs of Alzheimer's disease revisited. *Am J Public Health* 84:1261–1264.

Sano, M., Ernseto, C., Thomas, R. G., Klauber, M. R., Schafer, K., Grundman, M., woodbury, P., Growdon, J., Cotman, C. W., Preiffer, E., Schneider, L. S., Thal, L., and Members of the Alzheimer's Disease Cooperative Study. (1997). A controlled trial of selegeline, alpha-tocopherol, or both as treatment for Alzheimer's disease. *N Engl J Med* 336:1216–1222.

CHAPTER 20

Neurotrophic Factors, Gene Therapy, and Alzheimer's Disease

MARK H. TUSZYNSKI,*,† FRED H. GAGE,** ELLIOTT J. MUFSON,†† AND JEFFREY H. KORDOWER††

*Department of Neurosciences, University of California, San Diego, La Jolla, California, and †Department of Neurology, Veterans Affairs Medical Center, San Diego, California, **Laboratory of Genetics, Salk Institute for Biological Studies, La Jolla, California, and ††Research Center for Brain Repair and Department of Neurological Sciences, Rush Presbyterian St. Luke's Medical Center, Chicago, Illinois

Neurotrophic factors are a class of molecules that modulate several neuronal functions, including cell survival. Since the discovery approximately 10 years ago that nerve growth factor (NGF) prevents the degeneration and death of neurons in the adult rodent and primate brain, the use of neurotrophic factors as a potential means of preventing or reducing cell loss in nervous system disease has been the subject of intense study. The neuroprotective effects of NGF on basal forebrain cholinergic neurons have attracted particular attention, since cholinergic neurons are severely affected in Alzheimer's disease (AD). Indeed, the role of NGF in naturally modulating neuronal vulnerability in AD has quite possibly been the subject of more thorough study than any other growth factor in the context of human disease. The potential use of NGF to treat AD has emerged as a prototype of the balance between the beneficial effects of promoting neuronal survival, and the adverse consequences of eliciting adverse neuronal growth responses, that emerges when a nervous system disease is manipulated with a class of molecules as potent as the nervous system growth factors. A concept that has emerged from this work is that the accurately targeted and regionally restricted delivery of growth factors will be essential when attempting to treat central nervous system disease with these substances. Despite these concerns, NGF treatment remains an intriguingly promising means of preventing cholinergic neuronal degeneration in AD, and the development of effective technology for NGF delivery in the near future should permit the testing of its therapeutic potential in clinical trials.

I. INTRODUCTION

Neuroprotective strategies for Alzheimer's disease (AD) have entered a new era led by the expansion of our knowledge of neurotrophic factor (NTF) biology. For the first time, neurotrophic factors offer the possibility of *preventing* neuronal degeneration early in the course of AD, rather than merely compensating for cell loss after it has occurred. This potentially promising field of AD therapy is based upon a series of logical and consistent preclinical (animal) experiments in rodents and nonhuman primates. These studies demonstrate that neuronal degeneration can be prevented by neuronal growth factors such as nerve growth factor (NGF), fibroblast growth factor (FGF), and others. This chapter will present evidence that NGF is capable of preventing degeneration of cholinergic basal forebrain (CBF) neurons. This population of neurons is severely affected in AD and degeneration of CBF neurons correlates with the severity of cognitive decline in Alzheimer's patients. Therefore, rescue of cholinergic neurons may improve cognition. Although a large body of research indicates that NGF is a potentially powerful molecule for treating these vulnerable neurons in AD, safe and effective delivery of NGF to the brain is a major challenge. Gene therapy will be presented as a novel means of locally targeting NGF to degenerating neuronal populations without inducing adverse effects.

A. NGF as a Neurotrophic Factor

As reviewed in Chapter 5, a nervous system growth factor is a molecule that promotes neuronal survival, stimulates axon growth (Levi-Montalcini, 1987), and modulates neuronal plasticity (Thoenen, 1995). The first neurotrophic factor to be identified was NGF. Discovered by Levi-Montalcini in 1954, NGF was first found to influence developing sensory and sympathetic neurons of the peripheral neurons system (Levi-Montalcini *et al.*, 1954). For the next 30 years, it was generally believed that growth factors such as NGF played no role in the adult nervous system. Approximately 15 years ago, however, this perception dramatically changed following the discovery that NGF was present (Scott *et al.*, 1981; Sheldon and Reichardt, 1986; Whittemore *et al.*, 1986) and biologically active in the adult CNS (Hefti, 1986; Kromer, 1987; Gage *et al.*, 1988), as will be described in more detail in this chapter.

NGF is a pentameric protein that consists of two alpha, one beta, and two gamma subunits (see Levi-Montalcini, 1987). The active site of NGF is contained in the 118 amino acid beta subunit (MW 13,250 daltons; Angeletti *et al.*, 1973; Honegger and Lenoir, 1982) which binds to both low-affinity and specific, high-affinity neuronally based cell membrane receptors. The high-affinity NGF receptor, trkA, specifically binds NGF and mediates signal transduction through a tyrosine kinase-linked second messenger system. NGF binding to trkA is thought to mediate many of NGF's neuronal survival and axonal growth-promoting functions (Cordon-Cardo *et al.*, 1991; Alitalo *et al.*, 1982; Klein *et al.*, 1991; Ip *et al.*, 1993). The low-affinity receptor binds all members of the NGF family of neurotrophins including NGF, brain-derived neurotrophic factor (BDNF), neurotrophin-3 (NT-3), and neurotrophin-4/5 (NT-4/5) with similar affinity. This receptor is now termed the p75 neurotrophin ($p75^{NTF}$) receptor. The precise biological role of the $p75^{NTF}$ receptor has not been fully characterized, but roles have been reported in regulating apoptosis (programmed cell death) (Rabizadeh *et al.*, 1993; Yao and Cooper, 1995) and sphingomyelin metabolism (Dobrowsky *et al.*, 1994). $p75^{NTF}$ may also sequester NGF to the synaptic cleft to facilitate interaction with the trkA receptor, thus enhancing neurotrophin action (Johnson *et al.*, 1996).

B. NGF and the Cholinergic Basal Forebrain

A potential role for NGF in supporting neuronal survival and function in the *adult* central nervous system was first described approximately a decade and a half ago. First, it was discovered that NGF was retrogradely transported exclusively from the cerebral cortex and hippocampus in the adult brain to neurons within the cholinergic basal forebrain (Schwab *et al.*, 1979; Seiler and Schwab, 1984; Taniuchi and Johnson, 1985). Subsequently, NGF was found to be present in the adult brain (Scott *et al.*, 1981; Sheldon and Reichardt, 1986). These intriguing findings led to the hypothesis that NGF might exert some influence, and perhaps even promote survival, of adult cholinergic basal forebrain (CBF) neurons. Subsequently, in elegant and now classic experiments, three groups of investigators transected the fimbria–fornix, the main cholinergic projection from the medial septum to the hippocampus, and found that NGF completely rescued CBF neurons from injury-induced degeneration (Fig. 1) (Hefti, 1986; Kromer, 1987; Gage *et al.*, 1988). In the absence of NGF treatment after fimbria–fornix lesions, control animals showed a loss of choline acetyltransferase (ChAT) labeling in approximately 75% of CBF neurons. In contrast, animals that received injections or continuous infusions of NGF into the ventricular system of the brain at the time of the lesion showed virtually no loss of ChAT labeling in CBF neurons. These findings provided direct evidence that a single molecule, NGF, was capable of preventing what had heretofore been an irreversible loss of the cholinergic phenotype in neurons after injury. This not only firmly established a role for growth factors in the *adult* central nervous system, but it also opened a potential therapeutic window

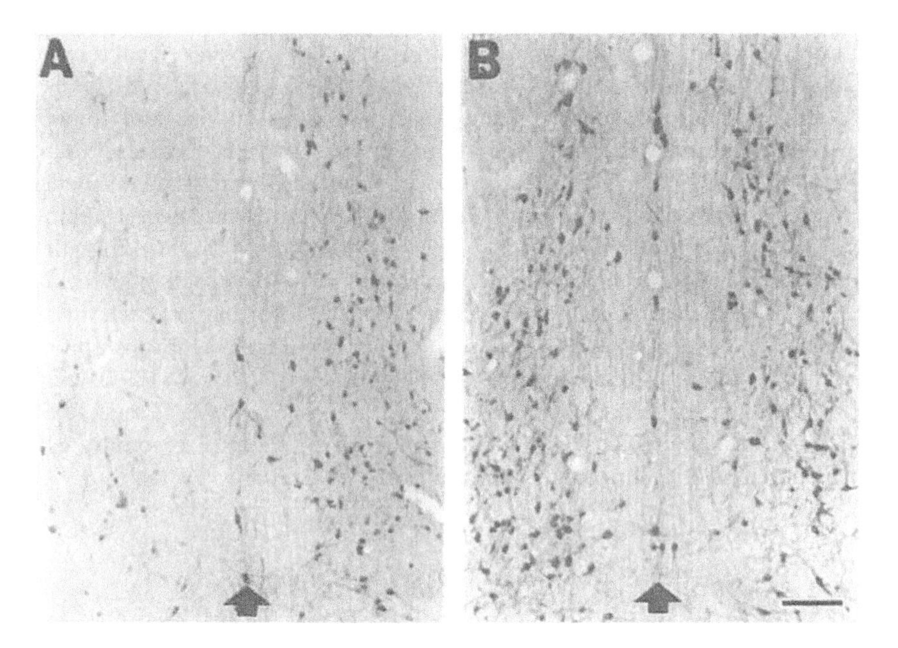

FIGURE 1 NGF infusions rescue lesioned basal forebrain cholinergic neurons in the adult mammalian brain. (A) The medial septal region of the basal forebrain cholinergic (BFC) system is pictured in the adult rat brain. Cholinergic axons have been cut on the left side, resulting in retrograde degeneration of neurons and loss of choline acetyltransferase (ChAT) immunolabeling. The intact (right) side of panel (A) contains a normal number and distribution of cholinergic neurons. Arrow indicates midline. (B) Intracerebroventricular infusions of nerve growth factor (NGF) at the time of the lesion completely prevent cholinergic neuronal degeneration, and a normal number and distribution of ChAT-labeled neurons remain on the lesioned (left) side. Scale bar = 100 μm in A and B.

for strategies aimed at *preventing* neuronal loss and ameliorating diminished ChAT expression in neurological disease.

Although these findings were groundbreaking, axotomy of cholinergic axons does not mimic the degenerative process seen in diseases such as AD. Interest in this approach was sustained, however, by the findings of Fischer and associates (1987), who demonstrated that NGF also prevented age-related *spontaneous* atrophy of CBF neurons. Thus NGF neuronal protection was not merely evident following acute experimental trauma to neurons; in addition, it prevented neuronal degeneration in the protracted biological process of aging. Importantly, this study also demonstrated that age-related cognitive impairments could be ameliorated by intraventricular infusions of NGF. Recovery of behavioral function based upon reconstruction of host-injured cholinergic systems was subsequently reported in adult rats with lesion-induced cholinergic neuronal degeneration (Tuszynski and Gage, 1995). These findings furthered the concept that NGF treatment may be able to improve cognitive dysfunction associated with the cholinergic deficits in AD.

Subsequent experiments in rats examined effects of NGF on the continuum of CBF neurons (i.e., septal-diagonal band complex and nucleus basalis of Meynert), which provides the major cholinergic input to the neocortex and hippocampus (Mesulam *et al.*, 1983, Mesulam *et al.*, 1983b; Liberini *et al.*, 1993; Dekker *et al.*, 1992). In these experiments, NGF was also found to ameliorate lesion-induced degeneration of CBF neurons and to improve behavioral outcomes on spatial memory tasks. These observations supported NGF's role as an important neurotrophic factor for the function of CBF neurons.

Most of these studies were performed in rat models of neuronal injury. To determine whether neurotrophic factor principles of neuronal protection were also relevant to primate systems, fornix lesions were performed in adult monkeys (Tuszynski *et al.*, 1990, 1991; Koliatsos *et al.*, 1990). Here too, transection of this projection to the hippocampus resulted in the retrograde degeneration of approximately 75% of CBF neurons. When monkeys received NGF infusions into the ventricular system, 80 to 100% of cholinergic neurons were rescued after fornix transection. The subsequent detection of NGF and its receptors in the human brain (see following discussion) further substantiated the biological significance of neurotrophins in human brain function.

NGF protection of CBF neurons is of particular interest in the context of AD. A number of neuronal populations degenerate in AD, including cholinergic, noradrenergic, serotonergic, and several peptidergic neurons (see Terry *et al.*, 1994). The loss of CBF neurons is particularly severe, often affecting 75 to 80% of this populations. Further, loss of CBF neurons has been correlated with the severity of synapse loss in the cortex, the density of amyloid plaques, the severity of clinical dementia, and the duration of disease (Coyle *et al.*, 1983; Bartus *et al.*, 1982; Perry *et al.*, 1978; Mufson *et al.*, 1989). Indeed of all of the neurochemical changes seen in AD, the cortical cholinergic deficit correlates best with cognitive dysfunction (Bierer *et al.*, 1995), suggesting that this neurochemical change is integral to the major symptomatology of the disease. For these reasons, therapeutic efforts in AD have often focused on augmenting CBF function using pharmacological approaches such as cholinesterase inhibitors. In general, trials employing cholinesterase inhibitor have been only modestly successful; pharmacokinetic characteristics of available anticholinesterase drugs as well as specificity and dosing factors limit the ability of pharmacological approaches to optimally modulate basal forebrain cholinergic function (Bartus *et al.*, 1990). NGF offers a novel and direct approach for targeting CBF degeneration in AD, presenting the opportunity for the first time to *prevent* cholinergic neuronal degeneration rather than to *compensate* for this degeneration after it has already occurred. To explore the possibility of NGF therapy for the cholinergic component of neuronal loss in AD, two issues require further exploration: (1) how is natural NGF biology/metabolism affected in the AD brain, and (2) how can NGF be delivered to the brain in AD?

II. ALTERATIONS IN NATURAL NGF BIOLOGY AND FUNCTION IN AD

The extensive literature linking NGF processes to basal forebrain function led to an early suggestion that impaired NGF trophic support underlies CBF degeneration in AD (Appel, 1981; Hefti *et al.*, 1986). Subsequent studies primarily examining NGF mRNA levels in cerebral cortex and the expression of protein and mRNA for the low-affinity $p75^{NTF}$ within the CBF failed to support this hypothesis (Goedert *et al.*, 1986; Kordower *et al.*, 1989; Ernfors *et al.*, 1990), and in fact cortical NGF synthesis was found to be unimpaired in AD (Goedert *et al.*, 1986). However, disturbances in NGF utilization or transport could occur at several potential levels in the brain in AD, contributing to cholinergic neuronal degeneration. Normally, NGF exerts trophic influences upon CBF neurons in a series of complex events that include (1) NGF synthesis, (2) *trk* receptor synthesis and anterograde transport, (3) NGF binding to its high-affinity *trk* receptor, (4) retrograde transport of the NGF–receptor complex from the cerebral cortex and hippocampus to CBF neurons, and (5) autophosphorylation of the NGF signal (see also Mufson and Kordower, 1997, in press; Tuszynski and Gage, 1994). Failure at any of these steps could cause basal forebrain degeneration *secondary* to impaired NGF trophic support. This secondary loss could result, directly or indirectly, from a primary neuropathological event in AD, such as β-amyloid-induced toxicity to neurons. There is now mounting evidence, summarized next, that NGF protein transport to CBF perikarya in AD is impaired, resulting in subnormal levels of NGF in cholinergic neurons and NGF accumulation in the cerebral cortex in AD.

A. NGF in the Primate CBF

NGF-like immunoreactivity has been visualized within the rodent basal forebrain by Conner and coworkers employing affinity-purified antibodies to mouse β-NGF (Connor *et al.*, 1992; Conner and Varon, 1992). Using this protocol, NGF-immunoreactive neurons were also visualized within CBF neurons and in the hippocampus of normal monkeys and aged humans (Mufson *et al.*, 1994). In the basal forebrain of both primate species, a granular NGF-like IR reaction product is present within neurons of the medial septum, the nucleus of the diagonal band and nucleus basalis of Meynert (regions Ch1, Ch2, and Ch4 in the classification of Mesulam; Fig. 2). NGF-like immunoreactivity exclusively colocalizes within $p75^{NTF}$ receptor-containing basal forebrain neurons. The intensity of NGF immunolabeling varies between cell bodies: many NGF-IR perikarya are highly immunoreactive in the CBF, whereas others show undetectable or minimal NGF immunoreactivity.

Because NGF gene expression is virtually nondetectable in CBF neurons (Whittemore *et al.*, 1986), NGF seen within these neurons is most likely packaged and retrogradely transported from its sites of production within cortex and hippocampus (Schwab *et al.*, 1979; Seiler and Schwab, 1984). Using a polyclonal antibody directed against β-NGF, Mufson and coworkers (1994) found that NGF-like immunoreactivity is mainly localized within the hilus of the dentate gyrus and within CA3 and CA2 hippocampal subfields of both humans and nonhuman primates (Fig. 3). Little NGF-like staining is seen in CA1. Within the hippocampal formation, NGF-like immunoreactivity is dense within the neuropil of stratum radiatum, intermediate in stratum oriens, and lightest in stratum pyramidale. NGF-like immunoreactivity is not found within granule or pyramidal cells of the dentate gyrus and hippocampal formation, respectively. It is likely that this neuropil staining represents immunoreactive

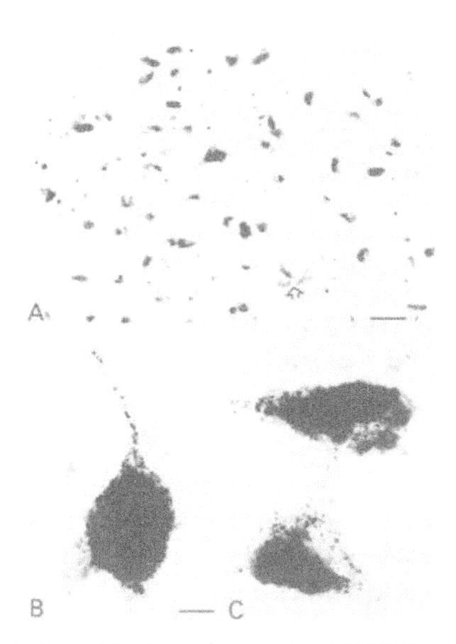

FIGURE 2 Nerve growth factor-like immunoreactivity (NGF-ir) within the human nucleus basalis (Ch4 anterolateral nucleus). (A) Numerous NGF-ir neurons exhibit differing levels of immunoreactivity. Open arrow indicates a cluster of lightly labeled neurons (B, C). Higher magnification photomicrographs of nucleus basalis NGF-ir neurons. Scale bar in A = 100 μm; B, C = 40 μm. (From Mufson *et al.* (1994). *J Comp Neurol* 341:507–519.)

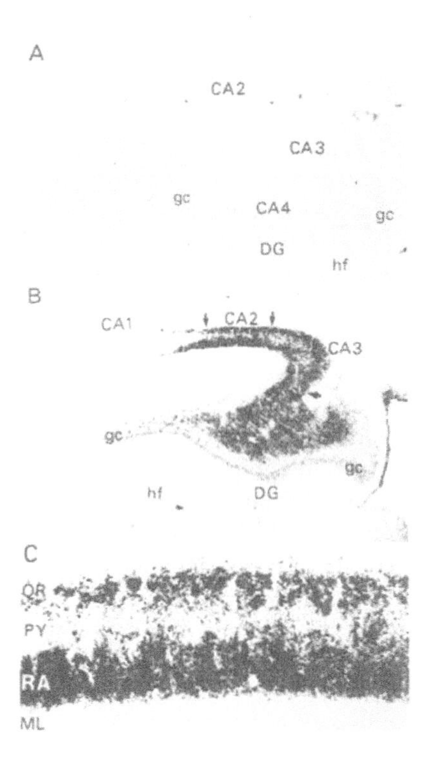

FIGURE 3 Photomicrographs of NGF-like immunoreactivity within the monkey hippocampal formation. (A) Hippocampal formation immunoreacted with anti-NGF antibodies preadsorbed to purified β-NGF results in a complete absence of hippocampal staining. (B) NGF-like immunoreactivity within the hippocampal formation reacted with anti-NGF antibodies. Note that the granular pattern of NGF-like immunoreactivity is restricted to the neuropil of the hilar (CA4) and CA3-CA2 hippocampal subfields. (C) High-power photomicrograph of CA2 as demarcated by arrows in B. Note that the trilaminar appearance of the reaction product is most intense in stratum radiatum (RA), moderate in stratum oriens (OR), and light in the stratum pyramidal (PY). Scale bar in A,B = 200 μm; C = 40 μm. (From Mufson *et al.* (1994). *J Comp Neurol* 341:507–519.)

processes originating from intrinsic hippocampal neurons because a similar staining pattern in rats has been found to emanate from intrinsic hippocampal neurons following colchicine administration (Connor *et al.*, 1992).

NGF expression has also been examined in the AD brain. Whereas virtually all CBF neurons in normal human aged brains exhibit NGF-IR (Mufson *et al.*, 1994), few of the remaining CBF neurons in AD brains display detectable NGF-IR (Mufson *et al.*, 1995). This reduction in NGF-IR in AD occurs despite persistent labeling for $p75^{NTR}$ in remaining neurons of the CBF (Kordower *et al.*, 1989; Ernfors *et al.*, 1990; Mufson *et al.*, 1995), suggesting a specific loss of transported NGF rather than a loss of all cellular markers from general cellular degeneration. Of the remaining CBF neurons in AD that exhibit NGF-IR, optical density measurements reveal a 33% reduction in the intensity of this labeling compared to age-matched control brains ($p < 0.001$) (Mufson *et al.*, 1995). These data support the hypothesis that there is a defect in the ability of CBF neurons to bind or retrogradely transport NGF in AD. This possibility is further supported by the finding that NGF protein accumulates in the neocortex in AD, suggesting an impairment in uptake or utilization by basal forebrain neurons. Indeed, NGF protein in AD is increased relative to age-matched controls within the superior frontal gyrus, superior temporal gyrus, parietal cortex, and occipital cortex (Crutcher *et al.*, 1993; Scott *et al.*, 1995).

Because NGF protein levels in AD are *increased* in the neocortex but *decreased* in the CBF, diminished NGF receptor expression might be causally related to reductions in NGF transport in AD. This is particularly important because NGF is retrogradely transported to CBF consumer neurons after binding to receptors on cholinergic terminals in the neocortex and hippocampus. Further, lesions of the basal forebrain reduce NGF uptake and retrograde transport from cholinergic terminals within the neocortex and hippocampus in rodent models, resulting in elevated NGF levels in these regions. Studies of low-affinity $p75^{NTR}$ NGF receptor levels in AD brains, however, revealed that virtually all magnocellular basal forebrain neurons in both normal aged individuals and even the most degenerated perikarya in AD express mRNA for $p75^{NTR}$ (Kordower *et al.*, 1989; Mufson *et al.*, 1989a,b; Mufson *et al.*, 1995). Further, relative levels of $p75^{NTR}$ mRNA are unchanged or even increased within the AD basal forebrain (Mufson *et al.*, 1996; Ernfors *et al.*, 1990). Thus, the $p75^{NTR}$ system is an unlikely candidate for mediating diminished NGF transport to CBF neurons. In contrast, studies of high-affinity trkA NGF receptor levels showed a *reduction* in trkA expression in AD brains relative to age-matched controls (Mufson *et al.*, 1996). Normally, virtually all primate basal forebrain magnocellular neurons express the high-affinity trkA receptor (Kordower *et al.*, 1994, Mufson *et al.*, 1997). Quantitative *in situ* hybridization and immunohistochemical analysis of trkA mRNA and protein in AD and age-matched controls revealed significant reductions in trkA markers within individual CBF neurons (Mufson *et al.*, 1996, 1997). Similar reductions in cholinergic basal forebrain trkA mRNA have recently been reported by Hirsch and coworkers using semiquantitative *in situ* hybridization (Boissiere *et al.*, 1997). Thus, a defect in trkA gene expression occurs in AD that may exacerbate or directly cause diminished NGF transport to cholinergic neurons of the basal forebrain. These findings lend support to the hypothesis that NGF could potentially reduce or prevent cholinergic neuronal degeneration if selectively delivered to the forebrain region in AD.

III. COMPLICATIONS OF NGF INFUSIONS TO THE BRAIN

Experiments described in the preceding section set the stage for trials of NGF therapy in AD. Indeed, large-scale clinical trials would likely be underway already were it not for the discovery two years ago that the traditional route of NGF administration in animal models, intracerebroventricular (ICV) infusions, results in unacceptable adverse effects. ICV NGF delivery floods the cerebrospinal fluid system with NGF, delivering high concentrations into the lateral, third and fourth ventricles, and into the subarachnoid space overlying the cerebral hemispheres, brain stem, and spinal cord. Following ICV NGF infusions, four notable complications in animals develop: (1) hypophagia caused by weight loss (Williams, 1991), (2) sprouting of sympathetic axons around the cerebral vasculature (Saffran and Crutcher, 1990; Menesini *et al.*, 1978), (3) migration and proliferation of Schwann cells in a thick pial layer that surrounds the brain stem and spinal cord (Winkler *et al.*, 1996), and (4) dense sprouting of sensory axons into the proliferating Schwann cell layer (Winkler *et al.*, 1996) (Fig. 4, see color insert). Although Schwann cell proliferation and sensory axon sprouting gradually regress following the discontinuation of NGF infusions, these findings severely limit the practicality of the ICV infusion approach for treating AD, a disease in which chronic long-term neurotrophin delivery is likely to be needed. Indeed, at least two patients received ICV NGF infusions in Sweden several years ago, and although reports on these patients are limited (Olson *et al.*, 1992), they developed pain syndromes that were likely related to Schwann cell and sensory axon growth.

Thus, whereas neurotrophins are a powerful means of preventing neuronal degeneration in animal models, their exposure to broad regions of the nervous system results in unacceptable adverse effects from nontargeted, neurotrophin-sensitive systems. Clearly, a specifically targeted, regionally restricted form of neurotrophin delivery to the CNS is needed.

IV. NGF DELIVERY TO THE AD BRAIN: GENE THERAPY

Relatively recent advances in molecular biology have introduced gene therapy as a potential means of delivering substances to the nervous system. Gene therapy offers the prospect of delivering neurotrophic factors directly into the brain parenchyma in a well-targeted, regionally restricted, long-term and potentially safe manner. Two approaches to gene therapy have been utilized to date in animal models and in human clinical trials: (1) *In vivo* gene therapy refers to the genetic alteration of host cells *in vivo,* using viral-derived vectors that directly insert DNA into a targeted brain region, and (2) *ex vivo* gene therapy refers to the genetic alteration of dividing cells in the culture dish, followed by the transplantation of these genetically altered cells to specific nervous system sites (see Miller, 1990; Suhr and Gage, 1993; Blau and Springer, 1995; Tuszynski, 1998). Each approach has unique benefits and drawbacks that are discussed in Chapter 9. In both approaches, genetically modified cells act as biological "minipumps" that can deliver substances such as NGF to precise brain targets.

Gene therapy strategies have been applied to the correlative animal models of AD discussed earlier: namely, rats and primates with injury-induced degeneration of BFC neurons after fornix lesions, and spontaneous age-related atrophy of CBF neurons. In these models, NGF gene therapy has rescued cholinergic neurons and, in some cases, improved cognition. In this section, gene therapy methods will be described briefly, followed by a discussion of how NGF delivered via these methods functions in animal models.

A. Gene Therapy: *Ex Vivo* Methods

Experiments performed to date in animal models of AD have primarily used *ex vivo* gene therapy because this method results in superior levels of transgene expression with little or no cellular toxicity compared to *in vivo* gene therapy methods (Miller, 1990; Rosenberg *et al.*, 1988; Tuszynski *et al.*, 1994, 1996; Chen and Gage, 1995; Tuszynski, 1998). Fibroblasts have been chosen as target cells to undergo *ex vivo* genetic modification in several experiments because fibroblasts (1) are easily obtained from skin biopsies, (2) divide in culture, (3) survive grafting to the nervous system for extended time periods, and (4) do not induce known adverse effects in the host brain.

In many experiments, cells have been genetically modified using retroviral vectors derived from Moloney murine leukemia virus (MLV). MLV is a retrovirus, containing RNA rather than DNA as the genetic coding information. For the MLV retrovirus to normally make functional proteins in a host cell, its viral RNA must be converted into DNA using the enzyme reverse transcriptase. Following reverse transcription, the viral DNA integrates into the host cell's genome, and the viral genes are expressed by powerful viral promoters that use normal host transcription machinery to produce large amounts of proteins. In gene therapy applications, several genes of the wild-type MLV retrovirus are excised and replaced with therapeutic genes, such as the human NGF gene. These new "gene therapy vector" particles are incapable of replication because wild-type genes needed for replication have been deleted. Typically, a second new gene is also placed into a transgene vector that makes the cells resistant to the antibiotic neomycin (Fig. 5); in this manner, cells that successfully incorporate the transgene can be *selected* from the cell culture by adding neomycin analogs, because all cells lacking the transgene will be killed by the antibiotic. Cells are subsequently grown for several cycles to amplify their numbers to amounts sufficient for transplantation to the brain.

In our studies, primary (nonimmortalized) fibroblasts that have been genetically modified to produce human NGF using these vectors secrete approximately 10 ng of human NGF/10^6 cells/day, whereas nonmodified fibroblasts produce no detectable NGF (Rosenberg *et al.*, 1988). This amount of NGF production from genetically modified cells exceeds physiological levels of NGF production in the adult brain by approximately 500-fold. Thus, prior to transplantation to the brain, host cells can be genetically modified *in vitro* to produce large amounts of human NGF.

B. Gene Therapy: Transplantation in Models of Cholinergic Degeneration

To test the ability of NGF gene therapy to rescue degenerating basal forebrain cholinergic neurons, primary fibroblasts genetically modified to produce and secrete human NGF were transplanted to the rat septal nucleus after fimbria–fornix lesions (Rosenberg *et al.*, 1988). Control animals received identical lesions and grafts of fibroblasts transduced to produce a nonneurotrophic

Neurotrophic Factor Vector

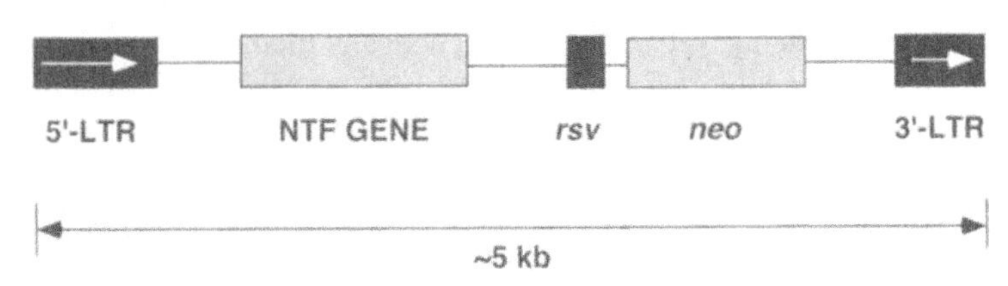

FIGURE 5 Prototypical vector design for gene therapy. The viral wild-type promoter (5′-long terminal repeat, or 5′-LTR) constitutively expresses the first transgene, in this case a neurotrophic factor (NTF) gene. This is followed by a second "internal" promoter from the SV40 virus (*rsv*) that constitutively expresses a second transgene for neomycin resistance (*neo*). Cells that contain the neomycin resistance gene will survive the addition of this antibiotic to the cell culture media; thus, cells that have successfully incorporated the vector can be *selected in vitro*. The transgene is completed by the wild-type 3′-long terminal repeat (3′-LTR) sequence. The entire construct is approximately 5 kilobases (5 kb) in size.

factor gene. Whereas the lesion caused degeneration of approximately 75% of cholinergic neurons in control animals, rats that received grafts of NGF-producing fibroblasts showed a loss of cellular labeling in only 5% of neurons. Thus, the transplantation of genetically modified cells to the brain was an effective means of preventing injury-induced cholinergic neuronal degeneration.

In subsequent experiments, NGF-producing fibroblasts were grafted to the nucleus basalis in rats with spontaneous age-related atrophy of basal forebrain cholinergic neurons and associated deficits in learning and memory. Animals that received NGF-producing cells showed reversal of mnemonic deficits and improvement of cholinergic neuronal morphology (Chen and Gage, 1995). In contrast, improvements were not observed in rats receiving control, non-NGF-secreting fibroblasts. Subsequently, other investigators replicated these findings using grafts of other genetically modified cell types (Martinez-Serrano *et al.*, 1995).

To determine whether gene therapy with human NGF was a practical means of protecting neurons in the larger primate brain, degeneration of basal forebrain cholinergic neurons in adult monkeys was induced by performing unilateral fornix transections. Monkeys then received transplants of autologous (self) fibroblasts genetically modified to produce human NGF (Tuszynski *et al.*, 1996; Fig. 6). Control subjects received lesions and grafts of fibroblasts that were not genetically modified. Transplants were placed intraparenchymally, directly into the septal region containing the degenerating cell bodies. The intraparenchymal placement of cells hypothetically achieved maximal neurotrophin delivery to degenerating neurons while shielding other potentially NGF-responsive neurons from neurotrophin exposure. One month after the lesion, control-lesioned monkeys showed degeneration of $73.8 \pm 5.0\%$ of basal forebrain cholinergic neurons, whereas monkeys that received NGF-secreting autologous genetically modified cell grafts showed an average loss of immunolabeling in only $38.3 \pm 8.9\%$ of cholinergic neurons ($p < 0.02$; Tuszynski *et al.*, 1996). The grafted monkey with the largest and most accurately targeted NGF-producing transplant showed a loss of only 8% of immunolabeled cholinergic neurons. Time points longer than one month after lesions are now being studied. To date, one monkey has been examined 8 months after lesion and grafting, and this subject shows rescue of 70% of medial septal neurons and persistent production of NGF protein. Because it is not known whether long-term delivery of NGF is required to sustain cholinergic neurons after fornix lesions, it is important to examine NGF protein production as a marker of sustained transgene expression *in vivo*. As assessed with an elisa protein assay, NGF expression has been sustained for at least 8 months in the primate brain. Importantly, monkeys with hNGF-producing intraparenchymal grafts do not experience weight loss, Schwann cell migration into the CNS, or abnormal sprouting of sympathetic and sensory systems for at least 8 months after genetically modified cell placement. Thus, gene therapy appears to provide a

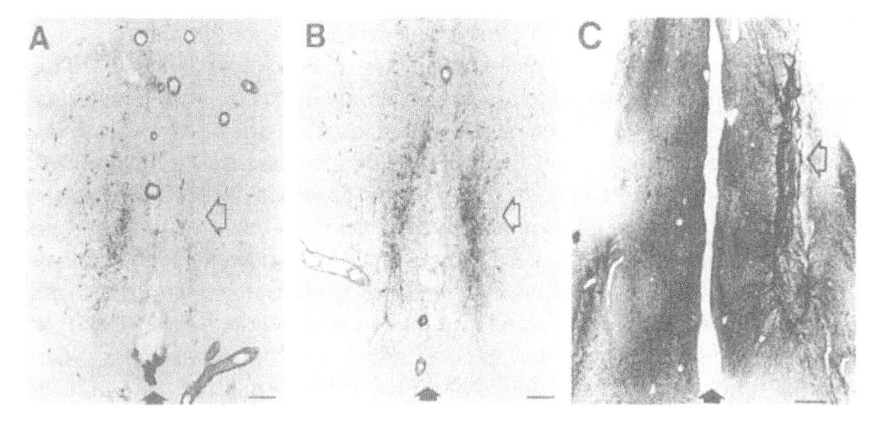

FIGURE 6 Photomicrograph illustrating effects of grafts of cells genetically modified to produce human nerve growth factor (hNGF) to the adult primate brain. (A) Following right-sided transections of the fornix, the number of cholinergic neurons immunolabeled for the p75 low-affinity neurotrophin receptor is reduced in the right medial septal region (open arrow). The left side, unaffected by the lesion, shows a normal number of p75-immunolabeled neurons. This monkey received an intraseptal graft of autologous control fibroblasts that were *not* genetically modified to produce NGF. Solid arrow indicates midline; scale bar in (A, B) = 100 μm. (B) In a monkey that has undergone a right-sided fornix transection and has received a graft of NGF-producing autologous fibroblasts, cholinergic neuronal degeneration is prevented on the right side of the medial septal nucleus (open arrow). (C) The NGF-secreting graft in the septal region is penetrated by cholinergic axons (acetyl cholinesterase stain), indicating tropic (growth) responses of adult primate cholinergic axons to an NGF source. g, graft. Open arrow indicates graft; solid arrow indicates midline. Scale bar = 62 μm. (From Tuszynski *et al.* (1996). *Gene therapy* 3:305–314.)

means of delivering NGF to the large primate brain that rescues cholinergic neurons without, at present, evidence of eliciting adverse effects.

To date, approximately 100 NGF-secreting grafts have been placed into 12 monkeys. In no case have grafted cells formed tumors, caused hydrocephalus, or resulted in other obvious adverse effects. No monkey has become ill as a result of the presence of an NGF-producing or control fibroblast graft. Thus, NGF delivery by transgenic approaches is a potentially practical means of delivering molecules to specific intraparenchymal sites in the brain. The feasibility of gene therapy for the treatment of human diseases such as Alzheimer's disease requires verification of the preceding efficacy and safety data in additional subjects. Further, questions remain to be resolved regarding the placement and number of NGF-secreting grafts that will be required in the Alzheimer's brain, the duration of gene expression beyond 8 months, the duration of neuronal protection after transgene expression ceases, and the feasibility of repeating the grafting procedure if downregulation of gene expression should occur.

V. ENCAPSULATED GRAFTS OF GENETICALLY MODIFIED NGF-SECRETING CELLS

An alternative method to *ex vivo* gene therapy using autologous host cells is *ex vivo* gene therapy using xenografted (different species) cells that are encapsulated in biopolymers to prevent graft rejection (Emerich *et al.*, 1994; Kordower *et al.*, 1994). The advantage of the encapsulated gene therapy approach is that banks of cells can be maintained that would be available for transplantation on short notice. Further, these cells are retrievable by removing implanted capsules, if problems related to the growth factor should be encountered.

To test the ability of encapsulated, genetically modified xenografts to rescue cholinergic neurons, baby hamster kidney (BHK) cells were genetically modified to secrete high levels of human NGF (hNGF): 21.4 $\pm$ 2.0 ng hNGF/d. These cells were then implanted into macaque monkeys to determine whether they would prevent lesion-induced degeneration of CBF neurons. Following polymer encapsulation, BHK cells were grafted into the lateral ventricle of four monkeys immediately following a unilateral transection/aspiration of the fornix. Three control monkeys received identical grafts, with the exception that the BHK cells were not genetically modified to secrete human NGF, thus differing by only the single gene construct. One monkey received a fornix transection only. All monkeys displayed complete transections of the fornix as revealed by a comprehensive loss of acetyl cholinesterase (AChE)-containing fibers within the hippocampus ipsilateral to the lesion. Control monkeys that were either nonimplanted or implanted with non-NGF secreting BHK cells displayed extensive losses of ChAT-ir (54%) and $p75^{NTR}$-ir (53%) neurons within the medial septum on the side ipsilateral to the lesion. Similar losses in ChAT-ir (19%) and $p75^{NTR}$-ir (20%) neurons were seen within the vertical limb of the diagonal band. In contrast, monkeys receiving BHK-hNGF grafts exhibited a significant ($p < .01$) amelioration of cholinergic neuronal loss within the septum (19% decrease for ChAT-ir and 20% decrease for $p75^{NTR}$-ir) and vertical limb of the diagonal band (7% loss for both markers). Furthermore, grafts of human NGF-secreting cells, but not control cells, induced a dense sprouting of cholinergic fibers within the septum that ramified against the ependymal lining of the ventricle adjacent to the transplant site. Examination of the capsules upon their removal from the animal prior to sacrifice revealed numerous healthy-appearing cells. Prior to transplantation, the capsules produced 21.4 + 2.0 ng hNGF/day. Following retrieval just prior to sacrifice, these capsules produced 8.5 + 1.2 ng hNGF/day. This level of NGF production was sufficient to induce differentiation of PC12 cells in culture, indicating that genetically modified hamster cells secrete biologically active human NGF.

A. Grafts of Encapsulated NGF-Secreting Cells in Aged Nonhuman Primates

AD is a disease that afflicts the elderly. Yet the vast majority of *in vivo* studies attempting to discover therapeutic strategies for AD patients employ young rodents or monkeys. Even rarer are studies employing aged nonhuman primates, a species that exhibits both behavioral and pathological sequelae similar to that seen in AD. To study the effects of cellular delivery of NGF upon degenerating CBF neurons in an aged primate, rhesus monkeys (ages 24–29 years) received unilateral fornix transections. Following sacrifice, all monkeys displayed numerous diffuse amyloid plaques in the temporal and limbic cortices. Three aged monkeys received intraventricular transplants of polymer-encapsulated baby hamster kidney (BHK) fibroblasts that had been genetically modified to secrete human nerve growth factor (hNGF). Three additional monkeys received grafts of identical cells that were not modified to secrete hNGF. Monkeys receiving the fornix transection and control grafts displayed extensive reductions in the number of ChAT-ir (68.3% ± 5.7) and $p75^{NTF}$-ir (53.0% ± 0.3) medial septal neurons ipsilateral to the lesion/graft. In contrast, monkeys receiving grafts of encapsulated hNGF-secreting cells displayed only a modest loss of ChAT-ir (16.5% + 8.3; range 0–36%) and $p75^{NTF}$-ir (14.7% + 0.4; range 7–22%) medial septal neurons. Additionally, all monkeys receiving the hNGF-secreting implants, but none receiving control implants, displayed robust sprouting of cholinergic fibers within the septum ipsilateral to the graft. In this study, capsules produced 44.7 ± 0.9 ng hNGF/day and at retrieval produced 9.6 ± 0.85 ng hNGF/day, a level that was again sufficient to differentiate PC12 cells in culture.

These data demonstrate that hNGF can provide trophic and tropic influences to degenerating CBF neurons in aged nonhuman primates. However, for this technology to be practical in the clinical realm, long-term gene expression is essential. In both the young and aged monkey studies, there was a significant reduction in hNGF expression from genetically modified encapsulated cells upon their retrieval from the brain at the time of sacrifice. This drop in expression is, in part, likely due to the optimal culture conditions in which the hNGF is analyzed preoperatively compared to the suboptimal environment in which the cells live *in vivo*. Studies in rodents demonstrate that reductions in hNGF expression routinely occur quickly following transplantation (Emerich *et al.*, unpublished data). However, following the initial drop-off, stable gene expression is achieved and biologically relevant levels of hNGF are secreted from encapsulated BHK cells for over a year post-transplantation (Winn *et al.*, 1996). Thus encapsulated, genetically modified cells may be able to provide long-term hNGF delivery.

B. *In Vivo* Gene Therapy Vectors

To date, most studies of gene therapy in the nervous system have used *ex vivo* methods. Reasons for this have been cited earlier, and include the fact that *ex vivo* vectors have generally been more extensively studied, and have exhibited superior levels and durations of gene expression in the CNS compared to *in vivo* gene therapy vectors. Nonetheless, *in vivo* approaches have certain advantages including less invasiveness and less potential risk of tumor formation in the host. Several *in vivo* vectors are currently under development, including adenovirus, adeno-associated virus (AAV), herpes virus (hsv), and retroviral vectors including lentivirus. Data are beginning to emerge that sustained *in vivo* gene expression with minimal toxicity can be achieved with *in vivo* gene delivery methods.

C. Regulating Gene Expression

One general disadvantage of gene therapy approaches at present is an inability to precisely control the amount or *dose* of the gene product delivered. Although the number or concentration of genetically modified cells injected into the host can be varied, ideally more precise control of molecular delivery in a fashion analogous to varying the dose of a drug is needed. Thus, efforts are underway to develop *regulatable* vectors that can turn on or off, or vary the production, of a gene product. Expression systems that are activated or repressed by tetracycline are one example of such a system (Gossen *et al.*, 1994; Weinmann *et al.*, 1994). Other regulators of gene expression, such as steroid-responsive systems, are also being developed (see Suhr and Gage, 1993).

In some cases it may also be desirable to completely stop expression of therapeutic genes from genetically modified cells. Discontinuation of transgene expression might be necessary if the biological function served by the transgene is no longer necessary, or if transgene expression actually becomes deleterious. "Suicide genes" such as thymidine kinase or the proapoptotic (programmed cell death-inducing) genes *p53* or *bax* could be inserted into transgene vectors to induce death in genetically modified cells (Ezzeddine *et al.*, 1991; Barba *et al.*, 1993). Studies of these approaches are ongoing.

VI. OTHER MEANS OF STIMULATING NGF FUNCTION IN THE BRAIN

Several means of delivering NGF or augmenting NGF function are being characterized that may deliver neurotrophins to the brain in a manner that is less invasive, albeit less targeted, than gene therapy. Small peptide analogs of NGF that putatively contain or mimic the active region of the NGF β-component have recently been developed. These molecules possess the hypothetical advantage that they would cross the blood–brain barrier because of their low molecular weight, eliminating the need for invasive brain delivery. *In vitro* peptide analogs have been demonstrated to mimic some NGF actions, including elicitation of neurite outgrowth from PC-12 cells. To date, the *in vivo* performance of peptide analogs in rescuing degenerating CBF neurons has been disappointing. The peptide analog approach suffers from the hypothetical drawback that it is a nontargeted approach, and thus might elicit growth responses from NGF-responsive nontargeted systems in AD, including Schwann cells, sensory axons, and sympathetic axons.

Another hypothetical means of increasing NGF availability in the AD brain is to augment its release. Some compounds have been developed that putatively enhance NGF release by mechanisms that remain to be identified. Some NGF-releasing agents are of sufficiently low molecular weight that they could cross the blood barrier after peripheral administration, again eliminating the need for invasive delivery to the CNS. On the other hand, this approach also suffers from the potential drawback that NGF release is nontargeted. Additionally, enhancing release of endogenous NGF from normal sites of synthesis may be ineffective because this neurotrophin may not be efficiently transported to CBF perikarya in AD (Mufson *et al.*, 1996).

A means of delivering NGF to the CNS has recently been reported in which NGF is attached to molecules that are actively transported into the CNS, e.g., linked to a transferrin receptor antibody (Friden *et al.*, 1993). This approach takes advantage of the fact that the iron-shuttling molecule transferrin enters the brain by active transport following its binding to the transferrin receptor. Recently, it has been demonstrated that the retrograde atrophy of nucleus basalis neurons that occurs following excitotoxic lesions of the cerebral cortex was prevented by systemic administration of NGF that was conjugated to an antibody against the transferrin receptor (Charles *et al.*, 1996). The specificity of this effect was illustrated by the fact that a nonconjugated mixture of NGF and the antibody to the transferrin receptor was without effect. This elegant approach eliminates the need for invasive CNS procedures to administer NGF, but also suffers from the drawback that its general delivery to the brain could elicit adverse effects from nontargeted structures (Williams, 1991; Winkler *et al.*, 1996).

VII. NGF AND β-AMYLOID EXPRESSION

For NGF to have clinical utility for the treatment of AD, it must protect cholinergic neurons without inducing deleterious effects that outweigh its potential benefits. One concern regarding NGF therapy is the induction or exacerbation of β-amyloid expression, a key component of plaque formation in AD. Previous studies in rodents demonstrate that NGF injections increase expression of β-amyloid precursor protein (β-APP) in the developing hamster brain (Mobley *et al.*, 1988). Because the β-amyloid protein itself is not normally deposited in the brain in most nonprimates, it could not be determined in the Mobley study whether upregulation of βAPP also resulted in increased formation of mature β-amyloid peptide. There are few animal models that display β-amyloid-containing plaques, thus it is difficult to study the *in vivo* effects of trophic factors such as NGF on amyloid expression and deposition. Transgenic animals have recently been generated that exhibit both β-amyloid-containing plaques (Quon *et al.*, 1991; Games *et al.*, 1995; Hsiao *et al.*, 1996) and cognitive dysfunction (Hsiao *et al.*, 1996). However, these animals are not yet generally available for study. Other animals that normally display β-amyloid containing plaque-like structures such as aged dogs or polar bears (Tekerian *et al.*, 1996) are also either unavailable or impractical for invasive study. In contrast, nonhuman primate brains contain β-amyloid-containing plaques as a normal consequence of aging that are similar in structure to those seen in aged humans and in patients with AD. The nonhuman primate studies described above provide the opportunity to determine whether grafts of hNGF secreting cells influence β-amyloid expression in a species that normally expresses it as a consequence of aging.

To determine whether NGF accelerates β-amyloid deposition in the brain, amyloid plaque formation was quantified in the brains of aged monkeys that had received grafts of cells genetically modified to produce and secrete NGF. NGF-secreting cells were placed either intracerebroventricularly and plaque density was measured two weeks later (Kordower *et al.*, 1997), or cells were grafted intraparenchymally and total plaque numbers were quantified three months later (Tuszynski *et al.*, 1998). Amyloid-containing plaques were identified using a specific antibody directed against amino acids 1 to 40 of the β-amyloid protein (Haass *et al.*, 1992). Findings were compared to subjects that received control grafts and to unoperated adult (but not aged) monkeys.

Adult non-aged monkeys never displayed β-amyloid-immunoreactive plaques. In aged monkeys, the density of β-amyloid-containing plaques was increased compared to young monkeys but plaque density was not significantly increased in NGF-treated aged subjects compared to aged controls (Kordower *et al.*, 1997). Similarly, aged monkeys that received intraparenchymal NGF-secreting fibroblast implants displayed β-amyloid plaques but did not show a significant increase in plaque number relative to control aged subjects that received uninfected fibroblast grafts (Tuszynski *et al.*, 1998). These data suggest that hNGF delivery to the adult and aged brain for up to three months does not upregulate expression of β-amyloid in aged primates.

VIII. CONCLUSION

NGF is a potent molecule that specifically rescues basal forebrain cholinergic neurons from both lesion-induced and age-related degeneration. The most common neurodegenerative disorder, Alzheimer's disease, is characterized by prominent degeneration of this population of cholinergic neurons, making it an attractive target for NGF therapy. However, traditional means of NGF delivery to the CNS, including intracerebroventricular infusions, induce unacceptable adverse effects. Other means of NGF delivery, or augmentation of NGF function in the CNS, are under development. Gene therapy offers a potential means of preventing the degeneration of basal forebrain cholinergic neurons while shielding non-targeted, NGF-responsive systems from abnormal stimulation and growth.

ACKNOWLEDGMENTS

Supported by the American Academy of Neurology and the National Institutes for Health AG10435 (MHT), NS25655 (JHK), AG 09466 (JHK and EJM) and AG14449 (EJM).

REFERENCES

Alitalo, K., Kurkinen, M., Virtanen, I., Mellstrom, K., and Vaheri, A. (1982). Deposition of basement membrane proteins in attachment and neurite formation of cultured murine C-1300 neuroblastoma cells. *J Cell Biochem* 18:25–36.

Angeletti, R. H., Hermodson, M. A., and Bradshaw, R. A. (1973). Amino acid sequences of mouse 2.5S nerve growth factor. II. Isolation and characterization of the thermolytic and peptic peptides and the complete covalent structure. *Biochem* 12:100–115.

Appel, S. H. (1981). A unifying hypothesis for the cause of amyotrophic lateral sclerosis, Parkinsonism, and Alzheimer's disease. *Ann Neurol* 10:499–505.

Barba, D., Hardin, J., Ray, J., and Gage, F. H. (1993). Thymidine kinase-mediated killing of rat brain tumors. *J Neurosurg* 79:729–735.

Barnett, J., Baecker, P., Routledge-Ward, C., *et al.* (1990). Human beta nerve growth factor obtained from a baculovirus expression system has potent *in vitro* and *in vivo* neurotrophic activity. *Exp Neurol* 110:11–24.

Bartus, R. T. (1990). Drugs to treat age-related neurodegenerative problems. The final frontier of medical science? *J Am Geriatrics Society* 38(6):680–695.

Bartus, R. T., Dean, R. L., Beer, C., and Lippa, A. S. (1982). The cholinergic hypothesis of geriatric memory dysfunction. *Science* 217:408–417.

Bierer, L. M., Haroutunian, V., Gabreil, S., Knott, P. J., Carlin, L. S., Purohit, D. P., Perl, D. P., Schmeidler, J., Kanof, P., and Davis, K. L. (1995). Neurochemical correlates of dementia in Alzheimer's disease: Relative importance of cholinergic deficits. *J Neurochem* 64:749–760.

Blau, H. M., and Springer, M. L. (1995). Gene therapy—a novel form of drug delivery. *N Engl J Med* 333:1204–1207.

Boissiere, F., Faucheux, B., Ruberg, M., Agid, Y., and Hirsch, E. C. (1997). Decreased TrkA gene expression in cholinergic neurons of the striatum and basal forebrain of patients with Alzheimer's disease. *Exp Neurol:* 245–252.

Charles, V., Mufson, E. J., Friden, P. M., Bartus, R. T., and Kordower, J. H. (1996). Atrophy of cholinergic basal forebrain neurons following excitotoxic cortical lesions is reversed by intravenous administration of an NGF conjugate. *Brain Res* 728:193–203.

Chen, K. S., and Gage, F. H. (1995). Somatic gene transfer of NGF to the aged brain: Behavioral and morphological amelioration. *J Neurosci* 15:2819–2825.

Conner, J. M., and Varon, S. (1992). Distribution of nerve growth factor-like immunoreactive neurons in the adult rat brain following colchicine treatment. *J Comp Neurol* 326:347–362.

Conner, J. M., Muir, D., Varon, S., Hagg, T., and Manthorpe, M. (1992). The localization of nerve growth factor-like immunoreactivity in the adult rat basal forebrain and hippocampal formation. *J Comp Neurol* 319:454–462.

Cordon-Cardo, C., Tapley, P., Jing, S., *et al.* (1991). The trk tyrosine protein kinase mediates the mitogenic properties of nerve growth factor and neurotrophin-3. *Cell* 66:173–183.

Coyle, J. T., Price, P. H., and Delong, M. R. (1983). Alzheimer's disease: A disorder of cortical cholinergic innervation. *Science* 219:1184–1189.

Crutcher, K. A., Scott, S. A., Liang, S., Everson, W. V., and Weingartner, J. (1993). Detection of NGF-like activity in the human brain tissue: Increased levels in Alzheimer's disease. *J Neurosci* 13:2540–2550.

Dekker, A. J., Gage, F. H., and Thal, L. J. (1992). Delayed treatment with nerve growth factor improves acquisition of a spatial task in rats with lesions of the nucleus basalis magnocellularis: Evaluation of the involvement of different neurotransmitter systems. *Neuroscience* 48:111–119.

Dobrowsky, R. T., Werner, M. H., Castellino, A. M., Chao, M. V., and Hannun, Y. A. (1994). Activation of the sphingomyelin cycle through the low-affinity neurotrophin receptor. *Science* 265:1596–1599.

Emerich, D. W., Winn, S., Harper, J., Hammang, J. P., Baetge, E. E., and Kordower, J. H. (1994). Implants of polymer-encapsulated human NGF-secreting cells in the nonhuman primate: Rescue and sprouting of degenerating cholinergic basal forebrain neurons. *J Comp Neurol* 349:148–164.

Ernfors, P., Lindefors, N., Chan-Palay, V., and Persson, H. (1990). Cholinergic neurons of the nucleus basalis express elevated levels of nerve growth factor receptor mRNA in senile dementia of the Alzheimer's type. *Dementia* 1:138–145.

Ezzeddine, Z. D., Martuza, R. L., Platika, D., *et al.* (1991). Selective killing of glioma cells in culture and *in vivo* by retrovirus transfer of the herpes simplex virus thymidine kinase. *New Biologist* 3:608–614.

Fischer, W., Bjorklund, A., Chen, K., and Gage F. H. (1991). NGF improves spatial memory in aged rodents as a function of age. *J Neurosci* 11:1889–1906.

Fischer, W., Wictorin, K., Bjorklund, A., Williams, L. R., Varon, S., and Gage, F. H. (1987). Amelioration of cholinergic neuron atrophy and spatial memory impairment in aged rats by nerve growth factor. *Nature* 329:65–68.

Friden, P. M., Walus, L. R., Watson, P., *et al.* (1993). Blood–brain barrier penetration and *in vivo* activity of an NGF conjugate. *Science* 259:373–377.

Gaffan, D., and Harrison, S. (1989). A comparison of the effects of fornix transection and sulcus principalis ablation upon spatial learning by monkeys. *Behav Brain Res* 31:207–220.

Gage, F. H., Armstrong, D. M., Williams, L. R., and Varon, S. (1988). Morphologic response of axotomized septal neurons to nerve growth factor. *J Comp Neurol* 269:147–155.

Games, D., Adams, D., Alessandrini, R., Barbour, R., Berthelette, P., Blackwell, C., Carr, T., Clemens, J., Donaldson, T., Gillespie, F., *et al.* (1995). Alzheimer-type neuropathology in transgenic mice overexpressing V717F beta-amyloid precursor protein [see comments]. *Nature* 3733(6514):523–527.

Goedert, M., Fine, A., Hunt, S. P., and Ulrich, A. (1986). NGF RNA in peripheral and rat central tissues and in the human central nervous system: Lesion effects in the rat brain and levels in Alzheimer's disease. *Mol Brain Res* 1:85–92.

Gossen, M., Bonin, A. L., Freundlieb, S., and Bujard, H. (1994). Inducible gene expression systems for higher eukaryotic cells. *Curr Opin Biotechnol* 5:516–520.

Haass, C., Scholssmacher, M. G., Hung, A. Y., Vigo-Pelfrey, C., Mellon, A., Ostaszewski, B. L., Lieberburg, I., Koo, E. H., Schenk, D., Teplow, D. B., and Selkoe, D. J. (1992). Amyloid β-peptide is produced by cultured cells during normal metabolism. *Nature* 359:322–327.

Hefti, F. (1986). Nerve growth factor (NGF) promotes survival of septal cholinergic neurons after fimbrial transection. *J Neurosci* 6:2155–2162.

Hefti, F., and Weiner, W. J. (1983). Is Alzheimer's disease caused by lack of nerve growth factor? *Ann Neurol* 13:109–110.

Hefti, F., Hartikka, J., Salviaterra, P., Weiner, W. J., and Mash, D. C. (1986). Localization of nerve growth factor receptors on cholinergic neurons of the human basal forebrain. *Neurosci Lett* 69:37–41.

Honegger, P., and Lenoir, D. (1982). Nerve growth factor (NGF) stimulation of cholinergic telencephalic neurons in aggregating cell cultures. *Dev Brain Res* 3:229–238.

Hsiao, K., Chapman, P., Nilsen, S., Eckman, C., Harigaya, Y., Younkin, S., Yang, F., and Cole, G. (1996). Correlative memory deficits, Aβ elevation, and amyloid plaques in transgenic mice. *Science* 274:99–102.

Ip, N. Y., Stitt, T. N., Tapley, P., *et al.* (1993). Similarities and differences in the way neurotrophins interact with the Trk receptors in neuronal and nonneuronal cells. *Neuron* 10:137–149.

Klein, R., Jing, S., Nanduri, V., O'Rourke, E., and Barabcid, M. (1991). The trk proto-oncogene encodes a receptor for nerve growth factor. *Cell* 65:189–197.

Koliatsos, V. E., Nauta, H. J., Clatterbuck, R. E., Holtzman, D. M., Mobley, W. C., and Price, D. L. (1990). Mouse nerve growth factor prevents degeneration of axotomized basal forebrain cholinergic neurons in the monkey. *J Neurosci* 10:3801–3813.

Kordower, J. H., and Mufson, E. J. (1989). NGF and Alzheimer's disease: Unfulfilled promise and untapped potential. *Neurobiol Aging* 10:543–544.

Kordower, J. H., Gash, D. M., Bothwell, M., Hersh, L. B., and Mufson, E. J. (1989). Nerve growth factor receptor and choline acetyltransferase remain colocalized in the nucleus basalis (Ch4) of Alzheimer's patients. *Neurobiol Aging* 10:287–294.

Kordower, J. H., Mufson, E. J., Fox, N., Martel, L., and Emerich, D. F. (1997). Cellular delivery of NGF does not alter the expression of β amyloid-immunoreactivity in young or aged nonhuman primates. *Exp Neurol* 145:586–591.

Kordower, J. H., Winn, S. R., Liu, Y.-T., *et al.* (1994). The aged monkey basal forebrain: Rescue and sprouting of axotomized basal forebrain neurons after grafts of encapsulated cells secreting human nerve growth factor. *PNAS* 91:10898–10902.

Kromer, L. F. (1987). Nerve growth factor treatment after brain injury prevents neuronal death. *Science* 235:214–216.

Levi-Montalcini, R. (1987). The nerve growth factor 35 years later. *Science* 237:1154–1162.

Levi-Montalcini, R., Meyer, H., and Hamburger, V. (1954). *In vitro* experiments on the effects of mouse sarcoma 180 and 37 on the spinal and sympathetic ganglia of the chick embryo. *Cancer Res* 14:49–57.

Liberini, P., Pioro, E. P., Maysinger, D., Ervin, F. R., and Cuello, A. C. (1993). Long-term protective effects of human recombinant nerve growth factor and monosialoganglioside GM1 treatment on primate nucleus basalis cholinergic neurons after neocortical infarction. *Neuroscience* 53:625–637.

Martinez-Serrano, A., Fischer, W., and Bjorklund, A. (1995). Reversal of age-dependent cognitive impairments and cholinergic neuron atrophy by NGF-secreting neural progenitors grafted to the basal forebrain. *Neuron* 15:473–484.

Menesini, M. G., Chen, J. S., and Levi-Montalcini, R. (1978). Sympathetic nerve fiber ingrowth in the central nervous system of neonatal rodent upon intracerebral nerve growth factor injections. *Arch Ital Biol* 116:53–84.

Mesulam, M. M., Mufson, E. J., Wainer, B. H., and Levey, A. I. (1983). Central cholinergic pathways in the rat: An overview based on an alternative nomenclature (Ch 1–Ch 6). *Neuroscience* 10:1185–201.

Mesulam, M. M., Mufson, E. J., Levey, A. I., and Wainer, B. H. (1983). Cholinergic innervation of cortex by the basal forebrain: cytochemistry and cortical connections of the septal area, diagonal band nuclei, nucleus basalis (substantia innominata), and hypothalamus in the rhesus monkey. *J Comp Neurol* 214:170–97.

Miller, D. A. (1990). Retrovirus packaging cells. *Human Gene Therapy* 1:5–14.

Mobley, W. C., Neve, R. L., Prusiner, S. B., and McKinley, M. P. (1988). Nerve growth factor induces gene expression for prion- and Alzheimer's beta-amyloid proteins. *PNAS* 85:9811–9815.

Mufson, E. J., Bothwell, M., Hersh, L. B., and Kordower, J. H. (1989a). Nerve growth factor receptor immunoreactive profiles in the normal aged human basal forebrain: Colocalization with cholinergic neurons. *J Comp Neurol* 285:196–217.

Mufson, E. J., Bothwell, M., and Kordower, J. H. (1989b). Loss of nerve growth factor receptor-containing neurons in Alzheimer's disease: A quantitative analysis across subregions of the basal forebrain. *Exp Neurol* 105:221–232.

Mufson, E. J., Connor, J. M., and Kordower, J. H. (1995). NGF and Alzheimer's disease: Defective retrograde transport to the nucleus basalis. *NeuroReport* 6:1063–1066.

Mufson, E. J., Connor, J., Varon, S., and Kordower, J. H. (1994). Nerve growth factor immunoreactivity in the basal forebrain and hippocampus in primates and Alzheimer's disease. *J Comp Neurol* 341:507–519.

Mufson, E. J., and Kordower, J. H. (1997). Nerve growth factor and its receptors in the primate forebrain: Alterations in Alzheimer's disease and potential use in experimental therapeutics. In: Mattson, M. P., ed., *Neuroprotective signal transduction*. Humana Press, pp. 23–59.

Mufson, E. J., and Kordower, J. H. (In press). Nerve growth factor in Alzheimer's disease. In: Morrison, J. H. and Peters, A. A., eds., *Cerebral cortex*. New York: Plenum Press.

Mufson, E. J., Lavine, N., Jaffar, S., Kordower, J. H., Quirion, R., and Saragovi, H. U. (1997). Reduction in p140-TrkA receptor protein within the nucleus basalis and cortex in Alzheimer's disease. *Exper Neurol* 146:91–103.

Mufson, E. J., Li, J.-M., Sobreviela, T., and Kordower, J. H. (1996). Decreased trkA gene expression within basal forebrain neurons in Alzheimer's disease. *NeuroReport* 7:25–29.

Olson, L., Nordberg, A., von Holst, H., *et al.* (1992). Nerve growth factor affects 11-C-nicotine binding, blood flow, EEG, and verbal episodic memory in an Alzheimer patient (Case report). *J Neural Trans* 4:79–95.

Perry, E. K., Tomlinson, B. E., Blessed, G., Bergmann, K., Gibson, P. H., and Perry, R. H. (1978). Correlation of cholinergic abnormalities with senile plaques and mental test scores in senile dementia. *Br Med J* 2:1457–1459.

Quon, D., Wang, Y., Catalano, R., Scardina, J. M., Murakami, K., Cordell, B. (1991). Formation of beta-amyloid protein deposits in brains of transgenic mice. *Nature* 352(6332):239–241.

Rabizadeh, S., Oh, J., Zhong, L. T., *et al.* (1993). Induction of apoptosis by the low-affinity NGF receptor. *Science* 261:345–348.

Rosenberg, M. B., Friedmann, T., Robertson, R. C., *et al.* (1988). Grafting genetically modified cells to the damaged brain: Restorative effects of NGF expression. *Science* 242:1575–1578.

Saffran, B. N., and Crutcher, K. A. (1990). NGF-induced remodeling of mature uninjured axon collaterals. *Brain Res* 525:11–20.

Schwab, M. E., Otten, U., Agid, Y., and Thoenen, H. (1979). Nerve growth factor (NGF) in the rat CNS: Absence of specific retrograde axonal transport and tyrosine hydroxylase induction in locus coeruleus and substantia nigra. *Brain Res* 168:473–483.

Scott, S. M., Tarris, R., Eveleth, D., Mansfield, H., Weichsel, M. E., and Fisher, D. A. (1981). Bioassay detection of mouse nerve growth factor (mNGF) in the brain of adult mice. *J Neurosci Res* 6:653–658.

Scott, S. A., Mufson, E. J., Weingartner, J. A., Skau, K. A., and Crutcher, K. A. (1995). Nerve growth factor in Alzheimer's disease: Increased levels throughout the brain coupled with declines in nucleus basalis. *J Neurosci* 15:6213–6221.

Seiler, M., and Schwab, M. E. (1984). Specific retrograde transport of nerve growth factor (NGF) from cortex to nucleus basalis in the rat. *Brain Res* 300:33–39.

Sheldon, D. L., and Reichardt, L. F. (1986). Studies on the expression of the beta-nerve growth factor (NGF) gene in the central nervous system; level and regional distribution of NGF mRNA suggest that NGF functions as a trophic factor for several distinct populations of neurons. *Proc Nat Acad Sci USA* 83:2714–2718.

Suhr, S., and Gage, F. H. (1993). Gene therapy for neurological disease. *Arch Neurol* 50:1252–1268.

Taniuchi, M., and Johnson, E. M. (1985). Characterization of the binding properties and retrograde axonal transport of monoclonal antibody directed against the rat nerve growth factor receptor. *J Cell Biol* 101:1100–1106.

Tekirian, T. L., Cole, G., Russell, M. J., Yang, F., Wekstein, D. R., Patel, E., Snowden, D. A., Marksberry, W. R., and Geddes, J. W. (1996). Carboxy terminal of β-a,yloid deposits in aged human, canine, and polar bear brain. *Neurobiology Aging* 17:249–257.

Terry, R. D., Katzman, R., and Bick, K. L. (1994) *Alzheimer disease.* New York:Raven Press.

Thoenen, H. (1995). Neurotrophins and neuronal plasticity. *Science* 270:593–598.

Tuszynski, M. H. (1998). Gene therapy for neurological disease. *Ann NY Acad Sci* 835:1–11.

Tuszynski, M. H., and Gage, F. H. (1994). Neurotrophic factors and neuronal loss: Potential relevance to Alzheimer's disease. In: Terry, R., Katzman, R., and Bick, K., eds., *Alzheimer's disease.* pp. New York: Raven Press, 405–418.

Tuszynski, M. H., and Gage, F. H. (1995). Bridging grafts and transient NGF infusions promote long-term CNS neuronal rescue and partial functional recovery. *PNAS* 92:4621–4625.

Tuszynski, M. H., Roberts, J., Senut, M. C., U., H.-S., and Gage, F. H. (1996). Gene therapy in the adult primate brain: Intraparenchymal grafts of cells genetically modified to produce nerve growth factor prevent cholinergic neuronal degeneration. *Gene Therapy* 3:305–314.

Tuszynski, M. H., Senut, M. C., Roberts, J., Ray, J., and Gage, F. H. (1994). Fibroblasts genetically modified to produce NGF promote sprouting in the adult primate brain. *Neurobiol Dis* 1:67–78.

Tuszynski, M. H., Smith, D. E., Roberts, J., McKay, H., and Mufson, E. (1998). Targeted chronic delivery of human NGF by gene therapy to the primate basal forebrain does not accelerate β-amyloid plaque deposition. *Abstr Soc Neurosci.*

Tuszynski, M. H., U., H. S., Amaral, D. G., and Gage, F. H. (1990). Nerve growth factor infusion in primate brain reduces lesion-induced cholinergic neuronal degeneration. *J Neurosci* 10:3604–3614.

Tuszynski, M. H., U., H. S., and Gage, F. H. (1991). Nerve human growth factor infusions prevent cholinergic neuronal degeneration in the adult primate brain. *Ann Neurol* 30:625–636.

Weinmann, P., Gossen, M., Hillen, W., Bujard, H., and Gatz, C. (1994). A chimeric transactivator allows tetracycline-responsive gene expression in whole plants. *Plant J* 5:559–569.

Whittemore, S. R., Ebendal, T., Larkfors, L., Olson, L., Seiger, A., Stromberg, I., and Persson, H. (1986). Development and regional expression of beta nerve growth factor messenger and protein in the rat central nervous system. *Proc Nat Acad Sci* 83:817–821.

Williams, L. R. (1991). Hypophagia is induced by intracerebroventricular administration of nerve growth factor. *Exp Neurol* 113:31–37.

Winkler, J., Ramirez, G. A., Kuhn, H. G., *et al.* (1996). Reversible induction of Schwann cell hyperplasia and sprouting of sensory and sympathetic neurites *in vivo* after continuous intracerebroventricular administration of nerve growth factor. *Ann Neurol* 40:128–139.

Winn, S. R., Linder, M. D., Lee, A., Haggett, G., Francis, J. M., and Emerich, D. F. (1996). Polymer-encapsulated genetically modified cells continue to secrete human nerve growth factor for over one year in rat ventricles: Behavioral and anatomical consequences. *Exper Neurol* 140:126–138.

Yao, R., and Cooper, G. M. (1995). Requirement for phosphatidylinositol-3 kinase in the prevention of apoptosis by nerve growth factor. *Science* 267:2003–2006.

CHAPTER 21

Trophic Factors and Cell Adhesion Molecules Can Drive Dysfunctional Plasticity and Senile Plaque Formation in Alzheimer's Disease through a Breakdown in Spatial and Temporal Regulation

CARL W. COTMAN AND BRIAN J. CUMMINGS
Institute for Brain Aging and Dementia, University of California, Irvine, Irvine, California

In this chapter, we discuss the role of trophic factors and cell adhesion molecules in brain plasticity and explore the hypothesis that these molecules contribute to the development of age-related pathology in Alzheimer's disease (AD). Although these molecules are well known to play a role in development and recovery of function after brain damage, ironically, they may also contribute to brain dysfunction in later life due to shifts in their regulation and organization that generate abnormal growth responses and subsequent degeneration. The first section examines the evidence for neuronal sprouting and plasticity in senile plaque formation and compares an animal model of injury-induced sprouting responses to the aberrant sprouting observed in AD. The second section explores if trophic molecules play a role in the process of plaque formation. Some of these molecules include bFGF, heparan sulfate proteoglycans, and the cell adhesion molecule, amyloid precursor protein (APP). We describe how β-amyloid and APP are involved in both attracting neurites and in their toxicity. Because much of the degeneration occurring within plaques is mediated by inflammation, the role of microglia and inflammation in response to cellular injury is briefly discussed. The chapter illustrates how common plasticity mechanisms may be activated by divergent insults, but converge into a positive feedback cascade that induces further degenera-

CNS Regeneration

tion, programmed cell death, and further inflammation. Finally, the potential therapeutic role for growth factors in modulating the progression of AD is examined. We conclude that early in the disease, trophic interventions may slow the progress of pathology, but caution that once sufficient damage has occurred, the continued administration of trophic support may exacerbate the disease process.

I. INTRODUCTION

The mechanisms that regulate the development and plasticity of the nervous system are traditionally viewed as separate from those that drive the evolution of brain pathology and disease. It is well known, for example, that neurotrophic factors are critical for the development, maintenance, and plasticity of the nervous system. And it is becoming increasingly clear that cell adhesion molecules also regulate neuronal development, maintenance, and plasticity. Together, trophic factors and adhesion molecules are the primary extracellular signaling systems controlling development and plasticity. Because these molecular systems are so important in brain function, it is likely that some of these mechanisms may become dysfunctional and contribute to brain pathology.

In this chapter, we will discuss the role of various trophic factors and cell adhesion molecules in brain plasticity in the adult nervous system and we will explore the hypothesis that these molecules contribute to the development of age-related pathology in Alzheimer's disease (AD); a process that is driven by a breakdown in the spatial segregation of molecular events. We have focused our discussion on fibroblast growth factor (bFGF) and cell adhesion molecules as illustrations of the principle. In the first section, we examine the evidence for neuronal sprouting and plasticity in senile plaque formation and compare an animal model of injury-induced sprouting responses to the aberrant sprouting observed in AD. We outline the key stages and cellular elements involved in the formation of senile plaques. In the second section, we ask whether or not trophic molecules may play a role in the process of plaque formation. We discuss several key molecules found within the plaque microenvironment and how they contribute to the progression of pathology. Some of these molecules include bFGF, heparan sulfate proteoglycans, and the cell adhesion molecule amyloid precursor protein (APP). We describe how β-amyloid and APP are involved in both attracting neurites and in their toxicity. Because much of the degeneration occurring within plaques is mediated by inflammation, the next section briefly discusses the role of microglia and inflammation in response to cellular injury. We illustrate how these common plasticity mechanisms may be activated by divergent insults, but converge into an unstoppable positive feedback cascade that induces further degeneration, programmed cell death, and further inflammation. Thus, the same convergent pathways can either serve to regulate growth and repair mechanisms or to participate in the destruction of the cell. Finally, we examine the potential therapeutic role for growth factors in modulating the progression of AD.

II. SPROUTING AND PLASTICITY IN SENILE PLAQUE FORMATION

A. Entorhinal Cell Loss Is One of the Initial Events in AD

In AD, there are three pathological hallmarks associated with the progressive degeneration of the brain—senile plaques, neurofibrillary tangles (NFTs), and neuronal cell loss. We suggest that in AD the senile plaques actually form local abnormal "micro-environments" that employ some of the same mechanisms that are used during normal growth and development. Detrimental consequences evolving from the brain's attempt to respond to injury may seem paradoxical, but pathology could occur if there is a breakdown of organizational and spatial relationships or dysregulation of developmental pathways.

Major neuronal loss occurs during AD in the temporal lobe, particularly in layers II and III of the entorhinal cortex (for a review, see Hyman *et al.*, 1986; Coleman and Flood, 1987). Neuronal loss and NFT are severe in layers II and III and, to a lesser, more variable extent, layer IV of the entorhinal cortex in AD (Hyman *et al.*, 1986). The layer II stellate cells of the entorhinal cortex form the perforant pathway and comprise 86% of the synapses within the outer three-fourths of the molecular layer of the dentate gyrus (Matthews *et al.*, 1976a). Degeneration of the entorhinal cortex may thus contribute to the functional disconnection of the hippocampus from the cerebral cortex.

B. Can Trophic Infusion Rescue Injured Perforant Path Neurons?

Infusion of nerve growth factor (NGF) has been demonstrated to ameliorate axotomy-induced cell loss within the septum and diagonal band of broca after fimbria–fornix transection (Hefti, 1986; Williams *et al.*, 1986; Gage *et al.*, 1988). However, NGF is relatively selective for peripheral sympathetic and sensory neurons and central nervous system cholinergic neurons (Barde, 1989; Hefti *et al.*, 1989). Fibroblast growth factor (bFGF) is among the most effective growth factors for supporting the survival and neuritic outgrowth of cultured neurons. Immature, developing neuronal cultures from both the hippocampus

(Mattson *et al.*, 1989; Walicke *et al.*, 1986) and the cerebral cortex (Morrison *et al.*, 1986; Walicke, 1988a) exhibit greatly enhanced survival in response to treatments with bFGF at concentrations as low as 1 picomolar. Like NGF, infusion of bFGF also prevents the loss of cholinergic neurons in the septum following axotomy (Anderson *et al.*, 1988).

Previous studies have shown that the majority of layer II stellate neurons utilize an excitatory amino acid as their neurotransmitter (White *et al.*, 1977; Mattson *et al.*, 1988). And, as discussed earlier, these neurons are one of the primary and initial cell groups to be lost in AD. In order to determine if bFGF acts on noncholinergic cortical neurons, a *in vivo* paradigm was developed to examine whether or not bFGF could spare layer II entorhinal stellate cells from axotomy-induced death or atrophy. A unilateral knife-cut of the medial entorhinal cortex fibers projecting to the dentate gyrus of the hippocampal formation via the perforant path led to retrograde cell loss in entorhinal cortex. Fourteen days after axotomy of the perforant path, layer II of medial entorhinal cortex showed a 28% decrease in large stellate neurons as well as many weakly stained, hollow cells compared to the nonlesioned side (Fig. 1). Layer IV neurons, however, which do not project via the perforant path, showed little detectable change in the number of cells ipsilateral to the knife-cut. Intraventricular infusion of bFGF over a period of 14 days reduced the 28% cell loss to less than 6% (Cummings *et al.*, 1992b). A more recent study using an

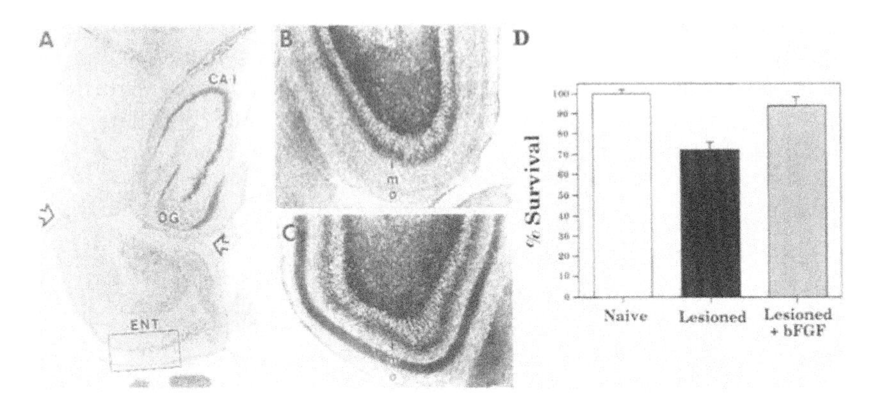

FIGURE 1 (A) Demonstration of a knife-cut to the perforant pathway in an adult rat. Arrows indicate the site of the lesion. Neurons in layer II of entorhinal cortex (shown by box) were counted 14 days after the lesion. (B and C) Acetylcholine esterase staining of the dentate gyrus contralateral (B) and ipsilateral (C) to the knife-cut demonstrates the reactive sprouting of cholinergic fibers within the middle molecular layer. (D) There is a 28% neuronal loss in layer II (black bar) compared to naive controls (white bar), however, intraventricular infusion of bFGF rescues the majority of these dying cells (gray bar) (adapted from Cummings *et al.*, 1992b).

ablation lesion and unbiased stereology techniques confirmed the protective effect, finding a 30% drop in total neuron number in layer II, which was prevented by FGF administration (Peterson *et al.*, 1996). Thus, bFGF is capable of preventing cortical neuronal loss and/or atrophy associated with retrograde degeneration of noncholinergic neurons following axotomy. It is natural to ask whether NGF or bFGF would be useful in treating AD. To address this question, however, we need to know more about how senile plaques form.

C. Entorhinal Cell Loss Initiates Compensatory Sprouting

In animal models, damage to the entorhinal cortex leads to hippocampal-dependent memory impairment, recovery from which correlates with sprouting of surviving entorhinal, commissural, or other undamaged cells (e.g., septal) that project to the deafferented dentate gyrus (Loesche and Steward, 1977; Scheff and Cotman, 1977). At the light microscopic level, this sprouting can be readily visualized by staining with acetylcholine esterase, a presynaptic marker for sprouting septal fibers. Sprouting of septal neurons occurs in the outer two-thirds of the molecular layer of the dentate gyrus (Cotman *et al.*, 1973) after an electrolytic lesion of the entorhinal cortex. Quantitative electron microscopy shows that sprouting occurs and new synapses form until pre-lesion synaptic density is restored (Matthews *et al.*, 1976a, 1976b). We have proposed that, in the early stages of AD, sprouting of surviving neurites may compensate for neuronal or synaptic loss. Partial neuronal loss within a circuit is offset by sprouting from the remaining cells (see, for example Anderson *et al.*, 1988).

After examining AD and control brains, we described a shift in the characteristic laminar pattern of AChE histochemistry in the hippocampi of AD brains as compared to nondemented controls (Geddes *et al.*, 1985). In control brains, intense bands of AChE staining appear in the supragranular and subgranular regions adjacent to the stratum granulosum, with a homogeneous, faint AChE pattern across both the inner and outer molecular layer of the dentate gyrus (DG). In AD brains, there was a further intensification in the middle to outer molecular layer, with a clearing in the inner layer (Geddes *et al.*, 1986). Additionally, the laminar distribution of the DG molecular layer changes with injury in rat, and plaque formation may coincide with the reorganization of the molecular layer due to deafferentation in AD. Neuritic plaques, therefore, are found where sprouting occurs. In select cases, even within the dentate gyrus, sprouting occurred in the molecular layer where few plaques existed. We have proposed that these sprouting fibers were attracted to early plaques and may contribute to their formation (Geddes *et al.*, 1986; Cotman and

Anderson, 1988; Cotman *et al.*, 1993a, 1993b). In the case of entorhinal lesion-induced degeneration and sprouting, the degenerating fibers are phagocytosed by microglia and astrocytes (Bechmann and Nitsch, 1997; Hailer *et al.*, 1997).

Most recently, Deller and associates (1996) have conducted a series of elegant studies tracing individual fibers sprouting into the dentate gyrus following unilateral entorhinal cortex lesions. Eight weeks after injury, extensive "tangle-like" axonal formations can be observed in the deafferented region. These axons originate from the contralateral entorhinal cortex and comprise the "crossed entorhino-dentate projection." In normal animals, the axons form primarily *en passant* boutons and exhibit only a few branches. Following injury, however, the crossed fibers sprout extensively and exhibit a significant increase in branch points and axon extensions (Deller *et al.*, 1996). Most striking, however, is the appearance of bulbous terminals and an aberrant sprouting reaction that results in focal areas of "axonal wandering," which appear identical to the aberrant regenerative sprouting seen in AD (Fig. 2, see color insert). See Deller and Frotscher (1997) for an extensive review.

Thus, experiments in animal models predict that entorhinal cell loss in AD should elicit a sprouting reaction in the hippocampus. Several anatomists who studied AD in the early 1900s had little doubt that the neuritic processes in plaques involved, in part, a regenerative phenomenon. As early as 1907, Fischer believed that "during the enlargements of the plaques, the fibrils [within them] spread out more and more and those that are closest show particular symptoms of proliferation, in the form of spindle-shaped thickenings, and rounded, club-like, multiply ramified sproutings which are arranged preferentially on the edge of the plaques" (Fischer, 1907). Ramon y Cajal (1928) and Bouman (1934) both proposed that plaques are not merely sites of degeneration but are actually areas of active neuritic involvement in and around the plaque. Cajal theorized: "It appears as if the sprouts had been attracted toward the region of the plaque under the influence of some special neurotropic substance" (Ramón y Cajal, 1928; Bouman, 1934).

D. Key Stages in Amyloid Accumulation and Plaque Formation

Senile plaques appear to develop and mature via a series of stages. The stages involve a progression from diffuse to primitive to neuritic plaque subtypes. Plaques are first identified as diffuse deposits of β-amyloid scattered throughout the neuropil. These deposits may originate from neurons because the amyloid precursor protein is overexpressed in neurons and neuritic processes surrounding and entering plaques (Cummings *et al.*, 1992a) and because, in the aged canine brain, amyloid accumulation occurs in neurons and plaques in the absence of other glial contributions (Cummings *et al.*, 1993b, 1996). At the early stages, the plaques are thioflavine negative, suggesting the absence of β-pleated sheet structures. They are also devoid of glial and neuritic involvement at this stage, because microglia and/or astrocytes are not concentrated in or around the plaque. As diffuse amyloid condenses into a primitive plaque, thioflavine staining develops and glial cells and neurites are drawn into the plaque microenvironment. The transition to β-sheet structure is crucial because amyloid has a distinct set of activities once this three-dimensional transition occurs (Pike *et al.*, 1993). As illustrated in the diagram (Fig. 3), it is possible that plaque biogenesis involves a cascade of molecular events: β-amyloid deposition, condensation, glial tropism, neuritic dystrophy, and inflammation (Cotman *et al.*, 1996b). These are mediated via a series of parallel and converging recursive

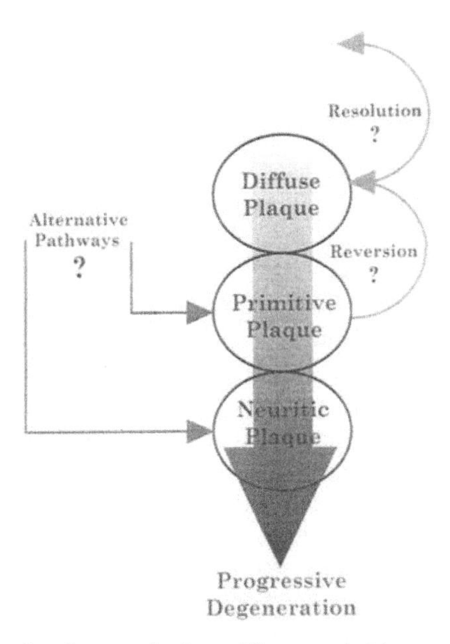

FIGURE 3 Plaques are thought to evolve from diffuse to primitive to neuritic subtypes based on evidence from nondemented aged individuals, the progression of pathology seen in Down's syndrome and in animal models. Plaques are thioflavine negative in the earliest stages, and are best detected with antibodies to β-amyloid. In the diffuse stage, plaques also contain APP and HSPG and many contain reactive microglia. Abnormal neurites are seen in primitive plaques and become more extensive in neuritic plaques. These later stage plaques contain trophic factors such as bFGF and a proliferation of reactive astrocytes as well. It is still unknown whether the brain is capable of resolving the earlier stages of β-amyloid deposition. As illustrated to the left, it is also possible that a subset of plaques do not progress through each of the stages, but rather are formed *de novo* by alternative mechanisms.

(positive feedback) molecular cascades. The following sections will outline these positive feedback cascades.

III. WHAT MOLECULES SUPPORT NEURITIC OUTGROWTH INTO PLAQUES?

Several mechanisms might trigger neuritic outgrowth into plaques. Regenerative growth, characterized by the enhanced viability and reactive sprouting of neurons surviving an insult, requires two fundamental components—neurotrophic factors and suitable substrates. The influence of these factors and their importance in hippocampal and cortical plasticity have been demonstrated in both tissue culture and animal models.

Neurite growth and degeneration coexist in AD. If sprouting is misdirected into plaques, a key issue becomes the identification of the specific molecules driving such growth. It would be predicted that the trophic and substrate elements detailed earlier play a critical role. The cellular microenvironment provides trophic support not only through growth factors but also by containing many different substrate components in the neuropil, including laminin, fibronectin, and glycosaminoglycans. Substrate factors are involved in a variety of processes essential to plastic and regenerative responses. The glycosaminoglycan heparan sulfate (HS) is one substrate factor that has been linked to plasticity. For example, neurons grown in culture on an HS substrate show dramatically increased axonal elongation. After 24 hours *in vitro,* the neurites of hippocampal neurons grown on a HS-coated tissue culture dish exhibit an approximately twofold increase in length in comparison to those grown on noncoated dishes.

Accumulating evidence suggests that bFGF and HS effectively promote plastic changes via independent actions as well as through interactions. These interactions are facilitated in part due to the strong binding affinity of bFGF for HS and the closely related glycose aminoglycan GAG, heparin (Baird *et al.*, 1988; Gospodarowicz *et al.*, 1984). As a result of this association, the biological activities of bFGF and HS may be reciprocally influenced. In fact, when bound to HS or heparin, bFGF has a larger effective radius from its source due to increased diffusion (Flaumenhaft *et al.*, 1990) and is protected from proteolytic degradation (Saksela *et al.*, 1988). In addition, evidence suggests that bFGF is internalized into cells only when it is bound to heparin (Ruoslahti and Yamaguchi, 1991). Thus, the activity of bFGF may require interaction with the substrate factors HS and heparin. Conversely, the activity of HS may require, or at least be enhanced by, association with bFGF. In support of this position, the *in vitro* neurite-outgrowth induced by bFGF is potentiated on HS and heparin substrates (Neufeld *et al.*, 1987; Walicke, 1988b).

A. Plaques Contain Fibroblast Growth Factor and Heparan Sulfate Glycosaminoglycans

Accordingly, we evaluated the possibility that bFGF and HS are present in or around senile plaques. Using AD brain tissue and aged matched controls, we found immunocytochemically that many plaques contain bFGF immunoreactivity (Gómez-Pinilla *et al.*, 1990). These bFGF-positive plaques are found both in the entorhinal cortex and along the middle of the DG molecular layer in a linear organization. However, not all plaques contained detectable levels of bFGF. In a follow-up study pairing bFGF immunocytochemistry with the detection of dystrophic neurites, we found that bFGF was primarily associated with the neuritic plaque subtype and not diffuse plaques, thereby supporting a relationship between bFGF and regenerative sprouting (Fig. 4, see color insert) (Cummings *et al.*, 1993a).

Others have also demonstrated bFGF-immunopositive plaques and suggested that whole brain levels of bFGF are elevated in AD compared to normal aged brains (Stopa *et al.*, 1990). Kato and colleagues (1991) demonstrated that exogenous bFGF can bind to plaques and tangles in sectioned AD tissue, and Perry and colleagues (1991) have shown that bFGF binding to NFTs is HS mediated. Transforming growth factor-β (TGFβ) is also detected within senile plaques (van der Wal *et al.*, 1993), and TGFβ has been shown to induce increases in the production of APP in microglia (Monning *et al.*, 1994, 1995). bFGF and IL-1 are produced by glia, and IL-1 can regulate the expression of APP (Goldgaber *et al.*, 1989; Donnelly *et al.*, 1990; Lahiri and Nall, 1995). It was originally thought that bFGF only induces the expression of APP mRNA in glia (Quon *et al.*, 1990; Gray and Patel, 1993), however, more recently, others have shown that treatment of neuroblastoma cells with bFGF results in a twofold increase in APP protein production as well (Ringheim *et al.*, 1997). bFGF and APP have been shown to colocalize within individual neurons, strengthening the hypotheses that injury begets more APP (Imaizumi *et al.*, 1993). Moreover, bFGF is capable of attenuating the toxicity of β-amyloid (Mark *et al.*, 1997).

B. Cell Adhesion Molecules Associated with Senile Plaques

Cell adhesion molecules have a well-documented role in developmental and regenerative growth responses in the CNS. Such mechanisms appear critical in the spatial and temporal responses for regrowth and remodeling of neuronal circuits. In the dentate gyrus after entorhinal lesions, for example, neuronal cell adhesion molecule (NCAM) is induced over the entire dendritic surface of the granule cells within the first day and then becomes restricted to the outer dendritic field in which new synapses are forming (Miller *et al.*, 1994). One adhesion molecule, intercellular adhesion molecule-1, is found within senile plaques (Frohman *et al.*, 1991; Akiyama *et al.*, 1993; Verbeek *et al.*, 1994) and may play a role in an inflammatory response to plaques (see the following discussion).

A more classical substrate molecule is heparan sulfate proteoglycan. Heparan sulfate proteoglycans are often found as an integral membrane component whose structure is three heparan sulfate (HS) side chains covalently linked to the core protein. HS is thought to play a role in cell adhesion (Laterra *et al.*, 1983; Gill *et al.*, 1986; Juliano, 1987), neural crest migration (Bronner-Fraser and Lallier, 1988), and neurite extension (Lander, 1982; Mathews and Patterson, 1983). In addition, HS may be involved in trophic factor regulation. For example, HS has been shown to alter the activity of nerve growth factor and fibroblast growth factor (bFGF) *in vitro* (Damon *et al.*, 1988; Lipton *et al.*, 1988).

An early study reported that AD tissue contained elevated levels of glycosaminoglycans when compared to normal aged brain (Suzuki *et al.*, 1955). Subsequently, histochemical techniques were used to demonstrate sulfated glycosaminoglycans and/or proteoglycans in the plaques, NFTs, and congophilic angiopathy of AD (Snow *et al.*, 1987, 1988). Until recently, the specific types of proteoglycans and glycosaminoglycans within the lesions of AD could not be determined. Using antibodies to the protein core of HSPG or the HS side chains of HSPG, Snow and co-workers reported the presence of HSPG core protein in plaques and cerebrovascular amyloid deposits and the presence of HS side chains in neurons, NFTs, plaques, glia, and the vessel walls of AD brain (Snow and Wight, 1989; Snow *et al.*, 1990). Perlmutter and co-workers (1990) have demonstrated that an antibody to HSPG core protein immunostained the basement membrane of capillaries in AD and control brains and plaques in AD brains. More recently, Snow and colleagues have gone on to identify and characterize a variety of glycosaminoglycans associated with AD lesions and their respective binding affinities to β-amyloid (Snow *et al.*, 1996; Castillo *et al.*, 1997).

In AD tissue, we found that the intensity of HS immunostaining in neurofibrillary tangles and senile plaques was more extensive than in control brains. In addition, within AD brains, the nuclei of select neurons and microglia were also immunopositive for HS, as were astrocytes (Fig. 4) (Su *et al.*, 1992). Some plaques in AD tissue also contained strong HS-positive neurites, which were not seen in controls. It is interesting to note that diffuse plaques, which are believed to be the earliest form of plaque, were immunostained by HS antibodies as well. Using double-labeling techniques or adjacent tissue sections, we have demonstrated that such diffuse HS-positive plaques were also immunostained by antibodies to β-amyloid protein, supporting the concept that both β-amyloid and HS are two components related to early plaque biogenesis (Snow and Kisilevsky, 1985). Thus, HS as well as bFGF may accumulate in concentrated areas associated with senile plaques, providing a microenvironment that could divert sprouting and plasticity processes away from potentially beneficial functions. Senile plaques contain other cell adhesion molecules in addition to heparan sulfate.

C. The Amyloid Precursor Protein Is a Cell Adhesion Molecule

Amyloid precursor protein (APP) is a central molecule in AD. In the normal brain, APP is thought to be involved in cell adhesion and stabilization of adhesion sites and synapses (see, for example, Coulson *et al.*, 1997). APP can interact with other cell adhesion proteins such as the extracellular matrix molecule laminin. APP is, in the strict definition of the term, a cell adhesion molecule. APP has heparin- and laminin-binding sites in the extracellular domain and APP appears capable of mediating cell-cell or cell-matrix adhesion. For example, in antisense APP-transfected cells, adhesion is reduced and can be reversed by addition of APP (LeBlanc *et al.*, 1992). Furthermore, APP has neurite growth-promoting activity and its secreted form has been reported to have a protective effect against neuronal excitotoxicity (Mattson *et al.*, 1993).

APP is localized to patches on the neuronal surface (Breen *et al.*, 1991; Storey *et al.*, 1996a, 1996b; Yamazaki *et al.*, 1997). Surface APP is not typically found over growth cones and, in fact, is reciprocally distributed with phalloidin staining for F-actin along the length of neurites. The segmental organization of extracellular APP does not seem to be restricted to any one neuronal type as neurons of different transmitter systems (i.e., GABA and nonGABA) and widely differing morphologies all appear similar. The segmental distribution of APP shows colocalization with focal adhesion patch components. There is a strong correspondence between the location of surface APP and β1-integrin. Talin also has a similar distribution (Storey *et al.*, 1996b). An interesting question is to determine if other cell surface adhesion molecules show segmental properties and whether these correspond to the distribution of focal adhesion kinases. Although poorly understood, APP production and processing become abnormal in AD. APP is upregulated early in AD and is found within normal neurons, dystrophic neurites, and senile plaques.

β-amyloid, which is derived from APP, also appears to have the ability to serve as an adhesion molecule. An antibody to β-amyloid as well as APP inhibits cell-substratum adhesion in several cell types (Chen and Yankner,

1991), possibly because of the presence of a laminin-like or other recognition sequences in the N-terminal domain of the peptide. Further, if β-amyloid is adsorbed to the surface of a culture dish, it will support the survival and outgrowth of neurons; similar results are obtained if soluble β-amyloid is presented to the cultures, which for several days show enhanced survival and neurite outgrowth (Whitson *et al.*, 1989, 1990). Over time, however, in this preparation the soluble amyloid assembles and clusters on the surface of the neurons and they degenerate (Pike *et al.*, 1991, 1992). Thus, the peptide stimulates growth but then causes degeneration.

D. β-Amyloid Initiates Neuronal Degeneration if It Assembles into β-Sheets

Over time β-amyloid assembles into fibrils and develops a B-sheet conformation. In this form it initiates apoptosis (Loo *et al.*, 1993). While soluble, β-amyloid is neurotrophic. When aggregated, it affects neurites and simulates, morphologically, the features of dystrophic neurites in and around senile plaques (Fig. 5, see color insert). At later times it initiates soma degeneration. It is possible that the initiation signal is due to an interaction with cell surface receptors and their cross-linking (Cotman *et al.*, 1996a). That is, the mechanism of β-amyloid-induced apoptosis involves the cross-linking of membrane receptors followed by receptor activation, which leads to aberrant signal transduction (Cotman *et al.*, 1996a; Cribbs *et al.*, 1996). Conversely, it should be noted that other plaque components, for example chondroitin sulfate, may also be capable of inhibiting normal neurite growth and inducing dystrophy (Canning *et al.*, 1993).

To test this hypothesis, a ligand is needed that (1) cross-links surface receptors and (2) can be modified so that binding is preserved but the cross-linking property is eliminated. The plant lectin Concanavalin A (Con A) has these properties and is a powerful stimulus of neuronal cell death. Like for Fas-induced cell death, the Con A needs to be multivalent to cross-link receptors; the divalent succinylated Con A that is unable to cross-link is inactive (Cribbs *et al.*, 1996) as are F_{ab} fragments of anti-Fas antibodies (Strasser, 1995). Furthermore, if Con A is bound to the surface of the cell culture dish it does not induce cell death but in fact stimulates process outgrowth. Con A-induced cell death shares many features of activation-induced cell death and β-amyloid-induced death, including the morphological characteristics and the rates of change and neuronal cell death (Cribbs *et al.*, 1996). Thus, β-amyloid and Con A provide two insults that appear to target receptors on the plasma membrane, cause their cross-linking, and therefore may represent a Fas-like mechanism of apoptosis in neurons. In the case of β-amyloid, it is likely that this cross-linking is irreversible. This of course does not define the nature of the surface components, because Con A and probably β-amyloid bind to many surface receptors, so the next question is whether specific molecules may also initiate apoptosis.

E. Cross-Linking of NCAM or APP Induces Cell Degeneration

To test if other, more relevant molecules can induce apoptosis in a similar fashion, NCAM antibodies were applied to neurons in a manner similar to the paradigm used in Con A and Fas receptor ligation experiments. Indeed, antibodies to NCAMs initiate neuronal degeneration that is apoptotic. Fab fragments are ineffective, as is adsorption of the antibody to the cell surface (Fig. 6) (Azizeh *et al.*, 1998). These results indicate that specific candidate CAMs are also capable of activating apoptotic pathways. Preliminary experiments indicate that cross-linking of APP may also activate apoptosis (Cotman and Cribbs, unpublished observations). Thus, an interesting hypothesis is that some cell adhesion molecules, such as NCAM or APP, can mediate multiple outcomes. Under normal circumstances, they may participate in neuronal growth and plasticity mechanisms. On the other hand, if a molecule interacts with them so that the structural organization in the membrane is not main-

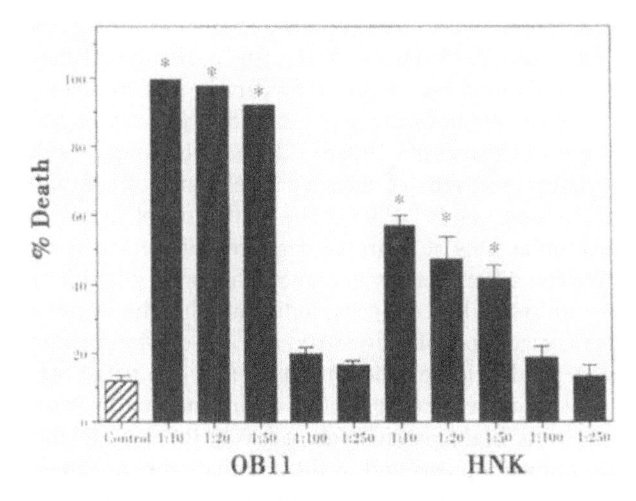

FIGURE 6 Dose-response for monoclonal anti-NCAM (OB11) and monoclonal anti-HNK-1/NCAM to induce programmed cell death in neuronal cultures. Cell death was determined after 24 hrs (taken from Azizeh *et al.*, 1998).

tained, growth is inhibited and degeneration may be induced. This type of mechanism may play a role in injury and in the evolution of senile plaque biogenesis in AD. Thus, it appears that the plaque is a microenvironment that can stimulate neurite growth and in the same locus place neuritic processes at risk for degeneration (Cotman *et al.*, 1993a, 1993b).

IV. THE MICROGLIAL/IMMUNE CONNECTION

Although the processes of tangle formation and plaque development primarily affect neurons, it has become increasingly clear that other mechanisms may also be involved, including apoptosis initiated by a variety of events and the recruitment of other cell types, such as astrocytes and microglia. In turn, pathogenesis may be driven by other, secondary factors that converge and add to the evolving cascades. One primary contributor is inflammation; both plaques and tangles initiate an injury response associated with inflammation. Normally, inflammation is an acute response after injury, but in the AD brain inflammation persists and evolves into a chronic condition.

To more fully understand the consequences of regenerative sprouting associated with plaques, we must also appreciate other cell types and signaling molecules present within plaques. In the past five years, researchers have focused much more closely on how the immune system participates in the pathogenesis of AD. Much of the groundwork for identifying this aspect of pathogenesis was initiated by McGeer and associates (McGeer *et al.*, 1987, 1989). Microglia, as local immunocompetent cells, are extremely sensitive to neuronal degeneration and become rapidly activated in the early stages of the disease (Griffin *et al.*, 1995; Sheng *et al.*, 1997). β-amyloid may serve as a mediator of the inflammatory response by driving and localizing a series of inflammatory events. We have proposed that this converts an acute response to a chronic one (Cotman *et al.*, 1996b). There is also ample evidence that β-amyloid is a potent activator of microglial cells and can induce microglial degeneration (Korotzer *et al.*, 1993). The activation of microglial cells and their accumulation at sites of neuronal degeneration obviously requires close interaction between different activated microglial cells, glial interactions with neurons, and with β-amyloid deposits, indicating that the expression of adhesion molecules as mediators of cell-matrix- and cell-cell-interactions is a necessary prerequisite in developing the typical signs of AD pathology.

Microglia have long been recognized as easily and rapidly activated phagocytic cells inside the brain parenchyma, actively involved in the removal of neurons undergoing apoptosis and in the elimination of synapses or neurites in the context of retrograde degeneration, a process that has been termed "synapse stripping" (Streit *et al.*, 1988; McGeer *et al.*, 1993). Activated microglial cells secrete pro-inflammatory cytokines and free radicals, similar to activated macrophages outside the brain. Furthermore, oxidative mechanisms seem to mediate changes in the state of microglial activation (Boje and Arora, 1992; Lee *et al.*, 1993; Heppner *et al.*, 1997), potentially contributing to a positive feedback loop of inflammation. Compounding the recruitment signals generated from injured cells or neurites, the primary component of plaques, β-amyloid, may contribute the recruitment of more microglia. *In vitro*, β-amyloid has been shown to be chemotaxic for microglia (Maeda *et al.*, 1997), and β-amyloid binds to C1q, a complement protein produced during inflammation (Jiang *et al.*, 1994). Furthermore, neuronal populations at risk in AD are immunopositive for C1q, and reactive microglia are observed surrounding these cells (Afagh *et al.*, 1996). In addition to the aggregation of microglial cells at sites of dendritic or axonal injury, microglia also establish direct contact with deposits of β-amyloid inside plaques and interact with β-amyloid via integrin adhesion molecules. *In vitro* experiments suggest that high concentrations of β-amyloid can even lead to degeneration of microglial cells themselves (Korotzer *et al.*, 1993). Further β-amyloid may act as a direct activator of microglial cells, as increases in the production of pro-inflammatory cytokines and free radicals occur in cells exposed to β-amyloid (Araujo and Cotman, 1992; Banati *et al.*, 1993; Haga *et al.*, 1993; Meda *et al.*, 1995; McDonald *et al.*, 1997). These findings suggest that inflammatory mechanisms may first respond to minor pathology and paradoxically become a key event in driving the progression of degeneration. Because β-amyloid has unique adhesion and self-assembly properties, it activates several signal transduction pathways and is a central driving force in the evolution of a chronic inflammation response.

A. The Vicious Circle of Microglial Activation

Following from the features outlined so far, it appears that the neuropathology observed in AD is perpetuated through a vicious feedback cascade of microglia activation. β-amyloid, as the central component of senile plaques in AD, directly and indirectly initiates activation of the brain-based components of the immune system, i.e., microglial cells, thereby causing disturbances in the composition of the specialized brain network of cytokines and growth factors (Owens *et al.*, 1994; Fabry *et al.*, 1994). This leads to the accumulation of neurotoxic substances, e.g., activated complement factors and free radicals. These and related events result in the brain equivalent of chronic inflammatory disease; the process sustains a vicious circle of primary neuronal damage, an increase in APP and β-amyloid production in response to injury, and subsequent local release of neurotoxic substances by activated microglia responding to the site; the result is further damage to as yet unaffected neurons (Fig. 7).

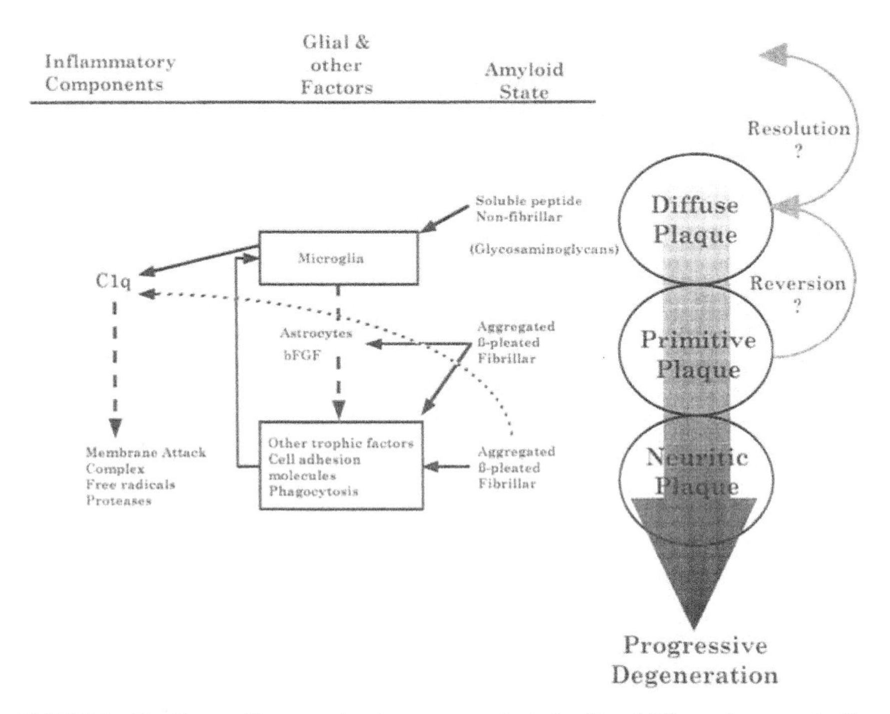

FIGURE 7 The diagram illustrates the time course and relationship of different factors and cell types in the progressive evolution of senile plaques and degeneration in AD. The initial condensation of β-amyloid within the neuropil (which is organized by proteoglycans) forms diffuse plaques. A substantial proportion of diffuse plaques contain reactive microglia, but do not yet contain astrocytes or complement. As additional β-amyloid collects and astrocytes are recruited to the region, dystrophic neurites are detectable, as are trophic factors and C1q. The disorganization of intra and extracellular components and mistiming of signaling events triggers further injury, recruitment of additional glia, and the cycle repeats and spreads.

V. A ROLE FOR TROPHIC FACTORS IN THE TREATMENT OF AD?

It should be clear from the preceding discussion that any therapeutic role for trophic factors in AD would have to be carefully timed with regard to the stage of the disease. If FGF or other trophic molecules were given to a patient in the moderate to severe stages of the disease, it is possible that these additional molecules may, in fact, exacerbate the evolution of plaque pathology and inflammation. We remind readers of just a few examples from the previous sections in this hypothetical scenario: a focal disturbance injures a group of neurons and triggers a local increase in bFGF. bFGF induces an increase in APP mRNA expression and protein production; bFGF also elevates the secretion of NGF. Microglia are recruited into the area and are activated by extracellular deposits of β-amyloid as they try to phagocytose the material. These activated microglia are responsive to bFGF and in turn secrete additional trophic factors and cytokines. Were this situation to stabilize and return to normal, the net result may be the loss of a few neurons and a successful glial response. However, mechanisms that tip the balance toward a greater glial and inflammatory response, which results in additional neuronal death, could inadvertently initiate a runaway feedback cascade and result in senile plaque formation.

Recalling the rescue of entorhinal cortex neurons after injury by bFGF in the previous section, it is possible that trophic therapy very early in the disease may be able to stave off the pathological progression of the disease for a time. The pre-alpha, or layer II, entorhinal cortex neurons are thought to be some of the earliest cells lost in the disease (Braak and Braak, 1991). These cells are located in a high order multimodal associational area and form the primary input into the hippocampus. Therapies that could delay the death of these cells should have a major impact on learning and memory in AD patients. In the rodent model, continuous micro infusion of bFGF was able to ameliorate axonal-induced cell death. However, this is only one type of cell death, and the question remains: how long after cells have begun to die will exogenous trophic molecule administration be effective? We do not know the answers to these basic questions in animal models. However, recent data from the ischemia field (Fisher *et al.*, 1995; Jiang *et al.*, 1996) support the hypothesis that post-trauma administration of trophic factors may ameliorate injury and actually enhance behavioral recovery.

The key question is how to accomplish a proper balance of bFGF in the brain in order to adjust the levels to protect stressed neurons vulnerable to further insults and to restore their health. Astrocytes are one of the main sources of bFGF in the brain and *in vitro* studies have shown that these cells can regulate bFGF production. For example, astrocytes upregulate FGF production in response to glutamate receptor activation (Pechan *et al.*, 1993). Furthermore, bFGF can induce specific glutamate receptors such as mGluR5 (Balazs *et al.*, 1997), which, in turn, increases the potential responsiveness of the cell and its ability to provide protective support to neurons. When neurons degenerate, this balance may require a signal to replace the trophic support that is lost. Small molecule pharmacology may be an effective means to regulate astrocyte function (e.g., via mGluR5). Although this approach has traditionally not been a target for therapeutic development, it should be one in the future. Although much needs to be defined and refined, it is likely that the regulation of neurotrophic factors in astrocytes will be crucial to the overall health of the aged brain. We stress that the issue is the homeostatic balance between too little and not enough trophic support. Perhaps early in the disease, strong trophic interventions may slow the progress of pathology, but once sufficient damage has occurred, the continued administration of trophic support may exacerbate the problem.

VI. CONCLUSION

In the aging and AD brain, neurons and their processes face a molecular battleground between opposing mechanisms—growth versus degeneration. The senile plaque is a classical focal point of these opposing mechanisms. Within senile plaques, there are both supportive and nonsupportive molecules, e.g., substrates, growth factors, cell adhesion molecules, and toxins that can masquerade as growth factors. APP has heparin- and laminin-binding sites in the extracellular domain, and APP appears capable of mediating cell-cell or cell-matrix adhesion. β-amyloid, which is derived from APP, appears to have the ability to serve as a pseudo cell-adhesion molecule. When β-amyloid assembles into fibrils and develops a β-sheet conformation, it induces in neuronal processes the morphological features of dystrophic neurites in and around senile plaques. As culture studies show, neurites are attracted to amyloid aggregates. These observations suggest a possible parallel between aggregation-related neurotoxicity to β-amyloid and neuritic changes associated with plaques. If so, it follows that cell adhesion-type interactions are one of the basic mechanisms in plaque formation. The plaque offers an attractive, permissive environment but also an unstable one that initiates cell surface receptor cross-linking and clustering and activates downstream signal transduction events that can cause degeneration. If the two surfaces maintain a proper alignment then growth proceeds, but if the matrix is unstable and can cross-link and cluster surface receptors, then the neurite or the entire cell degenerates. *In vitro*, β-amyloid in contact with the neuronal cell body initiates neuronal apoptosis. Neurons are probably most sensitive to such mechanisms, but there are also subpopulation responses. For example, GABA neurons *in vitro* are more resistant to β-amyloid than are glutamate neurons; and this appears to hold true within AD as well. The nature of the signal transduction processes may help to define the relative vulnerability mechanisms and in turn may locate key intervention points to afford more general protection to all cells. The activation of microglia and astrocytes within plaques initiates and further drives other molecular cascades, similar to normal injury responses, but their chronic nature promotes degeneration. The consequence of these convergent mechanisms is a microenvironment in which the neurite faces a molecular battleground between opposing mechanisms—growth versus degeneration and in which β-amyloid and other organizing factors drive molecular cascades that accumulate cell adhesion molecules and growth factors that perpetuate the evolution of the plaque. The insidious tragedy of AD is that as the pathology evolves is it compounded and rendered more devastating by microglia, which are the central part of a chronic inflammatory response that accompanies the disease.

REFERENCES

Afagh, A., Cummings, B. J., Cribbs, D., Cotman, C., and Tenner, A. (1996). Localization and cell association of C1q in Alzheimer's disease brain. *Exper Neurol* 138(1):22–32.

Akiyama, H., Kawamata, T., Yamada, T., Tooyama, I., Ishii, T., and McGeer, P. L. (1993). Expression of intercellular adhesion molecule (ICAM)-1 by a subset of astrocytes in Alzheimer disease and some other degenerative neurological disorders. *Acta Neuropathol Berl* 85:628–634.

Anderson, K. J., Dam, D., Lee, S., and Cotman, C. W. (1988). Basic fibroblast growth factor prevents death of lesioned cholinergic neurons *in vivo*. *Nature* 332(6162):360–361.

Araujo, D. M., and Cotman, C. W. (1992). Beta-amyloid stimulates glial cells *in vitro* to produce growth factors that accumulate in senile plaques in Alzheimer's disease. *Brain Res* 569:141–145.

Azizeh, B. Y., Cribbs, D. H., Kreng, V. M., and Cotman, C. W. (1998). Cross-linking of NCAM receptors on neurons induces programmed cell death. *Brain Research* 796:20–26.

Baird, A., Schubert, D., Ling, N., and Guillemin, R. (1988). Receptor and heparin binding domains of basic fibroblast growth factor. *Proc Natl Acad Sci USA* 85:2324–2328.

Balazs, R., Miller, S., Romano, C., de Vries, A., Chun, Y., and Cotman, C. W. (1997). Metabotropic glutamate receptor mGluR5 in astrocytes: Pharmacological properties and agonist regulation. *J Neurochem* 69(1):151–63.

Banati, R. B., Gehrmann, J., Czech, C., Monning, U., Jones, L. L., Konig, G., Beyreuther, K., and Kreutzberg, G. W. (1993). Early and rapid *de novo* synthesis of Alzheimer beta A4-amyloid precursor protein (APP) in activated microglia. *Glia* 9:199–210.

Barde, Y. A. (1989). Trophic factors and neuronal survival. *Neuron* 2:1525–1534.

Bechmann, I., and Nitsch, R. (1997). Astrocytes and microglial cells incorporate degenerating fibers following entorhinal lesion. A light, confocal, and electron microscopical study using a phagocytosis-dependent labeling technique. *Glia* 20:145–154.

Boje, K. M., and Arora, P. K. (1992). Microglial-produced nitric oxide and reactive nitrogen oxides mediate neuronal cell death. *Brain Res* 587:250–256.

Bouman, L. (1934). Senile plaques. *Brain* 57:128–142.

Braak, H., and Braak, E. (1991). Neuropathological stageing of Alzheimer-related changes. *Acta Neuropathol* 82:239–259.

Breen, K. C., Bruce, M., and Anderton, B. H. (1991). Beta amyloid precursor protein mediates neuronal cell-cell and cell-surface adhesion. *J Neurosci Res* 28(1):90–100.

Bronner-Fraser, M., and Lallier, T. (1988). A monoclonal antibody against a laminin-heparan sulfate proteoglycan complex perturbs cranial neural crest migration *in vivo*. *J Cell Biol* 106:1321–1329.

Canning, D. R., McKeon, R. J., DeWitt, D. A., Perry, G., Wujek, J. R., Frederickson, R. C., and Silver, J. (1993). β-Amyloid of Alzheimer's disease induces reactive gliosis that inhibits axonal outgrowth. *Exp Neurol* 124:289–98.

Castillo, G. M., Ngo, C., Cummings, J., Wight, T. N., and Snow, A. D. (1997). Perlecan binds to the beta-amyloid proteins (A beta) of Alzheimer's disease, accelerates A beta fibril formation, and maintains A beta fibril stability. *J Neurochem* 69(6):2452–2465.

Chen, M., and Yankner, B. A. (1991). An antibody to beta amyloid and the amyloid precursor protein inhibits cell-substratum adhesion in many mammalian cell types. *Neuroscience Letters* 125(2):223–226.

Coleman, P. D., and Flood, D. G. (1987). Neuron numbers and dendritic extent in normal aging and Alzheimer's disease. *Neurobiology of Aging* 8:521–545.

Cotman, C. W., and Anderson, K. J., eds. (1988). *Synaptic plasticity and functional stabilization in the hippocampal formation: Possible role in Alzheimer's disease. Advances in Neurology*. New York: Raven Press.

Cotman, C. W., Cribbs, D. H., and Anderson, A. J. (1996a). The β-amyloid model of Alzheimer's disease: Conformation, change, receptor cross-linking, and the initiation of apoptosis. In Wasco, W., and Tanzi, R. E., eds., *Molecular mechanisms of dementia*. Totowa, N.J.: Humana Press Inc., pp. 73–90.

Cotman, C. W., Cummings, B. J., and Pike, C. J. (1993a). Molecular cascades in adaptive versus pathological plasticity. *Neuroregeneration*. New York: Raven Press, pp. 217–240.

Cotman, C. W., Matthews, D., Taylor, D., and Lynch, G. (1973). Synaptic rearrangement in the dentate gyrus: Histochemical evidence of adjustments after lesions in immature and adult rats. *Proc Natl Acad Sci USA* 70:3473–3477.

Cotman, C. W., Pike, C. J., and Cummings, B. J. (1993b). Adaptive versus pathological plasticity: Possible contributions to age-related dementia. *Advances in Neurology*. New York: Raven Press, Ltd., pp. 35–45.

Cotman, C. W., Tenner, A. J., and Cummings, B. J. (1996b). β-amyloid converts an acute phase injury response to chronic injury responses. *Neurobiology of Aging* 15(5):723–731.

Coulson, E. J., Barrett, G. L., Storey, E., Bartlett, P. F., Beyreuther, K., and Masters, C. L. (1997). Down-regulation of the amyloid protein precursor of Alzheimer's disease by antisense oligonucleotides reduces neuronal adhesion to specific substrata. *Brain Res* 770(1–2):72–80.

Cribbs, D. H., Kreng, V. M., Anderson, A. J., and Cotman, C. W. (1996). Crosslinking of concanavalin A receptors on cortical neurons induces programmed cell death. *Neuroscience* 75(1):173–185.

Cummings, B. J., Head, E., Ruehl, W., Milgram, N., and Cotman, C. (1996). The canine as an animal model of human aging and dementia. *Neurobiol Aging* 17(2):259–268.

Cummings, B. J., Su, J. H., and Cotman, C. W. (1993a). Neuritic involvement within the bFGF immunopositive plaques of Alzheimer's disease. *Exper Neurol* 124:315–325.

Cummings, B. J., Su, J., Cotman, C., White, R., and Russell, M. (1993b). β-amyloid accumulation in aged canine brain: A model of early plaque formation in Alzheimer's disease. *Neurobiol Aging* 14(6):547–560.

Cummings, B. J., Su, J. H., Geddes, J. W., Van Nostrand, W., Wagner, S., Cunningham, D., and Cotman, C. W. (1992a). Aggregation of the amyloid precursor protein (APP) within degenerating neurons and dystrophic neurites in Alzheimer's Disease. *Neurosci* 48(4):763–777.

Cummings, B. J., Yee, G. J., and Cotman, C. W. (1992b). bFGF promotes entorhinal layer II cell survival after perforant path axotomy. *Brain Research* 591:271–276.

Damon, D., D'Amore, P., and Wagner, J. (1988). Sulfated glycosaminoglycans modify growth factor-induced neurite outgrowth in PC12 cells. *J Cell Phys* 135:293–300.

Deller, T., and Frotscher, M. (1997). Lesion-induced plasticity of central neurons: Sprouting of single fibres in the rat hippocampus after unilateral entorhinal cortex lesion. *Progress in Neurobiology* 53:687–727.

Deller, T., Frotscher, M., and Nitsch, R. (1996). Sprouting of crossed entorhinodentate fibers after a unilateral entorhinal lesion: Anterograde tracing of fiber reorganization with Phaseolus vulgaris-leucoagglutinin (PHAL). *J Comp Neurol* 365(1):42–55.

Donnelly, R. J., Friedhoff, A. J., Beer, B., Blume, A. J., and Vitek, M. P. (1990). Interleukin-1 stimulates the beta-amyloid precursor protein promoter. *Cell Mol Neurobiol* 10:485–495.

Fabry, Z., Raine, C. S., and Hart, M. N. (1994). Nervous tissue as an immune compartment: The dialect of the immune response in the CNS. *Immunol Today* 15:218–224.

Fischer, O. (1907). Miliary necrosis with nodular proliferation of the neurofibrils, a common change of the cerebral cortex in senile dementia. *Monatsschrift für Psychiatrie und Neurologie* 22:361–372.

Fisher, M., Meadows, M. E., Do, T., Weise, J., Trubetskoy, V., Charette, M., and Finklestein, S. P. (1995). Delayed treatment with intravenous basic fibroblast growth factor reduces infarct size following permanent focal cerebral ischemia in rats. *J Cereb Blood Flow Metab* 15(6):953–959.

Flaumenhaft, R., Moscatelli, D., and Rifkin, D. B. (1990). Heparin and heparan sulfate increase the radius of diffusion and action of basic fibroblast growth factor. *J Cell Biol* 111:1651–1659.

Frohman, E. M., Frohman, T. C., Gupta, S., de Fougerolles, A., and van den Noort, S. (1991). Expression of intercellular adhesion molecule 1 (ICAM-1) in Alzheimer's disease. *J Neurol Sci* 106:105–111.

Gage, F. H., Armstrong, D. M., Williams, L. R., and Varon, S. (1988). Morphological response of axotomized septal neurons to nerve growth factor. *J Comp Neurol* 269:147–155.

Geddes, J., Monaghan, D., Cotman, C., Lott, I., Kim, R., and Chui, H. (1985). Plasticity of hippocampal circuitry in Alzheimer's disease. *Science* 230:1179–1181.

Geddes, J. W., Anderson, K. J., and Cotman, C. W. (1986). Senile plaques as aberrant sprout-stimulating structures. *Expr Neurol* 94:767–776.

Gill, P., Silbert, C., and Silbert, J. (1986). Effects of heparan sulfate removal on attachment and reattachment of fibroblasts and endothelial cells. *Biochem* 25:405–410.

Goldgaber, D., Harris, H. W., Hla, T., Maciag, T., Donnelly, R. J., Jacobsen, J. S., Vitek, M. P., and Gajdusek, D. C. (1989). Interleukin 1 regulates synthesis of amyloid beta-protein precursor mRNA in human endothelial cells. *Proc Natl Acad Sci* 86:7606–7610.

Gómez-Pinilla, F., Cummings, B. J., and Cotman, C. W. (1990). Induction of basic fibroblast growth factor in Alzheimer's disease pathology, *NeuroReport* 1:211–214.

Gospodarowicz, D., Cheng, J., Lui, G., Baird, A., and Böhlent, P. (1984). Isolation of brain fibroblast growth factor by heparin-Sepharose affinity chromatography: Identity with pituitary fibroblast growth factor. *Proc Natl Acad Sci USA* 81:6963–6967.

Gray, C. W., and Patel, A. J. (1993). Regulation of beta-amyloid precursor protein isoform mRNAs by transforming growth factor-beta 1 and interleukin-1 beta in astrocytes. *Brain Res Mol Brain Res* 19(3):251–256.

Griffin, W. S., Sheng, J. G., Roberts, G. W., and Mrak, R. E. (1995). Interleukin-1 expression in different plaque types in Alzheimer's disease: Significance in plaque evolution. *J Neuropathol Exp Neurol* 54(2):276–281.

Haga, S., Ikeda, K., Sato, M., and Ishii, T. (1993). Synthetic Alzheimer amyloid beta/A4 peptides enhance production of complement C3 component by cultured microglial cells. *Brain Res* 601:88–94.

Hailer, N. P., Bechmann, I., Heizmann, S., and Nitsch, R. (1997). Adhesion molecule expression on phagocytic microglial cells following anterograde degeneration of perforant path axons. *Hippocampus* 7:341–349.

Hefti, F. (1986). Nerve growth factor (NGF) promotes survival of septal cholinergic neurons after fimbrial section. *J Neurosci* 6:2155–2162.

Hefti, F., Hartikka, J., and Knusel, B. (1989). Function of neurotrophic factors in the adult and aging brain and their possible use in the treatment of neurodegenerative disorders. *Neurobiol Aging* 10:515–533.

Heppner, F. L., Nitsch, R., and Hailer, N. P. (1997). Changes in microglial cell phenotype are mediated by oxidative mechanisms. (submitted).

Hyman, B. T., Van Hoesen, G. W., Kromer, L. J., and Damasio, A. R. (1986). Perforant pathway changes and the memory impairment of Alzheimer's disease. *Ann Neurol* 20:472–481.

Imaizumi, K., Iwata, H., Yoshida, S., Sun, G., Okumura, N., and Shiosaka, S. (1993). Coexistence of amyloid beta-protein precursor and basic fibroblast growth factor in single cells of the rat parietal cortex, hippocampus and basal magnocellular nucleus. *J Chem Neuroanat* 6(3):159–165.

Jiang, H., Burdick, D., Glabe, C. G., Cotman, C. W., and Tenner, A. J. (1994). β-amyloid activates complement by binding to a specific region of the collagen-like domain of the C1q A chain. *J Immunol* 152:5050–5059.

Jiang, N., Finklestein, S. P., Do, T., Caday, C. G., Charette, M., and Chopp, M. (1996). Delayed intravenous administration of basic fibroblast growth factor (bFGF) reduces infarct volume in a model of focal cerebral ischemia/reperfusion in the rat. *J Neurol Sci* 139(2):173–179.

Juliano, R. (1987). Membrane receptors for extracellular matrix macromolecules: Relationship to cell adhesion and tumor metastasis. *Biochim Biophys Acta* 907:261–278.

Kato, T, Sasaki, H, Katagiri, T, et al. (1991). The binding of basic fibroblast growth factor to Alzheimer's neurofibrillary tangles and senile plaques. *Neurosci Lett* 122:33–36.

Korotzer, A. R., Pike, C. J., and Cotman, C. W. (1993). β-Amyloid peptides induce degeneration of cultured rat microglia. *Brain Res* 624:121–125.

Lahiri, D. K., and Nall, C. (1995). Promoter activity of the gene encoding the beta-amyloid precursor protein is up-regulated by growth factors, phorbol ester, retinoic acid and interleukin-1. *Brain Res Mol Brain Res* 32(2):233–240.

Lander, A. D., Fujii, K., Gospodarowicz, D., and Reichardt, L. F. (1982). Characterization of a factor that promotes neurite outgrowth: Evidence linking activity to a heparan sulfate proteoglycan. *J Cell Biol* 94:574–585.

Laterra, J., Silbert, J., and Culp, L. (1983). Cell surface heparan sulfate mediates some adhesive responses to glycosaminoglycan-binding matrices, including fibronectin. *J Cell Biol* 96:112–123.

LeBlanc, A. C., Kovacs, D. M., Chen, H. Y., Villare, F., Tykocinski, M., Autilio-Gambetti, L., and Gambetti, P. (1992). Role of amyloid precursor protein (APP): Study with antisense transfection of human neuroblastoma cells. *Neuroscience Research* 31(4):635–645.

Lee, S. C., Liu, W., Dickson, D. W., Brosnan, C. F., and Berman, J. W. (1993). Cytokine production by human fetal microglia and astrocytes. Differential induction by lipopolysaccharide and IL-1 beta. *J. Immunol* 150:2659–2667.

Lipton, S., Wagner, J., Madison, R., and D'Amore, P. (1988). Acidic fibroblast growth factor enhances regeneration of processes by postnatal mammalian retinal ganglion cells in culture. *Proc Natl Acad Sci USA* 85:2388–2392.

Loesche, J., and Steward, O. (1977). Behavioral correlates of denervation and reinnervation of the hippocampal formation of the rat: Recovery of alternation performance following unilateral entorhinal cortex lesions. *Brain Res Bull* 2:31–39.

Loo, D. T., Copani, A., Pike, C. J., Whittemore, E. R., Walencewicz, A. J., and Cotman, C. W. (1993). Apoptosis is induced by beta-amyloid in cultured central nervous system neurons. *PNAS* 90:7951–7955.

Maeda, K., Nakai, M., Maeda, S., Kawamata, T., Yamaguchi, T., and Tanaka, C. (1997). Possible different mechanism between amyloid-beta (25–35)- and substance P-induced chemotaxis of murine microglia. *Gerontology* 43(Suppl. 1):11–15.

Mark, R. J., Keller, J. N., Kruman, I., and Mattson, M. P. (1997). Basic FGF attenuates amyloid beta-peptide-induced oxidative stress, mitochondrial dysfunction, and impairment of Na+/K+-ATPase activity in hippocampal neurons. *Brain Res* 756(1–2):205–214.

Mathews, W., and Patterson, P. (1983). The production of a monoclonal antibody that blocks the action of a neurite outgrowth-promoting factors. *Cold Springs Harb Symap Quant Biol* 48:625–637.

Matthews, D., Cotman, C., and Lynch, G. (1976a). An electron microscopic study of lesion-induced synaptogenesis in the dentate gyrus of the adult rat. I. Magnitude and time course of degeneration. *Brain Res* 115:1–21.

Matthews, D., Cotman, C., and Lynch, G. (1976b). An electron microscopic study of lesion-induced synaptogenesis in the dentate gyrus of the adult rat. II. Reappearance of morphologically normal synaptic contacts. *Brain Res* 115:23–41.

Mattson, M. P., Lee, R. E., Adams, M. E., Guthrie, P. B., and Kater, S. B. (1988). Interactions between entorhinal axons and target hippocampal neurons: A role for glutamate in the development of hippocampal circuitry. *Neuron* 1:865–876.

Mattson, M. P., Kumar, K. N., Wang, H., Cheng, B., and Michaelis, E. K. (1993). Basic bFGF regulates the expression of a functional 71 kDA NMDA receptor protein that mediates calcium influx and neurotoxicity in hippocampal neurons. *Journal of Neuroscience* 13(11):4575–4588.

McDonald, D. R., Brunden, K. R., and Landreth, G. E. (1997). Amyloid fibrils activate tyrosine kinase-dependent signaling and superoxide production in microglia. *J Neurosci* 17:2284–2294.

McGeer, P. L., Akiyama, H., Itagaki, S., and McGeer, E. G. (1989). Immune system response in Alzheimer's disease. *Can J Neurol Sci* 16:516–527.

McGeer, P. L., Itagaki, S., Tago, H., and McGeer, E. G. (1987). Reactive microglia in patients with senile dementia of the Alzheimer type are positive for the histocompatibility glycoprotein HLA-DR. *Neurosci Lett* 79:195–200.

McGeer, P. L., Kawamata, T., Walker, D. G., Akiyama, H., Tooyama, I, and McGeer, E. (1993). Microglia in degenerative neurological disease. *Glia* 7:84–92.

Meda, L., Cassatella, M. A., Szendrei, G. I., Otvos, L., Jr., Baron, P., Villalba, M., Ferrari, D., and Rossi, F. (1995). Activation of microglial cells by beta-amyloid protein and interferon-gamma. *Nature* 374:647–650.

Miller, P. D., Styren, S. D., Lagenaur, C. F., and DeKosky, S. T. (1994). Embryonic neural cell adhesion molecule (N-CAM) is elevated in the denervated rat dentate gyrus. *J Neurosci* 14:4217–4225.

Monning, U., Sandbrink, R., Banati, R. B., Masters, C. L., and Beyreuther, K. (1994). Transforming growth factor beta mediates increase of mature transmembrane amyloid precursor protein in microglial cells. *FEBS Lett* 342(3):267–272.

Monning, U., Sandbrink, R., Weidemann, A., Banati, R. B., Masters, C. L., and Beyreuther, K. (1995). Extracellular matrix influences the biogenesis of amyloid precursor protein in microglial cells. *J Biol Chem* 270(13):7104–7110.

Morrison, R. S., Sharma, A., De Vellis, J., and Brandshaw, R. A. (1986). Basic fibroblast growth factor supports the survival of cerebral cortical neurons in primary culture. *Proc Nat Acad Sci USA* 83:7537–7541.

Neufeld, G., Gospodarowicz, D., Dodge, L., and Fujii, D. K. (1987). Heparin modulation of the neurotropic effects of acidic and basic fibroblast growth factors and nerve growth factor on PC12 cells. 131:131–140.

Owens, T., Renno, T., Taupin, V., and Krakowski, M. (1994). Inflammatory cytokines in the brain: Does the CNS shape immune responses? *Immunol Today* 15:566–571.

Pechan, P. A., Chowdhury, K., Gerdes, W., and Seifert, W. (1993). Glutamate induces the growth factors NGF, bFGF, the receptor FGF-R1 and c-fos mRNA expression in rat astrocyte culture. *Neurosci Lett* 153(1):111–114.

Perlmutter, L. S., Chui, H. C., Saperia, D., and Athanikar, J. (1990). Microangiopathy and the colocalization of heparan sulfate proteoglycan with amyloid in senile plaques of Alzheimer's disease. *Brain Res* 508:13–19.

Perry, G., Siedlak, S. L., Richey, P., et al. (1991). Association of heparan sulfate proteoglycan with the neurofibrillary tangles of Alzheimer disease. *J Neurosci* 11:3679–3683.

Peterson, D. A., Lucidi-Phillipi, C. A., Murphy, D. P., Ray, J., and Gage, F. H. (1996). Fibroblast growth factor-2 protects entorhinal layer II glutamatergic neurons from axotomy-induced death. *J Neurosci* 16(3):886–898.

Pike, C. J., Walencewicz, A. J., Glabe, C. G., and Cotman, C. W. (1991). *In vitro* aging of β-amyloid protein causes peptide aggregation and neurotoxicity. *Brain Res* 563:311–314.

Pike, C. J., Cummings, B. J., and Cotman, C. W. (1992). β-amyloid induces neuritic dystrophy *in vitro:* Similarities with Alzheimer pathology. *NeuroReport* 3:769–772.

Pike, C. J., Burdick, D., Walencewicz, A., Glabe, C. G., and Cotman, C. W. (1993). Neurodegeneration induced by β-amyloid peptides *in vitro:* The role of peptide assembly state. *J. Neurosci* 13(4):1676–1687.

Quon, D., Catalano, R., and Cordell, B. (1990). Fibroblast growth factor induces beta-amyloid precursor mRNA in glial but not neuronal cultured cells. *Biochem Biophys Res* 167:96–102.

Ramón y Cajal, S., ed. (1928). *Degeneration and regeneration of the nervous system. History of Neuroscience.* New York: Oxford University Press.

Ringheim, G. E., Aschmies, S., and Petko, W. (1997). Additive effects of basic fibroblast growth factor and phorbol ester on beta-amyloid precursor protein expression and secretion. *Neurochem Int* 30(4–5):475–481.

Ruoslahti, E., and Yamaguchi, Y. (1991). Proteoglycans as modulators of growth factor activities. *Cell* 64:867–869.

Saksela, O., Moscatelli, D., Sommer, A., and Rifkin, D. (1988). Endothelial cell-derived heparan sulfate binds basic fibroblast growth factor and protects it from proteolytic degradation. *J Cell Bio* 107:743–751.

Scheff, S. W., and Cotman, C. W. (1977). Recovery of spontaneous alternation following lesions of the entorhinal cortex in adult rats: Possible correlation to axon sprouting. *Behavioral Biology* 21:286–293.

Sheng, J. G., Mrak, R. E., and Griffin, W. S. (1997). Neuritic plaque evolution in Alzheimer's disease is accompanied by transition of activated microglia from primed to enlarged to phagocytic forms. *Acta Neuropathol (Berl)* 94(1):1–5.

Snow, A. D., and Kisilevsky, R. (1985). Temporal relationship between glycosaminoglycan accumulation and amyloid deposition during experimental amyloidosis: A histochemical study. *Lab Invest* 53:37–44.

Snow, A. D., Willmer, J. P., and Kisilevsky, R. (1987). Sulfated glycosaminoglycans in Alzheimer's disease. *Hum Pathol* 18(5):506–510.

Snow, A. D., Mar, H., Nochlin, D., Kimata, K., Kato, M., Suzuki, S., Hassell, J., and Wight, T. N. (1988). The presence of heparan sulfate proteoglycans in the neuritic plaques and congophilic angiopathy in Alzheimer's disease. *Am J Pathol* 133(3):456–463.

Snow, A. D., and Wight, T. N. (1989). Proteoglycans in the pathogenesis of Alzheimer's disease and other Amyloidoses. *Neurobiol Aging* 10:481–497.

Snow, A. D., Mar, H., Nochlin, D., Sekiguchi, R. T., Kimata, K., Koike, Y., and Wight, T. N. (1990). Early accumulation of heparan sulfate in neurons and in the beta-amyloid protein-containing lesions of Alzheimer's disease and Down's syndrome. *Am J Pathol* 137:1253–1270.

Snow, A. D., Nochlin, D., Sekiguichi, R., and Carlson, S. S. (1996). Identification in immunolocalization of a new class of proteoglycan (keratan sulfate) to the neuritic plaques of Alzheimer's disease. *Exp Neurol* 138(2):305–317.

Stopa, E. G., Gonzalez, A., Chorsky, R., et al. (1990). Basic fibroblast growth factor in Alzheimer's disease. *Biochem Biophys Res Commun* 171:690–696.

Storey, E., Spurck, T., Pickett-Heaps, J., Beyreuther, K., and Masters, C. L. (1996a). The amyloid precursor protein of Alzheimer's disease is found on the surface of static but not activity motile portions of neurites. *Brain Research* 735(1):59–66.

Storey, E., Beyreuther, K., and Masters, C. L. (1996b). Alzheimer's disease amyloid precursor protein on the surface of cortical neurons in primary culture co-localizes with adhesion patch components. *Brain Research* 735(2):217–231.

Strasser, R. (1995). Life and death during lymphocyte development and function: evidence for two distinct killing mechanisms. *Current Opinion in Immunology* 7(2):228–234.

Streit, W. J., Graeber, M. B., and Kreutzberg, G. W. (1988). Functional plasticity of microglia: A review. *Glia* 1:301–307.

Su, J. H., Cummings, B. J., and Cotman, C. W. (1992). Localization of heparan sulfate glycosaminoglycan and proteoglycan core protein in aged brain and Alzheimer's disease. *Neurosci* 51:801–814.

Suzuki, K., Katzman, R., and Korey, S. (1955). Chemical studies on Alzheimer's disease. *J. Neuropathol Exp Neurol* 14:211–222.

van der Wal, E., Gómez-Pinilla, F., and Cotman, C. W. (1993). Transforming growth factor-β1 is in plaques in Alzheimer's and Down's pathologies. *NeuroReport* 4(1):69–72.

Verbeek, M. M., Otte Holler, I., Westphal, J. R., Wesseling, P., Ruiter, D. J., and de Waal, R. M. (1994). Accumulation of intercellular adhesion molecule-1 in senile plaques in brain tissue of patients with Alzheimer's disease. *Am J Pathol* 144:104–116.

Walicke, P. A. (1988a). Basic and acidic fibroblast growth factors have trophic effects on neurons from multiple CNS regions. *J Neurosci* 8:2618–2627.

Walicke, P. A. (1988b). Interactions between basic fibroblast growth factor (bFGF) and glycosoaminoglycans in promoting neurite outgrowth. *Exper Neurol* 102:144–148.

Walicke, P., Cowan, W., Ueno, N., Baird, A., and Guillemin, R. (1986). Fibroblast growth factor promotes survival of dissociated hippocampal neurons and enhances neurite extension. *Pro Natl Acad Sci USA* 83:3012–3016.

White, W. F., Nadler, J. V., Hamberger, A., and Cotman, C. W. (1977). Glutamate as transmitter of hippocampal perforant path. *Nature* 270:356–357.

Whitson, J. S., Glabe, C. G., Shintani, E., Abcar, A., and Cotman, C. W. (1990). β-amyloid protein promotes neuritic branching in hippocampal cultures. *Neurosci Lett* 110(3):319–324.

Whitson, J. S., Selkoe, D. J., and Cotman, C. W. (1989). Amyloid beta protein enhances the survival of hippocampal neurons *in vitro*. *Science* 243(4897):1488–1490.

Williams, L. R., Varon, S., Peterson, G. W., Wietorin, K., Fischer, W., Björklund, A., and Gage, F. H. (1986). Continuous infusion of nerve growth factor prevents basal forebrain neuronal death after fimbria-fornix transection. *Proc Natl Acad Sci USA* 83:9231–9235.

Yamazaki, T., Koo, E. H., and Selkoe, D. J. (1997). Cell surface amyloid beta-protein precursor colocalizes with beta 1 integrins at substrate contact sites in neural cells. *Neuroscience* 17(3):1004–1010.

CHAPTER 22

Cholinergic Basal Forebrain Grafts in Rodent and Nonhuman Primate Models of Alzheimer's Disease

JEFFREY H. KORDOWER AND TIMOTHY J. COLLIER

Research Center for Brain Repair and Department of Neurological Sciences, Rush Presbyterian St. Luke's Medical Center, Chicago, Illinois

Magnocellular neurons scattered heterogeneously throughout the mamalian basal forebrain provide virtually all of the cholinergic innervation to the hippocampus and cerebral cortex. This region has been linked to higher order cognitive function, particularly to attentional processes as well as processes related to learning and memory. Cholinergic basal forebrain neurons degenerate in many dementing illnesses such as Alzheimer's disease (Whitehouse *et al.*, 1982). The cholinergic deficit in Alzheimer's disease occurs by the time symptoms first manifest (Sims *et al.*, 1981) and correlates with the severity and duration of this disorder (Wilcox *et al.*, 1982; Mufson *et al.*, 1989). Thus novel therapeutic strategies addressing the cholinergic deficit may have significant clinical impact. One such strategy is the replacement of lost neurons via grafting of fetal cholinergic neurons. A body of evidence demonstrates that fetal basal forebrain grafts survive, innervate the cortex or hippocampus, form synaptic connections, are electrophysiologically active, and secrete acetylcholine. Most importantly, these grafts enhance cognitive function in many animal models of human cholinergic hypofunction including rats receiving fimbria–fornix transections or nucleus basalis lesions, aged rats receiving cholinergic neurotoxins, and genetic mutants. Fetal basal forebrain allografts can survive long-term in Rhesus monkeys and can attenuate the cognitive dysfunction seen in marmosets resulting from cholinergic denervation of the hippocampus or cerebral cortex.

I. INTRODUCTION

Surgical therapeutic approaches have made a resurgence in the field of neurodegenerative disease. For example, in Parkinson's disease, patients are undergoing precisely placed brain lesions (pallidotomies) or implantation of deep brain stimulators to control abnormal movements, and are undergoing transplantation of dopamine-secreting fetal cells to replace lost neuronal function. In animal models of Alzheimer's disease (AD), two surgical approaches have gained attention preclinically. One is a neuroprotective strategy, the infusion of trophic factors or the transplantation of trophic factor-secreting cells, which is intended to prevent the degeneration of cells destined to die in AD. This topic is addressed in chapter 20. The second is a replacement strategy in which fetal cholinergic basal forebrain (CBF) neurons are transplanted into the brain. The latter approach is intended to repopulate the brain with cholinergic neurons that are lost in AD. This chapter will review neuronal replacement strategies with a particular emphasis on their application in nonhuman primate models of AD.

II. FETAL GRAFTS OF CHOLINERGIC NEURONS

A. Clinical Rationale

Initial reports demonstrating reduced cholinergic markers within the cortex of AD patients (Davies and Maloney, 1976) led to intense interest in the source of this innervation, the CBF system. Cholinergic neurons within the basal forebrain undergo atrophy and eventually degenerate in AD (e.g., Whitehouse *et al.*, 1982; Doucette and Ball, 1987; Kordower *et al.*, 1989; Mufson *et al.*, 1989), making this system an attractive candidate for potential replacement therapies. There are a number of important concerns when considering which system to target with experimental therapeutic strategies. Is the system impaired when the symptoms first appear? Does the system continue to degenerate as the disease progresses? Does failure of the system in animal models mimic what is observed in humans? The answer to these types of questions furthers our interest in targeting the CBF with novel therapeutic interventions. The cortical cholinergic deficit in AD is present within one year of symptom onset (Sims *et al.*, 1981) and remains the neurotransmitter system that best correlates with the severity and duration of this illness (e.g., Wilcox *et al.*, 1982; Mufson *et al.*, 1989). Extensive animal literature also indicates that an intact CBF is needed for normal cognition (see reviews Bartus *et al.*, 1982; Dunnett and Fibiger, 1993; Wainer *et al.*, 1993). Lesions of the CBF system in rodents and monkeys induce long-lasting impairments of trial dependent (e.g., delayed nonmatch to sample) and trial independent (e.g., simple discrimination) memory tasks. In addition to lesion studies, pharmacological experiments indicate that an intact cholinergic system is necessary for normal cognition. Administration of cholinergic antagonists to rodents, monkeys, and humans consistently impairs short-term memory (see reviews Bartus *et al.*, 1982; Bartus, 1990; Dunnett and Fibiger 1993; Wainer *et al.*, 1993). Conversely, administration of cholinergic agonists to aged rodents and aged nonhuman primates enhances cognitive performance. These data support the concept that the cholinergic deficit mediates, in part, the memory deficits seen in AD.

For many years, research investigating the role of the CBF in higher order function was hampered by absence of a specific cholinergic neurotoxin. Recently, a selective neurotoxin has been created in which the low-affinity $p75^{NTR}$ receptor, which is selectively found on CBF neurons, is conjugated to the the ribosome-inactivating toxin saporin. Intraventricular or intraparenchymal injections of this toxin selectively destroy CBF neurons (Wiley *et al.*, 1991). Results from studies using this toxin are controversial. Some studies demonstrate that intraventricular injections of the I192-saporin toxin impair performance following water maze, passive avoidance (Leanza *et al.*, 1995), and radial arm maze (Dornan *et al.*, 1997; Shen *et al.*, 1996; Walsh *et al.*, 1996) testing. Others find little or no effect on these and other memory tasks following extensive, but specific, lesions using this neurotoxin (McMahan *et al.*, 1997; Baxter *et al.*, 1995; Torres *et al.*, 1994; Wenk *et al.*, 1994). Recently, Fine and coworkers (1997) have created lesions in the nonhuman primate CBF using a different version of this neurotoxin. At the highest dose, they found that these lesions induced deficits in learning a difficult visual discrimination task. No significant deficits were seen in retention of tasks learned prior to surgery or on simple tasks. Because nonselective lesions induce profound memory deficits, current thought is that degeneration of the CBF plus other noncholinergic regions are required for cognitive deficits to occur. Alternatively, the CBF may be more involved in attentional, rather than memory, processes (Voytko et al., 1996). However, attention is clearly an important parameter for normal cognition.

It is clear that the CBF is not the only region undergoing degeneration in AD. Pathological changes occur in regions across the neuraxis such as the locus coeruleus, raphae nucleus, and other brain regions that have diverse projections to regions of the neocortex that may be involved in memory. However, changes in these noncholinergic brain regions do not appear to correlate with the severity of cognitive dysfunction and thus may not be the highest priority for therapeutic consideration. Alternatively, brain areas such as layer II entorhinal cortex neurons, the cells of origin for the perforant path, likely contribute to the deficits seen in patients with AD. The perceived nonselectivity of degeneration (put more accurately, a selective vulnerability in multiple brain regions), has diminished enthusiasm for the cholinergic hypothesis of AD. Further, the failure of clinical trials using cholinomimetic agents or anticholinesterase inhibitors to substantially improve cognition in patients with AD has led some to abandon the concept that cholinergic systems play a role in the memory impairments seen in this disease. However, a detailed examination of this literature makes it difficult to reject the involvement of the cholinergic system in memory processes. As discussed by Bartus (1990), there are many explanations to account for the lack or more marked success in clinical trials that use cholinergic agonists or acetylcholinesterase inhibitors. Most of these drugs have poor pharmacokinetic profiles and their short half-life in the central nervous system may make it difficult to detect a drug effect. Additionally, nonhuman primates demonstrate a wide variability between subjects in the optimal effective dose of a memory-enhancing drug. This variability needs to be accounted for in clinical trials and drug effects in these trials need to be tested over wider dose ranges. Clearly, the extensive body of data linking CBF function with normal and abnormal cognitive processes cannot be ignored. Although the limitations in approaching AD from a single transmitter system perspective must be acknowledged, the preceding discussion highlights the potential that exists for improving cognition in patients with AD by site-specific enhancement of cholinergic forebrain neurotransmission.

Why use fetal neural CBF grafts? Numerous pharmalogical, trophic factor, and surgical treatment strategies have been attempted and proposed for the treatment of the cholinergic deficit in AD. Whereas Tacrine was the first drug approved for the treatment of AD, its effectiveness is modest. Why would transplants of cholinergic neurons be successful when other strategies have failed? First, fetal CBF grafts can provide a site-specific source of cholinergic activity. Second, grafts of fetal CBF neurons reinnervate target sites in an organotypic fashion. Fetal CBF grafts form synaptic connections with the host and restore relatively normal electrophysiological activity within the hippocampus. Third, fetal CBF grafts consistently improve memory function in (1) lesioned rodents, (2) aged rodents, and (3) lesioned nonhuman primates on a variety of cognitive tasks. We are currently extending these findings to aged nonhuman primates. If the latter studies demonstrate amelioration of morphological and functional consequences of aging, then several lines of evidence would indicate that fetal CBF grafts are effective in a logical series of correlational animal models of AD, including a primate species that displays both β-amyloid deposition and cognitive deficits that resemble some features of AD. This would support our contention that CBF grafting may be a useful strategy for the treatment of the cholinergic deficit of AD.

B. Rat Studies

Grafts of cholinergic neurons have been studied in three rodent models: (1) young adult rats with lesions of the fimbria–fornix or medial septum resulting in cholinergic denervation of the hippocampal formation, (2) young adult rats with lesions of basal forebrain cholinergic neurons in the nucleus basalis resulting in depletion of cholinergic innervation of neocortex, and (3) aged rats exhibiting a natural decline in cholinergic tone of both hippocampus and cortex. Experimental findings derived from each of these models will be discussed.

1. Cholinergic Grafts in Fimbria–Fornix Lesioned Young Adult Rats

Interruption of axons projecting from the septal area to the hippocampus via the fimbria–fornix in rats results in near-total depletion of cholinergic markers in the dorsal hippocampus (Bjorklund and Stenevi, 1977) and causes severe deficits in performance on learning and memory tasks (Kesner *et al.*, 1986; Olton *et al.*, 1978). It is in this model of cholinergic dysfunction that the most detailed information concerning the structural and functional correlates of transplanted fetal cholinergic neurons has been collected. Initial studies demonstrated that grafts of embryonic rat septal tissue survived transplantation both as solid grafts placed into an aspiration cavity adjacent to the hippocampus (Bjorklund and Stenevi, 1977), and as cell suspensions injected directly into the hippocampus (Bjorklund *et al.*, 1983b). An organotypic pattern of cholinergic reinnervation of the host hippocampus provided by grafted acetylcholine (ACh)-containing neurons was remarkably similar to the normal laminar distribution of nerve fibers staining positive for acetylcholinesterase (AChE) observed in intact animals. Although organotypic in organization, the density of this innervation decreased with increasing distance from the grafted cells. In the case of grafts comprised of blocks of embryonic septal tissue, levels of the cholinergic synthetic enzyme choline acetyltransferase (ChAT) were restored to 60 to 70% of control values within 1.75 mm of the graft, 30 to 45% of control at distances of 1.75 to 5.25 mm from the graft, and 25 to 30% of control at more distant levels (Bjorklund and Stenevi, 1977). In these studies, the lesion reduced ChAT levels to 5 to 10% of intact values. Cell suspension grafts placed at two sites within the hippocampus yielded even greater recovery of ChAT levels, increasing ChAT to 90 to 120% of control levels within 2.5 mm of grafts, and 60 to 70% of control at distances of 2.5 to 5.5 mm from grafts (Bjorklund *et al.*, 1983a). These values represent 3 to 6 months of graft survival and growth, with stability of these changes for at least 14 months after grafting.

The capacity for grafted ACh neurons to provide reinnervation of the hippocampus appears to emerge from an interaction between graft and host. Comparison of intrahippocampal implants of fetal rat tissue derived from five different regions of brain or spinal cord, but all containing rich populations of cholinergic neurons, clearly indicated that the most extensive reinnervation of host tissue was provided by fetal septal cholinergic neurons, the appropriate source of hippocampal cholinergic innervation (Nilsson *et al.*, 1988). This was true regardless of the degree of survival of grafted ACh neurons. Similarly, when cell suspension grafts of cholinergic neurons from fetal septum or nucleus basalis were implanted into the denervated hippocampus or frontal cortex, septal neurons provided the most extensive reinnervation of hippocampus, whereas nucleus basalis neurons provided the most extensive reinnervation of frontal cortex (Dunnett *et al.*, 1986). Thus, appropriate specificity of neuron-target relationships are maintained in transplanted cholinergic neurons.

The pattern of synaptic contacts provided by cholinergic grafts has been assessed with electron microscopy of material immunolabeled for ChAT (Clarke *et al.*, 1986a). Although graft-derived synaptic contacts were readily observed, the postsynaptic target of these contacts differed somewhat from the pattern observed in the intact, mature hippocampus. In rats receiving cell suspension grafts into the hippocampus, approximately 67% of all ChAT-positive contacts were on the soma of host hippocampal granule cells and CA1 pyramidal cells. Such contacts represent only 5% of all contacts in intact rats. Correspondingly, the normally prevalent contacts onto dendritic shafts and spines (>90% of all contacts) were reduced in grafted animals to 33% of all contacts. However, a second report is at odds with these findings. Anderson and co-workers (1986) reported that the vast majority of graft-derived contacts were on their normal dendritic targets, with small numbers of contacts on soma and axons. In addition, this report documents unstained, presumably host-derived, contacts onto grafted neurons. Thus, bidirectional synaptic interactions appear to exist between graft and host, but the details of the synaptic pattern remain in question.

Cholinergic grafts have been demonstrated to regulate muscarinic receptor binding in the host hippocampus (Joyce *et al.*, 1988). Following fimbria–fornix transection, binding of ligands specific for both M1 and M2 muscarinic receptors increases markedly. Following implantation of embryonic septal tissue, binding decreases by 29 to 47%, returning these values to near-normal. This downregulation of receptor binding suggests restoration of cholinergic tone by grafted neurons.

The functional state of graft-derived innervation has been assessed by measuring the synthesis of ^{14}C-acetylcholine from ^{14}C-glucose *in vitro* under conditions of low and high potassium (Bjorklund *et al.*, 1983a). Fimbria–fornix lesioned rats without grafts exhibited a 60 to 65% decrease in rate of acetylcholine synthesis under variable potassium conditions. These values were normalized in grafted animals. The ratio of ^{14}C-acetylcholine to ChAT in lesioned animals without grafts was elevated by 7 to 10-fold, reflecting increased turnover of acetylcholine in remaining terminals. This ratio was normalized in grafted rats. Regional glucose utilization also has been measured in this experimental paradigm utilizing ^{14}C-2-deoxyglucose autoradiography (Kelly *et al.*, 1985). Unilateral fimbria–fornix transection resulted in significant, long-lasting (6-month) decreases in glucose utilization in ipsilateral hippocampus and cingulate cortex. These deficits were significantly ameliorated by block grafts of septal tissue, and the extent of recovery of glucose utilization was correlated with the extent of graft-derived reinnervation of these structures.

Restoration of function provided by grafted acetylcholine neurons also has been demonstrated through utilizing electrophysiological techniques (Low *et al.*, 1982). Examination of block grafts of embryonic septal tissue at 7 months after implantation into a cavity adjacent to the hippocampus revealed that electrical stimulation of the graft evoked a field potential in the host hippocampus indicative of graft–host connections. Furthermore, electrical stimulation of the graft just prior to activation of the host angular bundle yielded heterosynaptic facilitation in the host hippocampus, an electrophysiological response characteristic of the intact septal-hippocampal pathway. Neither of these responses could be recorded in lesioned rats that did not receive grafts.

Similar results have been obtained from the study of slices of the hippocampus explanted from rats with cholinergic cell suspension grafts for study *in vitro* (Segal *et al.*, 1985a, 1985b). Electrical stimulation of the graft produced depolarization of host hippocampal CA1 neurons, associated with increased action potential discharge. These responses were blocked by administration of the cholinergic antagonist atropine, potentiated by the anticholinesterase drug physostigmine, and mimicked by local application of acetylcholine. No such responses were detectable in slices from lesioned rats that did not receive transplants.

A final electrophysiological study compared block grafts to cell suspension grafts for their capacity to restore the characteristic low-frequency rhythmic theta activity that is lost in the hippocampus after fimbria–fornix lesions (Buzsaki *et al.*, 1987). At 5 to 8 months after lesion and grafting, rats receiving block grafts of embryonic septal tissue exhibited a return of recognizable theta rhythm that was appropriately linked to certain behaviors and time-locked to theta in the intact contralateral hippocampus. Clearest theta was recorded from portions of the hippocampus receiving the densest graft-derived cholinergic innervation. Surprisingly, rats receiving suspension grafts into the host hippocampal parenchyma did not show restoration of theta rhythm despite having graft-derived cholinergic reinnervation that exceeded that observed in rats with block grafts. The authors speculated that block grafts may have had access to input from other host brain regions that the more isolated suspension grafts did not, or that block grafts also may serve as a physical bridge to encourage restoration of host septal-hippocampal connections severed by the lesion. Restoration of theta rhythm has been reported in other transplantation paradigms, however, including approaches using suspension grafts of cholinergic cells (Buzsaki *et al.*, 1987) or bridging grafts of fetal hippocampus placed into the lesion site to promote host cholinergic innervation of the hippocampus (Tuszynski *et al.*, 1990).

There is other evidence that graft-derived cholinergic innervation does not fully reproduce the function of the mature endogenous innervation. Cassel and colleagues (1991) demonstrated that 8 months after bilateral intrahippocampal grafts of septal cell suspensions in bilaterally fimbria–fornix transected host rats, grafts failed to improve lesion-induced deficits in ChAT activity, electrically evoked release of acetylcholine, or muscarinic-stimulated formation of inositol monophosphate from labeled choline. Thus, although grafted cholinergic neurons exhibit some encouraging features, they appear not to fully replace the lost host innervation.

If grafted cholinergic neurons are to be considered an experimental therapeutic option in human neurodegenerative diseases, the most significant assay of function may be the consequence of graft placement on cognitive tasks. As indicated earlier, cholinergic dysfunction has been linked to impaired performance on cognitive tasks in animals and humans (Drachman and Leavitt, 1974; Kesner *et al.*, 1986; Olton *et al.*, 1978). The specific deficit produced by cholinergic dysfunction is an area of active debate, ranging from effects on memory per se, to detection of cues–attentional mechanisms (Everitt *et al.*, 1987; Robbins *et al.*, 1989a, 1989b). This discussion will be restricted to graft-derived effects on performance of cognitive tasks, and will not address the neural-psychological mechanisms through which a change in performance is attained. Importantly, replacement of acetylcholine via neural grafts produces significant amelioration of deficits in behavioral performance exhibited by lesioned animals. The extent of graft-related recovery varies with the complexity of the task: simpler learning and memory paradigms show better recovery than more complicated tasks.

Good recovery of performance has been reported for rewarded alternation in a T-maze (Dunnett *et al.*, 1982). This task is run in pairs of trials, the first an information or training trial, the second a test trial. In this situation, rats must hold information regarding the location of a food reward in one of two possible goals. The correct response on the test trial is to enter the arm of the maze that did not contain food on the training trial. Intact rats learn this task rapidly, and perform at near 100% correct performance after one week of training. In contrast, fimbria–fornix lesioned rats perform at chance levels (50% correct) for extended periods of time (five weeks or more). Tests of rats that received either block grafts or cell suspension grafts rich in cholinergic neurons seven months previously revealed that these animals perform at chance levels initially, but improve over eight weeks of training to reach performance levels indistinguishable from intact animals (90% correct).

Behavior associated with acquisition and retention of spatial information in the Morris water maze also is improved in lesioned rats receiving cholinergic grafts (Nilsson *et al.*, 1987). The water maze requires rats to use spatial cues in the environment to map the location of a submerged escape platform in a circular pool of water. Animals are placed into the pool and allowed to swim-search until finding the platform. Over days of training, the animals learn the location of the platform and become increasingly efficient in swimming directly to it. Both long-term memory (platform in a constant location across trials) and short-term memory (platform in one of two locations on each trial) versions of this task have been studied in grafted rats. As with the T-maze task, intact animals learn this task quickly and perform with a high degree of accuracy, whereas fimbria–fornix transected animals are unable to acquire and retain the necessary spatial information. Rats with block grafts or cell suspension grafts of embryonic septal tissue exhibited significant recovery of performance on both versions of the task. In addition, improved performance in grafted rats was disrupted by pretreatment with the cholinergic antagonist atropine, suggesting that the grafts were influencing behavior via muscarinic cholinergic activity. A second study of water maze performance found a similar lesion effect of slower acquisition and poorer performance at the end of training, but did not detect a significant effect of cholinergic grafts (Segal *et al.*, 1989). Amelioration of behavioral deficits was obtained when animals were pretreated with a low dose of physostigmine, a drug that slows breakdown of acetylcholine provided by the graft, presumably boosting the efficacy of a relatively low concentration of transmitter.

Cholinergic grafts were less effective in augmenting performance of a more complicated task utilizing the radial eight-arm maze (Low *et al.*, 1982). Optimal performance on this task depends upon the rat's ability to learn and retain spatial information regarding the location of food rewards in eight runways radiating outward from a central start area. In this situation, maximum reward is attained by holding information concerning these eight goal locations, and retrieving each piece of food without re-entering locations from which food has already been retrieved. Intact rats perform well, whereas fimbria–fornix lesioned rats perform at chance levels. Rats that received block grafts of septal tissue months previously also exhibited deficits in performance of this task. However, treatment with the cholinesterase inhibitor physostigmine significantly enhanced the performance of grafted rats. Consistent with findings of the water maze study described previously, these results suggest that whereas graft-derived cholinergic innervation was not sufficient to significantly restore performance of this task, augmentation of graft cholinergic function by inhibiting degradation of acetylcholine can yield significant behavioral recovery. Similarly, study of rats with intrahippocampal cell suspension grafts of cholinergic tissue after medial septal lesions indicated that although there was a trend toward behavioral recovery in grafted rats, this improvement did not reach statistical significance by nine months posttransplantation (Pallage *et al.*, 1986). A third report indicates that cell suspension grafts can exhibit excessive growth that damages the host hippocampus, resulting in worsening of behavioral performance (Dalrymple-Alford *et al.*, 1988). Taken together, findings on graft-mediated improvement of cognitive behavior suggest that cholinergic grafts can provide significant amelioration of performance deficits in relatively simple learning and memory tasks, although grafted animals require more training than intact animals to achieve good performance. However, identical implants are much less effective in ameliorating deficits in performance of tasks sensitive to multiple mechanisms involved in cognition that require the animal to acquire, store, and retrieve information concerning multiple spatial locations.

2. Cholinergic Grafts in Nucleus Basalis-Lesioned Young Adult Rats

The large multipolar cholinergic neurons of the basal forebrain nucleus basalis magnocellularis provide an extensive cholinergic innervation of the cerebral cortex (Johnston *et al.*, 1979; Kievit and Kuypers, 1975). Degeneration of these neurons in humans is correlated with dementia severity (Perry *et al.*, 1978) and duration (Mufson *et al.*, 1989). Lesions of the nucleus basalis in rats lead to significant deficits in performance of multiple memory tasks (Hepler *et al.*, 1985; LoConte *et al.*, 1982). The primary model utilized in grafting studies is unilateral or bilateral excitotoxic lesion of nucleus basalis using ibotenic acid, quisqualic acid, or kainic acid. Delivery of these toxins yields extensive loss of large cholinergic neurons in the basal forebrain, a corresponding 60% decrease in ChAT activity, and significant loss of AChE-positive fibers in ipsilateral cortex (−40–60%), which persists for at least six months (Fine *et al.*, 1985a). Grafting studies have implanted cholinergic neurons into cortical regions exhibiting the greatest depletion of cholinergic innervation, the frontal and parietal cortex. At one month after implantation of embryonic septal-diagonal band cell suspension grafts into two locations in cortex, ChAT activity was increased by 20% in cortex adjacent to grafts (Fine *et al.*, 1985a). Clusters of cholinergic neurons were readily apparent in grafts, and graft-derived innervation of host cortex reproduced the laminar pattern of normal cholinergic innervation.

Ultrastructural analysis of the cortex surrounding cholinergic cell grafts indicated that grafted cells established synaptic contacts with host neurons

(Clarke and Dunnett, 1986). These synapses were symmetrical in form, like the normal innervation, but were shifted in their distribution from primarily dendritic contacts in intact animals to contacts on cell soma in grafted rats. Recall that similar shifts in the location of contacts were detected in a study of intrahippocampal grafts. The authors speculated that this shift may represent an immature pattern of synaptic contacts elaborated by grafted embryonic neurons.

Graft-derived innervation has been demonstrated to modulate cortical electroencephalographic (EEG) activity in a manner consistent with acetylcholine release (Vanderwolf *et al.*, 1990). Unilateral kainic acid lesions of the basal forebrain resulted in reduced occurrence of low-voltage fast activity and increased slow-wave activity in the cortical EEG during periods of waking immobility in rats. This lesion produced a 30 to 35% decrease in AChE fiber staining in the sensorimotor cortex from which the EEG was recorded. Following implantation of cholinergic suspension grafts within 1 mm of recording electrodes, slow-wave activity gradually decreased in the EEG, attaining a significant improvement beginning eight weeks postimplantation as compared to grafts of fetal hippocampal tissue or no graft at all. The effects of cholinergic grafts were blocked by administration of the cholinergic antagonist scopolamine. However, no correlation was found between graft volume or AChE fiber outgrowth and effects on EEG, suggesting to the authors that nonspecific acetylcholine release was involved.

Intracortical cholinergic grafts have been shown to influence stimulus-evoked metabolic activity of cortex (Jacobs *et al.*, 1994). In this study, lesion of the basal forebrain reduced AChE-positive fiber staining to 25 to 40% of normal, accompanied by a significant reduction in metabolic activity of the somatosensory cortex. Somatosensory cortex of rats contains cellular aggregates (i.e., cortical barrel fields), which functionally respond to stimulation of the vibrassae in part, by increasing their metabolic activity as measured by uptake of 2-deoxyglucose. Grafts of cholinergic neurons into cortex were associated with a gradient of AChE-positive fiber outgrowth—dense reinnervation of cortex proximal to the graft, less reinnervation distally. In those cortical barrels that were proximal to grafts, innervated by AChE-positive fibers at a density of at least 85% of normal, stimulation of the vibrassae yielded metabolic activity statistically indistinguishable from normal. Metabolic activity of cortical barrels outside the region of dense graft-derived innervation remained significantly decreased, as did barrels in cortex implanted with control fetal cortex transplants.

Rats receiving unilateral lesions of the nucleus basalis followed by intracortical cholinergic grafts or control grafts of embryonic hippocampal tissue have been assessed for behavioral changes three months after grafting (Dunnett *et al.*, 1985; Dunnett, 1987; Fine *et al.*, 1985b). Three tasks were examined:

(1) learning and memory for an inhibitory avoidance task, (2) a task in the Morris water maze, and (3) habituation of locomotion–sensorimotor reactivity. The inhibitory avoidance task trains the rat to inhibit its normal tendency to explore a particular area of a chamber by punishing entry with low-level footshock. Animals were trained to a criterion of not entering the location for 300 seconds, and were tested for recall of the situation five days later. During training, both groups of grafted rats required many more trials to reach the learning criterion than did intact control rats. Rats with nucleus basalis lesions alone were intermediate in performance. In contrast, during retention testing, rats with lesions alone and lesioned rats with control hippocampal grafts exhibited extremely poor retention of the task, whereas lesioned rats with cholinergic grafts and intact rats exhibited superior performance relative to hippocampal grafted controls. Thus, cholinergic grafts appeared to have no effect on defective learning of the avoidance task, but once learned, retention and recall of the situation was significantly improved in rats with cholinergic grafts suggesting that cortical grafts of cholinergic-rich tissue affect specific types of cognitive processes.

A similar result was obtained for learning and memory in the Morris water maze (Dunnet *et al.*, 1985). During training, lesioned rats and rats with lesions plus either cholinergic or hippocampal implants exhibited significantly longer latencies to locate the hidden platform. By the end of training, latencies to locate the platform were similar among all groups. On a retention test, subjects with lesions alone and lesioned animals with grafts of fetal hippocampal tissue were significantly impaired compared to unoperated controls and lesioned animals with cholinergic grafts. Again, deficits in acquisition of the spatial task appeared not to respond to cortical cholinergic implants, whereas retention of the spatial habit was significantly improved. Tests of locomotor behavior and sensorimotor responsiveness yielded mixed results. Cholinergic grafts did not reduce lesion-induced locomotor hyperactivity, but did improve some aspects of sensorimotor responsiveness (Dunnett *et al.*, 1985).

Rats with bilateral lesions of the nucleus basalis followed by bilateral implants of cholinergic neurons into cortex also have been evaluated for cognitive performance (Welner *et al.*, 1988). Animals were pretrained to a criterion of at least 75% accuracy for four consecutive days on a T-maze rewarded alternation task. Following pretraining, the basal forebrain was lesioned bilaterally and one week later rats received either (1) three cell suspension implants of cholinergic-rich tissue into the cortex of each hemisphere, (2) similar implants of control cortical tissue, (3) sham surgery, or (4) no surgery. Three months later, T-maze performance was retested. Lesioned rats were significantly impaired in performance of the alternation task and only animals receiving cholinergic grafts exhibited substantial improvement of performance.

A study by Hodges and colleagues (1991a) combined many of the factors examined individually in other studies. Bilateral ibotenic acid lesions of septum and nucleus basalis were performed, followed by cholinergic cell suspension grafts placed into cortex, hippocampus, both regions, or homotopically into the basal forebrain. Control implants of fetal hippocampal tissue also were performed. Rats were pretrained to perform "spatial" and "cue" tasks in the radial eight-arm maze. After lesions and 16 weeks of post-lesion testing, grafts were implanted. One week later, behavioral testing resumed and continued for 14 weeks. Lesions produced significant, long-lasting deficits in performance, particularly of the spatial version of the task. ChAT activity in cortex was reduced to about 60% of control levels, and to about 70% of control levels in hippocampus. Beginning 6 to 8 weeks after transplantation, only cholinergic grafts into cortex, hippocampus, or both sites resulted in significant improvement of behavioral performance, accompanied by restoration of ChAT activity to control levels. Grafting cholinergic neurons into both hippocampus and cortex did not produce additive effects on behavioral performance. Furthermore, cholinergic grafts into the basal forebrain were without effect and exhibited little fiber outgrowth into this part of the host brain. These findings support the view that specific cues exist in target brain regions that influence axon growth, and reinforce the concept that acetylcholine must be replaced in the targets of the intrinsic system in order to influence behavior. Study of the same animals after treatment with muscarinic and nicotinic agonist and antagonist drugs indicated that lesioned rats exhibited exaggerated behavioral responses to these agents, and that cholinergic grafts did not ameliorate this supersensitivity (Hodges, *et al.*, 1991b).

A final study examined performance of a more complicated two-lever operant task in which rats were required to learn and follow either a win-stay–lose-shift strategy, or a win-shift–lose-stay strategy (Sinden *et al.*, 1990). These tasks are "operant task" equivalents to delayed response and delayed alternation paradigms. Ibotenic acid lesions of nucleus basalis produced significant deficits in performance of these tasks at all inter-response delays, suggesting a general impairment of performance rather than a specific memory-related deficit. Rats were implanted with either cholinergic-rich grafts into cortex, grafts of hippocampal tissue into cortex, or sham surgery. At three months post-grafting, rats receiving cholinergic grafts exhibited no significant improvement of performance compared to control groups. However, cholinergic grafted rats did show improved retention of a simpler step-through inhibitory avoidance task.

Taken together, the evidence on behavior of rats with cholinergic cell implants into the cortex indicate that these grafts can significantly influence retention of learned information without affecting acquisition of information in relatively simple spatial learning and memory situations, but that performance of more complicated cognitive tasks is much less responsive to graft-related cholinergic replacement.

3. Cholinergic Grafts in Aged Rats

Cholinergic neurons projecting to hippocampus and cortex undergo functional decline with old age in rodents, and this change has been associated with memory impairment in aged rats (Bartus *et al.*, 1982; Lippa *et al.*, 1980). Initial morphological analyses of embryonic septal-diagonal band cell suspension grafts placed into the hippocampus of aged rats indicated that these grafts were similar in size to grafts in young rats, but appeared not to attain the largest graft volumes observed in young hosts (Gage *et al.*, 1983). Similarly, fiber outgrowth from cholinergic grafts in aged hosts was less extensive than that seen in young hosts. The laminar pattern of cholinergic innervation was reproduced in aged hosts, but the density of this graft-derived innervation was decreased in comparison to young hosts. Ultrastructural analysis of graft-host contacts indicated that cholinergic grafts in aged hosts accurately reproduced the pattern of contact seen in intact rats, with the majority of contacts residing on dendritic shafts and spines (Clarke *et al.*, 1986b). These results indicate that the aged rat brain remains an adequate environment for survival and growth of grafted cholinergic neurons. As noted by the authors, the decreased maximum size and neurite extension of cholinergic grafts in aged rats may be attributable to the fact that these animals have not experienced the extensive cholinergic denervation produced by experimental lesions, but rather display the partial denervation produced by normal aging. The results are consistent with evidence indicating that graft size and neurite outgrowth also are diminished in young adult animals that have not experienced denervating lesions prior to grafting (Gage and Bjorklund, 1986a). Thus, there was no clear indication that the aged brain, per se, negatively influenced the viability of cholinergic grafts.

Behavioral studies of aged animals demonstrate recovery of function after placement of cholinergic grafts. Aged rats screened for performance deficits on the Morris water maze task were implanted bilaterally with embryonic septal tissue rich in cholinergic neurons into three locations in the hippocampus (Gage *et al.*, 1984). Eight of 11 aged transplanted rats exhibited significant improvement of both acquisition and retention of the spatial task compared to unoperated aged behaviorally impaired rats. Although surviving grafts elaborating variable reinnervation of the hippocampus were found in all rats, there appeared to be no relationship between the extent of reinnervation and behavioral recovery; the three aged rats with cholinergic grafts that did not exhibit behavioral improvement did not have obviously less viable grafts. In a subsequent study, aged rats with cholinergic grafts were tested for their behavioral sensitivity to cholinergic drugs (Gage and Bjorklund, 1986b). As before, cholinergic grafts produced significant improvement in performance of the swim maze task. Pretreatment with the cholinesterase inhibitor physostigmine did not further augment performance, but treatment with the muscarinic antagonist atropine abolished the behavioral improvement seen in grafted aged rats.

Cholinergic grafts in aged rats also have been tested for effects on a type of spatial short-term memory (Dunnett *et al.*, 1988). Subjects were trained to press a lever for a food reward. On the first of a pair of trials, a single lever was presented in one of two possible locations, and the animal was rewarded for a lever-press. On trial 2, levers were presented in both locations, and reward followed pressing the lever in the location not presented on trial 1. Increasing memory requirement was introduced by increasing the delay between trials 1 and 2. Aged rats were significantly poorer at this task than young rats at all delays of four seconds or longer. Aged rats with cholinergic grafts either placed into the hippocampus or frontal and parietal cortex three months previously were significantly improved in their memory performance at delays of four, eight, and 12 seconds, but remained relatively impaired at delays of 16 and 24 seconds. Grafts into hippocampus were somewhat more effective than grafts into cortex. In combination with the water maze studies, these findings suggest that both deficient long-term and short-term memory performance can be significantly improved by cholinergic neuron grafts in aged rats. It should be noted, however, that total recovery of cognitive performance is rarely observed, and grafted aged rats often remain at least somewhat deficient as compared to intact young adult rats.

4. Summary of Evidence from Rat Studies

1. Transplantation of embryonic cholinergic neurons yields a very high percentage of subjects exhibiting surviving grafts. Whereas the presence of surviving grafted tissue is a high-probability event, estimates of the yield of surviving grafted cells relative to the number of neurons known to mature from these embryonic brain regions remain low, in the range of a few percent.

2. Implants of cholinergic neurons reinnervate hippocampal and cortical target regions in a pattern that is similar to that of the intact system. The density of reinnervation diminishes with distance from the grafts, with a significant decrease exhibited beyond a radius of approximately 2 mm. Grafted septal and basal forebrain cholinergic neurons maintain organotypic preference for reinnervating their normal targets, hippocampus, and cortex, respectively.

3. Ultrastructural studies indicate that grafted cholinergic neurons establish synaptic contacts with the host brain, but the exact location of these contacts

may be shifted somewhat in distribution from normal locations on dendritic shafts to more contacts with cell soma. It is unclear what effect, if any, this has on the physiology of the graft-host interaction. Host-derived contacts also have been observed on grafted cholinergic neurons, but the identity of these contacts has not been established.

4. Cholinergic grafts restore a variety of biochemical and physiological measures in close proximity to grafted neurons. Electrical stimulation of cholinergic grafts activate simple electrophysiological responses in hippocampus and cortex, but appear less able to influence complex responses such as the theta rhythm that is likely to require polysynaptic regulation.

5. Performance of relatively simple cognitive tasks involving learning and memory for spatial locations can be dramatically improved by cholinergic-rich transplants to hippocampus or cortex. As these tasks increase in complexity, involving information regarding multiple locations or complex response strategies, the efficacy of cholinergic grafts diminishes.

6. The aged brain is an environment that is compatible with graft survival and function. Comparable survival, reinnervation, and behavioral effects of cholinergic grafts in young and aged rats have been demonstrated.

III. FETAL NEURONAL GRAFTING STUDIES IN PRIMATES

A. Functional Studies

Prior to the initiation of clinical trials, especially for behaviors as complex as cognition, it is important that experimental therapeutic strategies demonstrate safety and efficacy in nonhuman primate models of the disease. Unlike Parkinson's disease, in which the MPTP-treated primate is an outstanding animal model, nonhuman primate models of AD are limited. For specifically modeling the cholinergic deficit in AD, lesion studies have been employed. Ridley and co-workers (1991) have pioneered the study of CBF transplants in monkeys with lesions of the basal forebrain system. Their initial investigations employed bilateral lesions of the fornix, thereby interrupting septohippocampal cholinergic projections. If should be noted that the septohippocampal system is less affected than the basalis-neocortical system in AD, especially in patients lacking the ApoE-4 genotype, and thus this lesion paradigm does not mimic the pattern of neural degeneration seen in most patients with AD. Still, given the more restricted target zone (hippocampus), and the numerous studies using this model system in rodents, the monkey septohippocampal system was a logical choice for initial fetal grafting studies.

In their first study (Ridley *et al.*, 1991), marmosets received bilateral transections of the fornix resulting in impaired function on a number of cognitive tasks including visuospatial conditional discriminations, visual conditional discriminations, and nonconditional spatial response tasks. Monkeys then received bilateral implants of cholinergic-rich fetal basal forebrain tissue into the hippocampus. Grafts resulted in a significant improvement in function on each of these tasks three to nine months posttransplantation. In contrast, lesioned animals that received fetal *hippocampal* rather than *septal* grafts remained impaired. Recipients of fetal hippocampal grafts displayed a severe reduction in AChE staining throughout the dentate gyrus and hippocampus, whereas recipients of septal grafts showed substantial restoration of AChE staining in these regions.

In a second study, marmosets with bilateral transections of the fornix were severely impaired on learning visuospatial conditional tasks presented in a Wisconsin General Test Apparatus (Ridley *et al.*, 1992). Bilateral transplantation of cholinergic-rich embryonic basal forebrain tissue into the hippocampus led to improvements on this task across a range of task difficulties. Administration of the direct cholinergic agonist pilocarpine to lesioned ungrafted animals immediately before testing also reduced their impairment, suggesting that the graft-associated recovery may have been mediated by acetylcholine release. Again, transection of the fornix produced a marked loss of acetylcholinesterase (AChE) staining confined to hippocampus and entorhinal cortex relative to controls. In all animals that received septal cell transplants, dense AChE-stained cellular masses of grafted tissue were seen bilaterally in the hippocampus and surrounding cortex, with fiber outgrowth from the graft into surrounding host tissue.

To explore the potential use of fetal CBF grafts in AD, a model system was employed (Ridley *et al.*, 1994), which more closely mimics the pattern of degeneration seen in AD. Rather than study graft effects in the septohippocampal system, marmosets with lesions of the nucleus basalis and grafts of fetal septal neurons to the cortex were studied. Three groups of marmosets were trained to perform a series of visual discrimination tasks using a Wisconsin General Test Apparatus. Two groups then received bilateral lesions of the basal nucleus of Meynert using the excitotoxin N-methyl-D-aspartate. Monkeys receiving these lesions were severely impaired on relearning in a visual discrimination test. One lesioned group then received grafts of acetylcholine-rich tissue dissected from the basal forebrain of fetal marmosets; grafts were placed bilaterally within the frontal and parietal association cortices. Three months later marmosets with lesions alone remained impaired on a number of retention and reversal tasks, whereas the transplanted animals were no longer significantly impaired. Histological examination of the brains indicated that all lesioned animals had sustained substantial loss of the cholinergic neurons of the basal nucleus of Meynert and that the lesion-alone animals showed marked loss of the cholinergic marker AChE in the dorsolateral and parietal cortex. All transplanted animals had surviving graft tissue as visualized by cresyl violet staining, dense AChE staining, and a limited number of p75 neurotrophin (p75-^{NTR})-immunoreactive neurons in the neocortex. Five of the six CBF-transplanted marmosets showed near complete restitution of AChE staining in frontal and parietal cortex. Examination of individual animal data showed that a single animal lacking neuroanatomical recovery also failed to exhibit significant behavioral recovery. The performance of the remaining transplanted animals was significantly improved relative to animals with lesions alone. There was a significant positive correlation between the degree of AChE staining and performance on tasks sensitive to frontal lobe damage. These results demonstrate that acetylcholine-rich tissue transplanted into the neocortex of primates with damage to the cholinergic projections to the neocortex can produce substantial restitution of function on specific tasks measuring memory function provided that an appropriate level of innervation between graft and host tissue is achieved.

B. Fetal CBF Allografts in Nonhuman Primates

1. Long-Term Survival, Cholinergic Innervation, and Effects of Cyclosporin

For CBF grafts to be useful for AD, they must survive, innervate, and function for an extended time period. We carried out a series of fetal allografts in rhesus monkeys to assess long-term graft survival and innervation of the large primate brain. Further, we examined whether graft viability was improved following immunosuppression with cyclosporin. Sixteen rhesus monkeys received unilateral lesions of the fornix to denervate the hippocampus of cholinergic inputs. Ten to 14 days later, monkeys received implants of solid monkey fetal basal forebrain neurons derived from a single donor into the hippocampus. All implants were between 35 and 50 days of gestation. This gestational age was chosen based upon the neurogenesis of the rhesus monkeys' CBF (Kordower and Rakic, 1990). Monkeys were sacrificed three, nine, 12, and 18 months following transplantation. Half of the monkeys received cyclosporin (15 mg/kg, im) daily beginning one day prior to the transplants and continuing until sacrifice. Viable grafts of CBF neurons were routinely observed upon histological examination. Importantly, large CBF grafts were observed in monkeys 12 and 18 months posttransplantation (Fig. 1). Grafted CBF neurons appeared healthy and exhibited magnocellular morphology appropriate to neurons of the CBF. Grafted neurons also displayed long multipolar neuritic processes emanating from the cell soma.

Numerous ChAT-ir, trk-ir, $p75^{NTR}$-ir, galanin-ir, and AChE-containing perikarya neurons could be observed within the grafted hippocampus surviving in an organotypic fashion (Fig. 1). Because these five proteins colocalize with CBF neurons, it was not surprising that similar numbers of neurons stained for each marker were seen within animals. Interestingly an exception to this finding was one monkey with a graft surviving for 18 months that displayed similar numbers of trk-ir, $p75^{NTR}$-ir, galanin-ir, and AChE grafted neurons but a marked diminution of ChAT-ir cells. The presence of trk receptors on grafted cholinergic neurons suggests that grafted neurons would be responsive to NGF exposure. This may be an important feature of these implants because the potential exists for the viability and neurite outgrowth from basal forebrain grafts to be augmented *in vivo* or *in vitro* by infusions of trophic factors such as NGF or the co-grafting of NGF-secreting cells.

Fetal CBF neurons were capable of providing extensive cholinergic innervation to the host hippocampus (Fig. 2). Beginning six to nine months posttransplantation, grafts provided innervation in an organotypic pattern that by 12 months was similar in magnitude to levels seen in the normal hippocampus. These data indicate that fetal CBF neurons can survive long-term grafting into the primate brain. Qualitative observations indicate that immunosuppression did not enhance graft viability. Thus, fetal CBF neurons are capable of providing a long-term, relatively normal regional cholinergic reinnervation in the young adult lesioned monkey. These data support the concept that fetal CBF grafts may be useful to treat the cholinergic deficit seen in AD.

It remains to be determined whether CBF grafts are capable of reversing age-related cognitive decline in the nonhuman primates in a manner similar to that seen in rodents. This point is not trivial because the type of trial-dependent cognitive deficits displayed by aged nonhuman primates are more similar to those observed in AD patients. For the most part, fornix lesions of the CBF system do not significantly affect trial-dependent memory processes in young monkeys, in which the correct response changes on each trial (e.g., delayed response, delayed alternation, or match to sample tasks). Rather, these lesions induce deficits in other types of trial-independent memory processes such as those responsible for simple visual discriminations (e.g., Gaffan *et al.*, 1989). Thus, the optimal model for examining the utility of fetal grafting for AD may be the aged nonhuman primate.

Whereas the above experimental data provides the groundwork to consider this approach clinically, a number of technical and ethical issues also need to be appreciated. Based upon experience with fetal nigral grafts for Parkinson's disease, multiple donors will be needed to sufficiently repopulate the brain with new basal forebrain cells. The low-pressure tissue-harvesting procedure

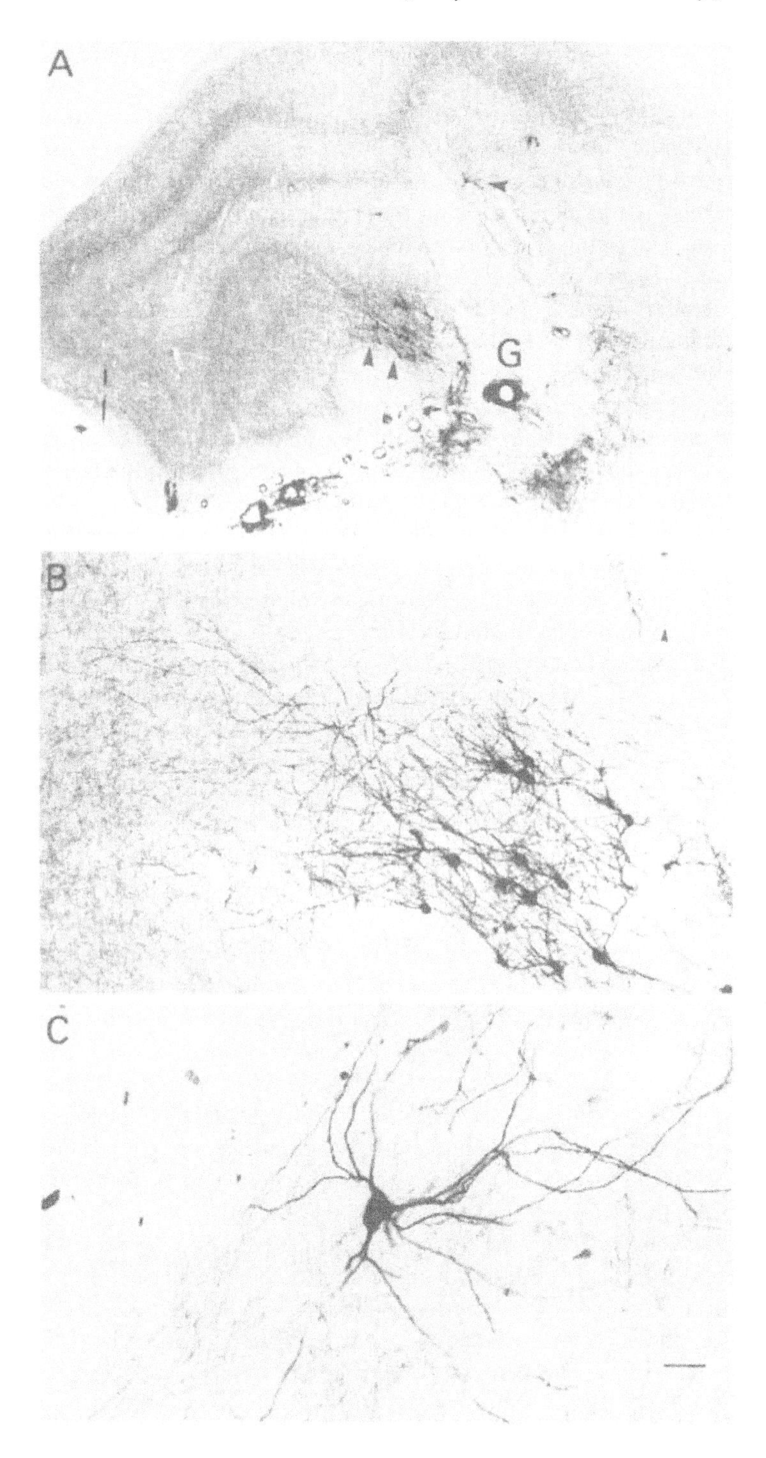
A
G
B
C

commonly used for tissue extraction to provide intact human midbrain for dopaminergic grafts may not be of use for the procurement of fetal basal forebrain because this region is usually extensively disrupted, making procurement extremely difficult. Even without this technical problem, the supply of tissue of the appropriate donor age is limited and procurement requirements for the 4,000,000 American suffering from AD would be staggering.

Another issue with regard to using CBF grafts is the optimal placement of cells in the host brain. AD selectively affects a particular subfield of the basal forebrain (Mufson *et al.*, 1989). We have reported that $p75^{NTR}$-containing neurons of the medial septum and vertical limb of the diagnonal band that projects to the hippocampal projection are minimally affected in AD relative to aged controls. In contrast, a significant loss of $p75^{NTR}$-immunoreactive neurons occurs in the nucleus basalis in AD that inversely correlated (−0.786) with disease duration. Moreover, all four subdivisions of the nucleus basalis are affected in AD cases with the anterolateral (76.4%), intermediate (62.1%), and posterior divisions (76.5%) demonstrating the greatest reduction in cholinergic basal forebrain neurons. Nissl-counterstained sections fail to reveal magnocellular neurons in the nucleus basalis, suggesting that reduction in immunopositive neurons reflects neuron death and not the failure of viable neurons to synthesize low-affinity neurotrophin receptors. These data indicate that cholinergic basal forebrain neurons that project to the amygdala, as well as the temporal, frontobasal, and frontodorsal cortices, are most affected in AD. Thus a neuronal repopulation strategy for AD would place CBF grafts within the cerebral cortex, a massive structure in the human brain, rather than in the smaller hippocampus. Whether this can be achieved successfully in AD remains to be determined.

Many challenges lie before us and much work needs to be done before even considering this strategy for clinical use. In addition, grafting strategies, especially those employing nonhuman primates, can help us understand the role of CBF in higher-order cognitive function.

FIGURE 1 (A) Low-power photomicrograph of a $p75^{NTR}$-immunostained section from a monkey that received a fornix transection followed 7 days later by a fetal CBF allograft (G). This monkey was sacrificed 18 months posttransplantation. Note that along the periphery of the graft are collections of grafted $p75^{NTR}$-ir neurons that give rise to an innervation pattern in this lesion monkey similar to what is normally seen *in situ*. (B) Higher-power photomicrograph from the region depicted with double arrowheads in panel A. These $p75^{NTR}$-ir magnocellular grafted neurons are extending processes into the host and then forming a terminal-like pattern. (C) High-power photomicrograph of a grafted neuron depicted with a single arrowhead in panel A. Note the classic magnocellular morphology of this grafted neuron with long multipolar processes. Scale bar in C represents the following magnifications: A = 400 μm; B = 85 μm; C = 40 μm.

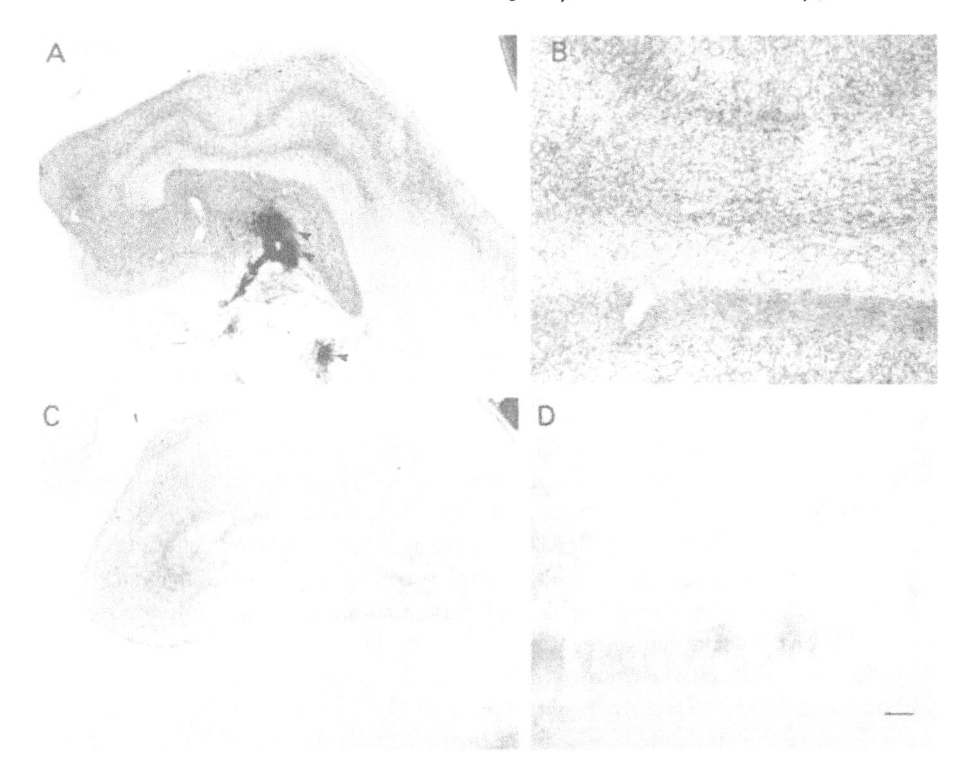

FIGURE 2 (A) Low- and (B) high-power photomicrographs of AChE-stained sections through the hippocampus of a monkey that received a fornix transection followed 7 days later by a fetal CBF allograft and was sacrificed 18 months posttransplantation. Note the dense staining of AChE-positive cholinergic neurons in the graft (arrowheads). (B) These cells gave rise to an extensive AChE-positive innervation pattern. (C and D) In contrast, low- and high-power photomicrographs from an unlesioned monkey that received a fornix transection revealed a comprehensive loss of AChE fibers within the hippocampus. Scale bar in D represents the following magnifications: A and C = 1200 μg; B and D = 125 μm.

REFERENCES

Anderson, K. J., Gibbs, R. B., Salvaterra, P. M., and Cotman, C. W. (1986). Ultrastructural characterization of identified cholinergic neurons transplanted to the hippocampal formation of the rat. *J Comp Neurol* 259:279–292.

Bartus, R. T. (1990). Drugs to treat age-related neurodegenerative problems. The final frontier of medical science? *Geriatric Bioscience* 38:680–695.

Bartus, R. T., Dean, R. L., Beer, B., and Lippa, A. S. (1982). The cholinergic hypothesis of geriatric memory dysfunction. *Science* 217:408–417.

Baxter, M. G., Bucci, D. J., Gorman, L. K., Wiley, R. G., and Gallagher, M. (1995). Selective immunotoxic lesions of basal forebrain cholinergic cells: Effects on learning and memory in rats. *Behav Neurosci* 109:714–722.

Bjorklund, A., Gage, F. H., Schmidt, R. H., Stenevi, U., and Dunnett, S. B. (1983a). Intracerebral grafting of neuronal cell suspensions. VII. Recovery of choline acetyltransferase activity and acetylcholine synthesis in the denervated hippocampus reinnervated by septal suspension implants. *Acta Physiol Scand Suppl* 522:59–66.

Bjorklund, A., Gage, F. H., Stenevi, U., and Dunnett, S. B. (1983b). Intracerebral grafting of neuronal cell suspensions. VI. Survival and growth of intrahippocampal implants of septal cell suspensions. *Acta Physiol Scand Suppl* 522:49–58.

Bjorklund, A., and Stenevi, U. (1977). Reformation of the severed septo-hippocampal cholinergic pathway in the adult rat by transplanted septal neurons. *Cell Tiss Res* 185:289–302.

Buzsaki, G., Gage, F. H., Czopf, J., and Bjorklund, A. (1987). Restoration of rhythmic slow activity in the subcortically denervated hippocampus by fetal CNS transplants. *Brain Res* 400:334–347.

Cassel, J. C., Jackisch, R., Duschek, M., Hornsperger, J. M., Richards, M. H., Kelche, C., Hertting, G., and Will, B. (1991). Long term effects of septohippocampal lesions and intrahippocampal grafts on acetylcholine concentration, muscarinic stimulated formation of inositol phospholipids and electrically evoked release of neurotransmitters in the rat hippocampus. *Exp Brain Res* 83:633–642.

Clarke, D. J., and Dunnett, S. B. (1986). Ultrastructural organization of choline-acetyltransferase immunoreactive fibers innervating the neocortex from embryonic ventral forebrain grafts. *J Comp Neurol* 250:192–205.

Clarke, D. J., Gage, F. H., and Bjorklund, A. (1986a). Formation of cholinergic synapses by intrahippocampal septal grafts as revealed by choline acetyltransferase immunocytochemistry. *Brain Res* 369:151–162.

Clarke, D. J., Gage, F. H., Nilsson, O. G., and Bjorklund, A. (1986b). Grafted septal neurons from cholinergic synaptic connections in the dentate gyrus of behaviorally impaired aged rats. *J Comp Neurol* 252:483–492.

Dalrymple-Alford, J. C., Kelche, C., Cassel, J. C., Toniolo, G., Pallage, V., and Will, B. E. (1988). Behavioral deficits after intrahippocampal fetal septal grafts in rats with selective fimbria–fornix lesions. *Exper Brain Res* 69:545–558.

Davies, P., and Maloney, A. F. J. (1976). Selective loss of central cholinergic neurons in Alzheimer's disease. *Lancet* 2:1403.

Dornan, W. A., McCampbell, A. R., Tinkler, G. P., Hickman, L. J., Bannon, A. W., Decker, M. W., and Gunther, K. L. (1997). Comparison of site specific injections into the basal forebrain on water maze performance in the male rat after immunolesioning with 192 IgG saporin. *Behav Brain Res* 86:181–189.

Doucette, R., and Ball, M. (1987). Left-right symmetry of neuronal cell counts in the nucleus basalis of control and Alzheimer's diseases brain. *Brain Res* 422:357–360.

Drachman, D. A., and Leavitt, J. (1974). Human memory and the cholinergic system. *Arch Neurol* 30:110–121.

Dunnett, S. B. (1987). Anatomical and behavioral consequences of cholinergic-rich grafts to the neocortex of rats with lesions of the nucleus basalis magnocellularis. *Annals NY Acad Sci* 495:415–429.

Dunnett, S. B., Badman, F., Rogers, D. C., Evenden, J. L., and Iversen, S. D. (1988). Cholinergic grafts in the neocortex or hippocampus of aged rats: Reduction of delay-dependent deficits in the delayed non-matching to position task. *Exper Neurol* 102:57–64.

Dunnett, S. B., and Fibiger, H. C. (1993). Role of forebrain cholinergic systems in learning memory: relevance to the cognitive deficits of aging and Alzheimer's dementia. *Prog Brain Res* 98:413–420.

Dunnett, S. B., Low, W. C., Iversen, S. D., Stenevi, U., and Bjorklund, A. (1982). Septal transplants restore maze learning in rats with fornix-fimbria lesions. *Brain Res* 251:335–348.

Dunnett, S. B., Toniolo, G., Find, A., Ryan, C. N., Bjorklund, A., and Iversen, S. D. (1985). Transplantation of embryonic ventral forebrain neurons to the neocortex of rats with lesions of nucleus basalis magnocellularis. II. Sensorimotor and learning impairments. *Neurosci* 16:787–797.

Dunnett, S. B., Whishaw, I. Q., Bunch, S. T., and Fine, A. (1986). Acetylcholine-rich neuronal grafts in the forebrain of rats: Effects of environmental enrichment, neonatal noradrenaline depletion, host transplantation site, and regional source of embryonic donor cells on graft size and acetylcholinesterase-positive fiber outgrowth. *Brain Res* 378:357–373.

Everitt, B. J., Robbins, T. W., Evenden, J. L., Marston, H. M., Jones, J. G., and Sirkia, T. E. (1987). The effects of excitiotoxic lesions of the substantia inominata, ventral and dorsal globus pallidus on the acquisition and retention of conditional visual dicrimination: Implications for the cholinergic hypothesis of learning and memory. *Neurosci* 22:441–470.

Fine, A., Dunnett, S. B., Bjorklund, A., Clarke, D. J., and Iversen, S. D. (1985a). Transplantation of embryonic ventral forebrain neurons to the neocortex of rats with lesions of nucleus basalis magnocellularis. I. Biochemical and anatomical observations. *Neurosci* 16:769–786.

Fine, A., Dunnett, S. B., Bjorklund, A., and Iversen, S. D. (1985b). Cholinergic ventral forebrain grafts into the neocortex improve passive avoidance memory in a rat model of Alzheimer disease. *PNAS USA* 82:5227–5230.

Fine, A., Hoyle, C., Maclean, C. J., Levatte, T. L., Baker, H. F., and Ridley, R. M. (1997). Learning impairments following injection of a selective cholinergic immunotoxin, ME20.4 IgG-saporin, into the basal nucleus of Meynert in monkeys. *Neuroscience* 81:331–343.

Gaffan, D., and Harrison, S. (1989). A comparison of the effects of fornix transection and sulcus principalis ablation upon spatial learning by monkeys. *Behav Brain Res* 31:207–220.

Gage, F. H., and Bjorklund, A. (1986a). Enhanced graft survival in the hippocampus following selective denervation. *Neurosci* 17:89–98.

Gage, F. H., and Bjorklund, A. (1986b). Cholinergic septal grafts into the hippocampal formation improve spatial learning and memory in aged rats by an atropine-sensitive mechanism. *J Neurosci* 6:2837–2847.

Gage, F. H., Bjorklund, A., Stenevi, U., and Dunnett, S. B. (1983). Intracerebral grafting of neuronal cell suspensions VII. Survival and growth of implants of nigral and septal cell suspensions in intact brains of aged rats. *Acta Physiol Scand* 522(Suppl.):67–75.

Gage, F. H., Bjorklund, A., Stenevi, U., Dunnett, S. B., and Kelly, P. A. T. (1984). Intrahippocampal septal grafts ameliorate learning impairments in aged rats. *Science* 225:533–536.

Hepler, D. G., Wenk, G. L., Cribbs, B. L., Olton, D. S., and Coyle, J. T. (1985). Memory impairment following basal forebrain lesions. *Brain Res* 346:8–14.

Hodges, H., Allen, Y., Kershaw, T., Lantos, P. L., Gray, J. A., and Sinden, J. (1991a). Effects of cholinergic-rich neural grafts on radial maze performance of rats after excitotoxic lesions of the forebrain cholinergic projection system. I. Amelioration of cognitive deficits by transplants into cortex and hippocampus but not into basal forebrain. *Neurosci* 45:587–607.

Hodges, H., Allen, Y., Sinden, J., Lantos, P. L., and Gray, J. A. (1991b). Effects of cholinergic-rich neural grafts on radial maze performance of rats after excitotoxic lesions of the forebrain cholinergic projection system. II. Cholinergic drugs as probes to investigate lesion-induced deficits and transplant-induced functional recovery. *Neurosci* 45:609–623.

Jacobs, S. E., Find, A., and Juliano, S. L. (1994). Cholinergic basal forebrain transplants restore diminished metabolic activity in the somatosensory cortex of rats with acetylcholine depletion. *J Neurosci* 14:697–711.

Johnston, M. V., McKinney, M., and Coyle, J. T. (1979). Evidence for a cholinergic projection to neocortex from neurons in basal forebrain. *PNAS USA* 76:5392–5396.

Joyce, J. N., Gibbs, B. R., Cottman, C. W., and Marshall, J. F. (1988). Regulation of acetylcholine muscarinic receptors by embryonic septal grafts showing cholinergic innervation of host hippocampus. *Prog Brain Res* 78:109–116.

Kelly, P. A. T., Gage, F. H., Ingvar, M., Lindvall, O., Stenevi, U., and Bjorklund, A. (1985). Functional reactivation of the deafferented hippocampus by embryonic septal grafts as assessed by measurements of local glucose utilization. *Exper Brain Res* 58:570–579.

Kesner, R. P., Crutcher, K. A., and Measom, M. O. (1986). Medial septal and nucleus basalis magnocellularis lesions produce order memory deficits in rats which mimic symptomatology of Alzheimer's disease. *Neurobiol Aging* 7:287–295.

Kievet, J., and Kuypers, H. G. J. M. (1975). Basal forebrain and hypothalamic connections to frontal and parietal cortex in the rhesus monkey. *Science* 187:660–662.

Kordower, J. H., and Rakic, P. (1990). Genesis of the magnocellular basal forebrain nuclei in Rhesus monkeys. *J Comp Neurol* 291:637–653.

Kordower, J. H., Gash, D. M., Bothwell, M., Hersh, L. B., and Mufson, E. J. (1989). Nerve growth factor receptor and choline acetyltransferase remain colocalized in the nucleus basalis (Ch4) of Alzheimer's patients. *Neurobiol Aging* 10:67–74.

Leanza, G., Nilsson, O. G., Wiley, R. G., and Bjorklund, A. (1995). Selective lesioning of the basal forebrain cholinergic system by intraventricular 192 IgG-saporin: Behavioral, biochemical and stereological studies in the rat. *Eur J Neurosci* 7:329–343.

Lippa, A. S., Pelham, R. W., Beer, B., Critchett, K. J., Dean, R. L., and Bartus, R. T. (1980). Brain cholinergic dysfunction and memory in aged rats. *Neurobiol Aging* 1:13–19.

LoConte, G., Bartolini, L., Casamenti, F., Marconcini-Pepeu, I., and Pepeau, G. (1982). Lesions of cholinergic forebrain nuclei: Changes in avoidance behavior and scopolamine actions. *Pharmacol Biochem Behav* 17:933–937.

Low, W. C., Lewis, P. R., Bunch, S. T., Dunnett, S. B., Thomas, S. R., Iversen, S. D., Bjorklund, A., and Stenevi, U. (1982). Function recovery following neural transplantation of embryonic septal nuclei in adult rats with septohippocampal lesions. *Nature* 300:260–262.

McMahan, R. W., Sobel, T. J., and Baxter, M. G. (1997). Selective immunolesions of hippocampal cholinergic input fail to impair spatial working memory. *Hippocampus* 7:130–136.

Mufson, E. J., Bothwell, M., and Kordower, J. H. (1989). Loss of nerve growth factor receptor-immunoreactive neurons in Alzheimer's disease: A quantitative analysis across subregions of the basal forebrain. *Exp Neurol* 105:221–232.

Nilsson, O. G., Clarke, D. J., Brundin, P., and Bjorklund, A. (1988). Comparison of growth and reinnervation properties of cholinergic neurons from different brain regions grafted to the hippocampus. *J. Comp Neurol* 268:204–222.

Nilsson, O. G., Shapiro, M. L., Gage, F. H., Olton, D. S., and Bjorklund, A. (1987). Spatial learning and memory following fimbria-fornix transection and grafting of fetal septal neurons to the hippocampus. *Exper Brain Res* 67:195–215.

Olton, D. S., Walker, J. A., and Gage, F. H. (1978). Hippocampal connections and spatial discrimination. *Brain Res* 139:295–308.

Pallage, V., Toniolo, G., Will, B., and Hefti, F. (1986). Long-term effects of nerve growth factor and neural transplants on behavior in rats with medial septal lesions. *Brain Res* 386:197–208.

Perry, E. K., Tomlinson, B. E., Blessed, G., Bergmann, K., Gibson, P. H., and Perry, R. H. (1978). Correlation of cholinergic abnormalities with senile plaques and mental test scores in senile dementia. *Brit Med J* 2:1457–1459.

Ridley, R. M., Baker, J. A., Baker, H. F., and Mclean, C. J. (1994). Restoration of cognitive abilities by cholinergic grafts in cortex of monkeys with lesions of the basal nucleus of Meynert. *Neuroscience* 63:653–666.

Ridley, R. M., Gribble, S., Clark, B., Baker, H. F., and Fine, A. (1992). Restoration of learning ability in fornix-transected monkeys after fetal basal forebrain but not fetal hippocampal tissue transplantation. *Neuroscience* 48:779–792.

Ridley, R. M., Thornley, H. D., Baker, H. F., and Fine, A. (1991). Cholinergic neural transplants into hippocampus restore learning ability in monkeys with fornix transections. *Expl Brain Res* 83:533–538.

Robbins, T. W., Everitt, B. J., Ryan, C. N., Marston, H. M., Jones, G. H., and Page, K. J. (1989a). Comparative effects of quisqualate and ibotenic acid-induced lesions of substantia innominata and globus pallidus on the acquisition of a conditioned visual discrimination: Differential effects on cholinergic mechanisms. *Neurosci* 28:337–352.

Robbins, T. W., Everitt, B. J., Marston, H. M., Wilkinson, J., Jones, G. H., and Page, K. J. (1989b). Comparative effects of ibotenic acid- and quisqualic acid-induced lesions of substantia innominata on attentional function in the rat: Further implications for the role of the cholinergic neurons of the nucleus basalis in cognitive processes. *Behav Brain Res* 35:221–241.

Segal, M., Bjorklund, A., and Gage, F. H. (1985a). Transplanted septal neurons make viable cholinergic synapses with a host hippocampus. *Brain Res* 336:302–307.

Segal, M., Bjorklund, A., and Gage, F. H. (1985b). Intracellular analysis of cholinergic synapses between grafted septal nuclei and host hippocampal pyramidal neurons. In: Bjorklund, A., and Stenevi, U., eds., *Neural grafting in the mammalian CNS*. Amsterdam, Elsevier, pp. 389–399.

Segal, M., Greenberger, V., and Pearl, E. (1989). Septal transplants ameliorate deficits and restore cholinergic functions in rats with a damaged septo-hippocampal connection. *Brain Res* 500:139–148.

Shen, J., Barnes, C. A., Wenk, G. L., and McNaughton, B. L. (1996). Differential effects of selective immunotoxic lesions of medial septal cholinergic cells on spatial working and reference memory. *Behav Neurosci* 110:1181–1186.

Sims, N. R., Bowen, D. M., Allen, S. J., Smith, C. C. T., Neary, D., Thomas, D. J., and Davidson, A. N. (1981). Presynaptic cholinergic dysfunction in patients with dementia. *Brain* 40:503–509.

Sinden, J. D., Allen, Y. S., Rawlins, J. N., and Gray, J. A. (1990). The effects of ibotenic acid lesions of the nucleus basalis and cholinergic-rich neural transplants on win-stay/lose-shift and win-shift/lose-stay performance in the rat. *Behav Brain Res* 36:229–249.

Torres, E. M., Perry, T. A., Blockland, A., Wilkinson, L. S., Wiley, R. G., Lappi, D. A., and Dunnet, S. B. (1994). Behavioural, histochemical and biochemical consequences of selective immunolesions in discrete regions of the basal forebrain cholinergic system. *Neuroscience* 63:95–122.

Tuszynski, M. H., Buzsaki, G., and Gage, F. H. (1990). NGF enhances graft-induced functional cholinergic reinnervation of the deafferented hippocampal formation. *Neuroscience* 36:33–44.

Vanderwolf, C. H., Fine, A., and Cooley, R. K. (1990). Intracortical grafts of embryonic basal forebrain tissue restore low voltage fast activity in rats with basal forebrain lesions. *Exper Brain Res* 81:426–432.

Voytko, M. L. (1996). Cognitive functions of the basal forebrain cholinergic system in monkeys: Memory or attention? *Behav Brain Res* 75:13–25.

Wainer, B. H., Steininger, T. L., Roback, J. D., Burke-Watson, M. A., Mufson, E. J., and Kordower, J. H. (1993). Ascending cholinergic pathways: Functional organization and implication disease models. *Prog Brain Res* 98:9–30.

Walsh, T. J., Herzog, C. D., Gandhi, C., Stackman, R. W., and Wiley, R. G. (1996). Injection of IgG 192-saporin into the medial septum produces cholinergic hypofunction and dose-dependent working memory deficits. *Brain Res* 726:69–79.

Welner, S. A., Dunnett, S. B., Salamone, J. D., MacLean, B., and Iversen, S. D. (1988). Transplantation of embryonic ventral forebrain grafts to the neocortex of rats with bilateral lesions of nucleus basalis magnocellularis ameliorates a lesion-induced deficit in spatial memory. *Brain Res* 463:192–197.

Wenk, G. L., Stoehr, J. D., Quintana, G., Mobley, S., and Wiley, R. G. (1994). Behavioral, biochemical, histological, and electrophysiological effects of 192 IgG-saporin injections into the basal forebrain of rats. *J Neurosci* 14:5986–5995.

Whitehouse, P. J., Price, D. L., Struble, R. G., Clark, A. W., Coyle, J. T., and DeLong, M. R. (1982). Alzheimer's disease and senile dementia: Loss of neurons in the basal forebrain. *Science* 215:1237–1239.

Wilcock, G. K., Eseri, M. M., Bowen, D. M., and Smith, C. T. (1982). Alzheimer's disease: Correlation of cortical choline acetyltransferase activity with the severity of dementia a histological abnormalities. *J Neurol Sci* 57:407–417.

Wiley, R. G., Oeltmann, T. N., and Lappi, D. A. (1991). Immunolesioning: Selective destructions of neurons using immunotoxin to rat NGF receptor. *Brain Res* 562:149–153.

CHAPTER 23

Neurotrophic Factors, Encapsulated Cells, and ALS

DIEGO BRAGUGLIA AND PATRICK AEBISCHER
Gene Therapy Center and Division of Surgical Research, CHUV, Lausanne University Medical School, Lausanne, Switzerland

Amyotrophic lateral sclerosis is a fatal neurodegenerative disease leading to paralysis and death within 3 to 5 years. Although ALS was described for the first time more than 100 years ago, the cause and mechanisms of the disease remain largely unknown. Two forms of ALS, a sporadic form accounting for 90% of cases and a familial form accounting for 10% of cases, have been identified. Approximately 20% of the familial cases correlate with a mutation in the Superoxide Dismutase 1 (SOD1) gene. To date, no therapeutic approaches have been able to significantly alter the course of the disease.

Neurotrophic factors hold promise for the treatment of neurodegenerative diseases. Demonstration of the neuroprotective effects of trophic factors in the central nervous system (CNS) and in various animal models of neurodegeneration has led to the development of strategies for the treatment of ALS. The presence of the blood–brain barrier remains, however, a major concern for the delivery of neurotrophic factors in the CNS. A technique involving the intrathecal implantation of polymer-encapsulated cell-lines genetically engineered to release neurotrophic factors provides a means to continuously deliver neurotrophic factors directly within the CNS, avoiding the numerous side effects observed following their systemic administration.

Among the motor neuron diseases (MND), amyotrophic lateral sclerosis (ALS) is the most common. Other MNDs include the clinical entities of progressive bulbar palsy (PBP), progressive muscular atrophy (PMA), and primary lateral sclerosis (PLS). ALS is a prototypic age-related disease, which typically appears in the middle of adult life, afflicts the motor neuron system, and leads to paralysis and death within 3 to 5 years (Williams and Windebank, 1991). ALS is diagnosed upon clinical features and/or neuropathological findings. These latter include loss and degeneration of the large anterior horn cells of the spinal cord and lower cranial motor nuclei of the brain stem, atrophy of striated muscles as a consequence of denervation, and degeneration of the upper motoneurons (i.e., Betz cells of the motor cortex). In spite of the lack of specific markers, cytoplasmic alterations of the lower motoneurons have been studied in detail and include a marked reduction in synaptophysin expression in the anterior horns of the spinal cord in ALS patients, which parallels the severity of the neuronal loss (Kato *et al.*, 1987), and the accumulation of neurofilaments in the proximal segment of anterior horn cell axons. This increase in neurofilament accumulation is considered to be one of the early cytoplasmic changes in ALS (Carpenter, 1968; Sobue *et al.*, 1990). The presence of ubiquitin-labelled skein-like structures in anterior horn cells in both sporadic and familial ALS constitutes another morphological marker of the disease (Carpenter, 1968; Leigh *et al.*, 1988, 1991; Sobue *et al.*, 1990; Suenage *et al.*, 1993). In CNS cells, the Golgi apparatus, in addition to its function in processing of the plasma membrane, lysosomal and secreted proteins also plays a key role in fast anterograde axonal transport of macromolecules. Motoneurons of ALS patients display a fragmented Golgi apparatus. This fragmentation is present in approximately 30% of diseased motoneurons, but in less than 1% of neurons from patients affected by other neurologic diseases. Thus, it is proposed that, in addition to synaptopysin decrease and neurofilament accumulation, fragmented Golgi apparatus could represent an additional marker and an early event in ALS-mediated neuronal degeneration (Mourelatos *et al.*, 1990, 1994; Gonatas *et al.*, 1992).

I. CAUSES OF ALS AND OTHER MOTOR NEURON DISEASES

From a pathological point of view, ALS is characterized by the atrophy and death of the affected motoneurons with a marked dissolution of the nuclei and cytoplasm. At least three different aspects of ALS must be considered—the mechanism leading to cell death, the selective loss of motor neurons, and the protracted course of motoneuron degeneration. Several hypothesis have been proposed to account for these features. The first one relies on an atypical polioviral infection, potentially explaining motoneuron selectivity. This hypothesis is unlikely because no polioviruses have been detected in ALS cerebrospinal fluids and brain samples. A second hypothesis postulates a role of metallotoxins in motoneuron pathology. To date no convincing results support this hypothesis. The observation that many patients suffering from sporadic ALS possess antibodies directed against their own distal motor neuron termini, in particular voltage-dependent Ca^{2+} channels, suggested the possibility of ALS being an autoimmune disease. However, an autoimmune reaction against voltage-dependent Ca^{2+} channels could not be demonstrated as the cause of ALS. These patients also do not benefit from an immunosuppressive therapy (Brown, 1995). A more likely hypothesis focuses upon toxicity caused by excitatory neurotransmitters—glutamate levels are abnormally elevated in sporadic ALS patients, and glutamate transport is impaired in synaptosomal preparations from ALS brain (Rothstein *et al.*, 1990, 1992). In the recent past, genetic investigations have identified two proteins that are defective in some ALS patients—superoxide dismutase (SOD) and neurofilament heavy chains (Brown, 1995).

The current belief is that motoneuron degeneration and death observed in ALS result from a complex interaction between oxidative stress, excitotoxic stimulation, and altered function of mitochondria and neurofilaments.

II. THE EXCITOTOXIC HYPOTHESIS

The observation that the dorsal root ganglia (DRG) of sensory neurons do not degenerate in ALS, despite sharing the property with motoneurons of uncommonly long axons, suggests that, in addition to cellular morphology and cytoskeletal function, additional factor(s) could act as critical determinants for motoneuron susceptibility in ALS. One difference between motoneurons and dorsal root ganglia neurons is the presence in the former of excitatory synapses. This observation, taken together with the detection of increased glutamate levels in the cerebrospinal fluid and impaired glutamate transport in ALS patients, suggests an important role of neurotransmitter-mediated toxicity in the pathogenesis of ALS. In normal circumstances, glutamate is released in a Ca^{2+}-dependent process in depolarized presynaptic termini. Glutamate can activate specific motoneuron receptors [NMDA (N-methyl-D-aspartate) and non-NMDA] that normally control sodium ion entry. The glutamate is then cleared from the synaptic and extracellular environment by specific transporters, localized to surrounding astrocytes and neuronal elements. Because elevated glutamate levels can be neurotoxic, an efficient maintenance of low extracellular glutamate concentration is necessary. Several lines of evidence suggest that inefficient glutamate transport leads to an excessive accumulation of this neurotransmitter and thus to neurotoxicity. In ALS, defective glutamate transport is implicated with the observed chronic loss of motoneurons (Rothstein *et al.*, 1992; Shaw *et al.*, 1994).

Four high-affinity glutamate transporters have been identified: EAAT-1 [excitatory amino acid transporter, (GLAST)] and EAAT-2 (GLT-1) are specific for astrocytes, EAAT-3 (EAAC-1) is specific for neurons, and EAAT-4 seems to be restricted to Purkinje cells (Rothstein, 1996). Oligo-antisense knock-out of the astroglial specific GLAST and GLT-1 glutamate transporters suggests key roles for these transporters in maintaining a low extracellular glutamate concentration (Rothstein, 1996). Astroglial glutamate transport accounts for approximately 80% of all glutamate transport, with more than 50% attributable to GLT-1.

Glutamate excitotoxicity is typically characterized by cellular swelling and vacuolization of mitochondria and endoplasmic reticulum. Glutamate exerts its toxic activity in two ways—the neurotoxic activity acts via a series of intracellular cascades, activated by excess calcium influx, which in turn leads to an activation of oxidant-generating enzymatic pathways including the formation of nitric oxide, peroxynitrite, hydroxyl radicals, and superoxide anions (Choi, 1988). Because the levels of the Ca^{2+}-binding protein calbindin D28K and parvalbumin are relatively low in motoneurons (Ince *et al.*, 1993) these cells may be particularly sensitive to elevated calcium levels. The glutamate excitotoxicity can also be exerted by the so-called weak excitotoxic pathway (Albin and Greenamyre, 1992), with normal concentration of glutamate becoming neurotoxic on metabolically compromised neurons.

The increased susceptibility of motoneurons to excitotoxicity may be related to the convergence of excitatory inputs, the consequence of neurofilament accumulation, and the metabolic requirements of a large cell. In addition, because motoneurons are among the most sensitive neuronal populations to non-NMDA mediated toxicity, the extracellular glutamate level may exert its toxicity via these calcium permeable non-NMDA glutamate receptors (Rothstein *et al.*, 1993).

III. SUPEROXIDE DISMUTASE AND ALS

Superoxide dismutase (SOD) dysfunction has been proposed to be linked to motoneuron degeneration observed in ALS. Sporadic forms of ALS (SALS) account for about 90% of all reported ALS cases (Emery and Holloway, 1982; Siddique *et al.*, 1989). In this population, the mean age of disease onset is 56 years and the mean disease duration is 3 years (Tandan and Bradley, 1985). Ten percent of ALS cases are familial and can be grouped in two forms—one inherited as an autosomal dominant marker (Dominant familial ALS, DFALS) and a more rare form inherited as an autosomal recessive form (Recessive Familial ALS, RFALS). Although sporadic and familial form of ALS are clinically indistinguishable, the average onset and the duration of the disease vary considerably between these two groups. For DFALS the onset is 46 years and the penetration of the disease is age related, with about 90% of patients displaying ALS by the age of 70 (Tandan and Bradley, 1985; Siddique, 1991). Mutations in the SOD1 gene (Deng *et al.*, 1993; Rosen *et al.*, 1993), located on the chromosome 21 (21q21) (Siddique *et al.*, 1991) have been found in about 20% of DFALS families. The genetic linkage for the other 80% cases of DFALS has not yet been established. RFALS is extremely rare and was reported in areas of high consanguinity. The age of onset is 12 years with a disease duration ranging between 5 and 20 years (Ben Hamida *et al.*, 1990).

In humans, three isoforms of superoxide dismutase protein have been identified: SOD1, also known as Cu-Zn-SOD1 or cytosolic COD; SOD2 mitochondrial or Mn-SOD; and SOD3, or extracellular SOD (Ec-SOD). These enzymes are encoded by three different genes located on chromosomes 21q21, 6q27, and 4p, respectively (Goner *et al.*, 1986). At least four nonfunctional copies of SOD1 pseudogenes are present on other chromosomes (Goner *et al.*, 1986). The SOD1 isoform is ubiquitously expressed and is particularly abundant in neurons (Pardo *et al.*, 1995). Besides its main enzymatic activity, superoxide dismutase also displays marginal peroxidase activity (Hodgson and Fridovich, 1975; Symonyan and Nalbandyan, 1972). Hydrogen peroxide, the product of dismutase activity, acts as an inhibitor of SOD1 peroxidase activity (Symonyan and Nalbandyan, 1972). The decreased activity in SOD1 enzymatic function may be attributed to (1) the instability of the dimers formed between a mutated and wild-type moiety or two mutated monomers, as predicted by random dimerization studies (Deng *et al.*, 1993), (2) to the reduction of the enzymatic activity itself due to a reduction of the amount of SOD1 present in the cells (Wiedau-Pazos *et al.*, 1996), or (3) to a reduced stability with the mutated SOD1 half-life being 7.5 to 20 hrs in comparison to 30 hrs for the wild-type SOD1 (Borchelt *et al.*, 1994). How could a mutated SOD1 produce a dominant trait? Several hypothesis are possible—(1) gain of function hypothesis for the dismutase or peroxidase activity as reported for the Gly93Ala and Ala4Val mutations (Wiedau-Pazos *et al.*, 1996), (2) gain of a new cell-toxic function or (3) loss of the normal SOD1 function due to a dominant negative effect of haplo insufficiency. However, *in vitro*, mutated SOD1 monomers have not been shown to have a dominant negative effect. Experiments performed with Drosophila SOD, however, demonstrate the contrary (Borchelt *et al.*, 1995; Phillips *et al.*, 1995).

A. A Mice Transgenic Model for Familial ALS (FALS)

Transgenic mice overexpressing either the wild-type or mutated forms of human SOD1 represent the most relevant animal for FALS (Gurney *et al.*, 1994; Ripps *et al.*, 1995; Wong *et al.*, 1995). These mice develop ALS-related clinical symptoms, including weakness and paralysis, and die at variable ages depending on the transgenic mouse line. For one mutation, Gly93Ala, the onset and age of death seems to be dependent on the number of copies integrated in the mouse genome (Dal Canto and Gurney, 1995). Mice with two copies develop ALS symptoms, later than mice harboring 18 to 36 copies, whereas mice overexpressing human wild-type SOD1 do not show any clinical symptoms for at least 2 years. The mutation Ala4Val was found in individuals with the most rapidly progressing DFALS. Interestingly, transgenic mice overexpressing this mutated protein do not develop the disease earlier than mice harboring other SOD1 mutations. In Ala4Val mice, the onset of the disease appears at about 700 days (Siddique *et al.*, 1996), compared to 140 days in the Gly93Ala transgenic mice (Gurney *et al.*, 1994). Pathological and histological studies confirm that motoneuron degeneration is the cause of the paralysis observed in SOD1 transgenic mice. Animals harboring a large number of mutated SOD1 gene-copies display vacuoles arising from rough endoplasmic reticulum and from mitochondria already in the early stage of the disease. In addition, the Golgi apparatus is fragmented. At a later stage, fibrillary deposits are observed in axons of degenerating neurons. Mice with lower levels of mutated SOD1 show fibrillary deposits in motoneurons resembling those observed in human ALS (Dal Canto and Gurney, 1995).

Observations made in transgenic mice overexpressing SOD1 human proteins would hardly be compatible with the loss of function hypothesis. On the other hand, a gain of a new toxic function is postulated. The hypothesized new toxic functions are the following: (1) increased production of free radicals, (2) new enzymatic function (i.e., the formation of nitronium-ion), or (3) a toxic-like function specifically deleterious to motoneurons. How these mutations can influence the age of disease onset but not its duration need further investigation.

IV. ALS and Neurofilaments

One of the common hallmarks in several motoneuron degenerative diseases, including sporadic and familial ALS, infantile spinal muscular atrophy, and hereditary sensory motor neuropathy, is the aberrant accumulation of neurofilaments (NF) in the affected motoneurons. The hypothesis of potential involvement of NF in these diseases received credit and experimental support from transgenic mice overexpressing mice NF-L or human NF-H (Cote *et al.*, 1993; Xu *et al.*, 1993). These animals develop loss of kinetic activity, muscular atrophy, and paralysis. Motoneurons of the anterior horn revealed enlarged perikarya with depleted rough endoplasmic reticulum and a mispositioned nucleus. The perikarya was filled with closely packed neurofilament bundles. These alterations are similar to those observed in sporadic and familial ALS (Hirano *et al.*, 1984a, 1984b; Hirano, 1988). In addition, a reduced level of NF undergoing axonal transport was also detected in hNF-H overexpressing mice (Collard *et al.*, 1995). However, the NF-L and NF-H mice models are not a very satisfying model for ALS because paralysis arising from neuronal failure and derived muscular atrophy is not associated with a significant motoneuron death even at the terminal stage of the disease. By expressing the mutated form of NF in mice, a more suggestive model for ALS-like diseases was created. Mutations affecting the NF-L subunit-conserved helical region were inserted in mice. These animals develop early abnormal gait and weakness in both upper and lower limbs. The abnormalities progress in severity and lead in most cases to death. Transgenic animals display a loss of motoneurons in the lumbar and cervical spinal cord. Distorted nuclei and microglial nodules, indicative of phagocytosis of degenerating cell bodies by glial cells, are often seen. Similarly to ALS, large neurofilaments-rich axons are lost, while smaller axons with less neurofilaments are spared (Kawamura *et al.*, 1981). These experiments suggest that mutated neurofilaments can be a primary cause in the selectivity of motoneuronal death. However, in spite of the identification of small deletions in repeated NF-H subunits in 5 of several hundred SALS patients, NF-mutations have not yet been identified in FALS individuals.

V. RATIONAL FOR USING NEUROTROPHIC FACTORS IN ALS

Several lines of evidence, including transgenic mice and *in vitro* and *in vivo* experiments, suggest that neurotrophic factors (NTF) play a critical role in supporting survival and differentiation of various neuronal population. Motoneurons respond to a variety of NTFs, namely CNTF, LIF, IGF-1, BDNF, NT3, NT4/5, GDNF, and Neurturin. Transgenic animals with targeted disruption of NTF genes and their receptors (Ernfors *et al.*, 1994; Frim *et al.*, 1994; Jones *et al.*, 1994; Klein *et al.*, 1993; Masu *et al.*, 1993; Smeyne *et al.*, 1994; Tojo *et al.*, 1995) have demonstrated and emphasized the NTF's physiological role in the development and survival of motoneurons. *In vitro,* survival of rat embryonic motoneurons is increased by a variety of NTFs, including CNTF, BDNF, NT3, NT4/5, and GDNF. In addition these factors have further been reported *in vivo* to reduce axotomy-induced motoneuron death (Magal *et al.*, 1993;

Sendtner *et al.*, 1990; Wong *et al.*, 1993; Zurn *et al.*, 1994; Zurn and Werren, 1994). Animal models for neurodegenerative diseases also provided additional evidence on the central role of NTFs in the biology of motoneurons (Ikeda *et al.*, 1995a, 1995b; Mitsumoto *et al.*, 1994; Sagot *et al.*, 1995; Sendtner *et al.*, 1992).

The NTF target-derived hypothesis states that developing neurons compete with each other for a limited supply of NTF provided by the innervation target. Only the successfully competing neurons will survive. The pleiotropic roles of NTFs can be summarized as follows: (1) NTF protects against injury and neuronal dysfunction and death, (2) NTFs action is not restricted to developing postmitotic neurons, (3) an individual neuronal population can respond to many different NTFs and a single NTF can act on several different neuronal population, and (4) NTF synthesis is temporally and spatially highly ordered. Up to now, however, there is no single example of a neurological disease caused by a deficiency of a NTF.

A. Neurotrophic Factors Acting on Motoneurons

Neurotropic factors affecting motoneurons can be divided into four families—the neurotrophins (NGF, BDNF, NT3, NT4/5), the neurokines (CNTF, CT-1, IL-6, LIF), the TGFβ-related family (GDNF, Neurturin), and the insulin-like growth factor family (IGF-1). Within the neurotrophin family all members share a significant structural homology and have structurally related receptors, the trk receptors. The binding of the different neurotrophins to their relative high-affinity trk receptors is modulated by a low-affinity receptor named $p75^{NTR}$. Three trk receptors have been characterized so far—trkA, trkB and trkC, which specifically interact with NGF, BDNF and NT4/5, and NT3, respectively (Yuen and Mobley, 1996). NGF was the first neurotrophic factor identified (Levi-Montalcini, 1987); a wide variety of neurons in the CNS and PNS respond to NGF. Antibody-mediated NGF-depletion or NGF or trkA gene disruptions are paralleled by severe loss of DRG sensory and sympathetic neurons (Aloe *et al.*, 1981; Crowley *et al.*, 1994; Smeyne *et al.*, 1994; Johnson *et al.*, 1980); however, NGF has no effect on motoneurons.

Brain-derived neurotrophic factor (BDNF), with its high-affinity receptor trkB, is widely expressed in both the developing and mature PNS and CNS (Lindsay, 1993). The importance of BDNF and trkB in the developing nervous system is highlighted by the heavy loss of cranial and DRG sensory neurons in mice lacking BDNF or trkB (Ernfors *et al.*, 1994; Jones *et al.*, 1994; Klein *et al.*, 1993). BDNF is a particularly attractive factor for ALS because it has been shown to act on motoneurons (Henderson *et al.*, 1993; Koliatsos *et al.*,

1993). BDNF is capable of rescuing motoneurons in various models, like the axotomy of facial nerve (Koliatsos *et al.*, 1993; Sendtner *et al.*, 1992) and of the sciatic nerve (Li *et al.*, 1994; Yan *et al.*, 1992), to prevent naturally occurring cell death of chick motoneurons *in vivo* (Oppenheim *et al.*, 1992). In adult motoneurons, BDNF attenuates the axotomy-induced reduction in ChAT activity (Yan *et al.*, 1994). In addition, BDNF is able to slow or even to stop in combination with CNTF (Mitsumoto *et al.*, 1994) the rate of decline in grip strength of the wobbler mouse, a model for motoneuron degeneration (Ikeda *et al.*, 1995a; Mitsumoto and Bradley, 1982; Mitsumoto *et al.*, 1994).

Neurotrophin 3 and its high-affinity trkC receptor are widely expressed both in the PNS and the CNS (Hohn *et al.*, 1990; Koliatsos *et al.*, 1993; Lamballe *et al.*, 1991). Due to the widespread expression of the trkC receptor, NT3 acts on largely different neuronal populations. NT3 has been shown to support the survival of motoneurons *in vitro* and during embryonic development. This suggests a potential utility of this neurotrophin in motoneuron degeneration as indicated by a 50% increase in the lifespan of pmn-pmn mice (Haase *et al.*, 1997), a mouse mutant characterized by a progressive motor neuronopathy. These mice develop weakness in the hindlimbs during the third week of life and die at 6 weeks of age (Schmalbruch *et al.*, 1991).

Of the numerous cytokines, CNTF, CT-1, LIF, and IL-6 have been described to act on motoneurons. CNTF is a 18 kD polypeptide (Sendtner *et al.*, 1994) that acts via a three component receptor composed of the CNTFR-α receptor, the transmembrane gp 130 and the leukemia inhibitory receptor β, LIFβ (Stahl and Yancopoulos, 1994). CNTF was initially described for its trophic action on chick embryonic parasympathetic ciliary ganglia. Since then, the survival of a variety of neuronal cell types, including motoneurons and sensory neurons (Sendtner *et al.*, 1994) was shown to be supported by CNTF (Arakawa *et al.*, 1990; Ip *et al.*, 1992; Magal *et al.*, 1993). *In vitro,* CNTF increases the survival, ChAT expression, and neurite outgrowth of purified embryonic motoneuron cultures (Wong *et al.*, 1993), whereas *in vivo* it rescues neonatal facial motoneurons from axotomy-induced cell death and it increases motoneuron survival and extends life expectancy in pmn-pmn mice (Sagot *et al.*, 1995; Sahenk *et al.*, 1994; Sendtner *et al.*, 1990; Tan *et al.*, 1996).

Cardiotrophin-1 (CT-1) was isolated because of its ability to induce cardiac myocyte hypertrophy (Pennica *et al.*, 1995). It is expressed at high levels in embryonic limb bud and is secreted by differentiated myotubes. CT-1 signals through heterodimer complexes formed by the LIFβ-gp130 proteins but does not bind to the CNTFR-α, suggesting the presence of a novel cytokine receptor α subunit, a putative CT-1Rα (Pennica *et al.*, 1996b). CT-1 expression patterns (mRNA detected in rat E7 embryos and in adult heart, skeletal muscle, liver, lung, kidney and weaker in testis and brain) and its range of activities in hematopoietic, neuronal, and developmental systems suggest that CT-1 may play a widespread important role in organ systems other than cardiac development and hypertrophy (Pennica *et al.*, 1996c). Confirming this hypothesis, CT-1 supports long-term *in vitro* survival of spinal motoneurons and was shown to protect *in vivo* neonatal lumbar motoneurons following a sciatic nerve axotomy. Inhibition of CT-1 action on motoneurons by PIPLC suggest that CT-1 may act through a GPI-linked component (Pennica *et al.*, 1996a).

B. TGFβ-Related Family

Glial cell line-derived neurotrophic factor (GDNF) has been shown to be a highly potent neurotrophic factor enhancing survival of various neuronal cells including motoneurons and substantia nigra dopaminergic neurons (Henderson *et al.*, 1994; Zurn *et al.*, 1994). GDNF-mediated signaling acts via a bipartite component receptor composed by the ret protein, the product of the c-ret protooncogene, and the GDNF-α receptor (GDNFR-α), a glycosyl-phospatydil-linked (GPI) protein (Durbec *et al.*, 1996; Treanor *et al.*, 1996; Trupp *et al.*, 1996). GDNF binds specifically to GDNFR-α and mediates the activation of the ret protein tyrosine kinase (PTK), inducing its autophosphorylation. The observation that the dose required to trigger GDNF-induced biological effects varies widely between different cell types suggests the existence of both high- and low-affinity GDNF receptors (Jing *et al.*, 1996). GDNFR-α does not need to be anchored to the cell membrane to induce ret PTK autophosphorylation (Jing *et al.*, 1996). GDNF-deficient mice completely lack the enteric nervous system (ENS), urethras, and kidneys, suggesting a key role of GDNF in organogenesis (Moore *et al.*, 1996; Pichel *et al.*, 1996; Sanchez *et al.*, 1996). In the pmn-pmn model, GDNF reduced the loss of facial motoneurons by 50% but was unable to influence the animals' life span nor their axonal degeneration (Sagot *et al.*, 1996).

Recently, the purification and characterization of neurturin, a protein closely related to GDNF, was reported (Kotzbauer *et al.*, 1996). Neurturin supports the survival of sympathetic neurons in culture. Like GDNF, neurturin can activate the MAP kinase signaling pathway in cultured sympathetic, DRG and nodose RG neurons. Neurturin expression varies considerably between neonatal and adult tissues, suggesting a role in the regulation of the development and maintenance of the central nervous system (Kotzbauer *et al.*, 1996). It also supports motoneuron survival *in vitro* and following sciatic nerve axotomy (A. Kato, personal communication).

VI. ENCAPSULATION TECHNOLOGY

Neuronal growth factors hold promise for the treatment of neurodegenerative diseases. The presence of the blood–brain barrier, however, represents a major hurdle for the delivery of neurotrophic factors to the CNS. A technique involving the intrathecal implantation of polymer encapsulated cell-lines genetically engineered to release neurotrophic factors provides a means to continuously deliver neurotrophic factors directly within the CNS.

In addition to direct exposure CNS neurons to the delivered neurotrophic factor, the use of encapsulated engineered cell-lines offers several other relevant advantages. Genetically engineered cell-lines, obtained by classical methods (Sambrook *et al.*, 1989), can be cloned by limited dilution. Clonal cell-lines can be characterized for parameters including neurotrophic factor secretion and stability, and *in vitro* and *in vivo* survival. Clonal cell-lines are amplified to generate a master cell bank (MCB), which is in turn amplified for the constitution of the working cell banks (WCB). MCB and WCB can be checked for the absence of adventitious agents (including bacteria, yeasts, mycoplasma, viruses, retroviruses). The chosen cell-line has to be exempt from transforming (tumorigenicity) activities and toxicity, as assessed on small (mice and rats) and large (monkeys and sheep) animals. The availability of both MCB and WCB will guarantee homogeneity and reproducibility in the device's preparation, ensuring that all patients will be implanted with an identical clonal cell-line. In addition, constant delivery will circumvent the rapid degradation of neurotrophic factors observed in pump-delivery administration (Penn *et al.*, 1997).

In vitro and *in vivo* experiments strongly suggest that administration of combinations of factors have additive and, in some combinations, synergistic effects on the survival of motoneurons. Second generations of genetically engineered cell-lines have been generated allowing the direct and constant delivery within the CNS of combinations of neurotrophic factors. These lines are currently being tested on *in vivo* models of motoneuron degeneration. A third generation of engineered cell-lines is under investigation and would provide *in vivo* modulation of gene expression. This system is based on transcriptional regulation driven by small, stable molecules (tetracycline and its derivatives), which diffuse across the blood–brain barrier (Gossen *et al.*, 1995). This system has already been reported in the generation of conditionally immortalized cell-lines (Ewald *et al.*, 1996) and in the control of the level of expression of Epo (Bohl *et al.*, 1997).

The encapsulation technique allows the transplantation of xenogeneic cells in the absence of pharmacological host immunosuppression (Fig. 1A and B). The immunoisolation is obtained by surrounding the cells by a selective semipermeable membrane barrier with a controlled pore size, allowing the inward diffusion of nutrients and the outward diffusion of cell-secreted bioactive factor(s). The immunoisolation approach was originally described in the mid-1970s for the transplantation of the islets of Langerhans (Chick *et al.*, 1975, 1977). Experiments examining its application to the nervous system

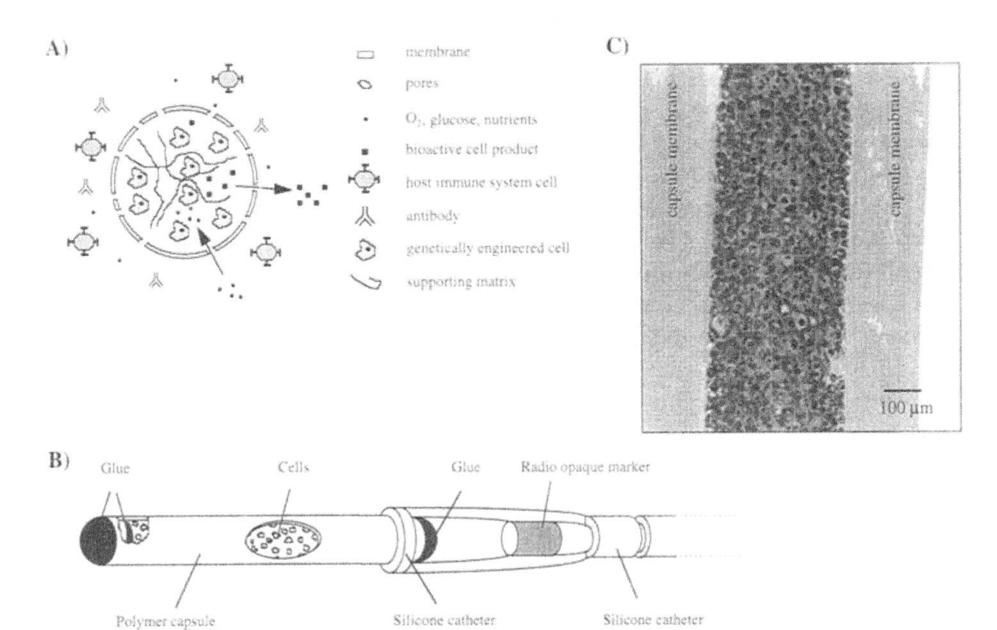

FIGURE 1 Principle of the immunoisolation by polymer encapsulation. (A) Genetically modified cells are embedded into a supporting matrix, usually collagen. The pores present in the membrane allow the diffusion of nutrients and cell-secreted bioactive products while protecting the xenogeneic cells from the host immune system. (B) Schematic drawing (not to scale) of the device employed in the human phase I ALS clinical trial (Aebischer *et al.*, 1996). The active part of the device containing the cells is 5 cm in length and 500 μm in diameter. Usually, between $7*10^5$ and 10^6 BHK cells engineered to secrete hCNTF are encapsulated. (C) Histological appearance of an explanted capsule from a patient 60 weeks postimplantation. Longitudinal section of the device is processed with hematoxylin and eosin to highlight viable BHK cells.

were initiated in the late 1980s. Xenogeneic cells can be encapsulated in polymer-based hollow fiber systems (Aebischer and Lysaght, 1995). The semi-permeable membrane serves as a critical immunoisolation barrier. Among the wide variety of membranes tested to date, two different types are particularly attractive. The first type of membrane is an anisotropic form prepared from thermoplastics in the form of films or hollow fibers. The shape is either a thin sandwich of discs sealed at the edges, or a cylinder sealed at both ends. The thickness of these macrofibers is usually 300 to 500 μm in diameter to avoid limited nutrient availability to cells located in the middle of the device. A second group of membranes includes films formed between two oppositely charged polyelectrolytes, commonly alginate and polylysine (Lim and Sun, 1980). Both type of membranes need to be semipermeable and biocompatible.

These parameters can be controlled by the fine-tuning of chemical composition and surface morphology to minimize fibrotic or other inflammatory responses, thus ensuring biocompatibility and by techniques of formulation and fabrication, control of "cut-off" parameters for diffusion of molecules of desired weight and size. Following the synthesis of the membranes, pores show a gaussian distribution in size. Formation of pores as small as 35 kD for asymmetric anisotropic membranes (i.e., PES Akso #11) and 70 kD for symmetric PAN-PVC T4 hollow fiber are obtained, allowing the diffusion of dissolved oxygen, nutrients, and bioactive cell secretions. Cells from the host immune system, immunoglobulins, and complement factors are excluded from the graft. In addition to physical properties inherent to membrane cut-off, composition, and thickness, an important parameter is the semisolid gel or matrix in which the cells are embedded (Zielinski and Aebischer, 1994). The encapsulation of dissociated cells in favorable biochemical and mechanical environment dramatically influences cell viability, such as by inhibiting cell clustering or gravimetric settling. The identification of the best matrix, membrane composition and membrane filling parameters depends widely on cell phenotype and growth characteristics and has to be adapted empirically for each cell line.

Transplantation of encapsulated cells provides a powerful approach for the continuous and local delivery of recombinant proteins. In addition, a shut-off mechanism is easily obtained by the retrieval of the device at the first appearance of adverse side effects. Although the mechanical characteristics of the implanted fibers make the possible inadvertent release of encapsulated cells unlikely, safety mechanisms are incorporated to ensure selected rejection of the implanted cells in the unlikely event of capsule breakage. The high immunogenicity of xenogeneic cells becomes an advantage; unencapsulated cross-species cell-lines are efficiently rejected even in immunoprivileged sites (Aebischer and Lysaght, 1995). As a further precaution, a suicide gene such as the herpes simplex virus thymidine kinase (HSV-TK) gene is incorporated into the engineered cell-line. Efficient killing of escaped cells can be achieved by the oral intake of ganciclovir, a chemical that is catabolised and becomes toxic only for cells expressing TK.

The encapsulated cell technology confers significant advantages in the treatment of neurological diseases. The presence of the highly selective blood–brain barrier (BBB) makes the delivery of systemically administered trophic factors to the CNS a challenging task. Intravenous administration of neurotrophic factors, even at low amounts, leads often to undersirable side effects due to the widespread distribution of their receptors outside the CNS. In addition, neurotrophic factors are quite unstable and their bioactivity decreases rapidly (Tan *et al.*, 1996). Due to the impermeability of the blood–brain barrier, the only mechanism for a systemically delivered factor to reach the CNS is by retrograde transport from neuron fibers.

A continuous delivery of neurotrophic factors by means of the immunoisolation technology was applied to various models of motoneuron degeneration, including axotomy-induced cell death and pmn-pmn mice. In addition, neurotrophic factors were successfully delivered by the same technology in the unilateral medial forebrain bundle (MFB) axotomy (Tseng *et al.*, 1997), a model for Parkinson's disease; in a quinolinic acid (QA) rodent and primate models for Huntington's disease (Emerich *et al.*, 1996, 1997); and to the lateral ventricle of fimbria–fornix-lesioned rats, a correlational model for Alzheimer's disease (Hoffman *et al.*, 1993, 1990).

A. Clinical Trials Using Encapsulated Genetically Modified Cells

After establishing the safety for device implantation and retrieval in various animal models, including sheep and monkeys, as well as having received the approval of the Ethical Committee of the Lausanne University Medical School and the Swiss Committee for Biological Safety, a clinical phase I trial for CNTF delivery was initiated in 12 ALS patients. It has demonstrated the feasibility and the safety of the encapsulated cell technology for the intrathecal constant delivery of hCNTF (Aebischer *et al.*, 1996). Patients were implanted with polymer capsules (Fig. 1B) containing genetically engineered baby hamster kidney (BHK) cells to produce hCNTF (between 0.5 and 1 μg/day). Nanogram levels of CNTF were detected in patients' cerebrospinal fluid (CSF) up to 60 weeks posttransplantation (Fig. 1C), whereas no CNTF was detectable prior to implantation. The intrathecal delivery of CNTF was not associated with any of the limiting adverse effects observed with systemic (intravenous) delivery of CNTF reported in clinical phase II trials undertaken by two private biotechnology companies (Lotz *et al.*, 1996; Penn *et al.*, 1997). All the implanted devices were intact and secreted CNTF at the time of explant. No evidence of tissue or cell adherence to the external surface was observed, confirming the excellent biocompatibility of the implant (Fig. 1C). All the patients have been evaluated by ALS clinical rating scores including Norris, TQNE, and FVC scores. Scores slopes indicate that the disease continues to progress. However, the small number of patients and the short observation periods (both pre- and posttransplantation) do not allow assessment of a potential slowing of the degenerative process.

Diffusion studies indicate that the bioactive secreted factor can diffuse within the CSF. Simultaneous CSF lumbar and cervical levels of CNTF were measured in one patient. The cervical concentration was 14.6% of the lumbar level of CNTF, demonstrating that all the spinal axis is exposed to the trophic factor after lumbar placement of the implant (Aebischer *et al.*, 1996). In addition, diffusion of tagged CNTF (CNTF-flag) released from capsules implanted in the striatum has been monitored in the rat (Fig. 2). CNTF-flag is found in the rat CSF 7, 14, and 30 days postimplantation, suggesting that CNTF is broadly distributed in the rat CNS. Technical limitations due to antibody sensitivity in the detection of CNTF-flag on brain histological sections did not allow for the determination of the diffusion kinetics.

In vitro and *in vivo* studies suggest that combinations of neurotrophic factors may have additive or in some cases synergistic effects (Zurn *et al.*, 1996). The possibility of mixing two different clonal cell lines, each secreting a factor, may encounter difficulties due to differing growth rates between the two clonal populations. The release of two different factors can be achieved by different methods. The most straightforward possibility is the implantation of two devices, one for each factor. A second approach deals with the generation of genetically modified cell-lines able to produce combination of factors. This second generation of secreting cell-lines can be generated either by a cotransfection of vectors expressing the desired combination of neurotrophic factors, or by transfecting a single vector able to express two factors (two monocistrons or one polycistronic transcription unit). BHK and C_2C_{12} cells were engineered to secrete combinations of factors by transfection with combinations of expression vectors. The effect of the combined factor administration was tested on purified spinal cord cultures (Zurn *et al.*, 1996). In all the combinations tested to date (CNTF-GDNF, CNTF-NT4/5, CT1-NT4/5) (Fig. 3 and data not shown), a marked synergistic effect on ChAT activity in purified spinal cord cultures was demonstrated. These results suggest that an enhanced effect on motoneuron survival and rescue after injury could be achieved by the application of combination of factors.

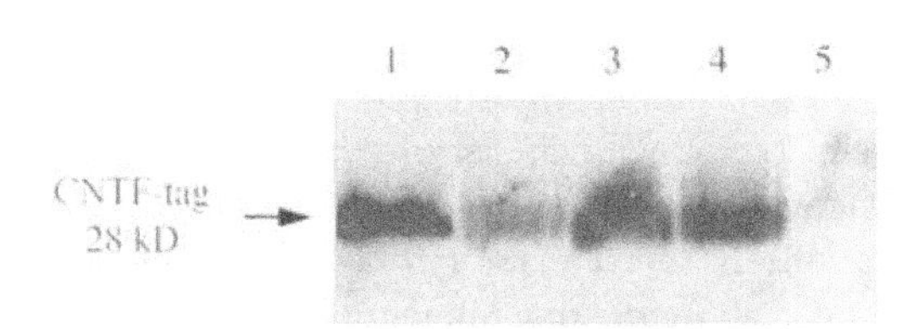

FIGURE 2 Diffusion of CNTF. Rats were implanted intrastriatally with BHK-secreting CNTF-flag (1.35 μg CNTF-flag/10^6 cells/24 hrs). The capsules contained at the loading $2*10^4$ cells. The cerebrospinal fluid (CSF) of rats sacrificed after 7, 14, and 30 day postimplant was collected and analyzed with M2 monoclonal antibody (Eastman Kodak Inc., New Haven). A Western-blot is shown. Lane 1: conditioned media from BHK transfected with CNTF-flag expression vector. Lane 2 to 4: CSF from rats implanted intrastriatally with BHK-secreting CNTF-flag for 7, 14, and 30 days, respectively. Lane 5: CSF from an animal implanted with native BHK cells. Equal amounts of proteins are loaded in each lane.

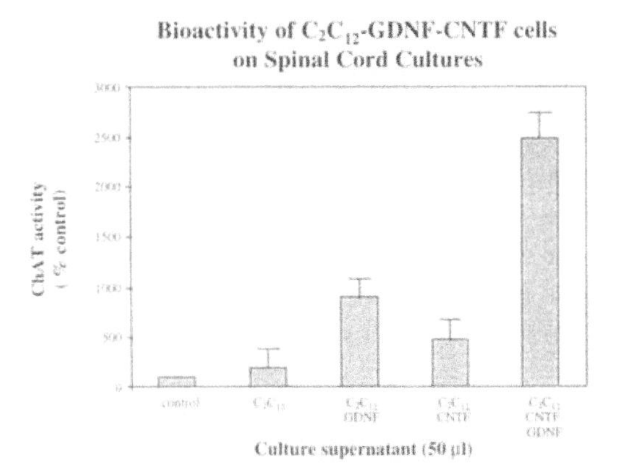

FIGURE 3 Synergistic effects of combined secretion of neurotrophic factors. ChAT activity was measured on purified motoneurons exposed to conditioned media as described (Zurn *et al.*, 1996). Bars indicate the relative increase in ChAT activity (in %) compared to control (100%).

VII. PERSPECTIVES

The encapsulation of xenogeneic modified cell-lines has the potential of overcoming many of the problems associated with systemic delivery and local administration of recombinant proteins. This technique minimizes the need for repeated injections and the effects of poor proteins stability. Although encapsulated cells were placed in the intrathecal region (Aebischer *et al.*, 1996), their clinical applications may not be restricted to the CNS. Encapsulated cell therapy may constitute a relevant advantage in diseases (i.e., diabetes, anemia, obesity) requiring a constant continuous delivery of pharmacological levels of bioactive recombinant proteins. Future implants may allow the release of recombinant proteins (i.e., insulin and erytropoietin) upon response to physiological stimuli, such as glucose concentration for insulin and hypoxia for erytropoietin. In addition, the availability of a shut-off mechanism represents a relevant advantage, allowing the interruption of the recombinant factor release by retrieval of the devices at the first appearance of side effects.

Further generations of genetically engineered cell-lines may allow the simultaneous administration of combinations of growth factors. In addition, the artificial *in vivo* modulation of transgene expression, based on the diffusion of small molecules, may allow the identification of ideal therapeutic windows for each factor.

REFERENCES

Aebischer, P., Schleup, M., Deglon, N., Joseph, J. M., Hirt, L., Heyd, B., Goddard, M., Hammang, J. P., Zurn, A. D., Kato, A. C., Regli, F., and Baetge, E. E. (1996). Intrathecal delivery of CNTF using encapsulated genetically modified xenogeneic cells in amyotrophic lateral sclerosis patients. *Nature Medicine* 2:696–699.

Aebischer, P., and Lysaght, M. J. (1995). Immunoisolation and cellular xenotransplantation. *Xeno* 3:43–48.

Albin, R. L., and Greenamyre, J. T. (1992). Alternative excitotoxic hypotheses. *Neurology* 42:733–738.

Aloe, L., Cozzari, C., Calissano, P., and Levi-Montalcini, R. (1981). Somatic and behavioral postnatal effects of fetal injections of nerve growth factor antibodies in the rat. *Nature* 291:413–415.

Arakawa, Y., Sendtner, M., and Thoenen, H. (1990). Survival effect of ciliary neurotrophic factor (CNTF) on chick embryonic motoneurons in culture: comparison with other neurotrophic factors and cytokines. *Journal of Neuroscience* 10:3507–3515.

Ben Hamida, M., Hentati, F., and Ben Hamida, C. (1990). Hereditary motor system diseases (chronic juvenile amyotrophic lateral sclerosis). Conditions combining a bilateral pyramidal syndrome with limb and bulbar amyotrophy. *Brain* 113:347–363.

Bohl, D., Naffakh, N., and Heard, J. M. (1997). Long-term control of erytropoietin secretion by doxycycline in mice transplanted with engineered primary myoblasts. *Nature Medicine* 3:299–305.

Borchelt, D. R., Guarnieri, M., Wong, P. C., Lee, M. K., Slunt, H. S., Xu, Z. S., Sisodia, S. S., Price, D. L., and Cleveland, D. W. (1995). Superoxide dismutase 1 subunits with mutations linked to familial amyotrophic lateral sclerosis do not affect wild-type subunit function. *Journal of Biological Chemistry* 270:3234–3238.

Borchelt, D. R., Lee, M. K., Slunt, H. S., Guarnieri, M., Xu, Z. S., Wong, P. C., Brown, R. H., Jr., Price, D. L., Sisodia, S. S., and Cleveland, D. W. (1994). Superoxide dismutase 1 with mutations linked to familial amyotrophic lateral sclerosis possesses significant activity. *Proceedings of the National Academy of Sciences of the United States of America* 91:8292–8296.

Brown, R. H., Jr. (1995). Amyotrophic lateral sclerosis: recent insights from genetics and transgenic mice. *Cell* 80:687–692.

Carpenter, S. (1968). Proximal axonal enlargement in motor neuron disease. *Neurology* 18:841–851.

Chick, W. L., Like, A. A., and Lauris, V. (1975). Beta cell culture on synthetic capillaries: an artificial endocrine pancreas. *Science* 187:847–849.

Chick, W. L., Perna, J. J., Lauris, V., Low, D., Galletti, P. M., Panol, G., Whittemore, A. D., Like, A. A., Colton, C. K., and Lysaght, M. J. (1977). Artificial pancreas using living beta cells:. effects of glucose homeostasis in diabetic rats. *Science* 197:780–782.

Choi, D. W. (1988). Glutamate neurotoxicity and diseases of the nervous system. *Neuron* 1:623–634.

Collard, J. F., Cote, F., and Julien, J. P. (1995). Defective axonal transport in a transgenic mouse model of amyotrophic lateral sclerosis. *Nature* 375:61–64.

Cote, F., Collard, J. F., and Julien, J. P. (1993). Progressive neuronopathy in transgenic mice expressing the human neurofilament heavy gene: A mouse model of amyotrophic lateral sclerosis. *Cell* 73:35–46.

Crowley, C., Spencer, S. D., Nishimura, M. C., Chen, K. S., Pitts-Meek, S., Armanini, M. P., Ling, L. H., MacMahon, S. B., Shelton, D. L., Levinson, A. D., *et al.* (1994). Mice lacking nerve growth factor display perinatal loss of sensory and sympathetic neurons yet develop basal forebrain cholinergic neurons. *Cell* 76:1001–1011.

Dal Canto, M. C., and Gurney, M. E. (1995). Neuropathological changes in two lines of mice carrying a transgene for mutant human Cu,Zn SOD, and in mice overexpressing wild type human SOD: A model of familial amyotrophic lateral sclerosis (FALS). *Brain Research* 676:25–40.

Deng, H. X., Hentati, A., Tainer, J. A., Iqbal, Z., Cayabyab, A., Hung, W. Y., Getzoff, E. D., Hu, P., Herzfeldt, B., Roos, R. P., *et al.* (1993). Amyotrophic lateral sclerosis and structural defects in Cu,Zn superoxide dismutase. *Science* 261:1047–1051.

Durbec, P., Marcos-Gutierrez, C. V., Kilkenny, C., Grigoriou, M., Wartiowaara, K., Suvanto, P., Smith, D., Ponder, B., Costantini, F., Saarma, M., *et al.* (1996). GDNF signalling through the Ret receptor tyrosine kinase. *Nature* 381:789–793.

Emerich, D. F., Linder, M. D., Winn, S. R., Chen, E.-Y., Frydell, B. R., and Kordower, J. H. (1996). Implants of encapsulated human CNTF-producing fibroblasts prevent behavioral deficits and striatal degeneration in a rodent model for Huntington's disease. *Journal of Neuroscience* 16:5168–5181.

Emerich, D. F., Winn, S. R., Hantraye, P. M., Peschanski, M., Chen, E.-Y., Chu, Y., McDermott, P., Baetge, E. E., and Kordower, J. H. (1996). Protective effect of encapsulated cell producing neurotrophic factor CNTF in a monkey model of Huntington's disease. *Nature* 386:395–399.

Emery, A. E. H., and Holloway, S., (1982). Familial motor neuron disease. In: Rowland, L. P., ed., *Human motor neuron disease.* New York: Raven Press, pp. 139–147.

Ernfors, P., Lee, K. F., and Jaenisch, R. (1994). Mice lacking brain-derived neurotrophic factor develop with sensory deficits. *Nature* 368:147–150.

Ernfors, P., Lee, K. F., Kucera, J., and Jaenisch, R. (1994). Lack of neurotrophin-3 leads to deficiencies in the peripheral nervous system and loss of limb proprioceptive afferents. *Cell* 77:503–512.

Ewald, E., Li, M., Efrat, S., Auer, G., Wall, R. J., Furth, P. A., and Hennighausen, L. (1996). Time-sensitive reversal of hyperplasia in transgenic mice expressing SV40 T antigen. *Science* 273:1384–1386.

Frim, D. M., Uhler, T. A., Galpern, W. R., Beal, M. F., Breakefield, X. O., and Isacson, O. (1994). Implanted fibroblasts genetically engineered to produce brain-derived neurotrophic factor prevent 1-methyl-4-phenylpyridinium toxicity to dopaminergic neurons in the rat. *Proceedings of the National Academy of Sciences of the United States of America* 91:5104–5108.

Gonatas, N. K., Stieber, A., Mourelatos, Z., Chen, Y., Gonatas, J. O., Appel, S. H., Hays, A. P., Hickey, W. F., and Hauw, J. J. (1992). Fragmentation of the Golgi apparatus of motoneurons in amyotrophic lateral sclerosis. *American Journal of Pathology* 140:731–737.

Goner, Y., Geiman-Hurvitz, J., Dafri, N., *et al.* (1986). The human Cu/Zn superoxide dismutase gene family: Architecture and expression of the chromosome 21-encoded functional gene and its processed pseudogenes. In: Rotilis, G., ed., *Superoxide and superoxide dismutase in chemistry, biology and medicine.* Amsterdam: Elsevier Science Publisher, Biochemical Division, pp. 247–255.

Gossen, M., Freundlieb, S., Bender, G., Müller, G., Hillen, W., and Bujard, H. (1995). Transcriptional activation by tetracyclines in mammalian cells. *Science* 268:1766–1769.

Gurney, M. E., Pu, H., Chiu, A. Y., Dal Canto, M. C., Polchow, C. Y., Alexander, D. D., Caliendo, J., Hentati, A., Kwon, Y. W., Deng, H. X., *et al.* (1994). Motor neuron degeneration in mice that express a human Cu,Zn superoxide dismutase. *Science* 264:1772–1775.

Haase, G., Kennel, P., Pettmann, B., Vigne, E., Akli, S., Revah, F., Schmalbruch, H., and Kahn, A. (1997). Adenovirus-mediated transfer of neurotrophin-3 gene into muscle reduces progressive motor neuropathy in mice. *Nature Medicine* 3:429–436.

Henderson, C. E., Camu, W., Mettling, C., Gouin, A., Poulsen, K., Karihaloo, M., Rullamas, J., Evans, T., McMahon, S. B., Armanini, M. P., *et al.* (1993). Neurotrophins promote motor neuron survival and are present in embryonic limb bud. *Nature* 363:266–270.

Henderson, C. E., Phillips, H. S., Pollock, R. A., Davies, A. M., Lemeulle, C., Armanini, M., Simmons, L., Moffet, B., Vandlen, R. A., Simpson, L. C., *et al.* (1994). GDNF: A potent survival factor for motoneurons present in peripheral nerve and muscles. *Science* 266:1062–1064.

Hirano, A. (1988). *Color atlas of pathology of the nervous system,* 2nd ed. New York: Igakushoin, p. 99.

Hirano, A., Donnenfeld, H., Sasaki, S., and Nakano, I. (1984a). Fine structural observations of neurofilamentous changes in amyotrophic lateral sclerosis. *Journal of Neuropathology & Experimental Neurology* 43:461–470.

Hirano, A., Nakano, I., Kurland, L. T., Mulder, D. W., Holley, P. W., and Saccomanno, G. (1984b). Fine structural study of neurofibrillary changes in a family with amyotrophic lateral sclerosis. *Journal of Neuropathology & Experimental Neurology* 43:471–480.

Hodgson, E. K., and Fridovich, L. (1975). The interaction of bovine erythrocyte superoxide dismutase with hydrogen peroxide: Inactivation of the enzyme. *Biochemistry* 14:5294–5299.

Hoffman, D., Breakefield, X. O., Short, M. P., and Aebischer, P. (1993). Transplantation of a polymer-encapsulated cell line genetically engineered to release NGF. *Experimental Neurology* 122:100–106.

Hoffman, D., Wahlberg, L., and Aebischer, P. (1990). NGF released from a polymer matrix prevents loss of ChAT expression in basal forebrain neurons following a fimbria-fornix lesion. *Experimental Neurology* 110:39–44.

Hohn, A., Leibrock, J., Bailey, K., and Barde, Y. A. (1990). Identification and characterization of a novel member of the nerve growth factor/brain-derived neurotrophic factor family. *Nature* 344:339–341.

Ikeda, K., Klinkosz, B., Greene, T., Cedarbaum, J. M., Wong, V., Lindsay, R. M., and Mitsumoto, H. (1995a). Effects of brain-derived neurotrophic factor on motor dysfunction in wobbler mouse motor neuron disease. *Annals of Neurology* 37:505–511.

Ikeda, K., Wong, V., Holmlund, T. H., Greene, T., Cedarbaum, J. M., Lindsay, R. M., and Mitsumoto, H. (1995b). Histometric effects of ciliary neurotrophic factor in wobbler mouse motor neuron disease. *Annals of Neurology* 37:47–54.

Ince, P., Stout, N., Shaw, P., Slade, J., Hunziker, W., Heizmann, C. W., and Baimbridge, K. G. (1993). Parvalbumin and calbindin D-28k in the human motor system and in motor neuron disease. *Neuropathology & Applied Neurobiology* 19:291–299.

Ip, N. Y., Nye, S. H., Boulton, T. G., Davis, S., Taga, T., Li, Y., Birren, S. J., Yasukawa, K., Kishimoto, T., Anderson, D. J., *et al.* (1992). CNTF and LIF act on neuronal cells via shared signaling pathways that involve the IL-6 signal transducing receptor component gp130. *Cell* 69:1121–1132.

Jing, S., Wen, D., Yu, Y., Holst, P. L., Luo, Y., Fang, M., Tamir, R., Antonio, L., Hu, Z., Cupples, R., Louis, J. C., Hu, S., Altrock, B. W., and Fox, G. M. (1996). GDNF-induced activation of the ret protein tyrosine kinase is mediated by GDNFR-alpha, a novel receptor for GDNF. *Cell* 85:1113–1124.

Johnson, E. M., Jr., Gorin, P. D., Brandeis, L. D., and Pearson, J. (1980). Dorsal root ganglion neurons are destroyed by exposure *in utero* to maternal antibody to nerve growth factor. *Science* 210:916–918.

Jones, K. R., Farinas, I., Backus, C., and Reichardt, L. F. (1994). Targeted disruption of the BDNF gene perturbs brain and sensory neuron development but not motor neuron development. *Cell* 76:989–999.

Kato, T., Hirano, A., and Donnenfeld, H. (1987). A Golgi study of the large anterior horn cells of the lumbar cords in normal spinal cords and in amyotrophic lateral sclerosis. *Acta Neuropathologica* 75:34–40.

Kawamura, Y., Dyck, P. J., Shimono, M., Okazaki, H., Tateishi, J., and Doi, H. (1981). Morphometric comparison of the vulnerability of peripheral motor and sensory neurons in amyotrophic lateral sclerosis. *Journal of Neuropathology & Eperimental Neurology* 40:667–675.

Klein, R., Smeyne, R. J., Wurst, W., Long, L. K., Auerbach, B. A., Joyner, A. L., and Barbacid, M. (1993). Targeted disruption of the trkB neurotrophin receptor gene results in nervous system lesions and neonatal death. *Cell* 75:113–122.

Koliatsos, V. E., Clatterbuck, R. E., Winslow, J. W., Cayouette, M. H., and Price, D. L. (1993). Evidence that brain-derived neurotrophic factor is a trophic factor for motoneurons *in vivo*. *Neuron* 10:359–367.

Kotzbauer, P. T., Lampe, P. A., Heuckeroth, R. O., Golden, J. P., Creedon, D. J., Johnson, E. M., Jr., and Milbrandt, J. (1996). Neurturin, a relative of glial-cell-line-derived neurotrophic factor. *Nature* 384:467–470.

Lamballe, F., Klein, R., and Barbacid, M. (1991). trkC, a new member of the trk family of tyrosine protein kinases, is a receptor for neurotrophin-3. *Cell* 66:967–979.

Leigh, P. N., Anderton, B. H., Dodson, A., Gallo, J. M., Swash, M., and Power, D. M. (1988). Ubiquitin deposits in anterior horn cells in motor neurone disease. *Neuroscience Letters* 93:197–203.

Leigh, P. N., Whitwell, H., Garofalo, O., Buller, J., Swash, M., Martin, J. E., Gallo, J. M., Weller, R. O., and Anderton, B. H. (1991). Ubiquitin-immunoreactive intraneuronal inclusions in amyotrophic lateral sclerosis. Morphology, distribution, and specificity. *Brain* 114:775–788.

Levi-Montalcini, R. (1987). The nerve growth factor 35 years later. *Science* 237:1154–1162.

Li, L., Oppenheim, R. W., Lei, M., and Houenou, L. J. (1994). Neurotrophic agents prevent motoneuron death following sciatic nerve section in the neonatal mouse. *Journal of Neurobiology* 25:759–766.

Lim, F., and Sun, A. M. (1980). Microencapsulated islets as bioartificial endocrine pancreas. *Science* 210:908–910.

Lindsay, R. M. (1993). Brain-derived neurotrophic factor: An NGF-related neurotrophin. In: Loughlin, J., E., and Fallon, J., H., eds., *Neurotrophic factors*. New York: Academic, pp. 257–284.

Lindsay, R. M., Wiegand, S. J., Altar, C. A., and DiStefano, P. S. (1994). Neurotrophic factors: From molecule to man. *Trends in Neurosciences* 17:182–190.

Lotz, B., Brooks, B., Sanjak, M., Weasler, C., Roelke, K., Parnell, J., Neville, H., Ringel, S., Brinkmann, J., Singh, K., Burns, D., Pestronk, A., Lopate, G., Florence, J., Blume, G., Eliott, J., Mitsumoto, H., Levin, K., Szirony, K., Caldwell, M., Bosch, P., Smith, B., Verheijde, J., Carr, S., Engel, W. K., Cedarbaum, J. M., *et al.* (1996). A double-blind placebo-controlled clinical trial of subcutaneous recombinant human ciliary neurotrophic factor (Rhcntf) in amyotrophic lateral sclerosis. *Neurology* 46:1244–1249.

Magal, E., Louis, J. C., Oudega, M., and Varon, S. (1993). CNTF promotes the survival of neonatal rat corticospinal neurons *in vitro*. *Neuroreport* 4:779–782.

Masu, Y., Wolf, E., Holtmann, B., Sendtner, M., Brem, G., and Thoenen, H. (1993). Disruption of the CNTF gene results in motor neuron degeneration. *Nature* 365:27–32.

Mitsumoto, H., and Bradley, W. G. (1982). Murine motor neuron disease (the wobbler mouse): Degeneration and regeneration of the lower motor neuron. *Brain* 105:811–834.

Mitsumoto, H., Ikeda, K., Klinkosz, B., Cedarbaum, J. M., Wong, V., and Lindsay, R. M. (1994). Arrest of motor neuron disease in wobbler mice cotreated with CNTF and BDNF. *Science* 1107–1110.

Moore, M. W., Klein, R. D., Farinas, I., Sauer, H., Armanini, M., Phillips, H., Reichardt, L. F., Ryan, A. M., Carver-Moore, K., and Rosenthal, A. (1996). Renal and neuronal abnormalities in mice lacking GDNF. *Nature* 382:76–79.

Mourelatos, Z., Adler, H., Hirano, A., Donnenfeld, H., Gonatas, J. O., and Gonatas, N. K. (1990). Fragmentation of the Golgi apparatus of motoneurons in amyotrophic lateral sclerosis revealed by organelle-specific antibodies. *Proceedings of the National Academy of Sciences of the United States of America* 87:4393–4395.

Mourelatos, Z., Hirano, A., Rosenquist, A. C., and Gonatas, N. K. (1994). Fragmentation of the Golgi apparatus of motoneurons in amyotrophic lateral sclerosis (ALS). Clinical studies in ALS of Guam and experimental studies in deafferented neurons and in beta,beta'-iminodipropionitrile axonopathy. *American Journal of Pathology* 144:1288–1300.

Oppenheim, R. W., Yin, Q. W., Prevette, D., and Yan, Q. (1992). Brain-derived neurotrophic factor rescues developing avian motoneurons from cell death. *Nature* 360:755–757.

Pardo, C. A., Xu, Z., Borchelt, D. R., Price, D. L., Sisodia, S. S., and Cleveland, D. W. (1995). Superoxide dismutase is an abundant component in cell bodies, dendrites, and axons of motoneurons and in a subset of other neurons. *Proceedings of the National Academy of Sciences of the United States of America* 92:954–958.

Penn, R. D., Krain, J. S., York, M. M., and Cedarbaum, J. M. (1997). Intrathecal ciliary neurotrophic factor delivery for the treatment of amyotrophic lateral sclerosis (phase I trial). *Neurosurgery* 40:94–99.

Pennica, D., Shaw, K. J., Swanson, T. A., Moore, M. W., Shelton, D. L., Zioncheck, K. A., Rosenthal, A., Taga, T., Paoni, N. F., and Wood, W. I. (1995). Cardiotrophin-1. Biological activities and binding to the leukemia inhibitory factor receptor/gp130 signaling complex. *Journal of Biological Chemistry* 270:10915–10922.

Pennica, D., Arce, V., Swanson, T. A., Vejsada, R., Pollock, R. A., Armanini, M., Dudley, K., Phillips, H. S., Rosenthal, A., Kato, A. C., and Henderson, C. E. (1996a). Cardiotrophin-1, a cytokine present in embryonic muscle, supports long-term survival of spinal montoneurons. *Neuron* 17:63–74.

Pennica, D., Swanson, T. A., Shaw, K. J., Kuang, W-J., Gray, C. L., Beatty, B. G., and Wood, W. I. (1996b). Human cardiotrophin-1: Protein and gene structure, biological and binding activities, and chromosomal localisation. *Cytokine* 8:183–189.

Pennica, D., Wood, W. I., and Chien, K. R. (1996c). Cardiotrophin-1: A multifunctional cytokine that signals via LIF receptor-gp 130 dependent pathways. *Cytokine and Growth Factor Reviews* 7:81–91.

Phillips, J. P., Tainer, J. A., Getzoff, E. D., Boulianne, G. L., Kirby, K., and Hilliker, A. J. (1995). Subunit-destabilizing mutations in Drosophila copper/zinc superoxide dismutase: Neuropathology and a model of dimer dysequilibrium. *Proceedings of the National Academy of Sciences of the United States of America* 92:8574–8578.

Pichel, J. G., Shen, L., Sheng, H. Z., Granholm, A. C., Drago, J., Grinberg, A., Lee, E. J., Huang, S. P., Saarma, M., Hoffer, B. J., Sariola, H., and Westphal, H. (1996). Defects in enteric innervation and kidney development in mice lacking GDNF. *Nature* 382:73–76.

Ripps, M. E., Huntley, G. W., Hof, P. R., Morrison, J. H., and Gordon, J. W. (1995). Transgenic mice expressing an altered murine superoxide dismutase gene provide an animal model of amyotrophic lateral sclerosis. *Proceedings of the National Academy of Sciences of the United States of America* 92:689–693.

Rosen, D. R., Siddique, T., Patterson, D., Figlewicz, D. A., Sapp, P., Hentati, A., Donaldson, D., Goto, J., O'Regan, J. P., Deng, H. X., *et al.* (1993). Mutations in Cu/Zn superoxide dismutase gene are associated with familial amyotrophic lateral sclerosis. *Nature* 362:59–62.

Rothstein, J. D. (1996). Excitotoxicity hypothesis. *Neurology* 47:19–25.

Rothstein, J. D., Jin, L., Dykes-Hoberg, M., and Kuncl, R. W. (1993). Chronic inhibition of glutamate uptake produces a model of slow neurotoxicity. *Proceedings of the National Academy of Sciences of the United States of America* 90:6591–6595.

Rothstein, J. D., Martin, L. J., and Kuncl, R. W. (1992). Decreased glutamate transport by the brain and spinal cord in amyotrophic lateral sclerosis. *New England Journal of Medicine* 326:1464–1468.

Rothstein, J. D., Tsai, G., Kuncl, R. W., Clawson, L., Cornblath, D. R., Drachman, D. B., Pestronk, A., Stauch, B. L., and Coyle, J. T. (1990). Abnormal excitatory amino acid metabolism in amyotrophic lateral sclerosis. *Annals of Neurology* 28:18–25.

Sagot, Y., Tan, S. A., Baetge, E., Schmalbruch, H., Kato, A. C., and Aebischer, P. (1995). Polymer encapsulated cell lines genetically engineered to release ciliary neurotrophic factor can slow down progessive motor neuronopathy in the mouse. *European Journal of Neuroscience* 7:1313–1322.

Sagot, Y., Tan, S. A., Hammang, J. P, Aebischer, P., and Kato, A. C. (1996). GDNF slows loss of motoneurons but not axonal degeneration or premature death of pmn/pmn mice. *Journal of Neuroscience* 16:2335–2341.

Sahenk, Z., Seharaseyon, J., and Mendell, J. R. (1994). CNTF potentiates peripheral nerve regeneration. *Brain Research* 655:246–250.

Sambrook, T., Fritsch, E. F., and Maniatis, T. (1989). *Molecular cloning. A laboratory manual.* Cold Spring Harbor, New York: Cold Spring Harbor.

Sanchez, M. P., Silos-Santiago, I., Frisen, J., He, B., Lira, S. A., and Barbacid, M. (1996). Renal agenesis and the absence of enteric neurons in mice lacking GDNF. *Nature* 382:70–73.

Schmalbruch, H., Jensen, H. J., Bjaerg, M., Kamieniecka, Z., and Kurland, L. (1991). A new mouse mutant with progressive motor neuronopathy. *J Neuropathol Exp Neurol* 50(3):192–204.

Sendtner, M., Carroll, P., Holtmann, B., Hughes, R. A., and Thoenen, H. (1994). Ciliary neurotrophic factor. *Journal of Neurobiology* 25:1436–1453.

Sendtner, M., Holtmann, B., Kolbeck, R., Thoenen, H., and Barde, Y. A. (1992). Brain-derived neurotrophic factor prevents the death of motoneurons in newborn rats after nerve section. *Nature* 360:757–759.

Sendtner, M., Kreutzberg, G. W., and Thoenen, H. (1990). Ciliary neurotrophic factor prevents the degeneration of motoneurons after axotomy. *Nature* 345:440–441.

Sendtner, M., Schmalbruch, H., Stockli, K. A., Carroll, P., Kreutzberg, G. W., and Thoenen, H. (1992). Ciliary neurotrophic factor prevents degeneration of motoneurons in mouse mutant progressive motor neuronopathy. *Nature* 358:502–504.

Shaw, P. J., Chinnery, R. M., and Ince, P. G. (1994). [3H]D-aspartate binding sites in the normal human spinal cord and changes in motor neuron disease: A quantitative autoradiographic study. *Brain Research* 655:195–201.

Siddique, T. (1991). Molecular genetics of familial amyotrophic lateral sclerosis. *Advances in Neurology* 56:227–231.

Siddique, T., Figlewicz, D. A., Pericak-Vance, M. A., Haines, J. L., Rouleau, G., Jeffers, A. J., Sapp, P., Hung, W. Y., Bebout, J., McKenna-Yasek, D., *et al.* (1991). Linkage of a gene causing familial amyotrophic lateral sclerosis to chromosome 21 and evidence of genetic-locus heterogeneity. *New England Journal of Medicine* 324:1381–1384.

Siddique, T., Nijhawan, D., and Hentati, A. (1996). Molecular genetic basis of familial ALS. *Neurology* 47:27–34.

Siddique, T., Pericak-Vance, M. A., Brooks, B. R., Roos, R. P., Hung, W. Y., Antel, J. P., Munsat, T. L., Phillips, K., Warner, K., Speer, M., *et al.* (1989). Linkage analysis in familial amyotrophic lateral sclerosis. *Neurology* 39:919–925.

Smeyne, R. J., Klein, R., Schnapp, A., Long, L. K., Bryant, S., Lewin, A., Lira, S. A., and Barbacid, M. (1994). Severe sensory and sympathetic neuropathies in mice carrying a disrupted Trk/NGF receptor gene. *Nature* 368:246–249.

Sobue, G., Hashizume, Y., Yasuda, T., Mukai, E., Kumagai, T., Mitsuma, T., and Trojanowski, J. Q. (1990). Phosphorylated high molecular weight neurofilament protein in lower motoneurons in amyotrophic lateral sclerosis and other neurodegenerative diseases involving ventral horn cells. *Acta Neuropathologica* 79:402–408.

Stahl, N., and Yancopoulos, G. D. (1994). The tripartite CNTF receptor complex: Activation and signaling involves components shared with other cytokines. *Journal of Neurobiology* 25:1454–1466.

Suenaga, T., Matsushima, H., Nakamura, S., Akiguchi, I., and Kimura, J. (1993). Ubiquitin-immunoreactive inclusions in anterior horn cells and hypoglossal neurons in a case with Joseph's disease. *Acta Neuropathologica* 85:341–344.

Symonyan, M. A., and Nalbandyan, R. M. (1972). Interaction of hydrogen peroxide with superoxide dismutase from erythrocytes. *FEBS Letters* 28:22–24.

Tan, S. A., Deglon, N., Zurn, A. D., Baetge, E. E., Bamber, B., Kato, A. C., and Aebischer, P. (1996). Rescue of motoneurons from axotomy-induced cell death by polymer encapsulated cells genetically engineered to release CNTF. *Cell Transplantation* 5:577–587.

Tandan, R., and Bradley, W. G. (1985). Amyotrophic lateral sclerosis: Part 1. Clinical features, pathology, and ethical issues in management. *Annals of Neurology* 18:271–280.

Tojo, H., Kaisho, Y., Nakata, M., Matsuoka, K., Kitagawa, M., Abe, T., Takami, K., Yamamoto, M., Shino, A., Igarashi, K., *et al.* (1995). Targeted disruption of the neurotrophin-3 gene with lacZ induces loss of trkC-positive neurons in sensory ganglia but not in spinal cords. *Brain Research* 669:163–175.

Treanor, J. J., Goodman, L., de Sauvage, F., Stone, D. M., Poulsen, K. T., Beck, C. D., Gray, C., Armanini, M. P., Pollock, R. A., Hefti, F., Phillips, H. S., Goddard, A., Moore, M. W., Buj-Bello, A., Davies, A. M., Asai, N., Takahashi, M., Vandlen, R., Henderson, C. E., and Rosenthal, A. (1996). Characterization of a multicomponent receptor for GDNF: *Nature* 382:80–83.

Trupp, M., Arenas, E., Fainzilber, M., Nilsson, A. S., Sieber, B. A., Grigoriou, M., Kilkenny, C., Salazar-Grueso, E., Pachnis, V., Arumae, U., *et al.* (1996). Functional receptor for GDNF encoded by the c-ret proto-oncogene. *Nature* 381:785–788.

Tseng, J. L., Beatge, E. E., Zurn, A. D., and Aebischer, P., (1997). GDNF reduces drug-induced rotational behavior after medial forebrain bundle transection by a mechanism not involving striatal dopamine. *Journal of Neuroscience* 17:325–333.

Wiedau-Pazos, M., Goto, J. J., Rabizadeh, S., Gralla, E. B., Roe, J. A., Lee, M. K., Valentine, J. S., and Bredesen, D. E. (1996). Altered reactivity of superoxide dismutase in familial amyotrophic lateral sclerosis. *Science* 271:515–518.

Williams, D. B., and Windebank, A. J. (1991). Motor neuron disease (amyotrophic lateral sclerosis). *Mayo Clinic Proceedings* 66:54–82.

Wong, P. C., Pardo, C. A., Borchelt, D. R., Lee, M. K., Copeland, N. G., Jenkins, N. A., Sisodia, S. S., Cleveland, D. W., and Price, D. L. (1995). An adverse property of a familial ALS-linked SOD1 mutation causes motor neuron disease characterized by vacuolar degeneration of mitochondria. *Neuron* 14:1105–1116.

Wong, V., Arriage, R., Ip, N. Y., and Lindsay, R. M. (1993). The neurotrophins BDNF, NT-3 and NT-4/5, but not NGF, up-regulate the cholinergic phenotype of developing motoneurons. *European Journal of Neuroscience* 5:466–474.

Xu, Z., Cork, L. C., Griffin, J. W., and Cleveland, D. W. (1993). Increased expression of neurofilament subunit NF-L produces morphological alterations that resemble the pathology of human motor neuron disease. *Cell* 73:23–33.

Yan, Q., Elliott, J., and Snider, W. D. (1992). Brain-derived neurotrophic factor rescues spinal motoneurons from axotomy-induced cell death. *Nature* 360:753–755.

Yan, Q., Matheson, C., Lopez, O. T., and Miller, J. A. (1994). The biological responses of axotomized adult motoneurons to brain-derived neurotrophic factor. *Journal of Neuroscience* 14:5281–5291.

Yuen, E. C., and Mobley, W. C. (1996). Therapeutic potential of neurotrophic factors for neurological disorders. *Annals of Neurology* 40:346–354.

Zielinski, B. A., and Aebischer, P. (1994). Chitosan as a matrix for mammalian cell encapsulation. *Biomaterials* 15:1049–1056.

Zurn, A. D., and Werren, F. (1994). Development of CNS cholinergic neurons *in vitro:* Selective effects of CNTF and LIF on neurons from mesencephalic cranial motor nuclei. *Developmental Biology* 163:309–315.

Zurn, A. D., Baetge, E. E., Hammang, J. P., Tan, S. A., and Aebischer, P. (1994). Glial cell line-derived neurotrophic factor (GDNF), a new neurotrophic factor for motoneurones. *Neuroreport* 6:113–118.

Zurn, A. D., Winkel, L., Menoud, A., Djabali, K., and Aebischer, P. (1996). Combined effects of GDNF, BDNF, and CNTF on motoneuron differentiation *in vitro. Journal of Neuroscience Research* 44:133–141.

CHAPTER 24

Spinal Cord Regeneration

Growth Factors, Inhibitory Factors and Gene Therapy

MARK H. TUSZYNSKI,*,† RAY GRILL,* AND ARMIN BLESCH*

*Department of Neurosciences, University of California, San Diego, La Jolla, California,
†Department of Neurology, Veterans Affairs Medical Center, San Diego, California

The dogma that axonal regeneration in the central nervous system cannot be accomplished has now been refuted. Several studies performed in the last two decades have convincingly demonstrated that a variety of approaches can promote axonal growth after injury. These approaches include the delivery of neurotrophic factors, the placement of conducive axonal growth scaffolds in the spinal cord, and the neutralization of inhibitory substances that may be upregulated after injury in the CNS. Indeed, several of these therapeutic approaches have generated functional recovery after spinal cord injury in adult mammals. Mounting evidence also suggests that axonal growth after chronic spinal cord injury is elicitable. Although these findings are impressive, several factors remain to be addressed before the translation of these approaches to human trials: (1) reports of functional recovery require replication by independent laboratories, (2) experiments are needed that simultaneously utilize more than one approach for promoting neural regeneration in the spinal cord in order to maximize axonal growth, and (3) studies should be undertaken in animal models larger than rodents before subjecting humans to the risk of manipulating already-compromised systems. The progress of the last two decades has gained much from experiments that aim to understand underlying mechanisms of basic neurobiology, and progress in this field will continue to benefit from this level of basic science research.

CNS Regeneration

I. INTRODUCTION

Recent reports of functional recovery resulting from reconstruction of host neural circuitry in adult mammals after spinal cord injury have dramatically altered perceptions regarding our ability to promote effective central nervous system axonal regeneration. First, Bregman and colleagues (1995) reported that neutralization of myelin-based axon growth inhibitors in the CNS promoted growth of corticospinal, coerulospinal, and raphaespinal axons, and could induce functional recovery. Second, Cheng and associates (1996) reported that peripheral nerve bridges inserted into the mechanically stabilized site of a complete spinal cord transection, together with delivery of acidic fibroblast growth factor (aFGF) in a fibrin glue around the injury site, would promote growth of corticospinal axons and several other supraspinal axonal populations, resulting in functional recovery. Third, our group recently reported that the localized delivery of neurotrophin-3 (NT-3) to spinal cord lesion sites using grafts of genetically modified cells could promote growth of corticospinal axons up to 8 mm past the injury site and result in functional recovery (Grill *et al.*, 1997b). These reports and other (Li *et al.*, 1997) share both common features and distinctions, and all require confirmation by other groups. Nonetheless, they challenge our traditional concepts regarding spinal cord regeneration and suggest that, armed with the proper tools (testable hypotheses and good animal models), axonal regrowth and amelioration of functional deficits after injury may be achievable.

In this chapter, findings from these three studies and other related to axonal regeneration will be reviewed. Neurotrophic factor effects on axonal regeneration will be described, together with the technique of gene transfer as a means of locally delivering growth factors to regions of spinal cord injury for promoting axonal growth. Because this book focuses on regeneration of the injured *adult* central nervous system, the considerable body of knowledge regarding neonatal CNS lesions will not be dealt with, but interested readers are referred to relevant papers (Himes *et al.*, 1994; Kunkel-Bagden and Bregman, 1990; Bregman, 1987; Diener and Bregman, 1998a, 1998b).

II. AXONAL REGENERATION AFTER SPINAL CORD INJURY

Early studies of axonal regeneration were conducted nearly 100 years ago. Tello (1911a and 1911b) grafted sciatic nerves into the rabbit brain and described axonal penetration into grafts for distances of up to 1 cm, findings that were to be confirmed with improved histochemical techniques 70 years later (Richardson *et al.*, 1980). Landmark studies by Ramon y Cajal in Spain in the 1920's lucidly reported the striking inability of the injured adult CNS to regenerate, a limitation attributed to an inhospitable CNS environment and a potential lack of neurotrophism (Ramon y Cajal, 1928). Numerous reports in the intervening years regularly reported "success" (Sugar and Gerard, 1940; Kao *et al.*, 1977; Clemente, 1958) or "failure" (Brown and McCough, 1947; Le Gros Clark, 1942) in attempts to promote axonal regeneration in the CNS. Yet in the final analysis these studies were in conclusive, suggesting but not adequately proving that injured adult mammalian CNS axons were capable of more than abortive growth or sprouting after injury (Clemente, 1964).

Reports by Aguayo and associates in the late 1970s and early 1980s proved that CNS axons could regenerate for long distances after injury if provided with suitable growth "terrains" or substrates (Richardson *et al.*, 1980; Keirstead *et al.*, 1989). When grafted to the lesioned spinal cord, brain stem, or optic nerve, peripheral nerve grafts supported regeneration of CNS axons over distances that at times exceeded the original length of the axon. However, axonal growth abruptly ceased upon re-encountering the CNS milieu at the end of the nerve graft. These findings indicated that adult CNS axons are capable of re-entering growth states after injury during adulthood, but that the environment of the injured CNS is nonpermissive to axonal growth. Subsequent work by many investigators has demonstrated several features of the injured CNS environment that inhibit axon growth, including glial reactions (McKeon *et al.*, 1991), inhibitory molecules of the extracellular matrix (Silver, 1994) and myelin-associated axonal growth inhibitors (Schwab, 1990; McKerracher *et al.*, 1994; Mukhopadhyay *et al.*, 1994; Filbin, 1995).

Experiments utilizing fetal neural transplants to the injured adult spinal cord also demonstrated that CNS axons were capable of regenerative growth after injury (Reier *et al.*, 1992a, 1992b; Bregman, 1987; Mori *et al.*, 1997; Himes *et al.*, 1994; Bregman *et al.*, 1993; Houle and Reier, 1989, 1988) and limited functional recovery (e.g., see Bregmann *et al.*, 1993; Stokes and Reier, 1992). In these experiments, axons of the adult host extended for short distances after injury, penetrating and innervating fetal transplants. Axons of dorsal root sensory origin extended and grew into suspension grafts of embryonic day 14 rat spinal cord (Houle and Reier, 1989). Serotonergic axons of putative raphaespinal origin and coerulospinal axons demonstrated growth up to the host–graft interface in adult spinal cord-injured rats that received fetal grafts, although axons demonstrated only minimal penetration into grafts compared to transplants into neonatal animals (Houle and Reier, 1988; Bregman *et al.*, 1993). Reciprocal innervation of the host spinal cord by transplanted neurons was also occasionally demonstrated, although the number of axons crossing the interface from host to graft and vice versa was relatively small.

Hypothesizing that Schwann cells were an important component of the peripheral nerve environment that supported axon growth, Bunge and coworkers grafted Schwann cells in artificial matrices to sites of complete spinal cord transection, demonstrating growth of spinal and supraspinal axons into this environment (Kuhlengel *et al.*, 1990; Duncan *et al.*, 1988; Xu *et al.*, 1995a, 1995b) (see Chapter 25). Schwann cells produce neurotrophic factors and extracellular matrix molecules that support axonal growth (Assouline *et al.*, 1987; Meyer *et al.*, 1992; Dobrea *et al.*, 1992; Henderson *et al.*, 1994; Masu *et al.*, 1993; Friedman *et al.*, 1992; Richardson *et al.*, 1980; Williams *et al.*, 1983; Mahut, 1972) and remyelination (Honmou *et al.*, 1996). This work provided further support for the concept that CNS axons, provided with a supportive environment, can mount substantial growth responses.

Although these modern studies unequivocally demonstrated that injured adult CNS axons were capable of growth, three important features were notably absent in these experiments: (1) Long-distance axonal growth in the CNS: axons extended only very short distances through adult CNS parenchyma, on the order of less than 1 mm. (2) Numbers of regenerating axons: the number of growing axons that extend after injury in the CNS, even in the presence of growth-supportive environments such as peripheral nerve grafts or Schwann cell grafts, is small. Generally 1% or less of the original population of axons affected by a lesion had been demonstrated to regenerate after experimental manipulation. (3) Function: functional recovery had yet to be convincingly demonstrated.

III. NEUTRALIZING INHIBITORY MOLECULES IN THE CNS

In 1990, a landmark study by Schnell and Schwab (1990) reported that long-distance axonal growth could be elicited in the lesioned adult CNS. In this study, axonal growth-inhibiting molecules present on CNS myelin were neutralized by delivering blocking antibodies produced by implanted hybridoma cells. However, the absolute number of regenerating axons did not appear to be enhanced compared to the control-injury condition; rather, the modest number of axons that normally extend for short (<1mm) distances after injury extended for far greater distances, on the order of 10 mm (Schnell and Schwab, 1990). In a follow-up study, the groups of Bregman and Schwab reported that the delivery of myelin-neutralizing antibodies to the injured adult spinal cord resulted in functional recovery (Bregman *et al.*, 1995). This effect was putatively associated with regeneration of the corticospinal projection, because ablation of the rat sensorimotor cortex reversed the behavioral recovery. In the same study, augmented growth of raphaespinal and coerulospinal axons was also described. These impressive functional results require replication by other laboratories. Future experiments should also examine the extent over which myelin-neutralizing anitbodies diffuse through the host CNS and penetrate the spinal cord. Because the delivery of myelin-neutralizing antibodies elicits the growth of only small numbers of axons, presumably more robust functional effects might be generated by recruiting growth from larger numbers of injured axons.

IV. BRIDGING THE INJURY SITES

In 1996, Cheng, Yihai, and Olson reported functional recovery in a different model of adult spinal cord injury (Cheng *et al.*, 1996). Adult rats underwent complete mid-thoracic spinal cord transections. Eighteen small peripheral nerve grafts were placed into the lesion cavity, adopting the "bridging" approach first used by Aguayo and co-workers 20 years earlier. Unlike the previous efforts of Aguayo's group, however, Cheng's study made three additional modifications: (1) the site of spinal injury was *stabilized* with wires and fibrin glue to hold the nerve grafts in place to reduce further damage to the cord and to prevent posttraumatic cyst formation; (2) the small nerve grafts were placed in such a manner as to channel the regenerating axons to grow through host *gray* matter rather than white matter, in order to avoid the inhibitory effects of myelin-associated molecules; and (3) *acidic fibroblast growth factor* (FGF-1) was provided to the injury site to stimulate the growth of injured axons. One year later, the investigators reported that rats receiving these manipulations showed dramatic partial recovery of locomotor function and regeneration of several supraspinal axonal systems including corticospinal, rubrospinal, reticulospinal, and raphaespinal systems.

Several features of this experimental design merit note. First, an extraordinary degree of technical skill was required to microsurgically place 18 small nerve grafts in a 3 mm-wide by 5 mm-long spinal cord cavity in such a manner as to precisely link-up specific proximal white matter pathways to distal adjacent gray matter regions, and maintain this complex orientation up to one year later. If such mechanical support is a fundamental mechanism contributing to regeneration in this experiment, as the data suggests, then it will be important to adopt this paradigm in future regeneration studies. Second, all of the lesioned spinal projections examined responded to FGF-1, an apparent global neurotrophic effect. Global effects of solitary growth factors have not been described previously, although the FGF receptor is broadly distributed and responses as broad as those described by the authors are hypothetically possible (see also Eckenstein *et al.*, 1991, 1994; Ferguson and Johnson, 1991; Emoto *et al.*, 1989; Gomez-Pinilla and Cotman, 1993). Third, it is not clear that axons regenerating in response to FGF-1 in the experiments of Cheng exhibited chemotropism, i.e., growth toward the highest concentration of growth factor. Instead, axons extended for some distance away from the presumed highest concentration of the growth factor, whereas previous reports had indicated that spinal cord axons extend in the direction of a growth factor (Conner and Varon, 1996; Hagg *et al.*, 1990b; Tuszynski *et al.*, 1994). This finding could result from nonchemotropic effects of FGF, including alteration of glial responses in the injured environment. A nonchemotropic mechanism might also account for the augmented growth of several different axonal populations. Alternatively, FGF might be biologically active for only brief time periods in this paradigm, diffusing from the glue for days or weeks after injury and then losing activity. In such a case, FGF-1 might initially induce axonal growth that is later sustained by other mechanisms. These questions need to be addressed in the Cheng model, and the findings require replication by other investigators. If these findings are indeed clearly verified independently, then this approach to spinal cord injury is adaptable to the treatment of human injury.

V. AUGMENTING AXONAL GROWTH: NEUROTROPHIC FACTOR GENE THERAPY AFTER SPINAL CORD INJURY

Recently we reported another approach to eliciting axonal growth and generating partial functional recovery after adult spinal cord injury—neurotrophic factor gene therapy.

A key problem in promoting regeneration after spinal cord injury has been eliciting growth from large numbers of injured axons. Neurotrophic factors are molecules with roles in several biological functions including cell survival, synaptogenesis, dendritic shaping, neurotransmitter release, and, importantly, in the context of spinal cord injury, *axonal outgrowth* (Levi-Montalcini, 1987; Thoenen, 1995). Often, axonal outgrowth in response to a neurotrophic factor occurs in the direction of the highest gradient of growth factor concentration, a *chemotropic* effect (Letourneau, 1978, 1983; Gundersen and Barrett, 1980). This chemotropic property suggests that growth factors should be delivered directly to a site of injury to optimally promote axonal growth and regeneration through a lesioned region. Trophic factor injections or continuous infusions with osmotic pumps have traditionally been used to examine neurotrophic factor effects *in vivo* in the CNS, but such infusions suffer from the drawback that they do not achieve *localized* high neurotrophin concentrations and are therefore not an optimal means of achieving directionally specific axonal growth.

Gene therapy is a tool for potentially accomplishing intraparenchymal, regionally specific, locally restricted, long-term, and safe delivery of growth factors to regions of neural injury or degeneration (Fig. 1) (Friedmann and Roblin, 1972; Blesch and Tuszynski, 1996; Schumacher *et al.*, 1991). Two forms of gene therapy exist: *ex vivo* gene transfer and *in vivo* gene transfer, as described in Chapter 9. Briefly, in *ex vivo* methods, cells from the host are obtained from a biopsy and established as *in vitro* cell cultures. The cells are then genetically modified *in vitro,* typically using viral-based vectors to introduce the novel genes into the cultured cells. If a gene encoding a special cell survival function (such as resistance to the drug neomycin) is included in the transgene vector, then cells containing and expressing the transgene can be selected *in vitro* (e.g., by adding neomycin to the growth medium). The optimal cellular producers of the transgene product are determined by molecular and cellular analyses, and optimal expressors are subcloned (chosen from the culture dish and replated) to maximize efficiency of gene expression from the cells. These selected cells are then grown to numbers sufficient for harvesting and *in vivo* grafting in host regions of neural trauma or neurodegeneration. By genetically modifying the host's own cells, there is no risk of immunological rejection of the grafted cells. In *in vivo* gene therapy methods, vectors capable of introducing novel genetic material are injected directly into the brain, eliminating the need for *in vitro* cycles of cell growth, selection, and biochemical–cellular characterization.

The advantage of *in vivo* gene therapy methods lies in the simplicity of the approach and the ability to target a specific intraparenchymal brain region. Drawbacks of *in vivo* gene therapy include (1) lack of long-term gene expression with vectors currently available, although more recent reports suggest that this limitation may have been overcome; (2) toxicity to host cells with some vectors currently used, including herpes virus; (3) activation of host immune responses with some vectors such as adenovirus; (4) the infection of a variety of cell types in the injected region, including neurons, glia, and vascular cells, with uncertain consequences of neurotrophin expression on normal cell metabolism and function; and (5) the inability to select cells that exhibit maximal gene expression, with subsequent compromise of gene expression.

Advantages of *ex vivo* gene therapy include the ability to select maximal expressors of the desired gene product *in vitro* prior to grafting to the host; the placement of cells into specific intraparenchymal spinal cord or brain sites; and the certainty that all cells transfected *in vitro* will express the transgene at the time that they are grafted *in vivo.* Disadvantages of *ex vivo* gene therapy include (1) the complexity of cell preparation and transfection techniques compared to *in vivo* gene therapy methods; (2) invasiveness of the procedure, such as the transplantation of cells into the nervous system; and (3) the risk of tumor formation or other deleterious effects of the grafted cells in the host. Disadvantages of both *ex vivo* and *in vivo* gene therapy methods at the present time include the inability to regulate the amount of transgene expression and

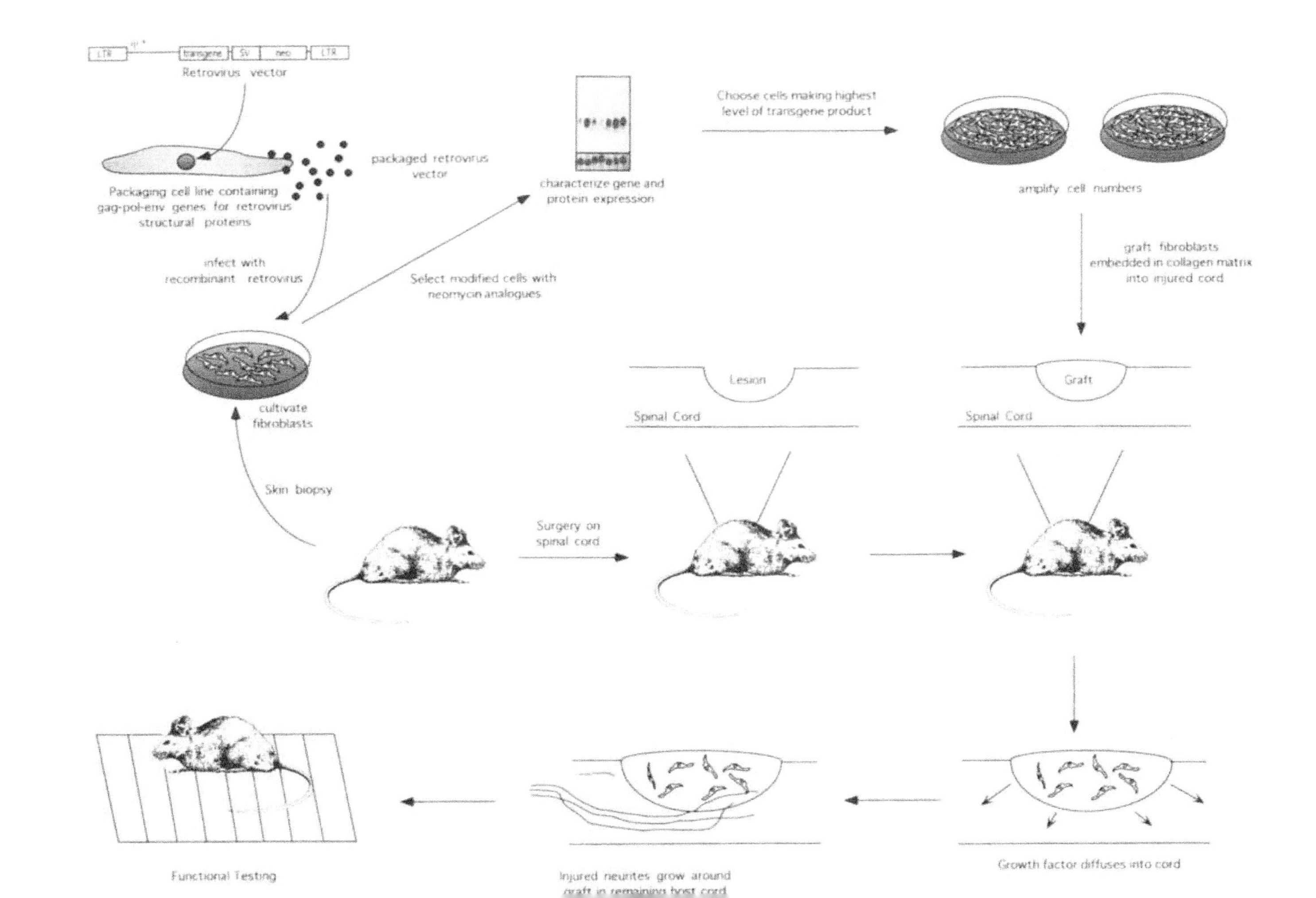
LTR
transgene
SV
neo
LTR
Retrovirus vector
Packaging cell line containing gag-pol-env genes for retrovirus structural proteins
packaged retrovirus vector
infect with recombinant retrovirus
cultivate fibroblasts
Select modified cells with neomycin analogues
characterize gene and protein expression
Choose cells making highest level of transgene product
amplify cell numbers
graft fibroblasts embedded in collagen matrix into injured cord
Lesion
Spinal Cord
Graft
Spinal Cord
Skin biopsy
Surgery on spinal cord
Growth factor diffuses into cord
Injured neurites grow around graft in remaining host cord
Functional Testing

to stop production of the transgene product, if needed. Both of the latter problems are potentially solvable using regulatable transgene vector systems (Gossen *et al.*, 1994).

We have applied *ex vivo* gene therapy techniques to the injured spinal cord as a means of targeting neuronal growth factors directly to sites of injury. Robust growth of specific classes of axons has been elicited, depending on the type of growth factor delivered. In the case of one growth factor, neurotrophin-3, functional recovery after injury has been detected. These studies are summarized next.

For these studies, we have used spinal cord "dorsal hemisection" lesions to model acute injury. Dorsal hemisection lesions (Fig. 2) are used because they are consistently reproducible, anatomically well defined, and result in specific functional deficits that can be assayed to examine functional recovery. Fibroblasts genetically modified to make a neurotrophic factor are embedded into a collagen matrix and grafted to the dorsal hemisection lesion cavity. Thus, axonal growth in response to the neurotrophic factor-secreting cell grafts can occur through three potential routes after a dorsal hemisection lesion: host

---

FIGURE 1 Schematic view of *ex vivo* gene therapy for spinal cord injury. *Top left:* Retroviral vectors derived from wild-type Moloney murine leukemia virus (*MLV*) are used to infect "packaging cell-lines" with a neurotrophin gene of interest. In the retrovirus vector, the transgene (in this case, a neurotrophic factor gene) is continually ("constitutively") expressed by a viral promoter in the long-terminal repeat (LTR) region of the vector. A ψ^+ sequence in the retroviral vector directs the "packaging" of vector particles into viral coats. The retrovirus vector also contains a gene (*neo*) that makes cells containing the transgene resistant to neomycin; cells that have successfully incorporated the transgene vector can therefore be selected *in vitro* by adding neomycin to the culture medium. The *neo* gene is constitutively expressed by a simian virus promoter (*SV*). When the retroviral vector is placed into a packaging cell-line, *gag*, *pol*, and *env* genes in the packaging cell-line result in amplification in the number of vector particles and the synthesis of viral coat proteins. The retrovirus vector lacks these essential genes, thus it is not capable of independent replication. The packaging cell-line then secretes the transgene particles within their viral coats; they are now capable of introducing the transgene into a new, target cell for gene therapy. In this case, primary fibroblasts obtained from rat skin biopsies are exposed to the retrovirus particles *in vitro*. The viral coats fuse to the fibroblast cell membrane, allowing entry of the retroviral vector into the cell. It travels to the fibroblast nucleus and integrates into the host genome, and expression of the transgene begins. *Top center:* Culture dishes containing genetically modified fibroblasts are grown in the presence of neomycin analogues to kill cells lacking the transgene. Remaining fibroblasts are assayed for production of the desired gene product, and those dishes ("bulk clones") producing the highest levels of transgene protein are amplified in number (top right). *Center:* Adult rats undergo mid-thoracic spinal cord dorsal hemisection lesions, and then receive transplants of genetically modified cells, embedded in collagen gels, into the lesion cavity. *Lower right:* Growth factors secreted from the genetically modified cells placed in the spinal cord diffuse from the grafted cells, eliciting growth of host spinal cord axons into or near the grafted region (see Figs. 3, 4, and 6). *Lower left:* Graft recipients undergo functional testing to determine whether axonal growth resulting from genetically modified cell implants induces functional recovery.

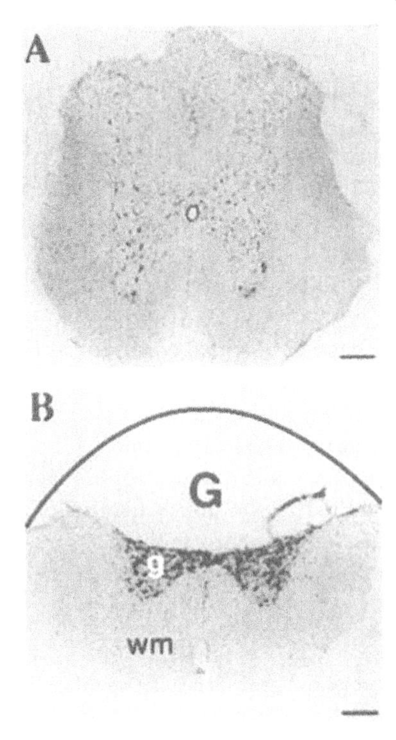

FIGURE 2 Spinal cord dorsal hemisection lesion model. Nissl-stained coronal sections illustrating (A) intact spinal cord and (B) dorsal hemisectioned spinal cord at the mid-thoracic level. G and elliptical line indicate region occupied by genetically modified cell graft in the lesion cavity; g, remaining gray matter; wm, remaining white matter. Scale bars = 400 μm (reproduced from Grill *et al.*, 1997b).

gray matter, host white matter, or the graft itself, which contains collagen, fibronectin, and a cellular substrate (see Fig. 2). This model correlates with human injury in that human injury is rarely anatomically complete, often leaving bridges of remaining host tissue. Shortcomings of this model include its failure to reproduce the mechanism of human injury (compression), and the need for caution in interpreting axonal "regeneration" (growth of transected axons) rather than sprouting (neurite elaboration from intact axons in response to injury of a neighboring system), because axons traveling in the ventral spinal cord are not transected. On balance, this model is extremely useful for examining injury-induced degeneration and axonal responses to injury (Bregman *et al.*, 1993, 1995; Schnell *et al.*, 1994; Schnell and Schwab, 1990; Kunkel-Bagden *et al.*, 1992; Grill *et al.*, 1997a, 1997b; Tuszynski *et al.*, 1996, 1997, 1998; Nakahara *et al.*, 1996).

A. Grafts of Cells Genetically Modified to Produce NGF to the Injured Spinal Cord

In the first series of studies to explore the effects of locally delivered neurotrophins on injured spinal cord axons, fibroblasts transduced to express human nerve growth factor (NGF) were transplanted to mid-thoracic acute lesion sites in adult rats (Fig. 3). Control animals received grafts of identical fibroblasts that were not genetically modified, or fibroblasts modified to express the reporter gene β-galactosidase. Two populations of injured axons responded to the NGF source—coerulospinal axons and primary nociceptive (pain-mediating) sensory axons. Primary nociceptive axons extended into NGF-secreting grafts within 7 days of injury, and densely and extensively innervated grafts by two weeks after injury. Sensory axons that penetrated the grafts extended from the dorsolateral fasciculus of the spinal cord, indicating that axons within spinal cord white matter can readily cross from white matter into a permissive growth terrain, the graft. Coerulospinal axons also extensively penetrated grafts, but only after more prolonged time periods of several weeks to months. By three to six months, coerulospinal axon growth into the grafts was complete. Coerulospinal axons originated from the locus coeruleus, indicating that a supraspinal axonal population was, like the local sensory axon population, capable of robust growth after injury. In roughly half of acutely injured animals, growth of choline acetyltransferase (ChAT)-labeled axons into grafts was also observed. Both primary motor axons and preganglionic sympathetic axons immunolabel for ChAT in the spinal cord, but based on the morphology and caliber of these axons, it is believed that the ChAt-labeled axons penetrating the grafts were of local motor origin.

The robust growth response from primary sensory axons was not surprising because these axons express both the low- and high-affinity forms of the NGF receptor, p75 and trkA, respectively. On the other hand, coerulospinal axons do not express either p75 or trkA, and local motor axons express only p75 after injury. Thus, the mechanism by which coerulospinal axons penetrated NGF-secreting grafts remains to be determined. NGF has been reported to induce secondary expression of BDNF in the spinal cord (Apfel *et al.*, 1996), a mechanism that could explain responses from local motor neurons. Similar NGF-induced expression of a coerulospinal growth factor might explain penetration of grafts by these axons.

Sensory and coerulospinal axons densely penetrated NGF-secreting grafts, and, not surprisingly, failed to sprout densely *outside* of the graft. Because NGF is believed to elicit chemotropic growth of axons (i.e., growth toward the highest concentration of NGF), axon growth beyond the graft may require both downregulation of NGF expression within the graft and NGF delivery

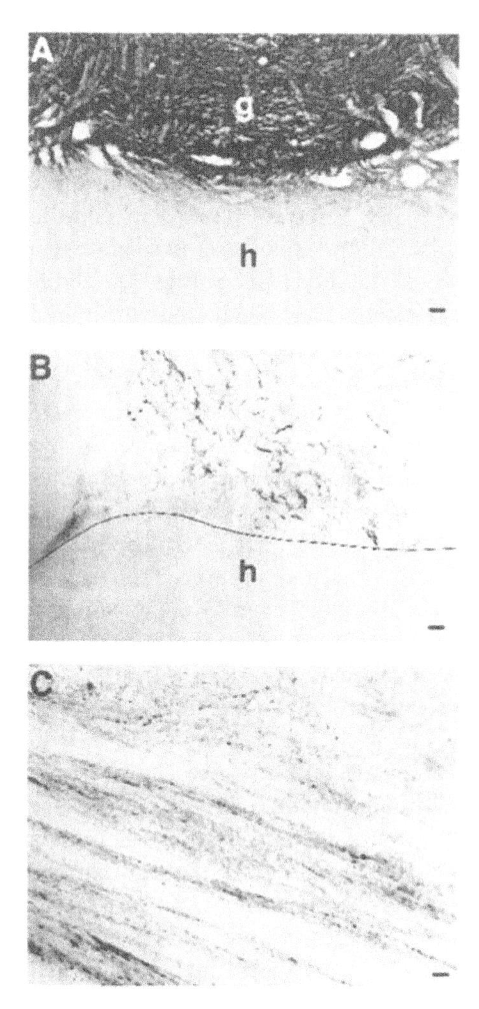

FIGURE 3 NGF-secreting cell grafts in the injured rat spinal cord. (A) Dense numbers of sensory axons (immunolabeled with an antibody to calcitonin gene-related peptide (CGRP), a marker of sensory axons and spinal motor neurons) penetrate an NGF-secreting graft one month after placement in the spinal cord lesion site. Motor neurons are present in the ventral spinal cord, but CGRP only labels the processes of sensory axons. (B) In contrast, in a control fibroblast graft that has been genetically modified to produce the reporter gene β-galactosidase, only a modest number of primary sensory axons have penetrated the graft one month after the placement of the transplant (g) in the spinal cord lesion site. (C) TH-labeled axons also densely penetrate NGF-secreting grafts (shown) but not control fibroblast grafts. These axons are putatively of coerulospinal origin because (1) they penetrate grafts directly from the spinal coerulospinal tract; (2) they are of fine caliber with multiple varicosities, resembling the morphology of other normal coerulospinal axons; and (3) the axons also immunolabel for dopamine beta-hydroxylase, excluding their origin from dopaminergic systems.

to spinal cord sites distant from the graft. Investigations currently in progress are examining these possibilities.

Functional studies in NGF-secreting graft recipients did not show evidence of motor recovery (Tuszynski *et al.*, 1997). Thus, NGF-secreting grafting studies have been extremely useful in terms of identifying the sensitivity of injured spinal cord axons to neurotrophins, but functional recovery is likely to require growth of axons beyond the lesions site and growth of additional axonal populations.

B. Grafts of Cells Genetically Modified to Produce NT-3 to the Injured Spinal Cord

Using the dorsal hemisection lesion model, we grafted cells genetically modified to produce human neurotrophin-3 (NT-3) to acutely injured spinal cords of adult rats (Grill *et al.*, 1997b). One and three months later, rats that received grafts of NT-3 secreting cells showed partial recovery of locomotor performance (Fig. 4a). Functional recovery was demonstrated on a grid task, in which rat ambulate over a mesh of plastic but must place their feet accurately on the components of the mesh to remove forward and to prevent their feet from falling through. Recipients of NT-3 secreting grafts, but not control-lesioned animals or animals with NGF-secreting grafts, showed significant amelioration of the motor deficit. Histological analysis revealed a significant increase in growth of the lesioned corticospinal projection at the level of the lesion and up to 8 mm distal to the injury site (Fig. 4b). Specific responses of other axon systems, including sensory, coerulospinal, raphaespinal, and local motor axons, were not observed.

C. Growth Substrates in NT-3 Grafts

Of note, the lesioned corticospinal axons responding to NT-3-secreting cell grafts in this experiment utilized the remaining ventral bridge of host *gray matter* as a growth substrate (Fig. 5) rather than extending into the *graft* or the remaining ventral bridge of host *white matter*. This pattern of axon growth is distinct from that observed in NGF-secreting grafts, in which sensory and coerulospinal axons directly penetrate the graft, presumably extending toward the highest concentration of NGF. Corticospinal axons on the other hand did not utilize the graft as a growth substrate despite the fact that axons clearly responded to the presence of an NT-3 secreting cell graft by extending for far greater distances than under control lesion–graft conditions. These findings suggest that either (1) the collagen–fibronectin substrate of the graft was

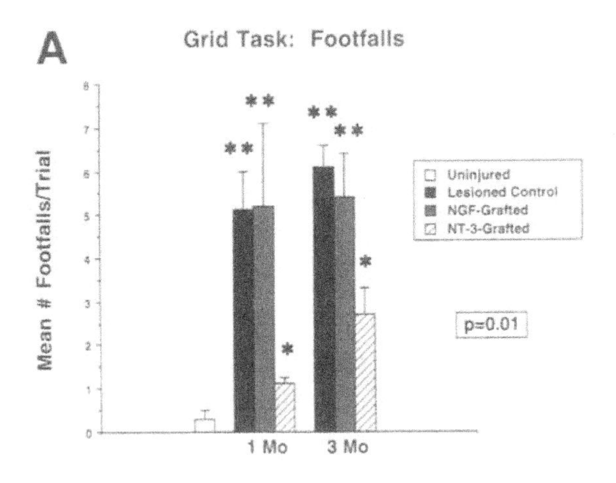

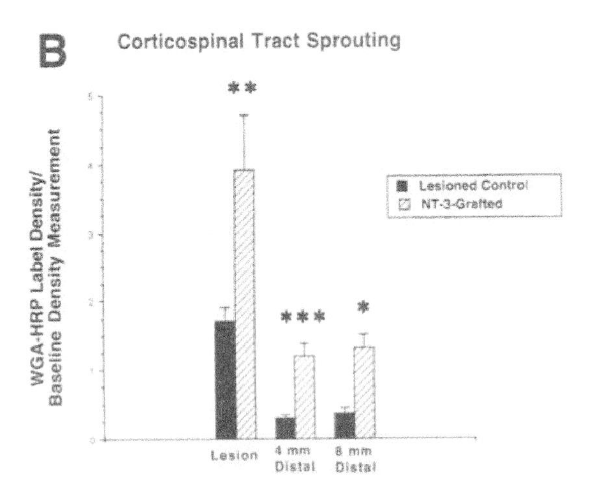

FIGURE 4 Functional recovery and enhanced corticospinal axonal growth in recipients of NT-3-secreting grafts. (A) Partial functional recovery is observed on a grid task-assessing sensorimotor integration in subjects grafted with NT-3-secreting fibroblasts compared with control- and NGF-grafted subjects (ANOVA, $p = 0.01$). NT-3-grafted subjects perform significantly better than control- or NGF-grafted subjects at one and three months post-grafting (*post hoc* Fischer's test), but NT-3-grafted subjects also differ significantly from nonlesioned subjects. **, Significant difference between NT3-grafted and nonlesioned animals; *, Significant difference from nonlesioned animals. Comparison of degree of recovery in NT-3 subjects at one and three months after lesion shows no significant difference (*post hoc* Fischer's test). (B) A significant increase in corticospinal axonal growth is evident in NT-3-grafted subjects compared to control, uninfected fibroblast-grafted animals at the lesion site and 4 mm and 8 mm distal to the caudal aspect of the lesion. $^{*}p < 0.05$; $^{**}p < 0.01$; $^{***}p < 0.005$ (reproduced from Grill *et al.*, 1997b).

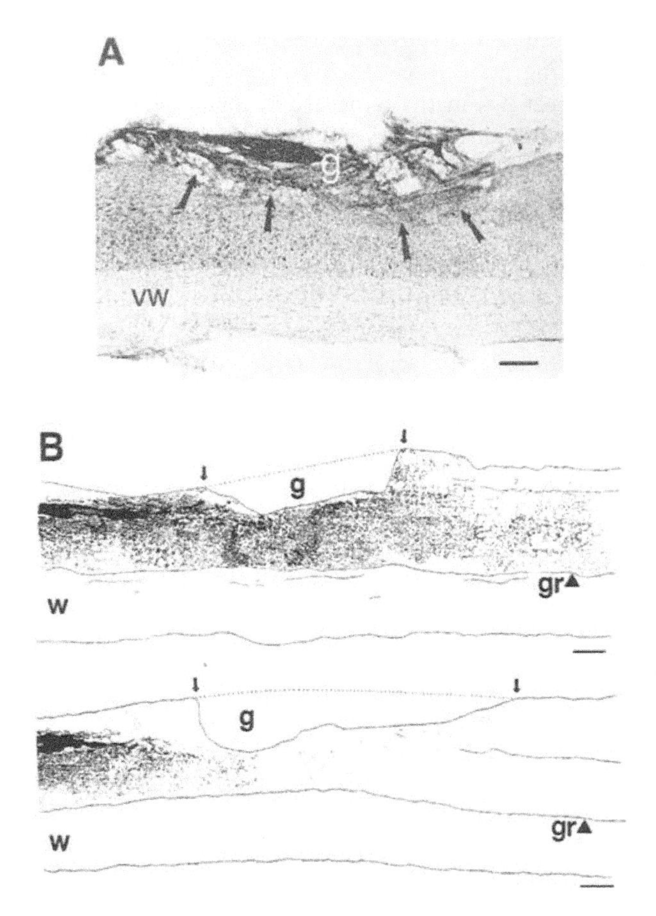

FIGURE 5 Morphology of NT-3-secreting cell grafts in the injured rat spinal cord. (A) Nissl-stained parasagittal section of NT3-secreting cell graft in dorsal hemisection spinal cord lesion cavity three months post-grafting. Graft (g) occupies lesion cavity. Arrows indicate graft/host interface. vw, ventral white matter. Scale bar = 100 μm. (B) Camera lucida drawings demonstrate enhanced growth of corticospinal axons exclusively through host gray matter rather than through the graft itself or through ventral host white matter. This pattern of growth is observed in both NT-3-grafted (top) and control-grafted (bottom) subjects, but is far greater in NT-3-grafted subjects. Lesion and graft (g) sites are outlined by dashed line and arrows. gr, remaining gray matter; w, white matter. Scale bar = 330 μm (reproduced from Grill *et al.*, 1997b).

nonpermissive to corticospinal axon growth, (2) inhibitory components of the extracellular environment at the host–graft interface prevented corticospinal axons from penetrating the graft, (3) host gray matter substrates more strongly attracted axons than graft or white matter substrates, or (4) the expression of a different growth-promoting substance was induced in the host gray matter by the NT-3-secreting graft. Interestingly, the fact that corticospinal axons extended through host gray matter rather than penetrating the graft itself made axonal growth for some distance beyond the lesion site possible: rather than becoming glued within the NT-3-secreting cell graft itself, axons grew for 8 mm through host gray matter distal to the lesion site, presumably responding to NT-3 that diffused from the graft or to some other factor in the host gray matter. Does this topography of axon growth indicate that nonchemotropic axonal extension occurred (growth that failed to follow the highest concentration of growth factor)? Presumably not, because axons in all likelihood followed the highest gradient of growth factor available to them *in a supportive growth substrate.* This hypothesis remains to be proven, however, by comparing NT-3 protein levels in the graft and host cord at successively greater distances from the graft. Thus, this experiment showed that the lesioned corticospinal tract responds to grafts of NT-3-producing cells, that these manipulations elicit functional recovery, and that distinct classes of axons possess differing affinities to similar extracellular growth milieus. The latter feature may be important for designing optimal strategies for promoting axonal regeneration.

D. Mechanisms of Functional Recovery with NT-3 Grafts

Axons in recipients of NT-3-secreting cell grafts extended approximately 8 mm caudal to the lesion site. An 8 mm growth distance is insufficient to reach motor neuron pools in the lumbosacral spinal cord, however, and *direct* reinnervation of the distal cord is therefore unlikely to account for the functional recovery observed. Rather, indirect mechanisms of graft-mediated functional recovery may be hypothesized: (1) Duffy and co-workers (1990) reported that functional recovery in the regenerated lizard spinal cord is mediated by formation of short polysynaptic projections from injured axons to caudal motor neuron pools, rather than by long-distance regrowth of axons. Further, stimulation of cortical motor neurons in intact cats (Kostyuk, Vasilenko, 1978) or after lesions of the corticospinal projection (Alstermark and Sasaki, 1985) generates excitation of spinal motor neurons through local propriospinal axon relays. Thus the partial functional recovery observed in this experiment may be mediated by excitation of distal motoneuron pools via regrown corticospinal projections propagating excitation through polysynaptic relays. (2) Corticospinal axons influence the intrinsic spinal locomotor pattern generator (Leonard and Goldberger, 1987; Grillner, 1986). Regrowing corticospinal projections in this experiment may have modulated the function of more rostrally located components of the spinal pattern generator, rather than directly reinnervating lumbar motoneurons. (3) Regrowing corticospinal axons may interact with other spinal projection systems at lower thoracic levels that modulate locomotor function. For example, sprouting of serotonergic systems can influence functional recovery in some lesion paradigms (Bregman *et al.*, 1993, 1995), and sprouting/regenerating corticospinal projections may interact with these motor system modulators (Thor *et al.*, 1993). Electrophysiological studies can address these possibilities.

Thus, this NT-3 study demonstrated that (1) NT-3 elicited the specific and robust growth of corticospinal motor axons, a important system for voluntary modulation of fine motor function; (2) animals that received NT-3-secreting grafts showed functional recovery; and (3) different classes of axons in the injured spinal cord exhibited differing affinities to the same growth substrates and environment. As with other studies reporting functional recovery in the spinal cord, the present functional findings also require replication by other investigators to fully accept their validity.

E. Studies in Chronic Spinal Cord Injury

The majority of the 200,000 people in the United States and 2 million people worldwide with spinal cord injuries are in chronic stages of injury. The relevance of regeneration strategies for promoting recovery of function in humans must therefore address chronic injury.

Although studies in chronic spinal cord injury models are relatively few in number, in general they support the concept that significant axonal growth is elicitable after chronic injury (Grill *et al.*, 1997a; Houle and Ye, 1997; Houle *et al.*, 1996; Stokes and Reier, 1992). The vast majority of neuronal cell bodies giving rise to spinal cord axons survive injury and their axons remain relatively close to the site of injury. Therefore, a favorable set of conditions for eliciting axonal regrowth potentially persists for many years after injury. Successful axonal regrowth after chronic injury requires the same factors that determine whether axons will regenerate after acute injury: (1) the expression of genes required to enter a new growth state and to synthesize cellular proteins necessary for axonal elongation, including neurofilaments, GAP-43 (Hagg *et al.*, 1990a), Tα1-tubulin (Kobayashi *et al.*, 1997), and possibly bcl-2 (Chen *et al.*, 1997); (2) the presence of a supportive extracellular matrix and cell adhesion molecules or recognition molecules such as L1, N-CAM, and others (Schachner and Martini, 1995; Zhang *et al.*, 1995b; Zhang *et al.*, 1995a; Goodman *et al.*, 1984; Edgar, 1989; Rutishauser *et al.*, 1988; Takeichi, 1988; Lagenauer and Lemmon, 1987; Rathjen *et al.*, 1987a; Mounard, 1988); (3) the expression of receptors to interact with extracellular factors that may promote axonal growth, including neurotrophin receptors and cell adhesion molecule receptors (Reichardt *et al.*, 1989; Turner and Flier, 1989); (4) the presence of guidance cues to direct growth toward appropriate targets (Harrelson and Goodman, 1988; Rathjen *et al.*, 1987b; Dodd *et al.*, 1988; Yamamoto *et al.*, 1986; Tisler *et al.*, 1981; Chuong and Edelman, 1984; Dodd and Jessell, 1988); (5) remyelination; and (6) the avoidance of inhibitory components found in injured extracellular matrix, CNS myelin, and possibly glia (Schwab, 1990; McKerracher *et al.*, 1994; Mukhopadhyay *et al.*, 1994; Filbin, 1995; Silver, 1994). In models of chronic injury using either fetal transplantation paradigms or neurotrophin infusions, some regrowth of injured axons has indeed been reported (Stokes and Reier, 1992; Houle and Reier, 1988; Houle, 1991; Houle and Ye, 1997; Houle *et al.*, 1996).

Recently we determined whether chronically injured axons of the adult spinal cord retained responsiveness to nerve growth factor-secreting genetically modified cells (Grill *et al.*, 1997a). Adult rats underwent spinal cord dorsal hemisection lesions (Grill *et al.*, 1997b). Three months later they received grafts of NGF-transduced cells to the chronic lesion cavity, or grafts of control, β-galactosidase-transduced fibroblasts. A further three months later (six months after the original injury), rats exhibited robust axonal growth that appeared to be equal in extent to responses observed after acute grafting studies (Fig. 6). As observed with grafts to the acutely injured spinal cord, primary sensory and supraspinal coerulospinal axons responded to NGF-secreting cells. We are in the process of determining whether NT-3-secreting cells elicit corticospinal axon growth after chronic injury, and the functional consequences of these interventions.

Thus, it appears that chronically injured axons of the adult mammalian spinal cord retain responsiveness to neurotrophic factors, and can mount substantial growth responses after injury. Whether growth in this model can occur for substantial distances, innervate distant targets, and elicit functional recovery remains to be determined.

VI. THE EXTRACELLULAR MATRIX AND SPINAL CORD INJURY

Interactions of injured axons with the extracellular matrix comprise an important determinant of axonal responses to injury. During development, cell-cell interactions and cell-substratum interactions influence the presence, strength, and direction of axonal growth (see Chapter 3) and (Purves and Lichtman, 1985; Baron van Evercooren *et al.*, 1982; Schinstine and Cornbrooks, 1990; Letourneau, 1975, 1983). After injury in the adult, peripheral axons readily regenerate, in part supported by interactions of injured axons with permissive components of the extracellular environment, the surfaces of other axons, and

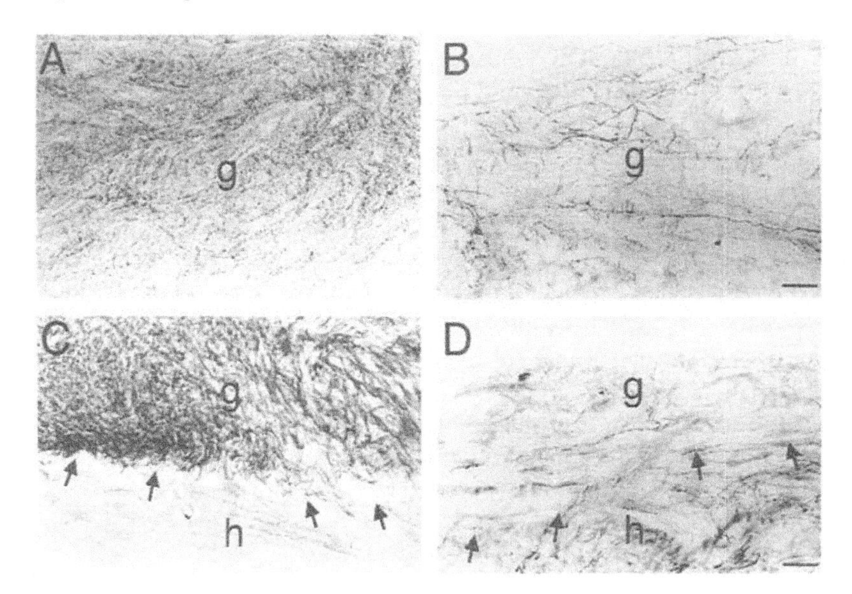

FIGURE 6 Neurotransmitter phenotype of penetrating axons within NGF-secreting cell grafts in the chronically injured rat spinal cord. (A) Coerulospinal axons identified by tyrosine hydroxylase (TH) immunolabeling robustly penetrate NGF-producing grafts. (B) Little TH immunolabeling is evident within control grafts. (C) Primary afferent sensory axons of the dorsolateral fasciculus identified by CGRP-immunolabeling densely penetrate NGF-grafted, chronically injured spinal cords. (D) Few CGRP-IR fibers are found in control-grafted subjects. All subjects were grafted three months after lesions and sacificed three months after grafting. g, Grafts; h, host tissue; arrows indicate graft–host interface. Scale bar = 150 μm (reproduced from Grill *et al.*, 1997a).

supportive features of Schwann cells. In the CNS, axonal regeneration after injury appears to be inhibited by features of CNS myelin (Schwab, 1990; Filbin, 1995; McKerracher *et al.*, 1994), the CNS extracellular matrix (Silver, 1994), and inhibitory components of astroglia (McKeon *et al.*, 1991). As reviewed earlier, the neutralization of myelin-associated growth inhibitors increases the distance over which injured axons will grow. As our experiments comparing NGF-secreting with NT-3-secreting cellular grafts have also shown, the nature of the growth milieu is important in determining growth of injured CNS axons, even when axons are isolated from inhibitory components of CNS white matter. Different types of spinal cord axons appear to exhibit differing affinities for identical CNS substrates, indicating that the identification of conducive growth substrates for each class of axon will be necessary to design an optimal regeneration environment.

VII. PROMOTING REGENERATION AFTER SPINAL CORD INJURY: A COMBINED APPROACH IS REQUIRED

From the preceding discussions, it is clear that no single factor accounts for the lack of axonal regeneration after spinal cord injury. Factors that stimulate axonal growth, such as nervous system growth factors, must be combined with the provision of conducive growth bridges and substrates. Optimally, to achieve long-distance regeneration, neutralization of myelin-associated inhibitors of CNS white matter might also be required. Thus, combined approaches that integrate several of these methods will likely optimize axonal regeneration. In this endeavor, as in most challenges that require the surmounting of formidable obstacles, cooperation amongst individuals with various areas of expertise will be the most fruitful route to reversing the human tragedy of spinal cord injury.

ACKNOWLEDGMENTS

Supported by the Hollfelder Foundation, American Paralysis Association, Paralyzed Veterans of America, Veterans Affairs, and the National Institute of Neurological Disorders and Stroke (NINDS).

REFERENCES

ALS CNTF Treatment Study Group. (1996). A double-blind placebo-controlled clinical trial of subcutaneous recombinant human ciliary neurotrophic factor (rHCNTF) in amyotrophic lateral sclerosis. *Neurology* 46:1244–1249.

Alstermark, B., and Sasaki, S. (1985). Integration in descending motor pathways controlling the forelimb in the cat. 13. Corticospinal effects in shoulder, elbow, wrist and digit motoneurones. *Exp Brain Res* 59:353–364.

Apfel, S. Y., Wright, D. E., Wiideman, A. M., Dormia, C., Snider, W. D., and Kessler, J. A. (1996). Nerve growth factor upregulates expression of brain-derived neurotrophic factor in the peripheral nervous system. *Molec Cell Neurosci* 7:134–142.

Assouline, J. G., Bosch, P., Lim, R., Kim, I. S., Jensen, R., and Pantazis, N. J. (1987). Rat astrocytes and Schwann cells in culture synthesize nerve growth factor-like neurite-promoting factors. *Dev Brain Res* 31:103–118.

Baron van Evercooren, A. H., Kleinman, H. K., Ohno, S., Marangos, J. P., Schwartz, J. P., and Dubois-Dalq, M. E. (1982). Nerve growth factor, laminin and fibronectin promote neurite growth in human fetal sensory ganglion cultures. *J Neurosci Res* 8:179–194.

Blesch, A., and Tuszynski, M. H. (1996). Gene therapy for neurological disease. *Clin Neurosci* 3:268–274.

Bregman, B. (1987). Spinal cord transplants permit the growth of serotonergic axons across the site of neonatal spinal cord transection. *Dev Brain Res* 34:265–279.

Bregman, B. S., Kunkel-Bagden, E., Reier, P. J., Dai, H. N., McAtee, M., and Gao, D. (1993). Recovery of function after spinal cord injury: Mechanisms underlying transplant-mediated recovery of function differ after spinal cord injury in newborn and adult rats. *Exp Neurol* 123:3–16.

Bregman, B. S., Kunkel-Bagden, E., Schnell, L., Dai, H. N., Gao, D., and Schwab, M. E. (1995). Recovery from spinal cord injury by antibodies to neurite growth inhibitors. *Nature* 378:498–501.

Brown, J. O., and McCough, G. P. (1947). Abortive regeneration of the transected spinal cord. *J Comp Neurol* 87:131–137.

Chen, D. F., Schneider, G. E., Martinou, J. C., and Tonegawa, S. (1997). Bcl-2 promotes regeneration of severed axons in mammalian CNS. *Nature* 385:434–439.

Cheng, H., Yihai, C., and Olson, L. (1996). Spinal cord repair in adult paraplegic rats: Partial restoration of hind limb function. *Science* 273:510–513.

Chuong, C. M., and Edelman, G. M. (1984). Alterations in neural cell adhesion molecules during development of different regions of the nervous system. *J Neurosci* 4:2354–2357.

Clemente, C. D. (1958). The regeneration of peripheral nerves inserted into the cerebral cortex and the healing of cerebral lesions. *J Comp Neurol* 109:123–151.

Clemente, C. D. (1964). Regeneration in the vertebrate central nervous system. *Int Rev Neurobiol* 6:257–301.

Conner, J. M., and Varon, S. (1996). Maintenance of sympathetic innervation into the hippocampal formation requires a continuous local availability of nerve growth factor. *Neuroscience* 72:933–945.

Diener, P. S., and Bregman, B. S. (1998a). Fetal spinal cord transplants support the development of target reaching and coordinated postural adjustments after neonatal spinal cord injury. *J Neurosci* 18:763–778.

Diener, P. S., and Bregman, B. S. (1998b). Fetal spinal cord transplants support growth of supraspinal cord segmental projections after cervical spinal cord hemisection in the neonatal rat. *J Neurosci* 18:779–793.

Dobrea, G. M., Unnerstall, J. R., and Rao, M. S. (1992). The expression of CNTF message and immunoreactivity in the central and peripheral nervous system of the rat. *Dev Brain Res* 66:209–219.

Dodd, J., and Jessell, T. M. (1988). Axon guidance and the patterning of neuronal projections in vertebrates. *Science* 242:692–699.

Dodd, J., Morton, S. B., Karagogeos, D., Yamamoto, M., and Jessell, T. M. (1988). Spatial regulation of axonal glycoprotein expression in subsets of embryonic spinal neurons. *Neuron* 1:105–116.

Duffy, M. T., Simpson, S. B., Liebich, D. R., and Davis, B. M. (1990). Origin of spinal cord axons in the lizard regenerated tail: Supernormal projections from local spinal neurons. *J Comp Neurol* 293:208–222.

Duncan, I. R., Hammang, J. P., Jackson, K. F., Wood, P. M., Bunge, R. P., and Langford, L. (1988). Transplantation of oligodendrocytes and Schwann cells into the spinal cord of the myelin-deficient rat. *J Neurocytol* 17:351–360.

Eckenstein, F. P., Shipley, G. D., and Nishi, R. (1991). Acidic and basic fibroblast growth factors in the nervous system: Distribution and differential alteration of levels after injury of central vs. peripheral nerve. *J Neurosci* 11:412–419.

Eckenstein, F. P., Andersson, C., Kuzis, K., and Woodward, W. R. (1994). Distribution of acidic and basic fibroblast growth factors in the mature, injured and developing rat nervous system. *Prog Brain Res* 103:55–64.

Edgar, D. (1989). Neuronal laminin receptors. *TINS* 12:248–251.

Emoto, N., Gonzales, A. M., Walicke, P., Wada, E., Simmons, D. M., Shimasaki, S., and Baird, A. (1989). Basic fibroblast growth factor (FGF) in the central nervous system: Identification of specific loci of basic FGF expression in the rat brain. *Growth Factors* 2:21–29.

Ferguson, I. A., and Johnson, E. M. (1991). Fibroblast growth factor receptor-bearing neurons in the CNS: Identification by receptor-mediated retrograde transport. *J Comp Neurol* 313:693–706.

Filbin, M. T. (1995). Myelin-associated glycoprotein: A role in myelination and in the inhibition of axonal regeneration? *Curr Opin Neurobiol* 5:588–595.

Friedman, B., Scherer, S. S., Rudge, J. S., Helgren, M., Morrisey, D., McClain, J., Wang, D. Y., Wiegand, S. J., Furth, M. E., Lindsay, R. M., *et al.* (1992). Regulation of ciliary neurotrophic factor expression in myelin-related schwann cells in vivo. *Neuron* 9:295–305.

Friedmann, T., and Roblin, R. (1972). Gene therapy for human genetic disease? *Science* 175:949–955.

Gomez-Pinilla, F., and Cotman, C. W. (1993). Distribution of fibroblast growth factor 5 mRNA in the rat brain: An *in situ* hybridization study. *Brain Res* 606:79–86.

Goodman, C. S., Bastiani, M. J., Doe, C. Q., du Lac, S., Helfand, S. L., Kuwada, K. Y., and Thomas, J. B. (1984). Cell recognition during neuronal development. *Science* 225:1271–1279.

Gossen, M., Bonin, A. L., Freundlieb, S., and Bujard, H. (1994). Inducible gene expression systems for higher eukaryotic cells. *Curr Opin Biotechnol* 5:516–520.

Grill, R., Blesch, A., Tuszynski, M. H. (1997a). Robust growth of chronically injured spinal cord axons induced by grafts of genetically modified NGF-secreting cells. *Exp Neurol* 148:444–452.

Grill, R., Murai, K., Blesch, A., Gage, F. H., and Tuszynski, M. H. (1997b). Cellular delivery of neurotrophin-3 promotes corticospinal axonal growth and partial functional recovery after spinal cord injury. *J Neurosci* 17:5560–5572.

Grillner, S. (1986). Locomotion in spinal vertebrates. Physiology and pharmacology. In: Goldberger, M. E., Gorio, A., and Murray, M., eds., *Development and plasticity of the mammalian spinal cord.* Padova: Liviana Press.

Gundersen, R. W., and Barrett, J. N. (1980). Characterization of the turning response of dorsal root neurites toward nerve growth factor. *J Cell Biol* 87:546–554.

Hagg, T., Vahlsing, H. L., Manthrope, M., and Varon, S. (1990a). Septohippocampal cholinergic axonal regeneration through peripheral nerve bridges: Quantification and temporal development. *Exp Neurol* 109:153–163.

Hagg, T., Vahlsing, L., Manthorpe, M., and Varon, S. (1990b). Nerve growth factor infusion into the denervated adult rat hippocampal formation promotes its cholinergic reinnervation. *J Neurosci* 10:3087–3092.

Harrelson, A. L., and Goodman, C. S. (1988). Growth cone guidance in insects: Fasciclin II is a member of the immunoglobin superfamily. *Science* 242:700–708.

Hendersen, C. E., Phillips, H. S., Pollock, R. A., Davies. A. M., Lemeulle, C., Armanini, M., Simpson, L. C., Moffet, B., Vandlen, R. A., Koliatsos, V. E., and Rosenthal, A. (1994). GDNF: A potent survival factor for motoneurons present in peripheral nerve and muscle. *Science* 266:1062–1064.

Henderson, J. T., Seniuk, N. A., and Roder, J. C. (1994). Localization of CNTF immunoreactivity to neurons and astrogia in the CNS. *Brain Res Mol Brain Res* 22:151–165.

Himes, B. T., Goldberger, M. E., and Tessler, A. (1994). Grafts of fetal central nervous system tissue rescue axotomized Clarke's nucleus neurons in adult and neonatal operates. *J Comp Neurol* 339:117–131.

Honmou, O., Felts, P. A., Waxman, S. G., and Kocsis, J. D. (1996). Restoration of normal conduction properties in demyelinated spinal cord axons in the adult rat by transplantation of exogenous Schwann cells. *J Neurosci* 16:3199–3208.

Houle, J. D. (1991). Demonstration of the potential for chronically injured neurons to regenerate axons into intraspinal peripheral nerve grafts. *Exp Neurol* 113:1–9.

Houle, J. D., and Reier, P. J. (1988). Transplantation of fetal spinal cord tissue into the chronically injured adult rat spinal cord. *J Comp Neurol* 269:535–547.

Houle, J. D., and Reier, P. J. (1989). Regrowth of calcitonin gene-related peptide (CGRP) immunoreactive axons from the chronically injured rat spinal cord into fetal spinal cord tissue transplants. *Neurosci Lett* 103:253–258.

Houle, J. D., and Ye, J.-H. (1997). Treatment of the chronically injured spinal cord with neurotrophic factors can promote axonal regeneration from supraspinal axons. *Exp Neurol* 143:70–81.

Houle, J. D., Ye, J.-H., and Kane, C. J. M. (1996). Axonal regeneration by chronically injured supraspinal neurons can be enhanced by exposure to insulin-like growth factor, basic fibroblast growth factor or transforming growth factor beta. *Restor Neurol Neurosci* 10:205–215.

Kao, C. C., Chang, L. W., and Bloodworth, J. M. B. (1977). Axonal regeneration across transected mammalian spinal cords: An electron microscopic study of delayed microsurgical grafting. *Exp Neurol* 54:591–615.

Keirstead, S. A., Rasminsky, M., Fukuda, Y., Carter, D., and Aguayo, A. J. (1989). Electrophysiologic responses in hamster superior colliculus evoked by regenerating retinal axons. *Science* 246:255–258.

Kobayashi, N. R., Fan, D.-P., Giehl, K. M., Bedard, A. M., Wiegand, S. J., and Tetzlaff, W. (1997). BDNF and NT-4/5 prevent atrophy or rat rubrospinal neurons after cervical axotomy, stimulate GAP-43 and Alpha-1-tubulin mRNA expression, and promote axonal regeneration. *J Neurosci* 17:9583–9595.

Kostyuk, P. G., and Vasilenko, D. A. (1978). Propriospinal neurones as a relay system for transmission of corticospinal influences. *J Physiol (Paris)* 74:247–250.

Kuhlengel, K. R., Bunge, M. B., Bunge, R. P., and Burton, H. (1990). Implantation of cultured sensory neurons and Schwann cells into lesioned spinal cord. II. Implant characteristics and examination of corticospinal tract growth. *J Comp Neurol* 293:74–91.

Kunkel-Bagden, E., and Bregman, B. S. (1990). Spinal cord transplants enhance the recovery of locomotor function after spinal cord injury at birth. *Exp Brain Res* 81:25–34.

Kunkel-Bagden, E., Dai, H. N., and Bregman, B. S. (1992). Recovery of function after spinal cord hemisection in newborn and adult rats: Differential effects on reflex and locomotor function. *Exp Neurol* 116:40–51.

Lagenauer, C., and Lemmon, V. (1987). An L1-like molecule, the 8D9 antigen, is a potent substrate for neurite extension. *Proc Natl Acad Sci U S A* 84:7753–7757.

Le Gros Clark, W. E. (1942). The problem of neuronal regeneration in the central nervous system. 1. The influence of spinal ganglia and nerve fragments grafted in the brain. *J Anatomy* 77:20–48.

Leonard, C. T., and Goldberger, M. E. (1987). Consequences of damage to the sensorimotor cortex in neonatal and adult cats. I. Sparing and recovery of function. *Dev Brain Res* 32:1–14.

Letourneau, P. (1975). Possible roles for cell-to-substratum adhesion in neuronal morphogenesis. *Dev Biol* 44:92–101.

Letourneau, P. C. (1978). Chemotaxic response of nerve fiber elongation to nerve growth factor. *Dev Biol* 66:183–196.

Letourneau, P. C. (1983). Axonal growth and guidance. *TINS* 1:451–455.

Levi-Montalcini, R. (1987). The nerve growth factor 35 years later. *Science* 237:1154–1162.

Li, Y., Field, P. M., and Raisman, G. (1997). Repair of adult rat corticospinal tract by transplants of olfactory ensheathing cells. *Science* 277:2000–2002.

Mahut, H. (1972). A selective spatial deficit in monkeys after transection of the fornix. *Neuropsychologia* 10:65–74.

Masu, Y., Wolf, E., Holtmann, B., Sendtner, M., Brem, G., and Thoenen, H. (1993). Disruption of the CNTF gene results in motor neuron degeneration. *Nature* 365:27–32.

McKeon, R. J., Schreiber, R. C., Rudge, J. S., and Silver, J. (1991). Reduction of neurite outgrowth in a model of glial scarring following CNS injury is correlated with expression of inhibitory molecules on reactive astrocytes. *J Neurosci* 11:3398–3411.

McKerracher, L., David, S., Jackson, D. L., Kottis, V., Dunn, R. J., and Braun, P. E. (1994). Identification of myelin-associated glycoprotein as a major myelin-derived inhibitor of neurite growth. *Neuron* 13:805–811.

Meyer, M., Matsuoka, I., Wetmore, C., Olson, L., and Thoenen, H. (1992). Enhanced synthesis of brain-derived neurotrophic factor in the lesioned peripheral nerve: Different mechanisms are responsible for the regulation of BDNF and NGF mRNA. *J Cell Biol* 119:45–54.

Mori, F., Himes, B. T., Kowada, M., Murray, M., and Tessler, A. (1997). Fetal spinal cord transplants rescue some axotomized rubrospinal neurons from retrograde cell death in adult rats. *Exp Neurol* 143:45–60.

Mounard, D. (1988). Cell-derived proteases and protease inhibitors as regulators of neurite outgrowth. *TINS* 11:541–544.

Mukhopadhyay, G., Doherty, P., Walsh, F. S., Crocker, P. R., and Filbin, M. T. (1994). A novel role for myelin-associated glycoprotein as an inhibitor of axonal regeneration. *Neuron* 13:757–767.

Nakahara, Y., Senut, M-C., Gage, F. H., and Tuszynski, M. H. (1996). Grafts of fibroblasts genetically modified to secrete NGF, BDNF, NT-3 or basic FGF elicit differential responses in the adult spinal cord. *Cell Transpl* 5:191–204.

Purves, D., and Lichtman, J. W. (1985). *Principles of neural development*. Sunderland, Mass: Sinauer.

Ramon y Cajal, S. (1928). *Degeneration and regeneration of the nervous system*. London: Oxford University Press.

Rathjen, F. G., Wolff, J. M., Chang, S., Bonhoeffer, F., and Raper, J. A. (1987a). Neurofascin: A novel chick cell surface glycoprotein involved in neurite-neurite interactions. *Cell* 52:841–849.

Rathjen, F. G., Wolff, J. M., Frank, R., Bonhoeffer, F., and Rutishauser, U. (1987b). Membrane glycoproteins involved in neurite fasciculation. *J Cell Biol* 104:343–353.

Reichardt, L. F., Bixby, J. L., Hall, D. E., Ignatius, M. J., Neugebauer, K. M., and Tomaselli, K. J. (1989). Integrins and cell adhesion molecules: Neuronal receptors that regulate axon growth on extracellular matrices and cell surfaces. *Dev Neurosci* 11:332–347.

Reier, P. J., Anderson, D. K., Thompson, F. J., and Stokes, B. T. (1992a). Neural tissue transplantation and CNS trauma: Anatomical and functional repair of the injured spinal cord. *J Neurotrauma* 9(Suppl. 1):S223–S248.

Reier, P. J., Stokes, B. T., Thompson, F. J., and Anderson, D. K. (1992b). Fetal cell grafts into resection and contusion/compression injuries of the rat and cat spinal cord. *Exp Neurol* 115:177–188.

Richardson, P. M., McGuiness, U. M., and Aguayo, A. J. (1980). Axons from CNS neurons regenerate into PNS grafts. *Nature* 284:264–265.

Rutishauser, U., Acheson, A., Hall, A. K., Mann, D. M., and Sunshine, J. (1988). The neural cell adhesion molecule (NCAM) as a regulator of cell-cell interactions. *Science* 240:53–57.

Schachner, M., and Martini, R. (1995). Glycans and the modulation of neural-recognition molecule function. *Trends Neurosci* 18:183–191.

Schinstine, M., and Cornbrooks, C. J. (1990). Axotomy enhances the outgrowth of neurites from embryonic rat septal-basal forebrain neurons on a laminin substratum. *Exp Neurol* 108:10–22.

Schnell, L., and Schwab, M. E. (1990). Axonal regeneration in the rat spinal cord produced by antibody against myelin-associated neurite growth inhibitors. *Nature* 343:269–272.

Schnell, L., Schneider, R., Kolbeck, R., Barde, Y. A., and Schwab, M. E. (1994). Neurotrophin-3 enhances sprouting of corticospinal tract during development and after adult spinal cord lesion. *Nature* 367:170–173.

Schumacher, J. M., Short, M. P., Hyman, B. T., Breakefield, X. O., and Isacson, O. (1991). Intracerebral implantation of nerve growth factor-producing (1991). Intracerebral implantation of nerve growth factor-producing fibroblasts protects striatum against neurotoxic levels of excitatory amino acids. *Neuroscience* 45:561–570.

Schwab, M. E. (1990). Myelin-associated inhibitors of neurite growth and regeneration in the CNS. *TINS* 13:452–456.

Silver, J. (1994). Inhibitory molecules in development and regeneration. *J Neurology* 242:S22–S24.

Stokes, B. T., and Reier, P. J. (1992). Fetal grafts alter chronic behavioral outcome after contusion damage to the adult rat spinal cord. *Exp Neurol* 116:1–12.

Sugar, O., and Gerard, R. W. (1940). Spinal cord regeneration on the rat. *J Neurophysiol* 3:1–19.

Takeichi, M. (1988). The cadherins: Cell-cell adhesion molecules controlling animal morphogenesis. *Development* 102:639–655.

Tello, F. (1911a). Un experimento sobre la influencia del neurotropismo en la regeneracion de la corteza cerebral. *Rev Clin Madr* 5:292–294.

Tello, F. (1911b). La influencia del neurotropismo en la regeneracion de las centros nerviosos. *Trab Lab Invest Biol Univ Madr* 9:123–159.

Thoenen, H. (1995). Neurotrophins and neuronal plasticity. *Science* 270:593–598.

Thor, K. B., Nickolaus, S., and Helke, C. J. (1993). Autoradiographic localization of 5-hydroxytryptamine 1A, 5-hydroxytryptamine 1B and 5-hydroxytryptamine 1C/2 binding sites in the rat spinal cord. *Neurosci* 55:235–252.

Tisler, G. D., Schneider, M. D., and Nirenberg, M. (1981). A topographic gradient of molecules in retina can be used to identify neuron positions. *Proc Natl Acad Sci U S A* 78:2145–2149.

Turner, D. C., and Flier, L. A. (1989). Receptor-mediated active adhesion to the substratum is required for neurite outgrowth. *Dev Neurosci* 11:300–312.

Tuszynski, M. H., Peterson, D. A., Ray, J., Baird, A., Nakahara, Y., and Gage, F. H. (1994). Fibroblasts genetically modified to produce nerve growth factor induce robust neuritic ingrowth after grafting to the spinal cord. *Exp Neurol* 126:1–14.

Tuszynski, M. H., Gabriel, K., Gage, F. H., Suhr, S., Meyer, S., and Rosetti, A. (1996). Nerve growth factor delivery by gene transfer induces differential outgrowth of sensory, motor and noradrenergic neurites after adult spinal cord injury. *Exp Neurol* 137:157–173.

Tuszynski, M. H., Murai, K., Blesch, A., Grill, R., and Miller, I. (1997). Functional characterization of NGF-secreting cell grafts to the acutely injured spinal cord. *Cell Transpl* 6:361–368.

Tuszynski, M. H., Weidner, N., McCormack, M., Miller, I., Powell, H., and Conner, J. (1998). Grafts of genetically modified Schwann cells to the spinal cord: Survival, axon growth and myelination. *Cell Transpl:* In press.

Williams, L. R., Longo, F. M., Powell, H. C., Lundborg, G., and Varon, S. (1983). Spatial-temporal progress of peripheral nerve regeneration within a silicone chamber: Parameters for a bioassay. *J Comp Neurol* 218:460–470.

Xu, X. M., Guenard, V., Kleitman, N., Aebischer, P., and Bunge, M. B. (1995a). A combination of BDNF and NT3 promotes supraspinal axonal regeneration in Schwann cell grafts in adult rat thoracic spinal cord. *Exp Neurol* 134:261–272.

Xu, X. M., Guenard, V., Kleitman, N., and Bunge, M. B. (1995b). Axonal regeneration into Schwann cell-seeded guidance channels grafted into transected adult rat spinal cord. *J Comp Neurol* 351:145–160.

Yamamoto M., Boyer, A. M., Crandall, J. E., Edwards, M., and Tanaka, H. (1986). Distribution of stage-specific neurite-associated proteins in the developing murine nervous system recognized by a monocloncal antibody. *J Neurosci* 6:3576–3594.

Zhang, Y., Campbell, G., Anderson, P. N., Martini, R., Schachner, M., and Lieberman, A. R. (1995a). Molecular basis of interactions between regenerating adult rat thalamic axons and Schwann cells in peripheral nerve grafts. II. Tenascin-C. *J Comp Neurol* 361:210–224.

Zhang, Y., Campbell, G., Anderson, P. N., Martini, R., Schachner, M., and Lieberman, A. R. (1995b). Molecular basis of interactions between regenerating adult rat thalamic axons and Schwann cells in peripheral nerve grafts. I. Neural cell adhesion molecules. *J Comp Neurol* 361:193–209.

CHAPTER 25

Neurotrophins and Neuroprotection Improve Axonal Regeneration into Schwann Cell Transplants Placed in Transected Adult Rat Spinal Cord

MARY BARTLETT BUNGE,*,† AND NAOMI KLEITMAN,*

*The Chambers Family Electron Microscopy Laboratory, The Miami Project to Cure Paralysis, Departments of *Neurological Surgery and †Cell Biology and Anatomy, University of Miami School of Medicine, Miami, Florida*

Can implanted cellular bridges enable axons to regrow across areas of injury in the adult mammalian spinal cord? We describe here studies demonstrating that bridges constructed of rat Schwann cells promote axonal regeneration across injury areas in adult rat spinal cord. Human Schwann cell constructs are effective in promoting axonal regeneration in the rat as well. In combination with rat Schwann cell transplants, infusion of the neurotrophins, brain-derived neurotrophic factor and neurotrophin-3 into the graft or the administration of a neuroprotective agent, methylprednisolone, improve the regenerative response, including the initiation of axonal regeneration from distant brain stem neurons. When Schwann cells, genetically modified to secrete human brain-derived neurotrophic factor, are transplanted into a complete transection site and as a trail in the distal spinal cord beyond, axons grow across the transection site and along the Schwann cell trail. Thus, it appears that a combination of strategies, among them cellular transplantation, neurotrophin administration, and neuroprotection, will be more successful than any single approach in promoting recovery after spinal cord injury.

As the first quarter century era of the National Institute for Neurological Disorders and Blindness drew to a close, R. P. Bunge posed the question of

CNS Regeneration

whether cellular transplantation could usefully correct dysfunction of the central nervous system (CNS) in animals and eventually humans (Bunge, 1975). More specifically, he stated that, "The possibility may now exist for the construction of a type of cellular bridge for use in vertebrates without the native capacity for . . . [regenerative] activity in response to injury." Although it did not seem practical at the time to construct bridges of neurons, supporting cells such as Schwann cells appeared likely to succeed as components of a cellular bridge. Purified populations of Schwann cells generated in culture had just become available (Wood and Bunge, 1975). It had been known for a long time that the terrain that regenerating peripheral axons preferentially follow *in situ* is composed of Schwann cells aligned in a connective tissue framework. Ramón y Cajal (1928) had concluded, from work demonstrating the capacity for regenerative CNS axonal growth into segments of peripheral nerve, that the Schwann cells served to nourish and orient the wandering axonal sprouts that contacted them. He further wrote that the deficiency in CNS regeneration was due not to the inability of central neurons to regrow fibers, but to an environment lacking in nutritive and specific orienting substances.

As the second era in NIH history began, the conclusion set forth by Ramón y Cajal (1928) was clearly substantiated by a landmark paper by Richardson and colleagues that appeared in 1980. A peripheral nerve segment placed across a gap in adult rat thoracic spinal cord engendered regrowth of propriospinal fibers from both transected stumps. The regenerated fibers that grew from these stumps were myelinated by the Schwann cells within the transplanted nerve segment. A more complex transection and peripheral nerve transplantation paradigm in adult rat spinal cord has been utilized recently, combining spinal column stabilization, precise placement of multiple fine peripheral nerves, and the addition of fibrin containing a growth factor (FGF1) (Cheng *et al.*, 1996). This combination approach led to the involvement of supraspinal neurons and some behavioral recovery not observed in earlier peripheral nerve-bridging work (Richardson *et al.*, 1980, 1984).

Schwann cells possess many attributes that encourage their use in CNS transplantation. It was shown in 1981 that purified populations of cultured rat Schwann cells remyelinated axons after transplantation into lysolecithin-demyelinated adult mouse spinal cord (Duncan *et al.*, 1981); these PNS cells readily functioned in the CNS environment. A substantial body of literature using *in vitro* preparations has since demonstrated that Schwann cells promote the growth of axons from central neurons (reviewed in Bunge and Hopkins, 1990). Schwann cells generate numerous neurotrophic factors, express cell adhesion molecules and integrins on their surfaces, and secrete a number of extracellular matrix components, such as laminin, which are known to be permissive for neurite growth (reviewed in Guénard *et al.*, 1993). Bridges of cultured Schwann cells have been shown to be effective in uniting transected stumps of thoracic spinal cord and promoting growth of fibers from both stumps across the graft (Xu *et al.*, 1997). These Schwann cell-bridging experiments will be described in this chapter. Recent work indicating that combining additional strategies with Schwann cell transplantation leads to a more effective outcome after transection will also be summarized.

I. THE SCHWANN CELL-BRIDGING MODEL

Cables of Schwann cells in polymer guidance channels (Guénard *et al.*, 1992) have been used as grafts to bridge the completely transected spinal cord in adult Fischer rats. The Schwann cells were combined with a basal lamina matrix, Matrigel, and drawn into a guidance channel. During storage in culture medium overnight, the Schwann cells in Matrigel developed into a cable smaller in diameter than the interior of the guidance channel, and the Schwann cells aligned along the longitudinal axis of the channel (as in Guénard *et al.*, 1992). The Schwann cells, taken from adult Fischer rat sciatic nerves, were 98% purified and used at a final density of 120×10^6 ml; a 10 mm-long Schwann cell cable contained about 6 million cells. The guidance channels, composed of polyacrylonitrile–polyvinylchloride, are semipermeable (50 kDa molecular weight cutoff), have a smooth interior that leads to the syneresis (shrinkage and alignment) effect just described, and have been used successfully in peripheral nerve transplantation paradigms (Aebischer *et al.*, 1988, 1990). They provide a controlled environment for the grafts, creating regions largely free of astrocytes and oligodendrocytes, and limit the invasion of mesenchymal cells. The transected stumps of the spinal cord are inserted about 1 mm into the ends of the channel to be contiguous with the ends of the Schwann cell–Matrigel graft.

We chose a transection model for a number of reasons. Following a transection lesion, the appearance of appropriately traced fibers beyond the graft is an unambiguous indication of regenerative growth. Moreover, rather than concentrating on one or a few tracts, it was our goal to determine which spinal and supraspinal tracts would ultimately prove capable of responding to Schwann cell grafts. Finally, in a study of 46 cases of injured human spinal cords, 35% of these exhibited complete anatomical discontinuity across the spinal cord (Bunge *et al.*, 1993, 1997), indicating that functionally significant numbers of clinical injuries would require a total repair strategy as utilized in our model system. Our grafts were 6 to 10 mm in length, depending upon whether the channel was open at both ends or distally capped, respectively. Methods of evaluation (Fig. 1) included counting myelinated axons in plastic embedded cross-sections of the graft, immunostaining of sagittally cryosectioned areas of the graft and of graft–host cord interface regions, and antero-

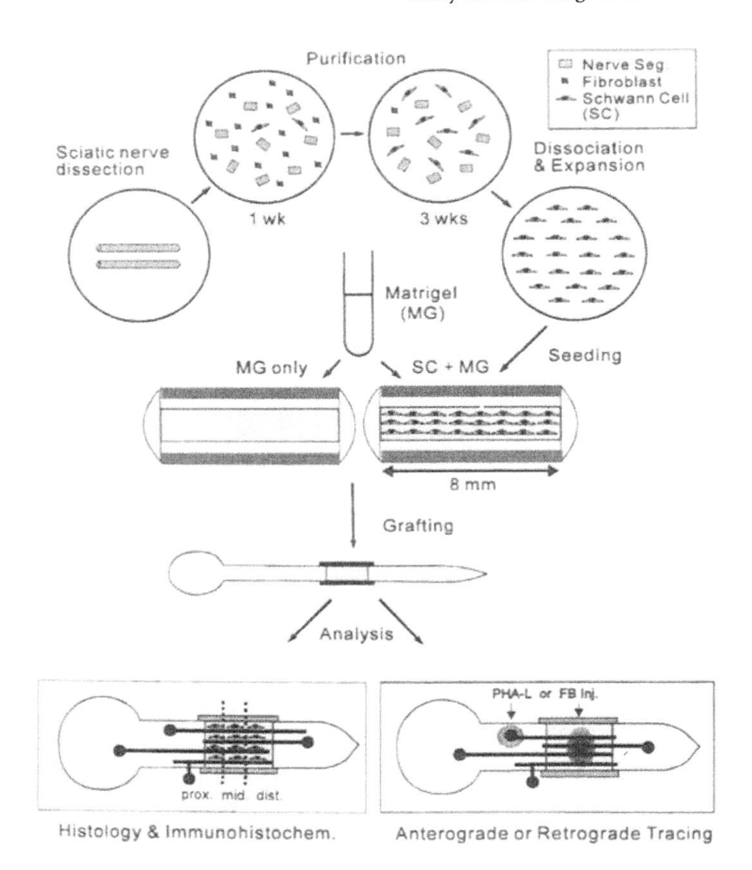

FIGURE 1 Schwann cells were purified and expanded in culture, suspended in Matrigel, and placed into guidance channels. After storage overnight, the channels were placed into spinal cord transected at T8 prior to removal of a 4 to 5 mm cord segment caudal to it. One month later, the tissues were examined histologically (from areas between the two dotted lines) and histochemically (in intervening areas plus adjacent cord). Some animals received an injection of Fast Blue into the Schwann cell cable to label neurons whose axons had regenerated into the graft. In other cases, an anterograde tracer, PHA-L or WGA-HRP, was injected into the cord 5 to 6 mm rostral to the graft to detect axonal growth along the cable and into the distal cord. (From Xu *et al.*, 1997; reprinted with permission of Chapman and Hall.)

grade and retrograde tracing through intact grafts to identify the origin of axons that regenerated into the grafts. The animals were maintained for 4 to 6 weeks. The transplantation paradigm just described was also employed to study the results of transplanting human Schwann cells into nude rats, which accept xenografts.

II. SCHWANN CELL BRIDGES UNITE THE TRANSECTED STUMPS

Schwann cell–Matrigel grafts, 6 mm in length, were transplanted inside guidance channels into a gap in the T8-9 spinal cord (Xu *et al.*, 1997). After one month, a bridge between the severed stumps had been formed (Fig. 2A, B), as observed from the gross and histological appearance and the ingrowth of propriospinal axons from both stumps. When the tracer Fast Blue was injected into the middle of the bridge (Fig. 3), propriospinal neurons (mean, 1064 $\pm$ 145 [$\pm$SEM]) were labeled as far away as C3 and S4. There was a mean of 1990 $\pm$ 594 myelinated axons and eight times as many unmyelinated by ensheathed axons at the graft midpoint (as resolved by electron microscopy). Myelinated and unmyelinated axons were not present in control Matrigel-only grafts (compare Fig. 2C and D). Earlier work with grafts in distally capped channels (Xu *et al.*, 1995a) or in collagen rolls rather than channels (Paino and Bunge, 1991; Paino *et al.*, 1994) had established that axonal growth into the grafts depends upon the presence of Schwann cells. Brain stem neurons were not retrogradely labeled from the graft; immunoreactive serotonergic and noradrenergic axons were observed only a short distance into the rostral end of the graft and did not reach the injection site of the tracer at the graft midpoint. Anterograde tracing with *Phaseolus vulgaris leucoagglutinin* (PHA-L) introduced above the graft demonstrated that axons extended into the graft but essentially did not leave its distal aspect. In sum, these bridging Schwann cell grafts supported regrowth of both ascending and descending axons from propriospinal neurons, regeneration of sensory axons, and limited regrowth of serotonergic and noradrenergic fibers from the rostral stump.

III. METHYLPREDNISOLONE IMPROVES THE EFFECTIVENESS OF SCHWANN CELL GRAFTS

In clinical studies, methylprednisolone has been shown to be neuroprotective if administration is begun within eight hours after spinal cord injury (Bracken *et al.*, 1990, 1997). We investigated whether axonal regrowth could be enhanced by methylprednisolone administration at the time of spinal cord transection and Schwann cell implantation (Chen *et al.*, 1996). Schwann cell grafts inside channels were prepared as described earlier. Methylprednisolone (30 mg/kg) or vehicle (control) was administered intravenously at five minutes, two hours, and four hours after transection. Thirty to 45 days later, the Schwann cell–methylprednisolone group exhibited large grafts in the channels and substantially more preservation of host cord tissue inside the rostral end of the

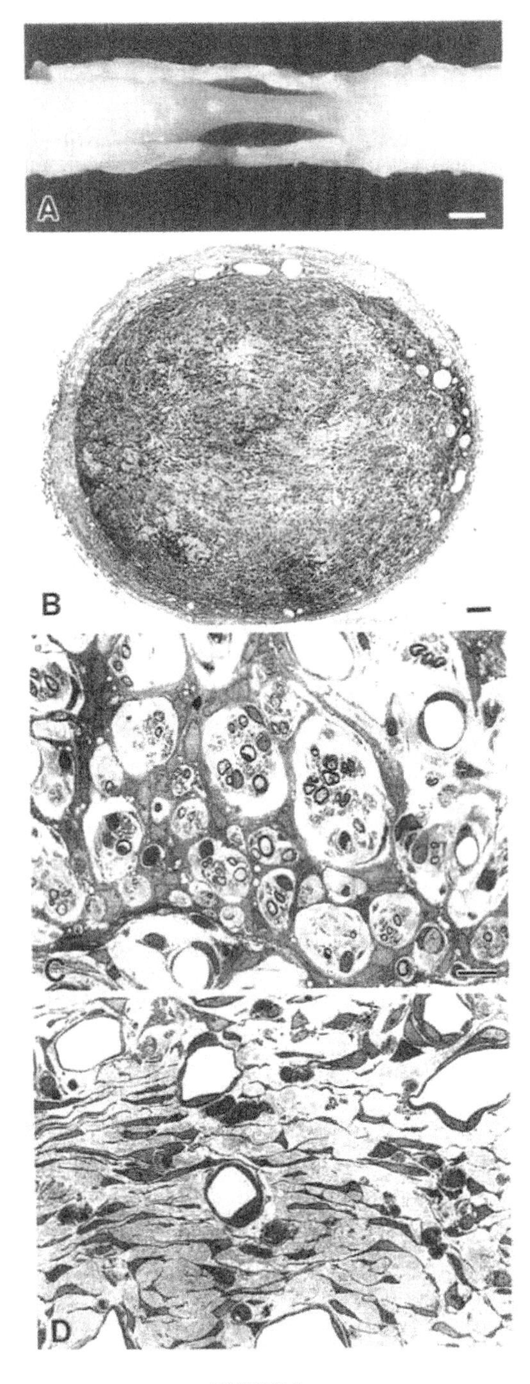

FIGURE 2

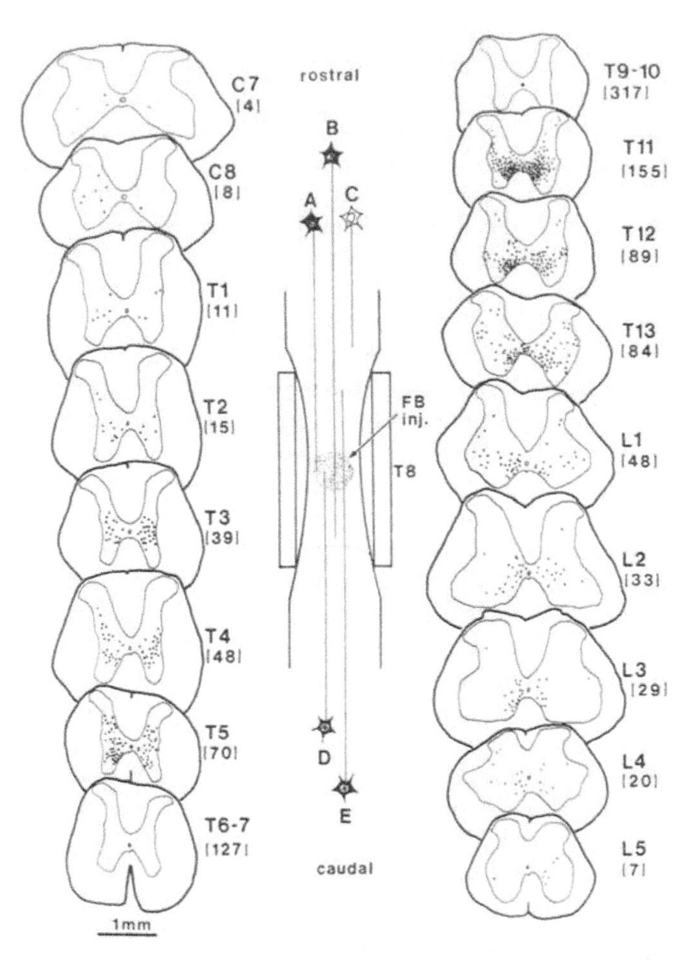

FIGURE 3 A drawing illustrating the injection of Fast Blue into the middle of a Schwann cell cable to label those neurons whose axons had grown to (A,D) or beyond (B,E) the site of injection for both rostral and caudal stumps. Also indicated (C) is the probability that many injured neurons did not regrow into the channel. The closer spinal cord segments contained more labelled cells, and most of the labelling was confined to lamina VII and the medial portion of lamina VIII, where most propriospinal neurons are located. The numbers in brackets are those for the labelled neurons observed at that distance from the graft in one of the cases studied. (From Xu *et al.*, 1997; reprinted with permission of Chapman and Hall.)

FIGURE 2 (A) The cable is well united to both stumps of the previously transected cord. The dorsal wall of the channel has been cut away to show the Schwann cell bridge. (B) Toluidine blue-stained cross-section of a similar bridge. Inside this cable, shown in C, are fascicles of myelinated and unmyelinated axons. (D) Myelinated axons are not found, however, in channels filled with Matrigel alone. Scale bars = 1 mm (A), 100 μm (B), and 10 μm (C and D). (From Xu *et al.*, 1997; reprinted with permission of Chapman and Hall.)

channel than in the group that did not receive the drug. In experiments done in distally capped channels, significantly more myelinated axons (1159 ± 308) were present at the 5 mm level in Schwann cell–methylprednisolone cases than in Schwann cell–vehicle-treated rats (355 ± 108), with again more unmyelinated than myelinated axons (approximately 4:1). In the Schwann cell–methylprednisolone group, unlike the Schwann cell–vehicle group, serotonergic and noradrenergic fibers were detected immunocytochemically 2.5 and 2.0 mm, respectively, into the graft. Fast Blue retrograde tracing confirmed that more cord neurons (1116 ± 113 versus 284 ± 88) and cord neurons more distant from the graft (C5 versus C8) responded by extending axons into the graft in the presence of methylprednisolone. Also, very significantly, supraspinal brain stem neurons extended axons into the graft only when methylprednisolone was administered (mean, 46 versus 0). These results indicate that methylprednisolone improves axonal regeneration for both spinal cord and brain stem neurons into thoracic Schwann cell grafts, possibly by reducing secondary host tissue loss adjacent to the graft.

When Schwann cell bridges open at both ends were tested in the presence or absence of methylprednisolone, myelinated axons were nearly tripled (3237 ± 2478), compared with open-ended channel grafts without methylprednisolone (1324 ± 342). Many more propriospinal neurons (2083 ± 321) responded to the graft in open-ended channels with methylprednisolone than without (1064 ± 145). Again, brain stem neurons responded in the presence of methylprednisolone (mean, 57) and serotonin and dopamine β hydroxylase (DBH)-immunoreactive axons were found 2.0 to 2.5 mm into the graft. Most significantly, following anterograde tracing with PHA-L injected 5 to 6 mm rostral to the graft, a modest number of fibers exited the distal host–graft interface to penetrate the distal cord.

IV. INFUSION OF NEUROTROPHINS INTO SCHWANN CELL GRAFTS IMPROVED REGENERATION

Recent experiments have shown that neurotrophins improve regeneration of axons in spinal cord; one example is the increased sprouting response of corticospinal axons following the injection of neurotrophin-3 (NT-3) into hemisected thoracic cord (Schnell *et al.*, 1994). Although Schwann cells produce many neurotrophic factors, the concentrations generated may not be adequate to elicit a major regenerative effort. We, therefore, tested the addition of neurotrophins to the Schwann cell bridges. Brain-derived neurotrophic factor (BDNF) and NT-3 were delivered together into the capped caudal end of the Schwann cell-filled channel by an Alzet minipump (12 μg/day/neurotrophin) for the first 14 days of the 30-day survival period. Controls received vehicle solution. One month later, a mean of 1523 ± 292 myelinated axons was present in Schwann cell–neurotrophin grafts, twice as many as in Schwann cell–vehicle grafts (882 ± 287) (Xu *et al.*, 1995b). In the Schwann cell–neurotrophin but not Schwann cell–vehicle graft, 5 mm or more from the rostral cord–graft interface, some nerve fibers were immunoreactive for serotonin, a neurotransmitter specific for raphe neurons. Fast Blue retrograde tracing for Schwann cell–neurotrophin grafts revealed labeled neurons in 10 different nuclei of the brain stem; 67% of the labeled neurons were in vestibular nuclei. The mean number of labeled brain stem neurons in the Schwann cell–neurotrophin group (92) contrasted sharply with the mean in the Schwann cell–vehicle group (6). More Fast Blue-labeled cells (967 ± 104) were also present throughout the rostal cord in Schwann cell–neurotrophin cases. These results clearly demonstrated that BDNF and NT-3 infusion enhances propriospinal axonal regeneration and, more significantly, promotes axonal regeneration of specific distant populations of brain stem neurons into grafts at the mid-thoracic level. Brain stem responses represented a particularly significant improvement in response to mid-thoracic grafts; peripheral nerve grafting at a similar cord level does not engender a regenerative response from brain stem neurons (Richardson *et al.*, 1984). Thus, regeneration of some neural populations distant from the spinal cord transection and implant can be enhanced by combinations of trophic factors and a favorable cellular substrate.

V. ADDITION OF INSULIN-LIKE GROWTH FACTOR-I AND PLATELET-DERIVED GROWTH FACTOR LESSENS REGENERATION

Because insulin-like growth factor-I (IGF-I) has been suggested to promote axonal regeneration in the PNS and this effect is enhanced by platelet-derived growth factor (PDGF) (Wells *et al.*, 1994), we tested the effect of IGF-I–PDGF in combination with Schwann cells in our transplantation paradigm (Oudega *et al.*, 1997). Channels filled with Matrigel, Schwann cells, and IGF-I–PDGF were transplanted; control animals received implants of only Matrigel and Schwann cells or only Matrigel and factors. Four weeks after implantation, electron microscopic analysis showed significant effects on both myelination and axonal regeneration. The addition of IGF-1–PDGF resulted in an increase in the ratio of myelinated to unmyelinated fibers from 1 : 7 in Schwann cell grafts to 1 : 3 in Schwann cell plus factors implants; myelin sheath thickness increased twofold, but there was a 36% decrease in the total number of myelinated axons in the Schwann cell implants. This finding and the altered myelinated : unmyelinated fiber ratio thus revealed that fiber regeneration overall was diminished up to 63% by IGF-I–PDGF. The fascicles were smaller than usual due to the paucity of small diameter, unmyelinated axons. In contrast to the overall decrease in axonal regeneration, the addition of IGF-I–PDGF promoted the growth of serotonergic and noradrenergic fibers up to 2 mm into the Schwann cell graft. Histological evaluation of the proximal spinal cord–graft interface region revealed more cavities in spinal tissue of animals with Schwann cell implants containing factors; the tissue cavitation may have impeded overall axonal penetration. In sum, even though this combination of factors enhanced myelination of axons that enter implants of Schwann cells and also allows the entry but not elongation of serotonergic or noradrenergic fibers, the overall regeneration of axons into Schwann cell grafts was diminished.

VI. SCHWANN CELLS GENETICALLY MODIFIED TO SECRETE HUMAN BDNF IMPROVE AXONAL REGENERATION

Not only is it imperative to obtain regrowth of appropriate axons into the graft, but it will be critical to promote growth *beyond* the graft into the distal cord to obtain functional recovery. Additional strategies will be required to enhance this axonal egress into the distal cord. Two possible strategies are to provide trails of Schwann cells in the host cord beyond the graft and to present neurotrophins beyond the graft as well, as suggested by the work of Brook and associates (1994) and Oudega and Hagg (1996), respectively. One way to deliver neurotrophins is to genetically engineer Schwann cells to secrete higher than normal levels of neurotrophins. The genetic engineering of cells to deliver neurotrophins in a cord transplantation paradigm has been pioneered by Tuszynski and colleagues (1994; Grill *et al.*, 1997). This work is reviewed elsewhere in this volume.

We have begun to study delivery of neurotrophins by Schwann cells genetically modified to secrete human BDNF (hBDNF) (Menei *et al.*, 1997). To initially assess the feasibility of this approach we used not the channel paradigm but complete transection at T8. The completeness of the transection was confirmed when elevating the distal stump to introduce the trail of Schwann cells. Three groups of animals were compared: (1) transected, (2) transected with normal Schwann cells transplanted, and (3) transected with hBDNF–Schwann cells transplanted. By means of a Hamilton syringe, 5×10^5 Schwann cells were deposited in a 5 mm-long trail in the gray matter beyond the transection and an equal number of Schwann cells were injected into the transection site, in which a clot had quickly formed. Schwann cells stained with Hoechst dye before transplantation were visualized in discrete trails for at least a month.

After 30 days in the transection-only animals, nerve fibers including serotonergic and noradrenergic axons did not cross the transection, in contrast to the groups with cellular trails; the most serotonergic and some noradrenergic axons were seen in the trails of hBDNF–Schwann cells. Fast Blue was injected at the distal end of the 5 mm-long Schwann cell trails to identify neurons that regenerated axons across the transection and along the length of the trail. When either control or hBDNF-secreting Schwann cells were grafted, neurons in the rostral cord, rostral sensory ganglia, and brain stem were labeled, whereas no labeled nerve cells were seen rostral to the transection when Schwann cells were not transplanted. More sensory and brain stem neurons, particularly in raphe and reticular formation nuclei, were labeled when hBDNF-secreting Schwann cells were transplanted; as many as 135 labeled brain stem neurons were detected in one case whereas no more than 22 were detected in control Schwann cell cases. Thus, hBDNF-secreting Schwann cells were more effective in promoting regeneration across a transection site and into the distal spinal cord than normal Schwann cells.

VII. AXONAL REGENERATION INTO HUMAN OR RAT SCHWANN GRAFTS PLACED IN NUDE RATS

Because of the potential for application of Schwann cell transplantation to spinal cord-injured persons, we have transplanted human Schwann cells into nude rats, which readily accept xenografts, and compared them with nude rat Schwann cell grafts (Guest *et al.*, 1996, 1997a, 1997b). Other than the difference in host and graft species, the experimental paradigm was the same as has been described earlier for Fischer rats. Using distally capped channels initially, we observed a mean of 2795 ± 772 myelinated axons in human Schwann cell–Matrigel grafts 35 days after grafting, a number greatly exceeding that found in Matrigel-only grafts (146 ± 71) or in the Fischer rat distally capped model (501 ± 83). Specific immunostaining demonstrated that myelin was formed by human Schwann cells. Following retrograde tracing with Fast Blue injected into human Schwann cell grafts, propriospinal (803 ± 58), sensory (287 ± 12), and brain stem (113 ± 14) neurons were labeled. When nude rat Schwann cell cables were grafted, similar numbers of retrogradely labeled propriospinal and supraspinal cells were obtained. Thus, the recruitment of supraspinal axons by nonengineered cells and substantial regeneration by propriospinal axons, substantially surpassing that seen in the Fischer rat model, was attributable to the regenerative capacity of the host species. Matrigel-only grafts induced scant sensory and propriospinal growth and no brain stem labeling, confirming the importance of Schwann cell presence. Regeneration of brain stem neurons into the human Schwann cell grafts was confirmed by the presence of serotonin- and DBH-immunostained fibers in the grafts. In addition, motoneuron processes regenerated into human Schwann cell grafts, as demonstrated by double labeling of lamina IX neurons by Fast Blue tracing and by diamidino yellow injected into paravertebral muscles.

When Schwann cell cables were grafted inside guidance channels open at both ends, human Schwann cells supported the ingrowth of propriospinal, sensory, motoneuronal, and brain stem neurons; however, only propriospinal and sensory neurons were observed to re-enter the host spinal cord. Using anterograde and retrograde tracers, we observed regeneration of propriospinal neurons up to 2.6 mm beyond the grafts, and estimated that 1% of the fibers that entered grafts re-entered the host spinal cord at 45 days after grafting. Three types of behavioral analysis were done. Animals with bridging grafts obtained modestly but significantly higher scores during open-field testing (8.2 ± 0.35 versus 6.8 ± 0.42; $p = 0.02$) and inclined plane testing (mean, 38.6 ± 0.54 versus 36.3 ± 0.53; $p < 0.01$) than animals with similar grafts in distally capped channels. A third type of test developed at The Miami Project (Broton *et al.*, 1996a) utilizes treadmill tests to perform a kinematic analysis of limb positions. When transected-only (Broton *et al.*, 1996b) and human Schwann cell-transplanted rats were evaluated in this way, the former exhibited rhythmic stepping 20% of the time and the latter exhibited stepping 60% of the time they were on the treadmill (J. Guest and J. Broton, unpublished observations). At times, both groups displayed alternate hindlimb stepping. Our interpretation is that the graft led to increased excitability in the distal stump although we do not yet know the basis for it; presumably, activity in the lumbar spinal cord stepping pattern-generating neurons was elevated. There was no evidence of fore- to hind-limb coordination.

In summary, human Schwann cells promote axonal regeneration and ensheathe and myelinate the regenerated fibers when transplanted into nude rat spinal cord. Axonal regeneration into Schwann cell grafts in the nude rat was superior to that in the Fischer rat model in a number of ways: more axonal growth occurred into the graft; brain stem neurons responded with no supplementation to the Schwann cell grafts or genetic engineering; and some outgrowth was seen from the graft into the distal cord. These differences may be attributable to immunological deficits in the nude rat that affect neuronal health or regenerative capacity after CNS injury.

VIII. CONCLUSION

R. P. Bunge's speculation that cellular therapy can promote regenerative growth in the damaged spinal cord is closer to reality now, near the end of the *second* 25-year period of history of what is now the National Institute of Neurological

Disorders and Stroke, than it was at its beginning. Schwann cells grafted into the spinal cord promote axonal regeneration in animal models; human Schwann cells have been harvested and shown to be as effective as rodent cells in transplantation experiments; and axonal regeneration, especially from distant supraspinal nuclei, can be augmented by neurotrophic factors and neuroprotection. Combinations of these experimental approaches, along with the use of inhibitor-blocking antibodies (Bregman *et al.*, 1995) or other factors, all seem readily amenable to clinical application. Another distinct advantage of therapies using Schwann cells is that they may be obtained from a peripheral nerve biopsy from a spinal cord-injured person, and expanded to adequate numbers in culture for autotransplantation to the area of injury. Furthermore, they may be genetically modified during the culture period. This is an advantage shared with engineered fibroblasts, but implanted Schwann cells may perform additional functions such as myelination. The rapid expansion of human Schwann cells is now technically feasible (Morrissey *et al.*, 1995; Levi *et al.*, 1995; Rutkowski *et al.*, 1995; Casella *et al.*, 1996; Li *et al.*, 1996). Recent modifications enable a 17,000-fold increase in human Schwann cell number within 30 to 35 days after harvest of nerve with a final purity of 90 to 95% (Casella *et al.*, 1997). A one centimeter piece of nerve can yield enough Schwann cells to fill a channel 12 meters long (assuming 2.6 mm inner diameter). Now that large numbers of these potent cells are available, the challenge before us is to determine the strategies that, in combination with Schwann cells, will be most effective in fostering the return of function after spinal cord injury.

ACKNOWLEDGMENTS

Experiments reviewed here were performed by Drs. X. M. Xu, A. Chen, V. Guénard, M. Oudega, P. Menei, C. Montero-Menei, and J. Guest. Expert help for these studies was provided by K. Akong, M. Bates, J.-P. Brunschwig, R. Camarena, E. Cuervo, A. Gomez, K. J. Klose, A. Rao, D. Santiago, C. Vargas, and A. Weber. Channels were provided by Dr. P. Aebischer (Universitaet de Lausanne); human nerve, by the Organ Procurement Team at the University of Miami School of Medicine; neurotrophins by Regeneron, Inc.; IGF-1 and PDGF by the Institute for Molecular Biology; and the ψBDNF 1B producer line by Dr. X. Breakefield (Harvard Medical School). R. Bunge and S. Whittemore offered valuable counsel. This work was supported by NIH grants NS28059 and NS09923, The Miami Project, and the Hollfelder, Rudin, and Heumann Foundations. Drs. Xu and Oudega were Werner Heumann Memorial International Scholars, Dr. Guest was a Fellow of the American Association of Neurological Surgeons, and Drs. Menei and Montero-Menei were funded by IRME to work in the Bunge Laboratory.

REFERENCES

Aebischer, P., Guènard, V., Winn, S. R., Valentini, R. F., and Galletti, P. M. (1988). Semi-permeable guidance channels allow peripheral nerve regeneration in the absence of a distal nerve stump. *Brain Res* 454:179–187.

Aebischer, P., Guènard, V., and Valentini, R. F. (1990). The morphology of regenerating peripheral nerves is modulated by the surface microgeometry of polymeric guidance channels. *Brain Res* 531:211–218.

Bracken, M. B., Shepard, M. J., Collins, W. F., Holford, T. R., Young, W., Baskin, D. S., Eisenberg, H. M., Flamm, E., Leo-Summers, L., Maroon, J., Marshall, L. F., Perot, P. L., Piepmeier, J., Sonntag, V. K. H., Wagner, F.C., Wilberger, J. E., and Winn, H. R. (1990). A randomized, controlled trial of methylprednisolone or naloxone in the treatment of acute spinal-cord injury. Results of the second National Acute Spinal Cord Injury Study. *N Engl J Med* 322:1405–1411.

Bracken, M. B., Shepard, M. J., Holford, T. R., Leo-Summers, L., Aldrich, E. F., Fazl, M., Fehlings, M., Herr, D. L., Hitchon, P. W., Marshall, L. F., Nockels, L. F., Pascale, V., Perot, P. L., Piepmeier, J., Sonntag, V. K. H., Wagner, F., Wilberger, J. E., Winn, H. R., and Young, W. (1997). Administration of methylprednisolone for 24 or 48 hours or tirilazad mesylate for 48 hours in the treatment of acute spinal cord injury. Results of the Third National Acute Spinal Cord Injury Randomized Controlled Trial. National Acute Spinal Cord Injury Study. *JAMA* 277:1597–1604.

Bregman, B. S., Kunkel-Bagden, E., Schnell, L., Dai, H. N., Gao, D., and Schwab, M. E. (1995). Recovery from spinal cord injury mediated by antibodies to neurite growth inhibitors. *Nature* 378:498–501.

Brook, G. A., Lawrence, J. M., Shah, B., and Raisman, G. (1994). Extrusion transplantation of Schwann cells into the adult rat thalamus induces directional host axon growth. *Exp Neurol* 126:31–43.

Broton, J. G., Nikolic, Z., Suys, S., and Calancie, B. (1996a). Kinematic analysis of limb position during quadrupedal locomotion in rats. *J Neurotrauma* 13:409–416.

Broton, J. G., Xu, X. M., Bunge, M. B., Lutton, S., Cuthbert, T., and Calancie, B. (1996b). Hindlimb movements of adult rats with transected spinal cords. *Soc Neurosci Abstr* 22:1096.

Bunge, R. P. (1975). Changing uses of nerve tissue culture 1950–1975. In: Tower, D. B., ed., *The nervous system, Vol. 1, The basic neurosciences*. New York: Raven Press, pp. 31–42.

Bunge, R. P., and Hopkins, J. M. (1990). The role of peripheral and central neuroglia in neural regeneration in vertebrates. *Sem Neurosci* 2:509–518.

Bunge, R. P., Puckett, W. R., Becerra, J. L., Marcillo, A., and Quencer, R. M. (1993). Observations on the pathology of human spinal cord injury. A review and classification of 22 new cases with details from a case of chronic cord compression with extensive focal demyelination. In: Seil, F. R., ed., *Advances in neurology*, Vol. 59. New York: Raven Press, Ltd., pp. 75–89.

Bunge, R. P., Puckett, W. R., and Hiester, E. D. (1997). Observations on the pathology of several types of human spinal cord injury, with emphasis on the astrocyte response in penetrating injuries. In: Seil, F. R., ed., *Advances in neurology, Vol. 72*. New York: Raven Press, Ltd., pp. 305–315.

Casella, G. T. B., Bunge, R. P., and Wood, P. M. (1996). Improved method for harvesting human Schwann cells from mature peripheral nerve and expansion *in vitro*. *Glia* 17:327–338.

Chen, A., Xu, X. M., Kleitman, N., and Bunge, M. B. (1996). Methylprednisolone administration improves axonal regeneration into Schwann cell grafts in transected adult rat thoracic spinal cord. *Exp Neurol* 138:261–276.

Cheng, H., Cao, Y., and Olson, L. (1996). Spinal cord repair in adult paraplegic rats: Partial restoration of hind limb function. *Science* 273:510–513.

Duncan, I. D., Aguayo, A. J., Bunge, R. P., and Wood, P. M. (1981). Transplantation of rat Schwann cells grown in tissue culture into the mouse spinal cord. *J Neurol Sci* 49:241–252.

Grill, R., Murai, K., Blesch, A., Gage, F. H., and Tuszynski, M. H. (1997). Cellular delivery of neurotrophin-3 promotes corticospinal axonal growth and partial functional recovery after spinal cord injury. *J Neurosci* 17:5560–5572.

Guènard, V., Kleitman, N., Morrissey, T. K., Bunge, R. P., and Aebischer, P. (1992). Syngeneic Schwann cells derived from adult nerves seeded in semi-permeable guidance channels enhance peripheral nerve regeneration. *J Neurosci* 12:3310–3320.

Guènard, V., Xu, X. M., and Bunge, M. B. (1993). The use of Schwann cell transplantation to foster central nervous system repair. *Sem Neurosci* 5:401–411.

Guest, J. D., Aebischer, P., Akong, K., Bunge, M. B., and Bunge, R. P. (1997b). Human Schwann cell transplants in transected nude rat spinal cord: Graft survival, axonal regeneration, and myelination. (submitted).

Guest, J. D., Kleitman, N., Aebischer, P., Bunge, M. B., and Bunge, R. P. (1996). Axonal regeneration into human Schwann cell grafts placed to span the transected spinal cord of the nude rat. *Soc Neurosci Abstr* 22:490.

Guest, J. D., Rao, A., Bunge, M. B., and Bunge, R. P. (1997a). The ability of human Schwann cell grafts to promote regeneration in the transected nude rat spinal cord. *Exp Neurol* (in press).

Levi, A. D. O., Bunge, R. P., Lofgren, J. A., Meima, L., Hefti, F., Nikolics, K., and Sliwkowski, M. X. (1995). The influence of heregulins on human Schwann cell proliferation. *J Neurosci* 15:1329–1340.

Li, R., Chen, J., Hammonds, G., Phillips, H., Armanini, M., Wood, P., Bunge, R., Godowski, P. J., Sliwkowski, M. X., and Mathes, J. P. (1996). Identification of Gas6 as a growth factor for human Schwann cells. *J Neurosci* 16:2012–2019.

Menei, P., Montero-Menei, C., Whittemore, S. R., Bunge, R. P., and Bunge, M. B. (1997). Schwann cells genetically modified to secrete human BDNF promote enhanced axonal regrowth across transected adult rat spinal cord. *Eur J Neurosci* (in press).

Morrissey, T. K., Levi, A. D. O., Neuijens, A. A., Sliwkowski, M. X., and Bunge, R. P. (1995). Axon-induced mitogenesis of human Schwann cells involves heregulin and $p185^{erbB2}$. *Proc Nat Acad Sci USA* 92:1431–1435.

Oudega, M., and Hagg, T. (1996). Nerve growth factor promotes regeneration of sensory axons into adult rat spinal cord. *Exp Neurol* 140:218–229.

Oudega, M., Xu, X. M., Guènard, V., Kleitman, N., and Bunge, M. B. (1997). A combination of insulin-like growth factor-I and platelet-derived growth factor enhances myelination but diminishes axonal regeneration into Schwann cell grafts in the adult rat spinal cord. *Glia* 19:247–258.

Paino, C. L., and Bunge, M. B. (1991). Induction of axon growth into Schwann cell implants grafted into lesioned adult rat spinal cord. *Exp Neurol* 114:254–257.

Paino, C. L., Fernandez-Valle, C., Bates, M. L., and Bunge, M. B. (1994). Regrowth of axons in lesioned adult rat spinal cord: Promotion by implants of cultured Schwann cells. *J Neurocytol* 23:433–452.

Ramón y Cajal, S. (1928). *Degeneration and regeneration of the nervous system,* May, R. M., trans. New York: Oxford University Press.

Richardson, P. M., Issa, V. M. K., and Aguayo, A. J. (1984). Regeneration of long spinal axons in the rat. *J Neurocytol* 13:165–182.

Richardson, P. M., McGuinness, U. M., and Aguayo, A. J. (1980). Axons from CNS neurones regenerate into PNS grafts. *Nature* 284:264–265.

Rutkowski, J. L., Kirk, C. J., Lerner, M. A., and Tennekoon, G. I. (1995). Purification and expansion of human Schwann cells *in vitro*. *Nature Med* 1:80–83.

Schnell, L., Schneider, R., Kolbeck, R., Barde, Y.-A., and Schwab, M. E. (1994). Neurotrophin-3 enhances sprouting of corticospinal tract during development and after adult spinal cord lesion. *Nature* 367:170–173.

Tuszynski, M. H., Peterson, D. A., Ray, J., Baird, A., Nakahara, Y., and Gage, F. H. (1994). Fibroblasts genetically modified to produce nerve growth factor induce robust neuritic ingrowth after grafting to the spinal cord. *Exp Neurol* 126:1–14.

Wells, M. R., Batter, D. K., Antoniades, H. N., Blunt, D. G., Weremowicz, J., Lynch, G. E., and Hansson, H. A. (1994). PDGF-BB and IGF-I in combination enhance axonal regeneration in a gap model of peripheral nerve injury. *Soc Neurosci Abstr* 20:1326.

Wood, P. M., and Bunge, R. P. (1975). Evidence that sensory axons are mitogenic for Schwann cells. *Nature* 256:662–664.

Xu, X. M., Chen, A., Guènard, V., Kleitman, N., and Bunge, M. B. (1997). Bridging Schwann cell transplants promote axonal regeneration from both the rostral and caudal stumps of transected adult rat spinal cord. *J Neurocytol* 26:1–16.

Xu, X. M., Guènard, V., Kleitman, N., and Bunge, M. B. (1995a). Axonal regeneration into Schwann cell-seeded guidance channels grafted into transected adult rat spinal cord. *J Comp Neurol* 351:145–160.

Xu, X. M., Guènard, V., Kleitman, N., Aebischer, P., and Bunge, M. B. (1995b). A combination of BDNF and NT-3 promotes supraspinal axonal regeneration into Schwann cell grafts in adult rat thoracic spinal cord. *Exp Neurol* 134:261–272.

CHAPTER 26

Designing Clinical Trials for Neural Repair

MICHAEL GRUNDMAN* AND MARK H. TUSZYNSKI*,†

*Department of Neurosciences, University of California, San Diego, La Jolla, California and
†Veterans Affairs Medical Center, San Diego, California

The design of clinical trials for neurodegenerative disease can be optimized if the objectives of the study are clearly defined. The rationale, safety, and efficacy of the putative therapeutic agent should be thoroughly evaluated in preclinical models prior to beginning a clinical development program. A good understanding of the mechanism of action may help in determining whether the design should incorporate maneuvers to test for symptomatic or disease-modifying properties. Other important issues to consider include the phase of the study, the choice of the primary outcome measure, and the estimated therapeutic effect size. The length of the trial, cost, number of subjects, clinical interpretation, and ultimately, its likelihood of success will depend on all of these factors.

I. INTRODUCTION

Steady and impressive advances in basic neuroscience have led over the past decade to the clinical transfer of several new potential approaches for treating nervous system disease. Clinical trials with several growth factors have been conducted, together with trials of fetal transplantation and several novel pharmacologic substances. Trials with these new agents have been conducted in neurodegenerative disorders, stroke, trauma, neuropathies, and CNS malignancy (see Table I). Yet several of these trials have yielded negative outcomes, highlighting the early and exploratory stage at which clinical transfer of these therapies is occuring. Clearly, two minimal criteria should be met when conducting clinical trials of novel neurologic therapeutics: (1) the approach must have been adequately tested in animal models of human disease and have been proven to show reasonable levels of efficacy and safety, preferably including data from large animal models, and (2) clinical trials must be rationally and logically formulated. Several of the preceding chapters focused on the former questions; the present chapter will focus on the appropriate design and performance of clinical trials. Particular emphasis will be placed on designing trials with sensitivity to potential outcomes of neural protective or regenerative strategies, because these are the focus of several new approaches emerging from basic neuroscience investigations. Readers interested in more general information on clinical trial design may consult texts by Porter and Schoenberg (1990) or Spilker (1991).

II. THE PHASES OF CLINICAL DRUG DEVELOPMENT

The objective of a clinical development program is to successfully test and market an effective therapeutic agent. The role of clinical trials is to determine the efficacy of a putative treatment under conditions that are likely to exist once the treatment is marketed. Such testing is traditionally conducted in three phases in the United States: Phase I, II, and III clinical trials.

Phase I studies determine the acute safety, tolerance, and pharmacokinetics of an agent in small samples of volunteers. Generally, the intent of a Phase I trial is to escalate the dose of a drug, or any therapy, to levels that produce clinical toxicity. In this manner, safe and well-tolerated treatment regimens can be planned for future trials. The agent's pharmacokinetic parameters, critical for planning dosing, such as the time to peak concentration and elimination half-life, are determined. Phase I trials are not typically designed to yield treatment efficacy data.

Phase II studies begin to examine efficacy across a range of potentially effective doses. Phase II studies also extend the safety profile of the drug to include short-term adverse side effects and further define the drug's pharmacokinetic properties. Phase II trials thereby yield data that permit optimal design of more extensive Phase III trials, providing information about optimal treatment doses and clinical parameters that should be assessed to determine whether the treatment is effective.

Phase III trials are designed to examine the efficacy of a therapeutic approach in a prospective and objective manner, sampling sufficiently large numbers of

TABLE I Clinical Neural Repair Trials

Substance	Class	Disease	Delivery route	Endpoint	Trial status	Outcome	Refer.
NGF	Neurotrophin	Alzheimer's Disease	Intraventricular	Cognition, PET	Halted	Negative	
		Parkinson's Disease*	Intraparenchymal	Motor Function	Halted	Negative	
		Diabetic Neuropathy	Subcutaneous	Sensorimotor, NCV/EMG	Phase III: In progress	? positive	
BDNF	Neurotrophin	Amyotrophic Lateral Sclerosis	Subcutaneous	Motor Function (ALS score)	In progress		
BDNF + CNTF	Neurotrophin	Amyotrophic Lateral Sclerosis	Subcutaneous	Motor Function (ALS score)	In progress		
CNTF	Neurotrophin	Amyotrophic Lateral Sclerosis	Subcutaneous Intrathecal	Motor Function (ALS score)	Complete	Negative	
GDNF	Neurotrophin	Parkinson's Disease	Intraventricular	Motor Function (UPDRS score) ?PET	In progress		
IGF-1 (Myotrophin)	Neurotrophin	Amyotrophic Lateral Sclerosis	Subcutaneous	Motor Function (ALS score)	Complete	Unclear	
FGF	Neurotrophin	Stroke	Subcutaneous	Overall Neurological Function	In progress		
Fetal Substantia Nigra	Fetal Graft	Parkinson's Disease	Intraparenchymal	Motor Function (UPDRS score) ?PET	In progress		
Fetal Striatum	Fetal Graft	Huntington's Disease	Intraparenchymal	Motor Function (UHDRS score) ?PET	In progress		
Fetal Spinal Cord	Fetal Graft	Syringomyelia (and spinal cord injury)	Intraparenchymal	Motor Function (ASIA score) MRI or cord	In progress		

*In combination with adrenal autograft
UPDRS: United Parkinson's Disease Rating Scale
UHDRS: United Huntington's Disease Rating Scale
ASIA: American Spinal Injury Association scale

patients to generate statistically significant outcomes. The design of Phase III trials depends on the outcome of Phase I and II trials, in which optimal drug dose and scheduling have been determined, and the appropriate clinical parameters for assessing effects of the treatment have, hopefully, been established. Phase III studies are usually large multicenter trials that are randomized and placebo controlled, with the treatment assignment unknown to the investigator or patient (double blind).

Phase III clinical trials will be designed differently, depending upon whether the goal of a therapeutic approach is to (1) ameliorate symptoms without influencing actual disease progression, (2) reduce the rate of disease progression, (3) delay disease onset, or (4) prevent disease. The design of a trial should therefore take into account the nature of the expected effect when testing regenerative therapies for neurological disease. For example, clinical trials of neurotrophic factors in neurodegenerative diseases such as Alzheimer's disease, Parkinson's disease, and ALS that are designed to slow disease progression will take longer than trials designed to demonstrate improvement of symptoms. A general goal of neurotrophic factor trials in neurodegenerative disorders is to reduce cell loss. For these agents, it may be reasonable to design therapeutic trials to detect a slowing in disease progression or that demonstrate a prolongation of time required to reach a defined functional endpoint. On the other hand, fetal-grafting strategies for Parkinson's disease or Huntington's disease can potentially restore dopamine levels in the striatum, and for clinical trials of restorative therapies of this sort, designs should be capable of detecting both symptom improvement and slowed progression. Clinical trials in spinal cord injury, head injury, or stroke might be designed to distinguish symptomatic improvements that are transient from those representing improved structural neural substrates.

Phase III trials share certain common elements. The hypothesis of the study must be defined explicitly. The study population must be chosen appropriately and an adequate sample size calculated to test the hypothesis. Treatment assignment must be made at random without the investigator's influence, because this might alter the outcome of the trial in favor of the active agent. Study endpoints should be defined that are objective and clinically meaningful, and assessments should be performed blind to treatment assignment.

Occasionally sponsors may choose to merge Phases II and III. This may involve the enrollment of large numbers of subjects to receive placebo, with other groups of patients receiving two or three widely spaced doses of the drug. The highest dose may be used as the primary comparison with placebo. Alternatively, the statistical significance level may be made more stringent than $p = 0.05$ to account for several planned comparisons.

III. GENERAL CLINICAL TRIAL DESIGNS

A frequent question that arises when considering the design of a clinical trial is whether to use a cross-over or parallel-group design. Both designs have advantages and disadvantages. For the reasons delineated below parallel group designs may be preferable for neurodegenerative diseases. Cross-over designs may be more appropriate for presumed symptomatic treatments in patients with nonprogressive illness such as chronic spinal cord injury.

A. Cross-over Design

The major appeal of cross-over studies is that each subject serves as his or her own control, which reduces the variance contributed by between-subject differences. In a simple two-period cross-over design, subjects are treated twice, once with the active agent and once with placebo (Stern *et al.*, 1987; Penn *et al.*, 1988; Gauthier *et al.*, 1990). Depending on whether the subject is assigned to treatment or placebo during the first period, the subject is assigned the alternative treatment during the second period. It is assumed that there are no carry-over effects (i.e., prolonged drug effects that affect outcomes after cross-over), withdrawal effects (deleterious effects resulting from drug withdrawal that impair function after cross-over), or treatment-by-period interactions (treatment response differs depending on the time period in which the drug is given). For this design to be valid, patients entering the second treatment period should return to the clinical state exhibited prior to the start of the first period. Unfortunately, period effects (the treatment effect is better during one period than another, regardless of which treatment is given) and order effects (the treatment response depends on the order in which the treatments are given) frequently occur, particularly if the disease progresses during the course of the trial. It is also possible that the investigator or subject may become unblinded at the time of cross-over. Patients who withdraw prematurely cause more problems in cross-over trials than in other trial designs because information from both treatment periods is needed to generate valid data. Cross-over trials are best suited for short symptomatic treatments in stable conditions. As neurodegenerative disorders are typically progressive, cross-over trials are frequently not ideal for these diseases. Patients with neurodegenerative disorders often require long time periods of observation to see a treatment effect, particularly if rate of progression is the primary outcome; this results in a strong likelihood that the patient will enter the second treatment period in a more advanced stage of the disease than the first.

B. Parallel-Group Designs

In parallel-group studies, participants are randomly divided into as many groups as there are treatments (Knapp *et al.*,, 1994; Becker *et al.*, 1996; Rogers and Friedhoff 1996; Thal *et al.*, 1996a; Rogers *et al.*, 1998). For example, if a study is designed to test the efficacy of a single dose of CNTF in reducing progression in ALS, then two parallel groups are studied: a CNTF treatment group and a control (placebo) group. This simplified study design has become utilized more commonly as the disadvantages of cross-over designs are more widely recognized. The analysis of parallel-group studies relies on fewer assumptions than that of cross-over designs. The main disadvantage of the parallel-group design is that the variability between subjects is greater than that within subjects, requiring more subjects for the same statistical power. It is also frequently assumed that the randomization process will balance important prognostic factors between the treatment groups (e.g., age), which does not always occur. To alleviate this problem, subjects may be stratified on important prognostic variables during the randomization. Similarly, during the analysis stage, treatment groups should be compared for potentially uneven distribution of prognostic factors, and appropriate analyses performed adjusting for these differences.

C. Enrichment Designs

Enrichment trials are designed in a manner to identify patients early in the trial that are potential drug-responders, and then to conduct more extensive controlled studies based upon this "enriched" population of patients compared to controls (Davis *et al.*, 1992; Thal *et al.*, 1996b). Studies are sometimes performed in this manner when it is likely that only a subset of patients will respond, but this subset cannot be determined prior to exposure to drug. For example, it might be assumed that only subjects with specific underlying pathophysiology or genetic susceptibility will respond; however, these cannot be distinguished from the general disease phenotype in a particular patient. Enrichment trials are performed in two stages. In the first stage, all participants are dose-titrated with the test substance in an attempt to identify subgroups of potential responders and to determine the optimal dosing for each subject. During the second stage, the subjects identified as potential responders in the first stage are randomized to active treatment or placebo. The nonresponding subjects from the first stage are dropped from the study. Advantages of this design include the ability to determine the most effective dose during the dose titration phase and to drop nonresponding subjects from the trial. A disadvantage of this design is that all patients are exposed to the drug in the first stage. This makes it difficult to determine true adverse event profiles if, for example, there is tolerance or increased sensitivity to the agent with repeated exposures. Carry-over effects may occur when patients are crossed over to placebo in the second stage. Finally, it is difficult to generalize the trial results from the pre-selected patients involved in the study to patients residing in the community.

IV. CHOOSING OUTCOME MEASURES TO FIT THE THERAPY AND THE DISEASE

The choice of a trial's primary and secondary outcome measures is an essential feature of designing clinical trials, especially when trying to assess the value of neuroregenerative therapies. The outcome measures must be chosen based on the expected effect of the therapeutic intervention.

Treatment with NGF in Alzheimer's disease might be predicted to improve cholinergic function. This would be expected to improve the cognitive symptoms of the disease, particularly with respect to spatial memory and attention. Therefore, a rational trial design for NGF in Alzheimer's disease might incorporate a general cognitive scale, such as the Alzheimer's Disease Assessment Scale (Rosen *et al.*, 1984) as well as more focused measures of spatial memory and attention.

In ALS studies, agents predicted to promote motor neuron survival such as ciliary neurotrophic factor would appropriately use measures of muscle strength as well as a functional measure, such as the ALS Functional Rating Scale (ALS CNTF Treatment Study Group 1996; The ALS CNTF Treatment Study or ACTS Phase I–II Study Group 1996). Similarly, studies of neuroregenerative therapies in trauma, such as spinal cord injury, should be designed to measure outcomes on tasks that are predicted to respond to the therapeutic manipulation. For example, a trial of NT-3 for spinal cord injury might focus on functions mediated by the corticospinal tract such as skilled motor tasks (digit function), and sensory proprioceptive tasks (see Chapter 24). A trial of a neuroprotective drug, such as a blocker of apoptosis or a blocker of excitatory amino acid receptors in spinal cord trauma might focus more broadly on several different types of motor and sensory functional outcomes (e.g., using the American Spinal Injury Association, or ASIA, scale).

Clinical trials with the conduction-enhancing drug 4-aminopyridine are currently in progress for chronic spinal cord injury. Because 4-AP is a general conduction enhancer that may benefit systems that are partially spared, the clinical trial design should focus on those systems that show partial function in individual patients. Thus, a cross-over trial design could be utilized that prescreens patients for the presence of partial injuries. Functional improvement would then be assessed in patients receiving both drug and placebo treatment, in a blinded fashion.

In summary, study outcomes, whether cognitive, motor, sensory, or functional in nature, should be chosen so that they are relevant to the disease under study and sensitive to the experimental intervention.

V. CONTINUOUS AND DISCRETE OUTCOME MEASURES

The outcomes traditionally chosen in Phase III clinical trials for neurodegenerative disorders have been a "change score" on a scale comparing an assessment performed on the last treatment visit and the patient's baseline. The scales are usually continuous (scales in which the outcome may take on any value within limits) or ordinal (scales based on rankings). In Alzheimer's disease trials, for example, both change in the ADAS-COG, a continuous scale between 0 and 70 (Rosen *et al.*, 1984), and a clinician's global impression of change, a 7-point scale that rates the patient's performance between "marked improvement" and "marked worsening" (Schneider *et al.*, 1997), have been used. Change scores on continuous measures generally have limited clinical interpretation. A specific change score on the ADAS-COG, for example, is not readily interpretable with respect to the magnitude of improvement that might be recognizable by a clinician. This presents some difficulty when trying to explain the potential benefit to a patient. Continuous measures may also suffer from lack of linearity over their entire spectrum. They are susceptible to ceiling effects (insensitivity to detect changes greater than the maximum score on the scale) and floor effects (insensitivity to detect changes below the minimum score on the scale), depending on the severity of the patient. This is a problem because it is frequently assumed that the same change score on a test reflects an equal change in performance. Another problem with continuous measurement scales in neurodegenerative diseases is that subjects may drop out of a clinical trial early or may expire, and cannot provide the correct change score needed for statistical analysis. This usually leads to the need for various imputation schemes to compensate for missing data.

Discrete clinical milestones, which are easily understood, may be used as study endpoints and are particularly useful in progressive neurodegenerative disorders, such as Alzheimer's disease, Parkinson's disease, ALS, and Huntington's disease. Such milestones or events may be analyzed using a survival analysis approach by computing the time to reach the endpoint of interest, and, when used in this fashion, are called time-to-event outcomes. Endpoints of this type include time to a clinical diagnosis, time to death or nursing home placement, and time to loss of independence in some activity of daily living (such as ambulation). Endpoints chosen for studies using a survival design should be clinically meaningful, well recognized, unlikely to be missed, and fixed (not changing) upon reaching them. They should occur consistently as the disease evolves and reflect the severity of the underlying pathology. Utilizing such endpoints has certain advantages compared to continuous measurement scales. Such endpoints may frequently be obtained without resorting to complex testing of the patient and may be retrieved in many cases without the patient even having to attend the clinic. This often helps to minimize the problem of missing data due to clinic visit dropouts. Moreover, the knowledge that a subject has survived up to a certain period of time without reaching the trial endpoint can be utilized in survival analyses even if the subject is removed (censored) prior to the end of the study. Current survival methods have the capacity to include important prognostic variables other than simply the treatment group in the analysis. The Cox proportional hazards model, for example, may be used to adjust for potential baseline imbalances in known prognostic variables. If possible, the trial should attempt to define endpoints that occur frequently enough during the period of follow-up so that results are not limited to only a small minority of the radomized population. In addition, trials using endpoints that occur only infrequently or after long observation periods require very large numbers of subjects in order to demonstrate a treatment effect. Survival approaches have been used successfully in neurodegenerative clinical trials for Alzheimer's disease (Sano *et al.*, 1996, 1997), ALS (Bensimon *et al.*, 1994; Lacomblez *et al.*, 1996), and Parkinson's Disease (The Parkinson's Study Group, 1993; LeWitt *et al.*, 1997).

Although the use of discrete outcomes and survival analysis has certain advantages alluded to here, this methodology may also have shortcomings. It may be difficult to find appropriate discrete endpoints. The relevance and validity of the outcome (e.g., death or institutionalization) may be questioned even if the outcome is clear enough (Drachman and Leber, 1997). Although disease severity is a contributing cause of death in approximately three-fourths of AD patients (Olichney *et al.*, 1995), some patients may die of myocardial infarction or cancer unrelated to their AD. Institutionalization generally correlates with disease stage, but may be prompted by factors not directly related to the patient's illness, such as caregiver characteristics, economic status, or the availability of a nearby nursing home. Outcomes more directly related to dementia severity such as conversion to a certain severity stage on the Clinical Dementia Rating Scale (Berg, 1988; Morris, 1993), require clinical experience and training. This type of outcome is thus harder to observe or retrieve without a clinical expert directly evaluating the patient, and is less readily understood by patients and other interested parties who are unfamiliar with the scale.

VI. TRIAL DESIGNS FOR SYMPTOMATIC IMPROVEMENT AND SLOWING DISEASE PROGRESSION

As noted earlier, therapeutic trials may aim to prevent disease, reduce the rate of progression, or ameliorate symptoms of the disease without altering its underlying structural pathology. This section will discuss how these various therapeutic aims can be incorporated into the study design.

Trials involving agents for which symptomatic improvement of the disease is the objective should examine a common and relevant core symptom of the disease. In Alzheimer's disease, a typical core symptom is memory loss; in spinal cord injury, it might be motor function or autonomic function. Generally, trials for symptomatic agents last 3 to 6 months. This duration is usually required in order to distinguish placebo and practice effects from the true benefit of the active treatment as well as to demonstrate that the improvement persists for a reasonable length of time. To guarantee that a therapeutic intervention produces a beneficial effect that is large enough to be recognized by the skilled clinician or patient, sensitive measures of change (e.g., neuropsychological tests in AD, isometric testing in ALS) may be supplemented with more clinical measures, e.g., clinical global assessments or functional scales measuring activities of daily living (ADL). In AD, some of the clinical global assessments that have been utilized include the Clinician's Interview Based Impression of Change (CIBIC) and the ADCS-Clinical Global Impression of Change (ADCS-CGIC) (Schneider and Olin, 1996; Schneider *et al.*, 1996). ADL scales that have been used include the Physical Self-Maintenance and Instrumental Activities of Daily Living Scales (Lawton and Brody, 1969; Lawton 1988), the Progressive Deterioration Scale (DeJong *et al.*, 1989), and the ADCS-ADL (Galasko *et al.*, 1997).

Phase III studies of symptomatic agents usually require several hundred subjects per treatment group depending on the disease, the outcome measure, and the assumptions involved in the power analysis. In planning a trial it is important to ensure that enough subjects are recruited to have adequate power. Pivotal trials are often designed with 90% power (i.e., allow only a 10% chance of the trial returning a false-negative result if the intervention truly works as expected). To this end, it is important to obtain an estimate of the mean rate of change and standard deviation for the placebo-treated subjects over the expected duration of the trial and an estimate of the mean difference in performance between the placebo and treatment groups that would be considered meaningful and important to detect. Studies reporting a significant symptomatic benefit in AD (Knapp *et al.*, 1994; Bodick *et al.*, 1997; Rogers *et al.*, 1998), for example, usually find a 1.4 to 2.0 point deterioration over six months on the cognitive measurement scale (the ADAS-COG) for the placebo-treated group, and a 2 to 5 point improvement relative to placebo for the active treatment. Based on the agents that have been approved for AD to date (e.g., tacrine and donepezil), a mean difference on the ADAS-COG between the active treatment and placebo group in this range will likely be the minimum difference necessary for future regulatory approval of symptomatic agents in AD.

Putative disease-modifying agents should demonstrate an effect on disease progression. Ideally a clinical trial would show (1) that the rate of decline (slope) is less in actively treated subjects (i.e., the longer the observation period, the greater the improvement of the active agent relative to placebo), and (2) that the therapeutic improvement persists long after its discontinuation or confers a sustained benefit when started earlier (Leber, 1996).

Symptomatic drugs typically result in a brief improvement in a particular measurement scale and then decline at the same rate as the placebo group. In contrast, disease-modifying drugs result in slower rates of decline than placebo when followed for an extended length of time (compare, for example, the pattern of decline during the "randomized phase" for symptomatic drugs in panels A and B of Fig. 1, to that of disease-modifying agents in panels C and D).

A number of clinical trial strategies have been suggested to test whether or not an agent has disease-modifying properties (Leber, 1996). These include adding a randomized "start" or "withdrawal" component to a clinical trial. In the "withdrawal" design (see panels A and C in Fig. 1), subjects on active drug at the conclusion of the trial are converted to placebo, while those receiving placebo at the trial's conclusion remain on placebo. Subjects taking the active drug during the trial would be predicted to lose efficacy over time and trend toward the placebo group after being placed on placebo, if the drug exerted only a symptomatic effect (panel A, Fig. 1). Subjects should maintain a therapeutic benefit if the agent resulted in a favorable structural change that did not revert to baseline immediately after the agent was discontinued (panel C, Fig. 1). By contrast, in the "start" design, one group of subjects is randomized to placebo first, followed by the active treatment, while the other group is randomized to the active treatment and remains on the active treatment throughout the entire trial. If the putative therapeutic agent has only symptomatic effects, the placebo-treated patients later placed on the active treatment should quickly approach the performance of those taking the active drug from the beginning of the trial (panel B, Fig. 1). On the other hand, if there is a true slowing in the rate of progression so that a structural change occurs in the actively treated subjects over the course of the trial, the placebo-treated patients who are newly placed on the active agent should not be able to catch up (panel D, Fig. 1).

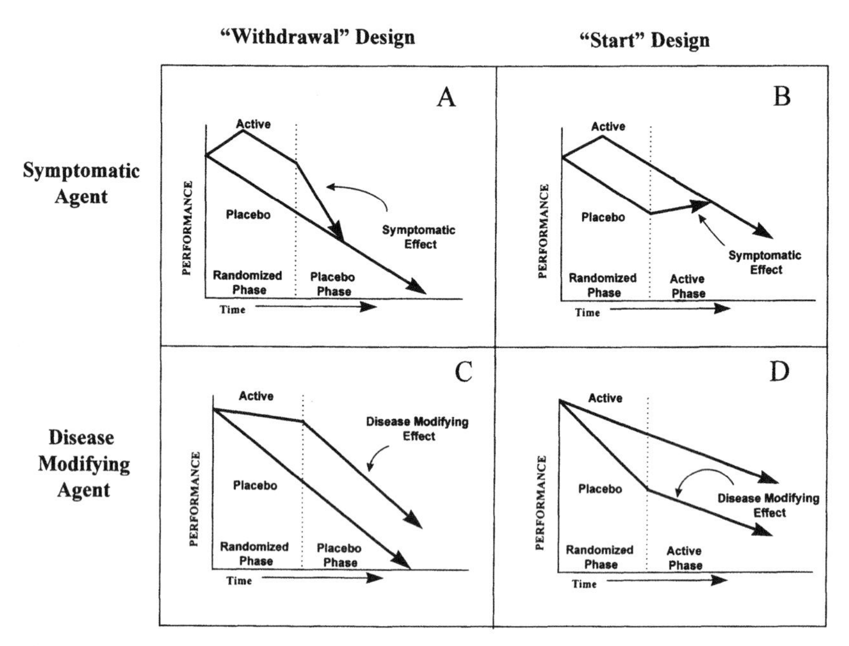

FIGURE 1 Expected patterns of decline using different trial designs for symptomatic and disease-modifying agents.

Clinical trials designed for interventions that slow the rate of disease progression take a longer time than symptomatic trials. Longer trials should, in principle, require fewer patients because, if the agent works as predicted, the treatment-effect size should increase over time in comparison to the variability of the test itself. On the other hand, longer trials are susceptible to a greater number of dropouts and therefore an estimate of the likely subject attrition rate during the trial is needed in order to include an adequate number of subjects. Importantly, many more subjects may be needed if it is intended to power adequately to demonstrate a persisting significant benefit using either the randomized "start" or "withdrawal" designs discussed above. It may be particularly difficult to demonstrate a significant effect on disease progression using these techniques if the agent has a combination of symptomatic and disease-modifying properties. If the agent demonstrates some persisting benefit but less than might be predicted if the agent were purely disease modifying, the detection of the smaller treatment difference would require an even larger number of subjects. Finally, subjects selected to participate in long trials should be healthy enough to remain in the study for its intended duration.

VII. NEURODEGENERATIVE DISEASE PREVENTION TRIALS

New potential therapeutic agents for preventing onset of neurodegenerative disorders are in the development stage, including neurotrophic factors, neuronal replacement strategies (fetal grafting and stem cell development), and antisense oligonucleotide approaches to blocking deleterious gene expression. It is likely that many neurodegenerative diseases are characterized by a preclinical period during which prevention of disease symptoms may be possible. Early pathologic changes in disorders such as Alzheimer's disease and Parkinson's disease may start years prior to the onset of clinical symptoms through genetic, traumatic, anoxic, or possibly other vulnerabilities. Aging may promote this process. Clinical symptoms appear when the pathology reaches a critical threshold. If correct, effective slowing of a disease in its early stages may yield superior efficacy compared to the current approach of treating symptoms later in the disease.

For example, trials to prevent Alzheimer's disease with gonadal steroids are currently underway. The Women's Health Initiative (WHI) is testing hormone replacement therapy (estrogen or estrogen and progesterone) in approximately 8,000 cognitively normal elderly women over a period of 6 to 9 years (McBee *et al.*, 1997). The Alzheimer's Disease Cooperative Study (ADCS) also has plans to initiate an antioxidant trial using high-dose vitamin E to determine if it can delay the diagnosis of Alzheimer's disease in subjects with mild cognitive impairment (Grundman and Thal, 1997). Support for vitamin E as a possible preventive agent for AD comes from a previously published trial by the ADCS, which found that Alzheimer's disease milestones could be delayed by antioxidants (Sano *et al.*, 1997). In the latter trial, AD patients with moderately severe disease were enrolled in a double-blind placebo-controlled trial in which they received either vitamin E (2,000 IU daily), selegiline (10 mg daily), the combination of selegiline and vitamin E, or placebo. The primary outcome was the time to reach one of the following endpoints: institutionalization, loss of basic activities of daily living, severe dementia (defined by a Clinical Dementia Rating of 3), or death. The results were analyzed using survival analysis. There were significant reductions in the risk of reaching the primary outcome for vitamin E, selegiline, or the combined treatment. It is possible that these neuroprotective agents will exhibit superior levels of efficacy if provided to patients earlier in the course of Alzheimer's disease.

VIII. COMBINATION STUDIES WITH DIFFERENT THERAPEUTIC AGENTS

More than one agent may be effective in treating a disease by acting through different mechanisms. In this case, the combined effect of both treatments may exceed that of either treatment alone. A symptomatic drug might be combined with a disease-modifying drug (e.g., a cholinesterase inhibitor and vitamin E), or two symptomatic drugs might be combined that work through different mechanisms (e.g., levodopa and a dopamine agonist). Alternatively, agents designed to slow the rate of progression through different mechanisms (e.g., vitamin E and a neurotrophic factor) might be used together. There is a greater need for trial designs of this sort as the number of potential neuroprotective–neuroregenerative therapeutic approaches reaching clinical testing from preclinical studies increases.

IX. CONCLUSION

As shown in the preceding chapters of this book, the treatment of neurological disease is on the threshold of substantial change. A number of therapeutic strategies, including neurotrophic factors, neural progenitor cells, and fetal cells, provide the potential to develop for the first time effective therapies to reduce disease progression or to promote neural repair. The proper design of clinical trials will be an essential requirement in the success of these efforts. The optimal design of a clinical trial will depend on both the nature of the neurological disease and the expected effect of the therapeutic intervention. Properly designed and conducted trials are becoming increasingly important as the number of neural repair strategies that are translated to the clinic steadily increases.

ACKNOWLEDGMENTS

Supported by the NIH (NINDS 1R01NS37083, NIA 2P01AG10435, NIA 5U01AG10483-06), American Academy of Neurology Research Foundation, Veterans Affairs, and the Hollfelder Foundation.

REFERENCES

ALS CNTF Treatment Study Group. (1996). A double-blind placebo-controlled clinical trial of subcutaneous recombinant human ciliary neurotrophic factor (rHCNTF) in amyotrophic lateral sclerosis. *Neurology* 46(5):1244–1249.

Becker, R. E., Colliver, J. A., *et al.* (1996). Double-blind, placebo-controlled study of metrifonate, an acetylcholinesterase inhibitor, for Alzheimer disease. *Alzheimer Dis Assoc Disord* 10(3):124–131.

Bensimon, G., Lacomblez, L., *et al.* (1994). A controlled trial of riluzole in amyotrophic lateral sclerosis. ALS/Riluzole Study Group. *N Engl J Med* 330(9):585–591.

Berg, L. (1988). Clinical Dementia Rating (CDR). *Psychopharmacol Bull* 24(4):637–639.

Bodick, N. C., Offen, W. W., *et al.* (1997). Effects of xanomeline, a selective muscarinic receptor agonist, on cognitive function and behavioral symptoms in Alzheimer disease. *Arch Neurol* 54(4):465–473.

Davis, K. L., Thal, L. J., *et al.* (1992). A double-blind, placebo-controlled multicenter study of tacrine for Alzheimer's disease. The Tacrine Collaborative Study Group. *N Engl J Med* 327(18):1253–1259.

DeJong, R., Osterlund, O. W., *et al.* (1989). Measurement of quality-of-life changes in patients with Alzheimer's disease. *Clin Ther* 11(4):545–554.

Drachman, D. A., and Leber, P. (1997). Treatment of Alzheimer's disease—searching for a breakthrough, settling for less. *N Engl J Med* 336(17):1245–1247.

Galasko, D., Bennett, D., *et al.* (1997). An inventory to assess activities of daily living for clinical trials in Alzheimer's disease. The Alzheimer's Disease Cooperative Study. *Alzheimer Dis Assoc Disord* 11(Suppl. 2):S33–39.

Gauthier, S., Bouchard, R., *et al.* (1990). Tetrahydroaminoacridine-lecithin combination treatment in patients with intermediate-stage Alzheimer's disease. Results of a Canadian double-blind, crossover, multicenter study. *N Engl J Med* 322(18):1272–1276.

Grundman, M., and Thal, L. J. (1997). Clinical trials to prevent Alzheimer's disease in a population at-risk. *Alzheimer's Disease: from Molecular Biology to Therapy*. R. E. Becker and E. Giacobini. Boston: *Birkhauser* 375–379.

Knapp, M. J., Knopman, D. S., *et al.* (1994). A 30-week randomized controlled trial of high-dose tacrine in patients with Alzheimer's disease. The Tacrine Study Group. *Jama* 271(13):985–991.

Lacomblez, L., Bensimon, G., *et al.* (1996). Dose-ranging study of riluzole in amyotrophic lateral sclerosis. Amyotrophic Lateral Sclerosis/Riluzole Study Group II. *Lancet* 347(9013):1425–1431.

Lawton, M. P. (1988). Scales to measure competence in everyday activities. *Psychopharmacol Bull* 24(4):609–614.

Lawton, M. P., and Brody, E. M. (1969). Assessment of older people: Self-maintaining and instrumental activities of daily living. *Gerontologist* 9(3):179–186.

Leber, P. (1996). Observations and suggestions on antidementia drug development. *Alzheimer Dis Assoc Disord* 10(Suppl. 1):31–35.

LeWitt, P., Oakes, D., *et al.* (1997). The need for levodopa as an end point of Parkinson's disease progression in a clinical trial of selegiline and alpha-tocopherol. Parkinson Study Group. *Mov Disord* 12(2):183–189.

McBee, W. L., Dailey, M. E., *et al.* (1997). Hormone replacement therapy and other potential treatments for dementias. *Endocrinol Metab Clin North Am* 26(2):329–345.

Morris, J. C. (1993). The Clinical Dementia Rating (CDR): Current version and scoring rules. *Neurology* 43(11):2412–2414.

Olichney, J. M., Hofstetter, C. R., *et al.* (1995). Death certificate reporting of dementia and mortality in an Alzheimer's disease research center cohort. *J Am Geriatr Soc* 43(8):890–893.

Penn, R. D., Martin, E. M., *et al.* (1988). Intraventricular bethanechol infusion for Alzheimer's disease: results of double-blind and escalating-dose trials. *Neurology* 38(2):219–222.

Porter, R. J., and Schoenberg, B. S. (1990). *Controlled clinical trials in neurological disease*. Boston: Kluwer Academic Publishers.

Rogers, S. L., Farlow, M. R., *et al.* (1998). A 24-week, double-blind, placebo-controlled trial of donepezil in patients with Alzheimer's disease. *Neurology* 50(January 1998):136–145.

Rogers, S. L., and Friedhoff, L. T. (1996). The efficacy and safety of donepezil in patients with Alzheimer's disease: Results of a US multicentre, randomized, double-blind, placebo-controlled trial. The Donepezil Study Group. *Dementia* 7(6):293–303.

Rosen, W. G., Mohs, R. C., *et al.* (1984). A new rating scale for Alzheimer's disease. *Am J Psychiatry* 141(11):1356–1364.

Sano, M., Ernesto, C., *et al.* (1996). Rationale and design of a multicenter study of selegiline and alpha-tocopherol in the treatment of Alzheimer disease using novel clinical outcomes. Alzheimer's Disease Cooperative Study. *Alzheimer Dis Assoc Disord* 10(3):132–140.

Sano, M., Ernesto, C., *et al.* (1997). A controlled trial of selegiline, alpha-tocopherol, or both as treatment for Alzheimer's disease. The Alzheimer's Disease Cooperative Study. *N Engl J Med* 336(17):1216–1222.

Schneider, L. S., and Olin, J. T. (1996). Clinical global impressions in Alzheimer's clinical trials. *Int Psychogeriatr* 8(2):277–288; discussion 288–290.

Schneider, L. S., Olin, J. T., *et al.* (1996). *Validity and reliability of the Alzheimer's disease cooperative study-clinical global impression of change. The Alzheimer's Disease Cooperative Study*. Boston: Birkhauser.

Schneider, L. S., Olin, J. T., *et al.* (1997). Validity and reliability of the Alzheimer's disease cooperative study-clinical global impression of change. The Alzheimer's Disease Cooperative Study. *Alzheimer Dis Assoc Disord* 11(Suppl. 2):S22–32.

Spilker, B. (1991). *Guide to clinical trials*. New York, N.Y., Raven Press, Ltd.

Stern, Y., Sano, M., *et al.* (1987). Effects of oral physostigmine in Alzheimer's disease. *Ann Neurol* 22(3):306–310.

Thal, L. J., Carta, A., *et al.* (1996). A 1-year multicenter placebo-controlled study of acetyl-L-carnitine in patients with Alzheimer's disease. *Neurology* 47(3):705–711.

Thal, L. J., Schwartz, G., *et al.* (1996). A multicenter double-blind study of controlled-release physostigmine for the treatment of symptoms secondary to Alzheimer's disease. Physostigmine Study Group. *Neurology* 47(6):1389–1395.

The ALS CNTF Treatment Study (ACTS) Phase I-II Study Group. (1996). The Amyotrophic Lateral Sclerosis Functional Rating Scale. Assessment of activities of daily living in patients with amyotrophic lateral sclerosis. *Arch Neurol* 53(2):141–147.

The Parkinson Study Group. (1993). Effects of tocopherol and deprenyl on the progression of disability in early Parkinson's disease. *N Engl J Med* 328(3):176–183.

INDEX

A

B

C

D

E

F

H

R

S

T

V

X

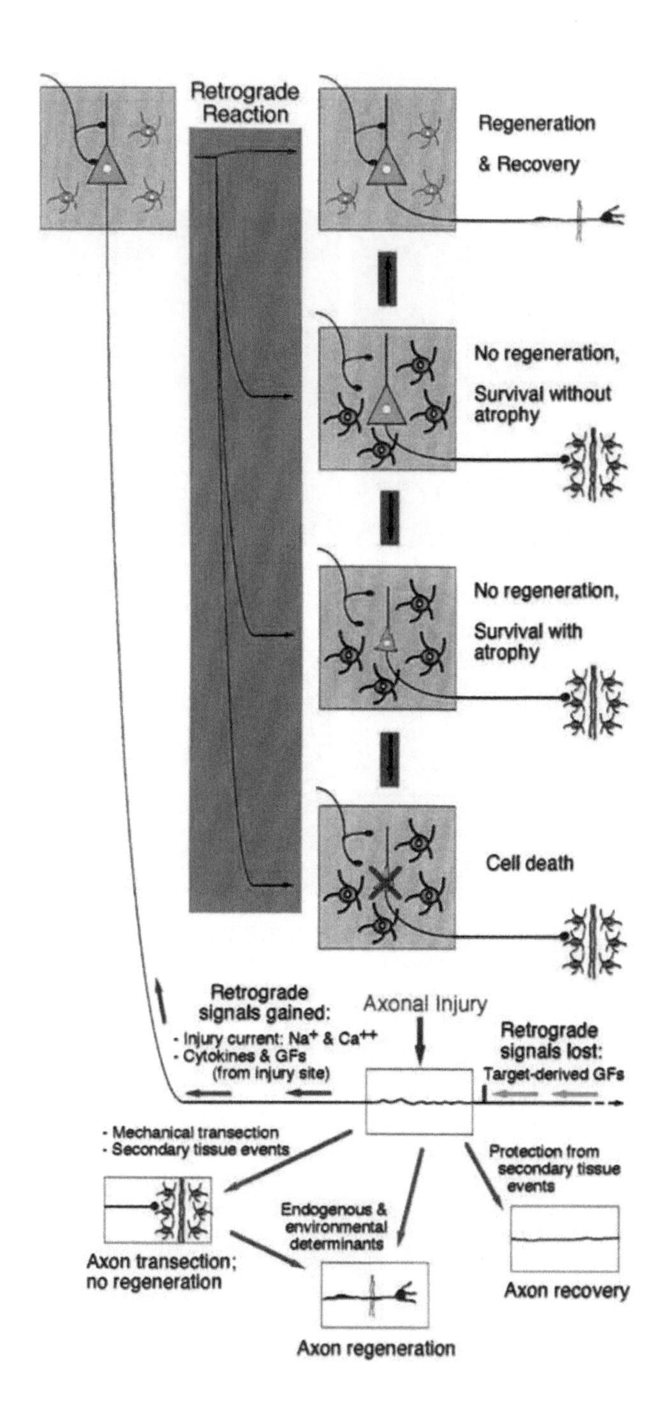

CHAPTER 1, FIGURE 1 Schematic summary of potential retrograde reactions of nerve cells to axonal injury and factors that may determine outcome.

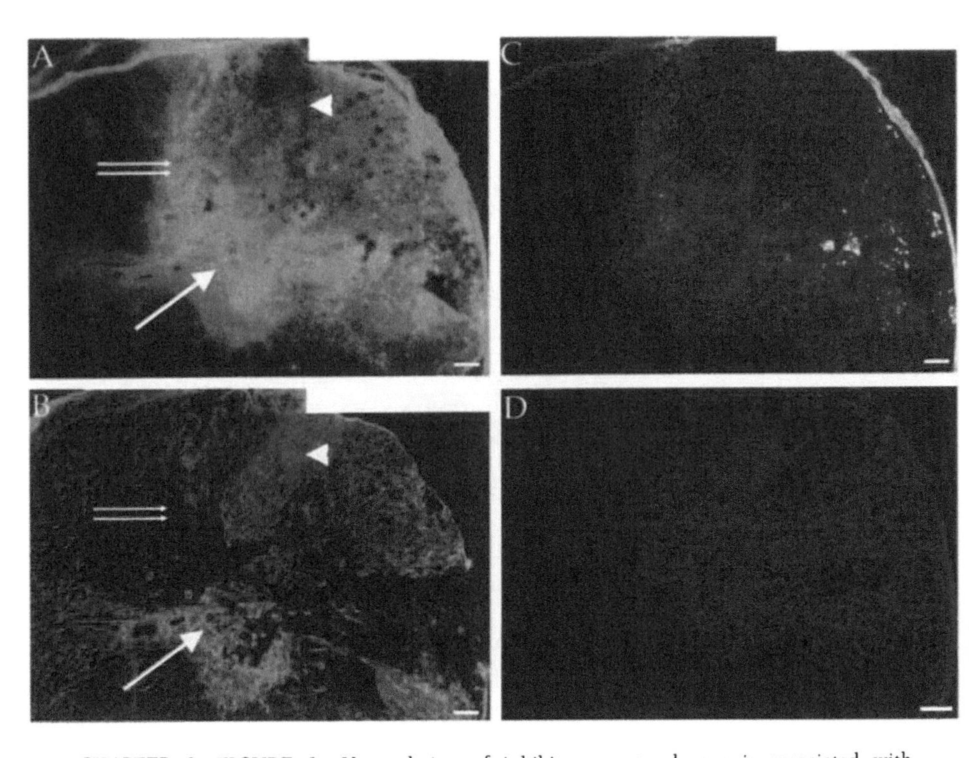

CHAPTER 3, FIGURE 1 Upregulation of inhibitory proteoglycans is associated with inflammation and breakdown of the blood brain barrier. (A–C) Representative cross section of a rat spinal cord hemicrush lesion one week after injury, triple stained for (A) chondroitin sulfate proteoglycans, (B) GFAP to visualize astrocytes, and (C) macrophages–microglial cells. (D) An adjacent section stained for extravasated immunoglobulins demonstrates areas with compromised blood brain barrier. Large arrow (A–B) indicates a region of astrocyte gliosis associated with increased levels of proteoglycan. Arrowhead (A–B) indicates a region of astrocyte gliosis not associated with increased levels of proteoglycan. Arrowhead (A–B) indicates a region of astrocyte gliosis not associated with increased proteoglycan staining. Double arrows (A–B) indicate an area with no GFAP positive astrocytes that demonstrates increased proteoglycan. Note the close association between the distribution of proteoglycan, inflammation, and blood brain barrier breakdown (A, C, and D) that is not consistently associated with regions of astrocyte gliosis (B), suggesting that extravasation of serum components and/or products secreted by inflammatory cells are triggering the production of inhibitory extracellular matrix molecules (see Fitch and Silver, 1997a). Scale bar, 100μ. (From Fitch and Silver, 1997a. *Experimental Neurology* 148: 587–603. Copyright © 1997 by Academic Press, used with permission)

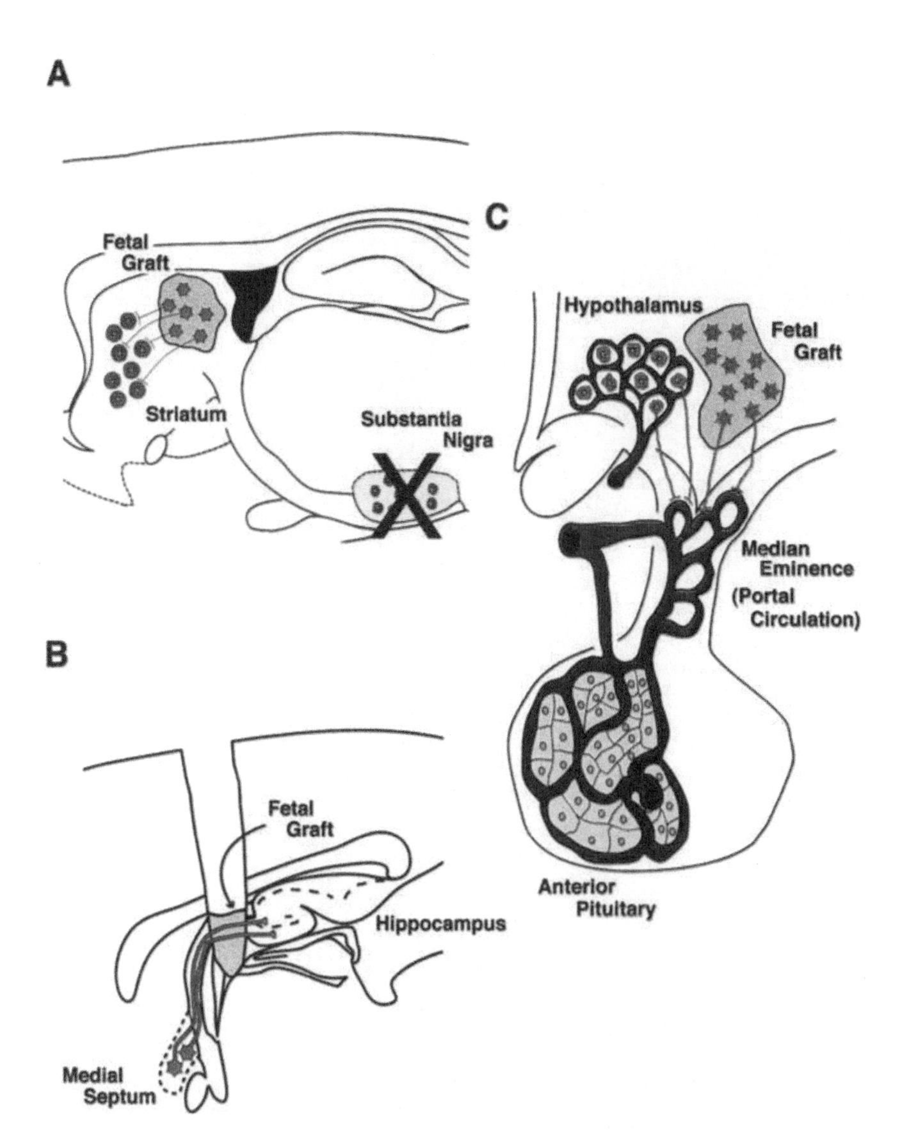

CHAPTER 6, FIGURE 1 Diagram illustrating potential applications of fetal nervous system grafting to treat neurological disease. (A) Fetal neural transplantation can be used to replace specific neural circuits. For example, fetal grafts placed into the striatum can replace degenerating neuronal circuitry projecting from the substantia nigra in models of Parkinson's disease. (B) Fetal grafts can also act as bridges to support the regeneration of *host* axons from one CNS location to another. For example, lesions of medial septal inputs to the hippocampus can regenerate through fetal grafts placed in the lesion cavity (see also Figs. 2–4). (C) Finally, fetal grafts can replace (or protect) endocrine function in association with host vasculature, as in this model of hypothalamic dysfunction. A fetal hypothalamic graft placed in the hypothalamus can compensate for dysfunctional host neurons by establishing projections onto host vasculature (neurohemal mechanism).

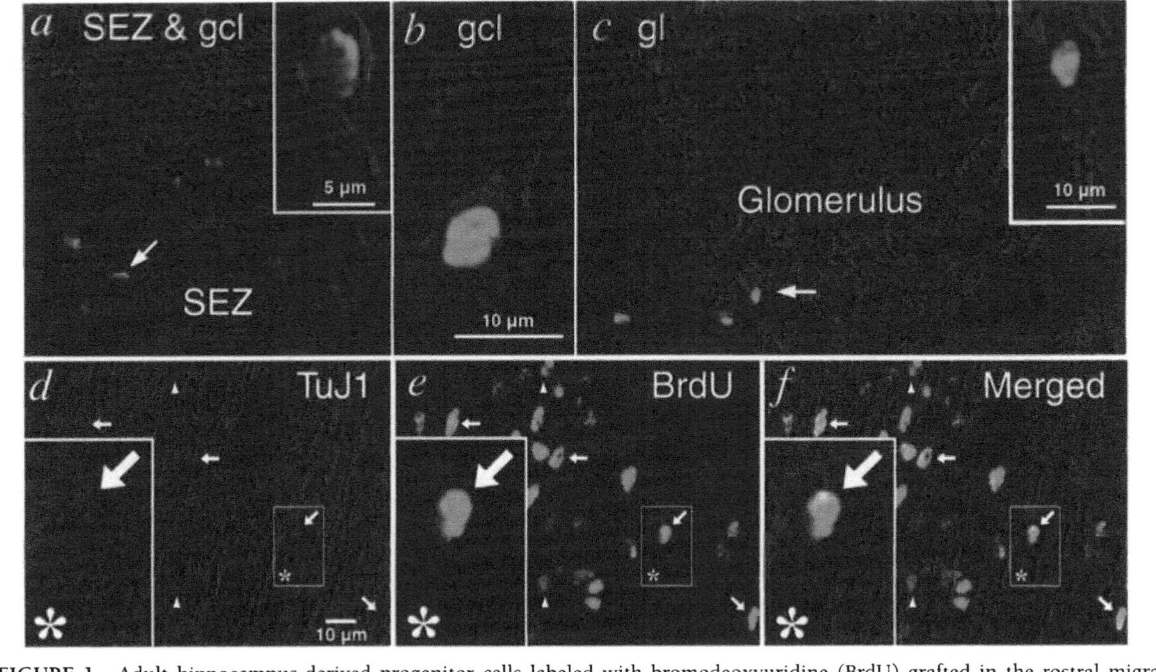

CHAPTER 7, FIGURE 1 Adult hippocampus-derived progenitor cells labeled with bromodeoxyuridine (BrdU) grafted in the rostral migratory pathway (RMP) of adult rat migrate from the site of injection. One week post-grafting, some $BrdU^+$ cells were found in association with polysialated form of neuron cell adhesion protein (PSA–NCAM) in subependymal zone (SEZ) and granule cell layer (gel) (a); and 8-weeks post-grafting in gel and glomeruli (gl) (b, c). Insets show the cells indicated by arrow in (a) and (c). Migrating cells also associate with TuJ1, a marker for immature neuroblasts, in RMP (d–f). Merged images of cells (f) showed that $BrdU^+$ cells are expressing TuJ1 (arrow). Association of the grafted cells with PSA–NCAM and TuJ1 indicate that they behave like endogenous progenitor cells that express TuJ1 and are associated with PSA–NCAM during their chain migration from subventricular zone (SVZ) to the olfactory bulb through RMP (Luskin *et al.*, 1997; Lois *et al.*, 1996). (From Suhonen *et al.*, (1996). *Nature* **383**: 624–627. Reprinted by permission from *Nature*.)

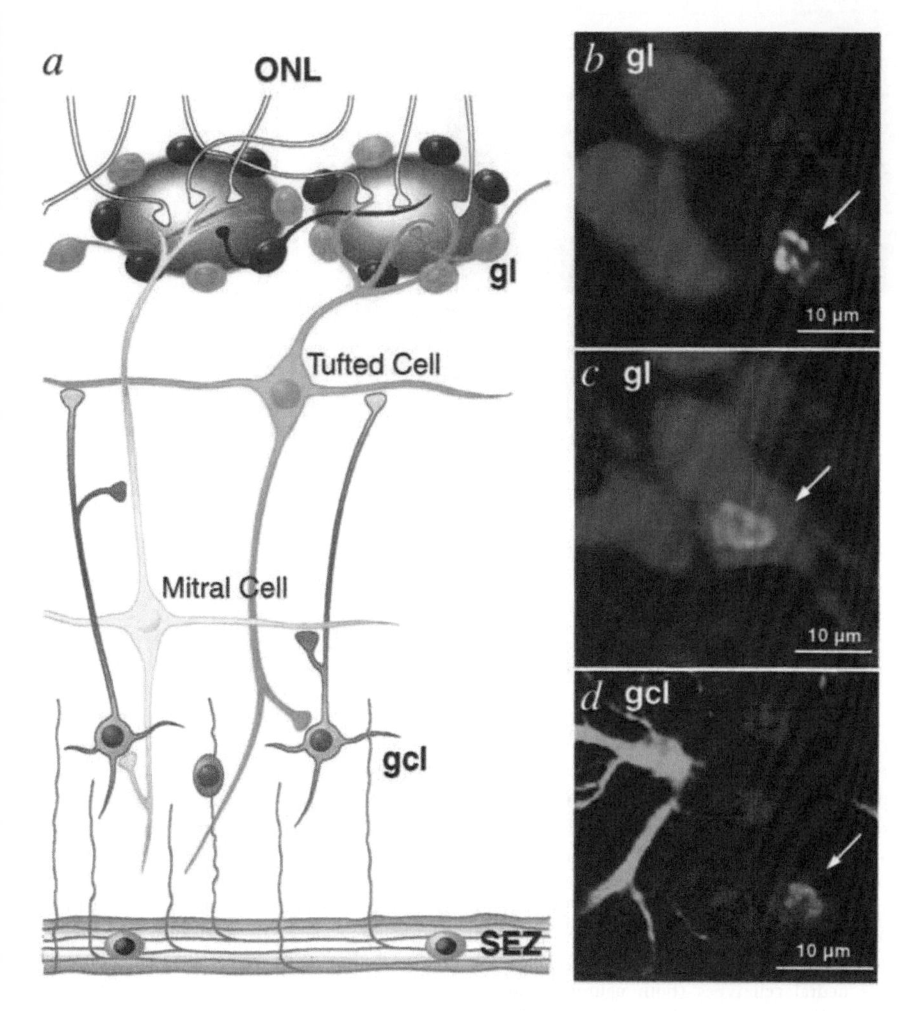

CHAPTER 7, FIGURE 2 Differentiation of grafted adult hippocampus-derived progenitor cells in olfactory bulb (OB). Schematic of adult rat OB is shown in (a). SEZ, subependymal zone; ONL, olfactory nerve layer. Phenotypes of grafted neurons 8-weeks post-transplantation (b–d). BrdU$^+$ progenitors (green) express calbindin (red; yellow pointed by arrow in b) and tyrosine hydroxyulase (TH) (blue; light blue pointed by arrow in c). TH$^+$ cells are not found in hippocampus. Presence of BrdU/Th-immunopositive (c) cells in gl indicates that grafted progenitors respond to local cues and differentiate into site-specific neuronal phenotypes. NeuN$^+$ neurons in gel (d). Yellow (pointed by arrow in d) indicates a grafted cell double-labeled for BrdU and NeuN. Turquoise cells are GFAP-expressing astrocytes. (From Suhonen *et al*, (1996). *Nature* 383: 624–627. Reprinted by permission from *Nature*.)

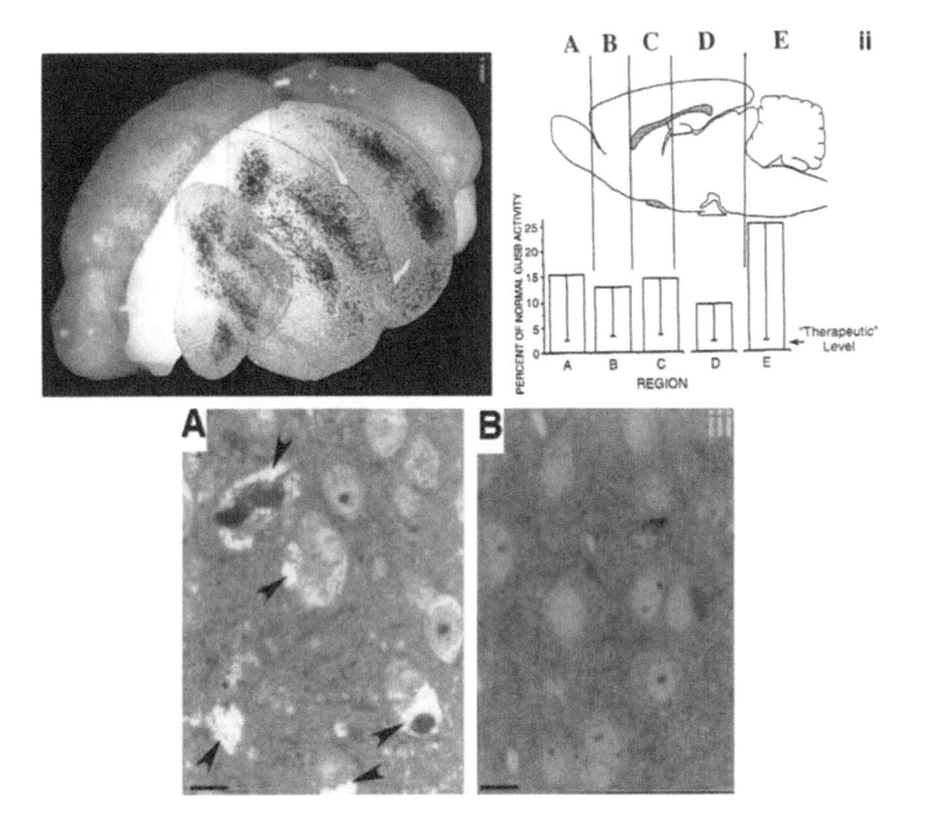

CHAPTER 8, FIGURE 1 Widespread engraftment of neural stem cells throughout the brain. (i) Computer reconstruction of the brain of a mature Mucopolysaccaridosis Type VII (MPS VII), β-glucuronidase (GUSB)-deficient mutant mouse after receiving a neonatal intraventricular transplant of neural stem cells expressing GUSB. Clone C17.2 neural stem cells, which express ~800 units of GUSB activity per mg protein *in vitro*, also express the *lacZ* reporter gene. They were implanted into the lateral ventricles of a newborn MPS VII mouse allowing them access, in part, to the subventricular germinal zone (SVZ). Donor-derived cells could be indentified *in vivo* by the Xgal histochemical reaction (blue precipitate) for expression of *lacZ*. The blue cells can be seen to have engrafted and integrated throughout the recipient mutant brain. They differentiate into the neural cell types (both neuronal and glial) appropriate to a given CNS region at the particular developmental period during which they were implanted. Cells implanted into the fetal ventricular system yield a similar distribution of integrated donor-derived cells. (ii) Widespread distribution of GUSB enzymatic activity throughout brains of MPS VII recipients of neural stem cell transplants, corresponding to the widespread distribution of donor-derived cells. Serial sections were collected from throughout the brains of MPS VII transplant recipients and quantitatively assayed for GUSB activity. Serial coronal slices were pooled to reflect the activity present within the regions demarcated in the schematic. The mean levels of GUSB enzyme activity for each region (n=17) are presented as the percentage of average normal levels for each region. GUSB activity co-localized with the distribution of the donor Xgal-positive (blue) cells (e.g., panel i). Untreated MPS VII mice show no GUSB activity biochemically or histochemically. Enzyme activity ≥ 2 to 5% of normal is corrective based on data from the liver and spleen. It is significant to note that, although these histograms illustrate the widespread distribution of GUSB, they could similarly reflect the transplant-mediated distribution of other diffusible (e.g., metabolic

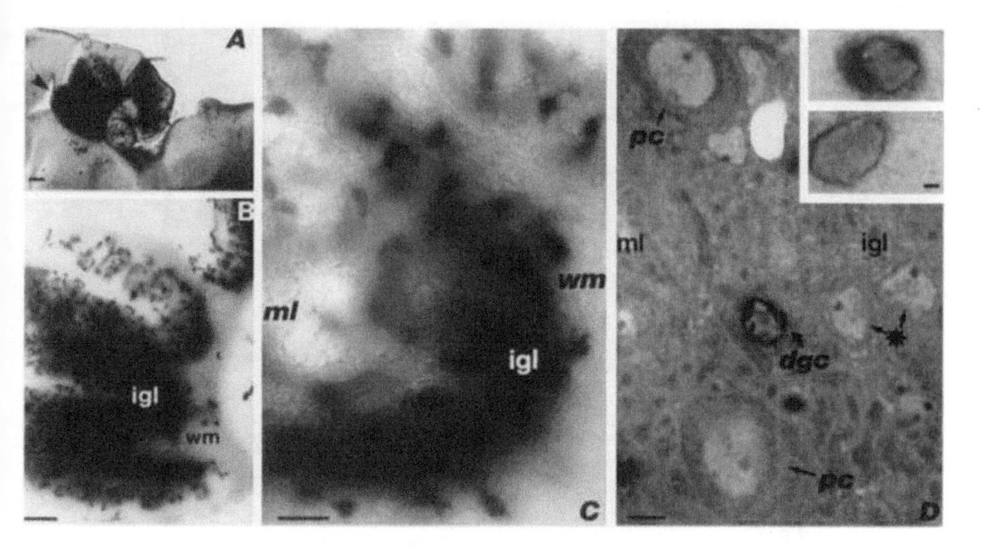

CHAPTER 8, FIGURE 3 Engrafted multipotent neural stem cells appear to have "shifted" their differentiation fate toward partially repopulating the granule neuron deficient anterior lobe in a mature *meander tail* mouse mutant cerebellum. Cells derived from *lacZ*-expressing clone C17.2 stem cells (implanted at birth into the external granular layer [EGL]) are visualized under brightfield following Xgal histochemistry, which confers a blue color to donor-derived cells. [A] Low-power overview of a representative parasagittal section through the mature cerebellum. There is dense engraftment in both the anterior lobe (AL) (arrowhead) and posterior lobe (PL). When compared to an unmanipulated, severely atrophic, granule neuron-deficient mutant AL, this transplanted AL gives the visual impression of having been "reconstituted" by donor cells. [B,C] Progressively higher-power views of the AL confirm that a relatively wide internal granular layer ("igl") is densely crowded with $Xgal^{+}$ cells. ("ml", molecular layer; "wm", tracts of white matter). [D] A 1 μm semi-thin section demonstrates that discrete donor-derived (blue) cells possessing definitive granule cell (GC) neuron size, morphology, and location ("dgc") were integrated within the neurophil of the host AL in the IGL. A donor-derived GC ("dgc") is evident, with its distinctive blue perinuclear ring, near a few, rare, unlabeled, residual host GCs (*). ("pc", Purkinje cell). [Insets] Higher magnification of two donor-derived GCs from a lightly counterstained semi-thin section. Although dense engraftment of clone C17.2 was similarly observed throughout the ALs and PLs of all transplanted mutant and unaffected mice, in the regions most deficient in GCs (i.e., the *mea* AL), a greater proportion of those donor-derived cells were noted to be GCs, as if there had been a "compensatory shift" in the differentiation of the progeny of this multipotent stem cell clone. This point is illustrated quantitatively in Fig. 3. Scale bars: (A) 250 μm; (B) 50 μm; (C) 10 μm; (D) 5 μm; (insets) 1 μm. (Figure reproduced with permission from Rosario *et al.*, 1997)

---

enzyme, neurotrophic agents, viral vectors) and non-diffusible (e.g., myelin, ECM) factors, as well as the distribution of "replacement" neural cells (see Table II). (iii) The efficacy of neural stem-cell-mediated CNS gene therapy: Decreased lysosomal storage in a treated MPS VII mouse brain at 8 months of age. (A) Extensive vacuolation representing distended lysosomes (arrowheads) in both neurons and glia in the cortex of an 8-month-old, untransplanted control MPS VII mouse. (B) Decrease in lysosomal storage in the neocortex of an MPS VII mouse treated at birth from a region analogous to the untreated control section (A). The other regions of this animal's brain showed a similar decrease in lysosomal storage compared to untreated, age-matched mutants in regions in which GUSB was expressed. Scale bars, 21μm. (Figure adapted with permission from Snyder *et al*, 1995.)

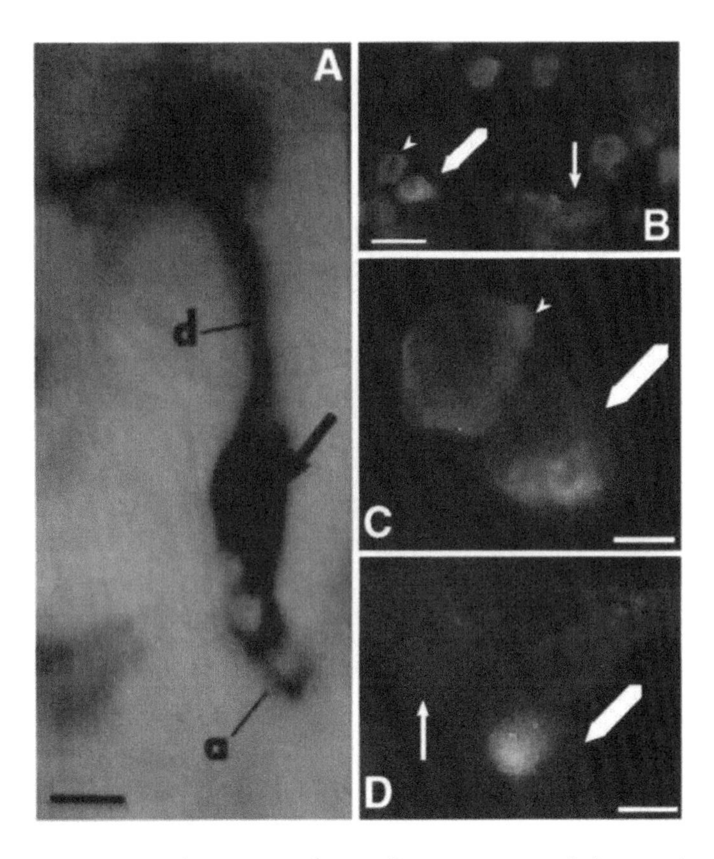

CHAPTER 8, FIGURE 5 Multipotent neural stem cells acquire a neuronal phenotype in regions of adult neocortex subjected to targeted *apoptotic* neuronal degeneration. They differentiate developmentally appropriately into only glia (or remain undifferentiated) in intact (or non-apoptotically lesioned) control postnatal mammalian neocortex, in which neurogenesis has normally ceased but gliogenesis persists. 15 ± 7% of engrafted cells (identified by their *lacZ* reporter gene expression) in regions of apoptotic neurodegeneration developed neuronal morphology, resembling pyramidal neurons within layer II/III 6 weeks following transplantation, at 12 weeks of age (A) and immunocytochemical properties consistent with a neuronal phenotype (B–D). [A] Donor-derived cell, which stains blue after following the Xgal histochemical reaction to detect *lacZ* reporter gene expression, with neuronal morphology (large arrow) under brightfield microscopy: Large somata (20–30μm diameter); characteristic large nucleus and prominent nucleolus; 300–600 μm presumptive apical dendrites("d"); presumptive axons ("a"). (The dark object at the upper end of the presumptive dendrite is another $Xgal^+$ cell out of the plane of focus). These features are readily confirmed by ultrastructural criteria under electron microscopy where the Xgal reaction product is electron dense. (Not shown here but pictured in Snyder *et al.*, 1997a.) Donor-derived cells have not only the above characteristics but are noted to contain cytoskeletal elements such as microtubules, to extend processes that contain neurofilaments and become myelinated, and to receive both axosomatic and axodendritic synaptic input. Donor-derived cells in control adult cortices (intact or non-apoptotically lesioned with kainic acid) were small (5–8 μm) and had only morphologic, ultrastructural, and immunocytochemical features of glia (astrocytes and oligodendrocytes). (Not shown here but pictured in Snyder *et al.*, 1997a.) (B–D) Immunocytochemical analysis. Donor-derived cells, identified by an anti-*lacZ* antibody, are also reacted with cell type-specific antibodies; immunoreactivity is visualized by

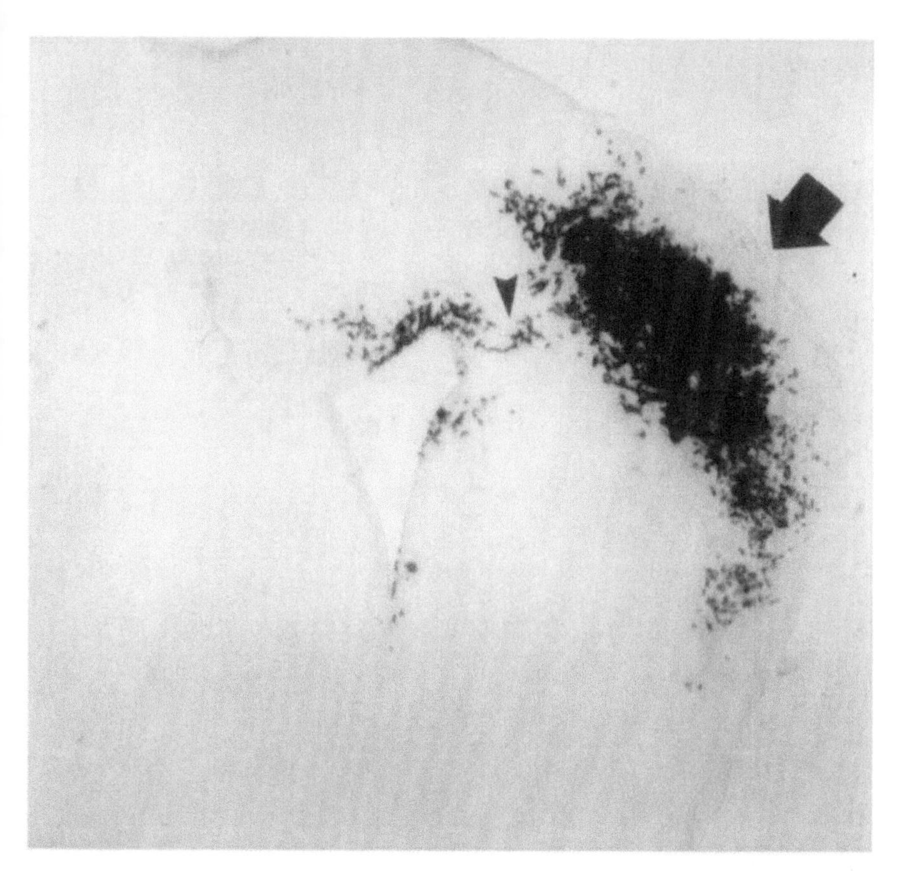

CHAPTER 8, FIGURE 6 Robust engraftment by donor neural stem cells within a representative coronal section through the ischemic region of a mouse brain subjected to unilateral asphyxial injury. This mouse was subjected to unilateral right-sided HI injury on P7, as described in the text. Three days later (at P10), the animal received a transplant of neural stem cells (clone C17.2) within the region of infarction. The animal was analyzed at maturity (15 days after transplantation) with Xgal histochemistry to identify *lacZ*-expressing donor-derived cell (which stain blue). Robust engraftment was evident within the ischemic area (arrow). Similar engraftment was evident throughout the hemisphere. Even cells which implanted outside the region of infarction (including in the contralateral hemisphere) appeared to migrate along the corpus callosum toward the ischemic area (arrowhead). The most exuberant engraftment was evident 3 to 7 days after HI. Immunocytochemical and ultrastructural analysis revealed that a subpopulation of donor-derived cells, especially those in the penumbra differentiated into neurons and oligodendroglia, the two neural cell types most characteristically damaged by HI and the cell types least likely to regenerate spontaneously in the postnatal brain.

---

fluorescent secondary antibodies. Donor-derived neurons (*lacZ*$^+$ by Texas Red) immunoreactive for NeuN (fluorescein, green), a marker for mature neurons. In (B) and at higher magnification in (C), *lacZ*$^+$ cell (large arrow) double-labels for NeuN. Other small *lacZ*$^+$ cells with nonneuronal morphology are NeuN$^-$ (small arrow). Remaining host neurons (NeuN$^+$) are *lacZ*$^-$ (arrowhead). (D) Donor-derived neuron (*lacZ*$^+$, NeuN$^+$; large arrow) adjacent to two NeuN$^-$ donor-drived cells (small arrow). Scale bars: (A) and (B) 25 μm; (C) and (D) 10μm. (Figure reproduced with permission from Snyder *et al.*, 1997a.)

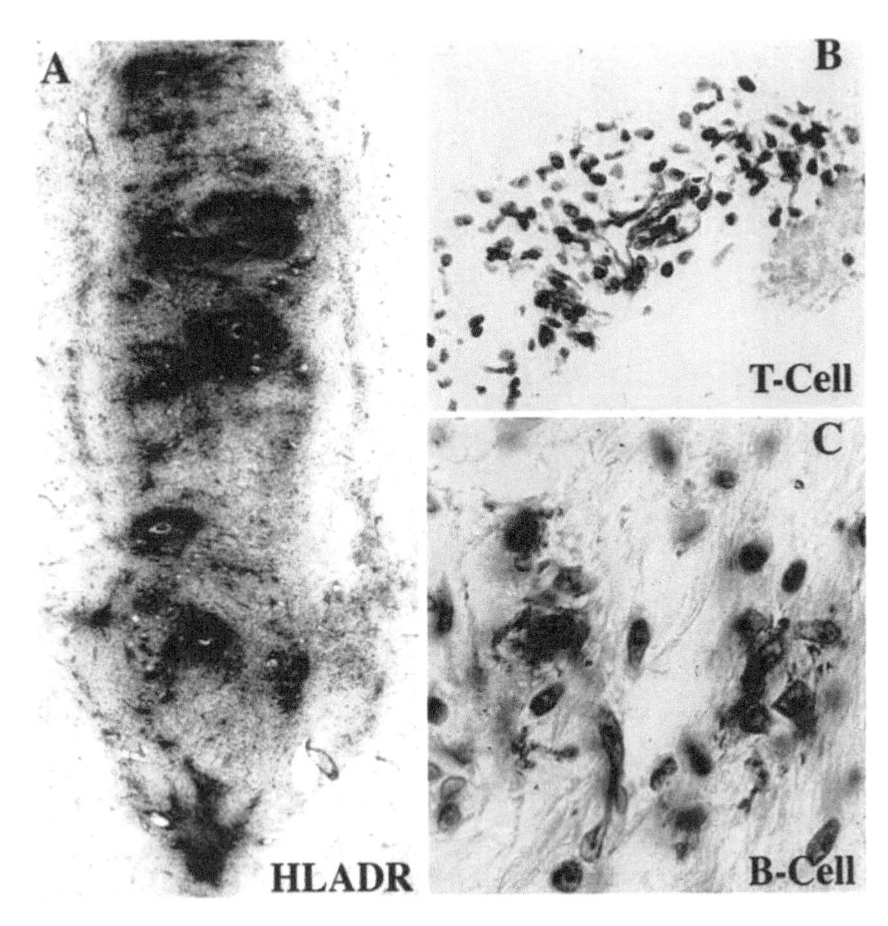

CHAPTER 14, FIGURE 1 (A) Low-power photomicrograph illustrating dense HLADR-ir staining within a human fetal nigral allograft. These grafts also contained a few (B) T cells and (C) B cells.

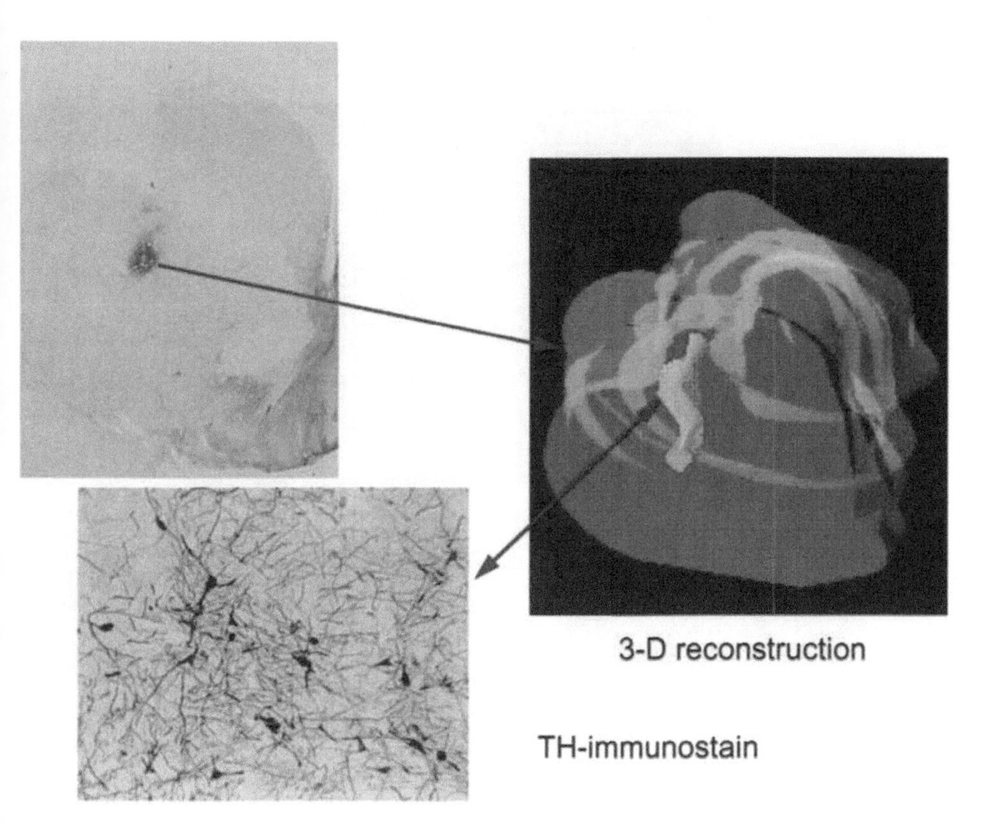

CHAPTER 16, FIGURE 4 Tyrosine-hydroxylase (TH) immunostaining of MPTP-lesioned monkey. In this animal, an AAV vector was introduced into the putamen by the convection-enhanced delivery system six weeks prior to necropsy (Bankiewicz *et al.*, 1996). Strong TH expression was detected in the resident striatal neurons and was spread by the infusion method, as shown in the 3-D reconstruction.

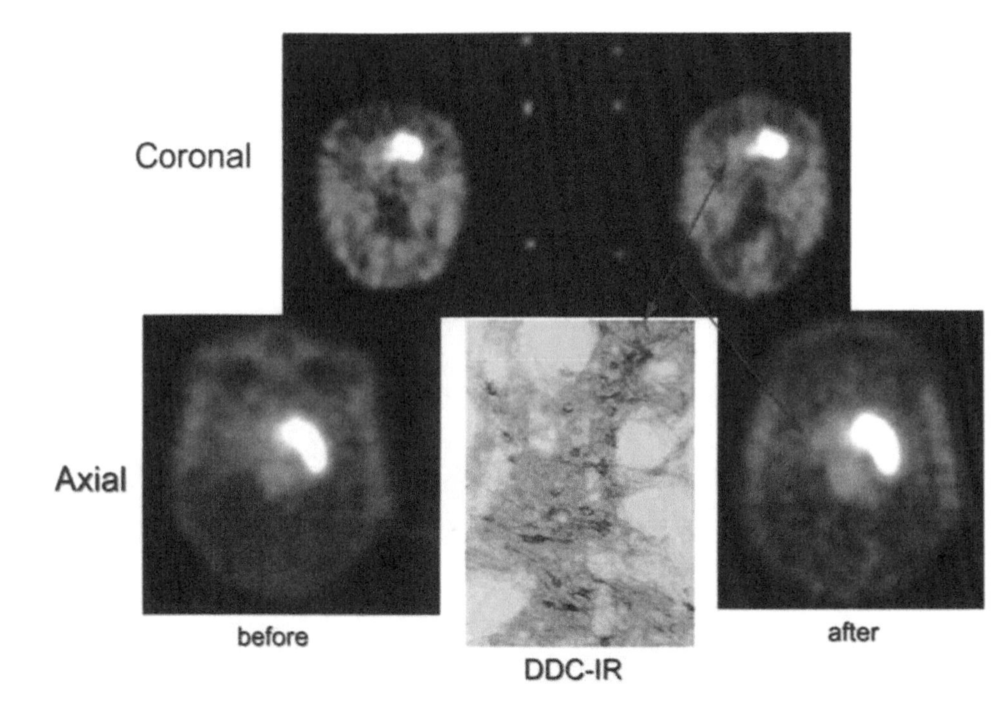

CHAPTER 16, FIGURE 5 Positron emission tomography (PET) images for a unilateral MPTP lesioned monkey using the aromatic amino acid decarboxylase (AADC) tracer, 6-[^{18}F]fluoro-L-*m*-tyrosine (FMT). Coronal and axial images were taken before (left) and six weeks after (right) the introduction of AAV–DDC. An area of increased FMT uptake corresponded to the area of AAV–DDC infusion. PET may provide an *in vivo* method for the detection of gene expression, extent of gene transfer, levels of gene expression, and monitoring gene regulation.

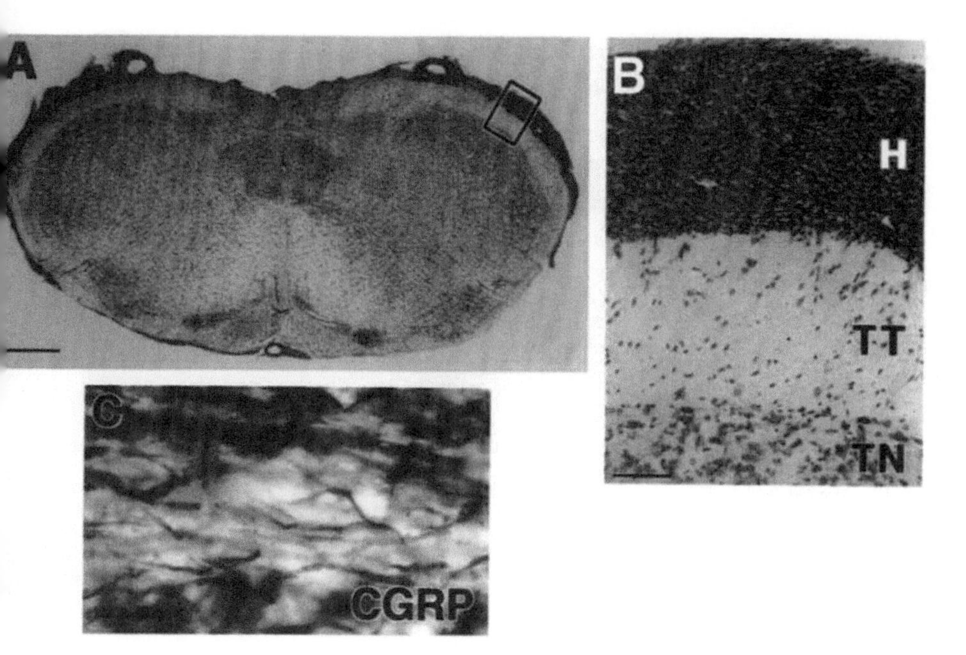

CHAPTER 20, FIGURE 4 Adverse consequences of intracerebroventricular infusions of NGF. (A) A layer of proliferating Schwann cells accumulates in the subpial region of the medulla, particularly on the dorsal surface. Ultrastructural analysis (not shown) and immunolabeling for p75 and S100 (not shown) verify the Schwann cell identity of the cells. Scale bar = 250 μm. Boxed region is detailed in (B). At higher magnification, the hypertrophic Schwann cell layer (H) is several layers thick. TT, Trigeminal tract; TN, trigeminal nucleus. Scale bar = 62 μm. (C) Immunocytochemical labeling for calcitonin gene-related peptide (CGRP), a marker for peripheral nervous system sensory axons, reveals dense sprouting of sensory axons into the proliferating Schwann cell layer. Sympathetic axons also sprout into the Schwann cell layer (not shown). Both nociceptive sensory and sympathetic axons are NGF-dependent systems for survival during nervous system development, thus their responsiveness to NGF in the adult is not surprising. Scale bar = 12 μm. (From Winkler *et al.* (1996), *Ann Neurol* 40: 128–139.)

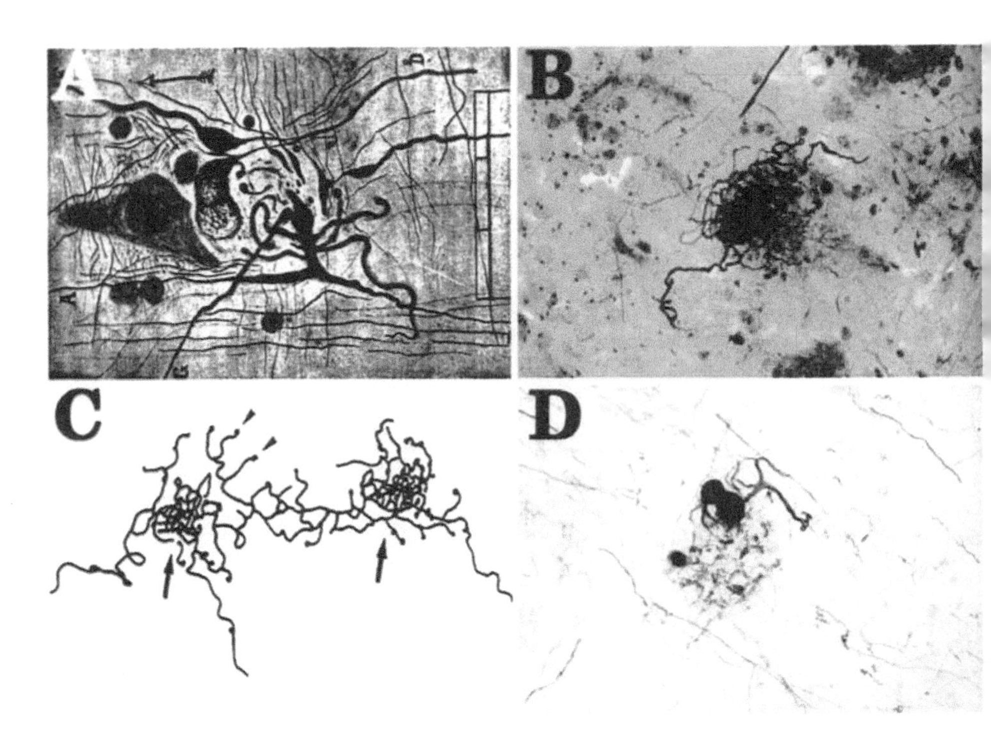

CHAPTER 21, FIGURE 2 (A) Earliest example of reactive sprouting and regrowth associated with a senile plaque in Alzheimer's disease. Even before the discovery of trophic factors, Cajal hypothesized that such a "trophic" molecule was present to trigger new growth (taken from Cajal, 1928). (B) Using the Bielschowsky silver stain that Alzheimer and Cajal employed, we can see the tortuous neuritic growth and dystrophic fibers associated with senile plaques. (C) In a rat entorhinal cortex lesion model, axons within the dentate gyrus undergo reactive sprouting that is morphologically similar to the dystrophic patterns seen in senile plaques (taken from Deller *et al.*, 1996). (D) Modern immunocytochemistry for β-amyloid (brown) and hyperphosphorylated neurofilament (blue) reveal swollen terminals and the redirection of neurite growth into a plaque (courtesy of Dr. Joseph H. Su).

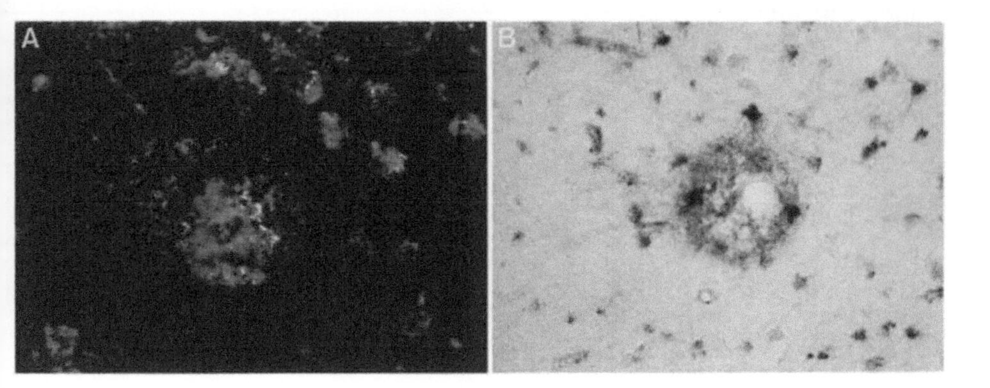

CHAPTER 21, FIGURE 4 (A) Double labeling for bFGF (green) and PHF-1 (red) reveals bFGF within a senile plaque (center) and in many cortical neurons. There are numerous dystrophic neurites in the periphery of the plaque (adapted from Cummings *et al.*, 1993a). (B) Antibodies for heparan sulfate proteoglycans detect epitopes within senile plaques and within astrocytes (adapted from Su *et al.*, 1992).

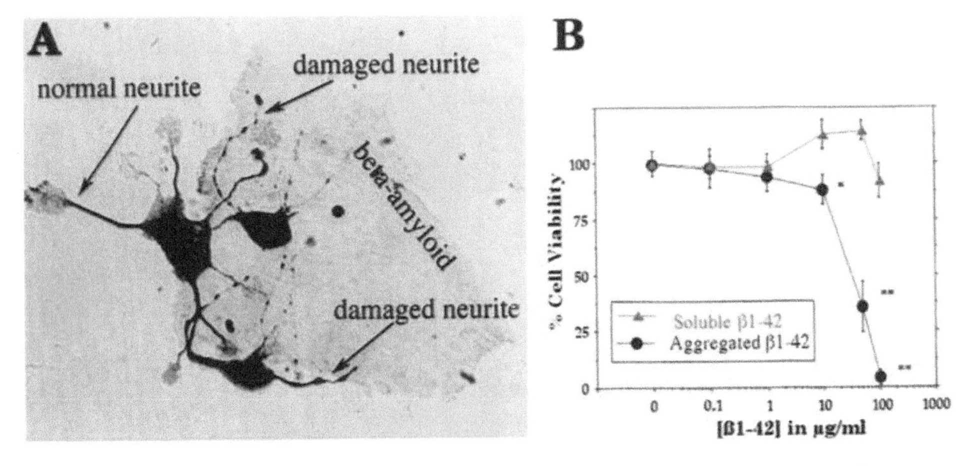

CHAPTER 21, FIGURE 5 (A) High-power photomicrograph of three neurons in culture contacting an aggregated core of β-amyloid peptide. Arrows denote beaded and dystrophic neurites that are in contact with the β-amyloid, while a neurite not in contact with the β-amyloid appears normal (adapted from Pike *et al.*, 1992). (B) The paradox is that β-amyloid is trophic to neurons when soluble, but is highly toxic when aggregated into a β-pleated sheet (adapted from Pike *et al.*, 1991).

Printed and bound by CPI Group (UK) Ltd, Croydon, CR0 4YY
08/05/2025
01865022-0005